SELECTED PAPERS

S. Chandrasekhar

SELECTED PAPERS

S. Chandrasekhar

*

VOLUME 2

Radiative Transfer and Negative Ion of Hydrogen

THE UNIVERSITY OF CHICAGO PRESS
CHICAGO AND LONDON

The University of Chicago Press, Chicago 60637
The University of Chicago Press, Ltd., London

Printed in the United States of America

98 97 96 95 94 93 92 91 90 89 5 4 3 2 1

 Acknowledgements appear on page 623.

Library of Congress Cataloging in Publication Data

Chandrasekhar, S. (Subrahmanyan), 1910–
Radiative transfer and negative ion of hydrogen.
p. cm.—(Selected papers of S. Chandrasekhar; v. 2)
1. Stars—Atmospheres. 2. Radiative transfer. 3. Hydrogen ions.
4. Anions. I. Title. II. Series: Chandrasekhar, S.
(Subrahmanyan), 1910– Selections. 1989; v. 2.
QB809.C47 1989
523.8′2—dc19 88-26111
ISBN 0-226-10092-8 (vol. 2) CIP
ISBN 0-226-10093-6 (vol. 2, pbk.)

∞ The paper used in this publication meets the minimum
requirements of the American National Standard for Information Sciences—
Permanence of Paper for Printed Library Materials, ANSI Z39.48-1984.

S. CHANDRASEKHAR is the Morton D. Hull Distinguished Service Professor
Emeritus in the Department of Astronomy and Astrophysics, the Department of
Physics, and the Enrico Fermi Institute at the University of Chicago.

Contents

2. *Radiative Transfer Allowing for Polarization of Scattering Light*

3. *Principles of Invariance: The H- and the X- and Y-Functions*

4. *Miscellaneous Problems*

5. *Review Articles*

PART TWO

Investigation on the Negative Ion of Hydrogen and of Two-Electron Atoms

A complete list of publications by S. Chandrasekhar will appear at the end of the final volume

Foreword

This second volume of Chandrasekhar's selected papers addresses that part of his prolific and diverse scientific career devoted to a mathematical analysis of the equations of radiative transfer. Although the papers in part 1 were published in a short time period, they had a dramatic and lasting influence on the field. Not only did he provide numerical results for comparison with observations, but he extended both the mathematical models and the mathematical techniques for solving them.

In the mid 1800s Maxwell completed the electromagnetic equations and, in applying them to light, made possible interpretations of natural light phenomena in mathematical terms. Feynman makes a play on words by stating in his first volume of *Lectures on Physics* that "Maxwell could say, 'let there be electricity and magnetism, and there is light.'"

By 1871 Rayleigh had used Maxwell's equations to explain the blue color of the sky. A more difficult problem was the polarization by the earth's atmosphere of unpolarized sunlight. Maxwell's equations predict polarization in scattering from particles, but multiple scattering must be accounted for to explain variations in the degree of polarization across the sky, as observed by Brewster and Babinet in the first half of the nineteenth century.

It was not until 1946 that Chandrasekhar gave a mathematical model which made it possible to do computations accounting for all orders of scattering in the study of polarization of the skylight. His mathematical and numerical analysis of the coupled system of linear Boltzmann equations for the Stokes vector and with the Rayleigh phase matrix was possible because of his previous work on the simpler models which account only for intensity, but not polarization, of radiation as presented in Schwarzschild's fundamental paper in 1906.

Schwarzschild's paper on radiative equilibrium was followed by contributions from Eddington, Hilbert, Jeans, Milne, Hopf, Ambartsumian, and others in the period 1906–1944. Radiative transfer problems were formulated both as integrodifferential equations and as functionals of associated integral equations. In a monograph in 1934, Hopf analyzed various integral equation formulations and applied the important mathematical technique of function factorization—the Wiener-Hopf method—to Milne's integral equation.

The present volume contains much of Chandrasekhar's analysis of various radiative transfer problems done primarily in the period 1944–50. The first section of part 1 contains papers published in the "linear period," when he applied the discrete ordinates technique suggested by results of Schuster, Schwarzschild, and Wick. Very accurate results were obtained by clever algebraic manipulations for several problems. In the 1950s Anselone proved convergence of these approximations, and later extended the techniques to a theory of "collectively compact operators" with applications to a large class of integral equations.

Chandrasekhar's approximation of solutions to boundary-value problems by finite superpositions of special solutions to the transfer equation was generalized by Case in the 1960s to the continuous superposition of distribution solutions of the transfer equation. Extensive applications have been made to problems in radiative transfer and neutron transport theory by Zweifel, Fertziger, Siewert, and many others, a recent application being Siewert's calculation of solutions to Chandrasekhar's "searchlight problem" of reflection of a pencil of radiation by a plane-parallel atmosphere.

Case's superposition of distributions can be cast in a more traditional mathematical framework by methods based on spectral theory or on Laplace transform methods and Wiener-Hopf factorizations. Solutions of boundary-value problems by any of these methods usually require the solving of singular integral equations, as done earlier by Case, Sobolev, and others in transfer problems. The Wiener-Hopf factorizations used here, and in other contexts, can often be expressed in terms of generalized Chandrasekhar H-functions or H-operators. An approach to rather general Wiener-Hopf integral equations and factorizations has been developed by Krein, Gohberg, and others in recent years.

The "nonlinear period" of Chandrasekhar's work, covered in section 3 of part 1, produced the expression of radiation intensities in terms of reflection and transmission functions, and these, in turn, as solutions to nonlinear integrodifferential equations of Riccati type. His derivation of these results by principles of invariance extended methods introduced by Ambartsumian. Chandrasekhar showed that reflection and transmission functions could be expressed in terms of simpler functions which also satisfy nonlinear equations—the simplest being the X- and Y-functions. This generalized a separation of variables given by Hopf for reflection from a gray, isotropically scattering halfspace by differentiation of the integral equation formulation. Anisotropic scattering and polarization problems lead to X- and Y-equations associated with Chandrasekhar's "pseudo-problems" in transfer theory. Existence and uniqueness questions for these equations were settled by Busbridge and Mullikin in the 1960s.

Chandrasekhar's representations of reflection and transmission functions in terms of X- and Y-functions made possible the generation of numerical results in a period when calculations were mechanical rather than electronic. Mathematically, it does not seem that progress is made by replacing a linear integro-differential equation by nonlinear integral equations. The progress was real, however, since two angular variables were separated in functions of three variables and optical thickness treated as a parameter in the nonlinear integral equations which could be solved by iterative methods on calculators of the 1940s. The transfer equation for the Stokes vector with the Rayleigh phase matrix in a finite atmosphere was solved by Chandrasekhar with a clever reduction of a vector problem to scalar *X*- and *Y*-functions, and results were related to the observations of Brewster and Babinet. These calculations for limited atmospheric thicknesses were extended in the 1960s by Sekera, Mullikin, and Kahle.

The interconnections of the linear transfer equation and associated nonlinear equations fit into a general framework identified by Bellman in the 1950s. Bellman, Wing, and others extended Chandrasekhar's and Ambartsumian's "invariance principles" to "invariant embedding," with applications to a variety of problems. Chandrasekhar's ideas are also generalized by Gohberg, Kailath, Casti, and others in the context of linear system theory, an area of research in which linear functional equations, Wiener-Hopf factorizations, bifurcation theory, and functional Riccati equations all play a role.

The portion of this volume devoted to radiative transfer problems shows the evolution and utility of Chandrasekhar's ideas in this subject. The continued development and application of his ideas attest to his mathematical insights.

Chandrasekhar speaks with fond recollections of his research in radiative transfer, from which he moved on to other research interests after publishing a book in 1950. Those who followed him in the study of radiative transfer owe him heartfelt thanks for his contributions to the subject, for his continued interest in the field while editor of the *Astrophysical Journal*, and for his kindness and encouragement in their own efforts to contribute to the subject of radiative transfer and to extensions of ideas generated within the subject.

T. W. MULLIKIN

Author's Note

The following remarks of a historical character are in the context of the last group of papers (in part 2) bearing on the continuous absorption coefficient of the negative ion of hydrogen (not included in Professor Mullikin's foreword).

The discovery (for it *was* a discovery) by Rupert Wildt in 1938, that the negative ion of hydrogen must be present in sufficient amounts to be the principal source of continuous absorption in the solar atmosphere and in the atmospheres of stars of spectral type later than A1, resolved an acute controversy that had persisted for several years during the thirties over the abundance of hydrogen relative to the metals. The resolution of the controversy can truly be described as a major event in modern astronomy. Nevertheless, Wildt's principal role in that event has largely been forgotten. This is neither the place nor the occasion to go into these historical matters. But one may obtain a flavor of the circumstances prevailing in those years (and an appreciation of the context in which the papers in part 2 were written) by the following translation of what A. Ünsold (one of the great pioneers in the theory of stellar atmospheres) has written in the second edition of his great treatise on *Physik der Sternatmosphären* (Springer Verlag, 1954, pp. 164–65).

> These ideas [relating to the determination of the abundances of the elements in stellar atmospheres] were incorporated in numerous calculations by H. N. Russell (1933), L. Biermann (1933), A. Ünsold (1935), and A. Pannekoek (1935). In these calculations, the following physical processes were taken into account: the "bound-free" and the "free-free" transitions of the electron in the hydrogen atom (H); and similar transitions of the outer [loosely bound] electrons in metals (M) evaluated more or less as in the case of hydrogen.
>
> The principal focus of the controversy was the abundance ratio, H : M, by numbers.
>
> H. N. Russell (1933) determined this ratio by the positions of the maxima of certain of the "spark lines" of the metals (by arguments by no means convincing) and by the observed intensities of several hydride bands in the solar spectrum. He derived a value for H : M in the range 1000 to 2000. A. Pannekoek (1935) took over into his calculations the ratio 1000 : 1.

A. Ünsold, on the other hand, considered the dependence on temperature of the discontinuity, $D = \Delta \log I$, at the limit (λ3700Å) of the Balmer series, particularly in stars with spectral types in the range *A* to *G*. If the atmospheres of the stars were purely of hydrogen, the discontinuity *D* must continuously *increase* with decreasing temperature, since the absorption from the second quantum state of hydrogen must decrease much less rapidly than the background absorption from the higher quantum states 3, 4, . . . The observed decrease of *D* below $T - 9000°K$ must then be accounted for by the enhancement, at lower temperatures, of the continuous absorption by the metals relative to that of hydrogen. [On this picture, *D* is directly related to the ratio of the number of hydrogen atoms in the second quantum state to the number of metal atoms contributing to the background absorption.] A quantitative evaluation of these effects led to a value H : M = 14-50—the spread in the ratio resulting only from the uncertainty in the quantal estimates of the absorption coefficient by the metals.

In 1938 R. Wildt successfully resolved these tangled contradictions by attributing to the negative ion of hydrogen—known at that time only to physical chemists and to "canal-ray" physicists—the principal contributory cause for continuous absorption in the atmospheres of the sun and the cooler stars, and showed, at the same time, how their presence required a much larger ratio for H : M. Quantal calculations by H. Bethe (1929) and E. A. Hylleraas (1930) had already shown [by variational methods] that the binding energy [i.e., the *electron affinity*] of a second electron to a neutral hydrogen atom must exceed 0.7 ev [as required by the adopted variational method]. The formation of H^- in the solar atmosphere, for example, is to be entirely expected, since neutral hydrogen atoms, in high concentration, are present together with electrons, provided by the metals of lower ionization potentials. The continuous absorption coefficient of H^- in the relevant spectral regions [in the visible and in the infrared] must also be high, since we are here concerned with its principal continuum. Quantal calculations of the continuous absorption coefficient of H^- by the bound-free transitions [derived from the ionization of H^- to neutral H] and by the free-free transitions [derived from the scattering of free electrons by neutral H] were carried out with ever increasing precision by C. K. Jen (1933), H. S. W. Massey, and D. R. Bates (1936), and finally by S. Chandrasekhar and F. H. Breen (1946) whose extensive tables provided the essential basis for the modern theory of the cooler stellar atmospheres.

For an account of the later developments on the physical aspects of the same problem, the reader may wish to consult D. R. Bates, *Other Men's Flowers* (*Physics Reports*, C, 35, no. 4 [January 1978], §5).

S. CHANDRASEKHAR

PART ONE

Radiative Transfer

1. *The Method of Discrete Ordinates*

Original page numbers appear in brackets at the bottoms of pages

1

ON THE RADIATIVE EQUILIBRIUM OF A STELLAR ATMOSPHERE. II

S. CHANDRASEKHAR
Yerkes Observatory
Received May 11, 1944

ABSTRACT

In this paper a new method is described for solving the various problems of radiative transfer in the theory of stellar atmospheres. The basic idea consists in expressing the integral (proportional to the density of radiation) which occurs in the equation of transfer as a sum according to Gauss's formula for numerical quadratures and replacing the integrodifferential equation by a system of linear equations. General solutions of this linear system can readily be written down, and this enables one to obtain solutions for the various problems to any desired degree of accuracy.

The method has been applied in detail to the problem considered in the earlier paper, and the first four approximations explicitly found. The corresponding laws of darkening have also been determined. An interesting by-product of the investigation is a new and entirely elementary proof of the exact Hopf-Bronstein relation, giving the boundary temperature in terms of the effective temperature.

1. *Introduction.*—In an earlier paper[1] it was shown how successively higher approximations to the solution of the equation of transfer

$$\mu \frac{dI}{d\tau} = I - \tfrac{1}{2} \int_{-1}^{+1} I \, d\mu \qquad (1)$$

can be obtained by expanding I in terms of spherical harmonics in the form

$$I(\tau, \mu) = \sum_{l=0}^{\infty} I_l(\tau) P_l(\mu) . \qquad (2)$$

More recently an important investigation by G. C. Wick[2] has come to the author's notice in which an alternative method for solving equations similar to (1) has been developed. It is the object of this paper to describe Wick's method in its astrophysical context and to show its particular adaptability for solving the standard problems of radiative transfer in the theory of stellar atmospheres.

2. *The outline of the method.*—Wick's basic idea consists in expressing the integral on the right-hand side of equation (1) as a sum, using for this purpose Gauss's formula for numerical quadratures.[3] Thus, denoting by $\mu_{-n}, \ldots, \mu_{-1}, \mu_1, \ldots, \mu_n$, and $\mu_{-i} = -\mu_i$ the $2n$ real roots of the Legendre polynomial $P_{2n}(\mu)$ of order $2n$, we can write, according to Gauss,

$$\int_{-1}^{+1} I(\tau, \mu) \, d\mu \simeq \sum_{j=-n}^{+n} a_j I(\tau, \mu_j) , \qquad (3)^{4}$$

[1] *Ap. J.*, **99**, 180, 1944. Referred to hereafter as "I."

[2] *Zs. f. Phys.*, **120**, 702, 1943.

[3] See, e.g., P. Frank and R. V. Misses, *Differentialgleichungen der Physik*, **1**, 394, New York, 1943.

[4] Note that the summation on the right-hand side does not include the term $j = 0$.

where the a_j's are certain weight factors. It may be further noted that

$$a_j = a_{-j}\,; \qquad \mu_j = -\mu_{-j}\,. \tag{4}$$

It is known that for a given number of subdivisions of the interval $(-1, +1)$ Gauss's choice of the points μ_j and the weights a_j yields the best value for the integral in the sense that for any *arbitrary* polynomial of degree $4n - 1$, the formula (3) is *exact*. In particular,

$$\sum_{j=1}^{n} a_j \mu_j^m = \int_0^1 \mu^m d\mu = \frac{1}{m+1}\,. \tag{5}$$

Accordingly, the representation of the integral as a finite sum of the form (3) can be made as accurate as may be desired by choosing n sufficiently large.

In the "nth approximation" we therefore replace equation (1) by the linear system of ordinary equation of order $2n$:

$$\mu_i \frac{dI_i}{d\tau} = I_i - \tfrac{1}{2}\Sigma a_j I_j \qquad (i = \pm 1, \ldots, \pm n)\,, \tag{6}$$

where, for the sake of brevity, we have written I_i for $I(\tau, \mu_i)$. Further, in equation (6) the summation over j is extended over all the positive and negative values.

3. *The general solution of the system of equations (6).*—We shall now find the different linearly independent solutions of the system (6) and later, by combining these, obtain the general solution.

First, we shall seek a solution of (6) of the form

$$I_i = g_i e^{-k\tau} \qquad (i = \pm 1, \ldots, \pm n)\,, \tag{7}$$

where the g_i's and k are constants, for the present unspecified. Introducing equation (7) into equation (6), we obtain the relation

$$g_i (1 + \mu_i k) = \tfrac{1}{2}\Sigma a_j g_j\,. \tag{8}$$

Hence,

$$g_i = \frac{\text{constant}}{1 + \mu_i k} \qquad (i = \pm 1, \ldots, \pm n)\,, \tag{9}$$

where the "constant" is independent of i. Substituting the foregoing form for g_i in equation (8), we obtain the following equation for k:

$$1 = \tfrac{1}{2}\Sigma \frac{a_j}{1 + \mu_j k}\,. \tag{10}$$

Remembering that $a_{-j} = a_j$ and $\mu_{-j} = -\mu_j$, we can re-write equation (10) as

$$1 = \sum_{j=1}^{n} \frac{a_j}{1 - \mu_j^2 k^2}\,. \tag{11}$$

It is thus seen that k^2 must satisfy an algebraic equation of order n. However, since

$$\sum_{j=1}^{n} a_j = 1 \tag{12}$$

(cf. eq. [5]), $k^2 = 0$ is a solution of equation (11). Accordingly, equation (10) has only $2n - 2$ distinct roots, which occur in pairs

$$\pm k_\alpha \qquad (\alpha = 1, \ldots, n-1)\,. \tag{13}$$

And, corresponding to these $2n-2$ roots, we have $2n-2$ independent solutions of equation (6). To complete the solution we notice that equation (6) admits of a solution of the form

$$I_i = b(\tau + q_i) \qquad (i = \pm 1, \ldots, \pm n), \quad (14)$$

where b is an arbitrary constant. For, inserting the form (14) for I_i in equation (6), we find that

$$\mu_i = q_i - \tfrac{1}{2}\Sigma a_j q_j ; \quad (15)$$

and this can be satisfied by

$$q_i = \mu_i + Q \qquad (i = \pm 1, \ldots, \pm n), \quad (16)$$

where Q is an arbitrary constant. Thus, the system (6) allows the solution

$$I_i = b(\tau + Q + \mu_i) \qquad (i = \pm 1, \ldots, \pm n), \quad (17)$$

where b and Q are two arbitrary constants.

Thus, combining solutions of the form (7) and (17), we have the general solution

$$I_i = b\left\{\sum_{a=1}^{n-1}\left[\frac{L_a e^{-k_a\tau}}{1+\mu_i k_a} + \frac{L_{-a} e^{+k_a\tau}}{1-\mu_i k_a}\right] + \mu_i + Q + \tau\right\}, \quad (18)$$

where b, $L_{\pm a}$, $(a = 1, \ldots, n-1)$, and Q are $2n$ arbitrary constants.

4. *The solution of equation (1) satisfying the necessary boundary conditions.*—For the astrophysical case under consideration the boundary conditions are[5] that none of the I_i's increase exponentially as $\tau \to \infty$ and that, further, there is no incident radiation on the surface $\tau = 0$. The first of these conditions implies that in the general solution (18) we omit all the terms in $\exp(+k_a\tau)$. Thus,

$$I_i = b\left\{\sum_{a=1}^{n-1}\frac{L_a e^{-k_a\tau}}{1+\mu_i k_a} + \mu_i + Q + \tau\right\} \qquad (i = \pm 1, \ldots, \pm n). \quad (19)$$

Next the nonexistence of any radiation incident on $\tau = 0$ requires that

$$I_{-i} = 0 \quad \text{at} \quad \tau = 0 \quad \text{and for} \quad i = 1, \ldots, n, \quad (20)$$

or, according to equations (4) and (19),

$$\sum_{a=1}^{n-1}\frac{L_a}{1-\mu_i k_a} + Q = \mu_i \qquad (i = 1, \ldots, n). \quad (21)$$

These are the n equations which determine the $(n-1)$ constants L_a and the further constant Q. The constant b is left arbitrary; and this, as we shall presently show, is related to the constant net flux in the atmosphere.

Now the net flux is defined in terms of F, where

$$F = 2\int_{-1}^{+1} I\mu\, d\mu. \quad (22)$$

Expressing the integral on the right-hand side as a sum over the $I_i\mu_i$'s according to Gauss's formula and using the solution (19) for the I_i's, we find

$$F = 2b\left\{\sum_{a=1}^{n-1} L_a e^{-k_a\tau}\sum_i \frac{a_i\mu_i}{1+\mu_i k_a} + \sum_i a_i\mu_i^2 + (Q+\tau)\sum_i a_i\mu_i\right\}. \quad (23)$$

[5] Cf. I, p. 182.

On the other hand (cf. eq. [5]),

$$\Sigma a_i \mu_i^2 = \tfrac{2}{3} \qquad \text{and} \qquad \Sigma a_i \mu_i = 0 \,. \tag{24}$$

Moreover,

$$\begin{aligned} \Sigma \frac{a_i \mu_i}{1+\mu_i k_\alpha} &= \frac{1}{k_\alpha} \Sigma a_i \left(1 - \frac{1}{1+\mu_i k_\alpha}\right) \\ &= \frac{1}{k_\alpha}\left(2 - \Sigma \frac{a_i}{1+\mu_i k_\alpha}\right), \end{aligned} \tag{25}$$

which vanishes according to equation (10). Hence,

$$F = \tfrac{4}{3} b = \text{constant} \,; \tag{26}$$

i.e., b is related to the constant net flux, as stated.

In terms of our solution for the I_i's we can obtain a convenient formula for J defined by

$$J = \tfrac{1}{2} \int_{-1}^{+1} I \, d\mu = \tfrac{1}{2} \Sigma a_i I_i \tag{27}$$

in our present approximation. We have

$$J = \tfrac{1}{2} b \left\{ \sum_{\alpha=1}^{n-1} L_\alpha e^{-k_\alpha \tau} \sum_i \frac{a_i}{1+\mu_i k_\alpha} + \sum_i a_i \mu_i + (Q+\tau) \sum_i a_i \right\} ; \tag{28}$$

or, according to equations (5), (10), (24), and (26), we have

$$J = \tfrac{3}{4} F \left(\tau + Q + \sum_{\alpha=1}^{n-1} L_\alpha e^{-k_\alpha \tau} \right) . \tag{29}$$

If we express J in its "normal" form (cf. I, eq. [47]),

$$J = \tfrac{3}{4} F \left(\tau + q[\tau] \right) , \tag{30}$$

we have

$$q(\tau) = Q + \sum_{\alpha=1}^{n-1} L_\alpha e^{-k_\alpha \tau} . \tag{31}$$

Finally, we may note that, according to the solution (29) for J, we have the law of darkening (cf. I, eq. [49]):

$$I(0, \mu) = \tfrac{3}{4} F \left(\mu + Q + \sum_{\alpha=1}^{n-1} \frac{L_\alpha}{1+k_\alpha \mu} \right) . \tag{32}$$

This completes the solution. The clear superiority of our present method over our earlier one of expanding I in terms of spherical harmonics is apparent. It is to be particularly noted that, in contrast to our earlier method, we can now write down the formal solution for any order of approximation quite generally.

5. *The first, second, third, and fourth approximations.*—We shall now obtain in their numerical forms the first four approximations to the solution for J and the corresponding laws of darkening.

i) *The first approximation.*—The first approximation is obtained by choosing $n = 1$, in which case

$$a_1 = a_{-1} = 1 \quad \text{and} \quad \mu_1 = -\mu_{-1} = \frac{1}{\sqrt{3}} . \tag{33}$$

There is no nonzero root for equation (10), and equation (21) now implies that

$$Q = \mu_1 = \frac{1}{\sqrt{3}} . \tag{34}$$

Accordingly,

$$q(\tau) = \frac{1}{\sqrt{3}} \tag{35}$$

and

$$I(0, \mu) = \tfrac{3}{4}F\left(\mu + \frac{1}{\sqrt{3}}\right) . \tag{36}$$

It is remarkable that in the very first approximation our method predicts a boundary value for $q(\tau)$ which is in *exact* agreement with the Hopf-Bronstein value (cf. I, eq. [73]). Actually, as we shall presently show (see § 6 below), this is identically the case in *all* approximations; and, consequently, we have here an essentially new and "elementary" proof of the Hopf-Bronstein relation.

One further remark about this first approximation may be made. The differential equations for I_1 and I_{-1} are

$$\left.\begin{aligned} \frac{1}{\sqrt{3}} \frac{dI_1}{d\tau} &= I_1 - \tfrac{1}{2}(I_1 + I_{-1}) , \\ -\frac{1}{\sqrt{3}} \frac{dI_{-1}}{d\tau} &= I_{-1} - \tfrac{1}{2}(I_1 + I_{-1}) , \end{aligned}\right\} \tag{37}$$

which are essentially the equations of Schwarzschild's first approximation.[6] However, the difference is that on the left-hand sides of the foregoing equations we now have $1/\sqrt{3}$ instead of the usual 1/2. It is interesting to speculate that, had Schwarzschild used our present systematic method, based on Gauss's formula, he might have discovered the exact boundary temperature some twenty-five years before Hopf and Bronstein!

ii) *The second approximation.*—To obtain the second approximation we have to choose for the μ_i's the zeros of $P_4(\mu)$ and the corresponding weight factors. We have[7]

$$\left.\begin{aligned} a_1 = a_{-1} &= 0.652145 ; \quad & \mu_1 = -\mu_{-1} &= 0.339981 , \\ a_2 = a_{-2} &= 0.347855 ; \quad & \mu_2 = -\mu_{-2} &= 0.861136 . \end{aligned}\right\} \tag{38}$$

Equation (11) reduces to

$$\mu_1^2 \mu_2^2 k^2 = a_1 \mu_1^2 + a_2 \mu_2^2 = \tfrac{1}{3} . \tag{39}$$

Hence,

$$k_1 = \frac{1}{\sqrt{3}\,\mu_1 \mu_2} = 1.972027 . \tag{40}$$

Solving for Q and L_1, we find

$$Q = 0.694025 ; \quad L_1 = -0.116675 . \tag{41}$$

[6] Cf. E. A. Milne, *Handb. d. Ap.*, **3,** No. 1, 114–116, Berlin: Springer, 1930.

[7] A. N. Lowan, N. Davids, and A. Levenson, *Bull. Amer. Math. Soc.*, **48,** 739, 1942.

Accordingly, on this approximation

$$q(\tau) = 0.694025 - 0.116675 e^{-1.97203\tau} \tag{42}$$

and

$$I(0, \mu) = \tfrac{3}{4}F\left(\mu + 0.694025 - \frac{0.116675}{1+1.97203\mu}\right). \tag{43}$$

iii) *The third approximation.*—We now have

$$\left.\begin{aligned} a_1 &= a_{-1} = 0.467914\ ; & \mu_1 &= -\mu_{-1} = 0.238619\ , \\ a_2 &= a_{-2} = 0.360762\ ; & \mu_2 &= -\mu_{-2} = 0.661209\ , \\ a_3 &= a_{-3} = 0.171324\ ; & \mu_3 &= -\mu_{-3} = 0.932470\ . \end{aligned}\right\} \tag{44}$$

The equation for k^2 reduces to

$$0.02164502k^4 - 0.254545k^2 + \tfrac{1}{3} = 0\ , \tag{45}$$

the positive roots of which are

$$k_1 = 3.202945 \quad \text{and} \quad k_2 = 1.225211\ . \tag{46}$$

Solving for Q, L_1, and L_2, we find

$$Q = 0.703899\ , \quad L_1 = -0.101245\ , \quad L_2 = -0.02530\ . \tag{47}$$

Hence,

$$q(\tau) = 0.703899 - 0.101245 e^{-3.20295\tau} - 0.02530 e^{-1.22521\tau} \tag{48}$$

and

$$I(0, \mu) = \tfrac{3}{4}F\left(\mu + 0.703899 - \frac{0.101245}{1+3.20295\mu} - \frac{0.02530}{1+1.22521\mu}\right). \tag{49}$$

iv) *The fourth approximation.*—To obtain the fourth approximation we have to choose for the μ_i's the roots of $P_8(\mu)$ and the corresponding weight factors. We have

$$\left.\begin{aligned} a_1 &= a_{-1} = 0.362684\ ; & \mu_1 &= -\mu_{-1} = 0.183435\ , \\ a_2 &= a_{-2} = 0.313707\ ; & \mu_2 &= -\mu_{-2} = 0.525532\ , \\ a_3 &= a_{-3} = 0.222381\ ; & \mu_3 &= -\mu_{-3} = 0.796666\ , \\ a_4 &= a_{-4} = 0.101229\ ; & \mu_4 &= -\mu_{-4} = 0.960290\ . \end{aligned}\right\} \tag{50}$$

The equation for k^2 reduces to

$$0.00543900k^6 - 0.1284982k^4 + 0.422222k^2 - \tfrac{1}{3} = 0\ , \tag{51}$$

the positive roots of which are

$$k_1 = 4.45808\ ; \quad k_2 = 1.59178\ ; \quad k_3 = 1.10319\ . \tag{52}$$

The constants Q, L_1, L_2, and L_3 were found to be

$$\left.\begin{aligned} Q &= 0.706920\ ; & L_1 &= -0.083921\ ; \\ L_2 &= -0.036187\ ; & L_3 &= -0.009461\ . \end{aligned}\right\} \tag{53}$$

Accordingly,

$$q(\tau) = 0.70692 - 0.08392 e^{-4.45808\tau} - 0.03619 e^{-1.59178\tau} - 0.00946 e^{-1.10319\tau} \tag{54}$$

and

$$I(0, \mu) = \tfrac{3}{4}F\left(\mu + 0.70692 - \frac{0.08392}{1+4.45808\mu} - \frac{0.03619}{1+1.59178\mu} - \frac{0.00946}{1+1.10319\mu}\right). \quad (55)$$

In Tables 1 and 2 we have tabulated the functions $q(\tau)$ and $I(0, \mu)/F$ according to our second, third, and fourth approximations. It is also seen that, in agreement with our

TABLE 1

THE FUNCTION $q(\tau)$ DERIVED ON THE BASIS OF THE SECOND, THIRD, AND FOURTH APPROXIMATIONS (EQS. [42], [48], AND [54])

τ	$q(\tau)$			τ	$q(\tau)$		
	Second Approximation	Third Approximation	Fourth Approximation		Second Approximation	Third Approximation	Fourth Approximation
0.00	0.5774	0.5774	0.5774	0.90	0.6743	0.6898	0.6933
.05	.5883	.5938	.5974	1.0	.6778	.6924	.6954
.10	.5982	.6080	.6139	1.2	.6831	.6959	.6987
.15	.6072	.6202	.6274	1.4	.6867	.6982	.7008
.20	.6154	.6307	.6386	1.6	.6891	.6997	.7024
.25	.6228	.6398	.6479	1.8	.6907	.7008	.7035
.30	.6295	.6477	.6557	2.0	.6918	.7016	.7044
.35	.6355	.6544	.6621	2.2	.6925	.7021	.7050
.40	.6410	.6603	.6676	2.4	.6930	.7025	.7055
.50	.6505	.6698	.6761	2.6	.6933	.7028	.7058
.60	.6583	.6770	.6823	2.8	.6936	.7031	.7064
.70	.6647	.6824	.6870	3.0	.6937	.7033	.7065
0.80	0.6699	0.6866	0.6905	∞	0.6940	0.7039	0.7069

TABLE 2

THE LAWS OF DARKENING GIVEN BY THE SECOND, THIRD AND FOURTH APPROXIMATIONS (EQS. [43], [49], AND [55])

μ	$I(0, \mu)/F$		
	Second Approximation	Third Approximation	Fourth Approximation
0.0	0.4330	0.4330	0.4330
0.1	0.5224	0.5285	0.5319
0.2	0.6078	0.6164	0.6205
0.3	0.6905	0.7003	0.7046
0.4	0.7716	0.7819	0.7861
0.5	0.8515	0.8620	0.8660
0.6	0.9304	0.9410	0.9449
0.7	1.0088	1.0193	1.0231
0.8	1.0866	1.0970	1.1007
0.9	1.1640	1.1743	1.1779
1.0	1.2411	1.2513	1.2548

earlier remarks, $q(0)$ agrees with the exact value $1/\sqrt{3}$ in all our approximations. (For a proof of this relation see § 6 below.) Moreover, a comparison of the law of darkening on our fourth approximation with the exact values given in I, Table 2, indicates that in this approximation we have reached an over-all accuracy of about one part in two hundred.

6. *A proof of the Hopf-Bronstein relation* J (0) = $(\sqrt{3}/4)$ F.—In the preceding section we have verified that $q(0)$ agrees with the Hopf-Bronstein value $1/\sqrt{3}$ in all the four approximations we have numerically worked out. We shall now show how this result can be demonstrated to be quite generally and rigorously true. In order to do this, we start by considering the function

$$S(\mu) = \sum_{a=1}^{n-1} \frac{L_a}{1-k_a\mu} + Q - \mu . \tag{56}$$

According to the boundary conditions (21),

$$S(\mu_i) = 0 \qquad (i = 1, \ldots, n) . \tag{57}$$

This fact enables us to determine $S(\mu)$ explicitly. For, by multiplying equation (56) by the function

$$R(\mu) = (1-k_1\mu)(1-k_2\mu) \ldots (1-k_{n-1}\mu) , \tag{58}$$

we obtain a polynomial of degree n in μ which vanishes for $\mu = \mu_i$, $i = 1, \ldots, n$. Accordingly, $S(\mu)R(\mu)$ cannot differ from the polynomial

$$P(\mu) = (\mu-\mu_1)(\mu-\mu_2) \ldots (\mu-\mu_n) \tag{59}$$

by more than a constant factor; and this factor can be determined by comparing the coefficients of the highest power of μ (namely, μ^n) in $P(\mu)$ and $S(\mu)R(\mu)$. In the former it is unity, while in the latter it is

$$(-1)^n k_1 k_2 \ldots k_{n-1} . \tag{60}$$

Hence,

$$S(\mu) = (-1)^n k_1 k_2 \ldots k_{n-1} \frac{P(\mu)}{R(\mu)} . \tag{61}$$ [8]

From equations (31), (56), and (61) we now conclude that

$$q(0) = \sum_{a=1}^{n-1} L_a + Q = S(0) = (-1)^n k_1 k_2 \ldots k_{n-1} \frac{P(0)}{R(0)} . \tag{62}$$

On the other hand, according to our definitions of the functions $P(\mu)$ and $R(\mu)$,

$$P(0) = (-1)^n \mu_1 \mu_2 \ldots \mu_n ; \qquad R(0) = 1 . \tag{63}$$

Hence,

$$q(0) = k_1 k_2 \ldots k_{n-1} \mu_1 \mu_2 \ldots \mu_n . \tag{64}$$

We shall now show how the quantity on the right-hand side of equation (64) can be explicitly evaluated from the equation for the roots $k_1, \ldots, k_{n-1}$. We have (eq. [11])

$$1 = \sum_{j=1}^{n} \frac{a_j}{1-k^2\mu_j^2} . \tag{65}$$

Multiplying this equation by the product of the factors $(1-k^2\mu_1^2), \ldots, (1-k^2\mu_n^2)$, we obtain

$$\left.\begin{aligned} (-1)^{n-1}\mu_1^2\mu_2^2 \ldots \mu_n^2 k^{2n} + \ldots - \sum_{i=1}^{n} a_i \left(\sum_{j=1}^{n} \mu_j^2 - \mu_i^2 \right) k^2 + \sum_{i=1}^{n} \mu_i^2 k^2 \\ + \sum_{i=1}^{n} a_i - 1 = 0 ; \end{aligned}\right\} \tag{66}$$

[8] This relation can be used for a direct evaluation of the constants L_a without going through a routine solution of linear equations (21).

or, using equation (5), we have

$$(-1)^{n-1}\mu_1^2\mu_2^2 \dots \mu_n^2 k^{2n-2} + \dots + \tfrac{1}{3} = 0 . \tag{67}$$

Hence the product of the roots $k_1^2 \dots k_{n-1}^2$ is given by

$$k_1^2 k_2^2 \dots k_{n-1}^2 = \frac{1}{3\mu_1^2\mu_2^2 \dots \mu_n^2} \tag{68}$$

or

$$k_1 k_2 \dots k_{n-1} = \frac{1}{\sqrt{3}\,\mu_1\mu_2 \dots \mu_n} . \tag{69}$$

Combining equations (64) and (69), we have

$$q(0) = \frac{1}{\sqrt{3}} , \tag{70}$$

a result which is thus seen to be true in all orders of approximation. We therefore conclude that equation (70) represents an exact relation.

Finally, it is to be noted that equations (30) and (70) imply that

$$J(0) = \frac{\sqrt{3}}{4} F , \tag{71}$$

which is the well-known relation of Hopf and Bronstein.

7. *Further applications of the method.*—Our analysis in the preceding sections has demonstrated the extreme simplicity with which solutions accurate to any desired extent can be obtained. But the usefulness of the method is by no means limited to the particular problem which has been considered. Indeed, the possible applications of the method are so numerous that it would hardly be possible to consider all of them within the limits of a single paper. We shall therefore content ourselves with a brief consideration of two further standard problems in the theory of radiative transfer, postponing to a later occasion the more detailed discussion of the various solutions.

i) *The radiative equilibrium of a planetary nebula.*[9]—As was first pointed out by Ambarzumian, the equation of transfer for the "ultraviolet" radiation (i.e., radiation beyond the head of the Lyman series) consistent with Zanstra's theory is

$$\mu \frac{dI}{d\tau} = I - \tfrac{1}{2} p \int_{-1}^{+1} I \, d\mu - \tfrac{1}{4} p S e^{-(\tau_1 - \tau)} , \tag{72}$$

where p is a certain factor less than unity, τ_1 the optical thickness of the nebula for the ultraviolet radiation, and πS the amount of ultraviolet radiant energy incident on each square centimeter of the inner surface of the nebula (i.e., at $\tau = \tau_1$).

Again, in equation (72), we approximate the integral occurring on the right-hand side by a sum according to Gauss's formula for numerical quadratures. And in this manner we replace equation (72) by the system of $2n$ linear equations

$$\mu_i \frac{dI_i}{d\tau} = I_i - \tfrac{1}{2} p \Sigma a_j I_j - \tfrac{1}{4} p S e^{-(\tau_1 - \tau)} \quad (i = \pm 1, \dots, \pm n) , \tag{73}$$

in the nth approximation. We shall now briefly indicate how the general solution of this linear system of equations can be obtained.

[9] For earlier discussions of this problem see V. A. Ambarzumian, *Pulkovo Obs. Bull.*, No. 13, and *M.N.*, **93**, 50, 1931; also S. Chandrasekhar, *Zs. f. Ap.*, **9**, 266, 1935.

Setting

$$I_i = S\,[\,g_i e^{-k\tau} + h_i e^{-(\tau_1-\tau)}\,] \qquad (i = \pm 1, \ldots, \pm n) \quad (74)$$

in equation (73) (where g_i, h_i, and k are constants), we find

$$g_i\,(1 + k\,\mu_i) = \tfrac{1}{2}p\Sigma a_j g_j \qquad (75)$$

and

$$h_i\,(1 - \mu_i) = \tfrac{1}{2}p\Sigma a_j h_j + \tfrac{1}{4}p\,. \qquad (76)$$

Equation (75) implies that

$$g_i = \frac{\text{constant}}{1 + \mu_i k} \qquad (i = \pm 1, \ldots, \pm n) \quad (77)$$

and that k is a root of the equation

$$1 = p \sum_{j=1}^{n} \frac{a_j}{1 - \mu_j^2 k^2}\,. \qquad (78)$$

Since $p < 1$, the foregoing equation admits of $2n$ distinct roots, which occur in pairs as

$$\pm k_\alpha \qquad (\alpha = 1, \ldots, n)\,. \quad (79)$$

Considering next equation (76), we observe that h_i must be expressible in the form

$$h_i = \frac{\beta}{1 - \mu_i} \qquad (i = \pm 1, \ldots, \pm n)\,, \quad (80)$$

where β is a constant which must, in turn, be so chosen that

$$\beta = \tfrac{1}{2}p\beta\Sigma \frac{a_j}{1 - \mu_j} + \tfrac{1}{4}p\,. \qquad (81)$$

Hence,

$$\beta = \tfrac{1}{4}p \frac{1}{1 - p \sum_{j=1}^{n} \frac{a_j}{1 - \mu_j^2}}\,. \qquad (82)$$

Accordingly, the general solution of equation (73) can be written in the form

$$I_i = S \left\{ \sum_{\alpha=1}^{n} \left[\frac{L_\alpha e^{-k_\alpha \tau}}{1 + \mu_i k_\alpha} + \frac{L_{-\alpha} e^{+k_\alpha \tau}}{1 - \mu_i k_\alpha} \right] + \frac{p\, e^{-(\tau_1 - \tau)}}{4\,(1 - \mu_i) \left(1 - p \sum_{j=1}^{n} \frac{a_j}{1 - \mu_j^2} \right)} \right\}, \quad (83)$$

where $L_{\pm\alpha}$, $\alpha = 1, \ldots, n$, are $2n$ constants of integration. For the problem under consideration the solution (83) becomes determinate when the boundary conditions at $\tau = \tau_1$ and at $\tau = 0$ are taken into account. These are

$$I_i = I_{-i} \quad \text{at} \quad \tau = \tau_1 \quad \text{for} \quad i = 1, \ldots, n \qquad (84)$$

and

$$I_{-i} = 0 \quad \text{at} \quad \tau = 0 \quad \text{for} \quad i = 1, \ldots, n\,. \qquad (85)$$

The conditions (84) arise from the geometry of the problem, as was first pointed out by Milne,[10] while the conditions (85) arise from the nonexistence of any radiation incident on $\tau = 0$. The explicit form which these conditions take can be readily written down in

[10] *Zs. f. Ap.*, **1**, 98, 1930.

terms of the general solution (83). We shall not continue with this discussion further, but it is apparent how solutions of any desired degree of accuracy can be obtained in this manner.

ii) *The standard case in the theory of the formation of absorption lines.*—In a standard notation[11] the equation of transfer appropriate for this problem is

$$\mu \frac{dI_\nu}{dt_\nu} = I_\nu - \frac{(1-\epsilon)\,\eta_\nu}{2(1+\eta_\nu)} \int_{-1}^{+1} I_\nu d\mu - \frac{1+\epsilon\eta_\nu}{1+\eta_\nu}(a_{\nu_0} + b_{\nu_0} t_\nu) \,. \tag{86}$$

In the "standard case" the ratio η_ν of the line to the continuous absorption coefficients is assumed to be constant. Suppressing the suffix ν and introducing the quantity

$$\lambda = \frac{1+\epsilon\eta}{1+\eta}\,, \tag{87}$$

we can re-write equation (86) more conveniently in the form

$$\mu \frac{dI}{dt} = I - \tfrac{1}{2}(1-\lambda) \int_{-1}^{+1} I d\mu - \lambda (a + bt) \,. \tag{88}$$

In the nth approximation we replace the foregoing equation by the system of linear equations

$$\mu_i \frac{dI_i}{dt} = I_i - \tfrac{1}{2}(1-\lambda)\Sigma a_j I_j - \lambda(a+bt) \qquad (i = \pm 1, \ldots, \pm n) \,. \tag{89}$$

Proceeding as before, we verify that the solution of this system appropriate for the problem on hand is

$$I_i = \sum_{a=1}^{n} \frac{L_a e^{-k_a t}}{1+\mu_i k_a} + (\mu_i b + a + bt) \qquad (i = \pm 1, \ldots, \pm n) \,, \tag{90}$$

where the k_a's are the n positive roots of the equation

$$1 = (1-\lambda) \sum_{j=1}^{n} \frac{a_j}{1-\mu_j^2 k^2} \tag{91}$$

and the L_a's ($a = 1, \ldots, n$) are the n constants of integration to be determined by the boundary conditions

$$\sum_{a=1}^{n} \frac{L_a}{1-\mu_i k_a} + a = \mu_i b \qquad (i = 1, \ldots, n) \,. \tag{92}$$

Again solutions to any desired degree of accuracy can be obtained.

In conclusion, we should further like to point out that the methods developed in this paper can readily be applied also to problems in which the scattering does not take place isotropically, as, for example, in the case of scattering by free electrons. But we postpone a discussion of this problem to a later occasion.

I am indebted to Miss Frances Herman, who carried out the numerical work involved in the preparation of Tables 1 and 2.

[11] See, e.g., B. Strömgren, *Ap. J.*, **86**, 1, 1937.

2

ON THE RADIATIVE EQUILIBRIUM OF A STELLAR ATMOSPHERE. V

S. Chandrasekhar

Yerkes Observatory

Received October 30, 1944

ABSTRACT

In this paper the methods developed in earlier papers are extended to solving the problem of radiative transfer in curved atmospheres, i.e., to solving the equation of transfer

$$\mu \frac{\partial I}{\partial r} + \frac{1-\mu^2}{r^2} \frac{\partial I}{\partial \mu} = -\kappa\rho I + \tfrac{1}{2}\kappa\rho \int_{-1}^{+1} I d\mu ,$$

where $\kappa\rho$ is a function only of r. After outlining a general method for replacing this partial integrodifferential equation by an equivalent system of $2n$ linear equations in the nth approximation, the most convenient forms of the equations for the first two approximations are found. The equations of the second approximation for the astrophysically important case $\kappa\rho \propto r^{-n}$ $(n > 1)$ are explicitly solved and found to involve quadratures over Bessel functions of purely imaginary arguments. For the case $n = 2$ the solutions have been found in their numerical forms. Finally, the physically interesting case of diffusion through a homogeneous sphere is also considered.

1. *Introduction.*—In the earlier papers of this series[1] an attempt has been made to present the solutions to the various plane problems of radiative transfer in the theory of stellar atmospheres in forms which would enable one to derive results to any desired degree of accuracy without much difficulty. In this paper we propose to extend this discussion to include the case of the radiative equilibrium of "extended atmospheres" in which it is not permissible to ignore the curvature of the outer layers. First approximations to the solution of this latter problem in curved atmospheres have been given by N. A. Kosirev[2] and the writer.[3] But all attempts to improve on the "first approximations" given by these writers have so far proved unsuccessful.[4] However, in view of the fact that the theory of extended atmospheres is finding increasing applications to a variety of practical problems,[5] it would appear worth while to re-examine the basic problem with a view toward developing systematic methods of approximation for obtaining solutions of higher accuracy. This is the object of this paper.

2. *The reduction of the equation of transfer to an equivalent system of* 2n *linear equations in the* nth *approximation.*—Let r denote the distance measured outward from the center of symmetry of the atmosphere and ϑ the angle measured from the positive direction of the radius vector. The equation of transfer which we have to deal with is

$$\cos\vartheta \frac{\partial I}{\partial r} - \frac{\sin\vartheta}{r} \frac{\partial I}{\partial \vartheta} = -\kappa\rho I + \tfrac{1}{2}\kappa\rho \int_0^\pi I(r, \vartheta) \sin\vartheta d\vartheta , \qquad (1)$$

where I, κ, and ρ have their usual meanings. Writing μ for $\cos\vartheta$, we can re-write equation (1) in the form

$$\mu \frac{\partial I}{\partial r} + \frac{1-\mu^2}{r} \frac{\partial I}{\partial \mu} = -\kappa\rho I + \tfrac{1}{2}\kappa\rho \int_{-1}^{+1} I(r, \mu) d\mu . \qquad (2)$$

[1] *Ap. J.*, **99**, 180; **100**, 76, 117, and 355, 1944. These papers will be referred to as "I," "II," "III," and "IV," respectively.

[2] *M.N.*, **94**, 430, 1934.

[3] S. Chandrasekhar, *M.N.*, **94**, 444, 1934; see also *Proc. Cambridge Phil. Soc.*, **31**, 390, 1935, and *Russian Astr. J.*, **11**, 550, 1934.

[4] Cf. L. Gratton, *Soc. astr. italiana*, **10**, 309, 1937.

[5] See F. L. Whipple and C. Payne-Gaposchkin, *Harvard Circ.*, No. 413, 1936, and C. Payne-Gaposchkin and Sergei Gaposchkin, *Ap. J.*, **101**, 56, 1945.

According to the ideas developed in papers II and III, we replace the integral which occurs on the right-hand side of this equation by a sum according to Gauss's formula for numerical quadratures. This permits the replacement of the integrodifferential equation (2) by a system of linear equations which in the nth approximation is

$$\mu_i \frac{dI_i}{dr} + \frac{1-\mu_i^2}{r}\left(\frac{\partial I}{\partial \mu}\right)_{\mu=\mu_i} = -\kappa \rho I_i + \tfrac{1}{2}\kappa\rho \Sigma a_j I_j \quad (i = \pm 1, \ldots, \pm n), \quad (3)$$

where we have used I_j to denote $I(r, \mu_j)$. Further, as in papers II, III, and IV, the μ_j's are the zeros of the Legendre polynomial $P_{2n}(\mu)$ and the a_j's are the appropriate Gaussian weights. It is at once seen that our present system of equations (cf. eq. [3]) differs in an essential way from those which occurred in our earlier studies on the plane problems; for equation (3) now involves $(\partial I/\partial\mu)_{\mu=\mu_i}$, and before we can proceed any further we must know the values which we are to assign to $\partial I/\partial\mu$ at the points of the Gaussian division μ_i in our present scheme of approximation. At first it might be supposed that the assignment of values to $\partial I/\partial\mu$ at $\mu = \mu_i$, $i = \pm 1, \ldots, \pm n$, is largely arbitrary, particularly when n is small. However, on consideration it appears that this assignment can be done in a satisfactory manner in only one way and, indeed, according to the following device:

Define the polynomials $Q_m(\mu)$ according to the formula

$$P_m(\mu) = -\frac{dQ_m}{d\mu} \qquad (m = 1, \ldots, 2n), \quad (4)$$

and adjust the constant of integration in Q_m by requiring that

$$Q_m = 0 \quad \text{for} \quad |\mu| = 1 . \qquad (5)$$

This can always be accomplished, since when m is odd, Q_m is even, and when m is even, the indefinite integral of $P_m(\mu)$ already contains $(1 - \mu^2)$ as a factor.[6] The first few of the polynomials $Q_m(\mu)$ are given in Table 1.

TABLE 1

m	$P_m(\mu)$	$Q_m(\mu)$	$\mathcal{Q}_m(\mu)$
1	μ	$\frac{1}{2}(1-\mu^2)$	$\frac{1}{2}$
2	$\frac{1}{2}(3\mu^2-1)$	$\frac{1}{2}\mu(1-\mu^2)$	$\frac{1}{2}\mu$
3	$\frac{1}{2}(5\mu^3-3\mu)$	$\frac{1}{8}(5\mu^2-1)(1-\mu^2)$	$\frac{1}{8}(5\mu^2-1)$
4	$\frac{1}{8}(35\mu^4-30\mu^2+3)$	$\frac{1}{8}\mu(7\mu^2-3)(1-\mu^2)$	$\frac{1}{8}\mu(7\mu^2-3)$
5	$\frac{1}{8}(63\mu^5-70\mu^3+15\mu)$	$\frac{1}{16}(21\mu^4-14\mu^2+1)(1-\mu^2)$	$\frac{1}{16}(21\mu^4-14\mu^2+1)$

Now by an integration by parts we arrive at the identity

$$\int_{-1}^{+1} Q_m(\mu)\frac{\partial I}{\partial\mu}\,d\mu = -\int_{-1}^{+1} I\,\frac{dQ_m}{d\mu}\,d\mu = \int_{-1}^{+1} IP_m(\mu)\,d\mu . \qquad (6)$$

Expressing the first and the last integrals in equation (6) as sums according to Gauss's formula, we have in the nth approximation

$$\Sigma a_i Q_m(\mu_i)\left(\frac{\partial I}{\partial\mu}\right)_{\mu=\mu_i} = \Sigma a_i I_i P_m(\mu_i) \quad (m = 1, \ldots, 2n). \quad (7)$$

Equation (7) provides us with exactly the right number of equations to express

[6] Actually, $Q_m(\mu) = [(P_{m-1} - P_{m+1})/(2\mu + 1)] +$ constant.

$(\partial I/\partial\mu)_{\mu=\mu_i}$, $i = \pm 1, \ldots, \pm n$, as linear combinations of I_i.[7] Accordingly, equations (3) and (7), together, provide the required reduction of the equation of transfer (2) to an equivalent system of linear equations for the I_i's in the nth approximation.

For purposes of practical solution it appears most convenient to combine equations (3) and (7) in the following manner:

Since we have arranged $Q_m(\mu)$ to be divisible by $1 - \mu^2$, we can clearly write

$$Q_m(\mu) = \mathcal{Q}_m(\mu)(1-\mu^2). \tag{8}$$

The first few of the polynomials $\mathcal{Q}_m(\mu)$, defined according to equation (8), are also listed in Table 1.

Now multiply equation (3) by $a_i\mathcal{Q}_m(\mu_i)$ and sum over all i's. We obtain

$$\left.\begin{aligned}\frac{d}{dr}\Sigma a_i\mu_i\mathcal{Q}_m(\mu_i)I_i + \frac{1}{r}\Sigma a_i(1-\mu_i^2)\mathcal{Q}_m(\mu_i)\left(\frac{\partial I}{\partial\mu}\right)_{\mu=\mu_i} &= -\kappa\rho\Sigma a_i\mathcal{Q}_m(\mu_i)I_i \\ +\tfrac{1}{2}\kappa\rho(\Sigma a_i I_i)[\Sigma a_i\mathcal{Q}_m(\mu_i)] \qquad & (m = 1, \ldots, 2n).\end{aligned}\right\} \tag{9}$$

But, according to equations (7) and (8),

$$\left.\begin{aligned}\Sigma a_i(1-\mu_i^2)\mathcal{Q}_m(\mu_i)\left(\frac{\partial I}{\partial\mu}\right)_{\mu=\mu_i} &= \Sigma a_i Q_m(\mu_i)\left(\frac{\partial I}{\partial\mu}\right)_{\mu=\mu_i} \\ &= \Sigma a_i P_m(\mu_i)I_i;\end{aligned}\right\} \tag{10}$$

and equation (9) reduces to

$$\left.\begin{aligned}\frac{d}{dr}\Sigma a_i\mu_i\mathcal{Q}_m(\mu_i)I_i + \frac{1}{r}\Sigma a_i P_m(\mu_i)I_i &= -\kappa\rho\Sigma a_i\mathcal{Q}_m(\mu_i)I_i \\ +\tfrac{1}{2}\kappa\rho(\Sigma a_i I_i)[\Sigma a_i\mathcal{Q}_m(\mu_i)] \qquad & (m = 1, \ldots, 2n),\end{aligned}\right\} \tag{11}$$

which is the required system of linear equations in the nth approximation.

Equation (11) for the case $m = 1$ admits of immediate integration. For, when $m = 1$,

$$P_1(\mu) = \mu \quad \text{and} \quad \mathcal{Q}_1(\mu) = \tfrac{1}{2}, \tag{12}$$

and equation (11) yields

$$\frac{1}{2}\frac{d}{dr}\Sigma a_i\mu_i I_i + \frac{1}{r}\Sigma a_i\mu_i I_i = 0. \tag{13}$$

Accordingly,

$$\Sigma a_i\mu_i I_i = \frac{1}{2}\frac{F_0}{r^2}, \tag{14}$$

where F_0 is a constant of integration. Equation (14) is the expression in our present approximation of the flux integral

$$F = 2\int_{-1}^{+1} I\mu\, d\mu = \frac{F_0}{r^2}, \tag{15}$$

which the equation of transfer (2) admits directly.

Again, since $\mathcal{Q}_m(\mu)$ is odd when m is even, equation (11) reduces for even values of m to the form

$$\left.\begin{aligned}\frac{d}{dr}\Sigma a_i\mu_i\mathcal{Q}_m(\mu_i)I_i + \frac{1}{r}\Sigma a_i P_m(\mu_i)I_i &= -\kappa\rho\Sigma a_i\mathcal{Q}_m(\mu_i)I_i \\ & (m = 2, 4, \ldots, 2n).\end{aligned}\right\} \tag{16}$$

[7] Essentially what eq. (7) allows is to determine in a "best possible way" the derivatives of a function in terms of its values at the points of the Gaussian division. This problem has apparently not been considered before.

For $m = 2n$ the foregoing equation further simplifies to

$$\frac{d}{dr}\Sigma a_i\mu_i\mathcal{Q}_{2n}(\mu_i)I_i = -\kappa\rho\Sigma a_i\mathcal{Q}_{2n}(\mu_i)I_i\,, \tag{17}$$

since the μ_i's are by definition the zeros of the polynomial $P_{2n}(\mu)$.

Finally, for further reference we may note here the explicit forms of equation (11) for the first few values of m. Using the definition of the polynomials P_m and $\mathcal{Q}_m$ given in Table 1, we find

$$\frac{d}{dr}\Sigma a_i\mu_i I_i + \frac{2}{r}\Sigma a_i\mu_i I_i = 0\,, \tag{18}$$

$$\frac{d}{dr}\Sigma a_i\mu_i^2 I_i + \frac{1}{r}\Sigma a_i(3\mu_i^2 - 1)I_i = -\kappa\rho\Sigma a_i\mu_i I_i\,, \tag{19}$$

$$\frac{d}{dr}\Sigma a_i\mu_i(5\mu_i^2 - 1)I_i + \frac{4}{r}\Sigma a_i\mu_i(5\mu_i^2 - 3)I_i = -\tfrac{5}{3}\kappa\rho\Sigma a_i(3\mu_i^2 - 1)I_i\,, \tag{20}$$

$$\frac{d}{dr}\Sigma a_i\mu_i^2(7\mu_i^2 - 3)I_i + \frac{1}{r}\Sigma a_i(35\mu_i^4 - 30\mu_i^2 + 3)I_i = -\kappa\rho\Sigma a_i\mu_i(7\mu_i^2 - 3)I_i\,, \tag{21}$$

et cetera.

3. *The first approximation.*—In the first approximation we consider equation (11) for $m = 1$ and 2 only, with (cf. II, eq. [33])

$$a_1 = a_{-1} = 1 \qquad \text{and} \qquad \mu_1 = -\mu_{-1} = \frac{1}{\sqrt{3}}\,. \tag{22}$$

We have (cf. eqs. [14] and [19])

$$I_1 - I_{-1} = \frac{\sqrt{3}}{2}\frac{F_0}{r^2} \tag{23}$$

and

$$\frac{1}{3}\frac{d}{dr}(I_1 + I_{-1}) = -\kappa\rho\frac{1}{\sqrt{3}}(I_1 - I_{-1})\,. \tag{24}$$

Combining equations (23) and (24), we have

$$\frac{d}{dr}(I_1 + I_{-1}) = -\frac{3}{2}\kappa\rho\frac{F_0}{r^2}\,. \tag{25}$$

Hence,

$$I_1 + I_{-1} = \frac{3}{2}F_0\int_r^R \frac{\kappa\rho\, dr}{r^2} + \text{constant}\,, \tag{26}$$

where $r = R$ defines the extent of the atmosphere. The constant of integration in equation (26) is determined by the condition that $I_{-1} = 0$ at $r = R$. In this manner we obtain for the source function J the solution

$$J = \tfrac{1}{2}(I_1 + I_{-1}) = \frac{\sqrt{3}}{4}\frac{F_0}{R^2} + \frac{3}{4}F_0\int_r^R \frac{\kappa\rho\, dr}{r^2}\,, \tag{27}$$

which is to be compared with the solution given earlier by Kosirev[8] and by Chandrasekhar.[9]

For an atmosphere which extends to infinity we should require that both I_1 and I_{-1} tend to zero as $r \to \infty$. In this case the solution for J reduces to

$$J = \frac{3}{4}F_0\int_r^\infty \frac{\kappa\rho\, dr}{r^2}\,, \tag{28}$$

[8] *M.N.*, **94,** 430, 1934. See particularly eq. (8) in this paper.

[9] *M.N.*, **94,** 444, 1934. See eqs. (49)–(51).

or, alternatively,

$$J=\frac{3}{4}F_0\int_0^{\tau}\frac{d\tau}{r^2}, \tag{29}$$

where τ denotes the radial optical thickness. It is in the form (29) that the solution to the problem of the extended atmospheres has been used in practice.

4. *The equations for the second approximation.*—In the second approximation we choose for the μ_i's the zeros of $P_4(\mu)$ and for the a_i's the appropriate Gaussian weights (II, eq. [38]). The equations which we have now to deal with are (cf. eqs. [14] and [19]-[21])

$$\Sigma a_i\mu_i I_i=\frac{1}{2}\frac{F_0}{r^2}, \tag{30}$$

$$\frac{d}{dr}\Sigma a_i\mu_i^2 I_i+\frac{1}{r}\Sigma a_i(3\mu_i^2-1)I_i=-\kappa\rho\Sigma a_i\mu_i I_i, \tag{31}$$

$$\frac{d}{dr}\Sigma a_i\mu_i(5\mu_i^2-1)I_i+\frac{4}{r}\Sigma a_i\mu_i(5\mu_i^2-3)I_i=-\tfrac{5}{3}\kappa\rho\Sigma a_i(3\mu_i^2-1)I_i, \tag{32}$$

and

$$\frac{d}{dr}\Sigma a_i\mu_i^2(7\mu_i^2-3)I_i=-\kappa\rho\Sigma a_i\mu_i(7\mu_i^2-3)I_i. \tag{33}$$

The foregoing equations can be written more compactly in terms of the quantities J, H, K, L, and M, defined as follows:

$$\left.\begin{aligned}&\tfrac{1}{2}\Sigma a_i I_i=J\,; && \tfrac{1}{2}\Sigma a_i\mu_i^3 I_i=L\,,\\ &\tfrac{1}{2}\Sigma a_i\mu_i I_i=H\,; && \tfrac{1}{2}\Sigma a_i\mu_i^4 I_i=M\,.\\ &\tfrac{1}{2}\Sigma a_i\mu_i^2 I_i=K\,; && \end{aligned}\right\} \tag{34}$$

We have

$$H=\frac{1}{4}\frac{F_0}{r^2}, \tag{34'}$$

$$\frac{dK}{dr}+\frac{1}{r}(3K-J)=-\kappa\rho H, \tag{35}$$

$$\frac{d}{dr}(5L-H)+\frac{4}{r}(5L-3H)=-\tfrac{5}{3}\kappa\rho(3K-J), \tag{36}$$

and

$$\frac{d}{dr}(7M-3K)=-\kappa\rho(7L-3H). \tag{37}$$

Equations (35)-(37) provide us with three equations for the four unknowns J, K, L, and M. However, in our present approximation we can express M linearly in terms of J and K; for, since the μ_i's ($i=\pm1, \pm2$) are now defined as the zeros of $P_4(\mu)$, we have identically

$$\Sigma a_i P_4(\mu_i)I_i\equiv 0, \tag{38}$$

or, substituting for $P_4(\mu)$, we have

$$\Sigma a_i(35\mu_i^4-30\mu_i^2+3)I_i\equiv 0. \tag{39}$$

In other words,

$$35M-30K+3J=0. \tag{40}$$

Accordingly,

$$7M - 3K = 3K - \tfrac{3}{5}J\,, \tag{41}$$

and equation (37) becomes

$$\frac{d}{dr}(3K - \tfrac{3}{5}J) = -\kappa\rho(7L - 3H)\,. \tag{42}$$

Equations (34)–(36) and (42), together with the relevant boundary conditions on the I_i's, make the problem determinate.

We now transform equations (35), (36), and (42) to more convenient forms. Letting X and Y stand for

$$X = 3K - J \quad \text{and} \quad Y = 5L - 3H\,, \tag{43}$$

we can re-write equation (35) as

$$\frac{dK}{dr} = -\frac{X}{r} - \frac{1}{4}\kappa\rho\frac{F_0}{r^2}, \tag{44}$$

where we have substituted for H according to equation (34′). Equation (44) can be formally integrated to give

$$K = -\int^r X\frac{dr}{r} - \frac{1}{4}F_0\int^r \frac{\kappa\rho\, dr}{r^2}\,. \tag{45}$$

Considering next equation (36), we obtain, after some minor reductions,

$$\frac{dY}{dr} + \frac{4}{r}Y = -\tfrac{5}{3}\kappa\rho X + \frac{F_0}{r^3}\,. \tag{46}$$

Again, since

$$3K - \tfrac{3}{5}J = \tfrac{6}{5}K + \tfrac{3}{5}(3K - J) = \tfrac{6}{5}K + \tfrac{3}{5}X\,, \tag{47}$$

equation (42) is clearly equivalent to

$$2\frac{dK}{dr} + \frac{dX}{dr} = -\tfrac{5}{3}\kappa\rho(7L - 3H)\,. \tag{48}$$

We can eliminate dK/dr from this equation by using equation (44). We obtain, in this manner,

$$\frac{dX}{dr} - \frac{2}{r}X = -\tfrac{7}{3}\kappa\rho Y\,. \tag{49}$$

Equations (46) and (49) provide us with a pair of simultaneous equations for X and Y. These equations can also be written in the forms

$$r^2\frac{d}{dr}\left(\frac{X}{r^2}\right) = -\tfrac{7}{3}\kappa\rho Y \tag{50}$$

and

$$\frac{1}{r^4}\frac{d}{dr}(r^4 Y) = -\tfrac{5}{3}\kappa\rho X + \frac{F_0}{r^3}\,. \tag{51}$$

From these equations we can readily eliminate X or Y to obtain a single second-order differential equation for either of them. We shall not perform this elimination at this stage, since to solve any of these equations we need a prior assumption concerning the dependence of $\kappa\rho$ on r.

5. *The solution in the second approximation for the case* $\kappa\rho \propto r^{-n}$, *when* $n > 1$.—In terms of the radial optical thickness defined by

$$d\tau = -\kappa\rho\, dr\,, \tag{52}$$

the equations of the second approximation derived in the preceding section (eqs. [45], [46], and [49]) become

$$K = \int^{\tau} X \frac{d\tau}{\kappa \rho r} + \frac{1}{4} F_0 \int^{\tau} \frac{d\tau}{r^2}, \tag{53}$$

$$\frac{dX}{d\tau} + \frac{2}{\kappa \rho r} X = \tfrac{7}{3} Y, \tag{54}$$

and

$$\frac{dY}{d\tau} - \frac{4}{\kappa \rho r} Y = \tfrac{5}{3} X - \frac{F_0}{\kappa \rho r^3}. \tag{55}$$

We shall now show how the foregoing equations can be solved in the case where $\kappa\rho$ is assumed to vary as some inverse power of r. Suppose, then, that

$$\kappa \rho = c r^{-n}, \tag{56}$$

where c is a constant. For a dependence of $\kappa\rho$ on r of this form and $n > 1$ we can define the optical depth τ measured from $r = \infty$ inward; for in that case the integral

$$\tau = \int_r^{\infty} \kappa \rho d r, \tag{57}$$

converges, and, in fact, we have

$$\tau = \frac{c}{n-1} \frac{1}{r^{n-1}} \qquad (n > 1). \tag{58}$$

For $n \leqslant 1$ the integral (57) diverges; but, as these cases have no astrophysical interest, we shall not consider them here.

From equations (56) and (58) we obtain the relations

$$\kappa \rho r = (n-1) \tau \tag{59}$$

and

$$\tau = \left(\frac{R}{r}\right)^{n-1}, \tag{60}$$

where R denotes the distance at which $\tau = 1$. Using these relations in the equations (53), (54), and (55) and measuring the various quantities J, H, K, L, X, and Y in units of F_0/R^2 (i.e., in units of the emergent flux at $r = R$), we obtain the equations

$$K = \frac{1}{n-1} \int_0^{\tau} X \frac{d\tau}{\tau} + \frac{n-1}{4(n+1)} \tau^{(n+1)/(n-1)}, \tag{61}$$

$$\frac{dX}{d\tau} + \frac{2}{(n-1)\tau} X = \tfrac{7}{3} Y, \tag{62}$$

and

$$\frac{dY}{d\tau} - \frac{4}{(n-1)\tau} Y = \tfrac{5}{3} X - \frac{1}{n-1} \tau^{(3-n)/(n-1)}. \tag{63}$$

It will be noticed that in the integral occurring in equation (61) we have set the lower limit for τ as zero. This is in accordance with the requirement that, since the atmosphere now extends to infinity, all the quantities must vanish at $r = \infty$, i.e., at $\tau = 0$.

Eliminating Y between equations (62) and (63), we find for X the differential equation

$$\frac{d^2X}{d\tau^2} - \frac{2}{(n-1)\tau} \frac{dX}{d\tau} - \frac{2(n+3)}{(n-1)^2 \tau^2} X = \tfrac{35}{9} X - \frac{7}{3(n-1)} \tau^{(3-n)/(n-1)}. \tag{64}$$

With the substitutions

$$z = q\tau \qquad (q = \sqrt{35}/3 = 1.9720), \tag{65}$$

and

$$X = \tau^{(n+1)/2(n-1)} f = q^{-(n+1)/2(n-1)} z^{(n+1)/2(n-1)} f(z), \tag{66}$$

we find that equation (64) can be brought to the form

$$z^2 \frac{d^2 f}{dz^2} + z \frac{df}{dz} - (z^2 + \nu^2) f = -k z^{\mu+1}, \tag{67}$$

where we have written

$$\nu = \frac{n+5}{2(n-1)}; \qquad \mu = \frac{3-n}{2(n-1)}, \tag{68}$$

and

$$k = \frac{7}{3(n-1)} q^{-(n+1)/2(n-1)}. \tag{69}$$

The accompanying short table (Table 2), giving the values of ν and μ for a few values of n, may be noted.

TABLE 2

n	ν	μ	n	ν	μ
1.25	12.5	3.5	3.0	2.0	0
1.50	6.5	1.5	4.0	1.5	−1/6
$1\frac{2}{3}$	5.0	1.0	7.0	1.0	−1/3
2.0	3.5	0.5	∞	1/2	−1/2

Before proceeding to the solution of equation (67) we may observe that, if ϕ be defined as the solution of

$$z^2 \frac{d^2\phi}{dz^2} + z \frac{d\phi}{dz} - (z^2 + \nu^2)\phi = -z^{\mu+1}, \tag{70}$$

we have, according to equations (66) and (69),

$$X = q^{-(n+1)/(n-1)} \frac{7}{3(n-1)} z^{(n+1)/2(n-1)} \phi(z). \tag{71}$$

Substituting this solution for ϕ in equation (61), we obtain

$$K = q^{-(n+1)/(n-1)} \left[\frac{7}{3(n-1)^2} \int_0^z z^{(3-n)/2(n-1)} \phi(z)\, dz + \frac{n-1}{4(n+1)} z^{(n+1)/(n-1)} \right]. \tag{72}$$

Moreover, since $3K - J = X$, we have for the source function J the solution

$$\left.\begin{aligned} J = q^{-(n+1)/(n-1)} &\left[\frac{7}{(n-1)^2} \int_0^z z^{(3-n)/2(n-1)} \phi(z)\, dz \right. \\ &\left. - \frac{7}{3(n-1)} z^{(n+1)/2(n-1)} \phi(z) + \frac{3(n-1)}{4(n+1)} z^{(n+1)/(n-1)} \right]. \end{aligned}\right\} \tag{73}$$

It now remains to solve equation (70).

First, it will be observed that the homogeneous part of equation (70) is simply Bessel's equation for a purely imaginary argument. The general solution of the homogeneous equation is, accordingly, known and can be expressed as a linear combination of the fundamental solutions $I_\nu(z)$ and $K_\nu(z)$.[10] The solution of the nonhomogeneous equation, therefore, can be found most conveniently by the method of the variation of the parameters.[11] Thus, writing

$$\phi = A(z) I_\nu(z) + B(z) K_\nu(z), \tag{74}$$

we determine the functions $A(z)$ and $B(z)$ by the equations

$$A'(z) I_\nu(z) + B'(z) K_\nu(z) = 0 \tag{75}$$

and

$$A'(z) I'_\nu(z) + B'(z) K'_\nu(z) = -z^{\mu-1}. \tag{76}$$

where we have used primes to denote differentiation with respect to the argument. Using the relation (Watson, p. 80, eq. [19])

$$I_\nu(z) K'_\nu(z) - I'_\nu(z) K_\nu(z) = -\frac{1}{z}, \tag{77}$$

we readily find from equation (75) and (76) that

$$A'(z) = -z^\mu K_\nu(z) \quad \text{and} \quad B'(z) = z^\mu I_\nu(z). \tag{78}$$

Hence,

$$A(z) = \int_z^{c_1} z^\mu K_\nu(z)\, dz \quad \text{and} \quad B(z) = \int_{c_2}^z z^\mu I_\nu(z)\, dz, \tag{79}$$

where c_1 and c_2 are constants unspecified for the present. The general solution of equation (70) can, accordingly, be expressed in the form

$$\phi = I_\nu(z) \int_z^{c_1} z^\mu K_\nu(z)\, dz + K_\nu(z) \int_{2}^z z^\mu I_\nu(z)\, dz. \tag{80}$$

For the problem on hand the arbitrary limits c_1 and c_2 in the general solution (80) are determined by the following considerations:

First, since none of the quantities must tend to infinity exponentially as $z \to \infty$, we must require that $c_1 = \infty$ in equation (80). This readily follows from the known asymptotic behaviors of $I_\nu(z)$ and $K_\nu(z)$ as $z \to \infty$ (cf. Watson, § 7.23, p. 202). Second, the vanishing of all the quantities as $z \to 0$ requires that (cf. eq. [71])

$$z^{(n+1)/2(n-1)} \phi(z) \to 0 \qquad (z \to 0). \tag{81}$$

But $K_\nu(z)$ diverges at the origin, and condition (81) can be met only by setting $c_2 = 0$. Thus the solution for ϕ appropriate for our problem is

$$\phi = I_\nu(z) \int_z^{\infty} z^\mu K_\nu(z)\, dz + K_\nu(z) \int_0^z z^\mu I_\nu(z)\, dz. \tag{82}$$

With this we have formally solved the equations of the second approximation for the case $\kappa\rho \propto r^{-n}$, where $n > 1$.

[10] We shall adopt, throughout, the definitions and notations of G. N. Watson in his *Treatise on the Theory of Bessel Functions*, Cambridge, England, 1922. In our further references to this work we shall simply refer to it as "Watson." In our particlar context see pp. 77–80.

[11] The treatment of Lommel's equation

$$z^2 \frac{d^2y}{dz^2} + z\frac{dy}{dz} + (z^2 - \nu^2)y = kz^{\mu+1}$$

in Watson, § 10.7, p. 345, is followed in our discussion of eq. (70).

6. *The numerical form of the solution for the case* $\kappa\rho \propto r^{-2}$.—As an example of the solution obtained in the preceding section, we shall consider the case $\kappa\rho \propto r^{-2}$. For this case the solutions for X, J, and K given in § 5 (eqs. [71]-[73]) become

$$\left.\begin{aligned} X &= q^{-3}\tfrac{7}{3}z^{3/2}\phi(z)\,, \\ K &= q^{-3}\left[\tfrac{7}{3}\int_0^z z^{1/2}\phi(z)\,dz + \tfrac{1}{12}z^3\right], \\ J &= q^{-3}\left[7\int_0^z z^{1/2}\phi(z)\,dz - \tfrac{7}{3}z^{3/2}\phi(z) + \tfrac{1}{4}z^3\right], \end{aligned}\right\} \tag{83}$$

where it might be recalled that

$$z = q\tau \quad \text{and} \quad q = \frac{\sqrt{35}}{3} = 1.972\,. \tag{84}$$

Moreover, when $n = 2$, $\nu = \frac{7}{2}$ and $\mu = \frac{1}{2}$ (cf. Table 2); and the solution (82) for ϕ takes the particular form

$$\phi = I_{7/2}(z)\int_z^\infty z^{1/2}K_{7/2}(z)\,dz + K_{7/2}(z)\int_0^z z^{1/2}I_{7/2}(z)\,dz\,. \tag{85}$$

The Bessel functions $I_{7/2}(z)$ and $K_{7/2}(z)$ are known explicitly, and we have (cf. Watson, p. 78)

$$I_{7/2}(z) = \left(\frac{2}{\pi z}\right)^{1/2}\left[\left(1+\frac{15}{z^2}\right)\cosh z - \left(\frac{6}{z}+\frac{15}{z^3}\right)\sinh z\right] \tag{86}$$

and

$$K_{7/2}(z) = \left(\frac{\pi}{2z}\right)^{1/2} e^{-z}\left(1+\frac{6}{z}+\frac{15}{z^2}+\frac{15}{z^3}\right). \tag{87}$$

Accordingly, we can re-write equation (85) in the form

$$\left.\begin{aligned} \phi = \frac{1}{z^{1/2}}\Bigg\{&\left[\left(1+\frac{15}{z^2}\right)\cosh z - \left(\frac{6}{z}+\frac{15}{z^3}\right)\sinh z\right]\int_z^\infty e^{-z}\left(1+\frac{6}{z}+\frac{15}{z^2}+\frac{15}{z^3}\right)dz \\ &+ e^{-z}\left(1+\frac{6}{z}+\frac{15}{z^2}+\frac{15}{z^3}\right)\int_0^z\left[\left(1+\frac{15}{z^2}\right)\cosh z - \left(\frac{6}{z}+\frac{15}{z^3}\right)\sinh z\right]dz\Bigg\}\,. \end{aligned}\right\} \tag{88}$$

It is seen that the first of the two integrals in equation (88) can be expressed in terms of known functions. We have

$$\int_z^\infty e^{-z}\left(1+\frac{6}{z}+\frac{15}{z^2}+\frac{15}{z^3}\right)dz = e^{-z}\left(1+\frac{15}{2z}+\frac{15}{2z^2}\right) - \tfrac{3}{2}Ei(z)\,, \tag{89}$$

where $Ei(z)$ stands for the exponential integral

$$Ei(z) = \int_1^\infty \frac{e^{-zt}}{t}\,dt\,. \tag{90}$$

The functions q^3X, q^3K, and q^3J, defined as in the foregoing paragraph, have been evaluated numerically for a range of values for z; and the results are given in Table 3.[12] For comparison we have also tabulated the function $z^3/4$, which is the solution for q^3J in the first approximation (cf. eq. [29]). It is seen that the second approximation introduces

[12] I am indebted to Miss Frances Herman for carrying out the necessary numerical work.

corrections to the extent of about 10 per cent. In view of this, it would be of interest further to examine the solution given in § 5 for other values of n. The case of $n = \frac{3}{2}$ would be of particular interest.

7. *Diffusion through a homogeneous sphere.*—The discussion of this case, while it has no special interest for astrophysics, is, however, likely to be of importance for problems of diffusion in physics.[13] But, apart from possible applications, the consideration of this

TABLE 3

z	q^3X	q^3K	q^3J	$\frac{1}{4}z^3$	z	q^3X	q^3K	q^3J	$\frac{1}{4}z^3$
0..	0	0	0	0	2.8..	2.5081	2.8719	6.1075	5.4880
0.1..	0.0002322	0.0001605	0.0002500	0.00025	2.9..	2.7048	3.1664	6.7944	6.0973
0.2..	0.0018377	0.0012825	0.0020097	0.00200	3.0..	2.9068	3.4791	7.5304	6.7500
0.3..	0.0061122	0.0043092	0.0068154	0.00675	3.1..	3.1137	3.8103	8.3172	7.4478
0.4..	0.014237	0.010158	0.016238	0.01600	3.2..	3.3254	4.1606	9.1564	8.1920
0.5..	0.027265	0.019718	0.031890	0.03125	3.3..	3.5414	4.5303	10.0494	8.9843
0.6..	0.046111	0.033844	0.055420	0.05400	3.4..	3.7617	4.9199	10.9979	9.8260
0.7..	0.071559	0.053357	0.088512	0.08575	3.5..	3.9858	5.3297	12.0033	10.7188
0.8..	0.10426	0.079047	0.13288	0.12800	3.6..	4.2136	5.7603	13.0671	11.6640
0.9..	0.14475	0.11167	0.19026	0.18225	3.7..	4.4448	6.2119	14.1910	12.6633
1.0..	0.19345	0.15195	0.26241	0.25000	3.8..	4.6792	6.6852	15.3763	13.7180
1.1..	0.25067	0.20059	0.35109	0.33275	3.9..	4.9165	7.1804	16.6246	14.8298
1.2..	0.31661	0.25824	0.45813	0.43200	4.0..	5.1565	7.6979	17.9373	16.0000
1.3..	0.39146	0.32557	0.58523	0.54925	4.1..	5.3991	8.2383	19.3159	17.2303
1.4..	0.47522	0.40317	0.73429	0.68600	4.2..	5.6440	8.8019	20.7618	18.5220
1.5..	0.56792	0.49165	0.90703	0.84375	4.3..	5.8911	9.3892	22.2766	19.8768
1.6..	0.66950	0.59158	1.10524	1.02400	4.4..	6.1401	10.0006	23.8617	21.2960
1.7..	0.77985	0.70352	1.33072	1.22825	4.5..	6.3909	10.6365	25.5184	22.7813
1.8..	0.89876	0.82800	1.58525	1.45800	4.6..	6.6434	11.2973	27.2484	24.3340
1.9..	1.0261	0.96556	1.8706	1.7148	4.7..	6.8975	11.9835	29.0529	25.9558
2.0..	1.1616	1.1167	2.1884	2.0000	4.8..	7.1528	12.6954	30.9334	27.6480
2.1..	1.3051	1.2819	2.5405	2.3153	4.9..	7.4095	13.4336	32.8914	29.4123
2.2..	1.4563	1.4616	2.9286	2.6620	5.0..	7.6672	14.1985	34.9283	31.2500
2.3..	1.6148	1.6564	3.3545	3.0418	5.1..	7.9260	14.9905	37.0454	33.1628
2.4..	1.7805	1.8667	3.8197	3.4560	5.2..	8.1856	15.8100	39.2443	35.1520
2.5..	1.9529	2.0930	4.3259	3.9063	5.3..	8.4461	16.6574	41.5263	37.2193
2.6..	2.1319	2.3356	4.8749	4.3940	5.4..	8.7072	17.5333	43.8928	39.3660
2.7..	2.3171	2.5951	5.4682	4.9208	5.5..	8.9690	18.4381	46.3453	41.5938

case is of definite interest, inasmuch as it provides the simplest illustration of the use of the equations of the second approximation obtained in § 4.

For a homogeneous sphere we naturally assume that

$$\kappa \rho = \text{constant} = \kappa_0 \, (\text{say}) . \tag{91}$$

As κ_0 is of dimensions (length)$^{-1}$ it is convenient to measure length in units of $1/\kappa_0$ and intensity in units of $F_0 \kappa_0^2$ (i.e., in units of the emergent flux at $r = 1/\kappa_0$). In these units, equations (45), (46), and (49) now reduce to the forms

$$K = -\int \frac{X}{r} \, d r + \frac{1}{4 r} , \tag{92}$$

$$\frac{d X}{d r} - \frac{2}{r} X = -\tfrac{7}{3} Y , \tag{93}$$

and

$$\frac{d Y}{d r} + \frac{4}{r} Y = -\tfrac{5}{3} X + \frac{1}{r^3} . \tag{94}$$

[13] Cf. W. Bothe, *Zs. f. Phys.*, **119**, 493, 1942.

Eliminating Y between equations (93) and (94), we obtain for X the differential equation

$$r^2 \frac{d^2X}{dr^2} + 2r\frac{dX}{dr} - 6X = \tfrac{35}{9} X r^2 - \frac{7}{3r}. \tag{95}$$

Making the substitutions

$$z = \frac{\sqrt{35}}{3} r \tag{96}$$

and

$$X = z^{-1/2} y , \tag{97}$$

we find that equation (95) is transformed to

$$z^2 \frac{d^2y}{dz^2} + z\frac{dy}{dz} - \left(z^2 + \tfrac{25}{4}\right) y = -\frac{7\sqrt{35}}{9} z^{-1/2}. \tag{98}$$

We verify that

$$y = \frac{7\sqrt{35}}{9} z^{-5/2} \tag{99}$$

represents a particular integral of equation (98); and, as the homogeneous part of this equation is Bessel's equation of order $\frac{5}{2}$ for a purely imaginary argument, we can write the general solution of equation (98) in the form

$$y = \frac{7\sqrt{35}}{9} z^{-5/2} + A I_{5/2}(z) + B K_{5/2}(z), \tag{100}$$

where A and B are two constants to be determined by the boundary conditions appropriate to the problem on hand. At this stage we can determine the constant B. For, as $z \to 0$,

$$K_{5/2}(z) \to 3\left(\frac{\pi}{2}\right)^{1/2} z^{-5/2} \qquad (z \to 0); \tag{101}$$

and this is seen to be too high an order for the singularity at the origin. But by choosing

$$B = -\left(\frac{2}{\pi}\right)^{1/2} \frac{7\sqrt{35}}{27}, \tag{102}$$

we can lower the order of the singularity at $z = 0$ by one to $z^{-3/2}$; for, with this choice of B the term in $z^{-5/2}$ in $K_{5/2}(z)$ cancels the particular integral (99). Accordingly, we write for ϕ the solution

$$y = \frac{7\sqrt{35}}{9} z^{-5/2} - \left(\frac{2}{\pi}\right)^{1/2} \frac{7\sqrt{35}}{27} K_{5/2}(z) + A I_{5/2}(z), \tag{103}$$

where A is a constant to be determined later by conditions at the boundary of the sphere.

Corresponding to the solution (103) for y, we have

$$X = \frac{7\sqrt{35}}{9} z^{-3} - \left(\frac{2}{\pi}\right)^{1/2} \frac{7\sqrt{35}}{27} z^{-1/2} K_{5/2}(z) + A z^{-1/2} I_{5/2}(z). \tag{104}$$

Equation (93) now enables us to determine Y; for, writing this equation in the form

$$Y = -\frac{3}{7} r^2 \frac{d}{dr}\left(\frac{X}{r^2}\right) = -\frac{\sqrt{35}}{7} z^2 \frac{d}{dz}\left(\frac{X}{z^2}\right) \tag{105}$$

and substituting for X according to equation (104), we find that

$$\left.\begin{aligned} Y &= 5L - 3H \\ &= \tfrac{175}{9} z^{-4} - \left(\frac{2}{\pi}\right)^{1/2} \tfrac{35}{27} z^{-1/2} K_{7/2}(z) - \frac{\sqrt{35}}{27} A\, z^{-1/2} I_{7/2}(z) . \end{aligned}\right\} \quad (106)$$

In obtaining the foregoing result we have made use of the recurrence relations satisfied by the Bessel functions (cf. Watson, p. 79, eq. [67]).

Again substituting for X in equation (92) according to equation (104), we similarly find that

$$\left.\begin{aligned} K = \frac{7\sqrt{35}}{27} z^{-3} - \left(\frac{2}{\pi}\right)^{1/2} \frac{7\sqrt{35}}{27} z^{-3/2} K_{3/2}(z) - A\, z^{-3/2} I_{3/2}(z) \\ + \frac{\sqrt{35}}{12 z} + C , \end{aligned}\right\} \quad (107)$$

where C is another constant of integration. Finally, remembering that $J = 3K - X$, we obtain for the source function J the solution

$$\left.\begin{aligned} J = \left(\frac{2}{\pi}\right)^{1/2} \frac{7\sqrt{35}}{27} z^{-3/2} [\, z K_{5/2}(z) - 3 K_{3/2}(z)\,] - A\, z^{-3/2} [\, z I_{5/2}(z) \\ + 3 I_{3/2}(z)\,] + \frac{\sqrt{35}}{4 z} + 3C . \end{aligned}\right\} \quad (108)$$

With this we have explicitly solved all the equations of the second approximation for the case under consideration. It only remains to determine the two constants of integrations A and C. These can be determined by the conditions at the boundary of the sphere, namely, that here both I_{-1} and I_{-2} must vanish. But we shall not continue here with the details of the elementary calculations necessary for the determination of these constants. We may, however, note that for the case of an *infinite* homogeneous sphere the constants A and C must vanish and that the solution for J reduces to

$$J = \left(\frac{2}{\pi}\right)^{1/2} \frac{7\sqrt{35}}{27} z^{-3/2} (z K_{5/2} - 3 K_{3/2}) + \frac{\sqrt{35}}{4 z} ; \quad (109)$$

or, substituting for $K_{5/2}$ and $K_{3/2}$ their known explicit forms and reverting to the original variable r, we find

$$J = \frac{3}{4 r} (1 + \tfrac{28}{27} e^{-1.972 r}) , \quad (110)$$

which is to be contrasted with what would be obtained on the first approximation, namely,

$$J = \frac{3}{4 r} \qquad \text{(first approximation)} . \quad (111)$$

It is seen that in going to the second approximation we introduce a "correction" term which amounts to 14.5 per cent at $r = 1$ and which further decreases rapidly for increasing r.

I wish to record my indebtedness to Drs. J. Sahade and C. U. Cesco for their careful revision of the manuscript.

3

ON THE RADIATIVE EQUILIBRIUM OF A STELLAR ATMOSPHERE. VII

S. CHANDRASEKHAR
Yerkes Observatory
Received March 9, 1945

ABSTRACT

In this paper the problem of the radiative equilibrium of a stellar atmosphere in local thermodynamic equilibrium is reconsidered with a view to examining in detail the effects of a continuous absorption coefficient, κ_ν, dependent on frequency. It is shown that when the material departs from grayness the corrections which have to be made to the temperature distribution are of two kinds: first, because the integrated Planck intensity B is not, in general, equal to the average integrated intensity J ($= \int\int I_\nu d\omega d\nu/4\pi$) and, second, because the energy density of the radiation does not have the gray atmospheric value. To evaluate the corrections arising from these two sources, a systematic method of approximation has been developed in which the first approximation is assumed to be given by the solution for a gray atmosphere with a κ_ν equal to a certain appropriately chosen mean absorption coefficient $\bar{\kappa}$. It is found that the best way of choosing $\bar{\kappa}$ is to define it as a straight mean of κ_ν, weighted according to the net monochromatic flux $F_\nu^{(1)}(\tau)$ of radiation of frequency ν in a gray atmosphere:

$$\bar{\kappa} = \int_0^\infty \kappa_\nu \frac{F_\nu^{(1)}}{F} d\nu \qquad (F = \text{integrated flux}) .$$

Moreover, in a second approximation,

$$B^{(2)} = J^{(2)} + \tfrac{1}{4}\int_0^\infty \left(\frac{\kappa_\nu}{\bar{\kappa}} - 1\right)\frac{dF_\nu^{(1)}}{d\tau} d\nu ,$$

where $J^{(2)}$ is the solution appropriate for J in this approximation. The solution for $J^{(2)}$ has, in turn, been obtained in an nth approximation.

To facilitate the use of the solutions in the higher approximations obtained in this paper, the monochromatic fluxes $F_\nu^{(1)}(\tau)$ and their derivatives $dF_\nu^{(1)}/d\tau$ have been evaluated for a range of values of τ and $\alpha = h\nu/kT_e$ (T_e denoting the effective temperature).

1. *Introduction.*—The solution to the problem of radiative transfer in an atmosphere in local thermodynamic equilibrium is fundamental to all investigations which are in any way related to the continuous spectrum of the stars. More particularly, the basic problem is one of solving the equation of transfer

$$\mu \frac{dI_\nu}{\rho dx} = -\kappa_\nu I_\nu + \kappa_\nu B_\nu , \tag{1}$$

under conditions of a constant net flux of radiation

$$\pi F = \pi \int^\infty F_\nu d\nu = 2\pi \int_0^\infty \int_{-1}^{+1} I_\nu \mu d\mu d\nu , \tag{1'}$$

in all the frequencies and where $B_\nu(T)$ is the Planck function for the temperature T at x. The complete solution to this problem depends only on the distribution of temperature in the atmosphere; for, once this is known, the source function $B_\nu(T)$ for the radiation of frequency ν is known at all points in the atmosphere and the determination of the intensity I_ν at any point and in any given direction is immediate (see eq. [25] below). However, when the continuous absorption coefficient κ_ν is allowed to be an arbitrary function of ν and x, the determination of the temperature distribution in the atmosphere is by no means a simple matter. Indeed, there exists at the present time no satisfactory attempt to solve the basic problem with any degree of generality. And what is already known in this connection[1] can be summarized quite simply.

[1] Cf., for what follows, A. Unsöld, *Physik der Sternatmosphären*, pp. 113–116, Berlin: Springer, 1938.

From the equation of transfer (1) we readily obtain

$$\frac{dK_\nu}{\rho dx} = -\tfrac{1}{4}\kappa_\nu F_\nu \,, \tag{2}$$

where, as usual, we have written

$$K_\nu = \frac{1}{2}\int_{-1}^{+1} I_\nu \mu^2 d\mu \,. \tag{3}$$

Defining the mean absorption coefficient $\bar{\kappa}$ by means of the relation

$$\frac{1}{\bar{\kappa}}\int_0^\infty \frac{dK_\nu}{\rho dx} d\nu = \int_0^\infty \frac{1}{\kappa_\nu}\frac{dK_\nu}{\rho dx} d\nu \,, \tag{4}$$

we can re-write equation (2) as an equation for integrated radiation in the form

$$\frac{dK}{d\tau} = \tfrac{1}{4}F \,, \tag{5}$$

where τ denotes the optical depth in the mean absorption coefficient $\bar{\kappa}$, measured from the boundary of the atmosphere normally inward. Remembering the constancy of the net integrated flux, we can integrate equation (5) to give

$$K = \tfrac{1}{4}F\tau + \text{constant} \,. \tag{6}$$

Using the conventional Milne-Eddington type of approximations, we readily derive from equation (6) the formula

$$J = \int_0^\infty J_\nu d\nu = \frac{1}{2}\int_0^\infty \int_{-1}^{+1} I_\nu d\mu d\nu = \tfrac{1}{2}F\left(1 + \tfrac{3}{2}\tau\right) . \tag{7}$$

But we *cannot*, in general, pass directly from this equation for J to an equation for the integrated Planck intensity B $(= \sigma T^4/\pi)$. For, according to the equation of transfer (1), the constancy of the net integrated flux implies only that

$$\int_0^\infty \kappa_\nu J_\nu d\nu = \int_0^\infty \kappa_\nu B_\nu d\nu \,; \tag{8}$$

and we cannot infer (except when the material is "gray") that $J = B$. In other words, the formula for the temperature distribution which is in common use, namely,

$$T^4 = \tfrac{1}{2}T_e^4\left(1 + \tfrac{3}{2}\tau\right) , \tag{9}$$

where τ is the optical depth measured in the "Rosseland mean"[2] absorption coefficient,

[2] In practice this is defined by

$$\frac{1}{\bar{\kappa}}\int_0^\infty \frac{dB_\nu}{dT} d\nu = \int_0^\infty \frac{1}{\kappa_\nu}\frac{dB_\nu}{dT} d\nu \,.$$

But this is not the same as the mean absorption coefficient defined by equation (4). It would therefore seem that the use of the conventional Rosseland mean absorption coefficient in the theory of stellar atmospheres is of questionable value. On the other hand, an alternative form of equation (4), namely,

$$\bar{\kappa} = \frac{1}{F}\int_0^\infty \kappa_\nu F_\nu d\nu \,,$$

would appear to provide a more satisfactory way of defining κ, as it has the further advantages of being a straight mean. Further remarks relating to this matter are made in § 5.

cannot, strictly speaking, be regarded as the solution of the fundamental equations in any well-defined scheme of approximation. Thus, while the use of equation (9) cannot be strictly defended, there does seem to be some evidence for the belief that the temperature distribution derived on the gray-body assumption provides a "first approximation" to the true distribution in some sense or another. For example, it is well known from the studies of Milne and Lindblad that the law of darkening in the different wave lengths over the solar disk and the intensity distribution in the continuous spectrum of the sun agree quite well with what can be predicted on the gray-body assumption. But attempts to improve on this agreement by allowing for a variation of κ_ν with ν on a "second approximation" generally indicate that the departures from grayness of the material is quite large.[3] It would, accordingly, appear that the solution for the temperature distribution on the gray-body assumption does provide a good first approximation even for substantial variations of κ_ν with ν. But, as to in what sense it is a first approximation can be understood only when we develop a systematic well-defined scheme of approximation and explicitly work out a second approximation. It is the object of this paper to provide such a scheme and to estimate the errors which are involved in the use of the solution (9).

2. *An iteration method for solving the equation of transfer for an atmosphere in local thermodynamic equilibrium and with a continuous absorption coefficient depending on the wave length.*—According to our remarks in the preceding section, it would appear that the temperature distribution in the atmosphere derived on the assumption of grayness of the material provides a first approximation. Assuming that this is the case, let κ be a certain mean absorption coefficient (undefined for the present) such that with $\kappa_\nu = \kappa =$ constant we obtain a first approximation. Let τ denote the optical depth measured in terms of this absorption coefficient. Further, let

$$\delta_\nu = \frac{\kappa_\nu}{\kappa} - 1 , \tag{10}$$

so that δ_ν is a measure of the departure from grayness of the material. With these definitions the equation of transfer (1) takes the form

$$\mu \frac{dI_\nu}{d\tau} = I_\nu - B_\nu + \delta_\nu (I_\nu - B_\nu) . \tag{11}$$

We suppose that this equation can be solved in two steps. First, we find the solution of the equation

$$\mu \frac{dI_\nu^{(1)}}{d\tau} = I_\nu^{(1)} - B_\nu^{(1)} \qquad [B_\nu^{(1)} = B_\nu (T^{(1)})] , \tag{12}$$

appropriate to the problem on hand, and use this solution in the term which occurs as the factor of δ_ν in equation (11). Thus, the second approximation will be given by the solution of

$$\mu \frac{dI_\nu^{(2)}}{d\tau} = I_\nu^{(2)} - B_\nu^{(2)} + \delta_\nu \mu \frac{dI_\nu^{(1)}}{d\tau} . \tag{13}$$

Formally, there is, of course, no difficulty in extending this method of iteration to obtain solutions to as high an approximation as may be needed. Thus, in the nth approximation we shall have

$$\mu \frac{dI_\nu^{(n)}}{d\tau} = I_\nu^{(n)} - B_\nu^{(n)} + \delta_\nu [I_\nu^{(n-1)} - B_\nu^{(n-1)}] . \tag{14}$$

[3] Cf. Unsöld, *op. cit.*, pp. 107–109 (see particularly Tables 24 and 26).

This method of iteration will obviously converge if δ_ν is sufficiently small. But the point to which attention may be drawn at this stage is that in practice the success of this method is not impaired even when δ_ν takes moderately large values of the order of 2 or 3 (see § 5 below).

In this paper we shall consider only the first two approximations.

In their integrated forms, equations (12) and (13) are

$$\mu \frac{dI^{(1)}}{d\tau} = I^{(1)} - B^{(1)} \tag{15}$$

and

$$\mu \frac{dI^{(2)}}{d\tau} = I^{(2)} - B^{(2)} + \mu \int_0^\infty \delta_\nu \frac{dI_\nu^{(1)}}{d\tau} d\nu . \tag{16}$$

The foregoing equations have, of course, to be solved under the conditions of a constant net integrated flux. In the first two approximations this condition requires, respectively, that

$$B^{(1)} = J^{(1)} \tag{17}$$

and

$$B^{(2)} = J^{(2)} + \frac{1}{2}\int_0^\infty \int_{-1}^{+1} \delta_\nu \frac{dI_\nu^{(1)}}{d\tau} \mu d\mu d\nu . \tag{18}$$

Equation (18) can be re-written in the form

$$B^{(2)} = J^{(2)} + \frac{1}{4}\int_0^\infty \delta_\nu \frac{dF_\nu^{(1)}}{d\tau} d\nu . \tag{19}$$

In other words, in the second approximation the integrated Planck intensity B differs from J by an amount which depends on the *nonconstancy* of the monochromatic fluxes F_ν. From general considerations we may expect (and this is confirmed by the calculations to be presented in § 5) that, unless δ_ν varies too widely over the relevant ranges of ν, the departures from constancy of the monochromatic fluxes F_ν will be of the second order of smallness. It is precisely on this account that the temperature distribution derived on the assumption of grayness of the material is as satisfactory as it has been found to be in practice.

We now proceed to the solutions of equations (15) and (16).

3. *The solution in the* (1, n) *approximation.*—Equations (15) and (17) together reduce to an equation of transfer of a standard type which has been treated in sufficient detail in earlier papers of this series.[4] Adopting in particular the scheme of approximation developed in paper II we can, in the nth approximation, write (II, eqs. [19] and [26])

$$I_i^{(1)} = \tfrac{3}{4}F \left\{ \sum_{a=1}^{n-1} \frac{L_a^{(1)}}{1+\mu_i k_a} e^{-k_a \tau} + \mu_i + Q^{(1)} + \tau \right\} \quad (i = \pm 1, \ldots, \pm n) , \tag{20}$$

where the n constants $L_a^{(1)}$, $(a = 1, \ldots, n-1)$, and $Q^{(1)}$ are determined by the equations (II, eq. [21])

$$\sum_{a=1}^{n-1} \frac{L_a^{(1)}}{1-\mu_i k_a} + Q^{(1)} - \mu_i = 0 \qquad (i = 1, \ldots, n) \tag{21}$$

[4] *Ap. J.*, **99**, 180, 1944, and **100**, 76, 1944. These papers will be referred to as "I" and "II," respectively.

and the k_a's are the $n-1$ distinct nonzero positive roots of the characteristic equation (II, eq. [10])

$$2=\sum_{i=-n}^{+n}\frac{a_i}{1+\mu_i k_a}. \tag{22}$$ [5]

On this approximation the solution for $J^{(1)}$ is (II, eq. [29])

$$J^{(1)}=B^{(1)}=\tfrac{3}{4}F\left(\tau+q\left[\tau\right]\right)=\tfrac{3}{4}F\left(\tau+Q^{(1)}+\sum_{a=1}^{n-1}L_a^{(1)}e^{-k_a\tau}\right). \tag{23}$$

The corresponding formula giving the temperature distribution is

$$(T^{(1)})^4=\tfrac{3}{4}T_e^4\left(\tau+Q^{(1)}+\sum_{a=1}^{n-1}L_a^{(1)}e^{-k_a\tau}\right). \tag{24}$$

Corresponding to the distribution of temperature (24), there is a determinate distribution of the intensities at various frequencies and at various levels. Thus, the intensity of the radiation of frequency ν at an optical depth τ and in a direction making an angle ϑ with the positive normal is given by

$$\left.\begin{aligned} I_\nu^{(1)}(\tau,\vartheta) &= \int_\tau^\infty e^{-(t-\tau)\sec\vartheta}B_\nu(T_t^{(1)})\sec\vartheta\,dt\quad(0<\vartheta<\pi/2)\\ &= -\int_0^\tau e^{-(t-\tau)\sec\vartheta}B_\nu(T_t^{(1)})\sec\vartheta\,dt\quad(\pi/2<\vartheta<\pi). \end{aligned}\right\} \tag{25}$$

In terms of this solution for $I_\nu^{(1)}(\tau,\vartheta)$ we can readily derive explicit formulae for quantities such as J_ν, F_ν, etc. In fact, it can be shown quite generally that

$$\left.\begin{aligned} \int_{-1}^{+1}I_\nu^{(1)}(\tau,\mu)\mu^j d\mu = \int_\tau^\infty B_\nu(T_t^{(1)})E_{j+1}(t-\tau)\,dt\\ +(-1)^j\int_0^\tau B_\nu(T_t^{(1)})E_{j+1}(\tau-t)\,dt. \end{aligned}\right\} \tag{26}$$

where $E_{j+1}(x)$ stands for the exponential integral

$$E_{j+1}(x)=\int_1^\infty\frac{e^{-xw}}{w^{j+1}}dw. \tag{27}$$

In our further work we shall find that we also need expressions for the quantities

$$\int_{-1}^{+1}\frac{dI_\nu^{(1)}}{d\tau}\mu^j d\mu. \tag{28}$$

From the equation of transfer (15) we find that

$$\int_{-1}^{+1}\frac{dI_\nu^{(1)}}{d\tau}\mu^j d\mu=\int_{-1}^{+1}I_\nu^{(1)}(\tau,\mu)\mu^{j-1}d\mu-\frac{2}{j}\epsilon_{j,\,\text{odd}}B_\nu(T_\tau^{(1)}), \tag{29}$$

where

$$\left.\begin{aligned} \epsilon_{j,\,\text{odd}} &= 1 \quad \text{if } j \text{ is odd}\\ &= 0 \quad \text{otherwise.} \end{aligned}\right\} \tag{30}$$

[5] In the summation on the right-hand side of this equation there is no term with $i = 0$.

Using equation (26), we can re-write equation (29) as

$$\left.\begin{aligned}\int_{-1}^{+1}\frac{dI_\nu^{(1)}}{d\tau}\mu^j d\mu &= \int_\tau^\infty B_\nu(T_t^{(1)})E_j(t-\tau)\,dt\\ &\quad+(-1)^{j-1}\int_0^t B_\nu(T_t^{(1)})E_j(\tau-t)\,dt-\frac{2}{j}\epsilon_{j,\,\mathrm{odd}}B_\nu(T_\tau^{(1)})\,.\end{aligned}\right\} \quad (31)$$

Equations (24), (25), (26), and (31) may be said to represent the solution to our problem in the (1, n) approximation referring to the fact that the equations of the first approximation have in turn been solved in an nth approximation.

4. *The solution of the equations in the* (2, n) *approximation.*—Turning next to the solution of the equations (16) and (18), we have to consider the equation

$$\mu\frac{dI^{(2)}}{d\tau}=I^{(2)}-\frac{1}{2}\int_{-1}^{+1}I^{(2)}d\mu+\mu\int_0^\infty\delta_\nu\frac{dI_\nu^{(1)}}{d\tau}d\nu-\frac{1}{2}\int_0^\infty\int_{-1}^{+1}\delta_\nu\mu\frac{dI_\nu^{(1)}}{d\tau}d\mu d\nu\,. \quad (32)$$

In solving this integrodifferential equation we shall follow the methods developed in the earlier papers of this series and replace the integrals which occur on the right-hand side of this equation by sums according to Gauss's formula for numerical quadratures. Thus, in the nth approximation, the equivalent systems of linear equations are

$$\left.\begin{aligned}\mu_i\frac{dI_i^{(2)}}{d\tau}&=I_i^{(2)}-\tfrac{1}{2}\Sigma a_j I_j^{(2)}+\mu_i\int_0^\infty\delta_\nu\frac{dI_{\nu,\,i}^{(1)}}{d\tau}d\nu\\ &\quad-\frac{1}{2}\int_0^\infty\delta_\nu\Sigma a_j\mu_j\frac{dI_{\nu,\,j}^{(1)}}{d\tau}d\nu\quad(i=\pm1,\ldots,\pm n)\,,\end{aligned}\right\} \quad (33)$$

where we have used $I_j^{(2)}$ and $I_{\nu,j}^{(1)}$ to denote, respectively, $I^{(2)}(\tau,\mu_j)$ and $I_\nu^{(1)}(\tau,\mu_j)$ and the rest of the symbols have the same meanings as in paper II.

The system of equations represented by (33) is most conveniently solved by the method of the variation of the parameters. Thus, since the homogeneous part of the system of equations (33) is of the same form as that considered in II, equation (6), we seek a solution of our present system of the form (cf. II, eqs. [18] and [26])

$$\left.\begin{aligned}I_i^{(2)}&=\tfrac{3}{4}F\left\{\sum_{a=1}^{n-1}\frac{e^{-k_a\tau}}{1+\mu_i k_a}L_a^{(2)}(\tau)\right.\\ &\quad\left.+\sum_{a=1}^{n-1}\frac{e^{+k_a\tau}}{1-\mu_i k_a}L_{-a}^{(2)}(\tau)+\mu_i+Q^{(2)}(\tau)+\tau\right\}\quad(i=\pm1,\ldots+n)\,,\end{aligned}\right\} \quad (34)$$

where, as we have indicated, $L_a^{(2)}$, $L_{-a}^{(2)}$, ($a=1,\ldots,n-1$), and $Q^{(2)}$ are all to be considered as functions of τ. It will be noticed that in writing the solution in this form we have treated as variable only ($2n-1$) of the $2n$ constants of integration which the general solution of the homogeneous system associated with equation (33) involves. This is, however, permissible in view of the fact that equation (33) admits the flux integral

$$\Sigma a_i\mu_i I_i^{(2)}=\text{constant}\ , \quad (35)$$

and we can arrange that this constant of integration has the same value as in the (1, n) approximation corresponding to the circumstance of a *given* constant net integrated flux.

Substituting for $I_i^{(2)}$ from equation (34) in equation (33), we obtain the variational equation

$$\left.\begin{aligned} &\tfrac{3}{4}F\mu_i\left\{\sum_{a=1}^{n-1}\frac{e^{-k_a\tau}}{1+\mu_i k_a}\frac{dL_a^{(2)}}{d\tau}+\sum_{a=1}^{n-1}\frac{e^{+k_a\tau}}{1-\mu_i k_a}\frac{dL_{-a}^{(2)}}{d\tau}+\frac{dQ^{(2)}}{d\tau}\right\} \\ &\quad=\mu_i\int_0^\infty \delta_\nu\frac{dI_{\nu,i}^{(1)}}{d\tau}d\nu-\tfrac{1}{2}\int_0^\infty \delta_\nu\Sigma a_j\mu_j\frac{dI_{\nu,j}^{(1)}}{d\tau}d\nu \quad (i=\pm 1,\ldots,\pm n)\,. \end{aligned}\right\} \tag{36}$$

Of the $2n$ equations represented in the foregoing equation, only $2n-1$ are linearly independent, since the equation derived from (36) by multiplying with a_i and summing over all i's is identically satisfied (cf. II, eq. [25]). The rank of the systems (36) is, accordingly, $(2n-1)$—in agreement with the fact that we have only $(2n-1)$ functions $L_a^{(2)}$, $L_{-a}^{(2)}$, $(a=1,\ldots,n-1)$, and $Q^{(2)}$ to determine.

The order of the system of equations (36) can be further reduced to $(2n-2)$ by a proper averaging of the continuous absorption coefficient κ_ν to yield a $\bar{\kappa}$ to be used in the first gray-body approximation. Thus, multiplying equation (36) by $a_i\mu_i$ and summing over all i's, we obtain

$$\tfrac{1}{2}F\frac{dQ^{(2)}}{d\tau}=\int_0^\infty \delta_\nu\Sigma a_i\mu_i^2\frac{dI_{\nu,i}^{(1)}}{d\tau}d\nu\,, \tag{37}$$

use having been made of the relations (II, eq. [24] and eq. [58] below)

$$\sum_i a_i\mu_i^2=\tfrac{2}{3} \qquad \text{and} \qquad \sum_i\frac{a_i\mu_i^2}{1+\mu_i k_a}=0\,. \tag{38}$$

Accordingly, if we arrange that

$$\int_0^\infty \delta_\nu\Sigma a_i\mu_i^2\frac{dI_{\nu,i}^{(1)}}{d\tau}d\nu=2\int_0^\infty \delta_\nu\frac{dK_\nu^{(1)}}{d\tau}d\nu=0\,, \tag{39}$$

we shall have the integral

$$Q^{(2)}=\text{constant}=Q^{(1)}+\Delta Q \qquad \text{(say)}\,. \tag{40}$$

But equation (39) implies that the mean absorption coefficient $\bar{\kappa}$ has to be defined according to

$$\bar{\kappa}\int_0^\infty\frac{dK_\nu^{(1)}}{d\tau}d\nu=\int_0^\infty\kappa_\nu\frac{dK_\nu^{(1)}}{d\tau}d\nu\,, \tag{41}$$

or, alternatively (cf. eqs. [2] and [5]),

$$\bar{\kappa}=\frac{1}{F}\int_0^\infty\kappa_\nu F_\nu^{(1)}d\nu\,; \tag{42}$$

for with this choice of $\bar{\kappa}$ the departures of κ_ν from $\bar{\kappa}$, when similarly averaged, will be zero:

$$\int_0^\infty\delta_\nu\frac{dK_\nu^{(1)}}{d\tau}d\nu=\frac{1}{4}\int_0^\infty\delta_\nu F_\nu^{(1)}d\nu\doteq 0\,. \tag{43}$$

Attention may be drawn here to a consequence of the foregoing method of averaging κ_ν for the important special case when δ_ν is independent of τ, i.e., when $\kappa_\nu/\bar{\kappa}$ is independent of depth. In this case we can re-write equation (19) in the form

$$B^{(2)}=J^{(2)}+\frac{1}{4}\frac{d}{d\tau}\int_0^\infty\delta_\nu F_\nu^{(1)}d\nu\,, \tag{44}$$

and equation (43) now implies that

$$B^{(2)} = J^{(2)} \qquad (\kappa_\nu/\bar{\kappa} \text{ independent of } \tau) . \quad (45)$$

With $\bar{\kappa}$ defined as in equation (42), the variational equations become

$$\left.\begin{aligned} &\tfrac{3}{4}F\mu_i\left\{\sum_{a=1}^{n-1}\frac{e^{-k_a\tau}}{1+\mu_i k_a}\frac{dL_a^{(2)}}{d\tau}+\sum_{a=1}^{n-1}\frac{e^{+k_a\tau}}{1-\mu_i k_a}\frac{dL_{-a}^{(2)}}{d\tau}\right\} \\ &= \mu_i\int_0^\infty \delta_\nu \frac{dI_{\nu,i}^{(1)}}{d\tau}d\nu-\frac{1}{2}\int_0^\infty \delta_\nu \Sigma a_j\mu_j\frac{dI_{\nu,j}^{(1)}}{d\tau}d\nu \quad (i=\pm 1, \ldots, \pm n) . \end{aligned}\right\} \quad (46)$$

Though $2n$ equations are represented in the foregoing equation, only $2n - 2$ of these are linearly independent, corresponding to the $(2n - 2)$ functions $L_a^{(2)}$ and $L_{-a}^{(2)}$, $(a = 1, \ldots, n - 1)$, which are to be determined.

In view of that fact that the rank of the system (46) is less than the number of equations, it appears that the most symmetrical way of treating the variational equations is the following:

Multiply equation (46) by $a_i\mu_i^{m-1}$, $(m = 1, \ldots, 2n)$, and sum over all i's. We obtain

$$\left.\begin{aligned} &\tfrac{3}{4}F\left\{\sum_{a=1}^{n-1}D_{m,a}e^{-k_a\tau}\frac{dL_a^{(2)}}{d\tau}+(-1)^m\sum_{a=1}^{n-1}D_{m,a}e^{+k_a\tau}\frac{dL_{-a}^{(2)}}{d\tau}\right\} \\ &\qquad = \bar{\delta}_m-\frac{\epsilon_{m,\,\text{odd}}}{m}\bar{\delta}_1 \quad (m=1, \ldots, 2n) , \end{aligned}\right\} \quad (47)$$

where we have written

$$D_{m,a}=\sum_i \frac{a_i\mu_i^m}{1+\mu_i k_a}=(-1)^m\sum_i\frac{a_i\mu_i^m}{1-\mu_i k_a} \quad (48)$$

and

$$\bar{\delta}_m=\int_0^\infty \delta_\nu \sum_i a_i\mu_i^m\frac{dI_{\nu,i}^{(1)}}{d\tau}d\nu \quad (49)$$

and where $\epsilon_{m,\,\text{odd}}$ has the same meaning as in equation (30). It may be noted that in deriving equation (47) use has been made of the relation

$$\sum_i a_i\mu_i^{m-1}=\frac{2}{m}\epsilon_{m,\,\text{odd}} \qquad (m=1, \ldots, 4n) . \quad (50)$$

There is a simple recursion formula which $D_{m,a}$ satisfies and which enables a direct evaluation of this quantity. We have

$$D_{m,a}=\frac{1}{k_a}\sum_i a_i\mu_i^{m-1}\left(1-\frac{1}{1+\mu_i k_a}\right), \quad (51)$$

or, using equation (50),

$$D_{m,a}=\frac{1}{k_a}\left(\frac{2}{m}\epsilon_{m,\,\text{odd}}-D_{m-1,a}\right) \qquad (m=1, \ldots, 4n) , \quad (52)$$

which is the required recursion formula. For odd, respectively even, values of m, the formula takes the forms

$$D_{2j-1,a}=\frac{1}{k_a}\left(\frac{2}{2j-1}-D_{2j-2,a}\right) \quad (53)$$

and

$$D_{2j,\,a} = -\frac{1}{k_a} D_{2j-1,\,a}\,. \tag{54}$$

Combining the relations (53) and (54), we have

$$D_{2j-1,\,a} = \frac{1}{k_a}\left(\frac{2}{2j-1} + \frac{1}{k_a} D_{2j-3,\,a}\right) \tag{55}$$

and

$$D_{2j,\,a} = -\frac{1}{k_a^2}\left(\frac{2}{2j-1} - D_{2j-2,\,a}\right). \tag{56}$$

On the other hand, in our present notation the equation for the characteristic roots k_a can be written as (cf. eq. [22])

$$D_{0,\,a} = 2 \qquad (a = 1, \ldots\ldots, n-1)\,. \tag{57}$$

From the recursion formula (52) we now conclude that

$$D_{2,\,a} = -\frac{1}{k_a} D_{1,\,a} = -\frac{1}{k_a^2}(2 - D_{0,\,a}) = 0 \tag{58}$$

and

$$D_{3,\,a} = \frac{2}{3k_a} \quad \text{and} \quad D_{4,\,a} = -\frac{2}{3k_a^2}\,. \tag{59}$$

The formula (55) will now enable us to determine $D_{m,\,a}$ successively for all odd values of m greater than 3; similarly, the expressions for the even values of m follow from equation (56). In this manner we find that

$$D_{2j-1,\,a} = \frac{2}{(2j-1)\,k_a} + \frac{2}{(2j-3)\,k_a^3} + \ldots\ldots + \frac{2}{3k_a^{2j-3}} \qquad (j = 2, \ldots\ldots, n) \tag{60}$$

and

$$D_{2j,\,a} = -\frac{2}{(2j-1)\,k_a^2} - \frac{2}{(2j-3)\,k_a^4} - \ldots\ldots - \frac{2}{3k_a^{2j-2}} \quad (j = 2, \ldots\ldots n)\,. \tag{61}$$ [6]

[6] In terms of these $D_{2j,a}$'s it is possible to write down an alternative form of the characteristic equation which, in contrast to equation (22), does not require an explicit knowledge of the Gaussian weights and divisions. For, if p_{2j}, $(j = 0, \ldots\ldots, n)$, are the coefficients of the polynomial $P_{2n}(\mu)$, so that

$$P_{2n}(\mu) = \sum_{j=0}^{n} p_{2j}\mu^{2j}\,,$$

it is evident that

$$\sum_{j=0}^{n} p_{2j}D_{2j,\,a} = \sum_i \frac{a_i}{1+\mu_i k_a}\left(\sum_{j=0}^{n} p_{2j}\mu_i^{2j}\right) \equiv 0 \qquad (a = 1, \ldots\ldots, n-1)\,,$$

since the μ_i's are, by definition, the zeros of $P_{2n}(\mu)$. Hence, with the definitions

$$-D_{2j} = \frac{2}{(2j-1)k^2} + \frac{2}{(2j-3)k^4} + \ldots\ldots + \frac{2}{3k^{2j-2}} \qquad (j = 2, \ldots\ldots, n)$$

and

$$D_0 = 2 \quad \text{and} \quad D_2 = 0\,,$$

the equation for the characteristic roots takes the form

$$\sum_{j=0}^{n} p_{2j}D_{2j} = 0\,.$$

Returning to equation (47), we first observe that, since D_1, D_2, and $\bar{\delta}_2$ are all zero (cf. eqs. [39] and [49]), then, of the $2n$ equations represented by equation (47), those for $m = 1$ and 2 are identically satisfied. Accordingly, we need consider equation (47) only for values of $m = 3, \ldots, 2n$. For odd, respectively even, values of m, equation (47) takes the forms

$$\tfrac{3}{4}F\sum_{a=1}^{n-1} D_{2j-1,\,a}\left(e^{-k_a\tau}\frac{dL_a^{(2)}}{d\tau}-e^{+k_a\tau}\frac{dL_{-a}^{(2)}}{d\tau}\right)=\bar{\delta}_{2j-1}-\frac{1}{2j-1}\,\bar{\delta}_1\quad (j=2,\ldots,n)\qquad(62)$$

and

$$\tfrac{3}{4}F\sum_{a=1}^{n-1} D_{2j,\,a}\left(e^{-k_a\tau}\frac{dL_a^{(2)}}{d\tau}+e^{+k_a\tau}\frac{dL_{-a}^{(2)}}{d\tau}\right)=\bar{\delta}_{2j}\qquad (j=2,\ldots,n)\,;\qquad(63)$$

or, substituting for $D_{2j-1,\,a}$ and $D_{2j,\,a}$ according to equations (60) and (61), we have

$$\left.\begin{aligned}\frac{3}{2}\sum_{a=1}^{n-1}\left\{\frac{1}{3k_a^{2j-3}}+\frac{1}{5k_a^{2j-5}}+\ldots+\frac{1}{(2j-1)\,k_a}\right\}X_a&=\bar{\delta}_{2j-1}-\frac{1}{2j-1}\,\bar{\delta}_1\\ (j&=2,\ldots,n)\end{aligned}\right\}\qquad(64)$$

and

$$\frac{3}{2}\sum_{a=1}^{n-1}\left\{\frac{1}{3k_a^{2j-2}}+\frac{1}{5k_a^{2j-4}}+\ldots+\frac{1}{(2j-1)\,k_a^2}\right\}Y_a=-\,\bar{\delta}_{2j}\quad (j=2,\ldots,n)\,,\qquad(65)$$

where, for the sake of brevity, we have written

$$\left.\begin{aligned}X_a&=F\left(e^{-k_a\tau}\frac{dL_a^{(2)}}{d\tau}-e^{+k_a\tau}\frac{dL_{-a}^{(2)}}{d\tau}\right),\\ Y_a&=F\left(e^{-k_a\tau}\frac{dL_a^{(2)}}{d\tau}+e^{+k_a\tau}\frac{dL_{-a}^{(2)}}{d\tau}\right).\end{aligned}\right\}\qquad(66)$$

The linear systems represented by equations (64) and (65) can be brought to forms more convenient for their solutions by the following procedure.

Considering, for example, equation (64), we find that the system of equations which this represents is

$$\left.\begin{aligned}\frac{3}{2}\sum_{a=1}^{n-1}\frac{1}{3k_a}X_a&=\bar{\delta}_3-\tfrac{1}{3}\bar{\delta}_1\,,\\ \frac{3}{2}\sum_{a=1}^{n-1}\left(\frac{1}{3k_a^3}+\frac{1}{5k_a}\right)X_a&=\bar{\delta}_5-\tfrac{1}{5}\bar{\delta}_1\,,\\ \frac{3}{2}\sum_{a=1}^{n-1}\left(\frac{1}{3k_a^5}+\frac{1}{5k_a^3}+\frac{1}{7k_a}\right)X_a&=\bar{\delta}_7-\tfrac{1}{7}\bar{\delta}_1\,,\\ \ldots\ldots\ldots&\ldots\ldots,\\ \frac{3}{2}\sum_{a=1}^{n-1}\left(\frac{1}{3k_a^{2n-3}}+\frac{1}{5k_a^{2n-5}}+\ldots+\frac{1}{2n-1}\,\frac{1}{k_a}\right)X_a&=\bar{\delta}_{2n-1}-\frac{1}{2n-1}\,\bar{\delta}_1\,.\end{aligned}\right\}\qquad(67)$$

The form of these equations suggests that we construct the following simpler system of equations:

$$\left.\begin{aligned}
&\frac{1}{2}\sum_{a=1}^{n-1}\frac{1}{k_a}X_a=\bar{\delta}_3-\tfrac{1}{3}\bar{\delta}_1=U_1\,,\\
&\frac{1}{2}\sum_{a=1}^{n-1}\frac{1}{k_a^3}X_a=\bar{\delta}_5-\tfrac{1}{5}\bar{\delta}_1-\tfrac{3}{5}U_1=U_2\,,\\
&\frac{1}{2}\sum_{a=1}^{n-1}\frac{1}{k_a^5}X_a=\bar{\delta}_7-\tfrac{1}{7}\bar{\delta}_1-\tfrac{3}{5}U_2-\tfrac{3}{7}U_1=U_3\,,\\
&\quad.\;.\;.\;.\;.\;.\;.\;.\;.\;.\;.\;.\;.\;.\;.\\
&\left.\begin{aligned}\frac{1}{2}\sum_{a=1}^{n-1}\frac{1}{k_a^{2j-3}}X_a&=\bar{\delta}_{2j-1}-\frac{1}{2j-1}\bar{\delta}_1-\tfrac{3}{5}U_{j-2}-\tfrac{3}{7}U_{j-3}-\;.\;.\;.\;.\;-\frac{3}{2j-1}U_1\\ &=U_{j-1}\end{aligned}\right\}\\
&\quad.\;,\\
&\left.\begin{aligned}\frac{1}{2}\sum_{a=1}^{n-1}\frac{1}{k_a^{2n-3}}X_a&=\bar{\delta}_{2n-1}-\frac{1}{2n-1}\bar{\delta}_1-\tfrac{3}{5}U_{n-2}-\tfrac{3}{7}U_{n-3}-\;.\;.\;.\;.\;-\frac{3}{2n-1}U_1\\ &=U_{n-1}\end{aligned}\right\}.
\end{aligned}\right\}\qquad(68)$$

In other words, we can reduce the systems of equations (67) to the simpler one

$$\sum_{a=1}^{n-1}\frac{1}{k_a^{2j-1}}X_a=2U_j\qquad(j=1,\;.\;.\;.\;.\;,n-1)\,,\qquad(69)$$

where the U_j's are defined as in equation (68).

Similarly, starting from equation (65), we can construct the simpler system

$$\sum_{a=1}^{n-1}\frac{1}{k_a^{2j}}Y_a=-2V_j\qquad(j=1,\;.\;.\;.\;.\;,n-1)\,.\qquad(70)$$

where the quantities V_j $(j = 1, \ldots, n-1)$ are defined by the recurrence relation

$$V_{j-1}=\bar{\delta}_{2j}-\tfrac{3}{5}V_{j-2}-\tfrac{3}{7}V_{j-3}-\;.\;.\;.\;.\;-\frac{3}{2j-1}V_1\quad(j=3,\;.\;.\;.\;.\;,n)\qquad(71)$$

and

$$V_1=\bar{\delta}_4\,.\qquad(72)$$

We shall now show how it is possible to write down the solution of the equations (69) and (70) quite generally.

Considering first equation (69), we find that the system of equations which this represents can be written down as a single vector equation

$$\boldsymbol{KX}=2\boldsymbol{U}\,,\qquad(73)$$

where $\boldsymbol{X}$ and $\boldsymbol{U}$ stand for the vectors

$$\boldsymbol{X}=(X_1,X_2,\;.\;.\;.\;.\;,X_{n-1})\quad\text{and}\quad\boldsymbol{U}=(U_1,\;.\;.\;.\;.\;,U_{n-1})\qquad(74)$$

and $\boldsymbol{K}$ for the matrix:

$$\boldsymbol{K}=\begin{pmatrix} \frac{1}{k_1} & \frac{1}{k_2} & \cdots & \frac{1}{k_{n-1}} \\ \frac{1}{k_1^3} & \frac{1}{k_2^3} & \cdots & \frac{1}{k_{n-1}^3} \\ \cdot & \cdot & \cdot & \cdot \, , \\ \frac{1}{k_1^{2n-3}} & \frac{1}{k_2^{2n-3}} & \cdots & \frac{1}{k_{n-1}^{2n-3}} \end{pmatrix} . \tag{75}$$

Equation (73) expresses a relation between $\boldsymbol{U}$ and a linear transform of $\boldsymbol{X}$. This relation can be inverted into a relation between $\boldsymbol{X}$ and the linear transform of $\boldsymbol{U}$ obtained with the inverse of the matrix $\boldsymbol{K}$.[7] Thus

$$\boldsymbol{X}=2\boldsymbol{K}^{-1}\boldsymbol{U}\,, \tag{76}$$

where $\boldsymbol{K}^{-1}$ denotes the inverse of $\boldsymbol{K}$. For the particular matrix $\boldsymbol{K}$ the inverse can be explicitly written down. But before we do this, it is first necessary to introduce certain definitions.

Consider the expansion of the product

$$\prod_{j\neq r}^{1,\,n-1}(x^2-k_j^2)=(x^2-k_1^2)\,(x^2-k_2^2)\ldots\ldots(x^2-k_{r-1}^2)\,(x^2-k_{r+1}^2)\ldots\ldots(x^2-k_{n-1}^2) \tag{77}$$

in descending powers of x^2. We have

$$\left.\begin{aligned}\prod_{j\neq r}^{1,\,n-1}(x^2-k_j^2) &= x^{2n-4}-x^{2n-6}\Bigl(\sum_{\mu_1\neq r}^{1,\,n-1}k_{\mu_1}^2\Bigr)+x^{2n-8}\Bigl(\sum_{\mu_1\neq\mu_2\neq r}^{1,\,n-1}k_{\mu_1}^2k_{\mu_2}^2\Bigr)\\ &\quad -x^{2n-10}\Bigl(\sum_{\mu_1\neq\mu_2\neq\mu_3\neq r}^{1,\,n-1}k_{\mu_1}^2k_{\mu_2}^2k_{\mu_3}^2\Bigr)+\ldots\ldots+(-1)^{n-2}\prod_{\mu\neq r}^{1,\,n-1}k_\mu^2\,;\end{aligned}\right\} \tag{78}$$

or, introducing the $(n-1)$ independent symmetric functions in the $(n-2)$ variables k_μ^2, $(\mu=1,\ldots\ldots,r-1,r+1,\ldots\ldots n-1)$,

$$\left.\begin{aligned} S_{0,\,r}&=1\,,\\ S_{1,\,r}&=-\sum_{\mu_1\neq r}^{1,\,n-1}k_{\mu_1}^2\,,\\ S_{2,\,r}&=+\sum_{\mu_1\neq\mu_2\neq r}^{1,\,n-1}k_{\mu_1}^2k_{\mu_2}^2\,,\\ &\cdots\cdots\cdots\\ S_{j,\,r}&=(-1)^j\sum_{\mu_1\neq\mu_2\neq\ldots\ldots\neq\mu_j\neq r}^{1,\,n-1}k_{\mu_1}^2k_{\mu_2}^2\ldots\ldots k_{\mu_j}^2\,,\\ &\cdots\cdots\cdots\cdots\cdots\,,\\ S_{n-2,\,r}&=(-1)^{n-2}k_1^2k_2^2\ldots\ldots k_{r-1}^2k_{r+1}^2\ldots\ldots k_{n-1}^2\,,\end{aligned}\right\} \tag{79}$$

[7] The existence of the inverse can be inferred from the linear independence of the equations (67), or more directly from the nonvanishing of the determinant of $\boldsymbol{K}$. The determinant of $\boldsymbol{K}$ can be written

we can re-write the expansion on the right-hand side of equation (78) in the form

$$\prod_{j\neq r}^{1,n-1}(x^2-k_j^2)=\sum_{\lambda=0}^{n-2}x^{2n-2\lambda-4}S_{\lambda,\,r}\,. \tag{80}$$

In terms of the functions $S_{\lambda,r}$, ($\lambda = 0, 1, \ldots, n-2$, and $r = 1, \ldots, n-1$), defined in this manner, the inverse of the matrix $\boldsymbol{K}$ can be written down. We have

$$\boldsymbol{K}^{-1}=\begin{pmatrix}
\frac{k_1^{2n-3}S_{0,1}}{\prod_{j\neq 1}(k_1^2-k_j^2)} & \frac{k_1^{2n-3}S_{1,1}}{\prod_{j\neq 1}(k_1^2-k_j^2)} & \cdots & \frac{k_1^{2n-3}S_{\lambda,1}}{\prod_{j\neq 1}(k_1^2-k_j^2)} & \cdots & \frac{k_1^{2n-3}S_{n-2,1}}{\prod_{j\neq 1}(k_1^2-k_j^2)}\\
\frac{k_2^{2n-3}S_{0,2}}{\prod_{j\neq 2}(k_2^2-k_j^2)} & \frac{k_2^{2n-3}S_{1,2}}{\prod_{j\neq 2}(k_2^2-k_j^2)} & \cdots & \frac{k_2^{2n-3}S_{\lambda,2}}{\prod_{j\neq 2}(k_2^2-k_j^2)} & \cdots & \frac{k_2^{2n-3}S_{n-2,2}}{\prod_{j\neq 2}(k_2^2-k_j^2)}\\
\cdot & \cdot & \cdot & \cdot & \cdot & \cdot\\
\frac{k_a^{2n-3}S_{0,a}}{\prod_{j\neq a}(k_a^2-k_j^2)} & \frac{k_a^{2n-3}S_{1,a}}{\prod_{j\neq a}(k_a^2-k_j^2)} & \cdots & \frac{k_a^{2n-3}S_{\lambda,a}}{\prod_{j\neq a}(k_a^2-k_j^2)} & \cdots & \frac{k_a^{2n-3}S_{n-2,a}}{\prod_{j\neq a}(k_a^2-k_j^2)}\\
\cdot & \cdot & \cdot & \cdot & \cdot & \cdot\\
\frac{k_{n-1}^{2n-3}S_{0,n-1}}{\prod_{j\neq n-1}(k_{n-1}^2-k_j^2)} & \frac{k_{n-1}^{2n-3}S_{1,n-1}}{\prod_{j\neq n-1}(k_{n-1}^2-k_j^2)} & \cdots & \frac{k_{n-1}^{2n-3}S_{\lambda,n-1}}{\prod_{j\neq n-1}(k_{n-1}^2-k_j^2)} & \cdots & \frac{k_{n-1}^{2n-3}S_{n-2,n-1}}{\prod_{j\neq n-1}(k_{n-1}^2-k_j^2)}
\end{pmatrix}. \tag{81}$$

That the foregoing matrix does, in fact, define the inverse of the matrix $\boldsymbol{K}$ can be verified directly. Thus,

$$\left.\begin{aligned}
(\boldsymbol{K}^{-1}\boldsymbol{K})_{l,\,m}&=\sum_{\lambda=0}^{n-2}\frac{k_l^{2n-3}S_{\lambda,\,l}}{\prod_{j\neq l}(k_l^2-k_j^2)}\,\frac{1}{k_m^{2\lambda+1}}\\
&=\left(\frac{k_l}{k_m}\right)^{2n-3}\frac{1}{\prod_{j\neq l}(k_l^2-k_j^2)}\sum_{\lambda=0}^{n-2}k_m^{2n-2\lambda-4}S_{\lambda,\,l}\,,
\end{aligned}\right\} \tag{82}$$

or, according to equation (80),

$$(\boldsymbol{K}^{-1}\boldsymbol{K})_{l,\,m}=\left(\frac{k_l}{k_m}\right)^{2n-3}\frac{1}{\prod_{j\neq l}(k_l^2-k_j^2)}\prod_{j\neq l}(k_m^2-k_j^2)\,. \tag{83}$$

---

down as it is related to the well-known Vandermonde determinant (cf. O. Perron, *Algebra*, Vol. 1, § 22, pp. 92–94, Leipzig: Gruyter, 1932). We have

$$|K|=\frac{1}{k_1k_2\ldots k_{n-1}}\prod_{i>j}^{1,n-1}\left(\frac{1}{k_i^2}-\frac{1}{k_j^2}\right). \tag{75$'$}$$

If $l \neq m$, the product which occurs in the numerator of the right-hand side of equation (83) includes the term $j = m$ and consequently vanishes. On the other hand, if $l = m$, the right-hand side of equation (83) clearly reduces to 1. Hence,

$$(\boldsymbol{K}^{-1}\boldsymbol{K})_{l,\,m} = 1 \text{ if } l = m \text{ and zero otherwise}\,, \tag{84}$$

proving the inverse relationship between the matrices (75) and (81).

Hence, with $\boldsymbol{K}^{-1}$ as defined in equation (81), equation (76) determines $\boldsymbol{X}$. In terms of the components X_a of $\boldsymbol{X}$ this equation becomes

$$\left. \begin{aligned} X_a = F\left(e^{-k_a\tau}\frac{dL_a^{(2)}}{d\tau} - e^{+k_a\tau}\frac{dL_{-a}^{(2)}}{d\tau}\right) = +\frac{2k_a^{2n-3}}{\prod_{j\neq a}(k_a^2-k_j^2)}\sum_{\lambda=0}^{n-2} S_{\lambda,\,a}U_{\lambda+1}(\tau) \\ (a = 1, \ldots, n-1)\,. \end{aligned} \right\} \tag{85}$$

Similarly, the solution of equation (70) is found to be

$$\left. \begin{aligned} Y_a = F\left(e^{-k_a\tau}\frac{dL_a^{(2)}}{d\tau} + e^{+k_a\tau}\frac{dL_{-a}^{(2)}}{d\tau}\right) = -\frac{2k_a^{2n-2}}{\prod_{j\neq a}(k_a^2-k_j^2)}\sum_{\lambda=0}^{n-2} S_{\lambda,\,a}V_{\lambda+1}(\tau) \\ (a = 1, \ldots, n-1)\,. \end{aligned} \right\} \tag{86}$$

From equations (85) and (86) we obtain

$$F\frac{dL_a^{(2)}}{d\tau} = +\frac{k_a^{2n-3}}{\prod_{j\neq a}(k_a^2-k_j^2)}\sum_{\lambda=0}^{n-2} S_{\lambda,\,a}\left[U_{\lambda+1}(\tau) - k_aV_{\lambda+1}(\tau)\right]e^{+k_a\tau} \tag{87}$$

and

$$F\frac{dL_{-a}^{(2)}}{d\tau} = -\frac{k_a^{2n-3}}{\prod_{j\neq a}(k_a^2-k_j^2)}\sum_{\lambda=0}^{n-2} S_{\lambda,\,a}\left[U_{\lambda+1}(\tau) + k_aV_{\lambda+1}(\tau)\right]e^{-k_a\tau}\,. \tag{88}$$

In their integrated forms the foregoing equations are

$$\left. \begin{aligned} FL_a^{(2)}(\tau) = \frac{k_a^{2n-3}}{\prod_{j\neq a}(k_a^2-k_j^2)}\sum_{\lambda=0}^{n-2} S_{\lambda,\,a}\int_0^\tau e^{+k_a\tau}\left[U_{\lambda+1}(\tau) - k_aV_{\lambda+1}(\tau)\right]d\tau \\ +F(L_a^{(1)} + \Delta L_a) \quad (a = 1, \ldots, n-1) \end{aligned} \right\} \tag{89}$$

and

$$\left. \begin{aligned} FL_{-a}^{(2)}(\tau) = \frac{k_a^{2n-3}}{\prod_{j\neq a}(k_a^2-k_j^2)}\sum_{\lambda=0}^{n-2} S_{\lambda,\,a}\int_\tau^\infty e^{-k_a\tau}\left[U_{\lambda+1}(\tau) + k_aV_{\lambda+1}(\tau)\right]d\tau \\ (a = 1, \ldots, n-1)\,, \end{aligned} \right\} \tag{90}$$

where in equation (89) ΔL_a, $(a = 1, \ldots, n-1)$, are $(n-1)$ constants of integration. However, it will be noticed that in integrating equation (88) in the form (90) we have made a particular choice of the constants of integrations so as to be compatible with the requirement that none of the quantities tend to infinity exponentially as $\tau \rightarrow \infty$ (cf. eq. [34]).

Finally, the constants of integration ΔL_a, $(a = 1, \ldots, n-1)$, and ΔQ (cf. eq. [40]) which occur in the solution of the variational equations are to be determined from the boundary conditions

$$I_{-i} = 0 \text{ at } \tau = 0 \text{ for } i = 1, \ldots, n. \tag{91}$$

Explicitly, these conditions reduce to (cf. II, eq. [21])

$$\sum_{a=1}^{n-1} \frac{L_a^{(1)} + \Delta L_a}{1 - \mu_i k_a} + Q^{(1)} + \Delta Q = \mu_i - \sum_{a=1}^{n-1} \frac{L_{-a}^{(2)}(0)}{1 + \mu_i k_a} \quad (i = 1, \ldots, n); \tag{92}$$

or, since the $L_a^{(1)}$'s and $Q^{(1)}$ satisfy the relation

$$\sum_{a=1}^{n-1} \frac{L_a^{(1)}}{1 - \mu_i k_a} + Q^{(1)} = \mu_i \qquad (i = 1, \ldots, n), \tag{93}$$

the equations which determine ΔL_a and ΔQ are

$$\sum_{a=1}^{n-1} \frac{\Delta L_a}{1 - \mu_i k_a} + \Delta Q = -\sum_{a=1}^{n-1} \frac{L_{-a}^{(2)}(0)}{1 + \mu_i k_a} \quad (i = 1, \ldots, n). \tag{94}$$ [8]

This completes the formal solution to the problem in the $(2, n)$ approximation.

5. *The solution in the* (2, 1) *approximation.*—The discussion of this case is of particular interest, as in this approximation

$$J^{(2)} = J^{(1)}, \tag{95}$$

and equation (19) becomes

$$B^{(2)} = J^{(1)} + \frac{1}{4} \int_0^\infty \delta_\nu \frac{dF_\nu^{(1)}}{d\tau} d\nu. \tag{96}$$

Substituting for $J^{(1)}$ its solution in the first approximation (II, eqs. [30] and [35]), we have

$$B^{(2)} = \tfrac{3}{4} F \left(\tau + \frac{1}{\sqrt{3}} + \frac{1}{3F} \int_0^\infty \delta_\nu \frac{dF_\nu^{(1)}}{d\tau} d\nu \right). \tag{97}$$

It may be recalled that in the foregoing equations the optical depth τ is evaluated in terms of the mean absorption coefficient $\bar{\kappa}$ defined by (cf. eq. [42])

$$\bar{\kappa} = \int_0^\infty \kappa_\nu \left(\frac{F_\nu^{(1)}}{F} \right) d\nu. \tag{98}$$

From equations (97) and (98) it is apparent that, when δ_ν is independent of depth,

$$B^{(2)} = B^{(1)} = \tfrac{3}{4} F \left(\tau + \frac{1}{\sqrt{3}} \right) \quad (\delta_\nu \text{ independent of } \tau). \tag{99}$$

[8] It is possible to write down explicitly the inverse of the linear transformation which appears on the left-hand side of this system (cf. *Ap. J.*, **101**, 320, 1945, eq. [8]). But, as the practical use to which these solutions can be put is limited, we shall not develop further these formalities here.

In other words, we have shown that *in a Milne-Eddington type of approximation the temperature distribution obtained on the gray-body assumption continues to be valid in a second approximation if the mean absorption coefficient* κ *(in terms of which* τ *has to be defined) is a straight average of* κ_ν*, weighted according to the net monochromatic flux of radiation of frequency* ν *in a gray atmosphere, and if, further,* $\kappa_\nu/\bar{\kappa}$ *is independent of depth.* This answers the question that we asked at the outset (§ 1), namely, as to the conditions under which an equation for the temperature distribution of the form (9) may be expected to hold even when the material departs from perfect grayness. We may draw particular attention to the fact that our answer has not provided any justification for averaging κ_ν according to the conventional method of Rosseland. Indeed, the Rosseland mean, as customarily defined by

$$\frac{1}{\bar{\kappa}}\int_0^\infty \frac{dB_\nu}{dT}\,d\nu = \int_0^\infty \frac{1}{\kappa_\nu}\frac{dB_\nu}{dT}\,d\nu\,, \tag{100}$$

weights κ_ν falsely and overemphasizes the violet region of the spectrum. On the other hand, our present discussion would rather suggest that in all preliminary considerations it would be preferable (and simpler) to average κ_ν directly, weighting with the Planck function corresponding to the effective temperature of the star. One further remark concerning the requirement of the constancy of $\kappa_\nu/\bar{\kappa}$ with depth may be made. With the negative ion of hydrogen firmly established as the principal source of absorption in stellar atmospheres of the solar and neighboring types, the requirement of the constancy of κ_ν/κ with depth is likely to be met fairly satisfactorily.

Returning to equations (97) and (98), we see that a practical use of these equations will require a knowledge of the monochromatic fluxes $F_\nu^{(1)}$ at various depths in a gray atmosphere.

Now, according to equation (26), we can write

$$F_\nu^{(1)}(\tau) = 2\int_\tau^\infty B_\nu(T_t^{(1)})\,E_2(t-\tau)\,dt - 2\int_0^\tau B_\nu(T_t^{(1)})\,E_2(\tau-t)\,dt\,, \tag{101}$$

where

$$T_t^{(1)} = T_e\left[\tfrac{3}{4}\left(t+q\left[t\right]\right)\right]^{1/4}. \tag{102}$$

Writing

$$\alpha = \frac{h\nu}{kT_e}, \tag{103}$$

we can express equation (101) in the form

$$\frac{F_\alpha^{(1)}(\tau)}{F} = \frac{30}{\pi^4}\,\frac{\alpha^3}{e^{1.23275\alpha}-1}\,f_\alpha(\tau)\,, \tag{104}$$

where

$$f_\alpha(\tau) = \left(e^{1.23275\alpha}-1\right)\left\{\int_\tau^\infty \frac{E_2(t-\tau)\,dt}{e^{\alpha/[3/4(t+q)]^{1/4}}-1} - \int_0^\tau \frac{E_2(\tau-t)\,dt}{e^{\alpha/[3/4(t+q)]^{1/4}}-1}\right\} \tag{105}$$

and

$$1.23275\ldots = \left(\frac{4}{\sqrt{3}}\right)^{1/4}. \tag{106}$$

Using for $q(\tau)$ the solution on the fourth approximation given in II, equation (54), Miss Frances Herman and the writer have numerically evaluated the function $f_\alpha(\tau)$ defined as in equation (105) for various values of α and τ. The results of the numerical integrations are summarized in Tables 1 and 2. In Table 1 we give the values of $f_\alpha(\tau)$, and in Table 2 these are converted according to equation (104) to give the monochromatic fluxes directly. And finally, in Table 3, we give values of $dF_a^{(1)}/d\tau$ obtained by direct numerical differentiation of the fluxes tabulated in Table 2.

TABLE 1

THE FUNCTION $f_a(\tau)$

τ	a = 0	1	2	3	4	6	8	10	12
0	0.601	0.682	0.803	0.982	1.236	2.10	3.82	7.42	15.2
0.1	.529	.619	.757	.960	1.249	2.23	4.20	8.32	17.3
0.2	.472	.568	.717	.938	1.257	2.35	4.58	9.24	19.4
0.3	.424	.524	.680	.915	1.259	2.46	4.94	10.19	21.7
0.4	.384	.485	.647	.893	1.257	2.56	5.29	11.15	24.1
0.5	.348	.450	.615	.869	1.251	2.64	5.62	12.11	26.6
0.6	.318	.420	.586	.847	1.244	2.72	5.95	13.08	29.1
0.7	.292	.393	.561	.826	1.236	2.79	6.26	14.06	31.8
0.8	.267	.368	.535	.804	1.224	2.85	6.56	15.03	34.6
0.9	.247	.346	.513	.784	1.214	2.90	6.84	16.00	37.4
1.0	.229	.326	.493	.765	1.203	2.95	7.12	16.96	40.3
1.2	.198	.292	.455	.729	1.177	3.04	7.64	18.85	46.2
1.4	.173	.264	.423	.696	1.152	3.10	8.11	20.70	52.2
1.6	.153	.241	.396	.666	1.126	3.16	8.54	22.49	58.4
1.8	.137	.221	.372	.639	1.101	3.20	8.94	24.23	64.5
2.0	0.124	0.204	0.351	0.613	1.076	3.23	9.30	25.90	70.7

TABLE 2

THE MONOCHROMATIC FLUXES $F_a^{(1)}(\tau)$ IN UNITS OF F

τ	a = 1	2	3	4	6	8	10	12
0	0.0864	0.1837	0.2074	0.1772	0.0856	0.0314	0.01013	0.00305
0.1	.0784	.1732	.2027	.1791	.0911	.0346	.01135	.00345
0.2	.0719	.1640	.1981	.1801	.0961	.0376	.01261	.00388
0.3	.0663	.1556	.1933	.1804	.1005	.0406	.01390	.00434
0.4	.0615	.1480	.1886	.1802	.1044	.0435	.01521	.00482
0.5	.0571	.1407	.1835	.1793	.1078	.0462	.01652	.00532
0.6	.0532	.1342	.1789	.1783	.1110	.0489	.01785	.00584
0.7	.0498	.1282	.1745	.1772	.1138	.0515	.01918	.00638
0.8	.0466	.1225	.1699	.1755	.1163	.0539	.02050	.00693
0.9	.0439	.1174	.1656	.1740	.1186	.0563	.02182	.00749
1.0	.0413	.1128	.1616	.1724	.1207	.0586	.02314	.00807
1.2	.0370	.1042	.1540	.1688	.1240	.0628	.02572	.00925
1.4	.0335	.0968	.1470	.1651	.1268	.0667	.02824	.01046
1.6	.0305	.0906	.1406	.1614	.1289	.0702	.03069	.01169
1.8	.0280	.0850	.1349	.1578	.1306	.0735	.03306	.01293
2.0	0.0259	0.0803	0.1295	0.1543	0.1319	0.0764	0.03534	0.01417

In Figure 1 we have illustrated the distribution in frequency of the net flux of radiation at various depths. It is of particular interest to observe how the redistribution in the frequencies takes place as we descend into the atmosphere.

With the data given in Tables 2 and 3 we should be able to use the solution (97) to determine the temperature distribution under a wide range of practical conditions. We shall return to such applications in later papers.

TABLE 3

THE DERIVATIVES OF THE MONOCHROMATIC FLUXES: $dF_\nu^{(1)}/d\tau$ IN UNITS OF F

τ	α = 1	2	3	4	6	8	10	12
0.......	−0.086	−0.11	−0.044	+0.024	+0.0585	+0.0323	+0.0120	+0.0039
0.1.....	.072	.098	.046	+ .015	.0523	.0311	.0124	.0044
0.2.....	.060	.088	.047	+ .007	.0467	.0302	.0128	.0048
0.3.....	.052	.080	.048	+ .001	.0417	.0293	.0130	.00465
0.4.....	.046	.074	.0485	− .005	.0367	.0281	.0131	.00490
0.5.....	.041	.068	.048	− .008	.0327	.0270	.0132	.00510
0.6.....	.036	.063	.046	− .010	.0295	.0260	.0133	.00530
0.7.....	.033	.058	.045	− .013	.0264	.0250	.0133	.00544
0.8.....	.030	.054	.043	− .015	.0240	.0242	.0132	.00558
0.9.....	.027	.050	.041	− .016	.0215	.0233	.0132	.00573
1.0.....	.024	.046	.040	− .017	.0193	.0222	.0131	.00583
1.2.....	.020	.040	.036	− .018	.0152	.0202	.0128	.00598
1.4.....	.016	.034	.033	− .0185	.0121	.0186	.0124	.00610
1.6.....	.014	.029	.030	− .0185	.0095	.0170	.0120	.00617
1.8.....	.011	.026	.028	− .018	.0075	.0156	.0116	.0062
2.0.....	−0.010	−0.022	−0.025	−0.017	+0.0060	+0.0142	+0.0112	+0.0062

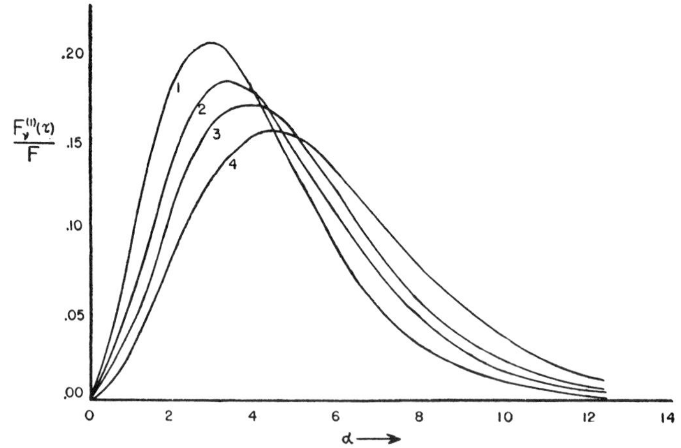

FIG. 1.—The frequency distribution of the net flux of radiation at various depths in a gray atmosphere. The abscissa measures the frequency (in units of kT_e/h), and the ordinate measures the flux (in units of the constant net integrated flux F). The curves *1*, *2*, *3*, and *4* refer to the depths $\tau=0$, 0.5, 1, and 2, respectively.

6. *The solutions for the temperature distribution in higher approximations.*—Turning next to solutions in higher approximations, we have (cf. eq. 34 and II, eqs. [28] and [29])

$$J^{(2)}=\tfrac{3}{4}F\left\{\tau+Q^{(2)}+\sum_{a=1}^{n-1}L_a^{(2)}(\tau)\,e^{-k_a\tau}+\sum_{a=1}^{n-1}L_{-a}^{(2)}(\tau)\,e^{+k_a\tau}\right\}; \qquad (107)$$

or, substituting for $Q^{(2)}$, $L_a^{(2)}$ and $L_{-a}^{(2)}$ according to equations (40), (89), and (90), we have

$$\left.\begin{aligned} J^{(2)} &= J^{(1)} + \tfrac{3}{4}F\left(\Delta Q + \sum_{a=1}^{n-1} \Delta L_a e^{-k_a\tau}\right) \\ &+ \frac{3}{4}\sum_{a=1}^{n-1} \frac{k_a^{2n-3}}{\prod_{j\neq a}(k_a^2 - k_j^2)} \sum_{\lambda=0}^{n-2} S_{\lambda,a}\Big\{ e^{-k_a\tau}\int_0^\tau e^{+k_a\tau}[U_{\lambda+1}(\tau) - k_a V_{\lambda+1}(\tau)]\,d\tau \\ &\qquad + e^{+k_a\tau}\int_\tau^\infty e^{-k_a\tau}[U_{\lambda+1}(\tau) + k_a V_{\lambda+1}(\tau)]\,d\tau \Big\}, \end{aligned}\right\} \tag{108}$$

where it may be recalled that the constants of integration ΔQ and ΔL_a, ($a = 1, \ldots,$ $n-1$), are to be determined according to equation (94). Corresponding to the foregoing solution for $J^{(2)}$, there is the temperature distribution

$$\frac{\sigma}{\pi}(T^{(2)})^4 = B^{(2)} = J^{(2)} + \frac{1}{4}\int_0^\infty \delta_\nu(\tau)\frac{dF_\nu^{(1)}}{d\tau}\,d\nu. \tag{109}$$

Equations (108) and (109) show that, when the material departs from being gray, the corrections which have to be made to the temperature distribution are of two kinds: first, because B is not, in general, equal to J, and, second, because the energy density of the radiation does not have the gray-atmosphere value. It is only under very special conditions (which we have already discussed in § 5) that these corrections vanish in a second approximation.

The solution (108) simplifies considerably for the case $n = 2$. For, in this (2, 2) approximation there is only one characteristic root (II, eq. [40]):

$$k_1 = \frac{\sqrt{35}}{3} = 1.97203; \tag{110}$$

and equation (108) becomes (cf. eqs. [68] and [72])

$$\left.\begin{aligned} J^{(2)} &= J^{(1)} + \tfrac{3}{4}F(\Delta Q + \Delta L_1 e^{-k_1\tau}) + \tfrac{3}{4}e^{-k_1\tau}\int_0^\tau e^{+k_1\tau}(\bar\delta_3 - \tfrac{1}{3}\bar\delta_1 - k_1\bar\delta_4)\,d(k_1\tau) \\ &\qquad + \tfrac{3}{4}e^{+k_1\tau}\int_\tau^\infty e^{-k_1\tau}(\bar\delta_3 - \tfrac{1}{3}\bar\delta_1 + k_1\bar\delta_4)\,d(k_1\tau), \end{aligned}\right\} \tag{111}$$

where it may be noted that (II, eqs. [30] and [42])

$$J^{(1)} = \tfrac{3}{4}F(\tau + 0.694025 - 0.116675e^{-k_1\tau}). \tag{112}$$

The equations which determine the constants of integrations ΔQ and ΔL_1 in the solution (111) are (cf. eq. [94])

$$\frac{\Delta L_1}{1 - \mu_1 k_1} + \Delta Q = -\frac{L_{-1}^{(2)}(0)}{1 + \mu_1 k_1} \tag{113}$$

and

$$\frac{\Delta L_1}{1 - \mu_2 k_1} + \Delta Q = -\frac{L_{-1}^{(2)}(0)}{1 + \mu_2 k_1}. \tag{114}$$

From these equations we find that

$$\Delta L_1 = \frac{(1 - \mu_1 k_1)(1 - \mu_2 k_1)}{(1 + \mu_1 k_1)(1 + \mu_2 k_1)} L_{-1}^{(2)}(0) \tag{115}$$

and

$$\Delta Q = -\frac{2}{(1 + \mu_1 k_1)(1 + \mu_2 k_1)} L_{-1}^{(2)}(0). \tag{116}$$

Substituting these values for ΔL_1 and ΔQ in equation (111), we obtain

$$\left.\begin{aligned}J^{(2)} &= J^{(1)}\\ &+\frac{3\left[(1-\mu_1 k_1)(1-\mu_2 k_1)e^{-k_1\tau}-2\right]}{4(1+\mu_1 k_1)(1+\mu_2 k_1)}\int_0^\infty e^{-k_1\tau}(\bar{\delta}_3-\tfrac{1}{3}\bar{\delta}_1+k_1\bar{\delta}_4)\,d(k_1\tau)\\ &+\tfrac{3}{4}e^{-k_1\tau}\int_0^\tau e^{+k_1\tau}(\bar{\delta}_3-\tfrac{1}{3}\bar{\delta}_1-k_1\bar{\delta}_4)\,d(k_1\tau)\\ &+\tfrac{3}{4}e^{+k_1\tau}\int_\tau^\infty e^{-k_1\tau}(\bar{\delta}_3-\tfrac{1}{3}\bar{\delta}_1+k_1\bar{\delta}_4)\,d(k_1\tau)\,.\end{aligned}\right\}\quad(117)$$

From equation (117) we find that for $\tau = 0$ (cf. II, eq. [71]),

$$\left.\begin{aligned}J^{(2)}(0) &= \frac{\sqrt{3}}{4}F+\frac{\sqrt{3}}{2}\,\frac{k_1}{(1+\mu_1 k_1)(1+\mu_2 k_1)}\\ &\qquad\times\int_0^\infty e^{-k_1\tau}(\bar{\delta}_3-\tfrac{1}{3}\bar{\delta}_1+k_1\bar{\delta}_4)\,d(k_1\tau)\,.\end{aligned}\right\}\quad(118)$$

The boundary temperature in the second approximation is therefore given by

$$\left.\begin{aligned}\frac{\sigma}{\pi}(T_0^{(2)})^4 &= \frac{\sigma}{\pi}(T_0^{(1)})^4+\frac{1}{4}\int_0^\infty\left(\delta_\nu\frac{dF_\nu^{(1)}}{d\tau}\right)_{\tau=0}d\nu\\ &+\frac{\sqrt{3}}{2}\,\frac{k_1}{(1+\mu_1 k_1)(1+\mu_2 k_1)}\int_0^\infty e^{-k_1\tau}(\bar{\delta}_3-\tfrac{1}{3}\bar{\delta}_1+k_1\bar{\delta}_4)\,d(k_1\tau)\,.\end{aligned}\right\}\quad(119)$$

It is seen that a practical use of the foregoing solutions in the (2, 2) approximation will require, in addition to the monochromatic fluxes $F_\nu^{(1)}$, a knowledge also of the quantities

$$M_\nu^{(1)}(\tau)=\frac{1}{2}\int_{-1}^{+1}\frac{dI_\nu^{(1)}}{d\tau}\mu^3 d\mu \quad\text{and}\quad N_\nu^{(1)}(\tau)=\frac{1}{2}\int_{-1}^{+1}\frac{dI_\nu^{(1)}}{d\tau}\mu^4 d\mu\,,\quad(120)$$

as these will be needed in the evaluation of $\bar{\delta}_3$ and $\bar{\delta}_4$. However, since the first approximation refers only to a gray atmosphere, $M_\nu^{(1)}$ and $N_\nu^{(1)}$ can be expressed in terms of $F_\nu^{(1)}$. For, from the equation of transfer appropriate to this case (eq. [12]), we can readily show that

$$\frac{dK_\nu^{(1)}}{d\tau}=\tfrac{1}{4}F_\nu^{(1)}\,;\qquad\frac{dM_\nu^{(1)}}{d\tau}=K_\nu^{(1)}(\tau)-\tfrac{1}{3}B_\nu^{(1)}(\tau)\,,\quad(121)$$

and

$$\frac{dN_\nu^{(1)}}{d\tau}=M_\nu^{(1)}(\tau)\,;\quad(122)$$

and, accordingly, $M_\nu^{(1)}(\tau)$ and $N_\nu^{(1)}(\tau)$ can be obtained successively by quadratures involving at each stage only known functions. Numerical work relating to these quantities with applications to practical problems will be given in later papers.

In conclusion, I wish to record my indebtedness to Miss Frances Herman for assistance with the very laborious numerical work which was involved in the preparation of Tables 1, 2, and 3.

4

ON THE RADIATIVE EQUILIBRIUM OF A STELLAR ATMOSPHERE. VIII

S. CHANDRASEKHAR
Yerkes Observatory
Received February 14, 1945

ABSTRACT

In this paper the methods which have been developed in the earlier papers of this series have been applied to the solution of the equation of transfer

$$\mu \frac{dI}{d\tau} = I(\tau, \mu) - B(\tau) \, ; \quad B = \frac{1}{2}\int_{-1}^{+1} I d\mu + \tfrac{1}{4} F e^{-\tau \sec \beta} \, ,$$

representing the radiative equilibrium of an atmosphere exposed to a parallel beam of radiation of flux πF per unit area normal to itself and incident at an angle β normal to the boundary of the atmosphere.

General solutions in the *nth* approximation have been found; and it is shown that, on the method of solution adopted, it is true in every approximation that

$$\frac{B(0)}{B(\infty)} = \frac{\sec \beta}{\sqrt{3}} \, .$$

It is further shown that the angular distribution of the "reflected" radiation for the problem on hand is related, in a simple way, with the law of darkening in an atmosphere characterized by a constant net flux and with no incident radiation. Thus, elementary and simple proofs have been obtained for the results first established by E. Hopf, using a special theory involving integral equations.

1. *Introduction.*—A problem in the theory of radiative transfer which has considerable practical interest is the one relating to the reflection effect in eclipsing binaries. This problem was first considered by A. S. Eddington[1] and later in greater detail by E. A. Milne.[2] As was shown particularly by Milne, the problem basic to the consideration of this effect is that of the radiative equilibrium of a semi-infinite atmosphere exposed to a parallel beam of radiation of flux πF per unit area normal to itself and incident at an angle β normal to the boundary of the atmosphere (see Fig. 1). In other words, we require to solve the equation of transfer (cf. Milne, *op. cit.*)

$$\mu \frac{dI}{d\tau} = I - \frac{1}{2}\int_{-1}^{+1} I d\mu - \tfrac{1}{4} F e^{-\tau \sec\beta} \, , \tag{1}$$

where I denotes the specific intensity of the integrated radiation derived from the material. The solution of the equation of transfer (1) was obtained by Milne in an approximation equivalent to that of Schwarzschild's and applied to the reflection problem. Later, certain exact results relating to the solution of equation (1) were proved by E. Hopf.[3] But apparently there has been no attempt to obtain solutions explicitly in approximations higher than those of Milne's. In this paper we propose to apply to the solution of equation (1) the methods which have been developed in the earlier papers of this series.[4] And, as we shall see, the application of these methods to the solution of equation (1) leads also to simple and elementary proofs of the exact results of Hopf.

[1] *M.N.*, **86,** 320, 1926.

[2] *M.N.*, **87,** 43, 1926; also *Handb. d. Ap.*, **3,** No. 1, 134–141, 1930.

[3] *Mathematical Problems of Radiative Equilibrium* (Cambridge Mathematical Tract No. 31), pp. 55–59, Cambridge, England, 1934.

[4] See particularly *Ap. J.*, **100,** 76, 1944. This paper will be referred to as "II." Familiarity with the methods and results of this paper will be assumed in the present investigation.

2. *The solution of equation (1) in the* n*th approximation.*—As in paper II, we replace the integral on the right-hand side of equation (1) by a sum according to Gauss's formula for numerical quadratures and obtain an equivalent system of linear equations which in the nth approximation is

$$\mu_i \frac{dI_i}{d\tau} = I_i - \tfrac{1}{2}\Sigma a_j I_j - \tfrac{1}{4}F e^{-\tau \sec\beta} \quad (i = \pm 1, \ldots, \pm n) . \quad (2)$$

where the various symbols have the same meaning as in II.

It is seen that the homogeneous system associated with equation (2) is the same as that considered in II, §§ 2-6. Accordingly, the complementary function for the solution of equation (2) is the same as the general solution (II, eq. [18]) of the homogeneous system. To complete the solution we need only find a particular integral. This can be found in the following manner.

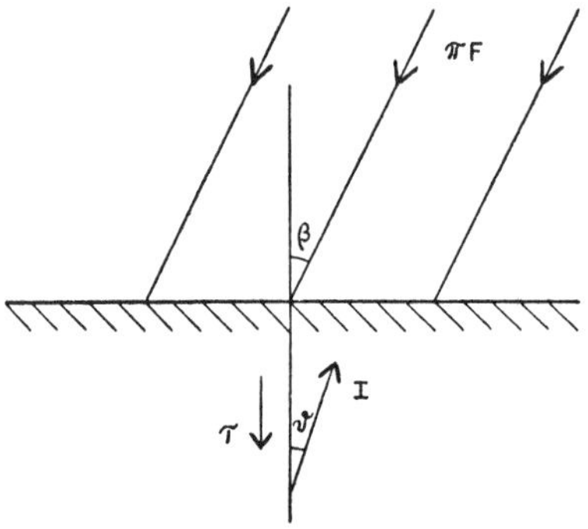

FIG. 1

Setting

$$I_i = \tfrac{1}{4}F h_i e^{-\tau \sec\beta} \quad (i = \pm 1, \ldots, \pm n) . \quad (3)$$

in equation (2) (the h_i's are certain constants unspecified for the present) we verify that we must have

$$h_i(1 + \mu_i \sec\beta) = \tfrac{1}{2}\Sigma a_j h_j + 1 \quad (i = \pm 1, \ldots, \pm n) . \quad (4)$$

Equation (4) implies that the constants h_i must be expressible in the form

$$h_i = \frac{\gamma}{1 + \mu_i \sec\beta} \quad (i = \pm 1, \ldots, \pm n) . \quad (5)$$

where the constant γ has to be determined by the condition (cf. eq. [4])

$$\gamma = \tfrac{1}{2}\gamma\Sigma \frac{a_i}{1 + \mu_i \sec\beta} + 1 . \quad (6)$$

In other words,

$$I_i = \tfrac{1}{4}F \frac{\gamma}{1 + \mu_i \sec\beta} e^{-\tau \sec\beta} \quad (i = \pm 1, \ldots, \pm n) \quad (7)$$

with

$$\gamma = \frac{1}{1 - \sum_{j=1}^{n} \frac{a_j}{1 - \mu_j^2 \sec^2\beta}} . \quad (8)$$

represents a particular integral of the equation (2). Adding to this particular integral the general solution of the homogeneous system which is compatible with the boundedness of the solution for $\tau \rightarrow \infty$, we have

$$I_i = \tfrac{1}{4}F\left\{\sum_{a=1}^{n-1}\frac{M_a e^{-k_a\tau}}{1+\mu_i k_a} + X + \frac{\gamma e^{-\tau\sec\beta}}{1+\mu_i \sec\beta}\right\} \quad (i = \pm 1, \ldots, \pm n) , \tag{9}$$

where M_a, $(a = 1, \ldots, n-1)$, and X are n constants of integration, to be determined by the boundary conditions at $\tau = 0$, and the k_a's $(a = 1, \ldots, n-1)$ are the $(n-1)$ distinct nonvanishing positive roots characteristic of the equation (cf. II, eq. [11])

$$1 = \sum_{j=1}^{n}\frac{a_j}{1-\mu_j^2 k^2} . \tag{10}$$

At $\tau = 0$ we have no radiation derived from the material. Accordingly, we should require that

$$I_{-i} = 0 \quad \text{at} \quad \tau = 0 \quad \text{and for} \quad i = 1, \ldots, n . \tag{11}$$

From equation (9) we now conclude that the equations which determine the M_a's and X are

$$\sum_{a=1}^{n-1}\frac{M_a}{1-\mu_i k_a} + X + \frac{\gamma}{1-\mu_i \sec\beta} = 0 \quad (i = 1, \ldots, n) . \tag{12}$$

In terms of the solution (9) for the I_i's we can obtain a convenient formula for J, defined by

$$J = \frac{1}{2}\int_{-1}^{+1} I\,d\mu = \tfrac{1}{2}\Sigma a_j I_j \tag{13}$$

in our present scheme of approximation. We have

$$J = \tfrac{1}{8}F\left\{\sum_{a=1}^{n-1} M_a e^{-k_a\tau}\sum_i \frac{a_i}{1+\mu_i k_a} + X\sum_i a_i + \gamma e^{-\tau\sec\beta}\sum_i \frac{a_i}{1+\mu_i \sec\beta}\right\} . \tag{14}$$

Using equations (6) and (10), we can reduce the foregoing equation to the form

$$J = \frac{1}{4}F\left\{\sum_{a=1}^{n-1} M_a e^{-k_a\tau} + X + (\gamma - 1)\, e^{-\tau\sec\beta}\right\} . \tag{15}$$

The temperature distribution in the atmosphere is given by (cf. eq. [1])

$$\frac{\sigma}{\pi}T^4 = B(\tau) = J + \tfrac{1}{4}F e^{-\tau\sec\beta} . \tag{16}$$

Substituting for J from equation (15), we obtain

$$B(\tau) = \tfrac{1}{4}F\left(\sum_{a=1}^{n-1} M_a e^{-k_a\tau} + X + \gamma e^{-\tau\sec\beta}\right) . \tag{17}$$

As $\tau \rightarrow 0$, respectively ∞, we have

$$B(0) = \tfrac{1}{4}F\left(\sum_{a=1}^{n-1} M_a + X + \gamma\right) \tag{18}$$

and

$$B(\infty) = \tfrac{1}{4}FX . \tag{19}$$

The law of darkening in the atmosphere corresponding to the temperature distribution (16) is readily found. We have

$$I(0, \mu) = \tfrac{1}{4}F\left\{\sum_{a=1}^{n-1} \frac{M_a}{1+k_a\mu} + X + \frac{\gamma}{1+\mu \sec \beta}\right\}. \tag{20}$$

This completes the formal solution. In the following section we shall show how the explicit forms of the solutions in the various approximations can be written down directly in terms of the respective solutions of the associated homogeneous equation

$$\mu \frac{dI}{d\tau} = I - \frac{1}{2}\int_{-1}^{+1} I d\mu, \tag{21}$$

given in paper II.

3. *The relation of the solution of equation (1) with that of the associated homogeneous equation (21).*—In the preceding section we have obtained the solutions for the temperature distribution and for the law of darkening for an atmosphere exposed to an incident parallel beam of radiation. The explicit numerical forms which these solutions will take in any particular approximation will depend on the solution of the linear equations for the constants of integration M_a, ($a = 1, \ldots, n-1$), and X (cf. eq. [12]). However, we shall now show how a straightforward solution of these equations can be avoided by expressing the constants M_a and X in terms of the solutions of the homogeneous equation (21) in the respective approximations. The analysis which follows is closely related to that of II, § 6.

Consider the function $G(\mu)$ defined by

$$G(\mu) = \sum_{a=1}^{n-1} \frac{M_a}{1-k_a\mu} + X + \frac{\gamma}{1-\mu \sec \beta}. \tag{22}$$

According to the boundary conditions (12),

$$G(\mu_i) = 0 \qquad (i = 1, \ldots, n). \tag{23}$$

This fact enables us to determine $G(\mu)$ explicitly. For, by multiplying equation (22) by the function (cf. II, eq. [58])

$$(1-\mu \sec \beta)R(\mu) = (1-\mu \sec \beta)\prod_{a=1}^{n-1}(1-k_a\mu), \tag{24}$$

we obtain a polynomial of degree n in μ which vanishes for $\mu = \mu_i$, $i = 1, \ldots, n$. The function $G(\mu)\,R(\mu)\,(1 - \mu \sec \beta)$ cannot, therefore, differ from the polynomial (cf. II, eq. [59])

$$P(\mu) = \prod_{j=1}^{n}(\mu - \mu_j) \tag{25}$$

by more than a constant factor, which can be determined by comparing the coefficients of the highest powers of μ in the two expressions. In the former the coefficient of μ^n is unity, while in the latter it is

$$(-1)^n k_1 k_2 \ldots k_{n-1} X \sec \beta. \tag{26}$$

Hence,

$$G(\mu) = (-1)^n k_1 k_2 \ldots k_{n-1} X \frac{\sec \beta}{1-\mu \sec \beta}\frac{P(\mu)}{R(\mu)}. \tag{27}$$

Comparing this with the function $S(\mu)$ introduced in II, equations (56) and (61), we see that

$$G(\mu) = X \frac{\sec\beta}{1-\mu\sec\beta} S(\mu) . \tag{28}$$

Several consequences follow from equations (27) and (28).

First, we observe that, according to equation (28),

$$G(0) = X\, S(0) \sec\beta. \tag{29}$$

But it has been shown in II that (cf. II, eqs. [62] and [70])

$$S(0) = \frac{1}{\sqrt{3}}. \tag{30}$$

Hence (cf. eq. [22])

$$G(0) = \sum_{a=1}^{n-1} M_a + X + \gamma = X \frac{\sec\beta}{\sqrt{3}}. \tag{31}$$

From equations (18), (19), and (31) it therefore follows that

$$\frac{B(0)}{B(\infty)} = \frac{\sec\beta}{\sqrt{3}}. \tag{32}$$

This is an exact result, since it is valid in all approximations. Relation (32) has been proved earlier by Hopf (*op. cit.*) but by entirely different methods of more advanced character.

By turning next to the constants M_a, it is apparent from equation (22) that

$$M_a = \lim_{\mu\to k_a^{-1}} (1-k_a\mu) G(\mu) \quad (a = 1, \ldots, n-1). \tag{33}$$

Using the explicit form of $G(\mu)$ (eq. [27]) in the foregoing equation, we find that

$$M_a = (-1)^n k_1 k_2 \ldots k_{n-1} X \frac{\sec\beta}{1-(\sec\beta/k_a)} \frac{P(\mu=k_a^{-1})}{R_a(\mu=k_a^{-1})} \quad (a = 1, \ldots, n-1), \tag{34}$$

where we have written

$$R_a(\mu) = \prod_{i\neq a}^{1,\, n-1} (1-k_i\mu) \quad (a = 1, \ldots, n-1). \tag{35}$$

On the other hand, the constants L_a in II, equation (19), are given by (cf. II, § 6)

$$L_a = (-1)^n k_1 k_2 \ldots k_{n-1} \frac{P(\mu=k_a^{-1})}{R_a(\mu=k_a^{-1})} \quad (a = 1 \ldots, n-1). \tag{36}$$

Hence,

$$M_a = X \frac{k_a}{k_a\cos\beta - 1} L_a \quad (a = 1, \ldots, n-1), \tag{37}$$

which relates the constants of integration for our present problem with those of the solution of the homogeneous equation (21) to which the L_a's refer.

Again, from equations (22) and (27) we infer that

$$\left.\begin{aligned}\gamma &= \lim_{\mu\to\cos\beta}(1-\mu\sec\beta)G(\mu)\\ &= (-1)^n k_1 k_2 \dots k_{n-1} X \sec\beta\,\frac{P(\cos\beta)}{R(\cos\beta)},\end{aligned}\right\} \tag{38}$$

or, more explicitly,

$$\gamma = (-1)^n k_1 k_2 \dots k_{n-1} X \sec\beta\,\frac{\prod_{j=1}^{n}(\cos\beta-\mu_j)}{\prod_{a=1}^{n-1}(1-k_a\cos\beta)}. \tag{39}$$

On the other hand, we know that (cf. eq. [8])

$$\gamma = \frac{1}{1-\sum_{j=1}^{n}\dfrac{a_j}{1-\mu_j^2\sec^2\beta}}. \tag{40}$$

Equations (39) and (40) therefore determine X. However, the evaluation of X is effected more conveniently by expressing the right-hand side of equation (40) differently, as follows:

Consider the function

$$T(x) = 1-\sum_{j=1}^{n}\frac{a_j}{1-\mu_j^2 x}. \tag{41}$$

Multiplying this equation by

$$\prod_{j=1}^{n}(1-\mu_j^2 x), \tag{42}$$

we obtain a polynomial of degree n in x, which clearly vanishes for

$$x=0 \quad\text{and}\quad x=k_a^2 \qquad (a=1,\dots,n-1). \tag{43}$$

Accordingly,

$$T(x)\prod_{j=1}^{n}(1-\mu_j^2 x) \tag{44}$$

cannot differ from

$$x\prod_{a=1}^{n-1}(x-k_a^2) \tag{45}$$

by more than a constant factor. The constant of proportionality can be determined by comparing the coefficients of the highest powers. In this manner we find that

$$T(x) = (-1)^n \mu_1^2\mu_2^2 \dots \mu_n^2\,\frac{x\prod_{a=1}^{n-1}(x-k_a^2)}{\prod_{j=1}^{n}(1-\mu_j^2 x)}. \tag{46}$$

But (cf. eqs. [40] and [41])

$$\gamma = \frac{1}{T(\sec^2 \beta)} . \tag{47}$$

Hence,

$$\gamma = (-1)^n \frac{\cos^2 \beta}{\mu_1^2 \mu_2^2 \ldots \mu_n^2} \frac{\prod_{j=1}^{n} (1 - \mu_j^2 \sec^2 \beta)}{\prod_{a=1}^{n-1} (\sec^2 \beta - k_a^2)} . \tag{48}$$

Comparing the two expressions for γ given by equations (39) and (48), we find that

$$X = \frac{\cos \beta}{\mu_1^2 \mu_2^2 \ldots \mu_n^2 k_1 k_2 \ldots k_{n-1}} \frac{\prod_{j=1}^{n} (\cos \beta + \mu_j)}{\prod_{a=1}^{n-1} (1 + k_a \cos \beta)} . \tag{49}$$

But (cf. II, eqs. [58], [59], and [61])

$$S(-\cos \beta) = k_1 k_2 \ldots k_{n-1} \frac{\prod_{j=1}^{n} (\cos \beta + \mu_j)}{\prod_{a=1}^{n-1} (1 + k_a \cos \beta)} . \tag{50}$$

Hence, combining equations (49) and (50) and remembering that (II, eq. [68])

$$\mu_1^2 \mu_2^2 \ldots \mu_n^2 k_1^2 k_2^2 \ldots k_{n-1}^2 = \tfrac{1}{3} , \tag{51}$$

we obtain

$$X = 3\, S(-\cos \beta) \cos \beta . \tag{52}$$

where it might be recalled that $S(\mu)$ determines the law of darkening for the standard problem considered in II (cf. II, eqs. [32] and [56]).

Substituting for X from equation (52) in equation (28), we have

$$G(\mu) = 3 \frac{\cos \beta}{\cos \beta - \mu} S(\mu) S(-\cos \beta) . \tag{53}$$

Now the law of darkening given by equation (20) can be expressed simply in terms of $G(\mu)$. We have (cf. eqs. [20] and [27])

$$I(0, \mu) = \tfrac{1}{4} F G(-\mu) . \tag{54}$$

Hence,

$$I(0, \mu) = \tfrac{3}{4} F \frac{\cos \beta}{\cos \beta + \mu} S(-\mu) S(-\cos \beta) , \tag{55}$$

which relates the angular distribution of the "reflected" radiation for the problem under consideration with the law of darkening in an atmosphere characterized by a constant net flux and with no incident radiation. This relation between these two problems expressed by equation (55) is an exact one, inasmuch as it is true in all orders of approximations. Our present derivation of equation (55) thus provides an alternative elementary proof of a result first established by Hopf by means of the theory of integral equations. It is remarkable that our present method of solving the equation of transfer should retain in every approximation all the known exact features of the problem.[5]

Applications of the theory presented here to problems of the reflection effect in eclipsing binaries will be found in a later paper.

[5] This is not the case in the method of approximation used by Milne. His solution (*M.N.*, **87**, 48, eq. [14], 1926) is not, for example, strictly compatible with the result (eq. [55]).

5

ON THE RADIATIVE EQUILIBRIUM OF A STELLAR ATMOSPHERE. XII

S. CHANDRASEKHAR
Yerkes Observatory
Received June 3, 1946

ABSTRACT

In this paper the theory of diffuse reflection, given in an earlier paper of this series, is further developed to allow for the scattering of radiation in accordance with a general phase function.

1. *Introduction.*—In an earlier paper[1] of this series the problem of diffuse reflection by a semi-infinite plane-parallel atmosphere has been considered, allowing for the anisotropy of the scattered radiation in accordance with the phase functions $\lambda(1 + x \cos \Theta)$ and $\frac{3}{4}(1 + \cos^2 \Theta)$, where Θ denotes the angle of scattering. In this paper we propose to extend this discussion to include all cases in which the phase function can be expressed as a series with a finite number[2] of terms in Legendre polynomials in $\cos \Theta$.

2. *The reduction of the equation of transfer.*—The equation of transfer appropriate to the problem of diffuse reflection of a parallel beam of radiation by a plane-parallel atmosphere has already been written down in Paper IX (eq. [2]). We shall now suppose that the phase function $p(\cos \Theta)$, which occurs in this equation, can be expressed as a finite series in the form

$$p(\cos\Theta) = \sum_{l=0}^{N} \varpi_l P_l(\cos\Theta) , \tag{1}$$

where ϖ_l, $l = 0, \ldots, N$ are a set of $N + 1$ constants and P_l denotes the Legendre polynomial of order l in the argument specified. For this phase function the equation of transfer is

$$\left.\begin{aligned}\cos\vartheta \frac{dI(\tau,\vartheta,\varphi)}{d\tau} = I(\tau,\vartheta,\varphi) - \frac{1}{4\pi}\int_0^{\pi}\int_0^{2\pi} I(\tau,\vartheta',\varphi')\sum_{l=0}^{N}\varpi_l \\ \times P_l(\cos\vartheta\cos\vartheta' + \sin\vartheta\sin\vartheta'\cos[\varphi'-\varphi])\sin\vartheta' d\vartheta' d\varphi' \\ - \tfrac{1}{4}F e^{-\tau\sec\beta}\sum_{l=0}^{N}\varpi_l P_l(-\cos\vartheta\cos\beta + \sin\vartheta\sin\beta\cos\varphi) .\end{aligned}\right\} \tag{2}$$

Expanding the Legendre functions in equation (2) in accordance with the addition theorem, we have

$$\left.\begin{aligned}\mu\frac{dI(\tau,\mu,\varphi)}{d\tau} = I(\tau,\mu,\varphi) - \frac{1}{4\pi}\int_{-1}^{+1}\int_0^{2\pi} I(\tau,\mu',\varphi')\sum_{l=0}^{N}\varpi_l \\ \times\Big[P_l(\mu)P_l(\mu') + 2\sum_{m=1}^{l}\frac{(l-m)!}{(l+m)!}P_l^m(\mu)P_l^m(\mu')\cos m(\varphi'-\varphi)\Big]d\mu' d\varphi' \\ - \tfrac{1}{4}F e^{-\tau\sec\beta}\sum_{l=0}^{N}(-1)^l\varpi_l\Big[P_l(\mu)P_l(\cos\beta) \\ + 2\sum_{m=1}^{l}(-1)^m\frac{(l-m)!}{(l+m)!}P_l^m(\mu)P_l^m(\cos\beta)\cos m\varphi\Big] ,\end{aligned}\right\} \tag{3}$$

where we have written μ and μ' for $\cos\vartheta$ and $\cos\vartheta'$, respectively.

[1] *Ap. J.*, **103**, 165, 1946. Referred to hereafter as "Paper IX."

[2] The necessity of this restriction to a finite number of terms will become apparent in § 3 (see eq. [7]).

The form of equation (3) suggests that we seek a solution of the form

$$I(\tau, \mu, \varphi) = I^{(0)}(\tau, \mu) + \sum_{m=1}^{N} I^{(m)}(\tau, \mu)\cos m\varphi\,, \tag{4}$$

where, as the notation indicates, $I^{(0)}$ and $I^{(m)}$ are functions of τ and μ only. With the substitution (4), equation (3) breaks up into the following equations:

$$\left.\begin{aligned} \mu\frac{dI^{(0)}}{d\tau} = I^{(0)} - \frac{1}{2}\int_{-1}^{+1} I^{(0)}(\tau, \mu')\sum_{l=0}^{N}\varpi_l P_l(\mu)P_l(\mu')\,d\mu' \\ -\tfrac{1}{4}Fe^{-\tau\sec\beta}\sum_{l=0}^{N}(-1)^l\varpi_l P_l(\mu)P_l(\cos\beta)\,, \end{aligned}\right\} \tag{5}$$

and

$$\left.\begin{aligned} \mu\frac{dI^{(m)}}{d\tau} = I^{(m)} - \frac{1}{2}\int_{-1}^{+1} I^{(m)}(\tau, \mu')\sum_{l=m}^{N}\varpi_l\frac{(l-m)!}{(l+m)!}P_l^m(\mu)P_l^m(\mu')\,d\mu' \\ -\tfrac{1}{2}Fe^{-\tau\sec\beta}\sum_{l=m}^{N}(-1)^{l+m}\varpi_l\frac{(l-m)!}{(l+m)!}P_l^m(\mu)P_l^m(\cos\beta) \\ (m = 1, \ldots, N)\,. \end{aligned}\right\} \tag{6}$$

We shall now show how these equations can be solved.

3. *The solution of equation (5) in the* n*th approximation.*—Following the general method developed in the earlier papers of this series, we replace the integral which occurs on the right-hand side of equation (5) by a sum according to Gauss's formula for numerical quadratures and reduce the integrodifferential equation to a system of linear equations. In the nth approximation the equivalent linear system is of order $2n$, corresponding to the Gaussian division of the interval $-1 \leqslant \mu \leqslant 1$ according to the zeros of $P_{2n}(\mu)$. It therefore follows that in our present context, in which the Legendre polynomials $P_l(\mu')$ with a maximum degree N occur under the integral sign, we must seek solutions in approximations n such that

$$4n - 1 > 2N\,. \tag{7}$$

Conversely, if we limit ourselves to solutions of the equation of transfer in the nth approximation, we are not entitled to include in an expansion of the phase function in Legendre polynomials terms of order higher than $2n - 1$.

Assuming, then, that condition (7) is fulfilled, the system of $2n$ linear equations, to which equation (5) is equivalent in the nth approximation, is

$$\left.\begin{aligned} \mu_i\frac{dI_i^{(0)}}{d\tau} = I_i^{(0)} - \frac{1}{2}\sum_j\sum_{l=0}^{N} a_j I_j^{(0)} P_l(\mu_j)\varpi_l P_l(\mu_i) \\ -\tfrac{1}{4}Fe^{-\tau\sec\beta}\sum_{l=0}^{N}(-1)^l\varpi_l P_l(\cos\beta)P_l(\mu_i) \qquad (i = \pm 1, \ldots, \pm n)\,, \end{aligned}\right\} \tag{8}$$

where the various symbols have their usual meanings.

In solving the system of equations represented by equation (8), we first seek the general solution of the associated homogeneous system

$$\mu_i\frac{dI_i}{d\tau} = I_i - \frac{1}{2}\sum_j\sum_{l=0}^{N} a_j I_j P_l(\mu_j)\varpi_l P_l(\mu_i) \qquad (i = \pm 1, \ldots, \pm n) \tag{9}$$

and then add to it a particular integral of the nonhomogeneous system.

To obtain the different linearly independent solutions of the system (9), we follow a procedure which has now become standard. Thus, setting

$$I_i = g_i e^{-k\tau} \qquad (i = \pm 1, \ldots, \pm n), \quad (10)$$

we find that the constants g_i must be expressible in the form

$$g_i = \frac{\sum_{l=0}^{N} \varpi_l \xi_l P_l(\mu_i)}{1+\mu_i k} \qquad (i = \pm 1, \ldots, \pm n), \quad (11)$$

where the ξ_l's $(l = 0, \ldots, N)$ are certain constants independent of i which have to be determined in conformity with the equation

$$\sum_{l=0}^{N} \varpi_l \xi_l P_l(\mu_i) = \sum_{l=0}^{N} \sum_{\lambda=0}^{N} \varpi_l P_l(\mu_i)\, D_{l,\lambda} \varpi_\lambda \xi_\lambda\,, \quad (12)$$

where

$$D_{l,\lambda} = \frac{1}{2} \sum_j \frac{a_j P_l(\mu_j) P_\lambda(\mu_j)}{1+\mu_j k}\,. \quad (13)$$

Since equation (12) must be valid for all i, we must require that

$$\xi_l = \sum_{\lambda=0}^{N} D_{l,\lambda} \varpi_\lambda \xi_\lambda \qquad (l = 0, \ldots, N). \quad (14)$$

We shall now show how equation (14) determines the ξ's apart from a constant of proportionality and leads also to the characteristic equation for k.

First, we may observe that $D_{l,\lambda}$ defined as in equation (13) satisfies a simple recursion formula. For, writing $D_{l,\lambda}$ in the form

$$D_{l,\lambda} = \frac{1}{2} \sum_j a_j P_l(\mu_j) P_\lambda(\mu_j) \left(1 - \frac{k\mu_j}{1+\mu_j k}\right) \quad (15)$$

and remembering that, since

$$l + \lambda \leqslant 2N < 4n - 1\,, \quad (16)$$

$$\left.\begin{aligned} \frac{1}{2} \sum_j a_j P_l(\mu_j) P_\lambda(\mu_j) &= \frac{1}{2} \int_{-1}^{+1} P_l(\mu) P_\lambda(\mu)\, d\mu \\ &= \frac{\delta_{l,\lambda}}{2l+1}\,, \end{aligned}\right\} \quad (17)$$

we have

$$\left.\begin{aligned} D_{l,\lambda} &= \frac{\delta_{l,\lambda}}{2l+1} - \tfrac{1}{2} k \sum_j \frac{a_j P_\lambda(\mu_j)}{1+\mu_j k} \mu_j P_l(\mu_j) \\ &= \frac{\delta_{l,\lambda}}{2l+1} - \frac{k}{2(2l+1)} \sum_j \frac{a_j P_\lambda(\mu_j) [(l+1) P_{l+1}(\mu_j) + l P_{l-1}(\mu_j)]}{1+\mu_j k} \\ &= \frac{\delta_{l,\lambda}}{2l+1} - \frac{k}{2l+1} [(l+1) D_{l+1,\lambda} + l D_{l-1,\lambda}]\,. \end{aligned}\right\} \quad (18)$$

In other words,

$$(2l+1) D_{l,\lambda} = \delta_{l,\lambda} - k [(l+1) D_{l+1,\lambda} + l D_{l-1,\lambda}]\,. \quad (19)$$

Using the recursion formula (19) in equation (14), we have

$$(2l+1)\,\xi_l=\sum_{\lambda=0}^{N}\{\delta_{l,\lambda}-k[(l+1)\,D_{l+1,\lambda}+lD_{l-1,\lambda}]\}\varpi_\lambda\xi_\lambda\,, \tag{20}$$

or, again using equation (14),

$$(2l+1)\,\xi_l=\varpi_l\xi_l-k[(l+1)\,\xi_{l+1}+l\xi_{l-1}]. \tag{21}$$

Alternatively, we can write

$$\xi_{l+1}=-\frac{2l+1-\varpi_l}{k(l+1)}\,\xi_l-\frac{l}{l+1}\,\xi_{l-1}\qquad(l=0,\ldots,N-1). \tag{22}$$

From equation (22) it follows that all the ξ's can be determined successively in terms of ξ_0. To remove this arbitrary constant of proportionality in the ξ's, we shall assume in our further work (as it will entail no loss of generality) that

$$\xi_0=1\,. \tag{23}$$

With this choice of ξ_0, the remaining ξ's are all uniquely determined by equation (22). Thus,

$$\xi_1=-\frac{1-\varpi_0}{k};\qquad \xi_2=\frac{(1-\varpi_0)\,(3-\varpi_1)}{2k^2}-\tfrac{1}{2};\qquad \text{etc.} \tag{24}$$

It is to be particularly noted that the ξ_l's ($l>0$) determined in this manner are, in general, functions of k. We shall accordingly write

$$\xi_l=\xi_l(k)\qquad(l=1,\ldots,N). \tag{25}$$

The characteristic equation for k now follows from equation (14) by considering the case $l=0$. Thus,

$$1=\sum_{\lambda=0}^{N}D_{0,\lambda}\varpi_\lambda\xi_\lambda\,, \tag{26}$$

or, more explicitly,

$$1=\frac{1}{2}\sum_j\frac{a_j\sum_{\lambda=0}^{N}\varpi_\lambda\xi_\lambda P_\lambda(\mu_j)}{1+\mu_jk}\,. \tag{27}$$

Equation (27) is of order n in k^2 and accordingly admits, in general, $2n$ distinct nonvanishing roots, which must occur in pairs, as

$$\pm k_\alpha\qquad(\alpha=1,\ldots,n). \tag{28}$$

However, an important exception arises when $\varpi_0=1$; for, in this case, $k=0$ is a root[3] and equation (27) admits of only $(2n-2)$ distinct nonvanishing roots. On the other hand, when $\varpi_0=1$, it may be readily verified that we also have the integral

$$I_i=b\left(\tau+\frac{\mu_i}{1-\frac{1}{3}\varpi_1}+Q\right)\qquad(i=\pm1,\ldots,n), \tag{29}$$

where b and Q are arbitrary constants. To avoid frequent repetition we shall suppose that $\varpi_0\neq1$ unless we explicitly state otherwise.

[3] Note that $\frac{1}{2}\sum_j\sum_{\lambda=0}^{N}a_j\varpi_\lambda\xi_\lambda P_\lambda(\mu_j)=\frac{1}{2}\sum_j a_j\varpi_0=1$.

Returning to equations (9) and (10), we have now shown that the homogeneous system (9) admits $2n$ linearly independent integrals of the form

$$I_i = \text{constant}\ \frac{\sum_{l=0}^{N} \varpi_l \xi_l (\pm k_a) P_l(\mu_i)}{1 \pm \mu_i k_a} e^{\mp k_a \tau} \qquad \begin{pmatrix} i = \pm 1, \ldots, \pm n \\ a = 1, \ldots, n \end{pmatrix}. \tag{30}$$

To complete the solution we need a particular integral of the nonhomogeneous system (8). Again this can be found by our standard procedure.

Setting

$$I_i^{(0)} = \tfrac{1}{4} F h_i e^{-\tau \sec \beta} \qquad (i = \pm 1, \ldots, \pm n), \tag{31}$$

we readily verify that the constants h_i must be expressible in the form (cf. eq. [11])

$$h_i = \frac{\sum_{l=0}^{N} \varpi_l \gamma_l P_l(\mu_i)}{1 + \mu_i \sec \beta} \qquad (i = \pm 1, \ldots, \pm n), \tag{32}$$

where the γ_l's ($l = 0, \ldots, N$) are certain constants to be determined in accordance with the equation (cf. eq. [14])

$$\gamma_l = \sum_{\lambda=0}^{N} E_{l,\lambda} \varpi_\lambda \gamma_\lambda + (-1)^l P_l(\cos \beta) \qquad (l = 0, \ldots, N) \tag{33}$$

where (cf. eq. [13])

$$E_{l,\lambda} = \frac{1}{2} \sum_j \frac{a_j P_l(\mu_j) P_\lambda(\mu_j)}{1 + \mu_j \sec \beta}. \tag{34}$$

The E_λ's satisfy a recursion formula similar to the $D_{l,\lambda}$'s. We have

$$(2l+1) E_{l,\lambda} = \delta_{l,\lambda} - \sec \beta \left[(l+1) E_{l+1,\lambda} + l E_{l-1,\lambda} \right]. \tag{35}$$

Using this formula in equation (33), we obtain

$$\left.\begin{aligned} (2l+1)\gamma_l = \varpi_l \gamma_l - \sec \beta \Big[\sum_{\lambda=0}^{N} \{ (l+1) E_{l+1,\lambda} + l E_{l-1,\lambda} \} \varpi_\lambda \gamma_\lambda \Big] \\ + (-1)^l (2l+1) P_l(\cos \beta) \end{aligned}\right\} \tag{36}$$

Again, using equation (33), we have

$$\left.\begin{aligned} (2l+1)\gamma_l = \varpi_l \gamma_l - \sec \beta \left[(l+1) \{ \gamma_{l+1} + (-1)^l P_{l+1}(\cos \beta) \} \right. \\ \left. + l \{ \gamma_{l-1} + (-1)^l P_{l-1}(\cos \beta) \} \right] + (-1)^l (2l+1) P_l(\cos \beta). \end{aligned}\right\} \tag{37}$$

Hence,

$$(2l+1)\gamma_l = \varpi_l \gamma_l - \sec \beta \left[(l+1) \gamma_{l+1} + l \gamma_{l-1} \right]. \tag{38}$$

It is therefore seen that the γ's satisfy a recursion formula of the same form as the ξ's (cf. eq. [21]). Accordingly, we may write

$$\gamma_l = \gamma_0 \xi_l (\sec \beta), \tag{39}$$

where γ_0 is a constant of proportionality. This can be determined from equation (33) by considering the case $l = 0$. We have

$$\gamma_0 = \gamma_0 \sum_{l=0}^{N} E_{0,\lambda} \varpi_\lambda \xi_\lambda (\sec \beta) + 1, \tag{40}$$

or

$$\gamma_0 = \frac{1}{1 - \sum_{\lambda=0}^{N} E_{0,\lambda} \varpi_\lambda \xi_\lambda (\sec \beta)} . \tag{41}$$

This is related to the characteristic equation (26) in a manner which enables us to write at once that (cf. paper IX, p. 171)

$$\gamma_0 = H^{(0)} (\cos \beta) H^{(0)} (-\cos \beta) , \tag{42}$$

where

$$H^{(0)} (\mu) = \frac{1}{\mu_1 \ldots \mu_n} \frac{\prod_{i=1}^{n} (\mu + \mu_i)}{\prod_{a=1}^{n} (1 + k_a \mu)} = \frac{(-1)^n}{\mu_1 \ldots \mu_n} \frac{P(-\mu)}{R^{(0)}(-\mu)} . \tag{43}$$

Thus equation (8) admits a particular integral of the form

$$I_i^{(0)} = \tfrac{1}{4} F \gamma_0 e^{-\tau \sec \beta} \frac{\sum_{l=0}^{N} \varpi_l \xi_l (\sec \beta) P_l (\mu_i)}{1 + \mu_i \sec \beta} \quad (i = \pm 1, \ldots, \pm n) . \tag{44}$$

Combining equations (30) and (44), we obtain the general solution of the nonhomogeneous system (8). We have

$$\left.\begin{aligned} I_i^{(0)} = \tfrac{1}{4} F \Bigg\{ & \sum_{a=1}^{n} \frac{M_a e^{-k_a \tau}}{1 + \mu_i k_a} \Big[\sum_{l=0}^{N} \varpi_l \xi_l (+k_a) P_l (\mu_i) \Big] \\ & + \sum_{a=1}^{n} \frac{M_{-a} e^{+k_a \tau}}{1 - \mu_i k_a} \Big[\sum_{l=0}^{N} \varpi_l \xi_l (-k_a) P_l (\mu_i) \Big] \\ & + \frac{\gamma_0 e^{-\tau \sec \beta}}{1 + \mu_i \sec \beta} \Big[\sum_{l=0}^{N} \varpi_l \xi_l (\sec \beta) P_l (\mu_i) \Big] \Bigg\} \quad (i = \pm 1, \ldots, \pm n) \end{aligned}\right\} \tag{45}$$

where $M_{\pm a}$ $(a = 1, \ldots, n)$ are the $2n$ constants of integration.

If we are dealing with a plane-parallel atmosphere of optical thickness τ_1, the boundary conditions which determine the constants $M_{\pm a}$ are

$$I_{-i}^{(0)} = 0 \quad \text{at} \quad \tau = 0 \quad \text{and} \quad i = 1, \ldots, n , \tag{46}$$

and

$$I_i^{(0)} = 0 \quad \text{at} \quad \tau = \tau_1 \quad \text{and} \quad i = 1, \ldots, n . \tag{47}$$

On the other hand, if we are dealing with diffuse reflection from a semi-infinite atmosphere, the boundedness of the solution for $\tau \to \infty$ requires that $M_{-a} = 0$ $(a = 1, \ldots, n)$. The solution then has the form

$$\left.\begin{aligned} I_i^{(0)} = \tfrac{1}{4} F \Bigg\{ & \sum_{a=1}^{n} \frac{M_a e^{-k_a \tau}}{1 + \mu_i k_a} \Big[\sum_{l=0}^{N} \varpi_l \xi_l (+k_a) P_l (\mu_i) \Big] \\ & + \frac{\gamma_0 e^{-\tau \sec \beta}}{1 + \mu_i \sec \beta} \Big[\sum_{l=0}^{N} \varpi_l \xi_l (\sec \beta) P_l (\mu_i) \Big] \Bigg\} \quad (i = \pm 1, \ldots, n) , \end{aligned}\right\} \tag{48}$$

and the constants M_a are now determined by the boundary condition at $\tau = 0$, namely, equations (46). Moreover, in this case, the angular distribution of the emergent radiation is given by

$$\left.\begin{aligned} I^{(0)}(0,\mu) = \tfrac{1}{4}F\Big\{\sum_{a=1}^{n}\frac{M_a}{1+\mu k_a}\Big[\sum_{l=0}^{N}\varpi_l\xi_l(k_a)P_l(\mu)\Big] \\ +\frac{\gamma_0}{1+\mu\sec\beta}\Big[\sum_{l=0}^{N}\varpi_l\xi_l(\sec\beta)P_l(\mu)\Big]\Big\}. \end{aligned}\right\} \quad (49)$$

This completes our discussion of equation (5).

4. *The solution of equation (6) in the* n*th approximation.*—Considering, next, equation (6), we have the following system of linear equations to which it is equivalent in the nth approximation:

$$\left.\begin{aligned} \mu_i\frac{dI_i^{(m)}}{d\tau} = I_i^{(m)} - \frac{1}{2}\sum_j\sum_{l=m}^{N}a_jI_j^{(m)}P_l^m(\mu_j)\varpi_l\frac{(l-m)!}{(l+m)!}P_l^m(\mu_i) \\ -\tfrac{1}{2}Fe^{-\tau\sec\beta}\sum_{l=m}^{N}(-1)^{l+m}\varpi_l\frac{(l-m)!}{(l+m)!}P_l^m(\mu_i)P_l^m(\cos\beta) \\ (i=\pm1,\ldots,\pm n;\ m=1,\ldots,N). \end{aligned}\right\} \quad (50)$$

The general solution of this system can be found by a procedure analogous to that which was adopted for the solution of the system (8).

Thus it can be shown that the system of homogeneous equations associated with equation (50) admits integrals of the form

$$I_i = \text{constant}\,\frac{\sum_{l=m}^{N}\varpi_l\frac{(l-m)!}{(l+m)!}\xi_l^mP_l^m(\mu_i)}{1+\mu_ik}\,e^{-k\tau} \quad (51)$$

where the ξ_l^m's are constants to be determined in accordance with the equation

$$\xi_l^m = \sum_{\lambda=m}^{N}D_{l,\lambda}^m\xi_\lambda^m\varpi_\lambda\frac{(\lambda-m)!}{(\lambda+m)!} \qquad (l=m,\ldots,N;\ m=1,\ldots,N), \quad (52)$$

where

$$D_{l,\lambda}^m = \frac{1}{2}\sum_j\frac{a_jP_l^m(\mu_j)P_\lambda^m(\mu_j)}{1+\mu_jk}. \quad (53)$$

Using the recursion formula

$$(2l+1)D_{l,\lambda}^m = \frac{(l+m)!}{(l-m)!}\delta_{l,\lambda} - k\left[(l-m+1)D_{l+1,\lambda}^m + (l+m)D_{l-1,\lambda}^m\right], \quad (54)$$

which the $D_{l,\lambda}^m$'s may be verified to satisfy, it can be readily deduced from equation (52) that

$$(2l+1)\xi_l^m = \varpi_l\xi_l^m - k\left[(l-m+1)\xi_{l+1}^m + (l+m)\xi_{l-1}^m\right], \quad (55)$$

or

$$\xi_{l+1}^m = -\frac{2l+1-\varpi_l}{k(l-m+1)}\xi_l^m - \frac{l+m}{l-m+1}\xi_{l-1}^m. \quad (56)$$

Equation (56) determines the ξ_l^m's apart from a constant of proportionality which we shall make determinate with the choice

$$\xi_m^m = 1\ . \tag{57}$$ [4]

Thus, we find from equation (56) that

$$\left.\begin{aligned} \xi_{m+1}^m &= -\frac{2m+1-\varpi_m}{k}\ ; \\ \xi_{m+2}^m &= \frac{(2m+1-\varpi_m)\,(2m+3-\varpi_{m+1})}{2k^2} - \frac{2m+1}{2}\ ; \\ &\text{etc.} \end{aligned}\right\} \tag{58}$$

The characteristic equation now follows from equation (52) by putting $l = m$. We have

$$1 = \sum_{\lambda=m}^{N} D_{m,\lambda}^m \xi_\lambda^m \varpi_\lambda \frac{(\lambda-m)\,!}{(\lambda+m)\,!}\ , \tag{59}$$

or, more explicitly,

$$1 = \frac{1}{2}\sum_j \frac{a_j \sum_{\lambda=m}^{N} \xi_\lambda^m \varpi_\lambda \frac{(\lambda-m)\,!}{(\lambda+m)\,!} P_m^m(\mu_j) P_\lambda^m(\mu_j)}{1+\mu_j k}\ . \tag{60}$$

Equation (60) is of order n in k^2 and admits, in general, $2n$ distinct nonvanishing roots, which must occur in pairs, as

$$\pm k_a^m \qquad (a = 1, \ldots, n\ ;\ \ m = 1, \ldots, N)\ . \tag{61}$$

The $2n$ linearly independent solutions of the homogeneous system associated with equation (50) are, therefore,

$$\left.\begin{aligned} I_i &= \text{constant}\ \frac{\sum_{l=m}^{N} \varpi_l \frac{(l-m)\,!}{(l+m)\,!} \xi_l^m (\pm k_a^m) P_l^m(\mu_i)}{1 \pm \mu_i k_a^m}\, e^{\mp k_a^m \tau} \\ &(a = 1, \ldots, n\ ;\ \ i = \pm 1, \ldots, \pm n\ ;\ \ m = 1, \ldots, N)\ . \end{aligned}\right\} \tag{62}$$

Similarly, it can be shown that equation (50) admits a particular integral of the form

$$\left.\begin{aligned} I_i^{(m)} &= \tfrac{1}{2} F e^{-\tau \sec\beta} \frac{\sum_{l=m}^{N} \varpi_l \frac{(l-m)\,!}{(l+m)\,!} \gamma_l^m P_l^m(\mu_i)}{1+\mu_i \sec\beta} \\ &(i = \pm 1, \ldots, \pm n\ ;\ \ m = 1, \ldots, N)\ , \end{aligned}\right\} \tag{63}$$

where the constants γ_l^m are to be determined in accordance with the equation

$$\gamma_l^m = \sum_{\lambda=m}^{N} E_{l,\lambda}^m \gamma_\lambda^m \varpi_\lambda \frac{(\lambda-m)\,!}{(\lambda+m)\,!} + (-1)^{l+m} P_l^m(\cos\beta)\ , \tag{64}$$

[4] It may be noted that, in applying equation (56) for the case $l = m$, we must put $\xi_{m-1}^m = 0$, as the ξ's are not defined for $l < m$.

where

$$E^m_{l,\lambda} = \frac{1}{2}\sum_j \frac{a_j P^m_l(\mu_j) P^m_\lambda(\mu_j)}{1+\mu_j \sec\beta} . \tag{65}$$

The $E^m_{l,\lambda}$'s defined in this manner satisfy the recursion formula

$$(2l+1)E^m_{l,\lambda} = \frac{(l+m)!}{(l-m)!}\delta_{l,\lambda} - \sec\beta\,[(l-m+1)E^m_{l+1,\lambda} + (l+m)E^m_{l-1,\lambda}] \; , \tag{66}$$

with which we can deduce from equation (64) that the γ^m_l's satisfy the relation

$$(2l+1)\gamma^m_l = \varpi_l \gamma^m_l - \sec\beta\,[(l-m+1)\gamma^m_{l+1} + (l+m)\gamma^m_{l-1}] . \tag{67}$$

This is of exactly the same form as the recursion formula satisfied by the ξ^m_l's. Hence, we may write

$$\gamma^m_l = \gamma^m_m \xi^m_l(\sec\beta) \qquad (l = m, \ldots . \; N), \tag{68}$$

where the constant of proportionality γ^m_m can be found from equation (64) by considering the case $l = m$. We find

$$\gamma^m_m = \frac{P^m_m(\cos\beta)}{1 - \sum_{\lambda=m}^{N} E^m_{m,\lambda}\xi^m_\lambda(\sec\beta)\,\varpi_\lambda \dfrac{(\lambda-m)!}{(\lambda+m)!}} . \tag{69}$$

The relation of this equation to the characteristic equation (59) is such that we can write

$$\gamma^m_m = \gamma^m_0 P^m_m(\cos\beta) , \tag{70}$$

where

$$\gamma^m_0 = H^{(m)}(\cos\beta)\,H^{(m)}(-\cos\beta) , \tag{71}$$

and

$$H^{(m)}(\mu) = \frac{1}{\mu_1 \ldots . \mu_n} \frac{\prod_{l=1}^{n}(\mu+\mu_i)}{\prod_{a=1}^{n}(1+k^m_a\mu)} = \frac{(-1)^n}{\mu_1 \ldots . \mu_n}\frac{P(-\mu)}{R^{(m)}(-\mu)} . \tag{72}$$

The particular integral (63) can therefore be written in the form

$$\left.\begin{aligned} I^{(m)}_i &= \tfrac{1}{2}F e^{-\tau\sec\beta}\gamma^m_0 P^m_m(\cos\beta)\frac{\sum_{l=m}^{N}\varpi_l \dfrac{(l-m)!}{(l+m)!}\xi^m_l(\sec\beta)P^m_l(\mu_i)}{1+\mu_i\sec\beta} \\ &(i = \pm 1, \ldots . , \pm n \; ; \;\; m = 1, \ldots . , N) . \end{aligned}\right\} \tag{73}$$

Finally, combining equations (62) and (63), we can write the general solution of equation (50) in the form

$$\left.\begin{aligned} I_i^{(m)} = \tfrac{1}{2}FP_m^m(\cos\beta)\Bigg\{&\sum_{a=1}^{n}\frac{M_a^m e^{-k_a^m\tau}}{1+\mu_i k_a^m}\Bigg[\sum_{l=m}^{N}\varpi_l\frac{(l-m)!}{(l+m)!}\xi_l^m(+k_a^m)P_l^m(\mu_i)\Bigg] \\ &+\sum_{a=1}^{n}\frac{M_{-a}^m e^{+k_a^m\tau}}{1-\mu_i k_a^m}\Bigg[\sum_{l-m}^{N}\varpi_l\frac{(l-m)!}{(l+m)!}\xi_l^m(-k_a^m)P_l^m(\mu_i)\Bigg] \\ &+\frac{\gamma_0^m e^{-\tau\sec\beta}}{1+\mu_i\sec\beta}\Bigg[\sum_{l=m}^{N}\varpi_l\frac{(l-m)!}{(l+m)!}\xi_l^m(\sec\beta)P_l^m(\mu_i)\Bigg]\Bigg\} \\ &(i=\pm1,\ldots,\pm n\,;\quad m=1,\ldots,N)\,. \end{aligned}\right\}\qquad(74)$$

The boundary conditions which determine the constants of integration $M_{\pm a}^m$ ($a = 1, \ldots, n$) are the same as for $I_i^{(0)}$ (cf. eqs. [46] and [47]). In particular, for the case of a semi-infinite atmosphere, the solution is

$$\left.\begin{aligned} I_i^{(m)} = \tfrac{1}{2}FP_m^m(\cos\beta)\Bigg\{&\sum_{a=1}^{n}\frac{M_a^m e^{-k_a^m\tau}}{1+\mu_i k_a^m}\Bigg[\sum_{l=m}^{N}\varpi_l\frac{(l-m)!}{(l+m)!}\xi_l^m(+k_a^m)P_l^m(\mu_i)\Bigg] \\ &+\frac{\gamma_0^m e^{-\tau\sec\beta}}{1+\mu_i\sec\beta}\Bigg[\sum_{l=m}^{N}\varpi_l\frac{(l-m)!}{(l+m)!}\xi_l^m(\sec\beta)P_l^m(\mu_i)\Bigg]\Bigg\} \\ &(i=\pm1,\ldots,\pm n\,;\quad m=1,\ldots,N)\,, \end{aligned}\right\}\qquad(75)$$

where the constants M_a^m ($a = 1, \ldots, n$) are to be found from the conditions

$$I_{-i}^{(m)} = 0 \quad\text{at}\quad \tau = 0 \quad\text{and for}\quad i = 1, \ldots, n\,. \qquad(76)$$

With this we have completed the formal solution of the equation of transfer. It is seen how solutions to any desired accuracy can always be found.

5. *Diffuse reflection in accordance with the phase function* $1 + \varpi_1 P_1(\cos\Theta) + \varpi_2 P_2(\cos\Theta)$.—As an illustration of the general theory developed in the preceding sections we shall consider the case of diffuse reflection by a semi-infinite plane-parallel atmosphere scattering radiation in accordance with the phase function

$$p(\cos\Theta) = 1 + \varpi_1 P_1(\cos\Theta) + \varpi_2 P_2(\cos\Theta)\,. \qquad(77)$$

For this phase function the solution has the form

$$I(\tau,\mu) = I^{(0)}(\tau,\mu) + I^{(1)}(\tau,\mu)\cos\varphi + I^{(2)}(\tau,\mu)\cos 2\varphi\,. \qquad(78)$$

Considering, first, the solution for $I^{(0)}(\tau, \mu)$, we have (cf. eq. [24])

$$\xi_0 = 1\,;\qquad \xi_1 = 0\,;\qquad\text{and}\qquad \xi_2 = -\tfrac{1}{2}\,. \qquad(79)$$

The corresponding characteristic equation is

$$1 = \frac{1}{2}\sum_j \frac{a_j\,[1-\tfrac{1}{2}\varpi_2 P_2(\mu_j)]}{1+\mu_j k}\,, \qquad(80)$$

or, somewhat differently,

$$1 = \tfrac{3}{4}\varpi_2\sum_{j=1}^{n}\left(\frac{4+\varpi_2}{3\varpi_2}-\mu_j^2\right)\frac{a_j}{1-\mu_j^2 k^2}\,. \qquad(81)$$

Equation (81) admits of only $(2n-2)$ distinct nonvanishing roots for k^2. We must accordingly make use of the integral (29) in writing the general solution for $I_i^{(0)}$.

The solution for $I_i^{(0)}$ has the form

$$I_i^{(0)}=\frac{3\varpi_2}{16}F\left[\left(\frac{4+\varpi_2}{3\varpi_2}-\mu_i^2\right)\sum_{a=1}^{n-1}\frac{M_a e^{-k_a\tau}}{1+\mu_i k_a}+X+\left(\frac{4+\varpi_2}{3\varpi_2}-\mu_i^2\right)\frac{\gamma_0 e^{-\tau\sec\beta}}{1+\mu_i\sec\beta}\right] \quad (i=\pm 1,\ldots,\pm n)\,. \tag{82}$$

The angular distribution of the emergent radiation $I^{(0)}(0,\mu)$ and the boundary conditions which determine the constants of integration M_a $(a=1,\ldots,n-1)$ and X can both be expressed in terms of the function

$$G(\mu)=\left(\frac{4+\varpi_2}{3\varpi_2}-\mu^2\right)\sum_{a=1}^{n-1}\frac{M_a}{1-\mu k_a}+X+\left(\frac{4+\varpi_2}{3\varpi_2}-\mu_i^2\right)\frac{\gamma_0}{1-\mu\sec\beta}\,. \tag{83}$$

Thus

$$I^{(0)}(0,\mu)=\frac{3\varpi_2}{16}FG(-\mu)\,, \tag{84}$$

and

$$G(\mu_i)=0 \qquad (i=1,\ldots,n)\,. \tag{85}$$

Comparing the foregoing solutions with those considered in paper IX, § 8 (see particularly eqs. [169]–[173]), it is apparent that in this case we can obtain a closed expression for $G(\mu)$ which does not explicitly involve the constants of integration. By an analysis similar to that in paper IX, § 8, we find that (cf. paper IX, eqs. [184], [188], and [189])

$$\begin{aligned}I^{(0)}(0,\mu)=\frac{3\varpi_2}{16}FH^{(0)}(\mu)\,H^{(0)}(\cos\beta)\frac{\cos\beta}{\cos\beta+\mu}\Bigg\{&\xi\left(\sqrt{\frac{4+\varpi_2}{3\varpi_2}}+\cos\beta\right)\\ \times\left(\sqrt{\frac{4+\varpi_2}{3\varpi_2}}+\mu\right)+(1-\xi)&\left(\sqrt{\frac{4+\varpi_2}{3\varpi_2}}-\cos\beta\right)\left(\sqrt{\frac{4+\varpi_2}{3\varpi_2}}-\mu\right)\Bigg\}\,,\end{aligned} \tag{86}$$

where

$$\xi=\frac{P\left(+\sqrt{\frac{4+\varpi_2}{3\varpi_2}}\right)R\left(-\sqrt{\frac{4+\varpi_2}{3\varpi_2}}\right)}{P\left(+\sqrt{\frac{4+\varpi_2}{3\varpi_2}}\right)R\left(-\sqrt{\frac{4+\varpi_2}{3\varpi_2}}\right)-P\left(-\sqrt{\frac{4+\varpi_2}{3\varpi_2}}\right)R\left(+\sqrt{\frac{4+\varpi_2}{3\varpi_2}}\right)}\,. \tag{87}$$

Considering, next, the solution for $I^{(1)}(\tau,\mu)$, we have

$$\xi_1^1=1 \quad \text{and} \quad \xi_2^1=-\frac{3-\varpi_1}{k}\,. \tag{88}$$

The characteristic equation (60) now becomes

$$1=\frac{1}{2}\sum_{j=1}^{n}\frac{a_j(1-\mu_j^2)}{1-\mu_j^2k^2}\left[\varpi_1+\varpi_2(3-\varpi_1)\mu_j^2\right]. \tag{89}$$

The corresponding solution for $I_i^{(1)}$ is

$$\begin{aligned}I_i^{(1)}=\frac{\varpi_1}{4}F(1-\mu_i^2)^{1/2}\sin\beta\Bigg\{&\sum_{a=1}^{n}\frac{M_a^1 e^{-k_a^1\tau}}{1+\mu_i k_a^1}\left[1-\frac{\varpi_2(3-\varpi_1)}{\varpi_1}\frac{\mu_i}{k_a^1}\right]\\ &+\frac{\gamma_0 e^{-\tau\sec\beta}}{1+\mu_i\sec\beta}\left[1-\frac{\varpi_2(3-\varpi_1)}{\varpi_1}\mu_i\cos\beta\right]\Bigg\} \quad (i=\pm 1,\ldots,\pm n)\,.\end{aligned} \tag{90}$$

From the similarity of this solution with equation (54) in paper IX, we conclude that, in this case also, closed expressions for $I^{(1)}(0, \mu)$ can be found which do not require an explicit solution for the constants of integration. We have (cf. paper IX, eq. [108])

$$\left.\begin{aligned} I^{(1)}(0, \mu) &= \frac{\varpi_1}{4} F (1-\mu^2)^{1/2} \sin\beta\, H^{(1)}(\mu)\, H^{(1)}(\cos\beta) \frac{\cos\beta}{\cos\beta+\mu} \\ &\times\left[1 - \frac{\varpi_2(3-\varpi_1)}{\varpi_1}\left\{\frac{\sigma}{\rho}(\cos\beta+\mu) + \mu\cos\beta\right\}\right], \end{aligned}\right\} \tag{91}$$

where σ and ρ are defined as in paper IX, equations (90) and (91), with $\varpi_2(3-\varpi_1)/\varpi_1$ replacing $x(1-\lambda)$.

Finally, considering the solution for $I^{(2)}(\tau, \mu)$, it is seen that the characteristic equation is (cf. paper IX, eq. [193])

$$1 = \tfrac{3}{8}\varpi_2 \sum_{j=1}^{n} \frac{a_j (1-\mu_j^2)^2}{1-\mu_j^2 k^2} \tag{92}$$

and, further, that the equation that has to be solved belongs to the class considered in paper IX, § 4.[5] We can accordingly write (cf. paper IX, eq. [191])

$$I^{(2)}(0, \mu) = \frac{3\varpi_2}{16} F \sin^2\beta \sin^2\vartheta\, H^{(2)}(\mu)\, H^{(2)}(\cos\beta) \frac{\cos\beta}{\cos\beta+\mu}. \tag{93}$$

It is therefore seen that in this case, as in the cases considered in paper IX, expressions for the angular distribution of the emergent radiation can be found which do not require an explicit solution of the constants of integration. Whether this is generally possible is not so easily answered, though it is a simple matter to write down a variety of special phase functions for which this can, in fact, be accomplished. We hope to return to these and related questions in the near future.

And, finally, there is the question of the relationship between our method (particularly in those cases in which we are able to eliminate the constants of integration) and Ambarzumian's,[6] in which integral equations are derived for the functions describing the angular distribution of the emergent radiation. An examination of this relationship leads to interesting developments in the solution of a large class of functional equations by a systematic method of approximation. But to go into these matters here will take us too far from the main objective of this paper.

[5] Indeed, this is generally true of the case $m = N$.

[6] *J. Phys. Acad. Sci. U.S.S.R.*, **8**, 64, 1944.

6

ON THE RADIATIVE EQUILIBRIUM OF A STELLAR ATMOSPHERE. XXI

S. CHANDRASEKHAR
Yerkes Observatory
Received June 20, 1947

ABSTRACT

In this paper the theory of diffuse reflection and transmission by a plane-parallel atmosphere of finite optical thickness is considered under conditions of (I) isotropic scattering with an albedo $\tilde{\omega}_0 \leqslant 1$, (II) scattering in accordance with Rayleigh's phase function, (III) scattering in accordance with the phase function $\lambda(1 + x \cos \Theta)$, and (IV) Rayleigh scattering with proper allowance for the polarization of the radiation field. In all cases considered, it has been possible to eliminate the constants of integration (which are twice as many as in the case of semi-infinite atmospheres) and express the solutions for the reflected and the transmitted radiations in closed forms in a general nth approximation. It is further shown how a pair of functions, $X(\mu)$ and $Y(\mu)$, depending only on the roots of a characteristic equation and the optical thickness of the atmosphere, play the same basic role in this theory as $H(\mu)$ does in the theory of semi-infinite atmospheres. The passage to the limit of infinite approximation and the determination of the exact laws of diffuse reflection and transmission are thus made possible.

1. *Introduction.*—In the earlier papers[1] of this series, the theory of the transfer of radiation in semi-infinite plane-parallel atmospheres has been developed to a point that it is possible to obtain by a definite algorism exact solutions for the various problems. But the corresponding theory for atmospheres of finite optical thicknesses is in a far less advanced stage. The difficulties confronting the development of this latter theory do not lie in the system of linear equations which replaces the equation of transfer in our scheme of approximation: they present, in fact, no problem which does not already require solution in the semi-infinite case.[2] The difficulties actually lie in the problem of eliminating the explicit appearance of the constants of integration in the solutions and expressing the angular distributions of the emergent radiations in terms of functions which involve only the roots of certain characteristic equations. This problem of the elimination of the constants and the reduction of the solution to the evaluation of a certain basic function (or a set of functions)[3] is of particular importance for passing to the limit of infinite approximation and obtaining the exact solutions. Thus, in analogy with the theory of semi-infinite atmospheres, we may expect that the angular distributions of the emergent radiations can be expressed in terms of certain functions which will be explicitly known in any finite approximation and which in the limit of infinite approximation will become solutions of functional equations of a standard form. We shall see that this reduction can be achieved in spite of the greater complexity of the problem arising from our present requirement of explicitly satisfying boundary conditions on both sides of the atmosphere and of obtaining solutions in closed forms for the expressions governing the radiation emergent from each of the two sides.

In this paper we shall take the first of the two principal steps required for the completion of the theory of radiative transfer in plane-parallel atmospheres of finite optical thicknesses. More particularly, in this paper, we shall carry out the elimination of the constants for the basic problem of diffuse reflection and transmission under a variety

[1] See particularly Papers XIV, XVI, XIX, and XX (*Ap. J.*, **105**, 164, 435, 1947; *ibid.*, **106**, 143, 145, 1947.

[2] Cf. Paper XII (eqs. [45] and [74]) (*ibid.*, **104**, 191, 1946).

[3] We shall see that, actually, a pair of functions defined in the interval (0, 1) is involved in the solutions for finite atmospheres.

of scattering conditions and show how, in each case considered, the solution can be expressed in terms of a single function defined in the interval $(-1, +1)$. The passage to the limit of infinite approximation and the exhibition of the relationship of these solutions[4] to the functional equations derived in Paper XVII[5] are postponed to a later paper.

I. ISOTROPIC SCATTERING WITH AN ALBEDO $\varpi_0 < 1$

2. *The expression of the boundary conditions and the emergent intensities in terms of two functions.*—As we shall see, the consideration of the problem of diffuse reflection and transmission by an atmosphere scattering radiation isotropically with an albedo $\varpi_0 < 1$ introduces us, in its simplest context, to a basic mathematical problem which is characteristic of this theory and which requires solution. Considering, then, the case of a plane-parallel atmosphere of a finite optical thickness, τ_1, on which is incident a parallel beam of radiation of net flux πF per unit area normal to itself, at an angle $\cos^{-1}\mu_0$ to the normal, we have[2]

$$I_i = \tfrac{1}{4}\varpi_0 F\left[\sum_{a=-n}^{+n} \frac{L_a e^{-k_a\tau}}{1+\mu_i k_a} + \frac{\gamma e^{-\tau/\mu_0}}{1+\mu_i/\mu_0}\right] \qquad (i = \pm 1, \ldots, \pm n), \quad (1)^{6}$$

for the solution of the intensities I_i in the nth approximation. In equation (1) the L_a's ($a = \pm 1, \ldots, \pm n$) are the $2n$ constants of integration, and the k_a's ($a = \pm 1, \ldots, \pm n$ and $k_{+a} = -k_{-a}$) are the $2n$ roots of the characteristic equation

$$1 = \varpi_0 \sum_{j=1}^{n} \frac{a_j}{1-k^2\mu_j^2}, \qquad (2)$$

which occurs in pairs ($k_{+a} = -k_{-a}$), and

$$\gamma = H(\mu_0)\,H(-\mu_0) = \frac{(-1)^n}{\mu_1^2 \cdots \mu_n^2}\,\frac{\prod_{i=1}^{n}(\mu_0^2-\mu_i^2)}{\prod_{a=1}^{n}(1-k_a^2\mu_0^2)}. \qquad (3)$$

The boundary conditions appropriate to our present problem are

$$I_{-i} = 0 \quad \text{at} \quad \tau = 0 \quad \text{and for} \quad i = 1, \ldots, n, \qquad (4)$$

and

$$I_{+i} = 0 \quad \text{at} \quad \tau = \tau_1 \quad \text{and for} \quad i = 1, \ldots, n. \qquad (5)$$

The equations which determine the $2n$ constants of integration are, therefore,

$$\sum_{a=-n}^{+n} \frac{L_a}{1-k_a\mu_i} + \frac{\gamma}{1-\mu_i/\mu_0} = 0 \qquad (i = 1, \ldots, n) \qquad (6)$$

and

$$\sum_{a=-n}^{+n} \frac{L_a e^{-k_a\tau_1}}{1+k_a\mu_i} + \frac{\gamma e^{-\tau_1/\mu_0}}{1+\mu_i/\mu_0} = 0 \qquad (i = 1, \ldots, n). \qquad (7)$$

[4] In the manner of Paper XIV. [5] *Ap. J.*, **106**, 441, 1947.

[6] In the summation over a in this equation, the term $a = 0$ is omitted. We shall adopt this convention throughout. It is, therefore, always to be understood that in all summations (or products) extended over a from $-n$ to $+n$ the term with $a = 0$ does not occur.

In terms of the functions

$$S(\mu) = \sum_{a=-n}^{+n} \frac{L_a}{1-k_a\mu} + \frac{\gamma}{1-\mu/\mu_0} \tag{8}$$

and

$$T(\mu) = \sum_{a=-n}^{+n} \frac{L_a e^{-k_a\tau_1}}{1+k_a\mu} + \frac{\gamma e^{-\tau_1/\mu_0}}{1+\mu/\mu_0}, \tag{9}$$

the boundary conditions can be expressed in the form

$$S(\mu_i) = T(\mu_i) = 0 \qquad (i = 1, \ldots, n). \tag{10}$$

In other words, the μ_i's $(i = 1, \ldots, n)$ are zeros of both S and T.

In analogy with the procedure adopted in the case of semi-infinite atmospheres, we must now try to express the reflected and the transmitted intensities, $I(0, \mu)$ and $I(\tau_1, -\mu)$, $(0 \leqslant \mu \leqslant 1)$, in terms of the same functions $S(\mu)$ and $T(\mu)$. It is, however, immediately apparent that $I(0, \mu)$ and $I(\tau_1, -\mu)$ cannot, simply, be proportional to $S(-\mu)$ and $T(-\mu)$, respectively, since these functions diverge for all those values of $\mu = k_a^{-1}$ for which $k_a^{-1} < 1$ $(a > 0)$; in nonconservative cases (such as the present) divergence from this source will occur for $(n - 1)$ values of μ in the interval $0 \leqslant \mu \leqslant 1$.[7] Consequently, a different procedure must be adopted for expressing the angular distributions of the reflected and the transmitted radiations. On consideration, it appears that the procedure which should be adopted is the following:

For the problem of diffuse reflection and transmission under consideration, the source function $\mathfrak{J}(\tau, \mu)$ is

$$\mathfrak{J}(\tau, \mu) = \tfrac{1}{2}\varpi_0 \int_{-1}^{+1} I(\tau, \mu')\, d\mu' + \tfrac{1}{4}\varpi_0 F e^{-\tau/\mu_0}, \tag{11}$$

or, in our scheme of approximation,

$$\mathfrak{J}(\tau, \mu) = \tfrac{1}{2}\varpi_0 \Sigma a_j I_j + \tfrac{1}{4}\varpi_0 F e^{-\tau/\mu_0}. \tag{12}$$

With the solution for I_i given by equation (1), the foregoing expression for $\mathfrak{J}(\tau, \mu)$ reduces to

$$\mathfrak{J}(\tau, \mu) = \tfrac{1}{4}\varpi_0 F \left[\sum_{a=-n}^{+n} L_a e^{-k_a\tau} + \gamma e^{-\tau/\mu_0} \right]. \tag{13}$$

Since, in general, the outward and the inward intensities, $I(\tau, +\mu)$ and $I(\tau, -\mu)$, $(0 < \mu < 1)$, at any level τ, are derivable from the source function in accordance with the equations

$$I(\tau, +\mu) = \int_{\tau}^{\tau_1} \mathfrak{J}(t, \mu)\, e^{-(t-\tau)/\mu} \frac{dt}{\mu} \tag{14}$$

and

$$I(\tau, -\mu) = \int_{0}^{\tau} \mathfrak{J}(t, -\mu)\, e^{-(\tau-t)/\mu} \frac{dt}{\mu}, \tag{15}$$

we find that in our particular case

$$\left.\begin{aligned} I(\tau, +\mu) = \tfrac{1}{4}\varpi_0 F e^{+\tau/\mu} \Bigg\{ \sum_{a=-n}^{+n} \frac{L_a}{1+k_a\mu} \left[e^{-\tau(1+k_a\mu)/\mu} - e^{-\tau_1(1+k_a\mu)/\mu} \right] \\ + \frac{\gamma}{1+\mu/\mu_0} \left[e^{-\tau(\mu_0+\mu)/\mu\mu_0} - e^{-\tau_1(\mu_0+\mu)/\mu\mu_0} \right] \Bigg\} \end{aligned}\right\} \tag{16}$$

[7] In addition, $T(-\mu)$ will also diverge for $\mu = \mu_0$.

and

$$I(\tau, -\mu) = \tfrac{1}{4}\varpi_0 F e^{-\tau/\mu}\left\{\sum_{a=-n}^{+n} \frac{L_a}{1-k_a\mu}\left[e^{\tau(1-k_a\mu)/\mu}-1\right] + \frac{\gamma}{1-\mu/\mu_0}\left[e^{\tau(\mu_0-\mu)/\mu\mu_0}-1\right]\right\}. \tag{17}$$

For the reflected and the transmitted intensities we therefore have

$$I(0, \mu) = \tfrac{1}{4}\varpi_0 F\left\{\sum_{a=-n}^{+n} \frac{L_a}{1+k_a\mu}\left[1-e^{-\tau_1(1+k_a\mu)/\mu}\right] + \frac{\gamma}{1+\mu/\mu_0}\left[1-e^{-\tau_1(\mu_0+\mu)/\mu\mu_0}\right]\right\} \tag{18}$$

and

$$I(\tau_1, -\mu) = \tfrac{1}{4}\varpi_0 F e^{-\tau_1/\mu}\left\{\sum_{a=-n}^{+n} \frac{L_a}{1-k_a\mu}\left[e^{\tau_1(1-k_a\mu)/\mu}-1\right] + \frac{\gamma}{1-\mu/\mu_0}\left[e^{\tau_1(\mu_0-\mu)/\mu\mu_0}-1\right]\right\}, \tag{19}$$

or, somewhat differently,

$$I(0, \mu) = \tfrac{1}{4}\varpi_0 F\left\{\sum_{a=-n}^{+n} \frac{L_a}{1+k_a\mu} + \frac{\gamma}{1+\mu/\mu_0} - e^{-\tau_1/\mu}\left[\sum_{a=-n}^{+n} \frac{L_a e^{-k_a\tau_1}}{1+k_a\mu} + \frac{\gamma e^{-\tau_1/\mu_0}}{1+\mu/\mu_0}\right]\right\}, \tag{20}$$

and

$$I(\tau_1, -\mu) = \tfrac{1}{4}\varpi_0 F\left\{\sum_{a=-n}^{+n} \frac{L_a e^{-k_a\tau_1}}{1-k_a\mu} + \frac{\gamma e^{-\tau_1/\mu_0}}{1-\mu/\mu_0} - e^{-\tau_1/\mu}\left[\sum_{a=-n}^{+n} \frac{L_a}{1-k_a\mu} + \frac{\gamma}{1-\mu/\mu_0}\right]\right\}. \tag{21}$$ [8]

According to our definitions of the functions $S(\mu)$ and $T(\mu)$, we can re-write the foregoing expressions for $I(0, +\mu)$ and $I(\tau_1, -\mu)$ in the forms

$$I(0, \mu) = \tfrac{1}{4}\varpi_0 F\left[S(-\mu) - e^{-\tau_1/\mu}T(\mu)\right] \tag{22}$$

and

$$I(\tau_1, -\mu) = \tfrac{1}{4}\varpi_0 F\left[T(-\mu) - e^{-\tau_1/\mu}S(\mu)\right]. \tag{23}$$ [9]

Thus, in the case of finite atmospheres, as in the case of semi-infinite atmospheres, there is a relationship of reciprocity between the equations which express the boundary conditions and the functions which describe the emergent radiations. The present relationship is naturally not so direct as the one encountered in the case of semi-infinite atmospheres. But it will appear that the relationship exemplified by equations (10), (22), and (23) is quite general and is precisely what is required to preserve the basic in-

[8] These expressions, do not, of course, diverge for any value of μ in the interval, $0 \leqslant \mu \leqslant 1$.

[9] Since $S(\mu_i) = T(\mu_i) = 0$,

$$I(0, \mu_i) = \tfrac{1}{4}\tilde{\omega}_0 F S(-\mu_i) \quad \text{and} \quad I(\tau_1, -\mu_i) = \tfrac{1}{4}\tilde{\omega}_0 F T(-\mu_i);$$

and this is in agreement with the solution (1) for the intensities at the points of the Gaussian division.

variances of the problem in all orders of approximation. And this last is, of course, an essential requirement for passing to the limit of infinite approximation on our method.

3. *The reduction to a problem in the theory of interpolation.*—In addition to the functions

$$P(\mu)=\prod_{i=1}^{n}(\mu-\mu_i) \quad \text{and} \quad R(\mu)=\prod_{a=1}^{n}(1-k_a\mu)\,, \tag{24}$$

which we have extensively used in the earlier papers, we shall now introduce the functions

$$W(\mu)=R(\mu)R(-\mu)=\prod_{a=-n}^{+n}(1-k_a\mu)=\prod_{a=1}^{n}(1-k_a^2\mu^2) \tag{25}$$

and

$$W_a(\mu)=\prod_{\substack{\beta=-n\\ \beta\neq a}}^{+n}(1-k_\beta\mu) \qquad (a=\pm 1\,,\ldots\ldots,\pm n)\,. \tag{26}$$

Identities which follow from definitions (25) and (26) and which we shall find useful, are

$$W(\mu)=W(-\mu) \tag{27}$$

and

$$W_a(\mu)=W_{-a}(-\mu) \qquad (a=1\,,\ldots\ldots,n)\,. \tag{28}$$

Now, from equations (8) and (9) it follows that

$$S(\mu)W(\mu)\left(1-\frac{\mu}{\mu_0}\right) \quad \text{and} \quad T(\mu)W(\mu)\left(1+\frac{\mu}{\mu_0}\right), \tag{29}$$

are polynomials of degree $2n$ in μ; and, according to equation (10), the μ_i's ($i=1, \ldots\ldots, n$) are zeros of both these polynomials. We may therefore write

$$S(\mu)=\frac{1}{\mu_1^2\ldots\ldots\mu_n^2}\frac{P(\mu)}{W(\mu)(1-\mu/\mu_0)}s(\mu) \tag{30}$$

and

$$T(\mu)=\frac{1}{\mu_1^2\ldots\ldots\mu_n^2}\frac{P(\mu)}{W(\mu)(1+\mu/\mu_0)}t(\mu)\,, \tag{31}$$ [10]

where $s(\mu)$ and $t(\mu)$ are polynomials of degree n in μ.

Two relations which follow immediately from equations (3), (8), (9), (30), and(31) are

$$\left.\begin{aligned}\gamma=\frac{1}{\mu_1^2\ldots\ldots\mu_n^2}\frac{P(\mu_0)P(-\mu_0)}{W(\mu_0)}&=\lim_{\mu\to\mu_0}\left(1-\frac{\mu}{\mu_0}\right)S(\mu)\\ &=\frac{1}{\mu_1^2\ldots\ldots\mu_n^2}\frac{P(\mu_0)}{W(\mu_0)}s(\mu_0)\,,\end{aligned}\right\} \tag{32}$$

and

$$\left.\begin{aligned}\gamma e^{-\tau_1/\mu_0}=\frac{e^{-\tau_1/\mu_0}}{\mu_1^2\ldots\ldots\mu_n^2}\frac{P(\mu_0)P(-\mu_0)}{W(\mu_0)}&=\lim_{\mu\to-\mu_0}\left(1+\frac{\mu}{\mu_0}\right)T(\mu)\\ &=\frac{1}{\mu_1^2\ldots\ldots\mu_n^2}\frac{P(-\mu_0)}{W(\mu_0)}t(-\mu_0)\,.\end{aligned}\right\} \tag{33}$$

[10] In eqs. (30) and (31) the factor $1/\mu_1^2\ldots\ldots\mu_n^2$ is introduced for reasons of convenience (see eqs. [32] and [33] below).

Hence

$$s(\mu_0) = P(-\mu_0)\,, \tag{34}$$

and

$$t(-\mu_0) = e^{-\tau_1/\mu_0} P(\mu_0)\,. \tag{35}$$

We next observe that, since (cf. eqs. [8] and [9])

$$L_a = \underset{\mu \to 1/k_a}{\text{limit}} (1 - k_a\mu) S(\mu) \qquad (a = \pm 1, \ldots, \pm n)\,, \tag{36}$$

and

$$L_a e^{-k_a\tau_1} = \underset{\mu \to -1/k_a}{\text{limit}} (1 + k_a\mu) T(\mu) \qquad (a = \pm 1, \ldots, \pm n)\,, \tag{37}$$

we must have

$$L_a = \frac{1}{\mu_1^2 \cdots \mu_n^2} \frac{P(1/k_a)}{W_a(1/k_a)(1 - 1/k_a\mu_0)} s(1/k_a) \qquad (a = \pm 1, \ldots, \pm n) \tag{38}$$

and (cf. eq. [28])

$$\left. \begin{aligned} L_a e^{-k_a\tau_1} &= \frac{1}{\mu_1^2 \cdots \mu_n^2} \frac{P(-1/k_a)}{W_a(1/k_a)(1 - 1/k_a\mu_0)} t(-1/k_a) \\ &\qquad (a = \pm 1, \ldots, \pm n)\,. \end{aligned} \right\} \tag{39}$$

Comparing equations (38) and (39), we conclude that

$$s(1/k_a) = e^{k_a\tau_1} \frac{P(-1/k_a)}{P(+1/k_a)} t(-1/k_a) \qquad (a = \pm 1, \ldots, \pm n)\,. \tag{40}$$

Re-writing this equation separately for the positive and the negative values of a, we have

$$s(1/k_a) = e^{k_a\tau_1} \frac{P(-1/k_a)}{P(+1/k_a)} t(-1/k_a) \qquad (a = 1, \ldots, n) \tag{41}$$

and

$$t(1/k_a) = e^{k_a\tau_1} \frac{P(-1/k_a)}{P(+1/k_a)} s(-1/k_a) \qquad (a = 1, \ldots, n)\,. \tag{42}$$

An immediate consequence of equations (41) and (42) is

$$s(1/k_a)\, s(-1/k_a) = t(1/k_a)\, t(-1/k_a) \qquad (a = \pm 1, \ldots, \pm n)\,. \tag{43}$$

From this equation it follows that

$$s(\mu)\, s(-\mu) - t(\mu)\, t(-\mu) \equiv \text{constant}\; W(\mu)\,, \tag{44}$$

since the quantity on the left-hand side is a polynomial of degree $2n$ in μ and vanishes for $\mu = \pm 1/k_a (a = 1, \ldots, n)$.

Next, writing

$$F(\mu) = s(\mu) + t(\mu) \qquad \text{and} \qquad G(\mu) = s(\mu) - t(\mu)\,, \tag{45}$$

we find, from equations (41) and (42),

$$F(1/k_a) = +\, e^{k_a\tau_1} \frac{P(-1/k_a)}{P(+1/k_a)} F(-1/k_a) \qquad (a = 1, \ldots, n) \tag{46}$$

and

$$G(1/k_a) = -\, e^{k_a\tau_1} \frac{P(-1/k_a)}{P(+1/k_a)} G(-1/k_a) \qquad (a = 1, \ldots, n)\,. \tag{47}$$

In § 5 we shall show how equations (46) and (47), together with equations (34) and (35), just suffice to determine the polynomials $F(\mu)$ and $G(\mu)$ uniquely. The problem, as we shall see, is essentially one in the theory of interpolation.

A more symmetrical way of writing equations (46) and (47) is

$$F(1/k_\alpha) = +\lambda_\alpha F(-1/k_\alpha) \qquad (\alpha = 1, \ldots, n) \tag{48}$$

and

$$G(1/k_\alpha) = -\lambda_\alpha G(-1/k_\alpha) \qquad (\alpha = 1, \ldots, n), \tag{49}$$

where

$$\lambda_\alpha = e^{k_\alpha \tau_1} \frac{P(-1/k_\alpha)}{P(+1/k_\alpha)}. \tag{50}$$

4. *The solution of the basic mathematical problem.*—The mathematical problem to which we reduced the solution of $S(\mu)$ and $T(\mu)$ in § 3 can be formulated as follows:

To determine two polynomials F(μ) *and* G(μ) *of degree* n *in* μ *such that*

$$F(x_\alpha) = +\lambda_\alpha F(-x_\alpha) \qquad (\alpha = 1, \ldots, n), \tag{51}$$

and

$$G(x_\alpha) = -\lambda_\alpha G(-x_\alpha) \qquad (\alpha = 1, \ldots, n), \tag{52}$$

where x_α, $\alpha = 1, \ldots, n$ *are* n *distinct values of the argument and* λ_α, $\alpha = 1, \ldots, n$, *are* n *assigned numbers, all different from one another.*

As we have already remarked, this problem is essentially one in the theory of interpolation. It does not, however, seem to have been considered in literature before. But it will appear that the problem is closely associated with the method of solution, in a finite approximation, of a class of simultaneous pairs of functional equations of which equations (119) and (120) of Paper XVII are typical. The problem would therefore appear to merit a closer investigation than we can afford in this paper. We shall, however, give explicit solutions for $F(\mu)$ and $G(\mu)$ which satisfy the required conditions.

First, it may be noted that $F(\mu)$ and $G(\mu)$ are related by the identity (cf. eq. [44])

$$F(\mu)G(-\mu) + F(-\mu)G(\mu) \equiv \text{constant} \prod_{\alpha=1}^{n} (x_\alpha^2 - \mu^2), \tag{53}$$

since the quantity on the right-hand side is a polynomial of degree $2n$ in μ and vanishes for $\mu = \pm x_\alpha (\alpha = 1, \ldots, n)$.

Next it should be observed that in general *the conditions stated determine* F(μ) *and* G(μ) *uniquely, apart from a constant factor of proportionality*. That this is the case can be seen by writing $F(\mu)$ (for example) in the form

$$F(\mu) = \sum_{m=0}^{n} a_m \mu^m \tag{54}$$

and noting that the conditions of the problem (eq. [51]) require that

$$\sum_{m=0}^{n} a_m [1 + (-1)^{m+1} \lambda_\alpha] x_\alpha^m = 0 \qquad (\alpha = 1, \ldots, n). \tag{55}$$

The $n + 1$ coefficients, $a_m (m = 0, \ldots, n)$, therefore satisfy a system of homogeneous linear equations of order n. Moreover, if the x_α's are all distinct and none of the λ_α's are equal to each other (as we have, indeed, assumed), the rank of the system (55) is also n. Consequently, the coefficients a_m are all uniquely determined apart from a constant factor of proportionality. The polynomial $F(\mu)$ is therefore also determined apart from a constant factor of proportionality. Similar remarks clearly apply to $G(\mu)$ also.

The arguments of the preceding paragraph further establish that *n is the lowest degree of a polynomial (not identically zero) which will satisfy* n *conditions of the form* (51) *or* (52). On the other hand, polynomials of degree higher than n can be readily constructed in terms of $F(\mu)$ and $G(\mu)$ which will satisfy conditions (51) or (52). For example,

$$a_0F(\mu) + a_1\mu G(\mu)\,, \tag{56}$$

where a_0 and a_1 are two arbitrary constants, is the most general polynomial of degree $(n+1)$ in μ which will satisfy the n conditions (51); for, according to equations (51) and (52),

$$\left.\begin{aligned} a_0F(x_a) + a_1x_aG(x_a) &= a_0\lambda_aF(-x_a) - a_1x_a\lambda_aG(-x_a) \\ &= \lambda_a[a_0F(\mu) + a_1\mu G(\mu)]_{\mu=-x_a} \qquad (a=1,\ldots,n)\,. \end{aligned}\right\} \tag{57}$$

Similarly,

$$b_0G(\mu) + b_1\mu F(\mu)\,, \tag{58}$$

where b_0 and b_1 are two arbitrary constants, is the most general polynomial of degree $(n+1)$ in μ which will satisfy conditions (52).

The foregoing observations can be readily extended to construct polynomials of any degree higher than n which will satisfy condition (51) or (52). We shall state the result in the form of the following theorem:

Theorem 1.—*The most general polynomials* $F^{(n+m)}(\mu)$ *and* $G^{(n+m)}(\mu)$ *of degree* $(n+m)$ *in* μ, $(m>0)$, *which will satisfy the conditions*

$$F^{(n+m)}(x_a) = +\lambda_aF^{(n+m)}(-x_a) \qquad (a=1,\ldots,n) \tag{59}$$

and

$$G^{(n+m)}(x_a) = -\lambda_aG^{(n+m)}(-x_a) \qquad (a=1,\ldots,n)\,, \tag{60}$$

are

$$F^{(n+m)}(\mu) = \sum_{l=0}^{m} a_l\mu^l[\epsilon_{l,\,\text{even}}F(\mu) + \epsilon_{l,\,\text{odd}}G(\mu)] \tag{61}$$

and

$$G^{(n+m)}(\mu) = \sum_{l=0}^{m} b_l\mu^l[\epsilon_{l,\,\text{odd}}F(\mu) + \epsilon_{l,\,\text{even}}G(\mu)]\,, \tag{62}$$

where a_l *and* $b_l (l=0,\ldots,n)$ *are arbitrary constants and*

$$\left.\begin{aligned} \epsilon_{l,\,\text{even}} &= 1 \text{ if } l \text{ is even} \\ &= 0 \text{ if } l \text{ is odd} \end{aligned}\right\} \text{ and } \left.\begin{aligned} \epsilon_{l,\,\text{odd}} &= 1 \text{ if } l \text{ is odd} \\ &= 0 \text{ if } l \text{ is even} \end{aligned}\right\}, \tag{63}$$

and $F(\mu)$ *and* $G(\mu)$ *are polynomials of degree* n *in* μ *which satisfy the same conditions as* $F^{(n+m)}(\mu)$ *and* $G^{(n+m)}(\mu)$, *respectively.*

This theorem suggests that *the polynomials* $F(\mu)$ *and* $G(\mu)$ *can be constructed by a process of induction;* for, if polynomials $F^{(n-1)}(\mu)$ and $G^{(n-1)}(\mu)$ of degree $(n-1)$ in μ which satisfy the $(n-1)$ conditions

$$F^{(n-1)}(x_a) = +\lambda_aF^{(n-1)}(-x_a) \qquad (a=1,\ldots,n-1) \tag{64}$$

and

$$G^{(n-1)}(x_a) = -\lambda_aG^{(n-1)}(-x_a) \qquad (a=1,\ldots,n-1) \tag{65}$$

are assumed known, then polynomials $F^{(n)}(\mu)$ and $G^{(n)}(\mu)$ of one higher degree satisfying conditions (51) and (52) appropriate to polynomials of degree n can be constructed. Thus, according to theorem 1,

$$F^{(n)}(\mu) = F^{(n-1)}(\mu) + a_1 \mu G^{(n-1)}(\mu) , \tag{66}$$

where a_1 is an arbitrary constant, will satisfy all the conditions (64) that are satisfied by $F^{(n-1)}(\mu)$. We therefore need to satisfy only the one additional condition,

$$F^{(n)}(x_n) = \lambda_n F^{(n)}(-x_n) . \tag{67}$$

This condition can be used to determine a_1. In this manner we find that, with the choice of

$$a_1 = -\frac{F^{(n-1)}(x_n) - \lambda_n F^{(n-1)}(-x_n)}{x_n [G^{(n-1)}(x_n) + \lambda_n G^{(n-1)}(-x_n)]} , \tag{68}$$

$F^{(n)}(\mu)$ defined as in equation (66) will satisfy all the required conditions. Similarly,

$$G^{(n)}(\mu) = G^{(n-1)}(\mu) + b_1 \mu F^{(n-1)}(\mu) , \tag{69}$$

where

$$b_1 = -\frac{G^{(n-1)}(x_n) + \lambda_n G^{(n-1)}(-x_n)}{x_n [F^{(n-1)}(x_n) - \lambda_n F^{(n-1)}(-x_n)]} , \tag{70}$$

will satisfy the n conditions (52). Thus polynomials of degree n satisfying the required number of conditions can be constructed if polynomials of one lower degree, each satisfying one less condition, are assumed known. On the other hand,

$$F^{(1)}(\mu) = (x_1 - \mu) + \lambda_1 (x_1 + \mu) \tag{71}$$

and

$$G^{(1)}(\mu) = (x_1 - \mu) - \lambda_1 (x_1 + \mu) \tag{72}$$

clearly satisfy the conditions appropriate to polynomials of degree 1. With this we have established that the solution to our problem can, in fact, be found by a process of induction.

While the construction by induction which we have outlined above solves our problem in principle, it is still unsatisfactory, in that the solution obtained by following the construction literally will be excessively complicated. This apparent complexity must, in part, be attributed to the fact that the manner of construction destroys the essential symmetry of the problem in the x_α's and the λ_α's. It would therefore seem that the construction of the polynomials $F(\mu)$ and $G(\mu)$ by a straightforward process of induction is not a significant approach to the problem. The method was therefore abandoned; and the somewhat indirect method that we shall now describe has at least the merit of providing explicit formulae for the functions $F(\mu)$ and $G(\mu)$ which preserve the symmetries of the problem.

From the form of the solution for the case $n = 2$, it appeared that $F(\mu)$ and $G(\mu)$ must be expressible as linear combinations of the 2^n polynomials of the form

$$\prod_{\alpha=1}^{n} (x_\alpha \pm \mu) , \tag{73}$$

where in each of the n factors we have either a plus or a minus sign in the parenthesis. Moreover, after some consideration, it also appeared that the coefficient of the term

$$\prod_{i=1}^{l} (x_{r_i} + \mu) \prod_{m=1}^{n-l} (x_{s_m} - \mu) , \tag{74}$$

where $r_1, \ldots, r_l$ and $s_1, \ldots, s_{n-l}$ are selections of l, respectively $n-l$, distinct integers from the set $(1, 2, \ldots, n)$, must be

$$\pm \lambda_{r_1} \ldots \lambda_{r_l} \frac{\prod_{m=1}^{n-l} \prod_{i=1}^{l} (x_{s_m} + x_{r_i})}{\prod_{m=1}^{n-l} \prod_{i=1}^{l} (x_{s_m} - x_{r_i})} . \tag{75}$$

But the decision regarding the sign turned out to be a rather more delicate matter. We shall, therefore, describe the method by which the decision was reached, as it will establish at the same time that the expression which we shall obtain does represent the solution to our problem.

According to our remarks in the preceding paragraph, we shall write

$$\left. \begin{aligned} F(\mu) = \sum_{2^n \text{ terms}} & \epsilon_{r_1 \ldots r_l} \lambda_{r_1} \ldots \lambda_{r_l} \\ & \times \frac{\prod_{m=1}^{n-l} \prod_{i=1}^{l} (x_{s_m} + x_{r_i})}{\prod_{m=1}^{n-l} \prod_{i=1}^{l} (x_{s_m} - x_{r_i})} \prod_{i=1}^{l} (x_{r_i} + \mu) \prod_{m=1}^{n-l} (x_{s_m} - \mu) , \end{aligned} \right\} \tag{76}$$

where

$$\epsilon^2_{r_1 \ldots r_l} = 1 \tag{77}$$

but is unspecified, otherwise, for the present.

It should be particularly noted that, in the summation on the right-hand side of equation (76), terms with the various factors $\lambda_{r_1} \ldots \lambda_{r_l}$ occur just exactly once.

Now if $F(\mu)$, as given by equation (76), represents a polynomial which satisfies the conditions of our problem, then we must have

$$F(x_{r_j}) = \lambda_{r_j} F(-x_{r_j}) . \tag{78}$$

According to equation (76), in $F(x_{r_j})$, there is *one and only one term* which occurs with the factor

$$\lambda_{r_1} \ldots \lambda_{r_j} \ldots \lambda_{r_l} , \tag{79}$$

and this arises, in fact, from the term

$$\left. \begin{aligned} & \epsilon_{r_1 \ldots r_j \ldots r_l} \lambda_{r_1} \ldots \lambda_{r_j} \ldots \lambda_{r_l} \frac{\prod_{m=1}^{n-l} \prod_{i=1}^{l} (x_{s_m} + x_{r_i})}{\prod_{m=1}^{n-l} \prod_{i=1}^{l} (x_{s_m} - x_{r_i})} \\ & \qquad \times \prod_{i=1}^{l} (x_{r_i} + \mu) \prod_{m=1}^{n-l} (x_{s_m} - \mu) , \end{aligned} \right\} \tag{80}$$

on the right-hand side of equation (76). The validity of equation (78) therefore requires that the term in $F(x_{r_j})$ arising from equation (80) must cancel the term in $F(-x_{r_j})$ which occurs with the factor

$$\lambda_{r_1} \dots \lambda_{r_{j-1}} \lambda_{r_{j+1}} \dots \lambda_{r_l}\,; \tag{81}$$

and the only term on the right-hand side of equation (76) which occurs with this factor is

$$\left.\begin{aligned} &\epsilon_{r_1 \dots r_{j-1} r_{j+1} \dots r_l} \lambda_{r_1} \dots \lambda_{r_{j-1}} \lambda_{r_{j+1}} \dots \lambda_{r_l} \\ &\times \frac{\prod_{m=1}^{n-l} \prod_{i=1, i\neq j}^{l} (x_{s_m}+x_{r_i}) \prod_{i=1, i\neq j}^{l} (x_{r_j}+x_{r_i})}{\prod_{m=1}^{n-l} \prod_{i=1, i\neq j}^{l} (x_{s_m}-x_{r_i}) \prod_{i=1, i\neq j}^{l} (x_{r_j}-x_{r_i})} \prod_{i=1, i\neq j}^{l} (x_{r_i}+\mu) \prod_{m=1}^{n-l} (x_{s_m}-\mu)(x_{r_j}-\mu)\,. \end{aligned}\right\} \tag{82}$$

Putting $\mu = x_{r_j}$ in equation (80) and $\mu = -x_{r_j}$ in equation (82), we have, respectively,

$$\epsilon_{r_1 \dots r_j \dots r_l} \lambda_{r_1} \dots \lambda_{r_j} \dots \lambda_{r_l} \frac{\prod_{m=1}^{n-l} \prod_{i=1}^{l} (x_{s_m}+x_{r_i})}{\prod_{m=1}^{n-l} \prod_{i=1, i\neq j}^{l} (x_{s_m}-x_{r_i})} \prod_{i=1}^{l} (x_{r_i}+x_{r_j}) \tag{83}$$

and

$$\left.\begin{aligned} &\epsilon_{r_1 \dots r_{j-1} r_{j+1} \dots r_l} \lambda_{r_1} \dots \lambda_{r_{j-1}} \lambda_{r_{j+1}} \dots \lambda_{r_l} \\ &\times \frac{\prod_{m=1}^{n-l} \prod_{i=1, i\neq j}^{l} (x_{s_m}+x_{r_i}) \prod_{i=1, i\neq j}^{l} (x_{r_j}+x_{r_i})}{\prod_{m=1}^{n-l} \prod_{i=1, i\neq j}^{l} (x_{s_m}-x_{r_i}) \prod_{i=1, i\neq j}^{l} (x_{r_j}-x_{r_i})} \prod_{i=1, i\neq j}^{l} (x_{r_i}-x_{r_j}) \prod_{m=1}^{n-l} (x_{s_m}+x_{r_j})(x_{r_j}+x_{r_j}) \\ &= \epsilon_{r_1 \dots r_{j-1} r_{j+1} \dots r_l} \lambda_{r_1} \dots \lambda_{r_{j-1}} \lambda_{r_{j+1}} \dots \lambda_{r_l} (-1)^{l-1} \\ &\qquad \times \frac{\prod_{m=1}^{n-l} \prod_{i=1}^{l} (x_{s_m}+x_{r_i})}{\prod_{m=1}^{n-l} \prod_{i=1, i\neq j}^{l} (x_{s_m}-x_{r_i})} \prod_{i=1}^{l} (x_{r_i}+x_{r_j})\,. \end{aligned}\right\} \tag{84}$$

Comparing (83) and (84), we observe that the validity of equation (78) requires only that

$$\epsilon_{r_1 \ldots\ldots r_l} = \epsilon_{r_1 \ldots\ldots r_{j-1} r_{j+1} \ldots\ldots r_l} (-1)^{l-1} . \tag{85}$$

Hence,

$$\epsilon_{r_1 \ldots\ldots r_l} = \epsilon_l = (-1)^{l-1} \epsilon_{l-1} . \tag{86}$$

Letting $\epsilon_n = 1$, we conclude from equation (86) that

$$\epsilon_n = +1 , \quad \epsilon_{n-1} = (-1)^{n-1} , \quad \epsilon_{n-2} = -1 , \quad \epsilon_{n-3} = (-1)^n , \quad \epsilon_{n-4} = +1 , \text{ etc.} \tag{87}$$

The ϵ's which occur in equation (76) can therefore be arranged in a sequence which we shall denote by $\epsilon_l^{(e)}$. With this choice of the ϵ's, equation (77) does, in fact, represent a solution for $F(\mu)$. (Any constant multiple of $F(\mu)$ will, of course, also be a solution.)

The solution for $G(\mu)$ can be constructed along similar lines. Thus $G(\mu)$ is also a linear combination of the 2^n polynomials (73), with, in fact, the same coefficients as $F(\mu)$ except for the ϵ-factor. And it can be verified that again ϵ depends only on the number of factors, l, in (73) which occurs with the positive sign in the parenthesis. However, in view of the minus sign in the conditions (52), we must now require that (cf. eq. [86])

$$\epsilon_l = \epsilon_{l-1} (-1)^l . \tag{88}$$

The ϵ's in the expansion for $G(\mu)$ therefore form the sequence

$$\epsilon_n = +1 . \quad \epsilon_{n-1} = (-1)^n , \quad \epsilon_{n-2} = -1 , \quad \epsilon_{n-3} = (-1)^{n-1} . \quad \epsilon_{n-4} = +1 , \text{ etc.}, \tag{89}$$

which we shall denote by $\epsilon_l^{(o)}$. We have thus established the following basic theorem:

Theorem 2.—The polynomials

$$F(\mu) = \sum_{2^n \text{ terms}} \epsilon_l^{(e)} \frac{\prod_{m=1}^{n-l} \prod_{i=1}^{l} (x_{s_m} + x_{r_i})}{\prod_{m=1}^{n-l} \prod_{i=1}^{l} (x_{s_m} - x_{r_i})} \prod_{i=1}^{l} \lambda_{r_i} (x_{r_i} + \mu) \prod_{m=1}^{n-l} (x_{s_m} - \mu) \tag{90}$$

and

$$G(\mu) = \sum_{2^n \text{ terms}} \epsilon_l^{(o)} \frac{\prod_{m=1}^{n-l} \prod_{i=1}^{l} (x_{s_m} + x_{r_i})}{\prod_{m=1}^{n-l} \prod_{i=1}^{l} (x_{s_m} - x_{r_i})} \prod_{i=1}^{l} \lambda_{r_i} (x_{r_i} + \mu) \prod_{m=1}^{n-l} (x_{s_m} - \mu) , \tag{91}$$

where $\epsilon_l^{(e)}$ *and* $\epsilon_l^{(o)}$ *denote the sequences*

$$\epsilon_l^{(e)} = +1 , \quad (-1)^{n-1} , \quad -1 , \quad (-1)^n , \quad +1 , \quad (-1)^{n-1} , \quad -1 , \quad (-1)^n , \ldots\ldots , \tag{92}$$

and

$$\epsilon_l^{(o)} = +1 , \quad (-1)^n , \quad -1 , \quad (-1)^{n-1} , \quad +1 , \quad (-1)^n , \quad -1 , \quad (-1)^{n-1} , \ldots\ldots , \tag{93}$$

satisfy the conditions

$$F(x_a) = \lambda_a F(-x_a) \quad \text{and} \quad G(x_a) = -\lambda_a G(-x_a) \qquad (a = 1 , \ldots\ldots , n) , \tag{94}$$

where x_a, $a = 1, \ldots, n$, *are* n *distinct values of the argument and* λ_a, $a = 1, \ldots, n$, *are* n *assigned numbers all different from one another. Any other polynomial of degree n which satisfies either of these conditions must be a simple numerical multiple of* $F(\mu)$ *or* $G(\mu)$ *as defined.*

By writing

$$x_a = \frac{1}{k_a} \qquad (a = 1, \ldots, n), \tag{95}$$

we can express the solutions (90) and (91) for $F(\mu)$ and $G(\mu)$ alternatively in the forms

$$F(\mu) = \sum_{2^n \text{ terms}} \epsilon_l^{(e)} \frac{\prod_{i=1}^{l}\prod_{m=1}^{n-l}(k_{r_i}+k_{s_m})}{\prod_{i=1}^{l}\prod_{m=1}^{n-l}(k_{r_i}-k_{s_m})} \prod_{i=1}^{l}(1+k_{r_i}\mu)\prod_{m=1}^{n-l}\frac{1}{\lambda_{s_m}}(1-k_{s_m}\mu) \tag{96}$$

and

$$G(\mu) = \sum_{2^n \text{ terms}} \epsilon_l^{(o)} \frac{\prod_{i=1}^{l}\prod_{m=1}^{n-l}(k_{r_i}+k_{s_m})}{\prod_{i=1}^{l}\prod_{m=1}^{n-l}(k_{r_i}-k_{s_m})} \prod_{i=1}^{l}(1+k_{r_i}\mu)\prod_{m=1}^{n-l}\frac{1}{\lambda_{s_m}}(1-k_{s_m}\mu). \tag{97}$$

Now, examining the sequences (92) and (93), we observe that the terms n, $n-2$, etc., agree, while the terms $n-1$, $n-3$, etc., are of opposite signs. We can, therefore, express $F(\mu)$ and $G(\mu)$ in the forms

$$F(\mu) = C_0(\mu) + C_1(\mu) \tag{98}$$

and

$$G(\mu) = C_0(\mu) - C_1(\mu), \tag{99}$$

where

$$C_0(\mu) = \sum_{2^{n-1}\text{ terms}}^{l=n,\,n-2,\,\ldots} \epsilon_l^{(0)} \frac{\prod_{i=1}^{l}\prod_{m=1}^{n-l}(k_{r_i}+k_{s_m})}{\prod_{i=1}^{l}\prod_{m=1}^{n-l}(k_{r_i}-k_{s_m})} \prod_{i=1}^{l}(1+k_{r_i}\mu)\prod_{m=1}^{n-l}\frac{1}{\lambda_{s_m}}(1-k_{s_m}\mu) \tag{100}$$

and

$$C_1(\mu) = (-1)^{n-1}\sum_{2^{n-1}\text{ term}}^{l=n-1;\,n-3,\,\ldots} \epsilon_l^{(1)} \frac{\prod_{i=1}^{l}\prod_{m=1}^{n-l}(k_{r_i}+k_{s_m})}{\prod_{i=1}^{l}\prod_{m=1}^{n-l}(k_{r_i}-k_{s_m})} \prod_{i=1}^{l}(1+k_{r_i}\mu)\prod_{m=1}^{n-l}\frac{1}{\lambda_{s_m}}(1-k_{s_m}\mu), \tag{101}$$

where

$$\left.\begin{aligned} \epsilon_l^{(0)} &= +1 \text{ for integers of the form } n-4l \\ &= -1 \text{ for integers of the form } n-4l-2 \\ &= 0 \text{ otherwise} \end{aligned}\right\} \tag{102}$$

and

$$\left.\begin{aligned}\epsilon_l^{(1)} &= +1 \text{ for integers of the form } n-4l-1\\ &= -1 \text{ for integers of the form } n-4l-3\\ &= 0 \text{ otherwise}\end{aligned}\right\} . \qquad (103)$$

In our future work we shall adopt equations (96)–(103) as our standard definitions of the various functions, and it will always have to be assumed that functions defined in this manner are meant unless something explicitly to the contrary is stated.

Finally, some properties of the functions $C_0(\mu)$ and $C_1(\mu)$ may be noted:

By virtue of equations (98) and (99), the identity (53) between $F(\mu)$ and $G(\mu)$ becomes, in our present notation (cf. eq. [25])

$$\left.\begin{aligned}C_0(\mu)C_0(-\mu)-C_1(\mu)C_1(-\mu) &= \text{constant}\prod_{a=1}^{n}(1-k_a^2\mu^2)\\ &= \text{constant } W(\mu) .\end{aligned}\right\} \qquad (104)$$

Since $W(0) = 1$, we can re-write equation (104) in the form

$$C_0(\mu)C_0(-\mu)-C_1(\mu)C_1(-\mu) = [C_0^2(0)-C_1^2(0)]\,W(\mu) , \qquad (105)$$

a relation which we shall find very useful in our further work.

Again, according to equations (51) and (52),

$$F(x_a)+G(x_a) = \lambda_a[F(-x_a)-G(-x_a)] , \qquad (106)$$

and

$$F(x_a)-G(x_a) = \lambda_a[F(-x_a)+G(-x_a)] . \qquad (107)$$

Expressing F and G as in equations (98) and (99), we find that the foregoing equations are equivalent, in our present notation, to

$$C_0(1/k_a) = \lambda_a C_1(-1/k_a) \qquad (a=1, \ldots, n) \qquad (108)$$

and

$$C_1(1/k_a) = \lambda_a C_0(-1/k_a) \qquad (a=1, \ldots, n) . \qquad (109)$$

Equations (108) and (109) will be formally equivalent to each other if (again formally!)

$$\lambda_a = \frac{1}{\lambda_{-a}} . \qquad (110)$$

For λ_a defined as in equation (50), this is *actually* the case.

5. *Completion of the solutions for* $s(\mu)$ *and* $t(\mu)$.—Returning, now, to the solution for $s(\mu)$ and $t(\mu)$ at the point where we left it in § 3, we conclude that $s(\mu) + t(\mu)$ and $s(\mu) - t(\mu)$ must be proportional, respectively, to $F(\mu)$ and $G(\mu)$ as we have defined them in equations (96) and (97), with λ_a having the particular value given by equation (50). Expressing $F(\mu)$ and $G(\mu)$ in terms of $C_0(\mu)$ and $C_1(\mu)$, as in equations (100) and (101), we can therefore write

$$s(\mu) = q_0 C_0(\mu) + q_1 C_1(\mu) \qquad (111)$$

and

$$t(\mu) = q_1 C_0(\mu) + q_0 C_1(\mu) , \qquad (112)$$

where q_0 and q_1 are two constants. To determine these constants we make use of equations (34) and (35), which require that

$$s(\mu_0) = q_0 C_0(\mu_0) + q_1 C_1(\mu_0) = P(-\mu_0) \tag{113}$$

and

$$t(-\mu_0) = q_0 C_1(-\mu_0) + q_1 C_0(-\mu_0) = P(\mu_0)\, e^{-\tau_1/\mu_0}\,. \tag{114}$$

Solving these equations for q_0 and q_1, we find

$$q_0 = \frac{P(-\mu_0)\,C_0(-\mu_0) - e^{-\tau_1/\mu_0} P(\mu_0)\,C_1(\mu_0)}{C_0(\mu_0)\,C_0(-\mu_0) - C_1(\mu_0)\,C_1(-\mu_0)} \tag{115}$$

and

$$q_1 = \frac{e^{-\tau_1/\mu_0} P(\mu_0)\,C_0(\mu_0) - P(-\mu_0)\,C_1(-\mu_0)}{C_0(\mu_0)\,C_0(-\mu_0) - C_1(\mu_0)\,C_1(-\mu_0)}\,. \tag{116}$$

The denominator in equations (115) and (116) can be simplified by using equation (105). We thus find that

$$q_0 = \frac{1}{[C_0^2(0) - C_1^2(0)]\,W(\mu_0)}\,[P(-\mu_0)\,C_0(-\mu_0) - e^{-\tau_1/\mu_0} P(\mu_0)\,C_1(\mu_0)] \tag{117}$$

and

$$q_1 = \frac{1}{[C_0^2(0) - C_1^2(0)]\,W(\mu_0)}\,[e^{-\tau_1/\mu_0} P(\mu_0)\,C_0(\mu_0) - P(-\mu_0)\,C_1(-\mu_0)]\,. \tag{118}$$

With this determination of the constants q_0 and q_1, we have completed the formal solution of our problem.

6. *The solution for the reflected and the transmitted radiations.*—With $s(\mu)$ and $t(\mu)$ given by equations (111) and (112), equations (30) and (31) for $S(\mu)$ and $T(\mu)$ take the forms

$$S(\mu) = \frac{1}{\mu_1^2 \cdots \mu_n^2}\,\frac{P(\mu)}{W(\mu)}\,\frac{\mu_0}{\mu_0 - \mu}\,[q_0 C_0(\mu) + q_1 C_1(\mu)] \tag{119}$$

and

$$T(\mu) = \frac{1}{\mu_1^2 \cdots \mu_n^2}\,\frac{P(\mu)}{W(\mu)}\,\frac{\mu_0}{\mu_0 + \mu}\,[q_1 C_0(\mu) + q_0 C_1(\mu)]\,. \tag{120}$$

Substituting for $S(\mu)$ and $T(\mu)$ from the foregoing equations in equations (22) and (23), we obtain, after some minor rearranging of the terms, the following expressions for the reflected and the transmitted intensities:

$$\left.\begin{aligned} I(0,\mu) = \frac{1}{4}\,\frac{\varpi_0 F}{\mu_1^2 \cdots \mu_n^2}\,\frac{1}{W(\mu)}\,[q_0\{P(-\mu)\,C_0(-\mu) - e^{-\tau_1/\mu} P(\mu)\,C_1(\mu)\} \\ - q_1\{e^{-\tau_1/\mu} P(\mu)\,C_0(\mu) - P(-\mu)\,C_1(-\mu)\}]\,\frac{\mu_0}{\mu_0 + \mu} \end{aligned}\right\} \tag{121}$$

and

$$\left.\begin{aligned} I(\tau_1,-\mu) = \frac{1}{4}\,\frac{\varpi_0 F}{\mu_1^2 \cdots \mu_n^2}\,\frac{1}{W(\mu)}\,[q_1\{P(-\mu)C_0(-\mu) - e^{-\tau_1/\mu} P(\mu)C_1(\mu)\} \\ - q_0\{e^{-\tau_1/\mu} P(\mu)C_0(\mu) - P(-\mu)C_1(-\mu)\}]\,\frac{\mu_0}{\mu_0 - \mu}\,. \end{aligned}\right\} \tag{122}$$

Substituting, next, for q_0 and q_1 according to equations (117) and (118), we have

$$\left.\begin{aligned} I(0,\mu)=&\frac{1}{4}\frac{\varpi_0 F}{\mu_1^2\ldots\mu_n^2}\frac{1}{[C_0^2(0)-C_1^2(0)]}\frac{1}{W(\mu)W(\mu_0)}\frac{\mu_0}{\mu_0+\mu}\\ &\times[\{P(-\mu)C_0(-\mu)-e^{-\tau_1/\mu}P(\mu)C_1(\mu)\}\{P(-\mu_0)C_0(-\mu_0)-e^{-\tau_1/\mu_0}P(\mu_0)C_1(\mu_0)\}\\ &-\{e^{-\tau_1/\mu}P(\mu)C_0(\mu)-P(-\mu)C_1(-\mu)\}\{e^{-\tau_1/\mu_0}P(\mu_0)C_0(\mu_0)-P(-\mu_0)C_1(-\mu_0)\}]\end{aligned}\right\}\quad(123)$$

and

$$\left.\begin{aligned} I(\tau_1,-\mu)=&\frac{1}{4}\frac{\varpi_0 F}{\mu_1^2\ldots\mu_n^2}\frac{1}{[C_0^2(0)-C_1^2(0)]}\frac{1}{W(\mu)W(\mu_0)}\frac{\mu_0}{\mu_0-\mu}\\ &\times[\{P(-\mu)C_0(-\mu)-e^{-\tau_1/\mu}P(\mu)C_1(\mu)\}\{e^{-\tau_1/\mu_0}P(\mu_0)C_0(\mu_0)-P(-\mu_0)C_1(-\mu_0)\}\\ &-\{e^{-\tau_1/\mu}P(\mu)C_0(\mu)-P(-\mu)C_1(-\mu)\}\{P(-\mu_0)C_0(-\mu_0)-e^{-\tau_1/\mu_0}P(\mu_0)C_1(\mu_0)\}].\end{aligned}\right\}\quad(124)$$

Now let

$$X(\mu)=\frac{(-1)^n}{\mu_1\ldots\mu_n}\frac{1}{[C_0^2(0)-C_1^2(0)]^{\frac{1}{2}}}\frac{1}{W(\mu)}[P(-\mu)C_0(-\mu)-e^{-\tau_1/\mu}P(\mu)C_1(\mu)]\quad(125)$$

and

$$Y(\mu)=\frac{(-1)^n}{\mu_1\ldots\mu_n}\frac{1}{[C_0^2(0)-C_1^2(0)]^{\frac{1}{2}}}\frac{1}{W(\mu)}[e^{-\tau_1/\mu}P(\mu)C_0(\mu)-P(-\mu)C_1(-\mu)].\quad(126)$$

It will appear that functions $X(\mu)$ and $Y(\mu)$, defined in this manner, play the same fundamental role in the theory of atmospheres of finite optical thicknesses as the function $H(\mu)$ did in the theory of semi-infinite atmospheres.

In terms of the functions $X(\mu)$ and $Y(\mu)$, equations (123) and (124) for the reflected and the transmitted intensities take the following simple forms:

$$I(0,\mu)=\tfrac{1}{4}\varpi_0 F\frac{\mu_0}{\mu_0+\mu}[X(\mu)X(\mu_0)-Y(\mu)Y(\mu_0)]\quad(127)$$

and

$$I(\tau_1,-\mu)=\tfrac{1}{4}\varpi_0 F\frac{\mu_0}{\mu_0-\mu}[X(\mu)Y(\mu_0)-Y(\mu)X(\mu_0)].\quad(128)$$

It will be seen that the solutions for $I(0, \mu)$ and $I(\tau_1, -\mu)$ given by equations (127) and (128) are of exactly the forms required by the functional equations satisfied by the scattering and the transmission functions (Paper XVII, § 6); they further bring into evidence Helmholtz' principle of reciprocity.

Finally, attention should be drawn to the fact that there is nothing in the analysis of the preceding sections which has depended on the k_a's being the roots of the particular characteristic equation (2) except that it has $2n$ roots which occur in pairs (i.e., $k_a = -k_{-a}$). The method of solution and the reduction to the basic problem considered in § 4 has depended only on the single circumstance of the solution for the intensities being of the form given by equations (1) and (3). Conversely, it follows that the expressions for the reflected and the transmitted radiations can always be brought to the forms (127) and (128), provided only that the intensities I_i at the points of the Gaussian division are given by equations of the general form of (1) and (3) and the k_a's are the roots of an equation of the form

$$1=2\sum_{j=1}^{n}\frac{a_j\Psi(\mu_j)}{1-k^2\mu_j^2},\quad(129)$$

where the characteristic function, $\Psi(\mu)$, is an even polynomial, satisfying the condition

$$\int_0^1 \Psi(\mu)\, d\mu < \tfrac{1}{2}\,. \tag{130}$$ [11]

An obvious corollary of this observation is that for all equations of transfer of the form considered in Paper IX, § 4, the solution for the reflected and the transmitted intensities can be reduced to the forms given by equations (127) and (128), in which the functions $X(\mu)$ and $Y(\mu)$ are defined according to equations (24), (25), (100), (101), (125), and (126).

II. ISOTROPIC SCATTERING WITH UNIT ALBEDO

7. *The reduction for the case* $\varpi_0 = 1$.—The solution for the case of isotropic scattering with unit albedo cannot be obtained by simply letting $\varpi_0 = 1$ in the equations of the preceding sections (§§ 5 and 6); for, in this case, the various functions become indeterminate because two of the characteristic roots become zero. While there can, of course, be no difficulty of principle in properly passing to the limit $\varpi_0 = 1$ with due regard to the indeterminateness we have mentioned, it appears simpler to treat this case separately.

When $\varpi_0 = 1$, the solution for the intensities I_i at the points of the Gaussian division is (cf. Paper VIII)

$$\left.\begin{aligned} I_i = \tfrac{1}{4}F\left[\sum_{a=-n+1}^{+n-1} \frac{L_a e^{-k_a\tau}}{1+k_a\mu_i} + L_0(\tau+\mu_i) + L_n + \frac{\gamma e^{-\tau/\mu_0}}{1+\mu_i/\mu_0}\right] \\ (i=\pm 1, \ldots, \pm n)\,, \end{aligned}\right\} \tag{131}$$

where the k_a's ($a = \pm 1, \ldots, \pm n \mp 1$, and $k_{+a} = -k_{-a}$) are the $(2n-2)$ distinct nonvanishing roots of the characteristic equation

$$1 = \sum_{j=1}^{n} \frac{a_j}{1-k^2\mu_j^2} \tag{132}$$

and the L_a's ($a = 0, \pm 1, \ldots, \pm n \mp 1, n$) are the $2n$ constants of integration and

$$\gamma = H(\mu_0)\, H(-\mu_0) = \frac{1}{\mu_1^2 \ldots \mu_n^2} \frac{P(\mu_0)\, P(-\mu_0)}{W(\mu_0)}\,. \tag{133}$$

In terms of the functions

$$S(\mu) = \sum_{a=-n+1}^{+n-1} \frac{L_a}{1-k_a\mu} - L_0\mu + L_n + \frac{\gamma}{1-\mu/\mu_0} \tag{134}$$

and

$$T(\mu) = \sum_{a=-n+1}^{+n-1} \frac{L_a e^{-k_a\tau_1}}{1+k_a\mu} + L_0(\tau_1+\mu) + L_n + \frac{\gamma e^{-\tau_1/\mu_0}}{1+\mu/\mu_0}, \tag{135}$$

the boundary conditions requiring

$$I_{-i} = 0 \text{ at } \tau = 0 \quad \text{and} \quad I_{+i} = 0 \text{ at } \tau = \tau_1 \quad \text{for} \quad i = 1, \ldots, n\,, \tag{136}$$

can be written as

$$S(\mu_i) = T(\mu_i) = 0 \qquad (i = 1, \ldots, n)\,. \tag{137}$$

[11] Notice the exclusion of the equality sign here. This means that we exclude conservative cases from this discussion (see Sec. II below).

And, as in the case $\varpi_0 < 1$ (§ 2), it can be shown that, in this case also, the reflected and the transmitted intensities can be expressed in the forms (cf. eqs. [22] and [23])

$$I(0, \mu) = \tfrac{1}{4}F[S(-\mu) - e^{-\tau_1/\mu}T(\mu)] \tag{138}$$

and

$$I(\tau_1, -\mu) = \tfrac{1}{4}F[T(-\mu) - e^{-\tau_1/\mu}S(\mu)]\,. \tag{139}$$

Returning to equations (134) and (135), we observe that, since the μ_i's ($i = 1, \ldots, n$) are zeros of $S(\mu)$ and $T(\mu)$, we can write (cf. eqs. [30], [31], and n. 10)

$$S(\mu) = \frac{1}{\mu_1^2 \ldots \mu_n^2} \frac{P(\mu)}{W(\mu)} \frac{1}{1-\mu/\mu_0} s(\mu) \tag{140}$$

and

$$T(\mu) = \frac{1}{\mu_1^2 \ldots \mu_n^2} \frac{P(\mu)}{W(\mu)} \frac{1}{1+\mu/\mu_0} t(\mu)\,, \tag{141}$$

where $s(\mu)$ and $t(\mu)$ are polynomials of degree n in μ.

From equations (133)–(135), (140), and (141), it readily follows that (cf. eqs. [32]–[35])

$$s(\mu_0) = P(-\mu_0) \qquad \text{and} \qquad t(-\mu_0) = e^{-\tau_1/\mu_0}P(\mu_0)\,. \tag{142}$$

But we now have only $(2n - 2)$ relations of the form (cf. eq. [40])

$$s(1/k_a) = e^{k_a\tau_1} \frac{P(-1/k_a)}{P(+1/k_a)} t(-1/k_a) \qquad (a = \pm 1, \ldots, \pm n \mp 1)\,. \tag{143}$$

Consequently, $s(\mu) + t(\mu)$ and $s(\mu) - t(\mu)$ satisfy only $(n - 1)$ (instead of n) relations of the forms (46) and (47), respectively. In accordance with theorems 1 and 2 (§ 4), we therefore conclude that, in the present case, $s(\mu) + t(\mu)$ must be a linear combination of $F(\mu)$ and $\mu G(\mu)$ with numerical coefficients, where $F(\mu)$ and $G(\mu)$ are polynomials of degree $n - 1$, defined in the manner of equations (50), (96), and (97) in terms of the $n - 1$ positive nonvanishing roots of equation (132).[12] Similarly, $s(\mu) - t(\mu)$ must be a linear combination of $G(\mu)$ and $\mu F(\mu)$ with numerical coefficients. We can therefore write (cf. eqs. [98] and [99])

$$s(\mu) = (p_0 + q_0\mu)C_0(\mu) + (p_1 - q_1\mu)C_1(\mu) \tag{144}$$

and

$$t(\mu) = (p_1 + q_1\mu)C_0(\mu) + (p_0 - q_0\mu)C_1(\mu)\,, \tag{145}$$

where p_0, q_0, p_1, and q_1 are certain constants. Equations (142) provide two relations between these four constants. Two further relations can be obtained in the following manner:

According to equations (134), (135), (140), and (141), we have

$$\left.\begin{aligned} \frac{1}{\mu_1^2 \ldots \mu_n^2} P(\mu)\, s(\mu) &= \left(1 - \frac{\mu}{\mu_0}\right) W(\mu)\, S(\mu) \\ &= (-1)^{n-1} k_1^2 \ldots k_{n-1}^2 \left[\frac{L_0}{\mu_0}\mu^{2n} - \left(L_0 + \frac{L_n}{\mu_0}\right)\mu^{2n-1} + \ldots\right] \end{aligned}\right\} \tag{146}$$

[12] We shall adopt this convention throughout. It is therefore always to be understood that W, W_a, F, G, C_0, C_1, X, and Y signify the functions defined as in eqs. (25), (26), (50), (96)–(103), (125), and (126) in terms of the positive nonvanishing roots of the particular characteristic equation which is appropriate in the context.

and

$$\left.\begin{aligned}\frac{1}{\mu_1\ldots\mu_n^2}P(\mu)\,t(\mu) &= \left(1+\frac{\mu}{\mu_0}\right)W(\mu)\,T(\mu).\\ &= (-1)^{n-1}k_1^2\ldots k_{n-1}^2\left[\frac{L_0}{\mu_0}\mu^{2n}+\left(L_0+\frac{L_n}{\mu_0}+\frac{L_0\tau_1}{\mu_0}\right)\mu^{2n-1}+\ldots\right].\end{aligned}\right\} \quad (147)$$

From equations (146) and (147) it is apparent that the coefficients of μ^n in $s(\mu)$ and $t(\mu)$ are the same. This requires that in equations (144) and (145)

$$q_0 = q_1 = a \text{ (say)}. \quad (148)$$

We can therefore write

$$s(\mu) = (p_0+a\mu)\,C_0(\mu) + (p_1-a\mu)\,C_1(\mu) \quad (149)$$

and

$$t(\mu) = (p_1+a\mu)\,C_0(\mu) + (p_0-a\mu)\,C_1(\mu). \quad (150)$$

With the foregoing forms for $s(\mu)$ and $t(\mu)$, it is readily verified that

$$\left.\begin{aligned}P(\mu)\,s(\mu) &= \mu^{2n}\{a[c_0^{(n-1)}-c_1^{(n-1)}]\}+\mu^{2n-1}\Big\{a[c_0^{(n-2)}-c_1^{(n-2)}]\\ &\quad +p_0c_0^{(n-1)}+p_1c_1^{(n-1)}-\Big(\sum_{i=1}^{n}\mu_i\Big)a[c_0^{(n-1)}-c_1^{(n-1)}]\Big\}+\ldots\end{aligned}\right\} \quad (151)$$

and

$$\left.\begin{aligned}P(\mu)\,t(\mu) &= \mu^{2n}\{a[c_0^{(n-1)}-c_1^{(n-1)}]\}+\mu^{2n-1}\Big\{a[c_0^{(n-2)}-c_1^{(n-2)}]\\ &\quad +p_1c_0^{(n-1)}+p_0c_1^{(n-1)}-\Big(\sum_{i=1}^{n}\mu_i\Big)a[c_0^{(n-1)}-c_1^{(n-1)}]\Big\}+\ldots,\end{aligned}\right\} \quad (152)$$

where $c_0^{(n-1)}$, $c_0^{(n-2)}$ and $c_1^{(n-1}$, $c_1^{(n-2)}$ are the coefficients of the highest and the next highest powers of μ in $C_0(\mu)$ and $C_1(\mu)$, respectively.

Comparing equations (146) and (147) with (151) and (152), we conclude that

$$\left.\begin{aligned}\frac{1}{c_0^{(n-1)}-c_1^{(n-1)}}\Big\{c_0^{(n-2)}-c_1^{(n-2)}+\frac{1}{a}[p_0c_0^{(n-1)}+p_1c_1^{(n-1)}]\Big\}-\sum_{i-1}^{n}\mu_i&\\ = -\left(\mu_0+\frac{L_n}{L_0}\right)&\end{aligned}\right\} \quad (153)$$

and

$$\left.\begin{aligned}\frac{1}{c_0^{(n-1)}-c_1^{(n-1)}}\Big\{c_0^{(n-2)}-c_1^{(n-2)}+\frac{1}{a}[p_1c_0^{(n-1)}+p_0c_1^{(n-1)}]\Big\}-\sum_{i=1}^{n}\mu_i&\\ = +\left(\mu_0+\frac{L_n}{L_0}+\tau_1\right).&\end{aligned}\right\} \quad (154)$$

Adding equations (153) and (154), we obtain

$$\frac{1}{c_0^{(n-1)}-c_1^{(n-1)}}\Big\{2[c_0^{(n-2)}-c_1^{(n-2)}]+\frac{p_0+p_1}{a}[c_0^{(n-1)}+c_1^{(n-1)}]\Big\}-2\sum_{i=1}^{n}\mu_i=\tau_1. \quad (155)$$

Hence

$$a = Q\,(p_0 + p_1)\,, \tag{156}$$

where

$$Q = \frac{c_0^{(n-1)} + c_1^{(n-1)}}{\left(\tau_1 + 2\sum_{i-1}^{n} \mu_i\right)[\,c_0^{(n-1)} - c_1^{(n-1)}] - 2\,[\,c_0^{(n-2)} - c_1^{(n-2)}]}\,. \tag{157}$$

Now, from equations (142), (149), and (150) we have

$$(p_0 + a\mu_0)\,C_0\,(\mu_0) + (p_1 - a\mu_0)\,C_1\,(\mu_0) = P\,(-\mu_0) \tag{158}$$

and

$$(p_0 + a\mu_0)\,C_1\,(-\mu_0) + (p_1 - a\mu_0)\,C_0\,(-\mu_0) = e^{-\tau_1/\mu_0}P\,(\mu_0)\,. \tag{159}$$

Solving these equations for $(p_0 + a\mu_0)$ and $(p_1 - a\mu_0)$, we have (cf. eqs. [113]–[118])

$$p_0 + a\mu_0 = \frac{1}{[C_0^2(0) - C_1^2(0)\,]W\,(\mu_0)}[P(-\mu_0)C_0(-\mu_0) - e^{-\tau_1/\mu_0}P(\mu_0)C_1(\mu_0)] \tag{160}$$

and

$$p_1 - a\mu_0 = \frac{1}{[C_0^2(0) - C_1^2(0)\,]W\,(\mu_0)}[\,e^{-\tau_1/\mu_0}P(\mu_0)C_0(\mu_0) - P(-\mu_0)C_1(-\mu_0)]\,. \tag{161}$$

And, finally, according to equations (156), (160), and (161)

$$\left.\begin{aligned} a = \frac{Q}{[C_0^2\,(0) - C_1^2\,(0)\,]\,W\,(\mu_0)}\,&[P\,(-\mu_0)\,C_0\,(-\mu_0) - e^{-\tau_1/\mu_0}P\,(\mu_0)\,C_1\,(\mu_0) \\ &+ e^{-\tau_1/\mu_0}P\,(\mu_0)\,C_0\,(\mu_0) - P\,(-\mu_0)\,C_1\,(-\mu_0)\,]\,. \end{aligned}\right\} \tag{162}$$

With this determination of the various constants, the solution of the formal problem is completed.

8. *The solution for the reflected and the transmitted radiations.*—The angular distributions of the reflected and the transmitted radiations are given by

$$I\,(0,\mu) = \frac{1}{4}\,\frac{F}{\mu_1^2 \ldots \mu_n^2}\,\frac{1}{W\,(\mu)}\,[P\,(-\mu)\,s\,(-\mu) - e^{-\tau_1/\mu}P\,(\mu)\,t\,(\mu)\,]\,\frac{\mu_0}{\mu_0 + \mu} \tag{163}$$

and

$$I\,(\tau_1, -\mu) = \frac{1}{4}\,\frac{F}{\mu_1^2 \ldots \mu_n^2}\,\frac{1}{W\,(\mu)}\,[P\,(-\mu)\,t(-\mu) - e^{-\tau_1/\mu}P\,(\mu)\,s\,(\mu)]\,\frac{\mu_0}{\mu_0 - \mu}\,. \tag{164}$$

On the other hand, according to equations (149) and (150),

$$\left.\begin{aligned} &P\,(-\mu)\,s\,(-\mu) - e^{-\tau_1/\mu}P\,(\mu)\,t\,(\mu) \\ &\quad = (p_0 - a\mu)\,[P\,(-\mu)\,C_0\,(-\mu) - e^{-\tau_1/\mu}P\,(\mu)\,C_1\,(\mu)] \\ &\qquad - (p_1 + a\mu)\,[\,e^{-\tau_1/\mu}P\,(\mu)\,C_0\,(\mu) - P\,(-\mu)\,C_1\,(-\mu)] \\ &\quad = (p_0 + a\mu_0)\,[P\,(-\mu)\,C_0\,(-\mu) - e^{-\tau_1/\mu}P\,(\mu)\,C_1\,(\mu)] \\ &\qquad - (p_1 - a\mu_0)\,[\,e^{-\tau_1/\mu}P\,(\mu)\,C_0\,(\mu) - P\,(-\mu)\,C_1\,(-\mu)] \\ &\qquad - a\,(\mu_0 + \mu)\,[P\,(-\mu)\,C_0\,(-\mu) - e^{-\tau_1/\mu}P\,(\mu)\,C_1\,(\mu) \\ &\qquad\qquad + e^{-\tau_1/\mu}P\,(\mu)\,C_0\,(\mu) - P\,(-\mu)\,C_1\,(-\mu)]\,. \end{aligned}\right\} \tag{165}$$

Similarly,

$$\left.\begin{aligned} P(-\mu)\,t(-\mu) - e^{-\tau_1/\mu}P(\mu)\,s(\mu) \\ = (p_1 - a\mu_0)\,[P(-\mu)\,C_0(-\mu) - e^{-\tau_1/\mu}P(\mu)\,C_1(\mu)] \\ - (p_0 + a\mu_0)\,[e^{-\tau_1/\mu}P(\mu)\,C_0(\mu) - P(-\mu)\,C_1(-\mu)] \\ + a(\mu_0 - \mu)\,[P(-\mu)\,C_0(-\mu) - e^{-\tau_1/\mu}P(\mu)\,C_1(\mu) \\ + e^{-\tau_1/\mu}P(\mu)\,C_0(\mu) - P(-\mu)\,C_1(-\mu)]\,. \end{aligned}\right\} \quad (166)$$

In the foregoing equations we can now substitute for $(p_0 + a\mu_0)$, $(p_1 - a\mu_0)$, and a in accordance with equations (160)–(162). In this manner we find

$$\left.\begin{aligned} I(0, \mu) = \tfrac{1}{4}F\frac{\mu_0}{\mu_0+\mu}\{X(\mu)\,X(\mu_0) - Y(\mu)\,Y(\mu_0) \\ - Q(\mu_0+\mu)\,[X(\mu)+Y(\mu)]\,[X(\mu_0)+Y(\mu_0)]\} \end{aligned}\right\} \quad (167)$$

and

$$\left.\begin{aligned} I(\tau_1, -\mu) = \tfrac{1}{4}F\frac{\mu_0}{\mu_0-\mu}\{X(\mu)\,Y(\mu_0) - Y(\mu)\,X(\mu_0) \\ + Q(\mu_0-\mu)\,[X(\mu)+Y(\mu)]\,[X(\mu_0)+Y(\mu_0)]\}\,. \end{aligned}\right\} \quad (168)$$

Equations (167) and (168) can be brought into forms analogous to solutions (127) and (128) for the case $\varpi_0 < 1$, if we introduce the functions

$$\psi(\mu) = X(\mu) - Q\mu\,[X(\mu) + Y(\mu)] \quad (169)$$

and

$$\phi(\mu) = Y(\mu) + Q\mu\,[X(\mu) + Y(\mu)]\,. \quad (170)$$

In terms of these functions, the angular distributions of the reflected and the transmitted radiations take the required forms:

$$I(0, \mu) = \tfrac{1}{4}F\frac{\mu_0}{\mu_0+\mu}\,[\psi(\mu)\,\psi(\mu_0) - \phi(\mu)\,\phi(\mu_0)] \quad (171)$$

and

$$I(\tau_1, -\mu) = \tfrac{1}{4}F\frac{\mu_0}{\mu_0-\mu}\,[\psi(\mu)\,\phi(\mu_0) - \phi(\mu)\,\psi(\mu_0)]\,. \quad (172)$$

III. SCATTERING IN ACCORDANCE WITH RAYLEIGH'S PHASE FUNCTION

9. *The azimuth independent term. The reduction of the solution.*—In solving the equation of transfer appropriately for the problem of diffuse reflection and transmission by a plane-parallel atmosphere scattering radiation in accordance with Rayleigh's phase function, we must distinguish three terms in the radiation field: an azimuth independent term, a term proportional to $\cos(\varphi - \varphi_0)$, and a term proportional to $\cos 2(\varphi - \varphi_0)$ (cf. Paper IX, eq. [152]). Considering, first, the azimuth independent term, we have the solution (cf. Papers III and IX, Sec. II)

$$\left.\begin{aligned} I_i = \tfrac{3}{32}F\Big[(3-\mu_i^2)\sum_{a=-n+1}^{+n-1}\frac{L_a e^{-k_a\tau}}{1+k_a\mu_i} + L_0(\tau+\mu_i) + L_n \\ + (3-\mu_i^2)\frac{\gamma e^{-\tau/\mu_0}}{1+\mu_i/\mu_0}\Big] \qquad (i = \pm 1, \ldots, \pm n)\,, \end{aligned}\right\} \quad (173)$$

where the k_a's ($a = \pm 1, \ldots, \pm n \mp 1$ and $k_{+a} = -k_{-a}$) are the $(2n-2)$ distinct nonvanishing roots of the characteristic equation

$$1 = \frac{3}{8}\sum_{j=1}^{n} \frac{a_j(3-\mu_j^2)}{1-k^2\mu_j^2}, \tag{174}$$

$$\gamma = H(\mu_0)\, H(-\mu_0), \tag{175}$$

and the rest of the symbols have their usual meanings.

In terms of the functions

$$S(\mu) = (3-\mu^2)\sum_{a=-n+1}^{+n-1} \frac{L_a}{1-k_a\mu} - L_0\mu + L_n + (3-\mu^2)\frac{\gamma}{1-\mu/\mu_0} \tag{176}$$

and

$$T(\mu) = (3-\mu^2)\sum_{a=-n+1}^{+n-1} \frac{L_a e^{-k_a\tau_1}}{1+k_a\mu} + L_0(\tau_1+\mu) + L_n + (3-\mu^2)\frac{\gamma e^{-\tau_1/\mu_0}}{1+\mu/\mu_0}, \tag{177}$$

the boundary conditions determining the constants of integration and the equations governing the reflected and the transmitted radiations can be written in the following forms:

$$S(\mu_i) = T(\mu_i) = 0 \qquad (i = 1, \ldots, n), \tag{178}$$

$$I(0, \mu) = \tfrac{3}{32}F\,[S(-\mu) - e^{-\tau_1/\mu}T(\mu)], \tag{179}$$

and

$$I(\tau_1, -\mu) = \tfrac{3}{32}F\,[T(-\mu) - e^{-\tau_1/\mu}S(\mu)]. \tag{180}$$

By virtue of the boundary conditions (eq. [178]), we can write

$$S(\mu) = \frac{1}{\mu_1^2 \ldots \mu_n^2}\,\frac{P(\mu)}{W(\mu)}\,\frac{1}{1-\mu/\mu_0}\, s(\mu) \tag{181}$$

an

$$T(\mu) = \frac{1}{\mu_1^2 \ldots \mu_n^2}\,\frac{P(\mu)}{W(\mu)}\,\frac{1}{1+\mu/\mu_0}\, t(\mu), \tag{182}$$

where $s(\mu)$ and $t(\mu)$ are polynomials of degree n in μ.

From equations (175)–(177), (181), and (182), it now follows that

$$s(\mu_0) = (3-\mu_0^2)\, P(-\mu_0) \tag{183}$$

and

$$t(-\mu_0) = (3-\mu_0^2)\, e^{-\tau_1/\mu_0} P(\mu_0). \tag{184}$$

And again from equations (176), (177), (181), and (182) we have

$$\left.\begin{aligned} \left(3-\frac{1}{k_a^2}\right) L_a = \frac{1}{\mu_1^2 \ldots \mu_n^2}\,\frac{P(+1/k_a)}{W_a(1/k_a)}\,\frac{1}{1-1/k_a\mu_0}\, s(1/k_a) \\ (a = \pm 1, \ldots, \pm n \mp 1) \end{aligned}\right\} \tag{185}$$

and

$$\left(3-\frac{1}{k_a^2}\right)L_a e^{-k_a\tau_1} = \frac{1}{\mu_1^2\ldots\mu_n^2}\,\frac{P(-1/k_a)}{W_a(1/k_a)}\,\frac{1}{1-1/k_a\mu_0}\,t(-1/k_a) \quad (a=\pm 1,\ldots,\pm n\mp 1)\,. \tag{186}$$

We therefore have the $2n-2$ relations

$$s(1/k_a) = e^{k_a\tau_1}\frac{P(-1/k_a)}{P(+1/k_a)}\,t(-1/k_a) \qquad (a=\pm 1,\ldots,\pm n\mp 1)\,. \tag{187}$$

Since, however, $s(\mu)$ and $t(\mu)$ are polynomials of degree n in μ, we conclude, in accordance with theorems 1 and 2 (§ 4), that $s(\mu)$ and $t(\mu)$ must be expressible in the forms (cf. eqs. [144] and [145])

$$s(\mu) = (p_0+q_0\mu)C_0(\mu) + (p_1-q_1\mu)C_1(\mu) \tag{188}$$

and

$$t(\mu) = (p_1+q_1\mu)C_0(\mu) + (p_0-q_0\mu)C_1(\mu)\,, \tag{189}$$

where p_0, q_0, p_1, and q_1 are certain constants to be determined and $C_0(\mu)$ and $C_1(\mu)$ are polynomials of degree $n-1$ in μ defined in the manner of equations (100) and (101) in terms of the $(n-1)$ positive nonvanishing roots of the equation (174) (see n. 12).

From equations (183), (184), (188), and (189) we now have

$$(p_0+q_0\mu_0)C_0(\mu_0) + (p_1-q_1\mu_0)C_1(\mu_0) = (3-\mu_0^2)P(-\mu_0) \tag{190}$$

and

$$(p_0+q_0\mu_0)C_1(-\mu_0) + (p_1-q_1\mu_0)C_0(-\mu_0) = (3-\mu_0^2)\,e^{-\tau_1/\mu_0}P(\mu_0)\,. \tag{191}$$

Solving these equations for $p_0+q_0\mu_0$ and $p_1-q_1\mu_0$, we have

$$p_0+q_0\mu_0 = \frac{3-\mu_0^2}{[C_0^2(0)-C_1^2(0)]W(\mu_0)}\,[P(-\mu_0)C_0(-\mu_0) - e^{-\tau_1/\mu_0}P(\mu_0)C_1(\mu_0)] \tag{192}$$

and

$$p_1-q_1\mu_0 = \frac{3-\mu_0^2}{[C_0^2(0)-C_1^2(0)]W(\mu_0)}\,[e^{-\tau_1/\mu_0}P(\mu_0)C_0(\mu_0) - P(-\mu_0)C_1(-\mu_0)]\,. \tag{193}$$

Next, combining equations (179)–(182), (188), and (189), we find, after some rearranging of the terms, that the reflected and the transmitted intensities can be expressed in the forms

$$\begin{aligned} I(0,\mu) = \tfrac{3}{32}\,\frac{F}{\mu_1^2\ldots\mu_n^2}\,\frac{1}{W(\mu)}\{ & (p_0+q_0\mu_0)[P(-\mu)C_0(-\mu) - e^{-\tau_1/\mu}P(\mu)C_1(\mu)] \\ & -(p_1-q_1\mu_0)[e^{-\tau_1/\mu}P(\mu)C_0(\mu) - P(-\mu)C_1(-\mu)] \\ & -q_0(\mu_0+\mu)[P(-\mu)C_0(-\mu) - e^{-\tau_1/\mu}P(\mu)C_1(\mu)] \\ & -q_1(\mu_0+\mu)[e^{-\tau_1/\mu}P(\mu)C_0(\mu) - P(-\mu)C_1(-\mu)]\}\,\frac{\mu_0}{\mu_0+\mu} \end{aligned} \tag{194}$$

and

$$
\left.
\begin{aligned}
I(\tau_1, -\mu) = \tfrac{3}{32}\frac{F}{\mu_1^2 \cdots\cdots \mu_n^2}\frac{1}{W(\mu)}\{&(p_1 - q_1\mu_0)\,[P(-\mu)\,C_0(-\mu) - e^{-\tau_1/\mu}P(\mu)\,C_1(\mu)] \\
&- (p_0 + q_0\mu_0)\,[e^{-\tau_1/\mu}P(\mu)\,C_0(\mu) - P(-\mu)\,C_1(-\mu)] \\
&+ q_1(\mu_0 - \mu)\,[P(-\mu)\,C_0(-\mu) - e^{-\tau_1/\mu}P(\mu)\,C_1(\mu)] \\
&+ q_0(\mu_0 - \mu)\,[e^{-\tau_1/\mu}P(\mu)\,C_0(\mu) - P(-\mu)\,C_1(-\mu)]\}\,\frac{\mu_0}{\mu_0 - \mu}\,.
\end{aligned}
\right\}
\qquad (195)
$$

It remains to determine the constants q_0 and q_1.

10. *The determination of the constants* q_0 *and* q_1.—Putting $\mu = +\sqrt{3}$, respectively, $-\sqrt{3}$, in equations (176), (177), (181), and (182) we have

$$
\left.
\begin{aligned}
\frac{1}{\mu_1^2 \cdots\cdots \mu_n^2}\frac{P(\pm\sqrt{3})}{W(\sqrt{3})}\frac{s(\pm\sqrt{3})}{1 \mp \dfrac{\sqrt{3}}{\mu_0}} &= S(\pm\sqrt{3}) \\
&= \mp\sqrt{3}L_0 + L_n
\end{aligned}
\right\}
\qquad (196)
$$

and

$$
\left.
\begin{aligned}
\frac{1}{\mu_1^2 \cdots\cdots \mu_n^2}\frac{P(\pm\sqrt{3})}{W(\sqrt{3})}\frac{t(\pm\sqrt{3})}{1 \pm \dfrac{\sqrt{3}}{\mu_0}} &= T(\pm\sqrt{3}) \\
&= (\tau_1 \pm \sqrt{3})\,L_0 + L_n\,.
\end{aligned}
\right\}
\qquad (197)
$$

The foregoing equations can be simplified by using the relation (Paper XIV, eq. [267])

$$
\frac{1}{\mu_1^2 \cdots\cdots \mu_n^2}\frac{P(+\sqrt{3})\,P(-\sqrt{3})}{W(\sqrt{3})} = -8\,. \qquad (198)
$$

We find

$$
-\frac{8}{1 \mp \dfrac{\sqrt{3}}{\mu_0}}\frac{s(\pm\sqrt{3})}{P(\mp\sqrt{3})} = \mp\sqrt{3}L_0 + L_n \qquad (199)
$$

and

$$
-\frac{8}{1 \pm \dfrac{\sqrt{3}}{\mu_0}}\frac{t(\pm\sqrt{3})}{P(\mp\sqrt{3})} = (\tau_1 \pm \sqrt{3})\,L_0 + L_n\,. \qquad (200)
$$

From equations (199) and (200) we obtain the following set of equations:

$$
\frac{8}{1 \mp \dfrac{\sqrt{3}}{\mu_0}}\left[\frac{s(\pm\sqrt{3})}{P(\mp\sqrt{3})} - \frac{t(\mp\sqrt{3})}{P(\pm\sqrt{3})}\right] = L_0\tau_1\,, \qquad (201)
$$

$$
-\frac{8}{1 + \dfrac{\sqrt{3}}{\mu_0}}\frac{t(+\sqrt{3})}{P(-\sqrt{3})} + \frac{8}{1 - \dfrac{\sqrt{3}}{\mu_0}}\frac{t(-\sqrt{3})}{P(+\sqrt{3})} = 2\sqrt{3}L_0\,, \qquad (202)
$$

and

$$
-\frac{8}{1 + \dfrac{\sqrt{3}}{\mu_0}}\frac{s(-\sqrt{3})}{P(+\sqrt{3})} + \frac{8}{1 - \dfrac{\sqrt{3}}{\mu_0}}\frac{s(+\sqrt{3})}{P(-\sqrt{3})} = 2\sqrt{3}L_0\,. \qquad (203)
$$

Now, eliminating L_0 from equations (201)–(203), we obtain the equations

$$\left(1+\frac{\sqrt{3}}{\mu_0}\right)\left[\frac{s(+\sqrt{3})}{P(-\sqrt{3})}-\frac{(t-\sqrt{3})}{(P+\sqrt{3})}\right]=\left(1-\frac{\sqrt{3}}{\mu_0}\right)\left[\frac{s(-\sqrt{3})}{P(+\sqrt{3})}-\frac{t(+\sqrt{3})}{P(-\sqrt{3})}\right] \quad (204)$$

and

$$\left.\begin{aligned} &2\sqrt{3}\left\{\left(1+\frac{\sqrt{3}}{\mu_0}\right)\left[\frac{s(+\sqrt{3})}{P(-\sqrt{3})}-\frac{t(-\sqrt{3})}{P(+\sqrt{3})}\right]+\left(1-\frac{\sqrt{3}}{\mu_0}\right)\left[\frac{s(-\sqrt{3})}{P(+\sqrt{3})}-\frac{t(+\sqrt{3})}{P(-\sqrt{3})}\right]\right\} \\ &=\tau_1\left\{\left(1+\frac{\sqrt{3}}{\mu_0}\right)\left[\frac{s(+\sqrt{3})}{P(-\sqrt{3})}+\frac{t(-\sqrt{3})}{P(+\sqrt{3})}\right]-\left(1-\frac{\sqrt{3}}{\mu_0}\right)\left[\frac{s(-\sqrt{3})}{P(+\sqrt{3})}+\frac{t(+\sqrt{3})}{P(-\sqrt{3})}\right]\right\} \\ &=\frac{\sqrt{3}}{2}\left(1-\frac{3}{\mu_0^2}\right)L_0\tau_1 . \end{aligned}\right\} \quad (205)$$

These equations can be further reduced to the forms

$$\left.\begin{aligned} &(\sqrt{3}+\mu_0)[s(+\sqrt{3})P(+\sqrt{3})-t(-\sqrt{3})P(-\sqrt{3})] \\ &\quad+(\sqrt{3}-\mu_0)[s(-\sqrt{3})P(-\sqrt{3})-t(+\sqrt{3})P(+\sqrt{3})]=0 \end{aligned}\right\} \quad (206)$$

and

$$\left.\begin{aligned} &2\sqrt{3}\{(\sqrt{3}+\mu_0)[s(+\sqrt{3})P(+\sqrt{3})-t(-\sqrt{3})P(-\sqrt{3})]-(\sqrt{3}-\mu_0) \\ &\times[s(-\sqrt{3})P(-\sqrt{3})-t(+\sqrt{3})P(+\sqrt{3})]\}=\tau_1\{(\sqrt{3}+\mu_0)[s(+\sqrt{3})P(+\sqrt{3}) \\ &+t(-\sqrt{3})P(-\sqrt{3})]+(\sqrt{3}-\mu_0)[s(-\sqrt{3})P(-\sqrt{3})+t(+\sqrt{3})P(+\sqrt{3})]\}. \end{aligned}\right\} \quad (207)$$

We now have to substitute for $s(\pm\sqrt{3})$ and $t(\pm\sqrt{3})$ in equations (206) and (207) according to equations (188) and (189). For this purpose, it is convenient to write $s(\pm\sqrt{3})$ and $t(\pm\sqrt{3})$ in the forms

$$\left.\begin{aligned} s(\pm\sqrt{3}) &= (p_0+q_0\mu_0)C_0(\pm\sqrt{3})+(p_1-q_1\mu_0)C_1(\pm\sqrt{3}) \\ &\quad+(\sqrt{3}\mp\mu_0)[\pm q_0C_0(\pm\sqrt{3})\mp q_1C_1(\pm\sqrt{3})] \end{aligned}\right\} \quad (208)$$

and

$$\left.\begin{aligned} t(\pm\sqrt{3}) &= (p_0+q_0\mu_0)C_1(\pm\sqrt{3})+(p_1-q_1\mu_0)C_0(\pm\sqrt{3}) \\ &\quad+(\sqrt{3}\pm\mu_0)[\mp q_0C_1(\pm\sqrt{3})\pm q_1C_0(\pm\sqrt{3})]. \end{aligned}\right\} \quad (209)$$

Substituting for $s(\pm\sqrt{3})$ and $t(\pm\sqrt{3})$ from equations (208) and (209) in equations (206) and (207), we find, after some lengthy but straightforward reductions, that

$$\left.\begin{aligned} &(p_0+q_0\mu_0)(\sqrt{3}w_1+w_2\mu_0)+(p_1-q_1\mu_0)(-\sqrt{3}w_1+w_2\mu_0) \\ &\quad+(3-\mu_0^2)w_2(q_0-q_1)=0 \end{aligned}\right\} \quad (210)$$

and

$$\left.\begin{aligned} &2\sqrt{3}\{(p_0+q_0\mu_0)(\sqrt{3}w_2+w_1\mu_0)+(p_1-q_1\mu_0)(\sqrt{3}w_2-w_1\mu_0) \\ &\quad+(3-\mu_0^2)w_1(q_0+q_1)\}=\tau_1\{(p_0+q_0\mu_0)(\sqrt{3}w_4+w_3\mu_0) \\ &\quad+(p_1-q_1\mu_0)(\sqrt{3}w_4-w_3\mu_0)+(3-\mu_0^2)w_3(q_0+q_1)\}, \end{aligned}\right\} \quad (211)$$

where, for the sake of brevity, we have written

$$\left.\begin{aligned}
w_1&=C_0(\sqrt{3})P(\sqrt{3})-C_1(\sqrt{3})P(\sqrt{3})+C_0(-\sqrt{3})P(-\sqrt{3})-C_1(-\sqrt{3})P(-\sqrt{3}),\\
w_2&=C_0(\sqrt{3})P(\sqrt{3})+C_1(\sqrt{3})P(\sqrt{3})-C_0(-\sqrt{3})P(-\sqrt{3})-C_1(-\sqrt{3})P(-\sqrt{3}),\\
w_3&=C_0(\sqrt{3})P(\sqrt{3})-C_1(\sqrt{3})P(\sqrt{3})-C_0(-\sqrt{3})P(-\sqrt{3})+C_1(-\sqrt{3})P(-\sqrt{3}),\\
w_4&=C_0(\sqrt{3})P(\sqrt{3})+C_1(\sqrt{3})P(\sqrt{3})+C_0(-\sqrt{3})P(-\sqrt{3})+C_1(-\sqrt{3})P(-\sqrt{3}).
\end{aligned}\right\} \tag{212}$$

Equations (210) and (211) can be re-written in the forms

$$\left.\begin{aligned}
(3-\mu_0^2)(q_0-q_1)+(p_0+q_0\mu_0)\left(\mu_0+\sqrt{3}\,\frac{w_1}{w_2}\right)&\\
+(p_1-q_1\mu_0)\left(\mu_0-\sqrt{3}\,\frac{w_1}{w_2}\right)&=0
\end{aligned}\right\} \tag{213}$$

and

$$\left.\begin{aligned}
(3-\mu_0^2)(q_0+q_1)+(p_0+q_0\mu_0)\left[\sqrt{3}\,\frac{w_4\tau_1-2\sqrt{3}w_2}{w_3\tau_1-2\sqrt{3}w_1}+\mu_0\right]&\\
+(p_1-q_1\mu_0)\left[\sqrt{3}\,\frac{w_4\tau_1-2\sqrt{3}w_2}{w_3\tau_1-2\sqrt{3}w_1}-\mu_0\right]&=0\,.
\end{aligned}\right\} \tag{214}$$

Solving these equations for q_0 and q_1, we have

$$(3-\mu_0^2)\,q_0=-[(c_1+\mu_0)(p_0+q_0\mu_0)+c_2(p_1-q_1\mu_0)] \tag{215}$$

and

$$(3-\mu_0^2)\,q_1=-[c_2(p_0+q_0\mu_0)+(c_1-\mu_0)(p_1-q_1\mu_0)], \tag{216}$$

where

$$c_1=\frac{\sqrt{3}}{2}\left[\frac{w_4\tau_1-2\sqrt{3}w_2}{w_3\tau_1-2\sqrt{3}w_1}+\frac{w_1}{w_2}\right] \tag{217}$$

and

$$c_2=\frac{\sqrt{3}}{2}\left[\frac{w_4\tau_1-2\sqrt{3}w_2}{w_3\tau_1-2\sqrt{3}w_1}-\frac{w_1}{w_2}\right]. \tag{218}$$

Since $(p_0+q_0\mu_0)$ and $(p_1-q_1\mu_0)$ have already been determined (eqs. [192] and [193]), the foregoing equations complete the formal solution for $S(\mu)$ and $T(\mu)$.

11. *The solution for the reflected and the transmitted radiations.*—Now, substituting for $(p_0+q_0\mu_0)$, $(p_1-q_1\mu_0)$, q_0, and q_1 according to equations (192), (193), (215), and (216) in equations (194) and (195) and introducing the functions $X(\mu)$ and $Y(\mu)$, defined as in equations (125) and (126), we have

$$\left.\begin{aligned}
I(0,\mu)=\tfrac{3}{32}F\,\frac{\mu_0}{\mu_0+\mu}\{&(3-\mu_0^2)[X(\mu)X(\mu_0)-Y(\mu)Y(\mu_0)]\\
&+(\mu_0+\mu)X(\mu)[(c_1+\mu_0)X(\mu_0)+c_2Y(\mu_0)]\\
&+(\mu_0+\mu)Y(\mu)[c_2X(\mu_0)+(c_1-\mu_0)Y(\mu_0)]\}
\end{aligned}\right\} \tag{219}$$

and

$$\left.\begin{aligned}
I(\tau_1,-\mu)=\tfrac{3}{32}F\,\frac{\mu_0}{\mu_0-\mu}\{&(3-\mu_0^2)[X(\mu)Y(\mu_0)-Y(\mu)X(\mu_0)]\\
&-(\mu_0-\mu)X(\mu)[c_2X(\mu_0)+(c_1-\mu_0)Y(\mu_0)]\\
&-(\mu_0-\mu)Y(\mu)[(c_1+\mu_0)X(\mu_0)+c_2Y(\mu_0)]\}\,.
\end{aligned}\right\} \tag{220}$$

After some rearranging, equations (219) and (220) can be brought to the forms

$$\left.\begin{aligned} I^{(0)}(0,\mu) = \tfrac{3}{32}F\frac{\mu_0}{\mu_0+\mu}\{ & X^{(0)}(\mu)\,X^{(0)}(\mu_0)\,[3+c_1(\mu_0+\mu)+\mu\mu_0] \\ & - Y^{(0)}(\mu)\,Y^{(0)}(\mu_0)\,[3-c_1(\mu_0+\mu)+\mu\mu_0] \\ & + c_2(\mu_0+\mu)\,[X^{(0)}(\mu)\,Y^{(0)}(\mu_0)+Y^{(0)}(\mu)\,X^{(0)}(\mu_0)]\} \end{aligned}\right\} \quad (221)$$

and

$$\left.\begin{aligned} I^{(0)}(\tau_1,-\mu) = \tfrac{3}{32}F\frac{\mu_0}{\mu_0-\mu}\{ & X^{(0)}(\mu)\,Y^{(0)}(\mu_0)\,[3-c_1(\mu_0-\mu)-\mu\mu_0] \\ & - Y^{(0)}(\mu)\,X^{(0)}(\mu_0)\,[3+c_1(\mu_0-\mu)-\mu\mu_0] \\ & - c_2(\mu_0-\mu)\,[X^{(0)}(\mu)\,X^{(0)}(\mu_0)+Y^{(0)}(\mu)\,Y^{(0)}(\mu_0)]\}\,. \end{aligned}\right\} \quad (222)$$

In equations (221) and (222) we have inserted a superscript "0" with the various functions to emphasize the fact that these equations represent only the solutions for the azimuth independent terms in the reflected and the transmitted intensities.

To complete the solution we need to find the remaining terms in the reflected and the transmitted intensities which are proportional to $\cos(\varphi-\varphi_0)$ and $\cos 2(\varphi-\varphi_0)$. The determination of these terms presents no difficulty, since the equations satisfied by $I^{(1)}(\tau,\mu)$ and $I^{(2)}(\tau,\mu)$ (cf. Paper IX, eqs. [154] and [155]) are of the standard form considered in Paper IX, § 4, and the analysis of Section I applies without any modifications. We can, therefore, write (cf. Paper IX, eqs. [190] and [194])

$$\left.\begin{aligned} I^{(1)}(0,\mu) &= -\tfrac{3}{8}F\mu\mu_0(1-\mu^2)^{\frac{1}{2}}(1-\mu_0^2)^{\frac{1}{2}}\frac{\mu_0}{\mu_0+\mu} \\ &\qquad \times [X^{(1)}(\mu)\,X^{(1)}(\mu_0)-Y^{(1)}(\mu)\,Y^{(1)}(\mu_0)]\,, \\ I^{(1)}(\tau_1,-\mu) &= +\tfrac{3}{8}F\mu\mu_0(1-\mu^2)^{\frac{1}{2}}(1-\mu_0^2)^{\frac{1}{2}}\frac{\mu_0}{\mu_0-\mu} \\ &\qquad \times [X^{(1)}(\mu)\,Y^{(1)}(\mu_0)-Y^{(1)}(\mu)\,X^{(1)}(\mu_0)]\,, \end{aligned}\right\} \quad (223)$$

and

$$\left.\begin{aligned} I^{(2)}(0,\mu) &= \tfrac{3}{32}F(1-\mu^2)(1-\mu_0^2)\frac{\mu_0}{\mu_0+\mu} \\ &\qquad \times [X^{(2)}(\mu)\,X^{(2)}(\mu_0)-Y^{(2)}(\mu)\,Y^{(2)}(\mu_0)]\,, \\ I^{(2)}(\tau_1,-\mu) &= \tfrac{3}{32}F(1-\mu^2)(1-\mu_0^2)\frac{\mu_0}{\mu_0-\mu} \\ &\qquad \times [X^{(2)}(\mu)\,Y^{(2)}(\mu_0)-Y^{(2)}(\mu)\,X^{(2)}(\mu_0)]\,, \end{aligned}\right\} \quad (224)$$

where $X^{(1)}$, $Y^{(1)}$ and $X^{(2)}$, $Y^{(2)}$ are defined in terms of the characteristic functions

$$\Psi^{(1)}(\mu) = \tfrac{3}{8}\mu^2(1-\mu^2) \qquad \text{and} \qquad \Psi^{(2)}(\mu) = \tfrac{3}{32}(1-\mu^2)^2\,. \quad (225)$$

In terms of the foregoing solutions (eqs. [221]–[224]), the reflected and the transmitted intensities are given by

$$\left.\begin{aligned} I(0,\mu,\varphi;\mu_0,\varphi_0) &= I^{(0)}(0,\mu)+I^{(1)}(0,\mu)\cos(\varphi-\varphi_0) \\ &\qquad + I^{(2)}(0,\mu)\cos 2(\varphi-\varphi_0) \end{aligned}\right\} \quad (226)$$

and

$$\left.\begin{aligned} I(\tau_1,-\mu,\varphi;\mu_0,\varphi_0) = I^{(0)}(\tau_1,-\mu) + I^{(1)}(\tau_1,-\mu)\cos(\varphi-\varphi_0) \\ + I^{(2)}(\tau_1,-\mu)\cos 2(\varphi-\varphi_0)\,. \end{aligned}\right\} \tag{227}$$

IV. SCATTERING IN ACCORDANCE WITH THE PHASE FUNCTION $\lambda(1 + x\cos\Theta)$

12. *The azimuth independent term.*—In the problem of diffuse reflection and transmission by a plane-parallel atmosphere scattering radiation in accordance with the phase function $\lambda(1 + x\cos\Theta)$ $(\lambda < 1, 1 \geqslant x \geqslant -1)$, we must distinguish between two terms in the radiation field: an azimuth independent term and a term proportional to $\cos(\varphi - \varphi_0)$ (cf. Paper IX, eq. [6]). Considering, first, the azimuth independent term, we can express the reflected and the transmitted radiations in terms of the functions

$$S(\mu) = \sum_{a=-n}^{+n} \frac{L_a[1 + x(1-\lambda)\mu/k_a]}{1 - k_a\mu} + \gamma \frac{[1 + x(1-\lambda)\mu\mu_0]}{1 - \mu/\mu_0} \tag{228}$$

and

$$\left.\begin{aligned} T(\mu) = \sum_{a=-n}^{+n} \frac{L_a[1 - x(1-\lambda)\mu/k_a]}{1 + k_a\mu} e^{-k_a\tau_1} \\ + \gamma e^{-\tau_1/\mu_0} \frac{[1 - x(1-\lambda)\mu\mu_0]}{1 + \mu/\mu_0}\,, \end{aligned}\right\} \tag{229}$$

where the k_a's $(a = \pm 1, \ldots, \pm n$ and $k_{+a} = -k_{-a})$ are the $2n$ roots of the characteristic equation

$$1 = \lambda \sum_{j=1}^{n} \frac{a_j[1 + x(1-\lambda)\mu_j^2]}{1 - k^2\mu_j^2}\,, \tag{230}$$

$$\gamma = H(\mu_0)H(-\mu_0)\,, \tag{231}$$

and the rest of the symbols have their usual meanings. We have

$$I(0,\mu) = \tfrac{1}{4}\lambda F[S(-\mu) - e^{-\tau_1/\mu}T(\mu)] \tag{232}$$

and

$$I(\tau_1,-\mu) = \tfrac{1}{4}\lambda F[T(-\mu) - e^{-\tau_1/\mu}S(\mu)]\,. \tag{233}$$

Moreover, the boundary conditions at $\tau = 0$ and $\tau = \tau_1$ require that

$$S(\mu_i) = T(\mu_i) = 0 \qquad (i = 1, \ldots, n)\,. \tag{234}$$

In view of these boundary conditions, we can write

$$S(\mu) = \frac{1}{\mu_1^2 \ldots \mu_n^2} \frac{P(\mu)}{W(\mu)} \frac{1}{1 - \mu/\mu_0} s(\mu) \tag{235}$$

and

$$T(\mu) = \frac{1}{\mu_1^2 \ldots \mu_n^2} \frac{P(\mu)}{W(\mu)} \frac{1}{1 + \mu/\mu_0} t(\mu)\,, \tag{236}$$

where $s(\mu)$ and $t(\mu)$ are polynomials of degree $n + 1$ in μ.

From equations (228), (229), (231), (235), and (236) it follows that

$$s(\mu_0) = [1 + x(1-\lambda)\mu_0^2]\,P(-\mu_0) \tag{237}$$

and

$$t(-\mu_0) = [1 + x(1-\lambda)\mu_0^2]\,e^{-\tau_1/\mu_0}P(\mu_0)\,. \tag{238}$$

And again, from equations (228), (229), (235), and (236), we have

$$L_a = \frac{1}{\mu_1^2 \ldots \mu_n^2}\,\frac{P(1/k_a)}{W_a(1/k_a)}\,\frac{1}{1-1/k_a\mu_0}\,\frac{s(1/k_a)}{1+x(1-\lambda)/k_a^2} \qquad (a = \pm 1, \ldots, \pm n) \tag{239}$$

and

$$L_a e^{-k_a\tau_1} = \frac{1}{\mu_1^2 \ldots \mu_n^2}\,\frac{P(-1/k_a)}{W_a(1/k_a)}\,\frac{1}{1-1/k_a\mu_0}\,\frac{t(-1/k_a)}{1+x(1-\lambda)/k_a^2} \qquad (a = \pm 1, \ldots, \pm n)\,. \tag{240}$$

We therefore have the $2n$ relations

$$s(1/k_a) = e^{k_a\tau_1}\frac{P(-1/k_a)}{P(+1/k_a)}\,t(-1/k_a) \qquad (a = \pm 1, \ldots, \pm n)\,. \tag{241}$$

However, since $s(\mu)$ and $t(\mu)$ are polynomials of degree $(n+1)$ in μ, we conclude, in accordance with theorems 1 and 2 (§ 4), that $s(\mu)$ and $t(\mu)$ must be expressible in the forms

$$s(\mu) = (p_0 + q_0\mu)C_0(\mu) + (p_1 - q_1\mu)C_1(\mu) \tag{242}$$

and

$$t(\mu) = (p_1 + q_1\mu)C_0(\mu) + (p_0 - q_0\mu)C_1(\mu)\,, \tag{243}$$

where p_0, q_0, p_1, and q_1 are certain constants and $C_0(\mu)$ and $C_1(\mu)$ are defined as in equations (50), (100), and (101) in terms of the n positive roots of equation (230).

From equations (237), (238), (242), and (243) it now follows that

$$p_0 + q_0\mu_0 = \frac{1 + x(1-\lambda)\mu_0^2}{[C_0^2(0) - C_1^2(0)]W(\mu_0)}\,[P(-\mu_0)C_0(-\mu_0) - e^{-\tau_1/\mu_0}P(\mu_0)C_1(\mu_0)] \tag{244}$$

and

$$p_1 - q_1\mu_0 = \frac{1 + x(1-\lambda)\mu_0^2}{[C_0^2(0) - C_1^2(0)]W(\mu_0)}\,[e^{-\tau_1/\mu_0}P(\mu_0)C_0(\mu_0) - P(-\mu_0)C_1(-\mu_0)]\,. \tag{245}$$

It remains to determine the constants q_0 and q_1.

13. *The determination of the constants* q_0 *and* q_1.—Putting $\mu = 0$ in equations (228), (229), (235), and (236), we obtain

$$S(0) = \sum_{a=-n}^{+n} L_a + \gamma = \frac{(-1)^n}{\mu_1 \ldots \mu_n}\,s(0) \tag{246}$$

and

$$T(0) = \sum_{a=-n}^{+n} L_a e^{-k_a\tau_1} + \gamma e^{-\tau_1/\mu_0} = \frac{(-1)^n}{\mu_1 \ldots \mu_n}\,t(0)\,. \tag{247}$$

Now, according to equation (239),

$$\left.\begin{aligned}\sum_{a=-n}^{+n} L_a &= \frac{1}{\mu_1^2 \ldots \mu_n^2} \sum_{a=-n}^{+n} \frac{P(1/k_a)\, s(1/k_a)}{W_a(1/k_a)(1-1/k_a\mu_0)[1+x(1-\lambda)/k_a^2]} \\ &= -\frac{\mu_0}{\mu_1^2 \ldots \mu_n^2 W(\mu_0)} \sum_{a=-n}^{+n} \frac{k_a P(1/k_a)\, s(1/k_a)}{W_a(1/k_a)[1+x(1-\lambda)/k_a^2]} W_a(\mu_0)\,.\end{aligned}\right\} \quad (248)$$

To evaluate the sum on the right-hand side, we introduce the function

$$f(z) = \sum_{a=-n}^{+n} \frac{k_a P(1/k_a)\, s(1/k_a)}{W_a(1/k_a)[1+x(1-\lambda)/k_a^2]} W_a(z) \quad (249)$$

and express ΣL_a in terms of it. Thus

$$\sum_{a=-n}^{+n} L_a = -\frac{\mu_0}{\mu_1^2 \ldots \mu_n^2} \frac{f(\mu_0)}{W(\mu_0)}\,. \quad (250)$$

As defined in equation (249), $f(z)$ is a polynomial of degree $(2n-1)$ in z and takes the values

$$\frac{k_a P(1/k_a)\, s(1/k_a)}{1+x(1-\lambda)/k_a^2} \quad (251)$$

for $z = 1/k_a$ ($a = \pm 1, \ldots, \pm n$). In other words,

$$\left.\begin{aligned} z[1+x(1-\lambda)z^2]\, f(z) - P(z)\, s(z) = 0 \qquad \text{for} \qquad & z = 1/k_a \\ \text{and} \qquad & a = \pm 1, \ldots, \pm n\,. \end{aligned}\right\} \quad (252)$$

Hence there must exist a relation of the form

$$z[1+x(1-\lambda)z^2]\, f(z) = P(z)\, s(z) + W(z)[A z^2 + B z + D] \quad (253)$$

where A, B, and D are certain constants to be determined.

Putting $z = 0$ in equation (253), we determine at once that

$$D = -(-1)^n \mu_1 \ldots \mu_n\, s(0)\,. \quad (254)$$

Next, putting $z = \pm i/\sqrt{x(1-\lambda)}$ in equation (253), we find

$$P(\pm i\zeta)\, s(\pm i\zeta) + W(i\zeta)[-A\zeta^2 \pm iB\zeta + D] = 0\,, \quad (255)$$

where, for the sake of brevity, we have written

$$\zeta = \frac{1}{\sqrt{x(1-\lambda)}}\,, \quad (256)$$

Solving equations (255) for A and B, we find

$$A = \frac{D}{\zeta^2} + \frac{1}{2\zeta^2 W(i\zeta)} [P(i\zeta)\, s(i\zeta) + P(-i\zeta)\, s(-i\zeta)] \quad (257)$$

and

$$B = \frac{i}{2\zeta W(i\zeta)} [P(i\zeta)\, s(i\zeta) - P(-i\zeta)\, s(-i\zeta)]\,. \quad (258)$$

Returning to equation (253) and setting $z = \mu_0$ and remembering that

$$s(\mu_0) = \left(1+\frac{\mu_0^2}{\zeta^2}\right) P(-\mu_0) \tag{259}$$

we obtain

$$f(\mu_0) = \frac{1}{\mu_0} P(\mu_0) P(-\mu_0) + \frac{W(\mu_0)}{\mu_0 [1+x(1-\lambda)\mu_0^2]} [A\mu_0^2 + B\mu_0 + D) . \tag{260}$$

Using the foregoing expression for $f(\mu_0)$ in equation (250) and inserting for A, B, and D their values given by equations (254), (257), and (258), we find, after some reductions, that

$$\left.\begin{aligned}\sum_{a=-n}^{+n} L_a = -\gamma + \frac{(-1)^n}{\mu_1 \cdots \mu_n} s(0) \\ -\frac{x(1-\lambda)\mu_0}{2\mu_1^2 \cdots \mu_n^2 [1+x(1-\lambda)\mu_0^2] W(i\zeta)} [\mu_0 \{P(i\zeta)s(i\zeta)+P(-i\zeta)s(-i\zeta)\} \\ +i\zeta\{P(i\zeta)s(i\zeta)-P(-i\zeta)s(-i\zeta)\}].\end{aligned}\right\} \tag{261}$$

Similarly,

$$\left.\begin{aligned}\sum_{a=-n}^{+n} L_a e^{-k_a\tau_1} = -\gamma e^{-\tau_1/\mu_0} + \frac{(-1)^n}{\mu_1 \cdots \mu_n} t(0) \\ +\frac{x(1-\lambda)\mu_0}{2\mu_1^2 \cdots \mu_n^2 [1+x(1-\lambda)\mu_0^2] W(i\zeta)} [-\mu_0 \{P(i\zeta)t(i\zeta)+P(-i\zeta)t(-i\zeta)\} \\ +i\zeta\{P(i\zeta)t(i\zeta)-P(-i\zeta)t(-i\zeta)\}].\end{aligned}\right\} \tag{262}$$

Equations (261) and (262) evaluate the required sums.

According to equations (246), (247), (261), and (262), we clearly have

$$(\mu_0+i\zeta) P(i\zeta) s(i\zeta) + (\mu_0-i\zeta) P(-i\zeta) s(-i\zeta) = 0 \tag{263}$$

and

$$(\mu_0-i\zeta) P(i\zeta) t(i\zeta) + (\mu_0+i\zeta) P(-i\zeta) t(-i\zeta) = 0 . \tag{264}$$

We now have to substitute for $s(\pm i\zeta)$ and $t(\pm i\zeta)$ in equations (263) and (264) according to equations (242) and (243). For this purpose it is convenient to write $s(\pm i\zeta)$ and $t(\pm i\zeta)$ in the forms

$$\left.\begin{aligned}s(\pm i\zeta) = (p_0+q_0\mu_0) C_0(\pm i\zeta) + (p_1-q_1\mu_0) C_1(\pm i\zeta) \\ + (\mu_0 \mp i\zeta)[-q_0 C_0(\pm i\zeta) + q_1 C_1(\pm i\zeta)]\end{aligned}\right\} \tag{265}$$

and

$$\left.\begin{aligned}t(\pm i\zeta) = (p_0+q_0\mu_0) C_1(\pm i\zeta) + (p_1-q_1\mu_0) C_0(\pm i\zeta) \\ + (\mu_0 \pm i\zeta)[-q_0 C_1(\pm i\zeta) + q_1 C_0(\pm i\zeta)] .\end{aligned}\right\} \tag{266}$$

Substituting from equations (265) and (266) in equations (263) and (264), we find, after some minor reductions, that

$$\left.\begin{aligned}(\mu_0^2+\zeta^2)(q_0\alpha_0 - q_1\alpha_1) = (p_0+q_0\mu_0)(\alpha_0\mu_0+\beta_0\zeta) \\ + (p_1-q_1\mu_0)(\alpha_1\mu_0+\beta_1\zeta)\end{aligned}\right\} \tag{267}$$

and

$$\left.\begin{aligned}(\mu_0^2+\zeta^2)(q_0\alpha_1-q_1\alpha_0) = {} & (p_0+q_0\mu_0)(\alpha_1\mu_0-\beta_1\zeta) \\ & +(p_1-q_1\mu_0)(\alpha_0\mu_0-\beta_0\zeta)\,,\end{aligned}\right\} \qquad (268)$$

where

$$\left.\begin{aligned}\alpha_0 &= P(i\zeta)C_0(i\zeta)+P(-i\zeta)C_0(-i\zeta)\,, \\ \alpha_1 &= P(i\zeta)C_1(i\zeta)+P(-i\zeta)C_1(-i\zeta)\,, \\ \beta_0 &= i\,[P(i\zeta)C_0(i\zeta)-P(-i\zeta)C_0(-i\zeta)]\,, \\ \beta_1 &= i\,[P(i\zeta)C_1(i\zeta)-P(-i\zeta)C_1(-i\zeta)]\,.\end{aligned}\right\} \qquad (269)$$

Defined in this manner, α_0, α_1, β_0, and β_1 are all real constants.

Solving equations (267) and (268) for q_0 and q_1, we find

$$q_0=\frac{x(1-\lambda)}{1+x(1-\lambda)\mu_0^2}[(c_1+\mu_0)(p_0+q_0\mu_0)+c_2(p_1-q_1\mu_0)] \qquad (270)$$

and

$$q_1=\frac{x(1-\lambda)}{1+x(1-\lambda)\mu_0^2}[c_2(p_0+q_0\mu_0)+(c_1-\mu_0)(p_1-q_1\mu_0)]\,, \qquad (271)$$

where

$$c_1=\frac{1}{\sqrt{x(1-\lambda)}}\,\frac{\alpha_0\beta_0+\alpha_1\beta_1}{\alpha_0^2-\alpha_1^2} \qquad (272)$$

and

$$c_2=\frac{1}{\sqrt{x(1-\lambda)}}\,\frac{\alpha_0\beta_1+\alpha_1\beta_0}{\alpha_0^2-\alpha_1^2}\,. \qquad (273)$$

With this, the determination of the constants is completed.

14. *The solutions for the reflected and the transmitted radiations.*—It is apparent that for $s(\mu)$ and $t(\mu)$ given by equations (242) and (243), the equations governing the angular distributions of the reflected and the transmitted radiations can be reduced to the forms (194) and (195) with $\lambda/4$ replacing "3/32." With the expressions for $(p_0+q_0\mu_0)$, $(p_1-q_1\mu_0)$, q_0, and q_1 given by equations (244), (245), (270), and (271), we therefore have

$$\left.\begin{aligned}I(0,\mu) = {} & \tfrac{1}{4}\lambda F\{[1+x(1-\lambda)\mu_0^2][X(\mu)X(\mu_0)-Y(\mu)Y(\mu_0)] \\ & -x(1-\lambda)(\mu_0+\mu)X(\mu)[(c_1+\mu_0)X(\mu_0)+c_2Y(\mu_0)] \\ & -x(1-\lambda)(\mu_0+\mu)Y(\mu)[c_2X(\mu_0)+(c_1-\mu_0)Y(\mu_0)]\}\frac{\mu_0}{\mu_0+\mu}\end{aligned}\right\} \qquad (274)$$

and

$$\left.\begin{aligned}I(\tau_1,-\mu) = {} & \tfrac{1}{4}\lambda F\{[1+x(1-\lambda)\mu_0^2][X(\mu)Y(\mu_0)-Y(\mu)X(\mu_0)] \\ & +x(1-\lambda)(\mu_0-\mu)X(\mu)[c_2X(\mu_0)+(c_1-\mu_0)Y(\mu_0)] \\ & +x(1-\lambda)(\mu_0-\mu)Y(\mu)[(c_1+\mu_0)X(\mu_0)+c_2Y(\mu_0)]\}\frac{\mu_0}{\mu_0-\mu}\,.\end{aligned}\right\} \qquad (275)$$

After some rearranging, equations (274) and (275) can be brought to the forms

$$\left.\begin{aligned} I^{(0)}(0,\mu)=\tfrac{1}{4}\lambda F\{X^{(0)}(\mu)X^{(0)}(\mu_0)[1-x(1-\lambda)c_1(\mu_0+\mu)-x(1-\lambda)\mu\mu_0] \\ -Y^{(0)}(\mu)Y^{(0)}(\mu_0)[1+x(1-\lambda)c_1(\mu_0+\mu)-x(1-\lambda)\mu\mu_0] \\ -x(1-\lambda)c_2(\mu_0+\mu)[X^{(0)}(\mu)Y^{(0)}(\mu_0)+Y^{(0)}(\mu)X^{(0)}(\mu_0)]\}\frac{\mu_0}{\mu_0+\mu} \end{aligned}\right\} \quad (276)$$

and

$$\left.\begin{aligned} I^{(0)}(\tau_1,-\mu)=\tfrac{1}{4}\lambda F\{X^{(0)}(\mu)Y^{(0)}(\mu_0)[1+x(1-\lambda)c_1(\mu_0-\mu)+x(1-\lambda)\mu\mu_0] \\ -Y^{(0)}(\mu)X^{(0)}(\mu_0)[1-x(1-\lambda)c_1(\mu_0-\mu)+x(1-\lambda)\mu\mu_0] \\ +x(1-\lambda)c_2(\mu_0-\mu)[X^{(0)}(\mu)X^{(0)}(\mu_0)+Y^{(0)}(\mu)Y^{(0)}(\mu_0)]\}\frac{\mu_0}{\mu_0-\mu}. \end{aligned}\right\} \quad (277)$$

In equations (276) and (277) we have inserted a superscript "0" to the various functions to emphasize the fact that these equations represent only the solutions for the azimuth independent terms in the reflected and the transmitted radiations.

To complete the solution we must find the remaining term in the reflected and the transmitted intensities which is proportional to cos $(\varphi-\varphi_0)$. The determination of this term presents no difficulty, since the equation satisfied by $I^{(1)}(\tau,\mu)$ (Paper IX, eq. [8]) is of the form for which the analysis of Section I is applicable without modifications. We therefore have

$$\left.\begin{aligned} I^{(1)}(0,\mu)=\tfrac{1}{4}x\lambda F(1-\mu^2)^{\frac{1}{2}}(1-\mu_0^2)^{\frac{1}{2}}\frac{\mu_0}{\mu_0+\mu} \\ \times[X^{(1)}(\mu)X^{(1)}(\mu_0)-Y^{(1)}(\mu)Y^{(1)}(\mu_0)] \end{aligned}\right\} \quad (278)$$

and

$$\left.\begin{aligned} I^{(1)}(\tau_1,-\mu)=\tfrac{1}{4}x\lambda F(1-\mu^2)^{\frac{1}{2}}(1-\mu_0^2)^{\frac{1}{2}}\frac{\mu_0}{\mu_0-\mu} \\ \times[X^{(1)}(\mu)Y^{(1)}(\mu_0)-Y^{(1)}(\mu)X^{(1)}(\mu_0)], \end{aligned}\right\} \quad (279)$$

where $X^{(1)}(\mu)$ and $Y^{(1)}(\mu)$ are defined in terms of the characteristic function

$$\Psi^{(1)}(\mu)=\tfrac{1}{4}x\lambda(1-\mu^2). \quad (280)$$

In terms of the foregoing solution, the reflected and the transmitted intensities are given by

$$I(0,\mu,\varphi;\mu_0,\varphi_0)=I^{(0)}(0,\mu)+I^{(1)}(0,\mu)\cos(\varphi-\varphi_0) \quad (281)$$

and

$$I(\tau_1,-\mu,\varphi;\mu_0;\varphi_0)=I^{(0)}(\tau_1,-\mu)+I^{(1)}(\tau_1,-\mu)\cos(\varphi-\varphi_0). \quad (282)$$

V. RAYLEIGH SCATTERING

15. *The formulation of the problem.*—In a proper treatment of problems involving multiple scattering we should take into account the fact that after the first scattering even natural light becomes partially polarized. In formulating the equations of transfer we must, therefore, include the polarization characteristics of the radiation and allow for the dependence of the scattered light on these characteristics. As we have shown in the earlier papers[13] of this series, this can best be accomplished by considering the in-

[13] See esp. Papers XI, XIII, XIV, and XV (*Ap. J.*, **104**, 110, 1946; **105**, 151, 424, 1947).

tensities I_l and I_r in two directions at right angles to each other in the plane of the electric and the magnetic vectors and the further quantities

$$U = (I_l - I_r) \tan 2\chi \quad \text{and} \quad V = (I_l - I_r) \sec 2\chi \tan 2\beta \, , \tag{283}$$

where χ denotes the inclination of the plane of polarization to the direction to which l refers and $-\pi/2 \leqslant \beta \leqslant +\pi/2$ is an angle whose tangent is equal to the ratio of the axes of the ellipse which characterizes the state of polarization. (The sign of β depends on whether the polarization is right-handed or left-handed.)

For the case of Rayleigh scattering, the equations of transfer for I_l, I_r,[14] U, and V have been explicitly formulated[15] and exactly solved[16] for transfer problems in semi-infinite plane-parallel atmospheres. In this Section we shall be concerned with the solutions of these same equations for the problem of diffuse reflection and transmission by a plane-parallel atmosphere of finite optical thickness.

16. *The solution for* V *and the azimuth dependent terms.*—Considering the general case of incidence of a parallel beam of partially elliptically polarized light on a plane-parallel atmosphere in the direction $(-\mu_0, \varphi_0)$, we may recall that the equations of transfer break up into seven independent sets of equations[17] when the intensities I_l, I_r, U, and V are expressed in the forms[18]

$$\left.\begin{aligned} I_l(\tau, \mu, \varphi) = I_l^{(0)}(\tau, \mu) + I_l^{(1)}(\tau, \mu)\cos(\varphi - \varphi_0) + I_l^{(-1)}(\tau, \mu)\sin(\varphi - \varphi_0) \\ + I_l^{(2)}(\tau, \mu)\cos 2(\varphi - \varphi_0) + I_l^{(-2)}(\tau, \mu)\sin 2(\varphi - \varphi_0)\, , \end{aligned}\right\} \tag{284}$$

$$I_r(\tau, \mu, \varphi) = I_r^{(0)}(\tau, \mu) + I_r^{(2)}(\tau, \mu)\cos 2(\varphi - \varphi_0) + I_r^{(-2)}(\tau, \mu)\sin 2(\varphi - \varphi_0)\, , \tag{285}$$

$$\left.\begin{aligned} U(\tau, \mu, \varphi) = U^{(1)}(\tau, \mu)\sin(\varphi - \varphi_0) + U^{(-1)}(\tau, \mu)\cos(\varphi - \varphi_0) \\ + U^{(2)}(\tau, \mu)\sin 2(\varphi - \varphi_0) + U^{(-2)}(\tau, \mu)\cos 2(\varphi - \varphi_0)\, , \end{aligned}\right\} \tag{286}$$

and

$$V(\tau, \mu, \varphi) = V^{(0)}(\tau, \mu) + V^{(1)}(\tau, \mu)\cos(\varphi - \varphi_0)\, . \tag{287}$$

Of the seven systems of equations which arise in this manner, the only one which requires any detailed consideration is the first, which governs the azimuth independent terms, $I_l^{(0)}(\tau, \mu)$ and $I_r^{(0)}(\tau, \mu)$. The other systems present no difficulties. Thus, Systems II–V (Paper XIII, pp. 154–55) governing the various azimuth dependent terms in I_l, I_r, and U admit the first integrals

$$I_l^{(1)}(\tau, \mu) = \mu U^{(1)}(\tau, \mu)\, ; \qquad I_l^{(-1)}(\tau, \mu) = -\mu U^{(-1)}(\tau, \mu)\, , \tag{288}$$

and

$$\left.\begin{aligned} I_l^{(2)}(\tau, \mu) &= -\mu^2 I_r^{(2)}(\tau, \mu) = +\tfrac{1}{2}\mu U^{(2)}(\tau, \mu)\, , \\ I_l^{(-2)}(\tau, \mu) &= -\mu^2 I_r^{(-2)}(\tau, \mu) = -\tfrac{1}{2}\mu U^{(-2)}(\tau, \mu)\, . \end{aligned}\right\} \tag{289}$$

Moreover, the resulting equations for $U^{(1)}(\tau, \mu)$, $U^{(-1)}(\tau, \mu)$, $I_r^{(2)}(\tau, \mu)$, and $I_r^{(-2)}(\tau, \mu)$ are all such that the analysis of Section I applies without any modifications. Conse-

[14] The symbols l and r now refer to the directions in the meridian plane and at right angles to it, respectively.

[15] See Papers XI (§ 5), XIII (§ 2), XIV (§ 2), and XV (eq. [74]).

[16] Cf. Paper XVI (*Ap. J.*, **105**, 425, 1947), where the functions and constants representing the exact solutions are tabulated.

[17] Paper XIII, Systems I–V (pp. 153–54) and Paper XV, eqs. (78) and (79).

[18] Paper XIII, eqs. (4)–(6) and Paper XV, eq. (77).

quently, we can write down at once the solutions for the corresponding reflected and transmitted intensities. We have (cf. Paper XIII, eqs. [47]–[50])

$$\left.\begin{aligned} I_l^{(1)}(0,\mu) &= +\mu U^{(1)}(0,\mu) \\ &= -\tfrac{3}{4}F_l\mu\mu_0(1-\mu^2)^{\frac{1}{2}}(1-\mu_0^2)^{\frac{1}{2}}[X^{(1)}(\mu)X^{(1)}(\mu_0)-Y^{(1)}(\mu)Y^{(1)}(\mu_0)]\frac{\mu_0}{\mu_0+\mu}, \end{aligned}\right\} \quad (290)$$

$$\left.\begin{aligned} I_l^{(-1)}(0,\mu) &= -\mu U^{(-1)}(0,\mu) \\ &= -\tfrac{3}{8}U_0\mu(1-\mu^2)^{\frac{1}{2}}(1-\mu_0^2)^{\frac{1}{2}}[X^{(1)}(\mu)X^{(1)}(\mu_0)-Y^{(1)}(\mu)Y^{(1)}(\mu_0)]\frac{\mu_0}{\mu_0+\mu}, \end{aligned}\right\} \quad (291)$$

$$\left.\begin{aligned} I_l^{(2)}(0,\mu) &= -\mu^2 I_r^{(2)}(0,\mu) = +\tfrac{1}{2}\mu U^{(2)}(0,\mu) \\ &= \tfrac{3}{16}(F_l\mu_0^2-F_r)\mu^2[X^{(2)}(\mu)X^{(2)}(\mu_0)-Y^{(2)}(\mu)Y^{(2)}(\mu_0)]\frac{\mu_0}{\mu_0+\mu}, \end{aligned}\right\} \quad (292)$$

$$\left.\begin{aligned} I_l^{(-2)}(0,\mu) &= -\mu^2 I_r^{(-2)}(0,\mu) = -\tfrac{1}{2}\mu U^{(-2)}(0,\mu) \\ &= \tfrac{3}{16}U_0\mu^2\mu_0[X^{(2)}(\mu)X^{(2)}(\mu_0)-Y^{(2)}(\mu)Y^{(2)}(\mu_0)]\frac{\mu_0}{\mu_0+\mu}, \end{aligned}\right\} \quad (293)$$

$$\left.\begin{aligned} I_l^{(1)}(\tau_1,-\mu) &= -\mu U^{(1)}(\tau_1,-\mu) \\ &= \tfrac{3}{4}F_l\mu\mu_0(1-\mu^2)^{\frac{1}{2}}(1-\mu_0^2)^{\frac{1}{2}}[X^{(1)}(\mu)Y^{(1)}(\mu_0)-Y^{(1)}(\mu)X^{(1)}(\mu_0)]\frac{\mu_0}{\mu_0-\mu}, \end{aligned}\right\} \quad (294)$$

$$\left.\begin{aligned} I_l^{(-1)}(\tau_1,-\mu) &= +\mu U^{(-1)}(\tau_1,-\mu) \\ &= \tfrac{3}{8}U_0\mu(1-\mu^2)^{\frac{1}{2}}(1-\mu_0^2)^{\frac{1}{2}}[X^{(1)}(\mu)Y^{(1)}(\mu_0)-Y^{(1)}(\mu)X^{(1)}(\mu_0)]\frac{\mu_0}{\mu_0-\mu}, \end{aligned}\right\} \quad (295)$$

$$\left.\begin{aligned} I_l^{(2)}(\tau_1,-\mu) &= -\mu^2 I_r^{(2)}(\tau_1,-\mu) = -\tfrac{1}{2}\mu U^{(2)}(\tau_1,-\mu) \\ &= \tfrac{3}{16}(F_l\mu_0^2-F_r)\mu^2[X^{(2)}(\mu)Y^{(2)}(\mu_0)-Y^{(2)}(\mu)X^{(2)}(\mu_0)]\frac{\mu_0}{\mu_0-\mu}, \end{aligned}\right\} \quad (296)$$

and

$$\left.\begin{aligned} I_l^{(-2)}(\tau_1,-\mu) &= -\mu^2 I_r^{(-2)}(\tau_1,-\mu) = +\tfrac{1}{2}\mu U^{(-2)}(\tau_1,-\mu) \\ &= \tfrac{3}{16}U_0\mu^2\mu_0[X^{(2)}(\mu)Y^{(2)}(\mu_0)-Y^{(2)}(\mu)X^{(2)}(\mu_0)]\frac{\mu_0}{\mu_0-\mu}, \end{aligned}\right\} \quad (297)$$

where $X^{(1)}(\mu)$, $Y^{(1)}(\mu)$, and $X^{(2)}(\mu)$, $Y^{(2)}(\mu)$, are defined as in equations (50), (100), (101), (125), and (126) in terms of the characteristic functions

$$\Psi^{(1)}(\mu) = \tfrac{3}{8}(1-\mu^2)(1+2\mu^2) \qquad \text{and} \qquad \Psi^{(2)}(\mu) = \tfrac{3}{16}(1+\mu^2)^2, \quad (298)$$

respectively.

Similarly, the equations for $V^{(0)}(\tau, \mu)$ and $V^{(1)}(\tau, \mu)$ also do not require any special consideration. The solutions for the reflected and the transmitted intensities are of standard form, and we have

$$V^{(0)}(0, \mu) = -\tfrac{3}{8} V_0 \mu \mu_0 [X_v(\mu) X_v(\mu_0) - Y_v(\mu) Y_v(\mu_0)] \frac{\mu_0}{\mu_0 + \mu}, \quad (299)$$

$$V^{(1)}(0, \mu) = +\tfrac{3}{8} V_0 (1-\mu^2)^{\frac{1}{2}} (1-\mu_0^2)^{\frac{1}{2}} [X_r(\mu) X_r(\mu_0) - Y_r(\mu) Y_r(\mu_0)] \frac{\mu_0}{\mu_0 + \mu}, \quad (300)$$

$$V^{(0)}(\tau_1, -\mu) = \tfrac{3}{8} V_0 \mu \mu_0 [X_v(\mu) Y_v(\mu_0) - Y_v(\mu) X_v(\mu_0)] \frac{\mu_0}{\mu_0 - \mu}, \quad (301)$$

and

$$V^{(1)}(\tau_1, -\mu) = \tfrac{3}{8} V_0 (1-\mu^2)^{\frac{1}{2}} (1-\mu_0^2)^{\frac{1}{2}} [X_r(\mu) Y_r(\mu_0) - Y_r(\mu) X_r(\mu_0)] \frac{\mu_0}{\mu_0 - \mu}, \quad (302)$$

where $X_r(\mu)$, $Y_r(\mu)$ and $X_v(\mu)$, $Y_v(\mu)$ are defined in terms of the characteristic functions

$$\Psi_r(\mu) = \tfrac{3}{8}(1-\mu^2) \quad \text{and} \quad \Psi_v(\mu) = \tfrac{3}{4}\mu^2 . \quad (303)$$

Finally, it may be noted that in equations (290)–(297) and (299)–(302) πF_l, πF_r, πU_0, and πV_0 represent the fluxes in the four components of the incident beam.

17. *The azimuth independent terms proportional to* F_l.—In the preceding section we have given the solutions for V and the various azimuth dependent terms in the expansions (eqs. [284]–[286]) for I_l, I_r, and U. The azimuth independent terms, $I_l^{(0)}$ and $I_r^{(0)}$, remain to be considered.

Now in the equations (Paper XIII, System I) governing $I_l^{(0)}$ and $I_r^{(0)}$, the inhomogeneous parts consist of two terms proportional, respectively, to F_l and F_r. The solutions for $I_l^{(0)}$ and $I_r^{(0)}$ can therefore be expressed in the forms

$$\left.\begin{aligned} I_l^{(0)}(\tau, \mu) &= I_{ll}^{(0)}(\tau, \mu) + I_{lr}^{(0)}(\tau, \mu) \\ \text{and} \quad I_r^{(0)}(\tau, \mu) &= I_{rl}^{(0)}(\tau, \mu) + I_{rr}^{(0)}(\tau, \mu) , \end{aligned}\right\} \quad (304)$$

where $I_{ll}^{(0)}$ and $I_{rl}^{(0)}$ are proportional to F_l and $I_{lr}^{(0)}$ and $I_{rr}^{(0)}$ are proportional to F_r. Considering, first, the terms proportional to F_l, we can express the corresponding terms in the reflected and the transmitted intensities in the following forms:

$$\left.\begin{aligned} I_{ll}^{(0)}(0, \mu) &= \tfrac{3}{16}[S_{ll}(-\mu) - e^{-\tau_1/\mu} T_{ll}(\mu)] F_l , \\ I_{ll}^{(0)}(\tau_1, -\mu) &= \tfrac{3}{16}[T_{ll}(-\mu) - e^{-\tau_1/\mu} S_{ll}(\mu)] F_l , \\ I_{rl}^{(0)}(0, \mu) &= \tfrac{3}{16}[S_{rl}(-\mu) - e^{-\tau_1/\mu} T_{rl}(\mu)] F_l , \\ \text{and} \quad I_{rl}^{(0)}(\tau_1, -\mu) &= \tfrac{3}{16}[T_{rl}(-\mu) - e^{-\tau_1/\mu} S_{rl}(\mu)] F_l , \end{aligned}\right\} \quad (305)$$

where (cf. Paper XIII, eqs. [12] and [13])

$$\left.\begin{aligned} S_{ll}(\mu) = (1-\mu^2)\sum_{\beta=-n+1}^{+n-1}\frac{L_\beta}{1-\kappa_\beta\mu} - L_0\mu + L_n \\ + \sum_{\alpha=-n}^{+n} M_\alpha(1+k_\alpha\mu) + 2(1-\mu^2)\frac{\Gamma}{1-\mu/\mu_0}, \end{aligned}\right\} \tag{306}$$

$$\left.\begin{aligned} T_{ll}(\mu) = (1-\mu^2)\sum_{\beta=-n+1}^{+n-1}\frac{L_\beta e^{-\kappa_\beta\tau_1}}{1+\kappa_\beta\mu} + L_0(\tau_1+\mu) + L_n \\ + \sum_{\alpha=-n}^{+n} M_\alpha(1-k_\alpha\mu)\, e^{-k_\alpha\tau_1} + 2(1-\mu^2)\frac{\Gamma e^{-\tau_1/\mu_0}}{1+\mu/\mu_0}, \end{aligned}\right\} \tag{307}$$

$$S_{rl}(\mu) = -L_0\mu + L_n - \sum_{\alpha=-n}^{+n}\frac{M_\alpha(k_\alpha^2-1)}{1-k_\alpha\mu}, \tag{308}$$

and

$$T_{rl}(\mu) = L_0(\tau_1+\mu) + L_n - \sum_{\alpha=-n}^{+n}\frac{M_\alpha(k_\alpha^2-1)\, e^{-k_\alpha\tau_1}}{1+k_\alpha\mu}. \tag{309}$$

In the foregoing equations the κ_β's $(\beta = \pm 1, \ldots, \pm n \mp 1)$ and k_α's $(\alpha = \pm 1, \ldots, \pm n)$ are the distinct nonvanishing roots of the characteristic equations

$$1 = \frac{3}{2}\sum_{j=1}^{n}\frac{a_j(1-\mu_j^2)}{1-\kappa^2\mu_j^2} \tag{310}$$

and

$$1 = \frac{3}{4}\sum_{j=1}^{n}\frac{a_j(1-\mu_j^2)}{1-k^2\mu_j^2}, \tag{311}$$

respectively, and

$$\Gamma = H_l(\mu_0)\, H_l(-\mu_0) = \frac{1}{\mu_1^2\ldots\mu_n^2}\,\frac{P(\mu_0)\,P(-\mu_0)}{\Omega(\mu_0)}, \tag{312}$$

$$\Omega(\mu) = \prod_{\beta=-n+1}^{+n-1}(1-\kappa_\beta\mu) = \prod_{\beta=1}^{n-1}(1-\kappa_\beta^2\mu^2), \tag{313}$$

and the rest of the symbols have their usual meanings.

The boundary conditions at $\tau = 0$ and $\tau = \tau_1$ require that

$$S_{ll}(\mu_i) = T_{ll}(\mu_i) = S_{rl}(\mu_i) = T_{rl}(\mu_i) = 0 \qquad (i = 1, \ldots, n). \tag{314}$$

By virtue of these boundary conditions we can write

$$S_{ll}(\mu)=\frac{1}{\mu_1^2\ldots\mu_n^2}\frac{P(\mu)}{\Omega(\mu)}\frac{1}{1-\mu/\mu_0}s_{ll}(\mu)\,, \tag{315}$$

$$T_{ll}(\mu)=\frac{1}{\mu_1^2\ldots\mu_n^2}\frac{P(\mu)}{\Omega(\mu)}\frac{1}{1+\mu/\mu_0}t_{ll}(\mu)\,, \tag{316}$$

$$S_{rl}(\mu)=\frac{1}{\mu_1^2\ldots\mu_n^2}\frac{P(\mu)}{W(\mu)}s_{rl}(\mu)\,, \tag{317}$$

and

$$T_{rl}(\mu)=\frac{1}{\mu_1^2\ldots\mu_n^2}\frac{P(\mu)}{W(\mu)}t_{rl}(\mu)\,, \tag{318}$$

where $s_{ll}(\mu)$ and $t_{ll}(\mu)$ are polynomials of degree n in μ and $s_{rl}(\mu)$ and $t_{rl}(\mu)$ are polynomials of degree $(n+1)$ in μ.

From equations (305), (306), (312), (315), and (316) it now follows that

$$s_{ll}(\mu_0)=2(1-\mu_0^2)P(-\mu_0)\qquad\text{and}\qquad t_{ll}(-\mu_0)=2(1-\mu_0^2)\,e^{-\tau_1/\mu_0}P(\mu_0)\,. \tag{319}$$

And again from equations (305), (306), (315), and (316) we have

$$\left.\begin{aligned}\left(1-\frac{1}{\kappa_\beta^2}\right)L_\beta&=\frac{1}{\mu_1^2\ldots\mu_n^2}\frac{P(1/\kappa_\beta)}{\Omega_\beta(1/\kappa_\beta)}\frac{1}{1-1/\kappa_\beta\mu_0}s_{ll}(1/\kappa_\beta)\\ &\qquad(\beta=\pm1,\ldots,\pm n\mp1)\end{aligned}\right\} \tag{320}$$

and

$$\left.\begin{aligned}\left(1-\frac{1}{\kappa_\beta^2}\right)L_\beta e^{-\kappa_\beta\tau_1}&=\frac{1}{\mu_1^2\ldots\mu_n^2}\frac{P(-1/\kappa_\beta)}{\Omega_\beta(1/\kappa_\beta)}\frac{1}{1-1/\kappa_\beta\mu_0}t_{ll}(-1/\kappa_\beta)\\ &\qquad(\beta=\pm1,\ldots,\pm n\mp1)\,.\end{aligned}\right\} \tag{321}$$

We therefore have $(2n-2)$ relations of the form

$$s_{ll}(1/\kappa_\beta)=e^{\kappa_\beta\tau_1}\frac{P(-1/\kappa_\beta)}{P(+1/\kappa_\beta)}t_{ll}(-1/\kappa_\beta)\qquad(\beta=\pm1,\ldots,\pm n\mp1)\,. \tag{322}$$

However, since $s_{ll}(\mu)$ and $t_{ll}(\mu)$ are polynomials of degree n in μ, we conclude in accordance with theorems 1 and 2 (§ 4) that (cf. eqs. [144] and [145])

$$s_{ll}(\mu)=(\varpi_0+\chi_0\mu)C_{0,l}(\mu)+(\varpi_1-\chi_1\mu)C_{1,l}(\mu) \tag{323}$$

and

$$t_{ll}(\mu)=(\varpi_1+\chi_1\mu)C_{0,l}(\mu)+(\varpi_0-\chi_0\mu)C_{1,l}(\mu) \tag{324}$$

where ϖ_0, ϖ_1, χ_0, and χ_1 are constants and $C_{0,l}$ and $C_{1,l}$ are polynomials of degree $(n-1)$ in μ defined in the manner of equations (50), (100), (101), (125), and (126) in terms of the roots κ_β $(\beta=1,\ldots,n-1)$ of equation (310).

From equations (319), (323), and (324) we now obtain

$$\left.\begin{aligned}\varpi_0+\chi_0\mu_0&=\frac{2(1-\mu_0^2)}{[C_{0,l}^2(0)-C_{1,l}^2(0)]\,\Omega(\mu_0)}\\ &\qquad\times[P(-\mu_0)C_{0,l}(-\mu_0)-e^{-\tau_1/\mu_0}P(\mu_0)C_{1,l}(\mu_0)]\end{aligned}\right\} \tag{325}$$

and

$$\left.\begin{aligned}\varpi_1 - \chi_1\mu_0 = &\frac{2(1-\mu_0^2)}{[C^2_{0,l}(0) - C^2_{1,l}(0)]\,\Omega(\mu_0)} \\ &\times [e^{-\tau_1/\mu_0}P(\mu_0)C_{0,l}(\mu_0) - P(-\mu_0)C_{1,l}(-\mu_0)]\,.\end{aligned}\right\} \quad (326)$$

Considering, next, equations (308), (309), (317), and (318), we have

$$M_a = -\frac{1}{\mu_1^2 \ldots \mu_n^2}\frac{P(1/k_a)}{W_a(1/k_a)(k_a^2-1)}\,s_{rl}(1/k_a) \quad (327)$$

and

$$M_a e^{-k_a\tau_1} = -\frac{1}{\mu_1^2 \ldots \mu_n^2}\frac{P(-1/k_a)}{W_a(1/k_a)(k_a^2-1)}\,t_{rl}(-1/k_a)\,. \quad (328)$$

We therefore have $2n$ relations of the form

$$s_{rl}(1/k_a) = e^{k_a\tau_1}\frac{P(-1/k_a)}{P(+1/k_a)}\,t_{rl}(-1/k_a) \qquad (a = \pm 1, \ldots, \pm n)\,. \quad (329)$$

It will be recalled that $s_{rl}(\mu)$ and $t_{rl}(\mu)$ are polynomials of degree $(n+1)$ in μ. However, it can be shown that $s_{rl}(\mu) + t_{rl}(\mu)$ is only of degree n; for, according to equations (308), (309), (317), and (318), we have

$$\left.\begin{aligned}\frac{1}{\mu_1^2 \ldots \mu_n^2}P(\mu)\,s_{rl}(\mu) &= W(\mu)\,S_{rl}(\mu) \\ &= (-1)^n k_1^2 \ldots k_n^2[-L_0\mu^{2n+1} + L_n\mu^{2n} + \ldots]\end{aligned}\right\} \quad (330)$$

and

$$\left.\begin{aligned}\frac{1}{\mu_1^2 \ldots \mu_n^2}P(\mu)\,t_{rl}(\mu) &= W(\mu)\,T_{rl}(\mu) \\ &= (-1)^n k_1^2 \ldots k_n^2[+L_0\mu^{2n+1} + (L_0\tau_1 + L_n)\,\mu^{2n} + \ldots]\,.\end{aligned}\right\} \quad (331)$$

And it is apparent from these equations that the coefficients of μ^{n+1} in $s_{rl}(\mu)$ and $t_{rl}(\mu)$ are of opposite signs. We may therefore write

$$s_{rl}(\mu) = (p_0 + a\mu)\,C_{0,r}(\mu) + (p_1 + a\mu)\,C_{1,r}(\mu) \quad (332)$$

and

$$t_{rl}(\mu) = (p_1 - a\mu)\,C_{0,r}(\mu) + (p_0 - a\mu)\,C_{1,r}(\mu)\,, \quad (333)$$

where p_0, p_1, and a are constants and $C_{0,r}$ and $C_{1,r}$ are defined in terms of the roots of equation (311).

With the forms (332) and (333) for $s_{rl}(\mu)$ and $t_{rl}(\mu)$ we can readily verify that

$$\left.\begin{aligned}P(\mu)\,s_{rl}(\mu) = &+a\,[c^{(n)}_{0,r} + c^{(n)}_{1,r}]\,\mu^{2n+1} + \Big\{a\,[c^{(n-1)}_{0,r} + c^{(n-1)}_{1,r}] \\ &+ p_0 c^{(n)}_{0,r} + p_1 c^{(n)}_{1,r} - \Big(\sum_{i=1}^{n}\mu_i\Big)\,a\,[c^{(n)}_{0,r} + c^{(n)}_{1,r}]\Big\}\,\mu^{2n} + \ldots\end{aligned}\right\} \quad (334)$$

and

$$P(\mu)\, t_{rl}(\mu) = -a\,[c_{0,r}^{(n)} + c_{1,r}^{(n)}]\,\mu^{2n+1} + \Big\{ -a\,[c_{0,r}^{(n-1)} + c_{1,r}^{(n-1)}] + p_1 c_{0,r}^{(n)} + p_0 c_{1,r}^{(n)} + \Big(\sum_{i=1}^{n}\mu_i\Big)\, a\,[c_{0,r}^{(n)} + c_{1,r}^{(n)}] \Big\}\,\mu^{2n} + \dots, \tag{335}$$

where $c_{0,r}^{(n)}$, $c_{1,r}^{(n)}$ and $c_{0,r}^{(n-1)}$, $c_{1,r}^{(n-1)}$ are the coefficients of the highest and the next highest powers of μ in $C_{0,r}$ and $C_{1,r}$.

From a comparison of equations (330) and (331) and (334) and (335) we now conclude that

$$-\sum_{i=1}^{n}\mu_i + \frac{1}{c_{0,r}^{(n)} + c_{1,r}^{(n)}}\Big\{ c_{0,r}^{(n-1)} + c_{1,r}^{(n-1)} + \frac{1}{a}\,[p_0 c_{0,r}^{(n)} + p_1 c_{1,r}^{(n)}] \Big\} = -\frac{L_n}{L_0} \tag{336}$$

and

$$-\sum_{i=1}^{n}\mu_i + \frac{1}{c_{0,r}^{(n)} + c_{1,r}^{(n)}}\Big\{ c_{0,r}^{(n-1)} + c_{1,r}^{(n-1)} - \frac{1}{a}\,[p_1 c_{0,r}^{(n)} + p_0 c_{1,r}^{(n)}] \Big\} = \frac{L_n}{L_0} + \tau_1\,. \tag{337}$$

From equations (336) and (337) we readily find that

$$a = Q\,(p_0 - p_1)\,, \tag{338}$$

where

$$Q = \frac{c_{0,r}^{(n)} - c_{1,r}^{(n)}}{\Big(\tau_1 + 2\sum_{i=1}^{n}\mu_i\Big)[c_{0,r}^{(n)} + c_{1,r}^{(n)}] - 2\,[c_{0,r}^{(n-1)} + c_{1,r}^{(n-1)}]}\,. \tag{339}$$

It will be observed that among the seven constants—ϖ_0, ϖ_1, χ_0, χ_1, p_0, p_1, and a—which occur in the solutions (323), (324), (332), and (333) for s_{ll}, t_{ll}, s_{rl}, and t_{rl}, we have so far obtained only three relations (eqs. [325], [326], and [338]). We therefore require four more relations to make the problem determinate. To obtain these additional relations we proceed in the following manner:

Putting $\mu = +1$, respectively, -1 in equations (306) and (307), we have

$$S_{ll}(\pm 1) = \mp L_0 + L_n + \sum_{\alpha=-n}^{+n} M_\alpha\,(1 \pm k_\alpha) \tag{340}$$

and

$$T_{ll}(\pm 1) = L_0\,(\tau_1 \pm 1) + L_n + \sum_{\alpha=-n}^{+n} M_\alpha\,(1 \mp k_\alpha)\, e^{-k_\alpha \tau_1}\,. \tag{341}$$

Similarly, putting $\mu = 0$ in equations (308) and (309), we have

$$S_{rl}(0) = L_n - \sum_{\alpha=-n}^{+n} M_\alpha\,(k_\alpha^2 - 1) \tag{342}$$

and

$$T_{rl}(0) = L_0\tau_1 + L_n - \sum_{\alpha=-n}^{+n} M_\alpha\,(k_\alpha^2 - 1)\, e^{-k_\alpha \tau_1}\,. \tag{343}$$

Combining equations (340)–(343) appropriately, we obtain the following relations:

$$\tfrac{1}{2}[S_{ll}(+1)+S_{ll}(-1)]-S_{rl}(0)=\sum_{a=-n}^{+n}M_a k_a^2, \tag{344}$$

$$\tfrac{1}{2}[T_{ll}(+1)+T_{ll}(-1)]-T_{rl}(0)=\sum_{a=-n}^{+n}M_a k_a^2 e^{-k_a\tau_1}, \tag{345}$$

$$\left.\begin{aligned}&\tfrac{1}{2}[S_{ll}(+1)-S_{ll}(-1)]+\tfrac{1}{2}[T_{ll}(+1)-T_{ll}(-1)]\\&\qquad=\sum_{a=-n}^{+n}M_a k_a-\sum_{a=-n}^{+n}M_a k_a e^{-k_a\tau_1},\end{aligned}\right\} \tag{346}$$

and

$$\left.\begin{aligned}&\tau_1\Big\{\tfrac{1}{2}[S_{ll}(+1)-S_{ll}(-1)]-\tfrac{1}{2}[T_{ll}(+1)-T_{ll}(-1)]\\&\quad-\sum_{a=-n}^{+n}M_a k_a-\sum_{a=-n}^{+n}M_a k_a e^{-k_a\tau_1}\Big\}=2\Big\{\tfrac{1}{2}[S_{ll}(+1)+S_{ll}(-1)]\\&\quad-\tfrac{1}{2}[T_{ll}(+1)+T_{ll}(-1)]-\sum_{a=-n}^{+n}M_a+\sum_{a=-n}^{+n}M_a e^{-k_a\tau_1}\Big\}=-2L_0\tau_1.\end{aligned}\right\} \tag{347}$$

We shall now show how the foregoing equations provide the required additional relations between the constants ϖ_0, ϖ_1, χ_0, χ_1, p_0, p_1, and a. However, we must first evaluate the summations

$$\sum_{a=-n}^{+n}M_a k_a^m \quad\text{and}\quad \sum_{a=-n}^{+n}M_a k_a^m e^{-k_a\tau_1} \qquad (m=0,\ 1,\ 2)\,, \tag{348}$$

which occur in these equations.

Considering the summation $\Sigma M_a k_a^m$, for example, we have (cf. eq. [327])

$$\sum_{a=-n}^{+n}M_a k_a^m=-\frac{1}{\mu_1^2\ldots\mu_n^2}\sum_{a=-n}^{+n}\frac{P(1/k_a)\,s_{rl}(1/k_a)}{k_a^{2-m}(1-k_a^{-2})\,W_a(1/k_a)}\,. \tag{349}$$

In terms of the function

$$f_m(x)=\sum_{a=-n}^{+n}\frac{P(1/k_a)\,s_{rl}(1/k_a)}{k_a^{2-m}(1-k_a^{-2})\,W_a(1/k_a)}\,W_a(x)\,, \tag{350}$$

we can therefore express $\Sigma M_a k_a^m$ as

$$\sum_{a=-n}^{+n}M_a k_a^m=-\frac{1}{\mu_1^2\ldots\mu_n^2}\,f_m(0)\,. \tag{351}$$

Now $f_m(x)$ is a polynomial of degree $(2n-1)$ in x, which takes the values

$$\frac{P(1/k_a)\,s_{rl}(1/k_a)}{k_a^{2-m}(1-k_a^{-2})} \tag{352}$$

for $x = 1/k_\alpha$ ($\alpha = \pm 1, \ldots, \pm n$). There must, accordingly, exist a relation of the form

$$(1-x^2)\, f_m(x) - x^{2-m} P(x)\, s_{rl}(x) = W(x)\, g_m(x) \qquad (m=0,1,2)\,, \qquad (353)$$

where $g_m(x)$ is a polynomial of degree $3-m$ in x. To determine $g_m(x)$ more explicitly, we must consider each case separately. We shall illustrate the procedure by considering the case $m = 0$.

For $m = 0$, equation (353) becomes

$$(1-x^2)\, f_0(x) - x^2 P(x)\, s_{rl}(x) = W(x)\,(Ax^3 + Bx^2 + Dx + E)\,, \qquad (354)$$

where A, B, D, and E are certain constants to be determined. The constants A and B follow directly from a comparison of the coefficients of x^{2n+3} and x^{2n+2} on either side of equation (354). We find (cf. eq. [334])

$$A = (-1)^{n+1} \frac{a\,[c^{(n)}_{0,r} + c^{(n)}_{1,r}]}{k_1^2 \ldots k_n^2} \qquad (355)$$

and

$$\left.\begin{aligned} B = \frac{(-1)^{n+1}}{k_1^2 \ldots k_n^2} \Big\{ & a\,[c^{(n-1)}_{0,r} + c^{(n-1)}_{1,r}] + p_0 c^{(n)}_{0,r} + p_1 c^{(n)}_{1,r} \\ & - \Big(\sum_{i=1}^{n} \mu_i\Big)\, a\,[c^{(n)}_{0,r} + c^{(n)}_{1,r}] \Big\}\,. \end{aligned}\right\} \qquad (356)$$

Next, putting $x = 1$, respectively -1, in equation (354), we have

$$-P(\pm 1)\, s_{rl}(\pm 1) = W(1)\,(\pm A + B \pm D + E)\,. \qquad (357)$$

From this equation it follows that

$$B + E = -\frac{1}{2W(1)}\,[P(+1)\, s_{rl}(+1) + P(-1)\, s_{rl}(-1)]\,. \qquad (358)$$

Hence

$$\left.\begin{aligned} f_0(0) = \frac{(-1)^n}{k_1^2 \ldots k_n^2} \Big\{ & a\,[c^{(n-1)}_{0,r} + c^{(n-1)}_{1,r}] + p_0 c^{(n)}_{0,r} + p_1 c^{(n)}_{1,r} - \Big(\sum_{i=1}^{n} \mu_i\Big) \\ & \times a\,[c^{(n)}_{0,r} + c^{(n)}_{1,r}] \Big\} - \frac{1}{2W(1)}\,[P(+1)\, s_{rl}(+1) + P(-1)\, s_{rl}(-1)]\,. \end{aligned}\right\} \qquad (359)$$

Inserting this value of $f_0(0)$ in equation (351) and making use of the relations (cf. Paper X, eq. [143] and Paper XI, n. 16)

$$k_1^2 \ldots k_n^2 \mu_1^2 \ldots \mu_n^2 = \tfrac{1}{2} \qquad (360)$$

and

$$\frac{1}{\mu_1^2 \ldots \mu_n^2} \frac{P(+1)\, P(-1)}{W(1)} = 4\,, \qquad (361)$$

we obtain

$$\left.\begin{aligned} \sum_{\alpha=-n}^{+n} M_\alpha = (-1)^{n+1} 2 \Big\{ & a\,[c^{(n-1)}_{0,r} + c^{(n-1)}_{1,r}] + p_0 c^{(n)}_{0,r} + p_1 c^{(n)}_{1,r} \\ & - \Big(\sum_{i=1}^{n} \mu_i\Big)\, a\,[c^{(n)}_{0,r} + c^{(n)}_{1,r}] \Big\} + 2\Big[\frac{s_{rl}(+1)}{P(-1)} + \frac{s_{rl}(-1)}{P(+1)}\Big]\,. \end{aligned}\right\} \qquad (362)$$

The other summations can be similarly evaluated. We find

$$\sum_{a=-n}^{+n} M_a k_a = (-1)^{n+1} 2a\,[c_{0,r}^{(n)} + c_{1,r}^{(n)}] + 2\left[\frac{s_{rl}(+1)}{P(-1)} - \frac{s_{rl}(-1)}{P(+1)}\right], \quad (363)$$

$$\sum_{a=-n}^{+n} M_a k_a^2 = (-1)^{n+1} \frac{s_{rl}(0)}{\mu_1 \cdots \mu_n} + 2\left[\frac{s_{rl}(+1)}{P(-1)} + \frac{s_{rl}(-1)}{P(+1)}\right], \quad (364)$$

$$\left.\begin{aligned} \sum_{a=-n}^{+n} M_a e^{-k_a\tau_1} = (-1)^{n+1} 2 \Big\{ &-a\,[c_{0,r}^{(n-1)} + c_{1,r}^{(n-1)}] + p_1 c_{0,r}^{(n)} + p_0 c_{1,r}^{(n)} \\ &+ \Big(\sum_{i=1}^{n} \mu_i\Big) a\,[c_{0,r}^{(n)} + c_{1,r}^{(n)}] \Big\} + 2\left[\frac{t_{rl}(+1)}{P(-1)} + \frac{t_{rl}(-1)}{P(+1)}\right], \end{aligned}\right\} \quad (365)$$

$$\sum_{a=-n}^{+n} M_a k_a e^{-k_a\tau_1} = (-1)^{n+1} 2a\,[c_{0,r}^{(n)} + c_{1,r}^{(n)}] - 2\left[\frac{t_{rl}(+1)}{P(-1)} - \frac{t_{rl}(-1)}{P(+1)}\right], \quad (366)$$

and

$$\sum_{a=-n}^{+n} M_a k_a^2 e^{-k_a\tau_1} = (-1)^{n+1} \frac{t_{rl}(0)}{\mu_1 \cdots \mu_n} + 2\left[\frac{t_{rl}(+1)}{P(-1)} + \frac{t_{rl}(-1)}{P(+1)}\right]. \quad (367)$$

A relation which follows from equations (362) and (365) may be noted here. We have

$$\left.\begin{aligned} \sum_{a=-n}^{+n} M_a - \sum_{a=-n}^{+n} M_a e^{-k_a\tau_1} = 2\left[\frac{s_{rl}(+1)}{P(-1)} + \frac{s_{rl}(-1)}{P(+1)}\right] - 2\left[\frac{t_{rl}(+1)}{P(-1)} + \frac{t_{rl}(-1)}{P(+1)}\right] \\ + (-1)^{n+1} 2 \Big\{ 2a[c_{0,r}^{(n-1)} + c_{1,r}^{(n-1)}] + (p_0 - p_1)[c_{0,r}^{(n)} - c_{1,r}^{(n)}] - 2\Big(\sum_{i=1}^{n} \mu_i\Big) a[c_{0,r}^{(n)} + c_{1,r}^{(n)}] \Big\}, \end{aligned}\right\} \quad (368)$$

or, substituting for $(p_0 - p_1)$ according to equations (338) and (339), we have

$$\left.\begin{aligned} \sum_{a=-n}^{+n} M_a - \sum_{a=-n}^{+n} M_a e^{-k_a\tau_1} &= (-1)^{n+1} 2a\,[c_{0,r}^{(n)} + c_{1,r}^{(n)}]\,\tau_1 \\ &+ 2\left[\frac{s_{rl}(+1)}{P(-1)} + \frac{s_{rl}(-1)}{P(+1)}\right] - 2\left[\frac{t_{rl}(+1)}{P(-1)} + \frac{t_{rl}(-1)}{P(+1)}\right]. \end{aligned}\right\} \quad (369)$$

Returning to equations (344)–(347), we first observe that (cf. eqs. [317] and [318])

$$S_{rl}(0) = \frac{(-1)^n}{\mu_1 \cdots \mu_n} s_{rl}(0) \qquad \text{and} \qquad T_{rl}(0) = \frac{(-1)^n}{\mu_1 \cdots \mu_n} t_{rl}(0). \quad (370)$$

From equations (344), (345), (364), (367), and (370) it now follows that

$$\tfrac{1}{2}[S_{ll}(+1) + S_{ll}(-1)] = 2\left[\frac{s_{rl}(+1)}{P(-1)} + \frac{s_{rl}(-1)}{P(+1)}\right] \quad (371)$$

and

$$\tfrac{1}{2}[T_{ll}(+1) + T_{ll}(-1)] = 2\left[\frac{t_{rl}(+1)}{P(-1)} + \frac{t_{rl}(-1)}{P(+1)}\right]. \quad (372)$$

Similarly, from equations (346), (363), and (366) we have

$$\left.\begin{aligned}&\tfrac{1}{2}[S_{ll}(+1)-S_{ll}(-1)]+\tfrac{1}{2}[T_{ll}(+1)-T_{ll}(-1)]\\&\qquad=2\left[\frac{s_{rl}(+1)}{P(-1)}-\frac{s_{rl}(-1)}{P(+1)}\right]+2\left[\frac{t_{rl}(+1)}{P(-1)}-\frac{t_{rl}(-1)}{P(+1)}\right].\end{aligned}\right\}\quad(373)$$

And, finally, from equations (347), (363), (366), and (369) we have

$$\left.\begin{aligned}&\tau_1\left\{\tfrac{1}{2}[S_{ll}(+1)-S_{ll}(-1)]-\tfrac{1}{2}[T_{ll}(+1)-T_{ll}(-1)]-2\left[\frac{s_{rl}(+1)}{P(-1)}-\frac{s_{rl}(-1)}{P(+1)}\right]\right.\\&\left.+2\left[\frac{t_{rl}(+1)}{P(-1)}-\frac{t_{rl}(-1)}{P(+1)}\right]\right\}=2\left\{\tfrac{1}{2}[S_{ll}(+1)+S_{ll}(-1)]-\tfrac{1}{2}[T_{ll}(+1)+T_{ll}(-1)]\right.\\&\qquad\left.-2\left[\frac{s_{rl}(+1)}{P(-1)}+\frac{s_{rl}(-1)}{P(+1)}\right]+2\left[\frac{t_{rl}(+1)}{P(-1)}+\frac{t_{rl}(-1)}{P(+1)}\right]\right\}.\end{aligned}\right\}\quad(374)$$

But, according to equations (371) and (372), the right-hand side of equation (374) vanishes ($\tau_1 \neq 0$). Hence

$$\left.\begin{aligned}&\tfrac{1}{2}[S_{ll}(+1)-S_{ll}(-1)]-\tfrac{1}{2}[T_{ll}(+1)-T_{ll}(-1)]\\&\qquad=2\left[\frac{s_{rl}(+1)}{P(-1)}-\frac{s_{rl}(-1)}{P(+1)}\right]-2\left[\frac{t_{rl}(+1)}{P(-1)}-\frac{t_{rl}(-1)}{P(+1)}\right].\end{aligned}\right\}\quad(375)$$

From equations (371), (372), (373), and (375) it is now apparent that

$$S_{ll}(\pm 1)=4\frac{s_{rl}(\pm 1)}{P(\mp 1)}\quad\text{and}\quad T_{ll}(\pm 1)=4\frac{t_{rl}(\pm 1)}{P(\mp 1)}.\qquad(376)$$

On the other hand, according to equations (315) and (316),

$$\left.\begin{aligned}S_{ll}(\pm 1)&=\frac{1}{\mu_1^2\cdots\cdot\mu_n^2}\frac{P(\pm 1)}{\Omega(1)}\frac{1}{1\mp\frac{1}{\mu_0}}s_{ll}(\pm 1),\\\text{and}\quad T_{ll}(\pm 1)&=\frac{1}{\mu_1^2\cdots\cdot\mu_n^2}\frac{P(\pm 1)}{\Omega(1)}\frac{1}{1\pm\frac{1}{\mu_0}}t_{ll}(\pm 1).\end{aligned}\right\}\quad(377)$$

The foregoing equations can be simplified by making use of the relation (Paper XI, n. 16)

$$\frac{1}{\mu_1^2\cdots\cdot\mu_n^2}\frac{P(+1)P(-1)}{\Omega(1)}=-2.\qquad(378)$$

We thus find

$$S_{ll}(\pm 1)=-\frac{2}{1\mp\frac{1}{\mu_0}}\frac{s_{ll}(\pm 1)}{P(\mp 1)}=4\frac{s_{rl}(\pm 1)}{P(\mp 1)}\qquad(379)$$

and

$$T_{ll}(\pm 1)=-\frac{2}{1\pm\frac{1}{\mu_0}}\frac{t_{ll}(\pm 1)}{P(\mp 1)}=4\frac{t_{rl}(\pm 1)}{P(\mp 1)}.\qquad(380)$$

Hence

$$\mu_0 s_{ll}(\pm 1) = \pm 2(1 \mp \mu_0)\, s_{rl}(\pm 1), \tag{381}$$

and

$$\mu_0 t_{ll}(\pm 1) = \mp 2(1 \pm \mu_0)\, t_{rl}(\pm 1), \tag{382}$$

or, alternatively,

$$(1 \pm \mu_0)\, s_{ll}(\pm 1) = \pm \frac{2}{\mu_0}(1-\mu_0^2)\, s_{rl}(\pm 1), \tag{383}$$

and

$$(1 \mp \mu_0)\, t_{ll}(\pm 1) = \mp \frac{2}{\mu_0}(1-\mu_0^2)\, t_{rl}(\pm 1). \tag{384}$$

We shall find it convenient to combine equations (383) and (384) in the following forms:

$$+(1+\mu_0)\, s_{ll}(+1) \mp (1-\mu_0)\, t_{ll}(+1) = \frac{2}{\mu_0}(1-\mu_0^2)\,[s_{rl}(+1) \pm t_{rl}(+1)] \tag{385}$$

and

$$-(1-\mu_0)\, s_{ll}(-1) \pm (1+\mu_0)\, t_{ll}(-1) = \frac{2}{\mu_0}(1-\mu_0^2)\,[s_{rl}(-1) \pm t_{rl}(-1)]. \tag{386}$$

We now have to substitute for $s_{ll}(\pm 1)$, $t_{ll}(\pm 1)$, $s_{rl}(\pm 1)$, and $t_{rl}(\pm 1)$ according to equations (323), (324), (332), and (333). For this purpose, we shall write these quantities in the following forms:

$$\left.\begin{aligned} s_{ll}(\pm 1) = {} & (\varpi_0 + \chi_0\mu_0)\, C_{0,\,l}(\pm 1) + (\varpi_1 - \chi_1\mu_0)\, C_{1,\,l}(\pm 1) \\ & + (1 \mp \mu_0)\,[\pm \chi_0 C_{0,\,l}(\pm 1) \mp \chi_1 C_{1,\,l}(\pm 1)], \end{aligned}\right\} \tag{387}$$

$$\left.\begin{aligned} t_{ll}(\pm 1) = {} & (\varpi_0 + \chi_0\mu_0)\, C_{1,\,l}(\pm 1) + (\varpi_1 - \chi_1\mu_0)\, C_{0,\,l}(\pm 1) \\ & + (1 \pm \mu_0)\,[\mp \chi_0 C_{1,\,l}(\pm 1) \pm \chi_1 C_{0,\,l}(\pm 1)], \end{aligned}\right\} \tag{388}$$

$$s_{rl}(\pm 1) = p_0 C_{0,\,r}(\pm 1) + p_1 C_{1,\,r}(\pm 1) \pm a\,[C_{0,\,r}(\pm 1) + C_{1,\,r}(\pm 1)], \tag{389}$$

and

$$t_{rl}(\pm 1) = p_0 C_{1,\,r}(\pm 1) + p_1 C_{0,\,r}(\pm 1) \mp a\,[C_{0,\,r}(\pm 1) + C_{1,\,r}(\pm 1)]. \tag{390}$$

Substituting from equations (387)–(390) in equations (385) and (386), we find, after some lengthy, but straightforward, reductions, that

$$\left.\begin{aligned} \frac{2}{\mu_0}(1-\mu_0^2)\, c_1(p_0+p_1) = {} & \mathfrak{A}_l(\gamma_2+\gamma_1\mu_0) + \mathfrak{B}_l(-\gamma_2+\gamma_1\mu_0) \\ & + (1-\mu_0^2)\,\gamma_1(\chi_0-\chi_1), \end{aligned}\right\} \tag{391}$$

$$\left.\begin{aligned} \frac{2}{\mu_0}(1-\mu_0^2)\, c_3(p_0+p_1) = {} & \mathfrak{A}_l(-\gamma_4+\gamma_3\mu_0) + \mathfrak{B}_l(\gamma_4+\gamma_3\mu_0) \\ & + (1-\mu_0^2)\,\gamma_3(\chi_0-\chi_1), \end{aligned}\right\} \tag{392}$$

$$\left.\begin{aligned} \frac{2}{\mu_0}(1-\mu_0^2)\,[c_2(p_0-p_1)+2a\,c_1] & = \frac{2}{\mu_0}(1-\mu_0^2)\,(c_2+2c_1 Q)\,(p_0-p_1) \\ & = \mathfrak{A}_l(\gamma_1+\gamma_2\mu_0) + \mathfrak{B}_l(\gamma_1-\gamma_2\mu_0) + (1-\mu_0^2)\,\gamma_2(\chi_0+\chi_1), \end{aligned}\right\} \tag{393}$$

and

$$\left.\begin{aligned}\frac{2}{\mu_0}(1-\mu_0^2)\left[c_4(p_0-p_1)-2ac_3\right] &= \frac{2}{\mu_0}(1-\mu_0^2)(c_4-2c_3Q)(p_0-p_1)\\ &= \mathfrak{A}_l(-\gamma_3+\gamma_4\mu_0)+\mathfrak{B}_l(-\gamma_3-\gamma_4\mu_0)+(1-\mu_0^2)\gamma_4(\chi_0+\chi_1)\,,\end{aligned}\right\}\quad(394)$$

where we have used the following abbreviations:

$$\mathfrak{A}_l=\varpi_0+\chi_0\mu_0 \qquad \text{and} \qquad \mathfrak{B}_l=\varpi_1-\chi_1\mu_0\,, \qquad (395)$$

and

$$\left.\begin{aligned}\gamma_1&=C_{0,\,l}(+1)+C_{1,\,l}(+1)\,; & c_1&=C_{0,\,r}(+1)+C_{1,\,r}(+1)\,,\\ \gamma_2&=C_{0,\,l}(+1)-C_{1,\,l}(+1)\,; & c_2&=C_{0,\,r}(+1)-C_{1,\,r}(+1)\,,\\ \gamma_3&=C_{0,\,l}(-1)+C_{1,\,l}(-1)\,; & c_3&=C_{0,\,r}(-1)+C_{1,\,r}(-1)\,,\\ \gamma_4&=C_{0,\,l}(-1)-C_{1,\,l}(-1)\,; & c_4&=C_{0,\,r}(-1)-C_{1,\,r}(-1)\,.\end{aligned}\right\}\quad(396)$$

Eliminating $(\chi_0-\chi_1)$ from equations (391) and (392) and $(\chi_0+\chi_1)$ from equations (393) and (394), we find

$$\frac{2}{\mu_0}(1-\mu_0^2)(p_0+p_1)=\frac{\gamma_1\gamma_4+\gamma_2\gamma_3}{c_3\gamma_1-c_1\gamma_3}(\mathfrak{B}_l-\mathfrak{A}_l) \qquad (397)$$

and

$$\frac{2}{\mu_0}(1-\mu_0^2)(p_0-p_1)=\frac{\gamma_1\gamma_4+\gamma_2\gamma_3}{(c_2\gamma_4-c_4\gamma_2)+2Q(c_1\gamma_4+c_3\gamma_2)}(\mathfrak{B}_l+\mathfrak{A}_l)\,. \qquad (398)$$

Similarly, eliminating (p_0+p_1) from equations (391) and (392) and (p_0-p_1) from equations (393) and (394), we find

$$(1-\mu_0^2)(\chi_1-\chi_0)=\mathfrak{A}_l\left[\mu_0+\frac{c_3\gamma_2+c_1\gamma_4}{c_3\gamma_1-c_1\gamma_3}\right]+\mathfrak{B}_l\left[\mu_0-\frac{c_3\gamma_2+c_1\gamma_4}{c_3\gamma_1-c_1\gamma_3}\right] \qquad (399)$$

and

$$\left.\begin{aligned}(1-\mu_0^2)(\chi_1+\chi_0)&=\mathfrak{A}_l\left[-\mu_0+\frac{c_4\gamma_1+c_2\gamma_3-2Q(c_3\gamma_1-c_1\gamma_3)}{c_2\gamma_4-c_4\gamma_2+2Q(c_1\gamma_4+c_3\gamma_2)}\right]\\ &\quad+\mathfrak{B}_l\left[\mu_0+\frac{c_4\gamma_1+c_2\gamma_3-2Q(c_3\gamma_1-c_1\gamma_3)}{c_2\gamma_4-c_4\gamma_2+2Q(c_1\gamma_4+c_3\gamma_2)}\right].\end{aligned}\right\}\quad(400)$$

Finally, solving equations (397) and (398) for p_0 and p_1 and equations (399) and (400) for χ_0 and χ_1, we find that these constants can be expressed in the following forms:

$$\frac{2}{\mu_0}(1-\mu_0^2)\,p_0=-\nu_2(\varpi_0+\chi_0\mu_0)+\nu_1(\varpi_1-\chi_1\mu_0)\,, \qquad (401)$$

$$\frac{2}{\mu_0}(1-\mu_0^2)\,p_1=-\nu_1(\varpi_0+\chi_0\mu_0)+\nu_2(\varpi_1-\chi_1\mu_0)\,, \qquad (402)$$

$$(1-\mu_0^2)\,\chi_0=-\,(\mu_0+\nu_4)(\varpi_0+\chi_0\mu_0)+\nu_3(\varpi_1-\chi_1\mu_0)\,, \qquad (403)$$

and

$$(1-\mu_0^2)\,\chi_1=+\nu_3(\varpi_0+\chi_0\mu_0)+(\mu_0-\nu_4)(\varpi_1-\chi_1\mu_0)\,, \qquad (404)$$

where

$$\nu_1 = \tfrac{1}{2}(\gamma_1\gamma_4 + \gamma_2\gamma_3)\left[\frac{1}{c_3\gamma_1 - c_1\gamma_3} + \frac{1}{c_2\gamma_4 - c_4\gamma_2 + 2Q(c_1\gamma_4 + c_3\gamma_2)}\right], \quad (405)$$

$$\nu_2 = \tfrac{1}{2}(\gamma_1\gamma_4 + \gamma_2\gamma_3)\left[\frac{1}{c_3\gamma_1 - c_1\gamma_3} - \frac{1}{c_2\gamma_4 - c_4\gamma_2 + 2Q(c_1\gamma_4 + c_3\gamma_2)}\right], \quad (406)$$

$$\nu_3 = \frac{1}{2}\left[\frac{c_3\gamma_2 + c_1\gamma_4}{c_3\gamma_1 - c_1\gamma_3} + \frac{c_4\gamma_1 + c_2\gamma_3 - 2Q(c_3\gamma_1 - c_1\gamma_3)}{c_2\gamma_4 - c_4\gamma_2 + 2Q(c_1\gamma_4 + c_3\gamma_2)}\right], \quad (407)$$

and

$$\nu_4 = \frac{1}{2}\left[\frac{c_3\gamma_2 + c_1\gamma_4}{c_3\gamma_1 - c_1\gamma_3} - \frac{c_4\gamma_1 + c_2\gamma_3 - 2Q(c_3\gamma_1 - c_1\gamma_3)}{c_2\gamma_4 - c_4\gamma_2 + 2Q(c_1\gamma_4 + c_3\gamma_2)}\right]. \quad (408)$$

Also, according to equations (338), (401), and (402) (or more directly from eq. [398]), we have

$$\frac{2}{\mu_0}(1 - \mu_0^2)\, a = Q(\nu_1 - \nu_2)\left[(\varpi_0 + \chi_0\mu_0) + (\varpi_1 - \chi_1\mu_0)\right]. \quad (409)$$

Since $(\varpi_0 + \chi_0\mu_0)$ and $(\varpi_1 - \chi_1\mu_0)$ are already known (eqs. [325] and [326]), the foregoing equations complete the solution of the formal problem.

Returning to equations (305) we first observe that, for $s_{ll}(\mu)$ and $t_{ll}(\mu)$ given by equations (323) and (324), the constants $\varpi_0 + \chi_0\mu_0$, $\varpi_1 - \chi_1\mu_0$, χ_0, and χ_1 must enter the equations governing the angular distributions of the reflected and the transmitted radiations in the component l in the manner of $p_0 + q_0\mu_0$, $p_1 - q_1\mu_0$, q_0, and q_1 in equations (194) and (195). With the expressions for $\varpi_0 + \chi_0\mu_0$, $\varpi_1 - \chi_1\mu_0$, χ_0, and χ_1 given by equations (325), (326), (403), and (404), we therefore have

$$\left.\begin{aligned} I_{ll}(0,\mu) = \tfrac{3}{16}F_l\frac{2\mu_0}{\mu_0+\mu}\{&(1-\mu_0^2)[X_l(\mu)X_l(\mu_0) - Y_l(\mu)Y_l(\mu_0)] \\ &-(\mu_0+\mu)X_l(\mu)[-(\mu_0+\nu_4)X_l(\mu_0) + \nu_3 Y_l(\mu_0)] \\ &-(\mu_0+\mu)Y_l(\mu)[\nu_3 X_l(\mu_0) + (\mu_0-\nu_4)Y_l(\mu_0)]\} \end{aligned}\right\} \quad (410)$$

and

$$\left.\begin{aligned} I_{ll}(\tau_1,-\mu) = \tfrac{3}{16}F_l\frac{2\mu_0}{\mu_0-\mu}\{&(1-\mu_0^2)[X_l(\mu)Y_l(\mu_0) - Y_l(\mu)X_l(\mu_0)] \\ &+(\mu_0-\mu)X_l(\mu)[\nu_3 X_l(\mu_0) + (\mu_0-\nu_4)Y_l(\mu_0)] \\ &+(\mu_0-\mu)Y_l(\mu)[-(\mu_0+\nu_4)X_l(\mu_0) + \nu_3 Y_l(\mu_0)]\}, \end{aligned}\right\} \quad (411)$$

where $X_l(\mu)$ amd $Y_l(\mu)$ are defined in terms of $C_{0,l}(\mu)$ and $C_{1,l}(\mu)$ (cf. eqs. [125] and [126]). Equations (410) and (411) can be reduced to the forms

$$\left.\begin{aligned} I_{ll}(0,\mu) = \tfrac{3}{16}F_l\frac{2\mu_0}{\mu_0+\mu}\{&X_l(\mu)X_l(\mu_0)[1+\nu_4(\mu_0+\mu)+\mu\mu_0] \\ &-Y_l(\mu)Y_l(\mu_0)[1-\nu_4(\mu_0+\mu)+\mu\mu_0] \\ &-\nu_3(\mu_0+\mu)[X_l(\mu)Y_l(\mu_0) + Y_l(\mu)X_l(\mu_0)]\} \end{aligned}\right\} \quad (412)$$

and

$$\left.\begin{aligned} I_{ll}(\tau_1,-\mu) = \tfrac{3}{16}F_l\frac{2\mu_0}{\mu_0-\mu}\{&X_l(\mu)Y_l(\mu_0)[1-\nu_4(\mu_0-\mu)-\mu\mu_0] \\ &-Y_l(\mu)X_l(\mu_0)[1+\nu_4(\mu_0-\mu)-\mu\mu_0] \\ &+\nu_3(\mu_0-\mu)[X_l(\mu)X_l(\mu_0) + Y_l(\mu)Y_l(\mu_0)]\}. \end{aligned}\right\} \quad (413)$$

Turning, next, to the reflected and the transmitted intensities in the component r, we have (cf. eqs. [305], [332], and [333])

$$\left.\begin{aligned} I_{rl}(0,\mu) &= \tfrac{3}{16}F_l\frac{(-1)^n}{\mu_1\ldots\mu_n}[C^2_{0,r}(0)-C^2_{1,r}(0)]^{\frac{1}{2}} \\ &\qquad\times[(p_0-a\mu)X_r(\mu)-(p_1-a\mu)Y_r(\mu)] \end{aligned}\right\} \quad (414)$$

and

$$\left.\begin{aligned} I_{rl}(\tau_1,-\mu) &= \tfrac{3}{16}F_l\frac{(-1)^n}{\mu_1\ldots\mu_n}[C^2_{0,r}(0)-C^2_{1,r}(0)]^{\frac{1}{2}} \\ &\qquad\times[(p_1+a\mu)X_r(\mu)-(p_0+a\mu)Y_r(\mu)], \end{aligned}\right\} \quad (415)$$

where $X_r(\mu)$ and $Y_r(\mu)$ are defined in terms of $C_{0,r}(\mu)$ and $C_{1,r}(\mu)$. Substituting for p_0, p_1, and a from equations (401), (402), and (409) in the foregoing equations, we find, after some minor rearrangement of the terms,

$$\left.\begin{aligned} I_{rl}(0,\mu) = \tfrac{3}{16}F_l\left[\frac{C^2_{0,r}(0)-C^2_{1,r}(0)}{C^2_{0,l}(0)-C^2_{1,l}(0)}\right]^{\frac{1}{2}} &\mu_0\{\nu_1[X_r(\mu)Y_l(\mu_0)+Y_r(\mu)X_l(\mu_0)] \\ &-\nu_2[X_r(\mu)X_l(\mu_0)+Y_r(\mu)Y_l(\mu_0)] \\ -Q(\nu_1-\nu_2)\mu[X_r(\mu)-Y_r(\mu)]&[X_l(\mu_0)+Y_l(\mu_0)]\} \end{aligned}\right\} \quad (416)$$

and

$$\left.\begin{aligned} I_{rl}(\tau_1,-\mu) = \tfrac{3}{16}F_l\left[\frac{C^2_{0,r}(0)-C^2_{1,r}(0)}{C^2_{0,l}(0)-C^2_{1,l}(0)}\right]^{\frac{1}{2}} &\mu_0\{\nu_2[X_r(\mu)Y_l(\mu_0)+Y_r(\mu)X_l(\mu_0)] \\ &-\nu_1[X_r(\mu)X_l(\mu_0)+Y_r(\mu)Y_l(\mu_0)] \\ +Q(\nu_1-\nu_2)\mu[X_r(\mu)-Y_r(\mu)]&[X_l(\mu_0)+Y_l(\mu_0)]\}. \end{aligned}\right\} \quad (417)$$

18. *The azimuth independent terms proportional to* F_r.—In the preceding section we completed the solution for the azimuth independent terms in the reflected and transmitted intensities which are proportional to F_l. We shall now consider the terms proportional to F_r.

The azimuth independent terms proportional to F_r in the reflected and transmitted intensities can be expressed in the forms

$$\left.\begin{aligned} I^{(0)}_{lr}(0,\mu) &= \tfrac{3}{16}[S_{lr}(-\mu)-e^{-\tau_1/\mu}T_{lr}(\mu)]F_r, \\ I^{(0)}_{lr}(\tau_1,-\mu) &= \tfrac{3}{16}[T_{lr}(-\mu)-e^{-\tau_1/\mu}S_{lr}(\mu)]F_r, \\ I^{(0)}_{rr}(0,\mu) &= \tfrac{3}{16}[S_{rr}(-\mu)-e^{-\tau_1/\mu}T_{rr}(\mu)]F_r, \\ \text{and}\quad I^{(0)}_{rr}(\tau_1,-\mu) &= \tfrac{3}{16}[T_{rr}(-\mu)-e^{-\tau_1/\mu}S_{rr}(\mu)]F_r, \end{aligned}\right\} \quad (418)$$

where (cf. Paper XIII, eqs. [33] and [34])

$$\left.\begin{aligned} S_{lr}(\mu) = (1-\mu^2)\sum_{\beta=-n+1}^{n-1} \frac{\mathfrak{L}_\beta}{1-\kappa_\beta\mu} - \mathfrak{L}_0\mu + \mathfrak{L}_n \\ + \sum_{\alpha=-n}^{+n} \mathfrak{M}_\alpha(1+k_\alpha\mu) - C\left(1+\frac{\mu}{\mu_0}\right)\mu_0^2, \end{aligned}\right\} \tag{419}$$

$$\left.\begin{aligned} T_{lr}(\mu) = (1-\mu^2)\sum_{\beta=-n+1}^{n-1} \frac{\mathfrak{L}_\beta e^{-\kappa_\beta\tau_1}}{1+\kappa_\beta\mu} + \mathfrak{L}_0(\tau_1+\mu) + \mathfrak{L}_n \\ + \sum_{\alpha=-n}^{+n} \mathfrak{M}_\alpha(1-k_\alpha\mu)e^{-k_\alpha\tau_1} - C\left(1-\frac{\mu}{\mu_0}\right)\mu_0^2 e^{-\tau_1/\mu_0}, \end{aligned}\right\} \tag{420}$$

$$S_{rr}(\mu) = -\mathfrak{L}_0\mu + \mathfrak{L}_n - \sum_{\alpha=-n}^{+n} \frac{\mathfrak{M}_\alpha(k_\alpha^2-1)}{1-k_\alpha\mu} + \frac{C(1-\mu_0^2)}{1+\mu/\mu_0}, \tag{421}$$

and

$$T_{rr}(\mu) = \mathfrak{L}_0(\tau_1+\mu) + \mathfrak{L}_n - \sum_{\alpha=-n}^{+n} \frac{\mathfrak{M}_\alpha(k_\alpha^2-1)e^{-k_\alpha\tau_1}}{1+k_\alpha\mu} + \frac{C(1-\mu_0^2)e^{-\tau_1/\mu_0}}{1+\mu/\mu_0}. \tag{422}$$

In the foregoing equations κ_β and k_α have the same meanings as in § 18,

$$C = H_r(\mu_0)H_r(-\mu_0) = \frac{1}{\mu_1^2\cdots\mu_n^2}\frac{P(\mu_0)P(-\mu_0)}{W(\mu_0)}; \tag{423}$$

further, we have used $\mathfrak{L}_\beta(\beta = 0, \pm 1, \ldots, \pm n \mp 1, n)$ and $\mathfrak{M}_\alpha$ $(\alpha = \pm 1, \ldots, \pm n)$ to denote the constants of integration to distinguish them from the L_β's and M_α's of the preceding section.

The boundary conditions, as usual, require that

$$S_{lr}(\mu_i) = T_{lr}(\mu_i) = S_{rr}(\mu_i) = T_{rr}(\mu_i) = 0 \qquad (i = 1, \ldots, n). \tag{424}$$

By virtue of these boundary conditions we can write

$$S_{lr}(\mu) = \frac{1}{\mu_1^2\cdots\mu_n^2}\frac{P(\mu)}{\Omega(\mu)}s_{lr}(\mu), \tag{425}$$

$$T_{lr}(\mu) = \frac{1}{\mu_1^2\cdots\mu_n^2}\frac{P(\mu)}{\Omega(\mu)}t_{lr}(\mu), \tag{426}$$

$$S_{rr}(\mu) = \frac{1}{\mu_1^2\cdots\mu_n^2}\frac{P(\mu)}{W(\mu)}\frac{1}{1-\mu/\mu_0}s_{rr}(\mu), \tag{427}$$

and

$$T_{rr}(\mu) = \frac{1}{\mu_1^2\cdots\mu_n^2}\frac{P(\mu)}{W(\mu)}\frac{1}{1+\mu/\mu_0}t_{rr}(\mu), \tag{428}$$

where $s_{lr}(\mu)$ and $t_{lr}(\mu)$ are polynomials of degree $n-1$ in μ and $s_{rr}(\mu)$ and $t_{rr}(\mu)$ are polynomials of degree $n+2$ in μ.

From equations (421), (422), (423), (427), and (428) it follows that

$$s_{rr}(\mu_0) = (1-\mu_0^2)\,P(-\mu_0) \qquad \text{and} \qquad t_{rr}(-\mu_0) = (1-\mu_0^2)\,e^{-\tau_1/\mu_0}P(\mu_0)\,. \tag{429}$$

From equations (419), (420), (425), and (426) we now have

$$\left(1-\frac{1}{\kappa_\beta^2}\right)\mathfrak{L}_\beta = \frac{1}{\mu_1^2\ldots\ldots\mu_n^2}\,\frac{P(1/\kappa_\beta)}{\Omega_\beta(1/\kappa_\beta)}\,s_{lr}(1/\kappa_\beta) \tag{430}$$

and

$$\left(1-\frac{1}{\kappa_\beta^2}\right)\mathfrak{L}_\beta e^{-\kappa_\beta\tau_1} = \frac{1}{\mu_1^2\ldots\ldots\mu_n^2}\,\frac{P(-1/\kappa_\beta)}{\Omega_\beta(1/\kappa_\beta)}\,t_{lr}(-1/\kappa_\beta)\,. \tag{431}$$

We therefore have $(2n-2)$ relations of the form

$$s_{lr}(1/\kappa_\beta) = e^{\kappa_\beta\tau_1}\frac{P(-1/\kappa_\beta)}{P(+1/\kappa_\beta)}\,t_{lr}(-1/\kappa_\beta) \qquad (\beta = \pm 1\,,\ldots\ldots,\pm n \mp 1)\,. \tag{432}$$

Since, however, $s_{lr}(\mu)$ and $t_{lr}(\mu)$ are polynomials of degree only $n-1$, we can conclude, in accordance with theorem 2, § 4, that

$$s_{lr}(\mu) = \varpi_0^* C_{0,\,l}(\mu) + \varpi_1^* C_{1,\,l}(\mu) \tag{433}$$

and

$$t_{lr}(\mu) = \varpi_1^* C_{0,\,l}(\mu) + \varpi_0^* C_{1,\,l}(\mu)\,, \tag{434}$$

where ϖ_0^* and ϖ_1^* are constants and $C_{0,l}(\mu)$ and $C_{1,l}(\mu)$ have the same meanings as in § 17.

Turning, next, to equations (421), (422), (425), and (426), we have

$$\left.\begin{aligned}\mathfrak{M}_a = -\frac{1}{\mu_1^2\ldots\ldots\mu_n^2}\,\frac{P(1/k_a)}{W_a(1/k_a)(k_a^2-1)}\,\frac{1}{1-1/k_a\mu_0}\,s_{rr}(1/k_a)&\\ (a = \pm 1\,,\ldots\ldots,\pm n)&\end{aligned}\right\} \tag{435}$$

and

$$\left.\begin{aligned}\mathfrak{M}_a e^{-k_a\tau_1} = -\frac{1}{\mu_1^2\ldots\ldots\mu_n^2}\,\frac{P(-1/k_a)}{W_a(1/k_a)(k_a^2-1)}\,\frac{1}{1-1/k_a\mu_0}\,t_{rr}(-1/k_a)&\\ (a = \pm 1\,,\ldots\ldots,\pm n)\,.&\end{aligned}\right\} \tag{436}$$

We therefore have $2n$ relations of the form

$$s_{rr}(1/k_a) = e^{k_a\tau_1}\frac{P(-1/k_a)}{P(+1/k_a)}\,t_{rr}(-1/k_a) \qquad (a = \pm 1\,,\ldots\ldots,\pm n)\,. \tag{437}$$

It will be recalled that $s_{rr}(\mu)$ and $t_{rr}(\mu)$ are polynomials of degree $(n+2)$ in μ. However, it can be shown that $s_{rr}(\mu) - t_{rr}(\mu)$ is only of degree $n+1$; for, according to equations (421), (422), (427), and (428),

$$\left.\begin{aligned}\frac{1}{\mu_1^2\ldots\ldots\mu_n^2}P(\mu)\,s_{rr}(\mu) &= \left(1-\frac{\mu}{\mu_0}\right)W(\mu)\,S_{rr}(\mu)\\ &= (-1)^n k_1^2\ldots\ldots k_n^2\left[\frac{\mathfrak{L}_0}{\mu_0}\,\mu^{2n+2} - \left(\mathfrak{L}_0 + \frac{\mathfrak{L}_n}{\mu_0}\right)\mu^{2n+1} + \ldots\ldots\right]\end{aligned}\right\} \tag{438}$$

and

$$\left.\begin{aligned}\frac{1}{\mu_1^2 \ldots . \mu_n^2} P(\mu)\, t_{rr}(\mu) &= \left(1+\frac{\mu}{\mu_0}\right) W(\mu)\, T_{rr}(\mu) \\ &= (-1)^n k_1^2 \ldots . k_n^2 \left[\frac{\mathfrak{L}_0}{\mu_0}\,\mu^{2n+2} + \left(\mathfrak{L}_0 + \frac{\mathfrak{L}_n + \mathfrak{L}_0 \tau_1}{\mu_0}\right)\mu^{n+1} + \ldots .\right].\end{aligned}\right\} \quad (439)$$

And it is apparent from these equations that the coefficients of μ^{n+2} in $s_{rr}(\mu)$ and $t_{rr}(\mu)$ are the same. We may therefore write

$$s_{rr}(\mu) = (p_0^* + q_0^* \mu + a^* \mu^2) C_{0,\,r}(\mu) + (p_1^* - q_1^* \mu + a^* \mu^2) C_{1,\,r}(\mu) \quad (440)$$

and

$$t_{rr}(\mu) = (p_1^* + q_1^* \mu + a^* \mu^2) C_{0,\,r}(\mu) + (p_0^* - q_0^* \mu + a^* \mu^2) C_{1,\,r}(\mu)\,, \quad (441)$$

where p_0^*, p_1^*, q_0^*, q_1^*, and a^* are constants and $C_{0,\,r}(\mu)$ and $C_{1,\,r}(\mu)$ again have the same meanings as in § 17.

For $s_{rr}(\mu)$ and $t_{rr}(\mu)$ given by equations (440) and (441) we can verify that

$$\left.\begin{aligned}P(\mu)\, s_{rr}(\mu) &= a^* [c_{0,\,r}^{(n)} + c_{1,\,r}^{(n)}]\,\mu^{2n+2} + \Big\{ a^* [c_{0,\,r}^{(n-1)} + c_{1,\,r}^{(n-1)}] \\ &\quad + q_0^* c_{0,\,r}^{(n)} - q_1^* c_{1,\,r}^{(n)} - \left(\sum_{i=1}^{n} \mu_i\right) a^* [c_{0,\,r}^{(n)} + c_{1,\,r}^{(n)}] \Big\}\, \mu^{2n+1} + \ldots .\end{aligned}\right\} \quad (442)$$

and

$$\left.\begin{aligned}P(\mu)\, t_{rr}(\mu) &= a^* [c_{0,\,r}^{(n)} + c_{1,\,r}^{(n)}]\,\mu^{2n+2} + \Big\{ a^* [c_{0,\,r}^{(n-1)} + c_{1,\,r}^{(n-1)}] \\ &\quad + q_1^* c_{0,\,r}^{(n)} - q_0^* c_{1,\,r}^{(n)} - \left(\sum_{i=1}^{n} \mu_i\right) a^* [c_{0,\,r}^{(n)} + c_{1,\,r}^{(n)}] \Big\}\, \mu^{2n+1} + \ldots .\,,\end{aligned}\right\} \quad (443)$$

where $c_{0,\,r}^{(n)}$, $c_{1,\,r}^{(n)}$ and $c_{0,\,r}^{(n-1)}$, $c_{1,\,r}^{(n-1)}$ are the coefficients of the highest and the next highest powers of μ in $C_{0,\,r}$ and $C_{1,\,r}$.

From a comparison of equations (438) and (439) and (442) and (443), we conclude that

$$\left.\begin{aligned}-\sum_{i=1}^{n} \mu_i + \frac{1}{c_{0,\,r}^{(n)} + c_{1,\,r}^{(n)}} \Big\{ c_{0,\,r}^{(n-1)} + c_{1,\,r}^{(n-1)} + \frac{1}{a^*} [q_0^* c_{0,\,r}^{(n)} - q_1^* c_{1,\,r}^{(n)}] \Big\} \\ = -\left(\mu_0 + \frac{\mathfrak{L}_n}{\mathfrak{L}_0}\right)\end{aligned}\right\} \quad (444)$$

and

$$\left.\begin{aligned}-\sum_{i=1}^{n} \mu_i + \frac{1}{c_{0,\,r}^{(n)} + c_{1,\,r}^{(n)}} \Big\{ c_{0,\,r}^{(n-1)} + c_{1,\,r}^{(n-1)} + \frac{1}{a^*} [q_1^* c_{0,\,r}^{(n)} - q_0^* c_{1,\,r}^{(n)}] \Big\} \\ = +\left(\mu_0 + \frac{\mathfrak{L}_n}{\mathfrak{L}_0} + \tau_1\right).\end{aligned}\right\} \quad (445)$$

From equations (444) and (445) we readily find that

$$a^* = Q\,(q_0^* + q_1^*)\,, \quad (446)$$

where Q has the same meaning as in § 17 (eq. [339]).

Finally, from equations (429), (440), and (441), we find

$$\left.\begin{aligned} p_0^* + q_0^*\mu_0 + a^*\mu_0^2 = &\frac{(1-\mu_0^2)}{[C_{0,r}^2(0) - C_{1,r}^2(0)]W(\mu_0)} \\ &\times [P(-\mu_0)C_{0,r}(-\mu_0) - e^{-\tau_1/\mu_0}P(\mu_0)C_{1,r}(\mu_0)] \end{aligned}\right\} \quad (447)$$

and

$$\left.\begin{aligned} p_1^* - q_1^*\mu_0 + a^*\mu_0^2 = &\frac{(1-\mu_0^2)}{[C_{0,r}^2(0) - C_{1,r}^2(0)]W(\mu_0)} \\ &\times [e^{-\tau_1/\mu_0}P(\mu_0)C_{0,r}(\mu_0) - P(-\mu_0)C_{1,r}(-\mu_0)]\,. \end{aligned}\right\} \quad (448)$$

Now it will be observed that, among the seven constants ϖ_0^*, ϖ_1^*, p_0^*, p_1^*, q_0^*, q_1^*, and a^* which we have introduced in the solutions for s_{lr}, t_{lr}, s_{rr}, and t_{rr}, we have so far only three relations (eqs. [446]–[448]). To obtain the four additional relations we proceed in the following manner:

Putting $\mu = +1$, respectively -1, in equations (419) and (420), we have

$$S_{lr}(\pm 1) = \mp \mathfrak{L}_0 + \mathfrak{L}_n + \sum_{a=-n}^{+n} \mathfrak{M}_a (1 \pm k_a) - C(\mu_0^2 \pm \mu_0) \quad (449)$$

and

$$T_{lr}(\pm 1) = \mathfrak{L}_0(\tau_1 \pm 1) + \mathfrak{L}_n + \sum_{a=-n}^{+n} \mathfrak{M}_a (1 \mp k_a)\, e^{-k_a\tau_1} - C(\mu_0^2 \mp \mu_0)\, e^{-\tau_1/\mu_0}\,. \quad (450)$$

Similarly, putting $\mu = 0$ in equations (421) and (422), we have

$$S_{rr}(0) = \mathfrak{L}_n - \sum_{a=-n}^{+n} \mathfrak{M}_a (k_a^2 - 1) + (1-\mu_0^2)C \quad (451)$$

and

$$T_{rr}(0) = \mathfrak{L}_0\tau_1 + \mathfrak{L}_n - \sum_{a=-n}^{+n} \mathfrak{M}_a (k_a^2 - 1)\, e^{-k_a\tau_1} + (1-\mu_0^2)C\, e^{-\tau_1/\mu_0}\,. \quad (452)$$

Combining equations (449)–(452) appropriately, we obtain the following relations:

$$\tfrac{1}{2}[S_{lr}(+1) + S_{lr}(-1)] - S_{rr}(0) = \sum_{a=-n}^{+n} \mathfrak{M}_a k_a^2 - C\,, \quad (453)$$

$$\tfrac{1}{2}[T_{lr}(+1) + T_{lr}(-1)] - T_{rr}(0) = \sum_{a=-n}^{+n} \mathfrak{M}_a k_a^2 e^{-k_a\tau_1} - C\, e^{-\tau_1/\mu_0}\,, \quad (454)$$

$$\left.\begin{aligned} &\tfrac{1}{2}[S_{lr}(+1) - S_{lr}(-1)] + \tfrac{1}{2}[T_{lr}(+1) - T_{lr}(-1)] \\ &\qquad = \sum_{a=-n}^{+n} \mathfrak{M}_a k_a - \sum_{a=-n}^{+n} \mathfrak{M}_a k_a\, e^{-k_a\tau_1} - C\mu_0(1 - e^{-\tau_1/\mu_0})\,, \end{aligned}\right\} \quad (455)$$

and

$$\left.\begin{aligned}\tau_1\Big\{\tfrac{1}{2}\,[S_{lr}(+1)-S_{lr}(-1)]-\tfrac{1}{2}\,[T_{lr}(+1)-T_{lr}(-1)]-\sum_{a=-n}^{+n}\mathfrak{M}_a k_a\\ -\sum_{a=-n}^{+n}\mathfrak{M}_a k_a e^{-k_a\tau_1}+C\mu_0(1+e^{-\tau_1/\mu_0})\Big\}=2\Big\{\tfrac{1}{2}\,[S_{lr}(+1)+S_{lr}(-1)]\\ -\tfrac{1}{2}[T_{lr}(+1)+T_{lr}(-1)]-\sum_{a=-n}^{+n}\mathfrak{M}_a+\sum_{a=-n}^{+n}\mathfrak{M}_a e^{-k_a\tau_1}+C\mu_0^2(1-e^{-\tau_1/\mu_0})\Big\}\\ =-2\mathfrak{L}_0\tau_1\,.\end{aligned}\right\}\quad(456)$$

We shall now indicate how the various summations which occur in the foregoing equations can be evaluated. Considering the summation $\Sigma\mathfrak{M}_a k_a^m$, for example, we have (cf. eq. [435])

$$\left.\begin{aligned}\sum_{a=-n}^{+n}\mathfrak{M}_a k_a^m&=\frac{\mu_0}{\mu_1^2\cdots\mu_n^2}\sum_{a=-n}^{+n}\frac{k_a^{m-1}P(1/k_a)\,s_{rr}(1/k_a)}{W_a(1/k_a)(1-k_a^{-2})(1-k_a\mu_0)}\\ &=\frac{\mu_0}{\mu_1^2\cdots\mu_n^2W(\mu_0)}\sum_{a=-n}^{+n}\frac{k_a^{m-1}P(1/k_a)\,s_{rr}(1/k_a)}{W_a(1/k_a)(1-k_a^{-2})}W_a(\mu_0)\,,\end{aligned}\right\}\quad(457)$$

or, in terms of the function

$$f_m(x)=\sum_{a=-n}^{+n}\frac{k_a^{m-1}P(1/k_a)\,s_{rr}(1/k_a)}{W_a(1/k_a)(1-k_a^{-2})}W_a(x)\,,\quad(458)$$

$$\sum_{a=-n}^{+n}\mathfrak{M}_a k_a^m=\frac{\mu_0}{\mu_1^2\cdots\mu_n^2W(\mu_0)}f_m(\mu_0)\,.\quad(459)$$

Now $f_m(x)$ is a polynomial of degree $(2n-1)$ in x, which takes the values

$$\frac{k_a^{m-1}}{(1-k_a^{-2})}P(1/k_a)\,s_{rr}(1/k_a)\quad(460)$$

for $x=1/k_a$ ($a=\pm1,\ldots,\pm n$). In other words, $(1-x^2)f_m(x)-x^{1-m}P(x)s_{rr}(x)$ vanishes for $x=1/k_a$ ($a=\pm1,\ldots,\pm n$); it must accordingly divide $W(x)$. This fact enables us to determine $f_m(x)$. To illustrate this, we shall consider the case $m=0$. In this case, there must exist a relation of the form

$$(1-x^2)\,f_0(x)=xP(x)\,s_{rr}(x)+W(x)\,(Ax^3+Bx^2+Dx+E)\,,\quad(461)$$

where A, B, D, and E are certain constants to be determined. The constants A and B follow from the fact that the coefficients of x^{2n+3} and x^{2n+2} on the right-hand side of equation (461) must vanish. We find (cf. eq. [442])

$$A=\frac{(-1)^{n+1}}{k_1^2\cdots k_n^2}a^*\,[c_{0,r}^{(n)}+c_{1,r}^{(n)}]\quad(462)$$

and

$$B = \frac{(-1)^{n+1}}{k_1^2 \dots k_n^2} \left\{ a^* [c_{0,r}^{(n-1)} + c_{1,r}^{(n-1)}] + q_0^* c_{0,r}^{(n)} - q_1^* c_{1,r}^{(n)} - \left(\sum_{i=1}^{n} \mu_i \right) a^* [c_{0,r}^{(n)} + c_{1,r}^{(n)}] \right\}. \tag{463}$$

Next, putting $x = +1$, respectively -1, in equation (461), we have

$$\pm P(\pm 1)\, s_{rr}(\pm 1) + W(1)(\pm A + B \pm D + E) = 0. \tag{464}$$

From this equation it readily follows that

$$B + E = -\frac{1}{2W(1)} [P(+1)\, s_{rr}(+1) - P(-1)\, s_{rr}(-1)] \tag{465}$$

and

$$A + D = -\frac{1}{2W(1)} [P(+1)\, s_{rr}(+1) + P(-1)\, s_{rr}(-1)]. \tag{466}$$

Returning to equation (461) and setting $x = \mu_0$ and remembering that

$$s_{rr}(\mu_0) = (1 - \mu_0^2) P(-\mu_0), \tag{467}$$

we have

$$f_0(\mu_0) = \mu_0 P(\mu_0) P(-\mu_0) - W(\mu_0) \left\{ A\mu_0 + B - \frac{(A+D)\mu_0 + B + E}{1 - \mu_0^2} \right\}. \tag{468}$$

Substituting for $f_0(\mu_0)$ according to equation (468) in equation (459) and making use of the relations (360), (361), (465), and (466), we obtain

$$\begin{aligned}\sum_{a=-n}^{+n} \mathfrak{M}_a = C\mu_0^2 + 2(-1)^n \Big\{ & a^* [c_{0,r}^{(n)} + c_{1,r}^{(n)}] \mu_0^2 + \Big[a^* [c_{0,r}^{(n-1)} + c_{1,r}^{(n-1)}] \\ & + q_0^* c_{0,r}^{(n)} - q_1^* c_{1,r}^{(n)} - \left(\sum_{i=1}^{n} \mu_i \right) a^* [c_{0,r}^{(n)} + c_{1,r}^{(n)}] \Big] \mu_0 \Big\} \\ & - \frac{2\mu_0}{1 - \mu_0^2} \left\{ \left[\frac{s_{rr}(+1)}{P(-1)} + \frac{s_{rr}(-1)}{P(+1)} \right] \mu_0 + \left[\frac{s_{rr}(+1)}{P(-1)} - \frac{s_{rr}(-1)}{P(+1)} \right] \right\}.\end{aligned} \tag{469}$$

The other summations can be similarly evaluated, and we find

$$\sum_{a=-n}^{+n} \mathfrak{M}_a k_a^2 = C + \frac{(-1)^{n+1}}{\mu_1 \dots \mu_n} s_{rr}(0) - \frac{2\mu_0}{1 - \mu_0^2} (\xi_1 \mu_0 + \xi_2), \tag{470}$$

$$\sum_{a=-n}^{+n} \mathfrak{M}_a k_a^2 e^{-k_a \tau_1} = C e^{-\tau_1/\mu_0} + \frac{(-1)^{n+1}}{\mu_1 \dots \mu_n} t_{rr}(0) - \frac{2\mu_0}{1 - \mu_0^2} (\eta_1 \mu_0 - \eta_2), \tag{471}$$

$$\sum_{a=-n}^{+n} \mathfrak{M}_a k_a = C\mu_0 + 2(-1)^n a^* [c_{0,r}^{(n)} + c_{1,r}^{(n)}] \mu_0 - \frac{2\mu_0}{1 - \mu_0^2} (\xi_2 \mu_0 + \xi_1), \tag{472}$$

$$\left.\begin{aligned}\sum_{a=-n}^{+n} \mathfrak{M}_a k_a e^{-k_a\tau_1} = C\mu_0 e^{-\tau_1/\mu_0} + 2(-1)^n a^* [c_{0,r}^{(n)} + c_{1,r}^{(n)}]\mu_0 \\ + \frac{2\mu_0}{1-\mu_0^2}(\eta_2\mu_0 - \eta_1),\end{aligned}\right\} \quad (473)$$

$$\left.\begin{aligned}\sum_{a=-n}^{+n} \mathfrak{M}_a = C\mu_0^2 + 2(-1)^n \Big\{ a^* [c_{0,r}^{(n)} + c_{1,r}^{(n)}]\mu_0^2 + \Big[a^* [c_{0,r}^{(n-1)} + c_{1,r}^{(n-1)}] \\ + q_0^* c_{0,r}^{(n)} - q_1^* c_{1,r}^{(n)} - \Big(\sum_{i=1}^{n}\mu_i\Big) a^*[c_{0,r}^{(n)} + c_{1,r}^{(n)}]\Big]\mu_0\Big\} - \frac{2\mu_0}{1-\mu_0^2}(\xi_1\mu_0 + \xi_2),\end{aligned}\right\} \quad (474)$$

and

$$\left.\begin{aligned}\sum_{a=-n}^{+n} \mathfrak{M}_a e^{-k_a\tau_1} = C\mu_0^2 e^{-\tau_1/\mu_0} + 2(-1)^n \Big\{ a^* [c_{0,r}^{(n)} + c_{1,r}^{(n)}]\mu_0^2 - \Big[a^* [c_{0,r}^{(n-1)} + c_{1,r}^{(n-1)}] \\ + q_1^* c_{0,r}^{(n)} - q_0^* c_{1,r}^{(n)} - \Big(\sum_{i=1}^{n}\mu_i\Big) a^* [c_{0,r}^{(n)} + c_{1,r}^{(n)}]\Big]\mu_0\Big\} - \frac{2\mu_0}{1-\mu_0^2}(\eta_1\mu_0 - \eta_2),\end{aligned}\right\} \quad (475)$$

where we have used the abbreviations

$$\left.\begin{aligned}\xi_1 &= \frac{s_{rr}(+1)}{P(-1)} + \frac{s_{rr}(-1)}{P(+1)}; \qquad \xi_2 = \frac{s_{rr}(+1)}{P(-1)} - \frac{s_{rr}(-1)}{P(+1)} \\ \text{and} \quad \eta_1 &= \frac{t_{rr}(+1)}{P(-1)} + \frac{t_{rr}(-1)}{P(+1)}; \qquad \eta_2 = \frac{t_{rr}(+1)}{P(-1)} - \frac{t_{rr}(-1)}{P(+1)}.\end{aligned}\right\} \quad (476)$$

A relation which follows from equations (474) and (475) may be noted here. We have

$$\left.\begin{aligned}\sum_{a=-n}^{+n} \mathfrak{M}_a - \sum_{a=-n}^{+n} \mathfrak{M}_a e^{-k_a\tau_1} - C\mu_0^2(1 - e^{-\tau_1/\mu_0}) \\ = 2(-1)^n \Big\{ 2a^* [c_{0,r}^{(n-1)} + c_{1,r}^{(n-1)}] + (q_0^* + q_1^*)[c_{0,r}^{(n)} - c_{1,r}^{(n)}] \\ - 2\Big(\sum_{i=1}^{n}\mu_i\Big) a^* [c_{0,r}^{(n)} + c_{1,r}^{(n)}]\Big\}\mu_0 - \frac{2\mu_0}{1-\mu_0^2}[(\xi_1\mu_0 + \xi_2) - (\eta_1\mu_0 - \eta_2)];\end{aligned}\right\} \quad (477)$$

or, substituting for $q_0^* + q_1^*$ according to equations (339) and (446), we have

$$\left.\begin{aligned}\sum_{a=-n}^{+n} \mathfrak{M}_a - \sum_{a=-n}^{+n} \mathfrak{M}_a e^{-k_a\tau_1} - C\mu_0^2(1 - e^{-\tau_1/\mu_0}) \\ = 2(-1)^n a^* \tau_1 [c_{0,r}^{(n)} + c_{1,r}^{(n)}]\mu_0 - \frac{2\mu_0}{1-\mu_0^2}[(\xi_1\mu_0 + \xi_2) - (\eta_1\mu_0 - \eta_2)].\end{aligned}\right\} \quad (478)$$

Returning to equations (453)–(456), we first observe that (cf. eqs. [427] and [428])

$$S_{rr}(0) = \frac{(-1)^n}{\mu_1 \ldots \mu_n} s_{rr}(0) \quad \text{and} \quad T_{rr}(0) = \frac{(-1)^n}{\mu_1 \ldots \mu_n} t_{rr}(0). \quad (479)$$

From equations (453), (454), (470), (471), and (479) it now follows that

$$\tfrac{1}{2}\left[S_{lr}(+1)+S_{lr}(-1)\right]=-\frac{2\mu_0}{1-\mu_0^2}(\xi_1\mu_0+\xi_2) \tag{480}$$

and

$$\tfrac{1}{2}\left[T_{lr}(+1)+T_{lr}(-1)\right]=-\frac{2\mu_0}{1-\mu_0^2}(\eta_1\mu_0-\eta_2)\,. \tag{481}$$

Similarly, from equations (455), (472), and (473) we have

$$\left.\begin{aligned}&\tfrac{1}{2}\left[S_{lr}(+1)-S_{lr}(-1)\right]+\tfrac{1}{2}\left[T_{lr}(+1)-T_{lr}(-1)\right]\\&\qquad=-\frac{2\mu_0}{1-\mu_0^2}\left[(\xi_2\mu_0+\xi_1)+(\eta_2\mu_0-\eta_1)\right].\end{aligned}\right\} \tag{482}$$

And, finally, from equations (456), (472), (473), and (478) we have

$$\left.\begin{aligned}&\tau_1\Big\{\tfrac{1}{2}\left[S_{lr}(+1)-S_{lr}(-1)\right]-\tfrac{1}{2}\left[T_{lr}(+1)-T_{lr}(-1)\right]\\&\quad+\frac{2\mu_0}{1-\mu_0^2}\left[(\xi_2\mu_0+\xi_1)-(\eta_2\mu_0-\eta_1)\right]\Big\}=2\Big\{\tfrac{1}{2}\left[S_{lr}(+1)+S_{lr}(-1)\right]\\&\quad-\tfrac{1}{2}\left[T_{lr}(+1)+T_{lr}(-1)\right]+\frac{2\mu_0}{1-\mu_0^2}\left[(\xi_1\mu_0+\xi_2)-(\eta_1\mu_0-\eta_2)\right]\Big\}.\end{aligned}\right\} \tag{483}$$

But, according to equations (480) and (481), the right-hand side of the foregoing equation vanishes. Hence

$$\left.\begin{aligned}&\tfrac{1}{2}\left[S_{lr}(+1)+S_{lr}(-1)\right]-\tfrac{1}{2}\left[T_{lr}(+1)+T_{lr}(-1)\right]\\&\qquad=-\frac{2\mu_0}{1-\mu_0^2}\left[(\xi_2\mu_0+\xi_1)-(\eta_2\mu_0-\eta_1)\right].\end{aligned}\right\} \tag{484}$$

From equations (480), (481), (482), and (484) it readily follows that

$$S_{lr}(+1)=-\frac{2\mu_0}{1-\mu_0^2}(\xi_1+\xi_2)(\mu_0+1)=-\frac{4\mu_0}{1-\mu_0}\frac{s_{rr}(+1)}{P(-1)}, \tag{485}$$

$$S_{lr}(-1)=-\frac{2\mu_0}{1-\mu_0^2}(\xi_1-\xi_2)(\mu_0-1)=+\frac{4\mu_0}{1+\mu_0}\frac{s_{rr}(-1)}{P(+1)}, \tag{486}$$

$$T_{lr}(+1)=-\frac{2\mu_0}{1-\mu_0^2}(\eta_1+\eta_2)(\mu_0-1)=+\frac{4\mu_0}{1+\mu_0}\frac{t_{rr}(+1)}{P(-1)}, \tag{487}$$

and

$$T_{lr}(-1)=-\frac{2\mu_0}{1-\mu_0^2}(\eta_1-\eta_2)(\mu_0+1)=-\frac{4\mu_0}{1-\mu_0}\frac{t_{rr}(-1)}{P(+1)}. \tag{488}$$

On the other hand, according to equations (378), (425), and (426),

$$S_{lr}(\pm 1) = \frac{1}{\mu_1^2 \cdots \mu_n^2} \frac{P(\pm 1)}{\Omega(1)} s_{lr}(\pm 1) = -2\frac{s_{lr}(\pm 1)}{P(\mp 1)}, \tag{489}$$

and

$$T_{lr}(\pm 1) = \frac{1}{\mu_1^2 \cdots \mu_n^2} \frac{P(\pm 1)}{\Omega(1)} t_{lr}(\pm 1) = -2\frac{t_{lr}(\pm 1)}{P(\mp 1)}. \tag{490}$$

We thus find

$$s_{lr}(\pm 1) = \pm\frac{2\mu_0}{1-\mu_0^2}(1\pm\mu_0)\, s_{rr}(\pm 1) \tag{491}$$

and

$$t_{lr}(\pm 1) = \mp\frac{2\mu_0}{1-\mu_0^2}(1\mp\mu_0)\, t_{rr}(\pm 1). \tag{492}$$

We now have to substitute for $s_{lr}(\pm 1)$, $t_{lr}(\pm 1)$, $s_{rr}(\pm 1)$, and $t_{rr}(\pm 1)$ according to equations (433), (434), (440), and (441). For this purpose it is convenient to write $s_{rr}(\pm 1)$ and $t_{rr}(\pm 1)$ in the forms

$$\left.\begin{aligned} s_{rr}(\pm 1) &= (p_0^* + q_0^*\mu_0 + a^*\mu_0^2)C_{0,r}(\pm 1) + (p_1^* - q_1^*\mu_0 + a^*\mu_0^2)C_{1,r}(\pm 1) \\ &+ (1\mp\mu_0)[\pm q_0^* C_{0,r}(\pm 1) \mp q_1^* C_{1,r}(\pm 1)] + a^*(1-\mu_0^2)[C_{0,r}(\pm 1) + C_{1,r}(\pm 1)] \end{aligned}\right\} \tag{493}$$

and

$$\left.\begin{aligned} t_{rr}(\pm 1) &= (p_0^* + q_0^*\mu_0 + a^*\mu_0^2)C_{1,r}(\pm 1) + (p_1^* - q_1^*\mu_0 + a^*\mu_0^2)C_{0,r}(\pm 1) \\ &+ (1\pm\mu_0)[\mp q_0^* C_{1,r}(\pm 1) \pm q_1^* C_{0,r}(\pm 1)] + a^*(1-\mu_0^2)[C_{0,r}(\pm 1) + C_{1,r}(\pm 1)]. \end{aligned}\right\} \tag{494}$$

We also have

$$s_{lr}(\pm 1) = \varpi_0^* C_{0,l}(\pm 1) + \varpi_1^* C_{1,l}(\pm 1) \tag{495}$$

and

$$t_{lr}(\pm 1) = \varpi_0^* C_{1,l}(\pm 1) + \varpi_1^* C_{0,l}(\pm 1). \tag{496}$$

Substituting from equations (493)–(496) in equations (491) and (492), we find, after some lengthy but straightforward reductions, that (cf. eq. [446])

$$\left.\begin{aligned} (\varpi_0^* + \varpi_1^*)\gamma_1 = \frac{2\mu_0}{1-\mu_0^2}[&\mathfrak{A}_r(\mu_0 c_1 + c_2) + \mathfrak{B}_r(\mu_0 c_1 - c_2) \\ &+ c_1(1-\mu_0^2)\{(q_0^* - q_1^*) + 2Q(q_0^* + q_1^*)\mu_0\}], \end{aligned}\right\} \tag{497}$$

$$\left.\begin{aligned} (\varpi_0^* + \varpi_1^*)\gamma_3 = \frac{2\mu_0}{1-\mu_0^2}[&\mathfrak{A}_r(\mu_0 c_3 - c_4) + \mathfrak{B}_r(\mu_0 c_3 + c_4) \\ &+ c_3(1-\mu_0^2)\{(q_0^* - q_1^*) + 2Q(q_0^* + q_1^*)\mu_0\}], \end{aligned}\right\} \tag{498}$$

$$\left.\begin{aligned} (\varpi_0^* - \varpi_1^*)\gamma_2 = \frac{2\mu_0}{1-\mu_0^2}[&\mathfrak{A}_r(\mu_0 c_2 + c_1) + \mathfrak{B}_r(-\mu_0 c_2 + c_1) \\ &+ (1-\mu_0^2)(c_2 + 2c_1 Q)(q_0^* + q_1^*)], \end{aligned}\right\} \tag{499}$$

and

$$(\varpi_0^* - \varpi_1^*)\,\gamma_4 = \frac{2\mu_0}{1-\mu_0^2}\,[\,\mathfrak{A}_r(\mu_0 c_4 - c_3) - \mathfrak{B}_r(\mu_0 c_4 + c_3) + (1-\mu_0^2)(c_4 - 2c_3Q)(q_0^* + q_1^*)\,]\,, \tag{500}$$

where γ_1, γ_2, γ_3, γ_4, and c_1, c_2, c_3, c_4, have the same meanings as in equations (396) and

$$\mathfrak{A}_r = p_0^* + q_0^*\mu_0 + a^*\mu_0^2 \qquad \text{and} \qquad \mathfrak{B}_r = p_1^* - q_1^*\mu_0 + a^*\mu_0^2\,. \tag{501}$$

Eliminating $[(q_0^* - q_1^*) + 2Q(q_0^* + q_1^*)]$ from equations (497) and (498) and $(q_0^* + q_1^*)$ from equations (499) and (500), we find

$$\varpi_1^* + \varpi_0^* = \frac{2\mu_0}{1-\mu_0^2}\,\frac{c_1c_4 + c_2c_3}{c_3\gamma_1 - c_1\gamma_3}\,(\mathfrak{A}_r - \mathfrak{B}_r) \tag{502}$$

and

$$\varpi_1^* - \varpi_0^* = \frac{2\mu_0}{1-\mu_0^2}\,\frac{c_1c_4 + c_2c_3}{(c_2\gamma_4 - c_4\gamma_2) + 2Q(c_1\gamma_4 + c_3\gamma_2)}\,[\,\mathfrak{A}_r(1 - 2Q\mu_0) + \mathfrak{B}_r(1 + 2Q\mu_0)\,]\,. \tag{503}$$

Similarly, eliminating $(\varpi_0^* + \varpi_1^*)$ from equations (497) and (498) and $(\varpi_0^* - \varpi_1^*)$ from equations (499) and (500), we find

$$(1-\mu_0^2)\,[\,(q_0^* - q_1^*) + 2Q(q_0^* + q_1^*)\,\mu_0\,] = \mathfrak{A}_r\left(\frac{c_2\gamma_3 + c_4\gamma_1}{c_3\gamma_1 - c_1\gamma_3} - \mu_0\right) - \mathfrak{B}_r\left(\frac{c_2\gamma_3 + c_4\gamma_1}{c_3\gamma_1 - c_1\gamma_3} + \mu_0\right) \tag{504}$$

and

$$(1-\mu_0^2)\left(1 + 2Q\,\frac{c_1\gamma_4 + c_3\gamma_2}{c_2\gamma_4 - c_4\gamma_2}\right)(q_0^* + q_1^*) = -\,\mathfrak{A}_r\left(\frac{c_1\gamma_4 + c_3\gamma_2}{c_2\gamma_4 - c_4\gamma_2} + \mu_0\right) - \mathfrak{B}_r\left(\frac{c_1\gamma_4 + c_3\gamma_2}{c_2\gamma_4 - c_4\gamma_2} - \mu_0\right). \tag{505}$$

Finally, solving equations (502) and (503) for ϖ_0^* and ϖ_1^* and equations (504) and (505) for q_0^* and q_1^*, we find that these constants can be expressed in the forms

$$\varpi_0^* = \frac{\mu_0}{1-\mu_0^2}\{\,\mathfrak{A}_r[u_2 + Q(u_1 - u_2)\mu_0] - \mathfrak{B}_r[u_1 + Q(u_1 - u_2)\mu_0]\,\}\,, \tag{506}$$

$$\varpi_1^* = \frac{\mu_0}{1-\mu_0^2}\{\,\mathfrak{A}_r[u_1 - Q(u_1 - u_2)\mu_0] - \mathfrak{B}_r[u_2 - Q(u_1 - u_2)\mu_0]\,\}\,, \tag{507}$$

$$(1-\mu_0^2)\,q_0^* = \mathfrak{A}_r(u_4 - u_5\mu_0 + u_6\mu_0^2) - \mathfrak{B}_r(u_3 + u_6\mu_0^2)\,, \tag{508}$$

and

$$(1-\mu_0^2)\,q_1^* = -\,\mathfrak{A}_r(u_3 + u_6\mu_0^2) + \mathfrak{B}_r(u_4 + u_5\mu_0 + u_6\mu_0^2)\,, \tag{509}$$

where

$$u_1 = (c_1c_4 + c_2c_3)\left[\frac{1}{c_3\gamma_1 - c_1\gamma_3} + \frac{1}{c_2\gamma_4 - c_4\gamma_2 + 2Q(c_1\gamma_4 + c_3\gamma_2)}\right], \tag{510}$$

$$u_2 = (c_1c_4 + c_2c_3)\left[\frac{1}{c_3\gamma_1 - c_1\gamma_3} - \frac{1}{c_2\gamma_4 - c_4\gamma_2 + 2Q(c_1\gamma_4 + c_3\gamma_2)}\right], \tag{511}$$

$$u_3 = \frac{1}{2}\left[\frac{c_2\gamma_3 + c_4\gamma_1}{c_3\gamma_1 - c_1\gamma_3} + \frac{c_1\gamma_4 + c_3\gamma_2}{c_2\gamma_4 - c_4\gamma_2 + 2Q(c_1\gamma_4 + c_3\gamma_2)}\right], \tag{512}$$

$$u_4 = \frac{1}{2}\left[\frac{c_2\gamma_3 + c_4\gamma_1}{c_3\gamma_1 - c_1\gamma_3} - \frac{c_1\gamma_4 + c_3\gamma_2}{c_2\gamma_4 - c_4\gamma_2 + 2Q(c_1\gamma_4 + c_3\gamma_2)}\right], \tag{513}$$

$$u_5 = \frac{c_2\gamma_4 - c_4\gamma_2}{c_2\gamma_4 - c_4\gamma_2 + 2Q(c_1\gamma_4 + c_3\gamma_2)}, \tag{514}$$

and

$$u_6 = Qu_5 \,. \tag{515}$$

Also, according to equations (446), (505), (514), and (515), we have

$$(1 - \mu_0^2)\, a^* = -u_6\left[\mathfrak{A}_r\left(\frac{c_1\gamma_4 + c_3\gamma_2}{c_2\gamma_4 - c_4\gamma_2} + \mu_0\right) + \mathfrak{B}_r\left(\frac{c_1\gamma_4 + c_3\gamma_2}{c_2\gamma_4 - c_4\gamma_2} - \mu_0\right)\right]. \tag{516}$$

It may be recalled that $\mathfrak{A}_r$ and $\mathfrak{B}_r$, which occur in the foregoing solution for the constants, are defined in equation (501). According to equations (447) and (448) we may, therefore, write

$$\mathfrak{A}_r = p_0^* + q_0^*\mu_0 + a^*\mu_0^2 = (-1)^n \mu_1 \ldots \mu_n \frac{(1 - \mu_0^2)\, X_r(\mu_0)}{[C_{0,r}^2(0) - C_{1,r}^2(0)]^{\frac{1}{2}}} \tag{517}$$

and

$$\mathfrak{B}_r = p_1^* - q_1^*\mu_0 + a^*\mu_0^2 = (-1)^n \mu_1 \ldots \mu_n \frac{(1 - \mu_0^2)\, Y_r(\mu_0)}{[C_{0,r}^2(0) - C_{1,r}^2(0)]^{\frac{1}{2}}}. \tag{518}$$

With this determination of the constants we have completed the solution of the formal problem.

We now return to equations (418) and the angular distributions of the reflected and the transmitted radiations.

It is first apparent that for $s_{lr}(\mu)$ and $t_{lr}(\mu)$ given by equations (433) and (434) the reflected and the transmitted intensities in the component l are of the forms

$$I_{lr}(0, \mu) = \tfrac{3}{16}F_r \frac{(-1)^n}{\mu_1 \ldots \mu_n}[C_{0,l}^2(0) - C_{1,l}^2(0)]^{\frac{1}{2}}[\varpi_0^* X_l(\mu) - \varpi_1^* Y_l(\mu)] \tag{519}$$

and

$$I_{lr}(\tau_1, -\mu) = \tfrac{3}{16}F_r \frac{(-1)^n}{\mu_1 \ldots \mu_n}[C_{0,l}^2(0) - C_{1,l}^2(0)]^{\frac{1}{2}}[\varpi_1^* X_l(\mu) - \varpi_0^* Y_l(\mu)]. \tag{520}$$

Substituting for ϖ_0^* and ϖ_1^* according to equations (506), (507), (517), and (518), we have

$$\left.\begin{aligned} I_{lr}(0, \mu) &= \tfrac{3}{16}F_r\left[\frac{C_{0,l}^2(0) - C_{1,l}^2(0)}{C_{0,r}^2(0) - C_{1,r}^2(0)}\right]^{\frac{1}{2}}\mu_0 \\ &\times \{X_l(\mu)[X_r(\mu_0)[\![u_2 + Q(u_1 - u_2)\mu_0]\!] - Y_r(\mu_0)[\![u_1 + Q(u_1 - u_2)\mu_0]\!]] \\ &- Y_l(\mu)[X_r(\mu_0)[\![u_1 - Q(u_1 - u_2)\mu_0]\!] - Y_r(\mu_0)[\![u_2 - Q(u_1 - u_2)\mu_0]\!]]\} \end{aligned}\right\} \tag{521}$$

and

$$\left.\begin{aligned} I_{lr}(\tau_1,-\mu) &= \tfrac{3}{16}F_r\left[\frac{C^2_{0,\,l}(0)-C^2_{1,\,l}(0)}{C^2_{0,\,r}(0)-C^2_{1,\,r}(0)}\right]^{\frac{1}{2}}\mu_0 \\ &\times\{X_l(\mu)[X_r(\mu_0)[\![u_1-Q(u_1-u_2)\mu_0]\!]-Y_r(\mu_0)[\![u_2-Q(u_1-u_2)\mu_0]\!]] \\ &-Y_l(\mu)[X_r(\mu_0)[\![u_2+Q(u_1-u_2)\mu_0]\!]-Y_r(\mu_0)[\![u_1+Q(u_1-u_2)\mu_0]\!]]\}. \end{aligned}\right\} \quad (522)$$

After some minor rearranging of the terms, equations (521) and (522) can be brought to the forms

$$\left.\begin{aligned} I_{lr}(0,\mu) &= \tfrac{3}{16}F_r\left[\frac{C^2_{0,\,l}(0)-C^2_{1,\,l}(0)}{C^2_{0,\,r}(0)-C^2_{1,\,r}(0)}\right]^{\frac{1}{2}}\mu_0 \\ &\times\{-u_1[X_l(\mu)Y_r(\mu_0)+Y_l(\mu)X_r(\mu_0)] \\ &+u_2[X_l(\mu)X_r(\mu_0)+Y_l(\mu)Y_r(\mu_0)] \\ &+Q(u_1-u_2)\mu_0[X_l(\mu)+Y_l(\mu)][X_r(\mu_0)-Y_r(\mu_0)]\} \end{aligned}\right\} \quad (523)$$

and

$$\left.\begin{aligned} I_{lr}(\tau_1,-\mu) &= \tfrac{3}{16}F_r\left[\frac{C^2_{0,\,l}(0)-C^2_{1,\,l}(0)}{C^2_{0,\,r}(0)-C^2_{1,\,r}(0)}\right]^{\frac{1}{2}}\mu_0 \\ &\times\{-u_2[X_l(\mu)Y_r(\mu_0)+Y_l(\mu)X_r(\mu_0)] \\ &+u_1[X_l(\mu)X_r(\mu_0)+Y_l(\mu)Y_r(\mu_0)] \\ &-Q(u_1-u_2)\mu_0[X_l(\mu)+Y_l(\mu)][X_r(\mu_0)-Y_r(\mu_0)]\}. \end{aligned}\right\} \quad (524)$$

Turning, next, to the reflected and the transmitted intensities in the component r, we have (cf. eqs. [418], [427], [428], [440], and [441])

$$\left.\begin{aligned} I_{rr}(0,\mu) &= \tfrac{3}{16}F_r\frac{(-1)^n}{\mu_1\ldots\mu_n}[C^2_{0,\,r}(0)-C^2_{1,\,r}(0)]^{\frac{1}{2}}\frac{\mu_0}{\mu_0+\mu} \\ &\times\{(p_0^*-q_0^*\mu+a^*\mu^2)X_r(\mu)-(p_1^*+q_1^*\mu+a^*\mu^2)Y_r(\mu)\} \end{aligned}\right\} \quad (525)$$

and

$$\left.\begin{aligned} I_{rr}(\tau_1,-\mu) &= \tfrac{3}{16}F_r\frac{(-1)^n}{\mu_1\ldots\mu_n}[C^2_{0,\,r}(0)-C^2_{1,\,r}(0)]^{\frac{1}{2}}\frac{\mu_0}{\mu_0-\mu} \\ &\times\{(p_1^*-q_1^*\mu+a^*\mu^2)X_r(\mu)-(p_0^*+q_0^*\mu+a^*\mu^2)Y_r(\mu)\}. \end{aligned}\right\} \quad (526)$$

We can re-write equations (525) and (526) in the forms

$$\left.\begin{aligned} I_{rr}(0,\mu) &= \tfrac{3}{16}F_r\frac{(-1)^n}{\mu_1\ldots\mu_n}[C^2_{0,\,r}(0)-C^2_{1,\,r}(0)]^{\frac{1}{2}}\frac{\mu_0}{\mu_0+\mu} \\ &\times\{(p_0^*+q_0^*\mu_0+a^*\mu_0^2)X_r(\mu)-(p_1^*-q_1^*\mu_0+a^*\mu_0^2)Y_r(\mu) \\ &-q_0^*(\mu_0+\mu)X_r(\mu)-q_1^*(\mu_0+\mu)Y_r(\mu)-a^*(\mu_0^2-\mu^2)[X_r(\mu)-Y_r(\mu)]\} \end{aligned}\right\} \quad (527)$$

and

$$I_{rr}(\tau_1,-\mu)=\tfrac{3}{16}F_r\frac{(-1)^n}{\mu_1\ldots\mu_n}[C_{0,r}^2(0)-C_{1,r}^2(0)]^{\frac{1}{2}}\frac{\mu_0}{\mu_0-\mu}$$
$$\times\{(p_1^*-q_1^*\mu_0+a^*\mu_0^2)X_r(\mu)-(p_0^*+q_0^*\mu_0+a^*\mu_0^2)Y_r(\mu)$$
$$+q_1^*(\mu_0-\mu)X_r(\mu)+q_0^*(\mu_0-\mu)Y_r(\mu)-a^*(\mu_0^2-\mu^2)[X_r(\mu)-Y_r(\mu)]\}. \quad (528)$$

Substituting for q_0^*, q_1^*, a^*, p_0^*, $+\ q_0^*\mu_0+a^*\mu_0^2$ and $p_1^*-q_1^*\mu_0+a^*\mu_0^2$ in accordance with equations (508), (509), and (516)–(518) in equations (527) and (528), we obtain

$$I_{rr}(0,\mu)=\tfrac{3}{16}F_r\frac{\mu_0}{\mu_0+\mu}\Big\{(1-\mu_0^2)[X_r(\mu)X_r(\mu_0)-Y_r(\mu)Y_r(\mu_0)]$$
$$-(\mu_0+\mu)X_r(\mu)[X_r(\mu_0)(u_4-u_5\mu_0+u_6\mu_0^2)-Y_r(\mu_0)(u_3+u_6\mu_0^2)]$$
$$-(\mu_0+\mu)Y_r(\mu)[-X_r(\mu_0)(u_3+u_6\mu_0^2)+Y_r(\mu_0)(u_4+u_5\mu_0+u_6\mu_0^2)]$$
$$-u_6(\mu_0^2-\mu^2)\left[-X_r(\mu_0)\left(\mu_0+\frac{c_1\gamma_4+c_3\gamma_2}{c_2\gamma_4-c_4\gamma_2}\right)+Y_r(\mu_0)\left(\mu_0-\frac{c_1\gamma_4+c_3\gamma_2}{c_2\gamma_4-c_4\gamma_2}\right)\right]$$
$$\times[X_r(\mu)-Y_r(\mu)]\Big\}. \quad (529)$$

and

$$I_{rr}(\tau_1,-\mu)=\tfrac{3}{16}F_r\frac{\mu_0}{\mu_0-\mu}\Big\{(1-\mu_0^2)[X_r(\mu)Y_r(\mu_0)-Y_r(\mu)X_r(\mu_0)]$$
$$+(\mu_0-\mu)X_r(\mu)[-X_r(\mu_0)(u_3+u_6\mu_0^2)+Y_r(\mu_0)(u_4+u_5\mu_0+u_6\mu_0^2)]$$
$$+(\mu_0-\mu)Y_r(\mu)[X_r(\mu_0)(u_4-u_5\mu_0+u_6\mu_0^2)-Y_r(\mu_0)(u_3+u_6\mu_0^2)]$$
$$-u_6(\mu_0^2-\mu^2)\left[-X_r(\mu_0)\left(\mu_0+\frac{c_1\gamma_4+c_3\gamma_2}{c_2\gamma_4-c_4\gamma_2}\right)+Y_r(\mu_0)\left(\mu_0-\frac{c_1\gamma_4+c_3\gamma_2}{c_2\gamma_4-c_4\gamma_2}\right)\right]$$
$$\times[X_r(\mu)-Y_r(\mu)]\Big\}. \quad (530)$$

After some lengthy reductions, the foregoing equations can be brought to the forms

$$I_{rr}(0,\mu)=\tfrac{3}{16}F_r\frac{\mu_0}{\mu_0+\mu}\{X_r(\mu)X_r(\mu_0)[1-u_4(\mu_0+\mu)+u_5\mu\mu_0]$$
$$-Y_r(\mu)Y_r(\mu_0)[1+u_4(\mu_0+\mu)+u_5\mu\mu_0]$$
$$+u_3(\mu_0+\mu)[X_r(\mu)Y_r(\mu_0)+Y_r(\mu)X_r(\mu_0)]$$
$$-Qu_5\mu\mu_0(\mu_0+\mu)[X_r(\mu)-Y_r(\mu)][X_r(\mu_0)-Y_r(\mu_0)]$$
$$-Q(u_3-u_4)\mu^2[X_r(\mu)-Y_r(\mu)][X_r(\mu_0)+Y_r(\mu_0)]$$
$$-Q(u_3-u_4)\mu_0^2[X_r(\mu)+Y_r(\mu)][X_r(\mu_0)-Y_r(\mu_0)]\} \quad (531)$$

and

$$\left.\begin{aligned} I_{rr}(\tau_1, -\mu) = \tfrac{3}{16}F_r \frac{\mu_0}{\mu_0-\mu}\{ & X_r(\mu)\, Y_r(\mu_0)\,[1 + u_4(\mu_0+\mu) - u_5\mu\mu_0] \\ & - Y_r(\mu)\, X_r(\mu_0)\,[1 - u_4(\mu_0-\mu) - u_5\mu\mu_0] \\ & - u_3(\mu_0-\mu)\,[X_r(\mu)\, X_r(\mu_0) + Y_r(\mu)\, Y_r(\mu_0)] \\ & + Qu_5\mu\mu_0(\mu_0-\mu)\,[X_r(\mu) - Y_r(\mu)]\,[X_r(\mu_0) - Y_r(\mu_0)] \\ & - Q(u_3-u_4)\,\mu^2\,[X_r(\mu) - Y_r(\mu)]\,[X_r(\mu_0) + Y_r(\mu_0)] \\ & + Q(u_3-u_4)\,\mu_0^2\,[X_r(\mu) + Y_r(\mu)]\,[X_r(\mu_0) - Y_r(\mu_0)]\}\,. \end{aligned}\right\} \quad (532)$$

19. *The scattering and the transmission matrices. The reciprocity principle.*—The solutions for the various terms given in the preceding sections can be combined to give the complete distribution of the reflected and the transmitted radiations. The resulting laws of diffuse reflection and transmission can be expressed in terms of a scattering (**S**) and a transmission (**T**) matrix in the forms (cf. Paper XIII, § 4; Paper XIV, § 2; and Paper XVI, Sec. II)

$$\boldsymbol{I}(0;\, \mu, \varphi;\, \mu_0, \varphi_0) = \frac{3}{16\mu} Q\boldsymbol{S}(\mu, \varphi;\, \mu_0, \varphi_0)\,\boldsymbol{F} \quad (533)$$

and

$$\boldsymbol{I}(\tau_1;\, -\mu, \varphi;\, \mu_0, \varphi_0) = \frac{3}{16\mu} Q\boldsymbol{T}(\mu, \varphi;\, \mu_0, \varphi_0)\,\boldsymbol{F}\,, \quad (534)$$

where

$$\boldsymbol{I} = (I_l, I_r, U) \quad \text{and} \quad \boldsymbol{F} = (F_l, F_r, U_0) \quad (535)$$

and

$$Q = \begin{pmatrix} 1 & 0 & 0 \\ 0 & 1 & 0 \\ 0 & 0 & 2 \end{pmatrix}. \quad (536)$$

In accordance with equations (284)–(287) and the solutions obtained in the preceding sections, we can write

$$\left.\begin{aligned} \boldsymbol{S}(\mu, \varphi;\, \mu_0, \varphi_0) = \boldsymbol{S}^{(0)}(\mu;\, \mu_0) + (1-\mu^2)^{\frac{1}{2}}(1-\mu_0^2)^{\frac{1}{2}}\boldsymbol{S}^{(1)}(\mu, \varphi;\, \mu_0, \varphi_0) \\ + \boldsymbol{S}^{(2)}(\mu, \varphi;\, \mu_0, \varphi_0) \end{aligned}\right\} \quad (537)$$

and

$$\left.\begin{aligned} \boldsymbol{T}(\mu, \varphi;\, \mu_0, \varphi_0) = \boldsymbol{T}^{(0)}(\mu;\, \mu_0) + (1-\mu^2)^{\frac{1}{2}}(1-\mu_0^2)^{\frac{1}{2}}\boldsymbol{T}^{(1)}(\mu, \varphi;\, \mu_0, \varphi_0) \\ + \boldsymbol{T}^{(2)}(\mu, \varphi;\, \mu_0, \varphi_0)\,, \end{aligned}\right\} \quad (538)$$

where

$$\left.\begin{aligned} \left(\frac{1}{\mu}+\frac{1}{\mu_0}\right) S_{11}^{(0)}(\mu;\, \mu_0) = 2\{ & X_l(\mu)\, X_l(\mu_0)\,[1 + \nu_4(\mu_0+\mu) + \mu\mu_0] \\ & - Y_l(\mu)\, Y_l(\mu_0)\,[1 - \nu_4(\mu_0+\mu) + \mu\mu_0] \\ & - \nu_3(\mu_0+\mu)\,[X_l(\mu)\, Y_l(\mu_0) + Y_l(\mu)\, X_l(\mu_0)]\}\,, \end{aligned}\right\} \quad (539)$$

$$\left.\begin{aligned}\left(\frac{1}{\mu}+\frac{1}{\mu_0}\right) S_{12}^{(0)}(\mu;\mu_0) = \left[\frac{C_{0,l}^2(0)-C_{1,l}^2(0)}{C_{0,r}^2(0)-C_{1,r}^2(0)}\right]^{\frac{1}{2}} (\mu_0+\mu) \\ \times\{-u_1[X_l(\mu)\,Y_r(\mu_0)+Y_l(\mu)\,X_r(\mu_0)] \\ +u_2[X_l(\mu)\,X_r(\mu_0)+Y_l(\mu)\,Y_r(\mu_0)] \\ +Q(u_1-u_2)\mu_0[X_l(\mu)+Y_l(\mu)][X_r(\mu_0)-Y_r(\mu_0)]\},\end{aligned}\right\} \quad (540)$$

$$\left.\begin{aligned}\left(\frac{1}{\mu}+\frac{1}{\mu_0}\right) S_{21}^{(0)}(\mu;\mu_0) = \left[\frac{C_{0,r}^2(0)-C_{1,r}^2(0)}{C_{0,l}^2(0)-C_{1,l}^2(0)}\right]^{\frac{1}{2}} (\mu_0+\mu) \\ \times\{\nu_1[X_r(\mu)\,Y_l(\mu_0)+Y_r(\mu)\,X_l(\mu_0)] \\ -\nu_2[X_r(\mu)\,X_l(\mu_0)+Y_r(\mu)\,Y_l(\mu_0)] \\ -Q(\nu_1-\nu_2)\mu[X_r(\mu)-Y_r(\mu)][X_l(\mu_0)+Y_l(\mu_0)]\},\end{aligned}\right\} \quad (541)$$

$$\left.\begin{aligned}\left(\frac{1}{\mu}+\frac{1}{\mu_0}\right) S_{22}^{(0)}(\mu;\mu_0) = X_r(\mu)\,X_r(\mu_0)[1-u_4(\mu_0+\mu)+u_5\mu\mu_0] \\ -Y_r(\mu)\,Y_r(\mu_0)[1+u_4(\mu_0+\mu)+u_5\mu\mu_0] \\ +u_3(\mu_0+\mu)[X_r(\mu)\,Y_r(\mu_0)+Y_r(\mu)\,X_r(\mu_0)] \\ -Qu_5\mu\mu_0(\mu_0+\mu)[X_r(\mu)-Y_r(\mu)][X_r(\mu_0)-Y_r(\mu_0)] \\ -Q(u_3-u_4)\{\mu^2[X_r(\mu)-Y_r(\mu)][X_r(\mu_0)+Y_r(\mu_0)] \\ +\mu_0^2[X_r(\mu)+Y_r(\mu)][X_r(\mu_0)-Y_r(\mu_0)]\};\end{aligned}\right\} \quad (542)$$

$$\left.\begin{aligned}\left(\frac{1}{\mu}-\frac{1}{\mu_0}\right) T_{11}^{(0)}(\mu;\mu_0) = 2\{X_l(\mu)\,Y_l(\mu_0)[1-\nu_4(\mu_0-\mu)-\mu\mu_0] \\ -Y_l(\mu)\,X_l(\mu_0)[1+\nu_4(\mu_0-\mu)-\mu\mu_0] \\ +\nu_3(\mu_0-\mu)[X_l(\mu)\,X_l(\mu_0)+Y_l(\mu)\,Y_l(\mu_0)]\},\end{aligned}\right\} \quad (543)$$

$$\left.\begin{aligned}\left(\frac{1}{\mu}-\frac{1}{\mu_0}\right) T_{12}^{(0)}(\mu;\mu_0) = \left[\frac{C_{0,l}^2(0)-C_{1,l}^2(0)}{C_{0,r}^2(0)-C_{1,r}^2(0)}\right]^{\frac{1}{2}} (\mu_0-\mu) \\ \times\{-u_2[X_l(\mu)\,Y_r(\mu_0)+Y_l(\mu)\,X_r(\mu_0)] \\ +u_1[X_l(\mu)\,X_r(\mu_0)+Y_l(\mu)\,Y_r(\mu_0)] \\ -Q(u_1-u_2)\mu_0[X_l(\mu)+Y_l(\mu)][X_r(\mu_0)-Y_r(\mu_0)]\},\end{aligned}\right\} \quad (544)$$

$$\left.\begin{aligned}\left(\frac{1}{\mu}-\frac{1}{\mu_0}\right) T_{21}^{(0)}(\mu;\mu_0) = \left[\frac{C_{0,r}^2(0)-C_{1,r}^2(0)}{C_{0,l}^2(0)-C_{1,l}^2(0)}\right]^{\frac{1}{2}} (\mu_0-\mu) \\ \times\{\nu_2[X_r(\mu)\,Y_l(\mu_0)+Y_r(\mu)\,X_l(\mu_0)] \\ -\nu_1[X_r(\mu)\,X_l(\mu_0)+Y_r(\mu)\,Y_l(\mu_0)] \\ +Q(\nu_1-\nu_2)\mu[X_r(\mu)-Y_r(\mu)][X_l(\mu_0)+Y_l(\mu_0)]\},\end{aligned}\right\} \quad (545)$$

$$\left(\frac{1}{\mu}-\frac{1}{\mu_0}\right)T_{22}^{(0)}(\mu;\mu_0) = X_r(\mu)\,Y_r(\mu_0)\,[1+u_4(\mu_0-\mu)-u_5\mu\mu_0] - Y_r(\mu)\,X_r(\mu_0)\,[1-u_4(\mu_0-\mu)-u_5\mu\mu_0] - u_3(\mu_0-\mu)\,[X_r(\mu)\,X_r(\mu_0)+Y_r(\mu)\,Y_r(\mu_0)] + Qu_5\mu\mu_0(\mu_0-\mu)\,[X_r(\mu)-Y_r(\mu)]\,[X_r(\mu_0)-Y_r(\mu_0)] - Q(u_3-u_4)\,\{\mu^2[X_r(\mu)-Y_r(\mu)]\,[X_r(\mu_0)+Y_r(\mu_0)] - \mu_0^2[X_r(\mu)+Y_r(\mu)]\,[X_r(\mu_0)-Y_r(\mu_0)]\}; \tag{546}$$

$$\begin{aligned} S_{31}^{(0)} &= S_{13}^{(0)} = S_{32}^{(0)} = S_{23}^{(0)} = S_{33}^{(0)} = 0\,, \\ T_{31}^{(0)} &= T_{13}^{(0)} = T_{32}^{(0)} = T_{23}^{(0)} = T_{33}^{(0)} = 0\,; \end{aligned} \tag{547}$$

$$\left(\frac{1}{\mu}+\frac{1}{\mu_0}\right)\boldsymbol{S}^{(1)}(\mu,\varphi;\mu_0,\varphi_0) = [X^{(1)}(\mu)\,X^{(1)}(\mu_0) - Y^{(1)}(\mu)\,Y^{(1)}(\mu_0)] \times\begin{pmatrix} -4\mu\mu_0\cos(\varphi-\varphi_0) & 0 & -2\mu\sin(\varphi-\varphi_0) \\ 0 & 0 & 0 \\ -2\mu_0\sin(\varphi-\varphi_0) & 0 & \cos(\varphi-\varphi_0) \end{pmatrix}; \tag{548}$$

$$\left(\frac{1}{\mu}-\frac{1}{\mu_0}\right)\boldsymbol{T}^{(1)}(\mu,\varphi;\mu_0,\varphi_0) = [X^{(1)}(\mu)\,Y^{(1)}(\mu_0) - Y^{(1)}(\mu)\,X^{(1)}(\mu_0)] \times\begin{pmatrix} 4\mu\mu_0\cos(\varphi-\varphi_0) & 0 & 2\mu\sin(\varphi-\varphi_0) \\ 0 & 0 & 0 \\ -2\mu_0\sin(\varphi-\varphi_0) & 0 & \cos(\varphi-\varphi_0) \end{pmatrix}; \tag{549}$$

$$\left(\frac{1}{\mu}+\frac{1}{\mu_0}\right)\boldsymbol{S}^{(2)}(\mu,\varphi;\mu_0,\varphi_0) = [X^{(2)}(\mu)\,X^{(2)}(\mu_0) - Y^{(2)}(\mu)\,Y^{(2)}(\mu_0)] \times\begin{pmatrix} \mu^2\mu_0^2\cos 2(\varphi-\varphi_0) & -\mu^2\cos 2(\varphi-\varphi_0) & \mu^2\mu_0\sin 2(\varphi-\varphi_0) \\ -\mu_0^2\cos 2(\varphi-\varphi_0) & \cos 2(\varphi-\varphi_0) & -\mu_0\sin 2(\varphi-\varphi_0) \\ \mu\mu_0^2\sin 2(\varphi-\varphi_0) & -\mu\sin 2(\varphi-\varphi_0) & -\mu\mu_0\cos 2(\varphi-\varphi_0) \end{pmatrix}; \tag{550}$$

and

$$\left(\frac{1}{\mu}-\frac{1}{\mu_0}\right)\boldsymbol{T}^{(2)}(\mu,\varphi;\mu_0,\varphi_0) = [X^{(2)}(\mu)\,Y^{(2)}(\mu_0) - Y^{(2)}(\mu)\,X^{(2)}(\mu_0)] \times\begin{pmatrix} \mu^2\mu_0^2\cos 2(\varphi-\varphi_0) & -\mu^2\cos 2(\varphi-\varphi_0) & \mu^2\mu_0\sin 2(\varphi-\varphi_0) \\ -\mu_0^2\cos 2(\varphi-\varphi_0) & \cos 2(\varphi-\varphi_0) & -\mu_0\sin 2(\varphi-\varphi_0) \\ -\mu\mu_0^2\sin 2(\varphi-\varphi_0) & \mu\sin 2(\varphi-\varphi_0) & \mu\mu_0\cos 2(\varphi-\varphi_0) \end{pmatrix}. \tag{551}$$

Now, Helmholtz' principle of reciprocity as reformulated in Paper XIII (§ 4) requires that the scattering and the transmission matrices have the following property for transposition:

$$\left.\begin{aligned} \boldsymbol{S}(\mu, \varphi; \mu_0, \varphi_0) &= \tilde{\boldsymbol{S}}(\mu_0, \varphi; \mu, \varphi_0) \\ \text{and} \qquad \boldsymbol{T}(\mu, \varphi; \mu_0, \varphi_0) &= \tilde{\boldsymbol{T}}(\mu_0, \varphi_0; \mu, \varphi)\,. \end{aligned}\right\} \tag{552}$$

From equations (548)–(551) it is evident that the matrices $\boldsymbol{S}^{(1)}$, $\boldsymbol{S}^{(2)}$, $\boldsymbol{T}^{(1)}$, and $\boldsymbol{T}^{(2)}$ have the required symmetries. But it is not at once apparent that the matrices $\boldsymbol{S}^{(0)}$ and $\boldsymbol{T}^{(0)}$ are in conformity with the reciprocity principle; for, though the diagonal elements of $\boldsymbol{S}^{(0)}$ and $\boldsymbol{T}^{(0)}$ clearly satisfy the necessary conditions, for the nondiagonal elements the validity of the principle requires that (cf. eqs. [540] and [541] or [543] and [544])

$$\left[\frac{C_{0,r}^2(0) - C_{1,r}^2(0)}{C_{0,l}^2(0) - C_{1,l}^2(0)}\right]^{\frac{1}{2}} \nu_i = -\left[\frac{C_{0,l}^2(0) - C_{1,l}^2(0)}{C_{0,r}^2(0) - C_{1,r}^2(0)}\right]^{\frac{1}{2}} u_i \qquad (i = 1, 2)\,. \tag{553}$$

From equations (405), (406), (510), and (511) defining the constants ν_i amd u_i, it is seen that the condition (553) is equivalent to

$$\frac{1}{2}\left[\frac{C_{0,r}^2(0) - C_{1,r}^2(0)}{C_{0,l}^2(0) - C_{1,l}^2(0)}\right]^{\frac{1}{2}} (\gamma_1\gamma_4 + \gamma_2\gamma_3) = -\left[\frac{C_{0,l}^2(0) - C_{1,l}^2(0)}{C_{0,r}^2(0) - C_{1,r}^2(0)}\right]^{\frac{1}{2}} (c_1 c_4 + c_2 c_3) \tag{554}$$

or

$$\frac{\gamma_1\gamma_4 + \gamma_2\gamma_3}{c_1 c_4 + c_2 c_3} = -2\,\frac{C_{0,l}^2(0) - C_{1,l}^2(0)}{C_{0,r}^2(0) - C_{1,r}^2(0)}\,. \tag{555}$$

On the other hand, from the definitions of the various constants γ_i (eq. [396]), it readily follows that

$$\gamma_1\gamma_4 + \gamma_2\gamma_3 = 2\,[C_{0,l}(+1)\,C_{0,l}(-1) - C_{1,l}(+1)\,C_{1,l}(-1)]\,; \tag{556}$$

or, using the identity (105) satisfied by the C-functions in general, we have

$$\gamma_1\gamma_4 + \gamma_2\gamma_3 = 2\,[C_{0,l}^2(0) - C_{1,l}^2(0)]\,\Omega(1)\,. \tag{557}$$

Similarly,

$$c_1 c_4 + c_2 c_3 = 2\,[C_{0,r}^2(0) - C_{1,r}^2(0)]\,W(1)\,. \tag{558}$$

Hence (cf. Paper XI, eq. [133]),

$$\frac{\gamma_1\gamma_4 + \gamma_2\gamma_3}{c_1 c_4 + c_2 c_3} = \frac{C_{0,l}^2(0) - C_{1,l}^2(0)}{C_{0,r}^2(0) - C_{1,r}^2(0)}\,\frac{\Omega(1)}{W(1)} = -2\,\frac{C_{0,l}^2(0) - C_{1,l}^2(0)}{C_{0,r}^2(0) - C_{1,r}^2(0)}\,, \tag{559}$$

in agreement with equation (555). Thus our solution for $\boldsymbol{S}$ and $\boldsymbol{T}$ is in conformity with the requirements of the reciprocity principle.

2. *Radiative Transfer Allowing for Polarization of Scattering Light*

7

ON THE RADIATIVE EQUILIBRIUM OF A STELLAR ATMOSPHERE. X

S. CHANDRASEKHAR
Yerkes Observatory
Received February 12, 1946

ABSTRACT

In this paper a detailed theory of the radiative equilibrium of an atmosphere in which the Thomson scattering by free electrons governs the transfer of radiation is developed. In particular, allowance has been made for the polarization of the scattered radiation; and the equations of transfer for the intensities I_l and I_r, referring, respectively, to the two states of polarization in which the electric vector vibrates in the meridian plane and at right angles to it, are separately formulated. The equations of transfer are found to be

$$\mu \frac{dI_l}{d\tau} = I_l - \frac{3}{8}\left\{2\int_{-1}^{+1} I_l(\tau, \mu')(1-\mu'^2)d\mu' + \mu^2\int_{-1}^{+1} I_l(\tau, \mu')(3\mu'^2-2)d\mu' + \mu^2\int_{-1}^{+1} I_r(\tau, \mu')d\mu'\right\}$$

and

$$\mu \frac{dI_r}{d\tau} = I_r - \frac{3}{8}\left\{\int_{-1}^{+1} I_r(\tau, \mu')d\mu' + \int_{-1}^{+1} I_l(\tau, \mu')\mu'^2 d\mu'\right\}.$$

These equations have been solved in a general *n*th approximation, and their explicit numerical forms have been found in the second and the third approximations.

It is found that the theory predicts different laws of darkening for the two states of polarization distinguished by I_l and I_r. The emergent radiation is therefore polarized, and it is further predicted that the degree of polarization must vary from zero at the center of the disk to 11 per cent at the limb.

1. *Introduction.*—It is now generally recognized that the Thomson scattering by free electrons must play an important role in the transfer of radiation in the atmospheres of the early-type stars.[1] Thus, it has been suggested by J. L. Greenstein[2] that the absolute-magnitude effect among the early-types stars shown by the discontinuity at the head of the Balmer series is probably to be attributed to the increased importance of electron scattering as we go to the more luminous stars. But it does not seem to have been observed so far that, if electron scattering is as major a factor as there appears to be evidence for, then there is an associated effect which should be detectable, namely, the polarization of the continuous radiation. It is the object of this paper to analyze this effect theoretically and to show that it is within the possibilities of detection. The detailed theory of radiative transfer in an atmosphere in which electron scattering plays the principal part, developed in this paper, predicts a degree of polarization to the extent of 11 per cent at the limb. In practice, this effect may be partially masked by other sources of continuous opacity; but it would seem, from all the available evidence, that the effect is probably present in detectable amounts in the early-type stars. Moreover, it would appear that the most favorable conditions under which the phenomenon could be observed are during the phases close to the primary minimum in an eclipsing binary, one component of which is an early-type star.

On the theoretical side, the problem discussed in this paper provides the first example in which the equations of transfer for the two states of polarization have been explicitly formulated and solved.[3]

[1] A. Unsöld, *Zs. f. Ap.*, **21**, 229, 1942; M. Rudkjøbing, *Zs. f. Ap.*, **21**, 254, 1942.

[2] *Ap. J.*, **95**, 299, 1942.

[3] The method of solution adopted is those of the earlier papers of this series. Familiarity with papers II and III (*Ap. J.*, **100**, 76, 117, 1944) is necessary to follow the analysis of this paper.

2. *The equations of transfer for the two components of polarization in an atmosphere in which the scattering by free electrons governs the transfer of radiation.*—We shall consider the radiative equilibrium of a semi-infinite plane-parallel atmosphere with a constant net flux of radiation and in which the transfer of radiation takes place in accordance with Thomson's laws of scattering by free electrons. It is apparent that under these conditions we can characterize the field of radiation by the intensities $I_l(z, \vartheta)$ and $I_r(z, \vartheta)$ at height z and inclined at an angle ϑ to the positive normal and referring to the states of polarization in which the electric vectors vibrate along the principal meridian and at right angles to it, respectively. And we require to write down the equations of transfer for the two components separately. For this purpose we shall first formulate the laws of scattering in the form we shall need them.

According to Thomson's classical theory[4] of the scattering by free electrons, the amount of radiation (initially unpolarized) which is scattered (per unit time) in a direction inclined at an angle Θ to the direction of incidence, and confined to an element of solid angle $d\omega'$ and per free electron, is given by

$$\frac{8\pi}{3}\frac{e^4}{m^2c^4} I\left\{\tfrac{3}{4}(1+\cos^2\Theta)\frac{d\omega'}{4\pi}\right\}, \tag{1}$$

where I denotes the intensity[5] of the incident radiation, e the charge of the electron, m its mass, and c the velocity of light. Moreover, the scattered radiation is polarized with the direction of the electric vector perpendicular to the *plane of scattering*.[6] For our purposes, however, we need a more detailed formulation of the law of scattering which will allow us to take into account a partial polarization of the incident light. We shall formulate these more detailed laws in the following manner:

Let $I_\perp$ denote the intensity of radiation plane-polarized with the electric vector perpendicular to the plane of scattering. Then the amount of radiation scattered in a direction inclined at an angle Θ to direction of $I_\perp$ and confined to an element of solid angle $d\omega'$ and per free electron is

$$\frac{8\pi}{3}\frac{e^4}{m^2c^4} I_\perp\left(\frac{3}{2}\frac{d\omega'}{4\pi}\right). \tag{2}$$

The scattered radiation is also polarized with the electric vector perpendicular to the plane of scattering. On the other hand, if the incident light is polarized with the electric vector parallel to the plane of scattering, then the scattered radiation is also polarized in the same way, but its amount is now given by (cf. eq.[2])

$$\frac{8\pi}{3}\frac{e^4}{m^2c^4} I_{||}\left(\frac{3}{2}\cos^2\Theta\frac{d\omega'}{4\pi}\right), \tag{3}$$

where $I_{||}$ denotes the intensity of the incident polarized radiation. More generally, if the incident light is plane-polarized with its electric vector inclined to the plane of scattering by an angle α, then the amount of scattered radiation (under the same circumstances to which equations [2] and [3] refer) is given by

$$\frac{8\pi}{3}\frac{e^4}{m^2c^4} I\left\{\tfrac{3}{2}(\sin\alpha+\cos\alpha\cos\Theta)^2\frac{d\omega'}{4\pi}\right\}, \tag{4}$$

[4] Cf. A. H. Compton and S. K. Allison, *X-Rays in Theory and Experiment*, pp. 117–119, New York: D. Van Nostrand, 1935.

[5] Since the Thomson scattering coefficient is independent of wave length, we can directly consider the intensity I integrated over all the frequencies.

[6] This is the plane which contains the directions of the incident and the scattered light.

while its plane of polarization is such that the electric vector is inclined to the plane of scattering by an angle

$$\beta = \tan^{-1}(\tan\alpha \sec\Theta)\,. \tag{5}$$

Thus there is, in general, a "turning" of the plane of polarization. We shall now show how, with the laws of scattering formulated in this manner, we can obtain the equations of transfer for the two components of polarization distinguished by I_l and I_r.

It is first evident that the equations of transfer must be expressible in the forms

$$\cos\vartheta \frac{dI_l}{d\tau} = I_l - \mathfrak{J}_{l,l}(\tau,\vartheta) - \mathfrak{J}_{r,l}(\tau,\vartheta) \tag{6}$$

and

$$\cos\vartheta \frac{dI_r}{d\tau} = I_r - \mathfrak{J}_{l,r}(\tau,\vartheta) - \mathfrak{J}_{r,r}(\tau,\vartheta)\,, \tag{7}$$

where τ denotes the optical depth, measured in terms of the Thomson scattering coefficient

$$\sigma = \frac{8\pi}{3}\frac{e^4}{m^2c^4}N_e\,. \tag{8}$$

with N_e denoting the number of electrons per unit mass. Further, in equation (6), $\mathfrak{J}_{l,l}$ and $\mathfrak{J}_{r,l}$ are the contributions to the source function for the radiation in a particular direction and polarized with the electric vector in the meridian plane, arising from the scattering from all other directions of radiations polarized with the electric vectors parallel, respectively, perpendicular to the appropriate meridian planes. Similarly, in equation (7), $\mathfrak{J}_{l,r}$ and $\mathfrak{J}_{r,r}$ are the contributions to the source function for the radiation in a particular direction polarized with the electric vector perpendicular to the meridian plane, arising from the scattering from all other directions of radiations polarized with the electric vectors parallel, respectively, perpendicular to the appropriate meridian planes.

To evaluate $\mathfrak{J}_{l,l}(\tau,\vartheta)$ and $\mathfrak{J}_{l,r}(\tau,\vartheta)$, we consider the contributions to these source functions for the radiation in the direction $(\vartheta, 0)$, say, arising from the scattering of radiation of intensity $I_l(\tau,\vartheta')$ in the direction (ϑ', φ') and polarized with the electric vector in the meridian plane through (ϑ', φ'). Let ξ_l denote a quantity proportional to the amplitude of the electric vector such that

$$I_l = \xi_l^2\,. \tag{9}$$

(We shall refer, quite generally, to ξ's defined in this manner as simply the *amplitude*.) The components of the amplitude that are parallel, respectively, perpendicular to the plane of scattering, are

$$\xi_l \cos i_1 \qquad \text{and} \qquad \xi_l \sin i_1\,, \tag{10}$$

where i_1 denotes the angle between the meridian plane OZP_1 through $P_1 = (\vartheta', \varphi')$ and the plane of scattering OP_1P_2 ($P_2 = [\vartheta, 0]$) (see Fig. 1). When this radiation is scattered into the direction OP_2, the components of the scattered amplitude that are parallel, respectively, perpendicular to the plane OP_1P_2, are proportional to

$$\xi_l \cos i_1 \cos\Theta \qquad \text{and} \qquad \xi_l \sin i_1\,, \tag{11}$$

while the amplitudes in the meridian plane OZX through P_2, and at right angles to it, are, respectively, proportional to

$$\xi_l(\sin i_1 \sin i_2 - \cos i_1 \cos i_2 \cos\Theta) \tag{12}$$

and

$$-\xi_l(\sin i_1 \cos i_2 + \sin i_2 \cos i_1 \cos\Theta)\,. \tag{13}$$

where i_2 denotes the angle between the planes OZX and OP_1P_2. The contributions to the source functions $\mathfrak{J}_{l,l}(z, \vartheta, 0)$ $(= \mathfrak{J}_{l,l}(z, \vartheta))$ and $\mathfrak{J}_{l,r}(z, \vartheta, 0)$ $(= \mathfrak{J}_{l,r}(z, \vartheta))$, arising from the scattering of the radiation $I_l(z, \vartheta', \varphi')$ $(= I_l(z, \vartheta'))$ in the direction (ϑ', φ') and confined to an element of solid angle $d\omega'$ are, therefore,

$$d\mathfrak{J}_{l,l} = \tfrac{3}{2} I_l(\tau, \vartheta')(\sin i_1 \sin i_2 - \cos i_1 \cos i_2 \cos \Theta)^2 \frac{d\omega'}{4\pi} \tag{14}$$

and

$$d\mathfrak{J}_{l,r} = \tfrac{3}{2} I_l(\tau, \vartheta')(\sin i_1 \cos i_2 + \sin i_2 \cos i_1 \cos \Theta)^2 \frac{d\omega'}{4\pi}. \tag{15}$$

Hence,

$$\mathfrak{J}_{l,l} = \frac{3}{8\pi} \int_0^\pi \int_0^{2\pi} I_l(\tau, \vartheta')(\sin i_1 \sin i_2 - \cos i_1 \cos i_2 \cos \Theta)^2 \sin \vartheta' d\vartheta' d\varphi' \tag{16}$$

and

$$\mathfrak{J}_{l,r} = \frac{3}{8\pi} \int_0^\pi \int_0^{2\pi} I_l(\tau, \vartheta')(\sin i_1 \cos i_2 + \sin i_2 \cos i_1 \cos \Theta)^2 \sin \vartheta' d\vartheta' d\varphi'. \tag{17}$$

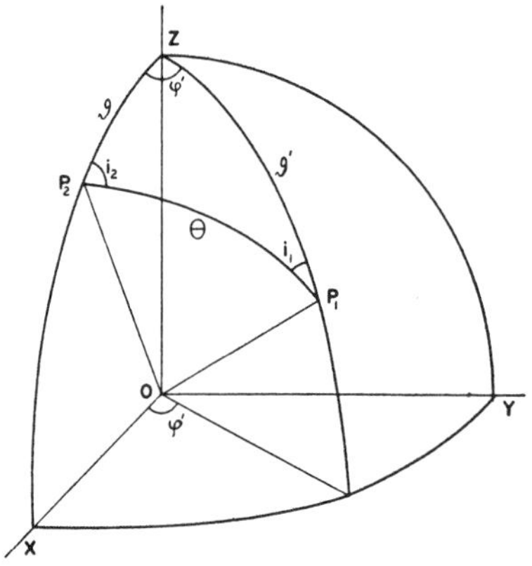

FIG. 1

On the other hand, from the spherical triangle ZP_1P_2 we have

$$\sin \varphi' \cos \vartheta' = \cos i_2 \sin i_1 + \sin i_2 \cos i_1 \cos \Theta\,. \tag{18}$$

Equation (17) accordingly reduces to the form

$$\mathfrak{J}_{l,r} = \frac{3}{8\pi} \int_0^\pi \int_0^{2\pi} I_l(\tau, \vartheta') \sin^2 \varphi' \cos^2 \vartheta' \sin \vartheta' d\vartheta' d\varphi'; \tag{19}$$

or, in view of the axial symmetry of the radiation field about the z-axis, in our problem, we can write

$$\mathfrak{J}_{l,r}(\tau, \mu) = \frac{3}{8} \int_{-1}^{+1} I_l(\tau, \mu') \mu'^2 d\mu', \tag{20}$$

where we have used μ and μ' to denote $\cos \vartheta$ and $\cos \vartheta'$, respectively.

From equations (16) and (17) we have

$$\mathfrak{J}_{l,l} + \mathfrak{J}_{l,r} = \frac{3}{8\pi} \int_0^\pi \int_0^{2\pi} I_l(\tau, \vartheta')(\sin^2 i_1 + \cos^2 i_1 \cos^2 \Theta) \sin \vartheta' d\vartheta' d\varphi'; \tag{21}$$

or, since

$$\left.\begin{aligned}\sin^2 i_1 + \cos^2 i_1 \cos^2 \Theta &= 1 - \cos^2 i_1 \sin^2 \Theta \\ &= 1 - (\cos\vartheta \sin\vartheta' - \sin\vartheta \cos\vartheta' \cos\varphi')^2 ,\end{aligned}\right\} \tag{22}$$

we have

$$\mathfrak{J}_{l,l} + \mathfrak{J}_{l,r} = \frac{3}{8}\int_{-1}^{+1} I_l(\tau, \mu')\,[2 - \mu'^2 + \mu^2(3\mu'^2 - 2)]\,d\mu' . \tag{23}$$

From equations (20) and (23) we now obtain

$$\mathfrak{J}_{l,l}(\tau, \mu) = \frac{3}{8}\int_{-1}^{+1} I_l(\tau, \mu')\,[2(1 - \mu'^2) + \mu^2(3\mu'^2 - 2)]\,d\mu' . \tag{24}$$

To determine the source functions $\mathfrak{J}_{r,l}$ and $\mathfrak{J}_{r,r}$, we proceed along similar lines, considering radiation of intensity $I_r(\tau, \vartheta', \varphi')$ $(= I_r(\tau, \vartheta'))$ in the direction (ϑ', φ') and polarized with the electric vector perpendicular to the meridian plane OZP_1 and evaluating its contribution to the source function for the radiation in the direction $(\vartheta, 0)$. Let ξ_r denote the amplitude corresponding to the intensity $I_r(\tau, \vartheta', \varphi')$. Its components, parallel, respectively, perpendicular to the scattering plane, are

$$\xi_r \sin i_1 \qquad \text{and} \qquad \xi_r \cos i_1 . \tag{25}$$

When this radiation is scattered into the direction OP_2, the components of the scattered amplitude, parallel, respectively, perpendicular to the plane OP_1P_2, are proportional to

$$\xi_r \sin i_1 \cos\Theta \qquad \text{and} \qquad \xi_r \cos i_1 . \tag{26}$$

The amplitudes in the meridian plane OZX and at right angles to it are, respectively, proportional to

$$\xi_r (\sin i_1 \cos i_2 \cos\Theta + \cos i_1 \sin i_2) = \xi_r \sin\varphi' \cos\vartheta \tag{27}$$

and

$$\xi_r (\sin i_1 \sin i_2 \cos\Theta - \cos i_1 \cos i_2) = \xi_r \cos\varphi' . \tag{28}$$

Accordingly, the contributions to the source functions $\mathfrak{J}_{r,l}(\tau, \vartheta, 0)$ and $\mathfrak{J}_{r,r}(\tau, \vartheta, 0)$, arising from the scattering of the radiation $I_r(\tau, \vartheta', \varphi')$ in the direction (ϑ', φ') and confined to an element of solid angle $d\omega'$, are given by

$$d\mathfrak{J}_{r,l} = \tfrac{3}{2} I_r(\tau, \vartheta')(\sin i_1 \cos i_2 \cos\Theta + \cos i_1 \sin i_2)^2 \frac{d\omega'}{4\pi} \tag{29}$$

and

$$d\mathfrak{J}_{r,r} = \tfrac{3}{2} I_r(\tau, \vartheta') \cos^2\varphi' \frac{d\omega'}{4\pi} . \tag{30}$$

Hence,

$$\mathfrak{J}_{r,r} = \frac{3}{8\pi}\int_0^{\pi}\int_0^{2\pi} I_r(\tau, \vartheta') \cos^2\varphi' \sin\vartheta' d\vartheta' d\varphi' , \tag{31}$$

or

$$\mathfrak{J}_{r,r}(\tau, \mu) = \frac{3}{8}\int_{-1}^{+1} I_r(\tau, \mu')\,d\mu' . \tag{32}$$

On the other hand (cf. eqs. [27] and [28]),

$$\mathfrak{J}_{r,r} + \mathfrak{J}_{r,l} = \frac{3}{8\pi}\int_0^{\pi}\int_0^{2\pi} I_r(\tau, \vartheta')(\sin^2 i_1 \cos^2\Theta + \cos^2 i_1) \sin\vartheta' d\vartheta' d\varphi' ; \tag{33}$$

or, since

$$\sin^2 i_1 \cos^2\Theta + \cos^2 i_1 = 1 - \sin^2 i_1 \sin^2\Theta = 1 - \sin^2\vartheta \sin^2\varphi' . \tag{34}$$

we have

$$\mathfrak{J}_{r,r}+\mathfrak{J}_{r,l}=\frac{3}{8}\int_{-1}^{+1} I_r(\tau,\mu')(1+\mu^2)\,d\mu' . \tag{35}$$

From equations (32) and (35) we now obtain

$$\mathfrak{J}_{r,l}(\tau,\mu) = \tfrac{3}{8}\mu^2 \int_{-1}^{+1} I_r(\tau,\mu')\,d\mu' . \tag{36}$$

Combining equations (6), (24), and (36) and similarly equations (7), (20), and (32), we obtain

$$\left.\begin{aligned}\mu\frac{dI_l}{d\tau}=I_l-\tfrac{3}{8}\Big\{2\int_{-1}^{+1} I_l(\tau,\mu')(1-\mu'^2)\,d\mu'+\mu^2\int_{-1}^{+1} I_l(\tau,\mu')(3\mu'^2-2)\,d\mu'\\ +\mu^2\int_{-1}^{+1} I_r(\tau,\mu')\,d\mu'\Big\}\end{aligned}\right\} \tag{37}$$

and

$$\mu\frac{dI_r}{d\tau}=I_r-\tfrac{3}{8}\Big\{\int_{-1}^{+1} I_r(\tau,\mu')\,d\mu'+\int_{-1}^{+1} I_l(\tau,\mu')\mu'^2 d\mu'\Big\}, \tag{38}$$

which are the required equations of transfer for I_l and I_r.

We may note here that, for radiation initially unpolarized, the source functions for radiations polarized in the two ways can be obtained from equations (20), (24), (32), and (36) by setting $I_l = I_r = \frac{1}{2}I$. Thus,

$$\mathfrak{J}_l=\frac{3}{16}\Big\{2\int_{-1}^{+1} I(\tau,\mu')(1-\mu'^2)\,d\mu'+\mu^2\int_{-1}^{+1} I(\tau,\mu')(3\mu'^2-1)\,d\mu'\Big\} \tag{39}$$

and

$$\mathfrak{J}_r=\frac{3}{16}\int_{-1}^{+1} I(\tau,\mu')(1+\mu'^2)\,d\mu', \tag{40}$$

which agree with A. Schuster's well-known formulae.[7]

In terms of the quantities J and K, defined in the usual manner (cf. paper III, eq. [6]), we can re-write the equations of transfer (37) and (38) in the following forms:

$$\mu\frac{dI_l}{d\tau}=I_l-\tfrac{3}{4}\Big\{2(J_l-K_l)+\mu^2(3K_l-2J_l+J_r)\Big\} \tag{41}$$

and

$$\mu\frac{dI_r}{d\tau}=I_r-\tfrac{3}{4}(J_r+K_l) . \tag{42}$$

From the equations of transfer in the foregoing forms we can readily establish the flux integral

$$F_l+F_r=F=2\int_{-1}^{+1}[I_l(\tau,\mu)+I_r(\tau,\mu)]\,\mu d\mu=\text{constant} \tag{43}$$

and the "K-integral"

$$K_l+K_r=\tfrac{1}{4}F(\tau+Q) \tag{44}$$

where Q is a constant.

3. *The general solution of the equations of transfer in the* n*th approximation.*—In solving equations (37) and (38) we shall follow the method developed in the earlier papers of this series and replace the various integrals which occur on the right-hand sides of the equations by sums according to Gauss's formula for numerical quadratures. Thus, in

[7] *M.N.*, **40**, 35, 1879; see also M. Minnaert, *Zs. f. Ap.*, **1**, 209, 1930, and H. Zanstra, *M.N.*, **101**, 250, 1941.

the nth approximation, equations (37) and (38) are replaced by the following system of $4n$ linear equations:

$$\left.\begin{array}{r}\mu_i \dfrac{dI_{l,i}}{d\tau} = I_{l,i} - \tfrac{3}{8}\left[2\Sigma a_j I_{l,j}(1-\mu_j^2) + \mu_i^2\{\Sigma a_j I_{l,j}(3\mu_j^2 - 2) + \Sigma a_j I_{r,j}\}\right] \\ (i = \pm 1, \ldots, \pm n)\end{array}\right\} \quad (45)$$

and

$$\mu_i \frac{dI_{r,i}}{d\tau} = I_{r,i} - \tfrac{3}{8}(\Sigma a_j I_{r,j} + \Sigma a_j I_{l,j}\mu_j^2) \qquad (i = \pm 1, \ldots, \pm n), \quad (46)$$

where the μ_i's ($i = \pm 1, \ldots, \pm n$) are the zeros of the Legendre polynomial of order $2n$ and the a_i's are the appropriate Gaussian weights. Further, in equations (45) and (46) we have written $I_{l,i}$ and $I_{r,i}$ for $I_l(\tau, \mu_i)$ and $I_r(\tau, \mu_i)$, respectively.

We shall now find the different linearly independent solutions of equations (45) and (46) and later, by combining these, obtain the general solution.

First, we seek a solution of equations (45) and (46) of the form

$$I_{l,i} = g_i e^{-k\tau} \quad \text{and} \quad I_{r,i} = h_i e^{-k\tau} \qquad (i = \pm 1, \ldots, \pm n), \quad (47)$$

where the g_i's, h_i's, and k are constants, for the present unspecified. Substituting the foregoing forms for $I_{l,i}$ and $I_{r,i}$ in equations (45) and (46), we obtain

$$g_i(1 + \mu_i k) = \tfrac{3}{8}\left[2\Sigma a_j g_j(1 - \mu_j^2) + \mu_i^2\{\Sigma a_j g_j(3\mu_j^2 - 2) + \Sigma a_j h_j\}\right] \quad (48)$$

and

$$h_i(1 + \mu_i k) = \tfrac{3}{8}\left[\Sigma a_j g_j \mu_j^2 + \Sigma a_j h_j\right]. \quad (49)$$

Equations (48) and (49) imply that g_i and h_i must be expressible in the forms

$$g_i = \frac{\alpha\mu_i^2 + \beta}{1 + \mu_i k} \quad \text{and} \quad h_i = \frac{\gamma}{1 + \mu_i k} \qquad (i = \pm 1, \ldots, \pm n), \quad (50)$$

where α, β, and γ are certain constants, independent of i. Inserting the solution (50) back into equations (48) and (49), we find

$$\left.\begin{array}{l}\alpha\mu_i^2 + \beta = \tfrac{3}{8}\left[2\{\alpha(D_2 - D_4) + \beta(D_0 - D_2)\} + \mu_i^2\{\alpha(3D_4 - 2D_2)\right. \\ \qquad\qquad\qquad\qquad\qquad\qquad \left. + \beta(3D_2 - 2D_0) + \gamma D_0\}\right]\end{array}\right\} \quad (51)$$

and

$$\gamma = \tfrac{3}{8}(\alpha D_4 + \beta D_2 + \gamma D_0), \quad (52)$$

where we have introduced the quantities D_0, D_2, and D_4, defined according to the formula

$$D_m = \Sigma \frac{a_j \mu_j^m}{1 + \mu_j k}. \quad (53)$$

Since equations (51) and (52) are valid for all i, we must require that

$$\tfrac{8}{3}\alpha = \alpha(3D_4 - 2D_2) + \beta(3D_2 - 2D_0) + \gamma D_0, \quad (54)$$

$$\tfrac{4}{3}\beta = \alpha(D_2 - D_4) + \beta(D_0 - D_2), \quad (55)$$

and

$$\tfrac{8}{3}\gamma = \alpha D_4 + \beta D_2 + \gamma D_0. \quad (56)$$

It is seen that equations (54), (55), and (56), together, represent a system of homogeneous linear equations for α, β, and γ. The determinant of this system must, therefore, be required to vanish. Thus,

$$\begin{vmatrix} 3D_4 - 2D_2 - \frac{8}{3} & 3D_2 - 2D_0 & D_0 \\ D_2 - D_4 & D_0 - D_2 - \frac{4}{3} & 0 \\ D_4 & D_2 & D_0 - \frac{8}{3} \end{vmatrix} = 0 . \tag{57}$$

Expanding this determinant by the elements of the last column, we find, after some minor reductions, that

$$D_0(\tfrac{4}{3} - D_0 + 2D_2 - D_4) - (\tfrac{32}{9} - \tfrac{8}{3}D_0 + \tfrac{16}{3}D_2 - 4D_4 + D_0D_4 - D_2^2) = 0 . \tag{58}$$

To simplify equation (58) still further, we must make use of certain relations which can be derived from the recursion formulae[8]

$$D_{2m} = \frac{1}{k^2}\left(D_{2m-2} - \frac{2}{2m-1}\right) \tag{59}$$

and

$$D_{2m-1} = -kD_{2m}, \tag{60}$$

which the D's satisfy. From equation (59) we infer, in particular, that

$$D_2 = \frac{1}{k^2}(D_0 - 2) \tag{61}$$

and

$$D_4 = \frac{1}{k^2}(D_2 - \tfrac{2}{3}) = \frac{1}{k^4}(D_0 - 2) - \frac{2}{3k^2}. \tag{62}$$

A further relation which follows from equations (61) and (62) is

$$D_2(D_2 - \tfrac{2}{3}) = D_4(D_0 - 2), \tag{63}$$

or

$$D_0D_4 - D_2^2 = 2D_4 - \tfrac{2}{3}D_2 . \tag{64}$$

By using equation (64), equation (58) reduces to

$$D_0(\tfrac{4}{3} - D_0 + 2D_2 - D_4) - (\tfrac{32}{9} - \tfrac{8}{3}D_0 + \tfrac{14}{3}D_2 - 2D_4) = 0 . \tag{65}$$

Now, substituting for D_2 and D_4 according to equations (61) and (62) in terms of D_0 in equation (65), we find, after some simplification, that

$$\frac{1}{k^4}(D_0 - 2)^2 - \frac{2}{k^2}(D_0 - 2)^2 + D_0^2 - 4D_0 + \tfrac{32}{9} = 0 . \tag{66}$$

Again using equations (61) and (62), we can re-write the foregoing equation in the form

$$D_2^2 - 2D_2(D_0 - 2) + D_0^2 - 4D_0 + \tfrac{32}{9} = 0 \tag{67}$$

or

$$(D_0 - D_2)^2 - 4(D_0 - D_2) + \tfrac{32}{9} = 0 . \tag{68}$$

[8] Cf. *Ap. J.*, **101**, 328, 1945 (eqs. [54] and [56]).

Equation (68) is equivalent to

$$(D_0 - D_2 - \tfrac{8}{3})(D_0 - D_2 - \tfrac{4}{3}) = 0 . \tag{69}$$

In other words, either

$$D_0 - D_2 = \Sigma \frac{a_j(1-\mu_j^2)}{1+\mu_j k} = \frac{8}{3} \qquad \text{(case 1)} \tag{70}$$

or

$$D_0 - D_2 = \Sigma \frac{a_j(1-\mu_j^2)}{1+\mu_j k} = \frac{4}{3} \qquad \text{(case 2)} . \tag{71}$$

Equations (70) and (71) can be written alternatively in the forms

$$\sum_{j=1}^{n} \frac{a_j(1-\mu_j^2)}{1-\mu_j^2 k^2} = \frac{4}{3} \qquad \text{(case 1)} \tag{72}$$

and

$$\sum_{j=1}^{n} \frac{a_j(1-\mu_j^2)}{1-\mu_j^2 k^2} = \frac{2}{3} \qquad \text{(case 2)} . \tag{73}$$

And k^2 must be a root of either of the two foregoing equations.

Equation (72) is of order n in k^2 and admits $2n$ distinct nonvanishing roots for k which occur in pairs as

$$\pm k_a \qquad (a = 1, \ldots, n) . \tag{74}$$

On the other hand, equation (73), though of order $2n$ in k^2, admits of only $(2n - 2)$ distinct nonvanishing roots for k, since $k^2 = 0$ is a root.[9] However, these $2n - 2$ roots also occur in pairs, which we shall denote by

$$\pm \kappa_\beta \qquad (\beta = 1 \ldots, n-1) , \tag{75}$$

to distinguish them from the roots of equation (72).

Case 1: k^2 *a root of equation* (72).—In this case, $D_0 - D_2 = 8/3$; and from equations (61) and (62) we readily find that

$$D_0 = \frac{2}{3}\frac{4k^2-3}{k^2-1}; \qquad D_2 = \frac{2}{3(k^2-1)}; \qquad D_4 = \frac{2}{3k^2}\frac{2-k^2}{k^2-1} . \tag{76}$$

With these values for D_0, D_2, and D_4, equations (55) and (56) lead to

$$\alpha = -k^2\beta \tag{77}$$

and

$$\gamma = -(k^2-1)\beta . \tag{78}$$

Accordingly, equations (45) and (46) allow $2n$ linearly independent integrals of the form

$$\left.\begin{aligned} I_{l,i} &= \text{constant}\,(1 \mp k_a\mu_i)\,e^{\mp k_a\tau} \\ I_{r,i} &= -\,\text{constant}\,\frac{k_a^2-1}{1 \pm k_a\mu_i}\,e^{\mp k_a\tau} \end{aligned}\right\}, \qquad \left\{\begin{aligned} &(i = \pm 1, \ldots, \pm n) \\ &(a = 1, \ldots, n) \end{aligned}\right\} . \tag{79}$$

[9] Note that $\sum_{j=1}^{n} a_j(1-\mu_j^2) = \frac{2}{3}$.

Case 2: κ^2 a root of equation (73).—In this case, $D_0 - D_2 = 4/3$; and equations (61) and (62) now give

$$D_0 = \frac{2}{3}\frac{2\kappa^2 - 3}{\kappa^2 - 1}; \qquad D_2 = D_4 = -\frac{2}{3(\kappa^2 - 1)}. \tag{80}$$

With these values of D_0, D_2, and D_4 it is seen that equation (55) is satisfied identically, while the consideration of equations (54) and (56) leads to the result that

$$a = -\beta \quad \text{and} \quad \gamma = 0 . \tag{81}$$

Accordingly, equations (45) and (46) admit of $(2n - 2)$ linearly independent integrals of the form

$$I_{l,i} = \text{constant}\,\frac{1 - \mu_i^2}{1 \pm \mu_i \kappa_\beta}\, e^{\mp\kappa_\beta\tau} \qquad \left\{ \begin{array}{l} (i = \pm 1, \ldots, \pm n) \\ (\beta = 1, \ldots, n-1) \end{array} \right\} \tag{82}$$

and

$$I_{r,i} \equiv 0 \qquad (i = \pm 1, \ldots, \pm n). \tag{83}$$

To complete the solution, we verify that equations (45) and (46) also admit the solution

$$I_{l,i} = I_{r,i} = b(\tau + \mu_i + Q) \qquad (i = \pm 1, \ldots, \pm n), \tag{84}$$

where b and Q are arbitrary constants.

Combining the solutions (79), (82), (83), and (84), we observe that the general solution of the system of equations (45) and (46) can be written in the forms

$$\left.\begin{aligned} I_{l,i} = b\Big\{\tau + \mu_i + Q + (1 - \mu_i^2)\sum_{\beta=1}^{n-1}\left(\frac{L_\beta e^{-\kappa_\beta\tau}}{1 + \mu_i\kappa_\beta} + \frac{L_{-\beta} e^{+\kappa_\beta\tau}}{1 - \mu_i\kappa_\beta}\right) \\ + \sum_{a=1}^{n} M_a(1 - k_a\mu_i)\, e^{-k_a\tau} + \sum_{a=1}^{n-1} M_{-a}(1 + k_a\mu_i)\, e^{+k_a\tau}\Big\} \\ (i = \pm 1, \ldots, \pm n) \end{aligned}\right\} \tag{85}$$

and

$$\left.\begin{aligned} I_{r,i} = b\Big\{\tau + \mu_i + Q - \sum_{a=1}^{n} M_a \frac{(k_a^2 - 1)}{1 + \mu_i k_a}\, e^{-k_a\tau} - \sum_{a=1}^{n} M_{-a}\frac{(k_a^2 - 1)}{1 - \mu_i k_a}\, e^{+k_a\tau}\Big\} \\ (i = \pm 1, \ldots, \pm n), \end{aligned}\right\} \tag{86}$$

where $L_{\pm\beta}$ $(\beta = 1, \ldots, n-1)$, $M_{\pm a}$ $(a = 1, \ldots, n)$, b, and Q are the $4n$ constants of integration.

4. *The solution satisfying the necessary boundary conditions.*—The boundary conditions for the astrophysical problem on hand are that none of the I_i's tend to infinity exponentially as $\tau \to \infty$ and that there is no radiation incident on the surface $\tau = 0$. The first of these conditions implies that in the general solution (85) and (86) we omit all terms in $\exp(+k_a\tau)$ and $\exp(+\kappa_\beta\tau)$. We are thus left with

$$\left.\begin{aligned} I_{l,i} = b\Big\{\tau + \mu_i + Q + (1 - \mu_i^2)\sum_{\beta=1}^{n-1}\frac{L_\beta e^{-\kappa_\beta\tau}}{1 + \mu_i\kappa_\beta} + \sum_{a=1}^{n} M_a(1 - k_a\mu_i)\, e^{-k_a\tau}\Big\} \\ (i = \pm 1, \ldots, \pm n) \end{aligned}\right\} \tag{87}$$

and

$$I_{r,i} = b\Big\{\tau + \mu_i + Q - \sum_{a=1}^{n}\frac{M_a(k_a^2 - 1)}{1 + \mu_i k_a}\, e^{-k_a\tau}\Big\} \qquad (i = \pm 1, \ldots, \pm n). \tag{88}$$

Next, the nonexistence of any radiation incident on $\tau = 0$ requires that

$$I_{l,i} = I_{r,i} = 0 \quad \text{at} \quad \tau = 0 \quad \text{and for} \quad i = -1, \ldots, -n; \tag{89}$$

or, according to equations (86) and (87),

$$(1-\mu_i^2)\sum_{\beta=1}^{n-1}\frac{L_\beta}{1-\mu_i\kappa_\beta}+\sum_{\alpha=1}^{n}M_\alpha(1+k_\alpha\mu_i)-\mu_i+Q=0 \qquad (i=1,\ldots,n) \tag{90}$$

and

$$\sum_{\alpha=1}^{n}\frac{M_\alpha(k_\alpha^2-1)}{1-\mu_i k_\alpha}+\mu_i-Q=0 \qquad (i=1,\ldots,n). \tag{91}$$

These are the $2n$ equations which determine the $2n$ constants L_β ($\beta = 1, \ldots, n-1$), M_α ($\alpha = 1, \ldots = n$), and Q.[10] The constant b is left arbitrary and is related to the constant net flux of radiation in the atmosphere.

For, defining the net flux in terms of F_l and F_r where

$$F_q = 2\int_{-1}^{+1} I_q \mu\, d\mu \simeq 2\Sigma a_i I_{q,i}\mu_i \qquad (q = l, r), \tag{92}$$

we have, according to equations (87) and (88),

$$F_l = 2b\left\{\frac{2}{3}+\sum_{\beta=1}^{n-1}L_\beta[D_1(\kappa_\beta)-D_3(\kappa_\beta)]e^{-\kappa_\beta\tau}-\frac{2}{3}\sum_{\alpha=1}^{n}M_\alpha k_\alpha e^{-k_\alpha\tau}\right\} \tag{93}$$

and

$$F_r = 2b\left\{\frac{2}{3}-\sum_{\alpha=1}^{n}M_\alpha(k_\alpha^2-1)D_1(k_\alpha)e^{-k_\alpha\tau}\right\}. \tag{94}$$

On the other hand, from equations (60), (76), and (80) we conclude that

$$D_1(k_\alpha) = -k_\alpha D_2(k_\alpha) = -\frac{2}{3}\frac{k_\alpha}{k_\alpha^2-1} \tag{95}$$

and

$$D_1(\kappa_\beta)-D_3(\kappa_\beta) = -\kappa_\beta[D_2(\kappa_\beta)-D_4(\kappa_\beta)] = 0\,. \tag{96}$$

Hence,

$$F_l = \tfrac{4}{3}b\left(1-\sum_{\alpha=1}^{n}M_\alpha k_\alpha e^{-k_\alpha\tau}\right) \tag{97}$$

and

$$F_r = \tfrac{4}{3}b\left(1+\sum_{\alpha=1}^{n}M_\alpha k_\alpha e^{-k_\alpha\tau}\right). \tag{98}$$

From the two preceding equations we infer the constancy of the net flux. More particularly,

$$F = F_l + F_r = \tfrac{8}{3}b = \text{constant}\,. \tag{99}$$

[10] Adding equations (90) and (91), we obtain the equation

$$\sum_{\alpha=1}^{n}\frac{M_\alpha k_\alpha^2}{1-\mu_i k_\alpha}+\sum_{\beta=1}^{n}\frac{L_\beta}{1-\mu_i\kappa_\beta}=0 \qquad (i=1,\ldots,n),$$

which together with equation (91) is more convenient for the practical determination of these constants.

We can, therefore, re-write the solution (85) and (86) in the form

$$I_{l,i}=\tfrac{3}{8}F\left\{\tau+\mu_i+Q+(1-\mu_i^2)\sum_{\beta=1}^{n-1}\frac{L_\beta e^{-\kappa_\beta\tau}}{1+\mu_i\kappa_\beta}+\sum_{\alpha=1}^{n}M_\alpha(1-k_\alpha\mu_i)\,e^{-k_\alpha\tau}\right\} \quad (i=\pm1,\ldots,\pm n) \tag{100}$$

and

$$I_{r,i}=\tfrac{3}{8}F\left\{\tau+\mu_i+Q-\sum_{\alpha=1}^{n}\frac{M_\alpha(k_\alpha^2-1)}{1+\mu_i k_\alpha}\,e^{-k_\alpha\tau}\right\} \quad (i=\pm1,\ldots,\pm n). \tag{101}$$

In terms of the foregoing solutions for $I_{l,i}$ and $I_{r,i}$ we can readily establish the following formulae

$$J_l=\tfrac{1}{2}\Sigma a_j I_{l,j}=\tfrac{3}{8}F\left(\tau+Q+\frac{2}{3}\sum_{\beta=1}^{n-1}L_\beta e^{-\kappa_\beta\tau}+\sum_{\alpha=1}^{n}M_\alpha e^{-k_\alpha\tau}\right), \tag{102}$$

$$K_l=\tfrac{1}{2}\Sigma a_j I_{l,j}\mu_j^2=\tfrac{1}{8}F\left(\tau+Q+\sum_{\alpha=1}^{n}M_\alpha e^{-k_\alpha\tau}\right), \tag{103}$$

$$J_r=\tfrac{3}{8}F\left(\tau+Q-\frac{1}{3}\sum_{\alpha=1}^{n}M_\alpha[4k_\alpha^2-3]\,e^{-k_\alpha\tau}\right), \tag{104}$$

and

$$K_r=\tfrac{1}{8}F\left(\tau+Q-\sum_{\alpha=1}^{n}M_\alpha e^{-k_\alpha\tau}\right). \tag{105}$$

Now the source functions $\mathfrak{J}_l(\tau,\mu)$ and $\mathfrak{J}_r(\tau,\mu)$ for I_l and I_r are (cf., eqs. [41] and [42])

$$\mathfrak{J}_l(\tau,\mu)=\tfrac{3}{4}[2(J_l-K_l)+\mu^2(3K_l-2J_l+J_r)] \tag{106}$$

and

$$\mathfrak{J}_r(\tau,\mu)=\tfrac{3}{4}(J_r+K_l); \tag{107}$$

or, substituting for J_l, K_l, J_r, and K_r from equations (102)–(105), we find

$$\mathfrak{J}_l(\tau,\mu)=\tfrac{3}{8}F\left\{\tau+Q+(1-\mu^2)\sum_{\beta=1}^{n-1}L_\beta e^{-\kappa_\beta\tau}+\sum_{\alpha=1}^{n}M_\alpha(1-k_\alpha^2\mu^2)\,e^{-k_\alpha\tau}\right\} \tag{108}$$

and

$$\mathfrak{J}_r(\tau,\mu)=\tfrac{3}{8}F\left\{\tau+Q-\sum_{\alpha=1}^{n}M_\alpha(k_\alpha^2-1)\,e^{-k_\alpha\tau}\right\}. \tag{109}$$

With the explicit forms for the source functions now found, we can readily obtain formulae for the intensity distribution of the emergent radiation. For, since quite generally

$$I(0,\mu)=\int_0^\infty \mathfrak{J}(\tau,\mu)\,e^{-\tau/\mu}\frac{d\tau}{\mu}, \tag{110}$$

we have, in our present case,

$$I_l(0,\mu)=\tfrac{3}{8}F\left\{\mu+Q+(1-\mu^2)\sum_{\beta=1}^{n-1}\frac{L_\beta}{1+\kappa_\beta\mu}+\sum_{\alpha=1}^{n}M_\alpha(1-k_\alpha\mu)\right\} \tag{111}$$

and

$$I_r(0, \mu) = \tfrac{3}{8}F\left\{\mu + Q - \sum_{a=1}^{n}\frac{M_a(k_a^2 - 1)}{1 + k_a\mu}\right\}. \tag{112}$$

It is to be particularly noted that the foregoing expressions for $I_l(0, \mu)$ and $I_r(0, \mu)$ agree with the solution (100) and (101) for $\tau = 0$ at the points of the Gaussian division.

This completes the solution of the problem in the nth approximation.

5. *The numerical forms of the solution in the second and the third approximations; the degree of polarization of the emergent radiation.*—The consideration of the solution obtained in the preceding section in the first approximation is of no special interest except possibly to emphasize the importance of having a method which gives solutions to any desired degree of accuracy. For, in the first approximation ($n = 1$) there are no L's; and, though there is an M, this is also seen to be zero; and the only nonvanishing constant of integration is Q, which has the value $1/\sqrt{3}$. Accordingly, in this approximation (cf. II, eq. [35], and III eq. [72])

$$I_l(0, \mu) = I_r(0, \mu) = \tfrac{3}{8}F\left(\mu + \frac{1}{\sqrt{3}}\right). \tag{113}$$

It is, therefore, seen that the first approximation is far too crude to disclose the essentially finer features of our problem. We therefore proceed to the higher approximations.

i) *Second approximation.*—In this approximation it is found that

$$\kappa_1 = \sqrt{\tfrac{7}{3}}\ ; \qquad k_1 = \sqrt{5}; \qquad \text{and} \qquad k_2 = \sqrt{\tfrac{7}{6}}\ . \tag{114}$$

Further, the constants L_1, M_1, M_2, and Q have the values

$$\left.\begin{aligned} L_1 = -0.19265\ ; \quad M_1 = +0.021830\ ; \quad M_2 &= -0.029516\ ; \\ Q &= +0.69638\ ; \end{aligned}\right\} \tag{115}$$

and the laws of darkening for the emergent radiation in the two states of polarization take the forms

$$\left.\begin{aligned} I_l(0, \mu) = \tfrac{3}{8}F\Big\{\mu + 0.69638 - (1 - \mu^2)\frac{0.19265}{1 + 1.5275\mu} \\ + 0.021830(1 - 2.23607\mu) - 0.029516(1 - 1.08012\mu)\Big\} \end{aligned}\right\} \tag{116}$$

and

$$I_r(0, \mu) = \tfrac{3}{8}F\left\{\mu + 0.69638 - \frac{0.0873215}{1 + 2.23607\mu} + \frac{0.0049193}{1 + 1.08012\mu}\right\}. \tag{117}$$

Values of $I_l(0, \mu)$ and $I_r(0, \mu)$ obtained[11] from the preceding formulae are given in Table 1.

ii) *Third approximation.*—In this approximation the characteristic equations for κ^2 and k^2 are

$$\kappa^4 - 8.64\kappa^2 + 9.24 = 0 \tag{118}$$

and

$$k^6 - 14.82k^4 + 36.12k^2 - 23.1 = 0\ . \tag{119}$$

The characteristic roots are

$$\left.\begin{aligned} \kappa_1 = 2.718381\ ; \quad \kappa_2 = 1.118216\ ; \quad k_1 &= 3.458589\ ; \\ k_2 = 1.327570\ ; \quad \text{and} \quad k_3 &= 1.046766\ . \end{aligned}\right\} \tag{120}$$

[11] I am indebted to Mrs. Frances H. Breen for assistance with these calculations.

The constants of integration L_1, L_2, Q, M_1, M_2, and M_3 are found to have the values

$$\left.\begin{array}{lll} L_1 = -0.1402646\,; & L_2 = -0.06791696\,; & Q = +0.705927\,, \\ M_1 = +0.00718392\,; & M_2 = +0.01861255\,; & M_3 = -0.0328664\,; \end{array}\right\} \quad (121)$$

and the laws of darkening for the emergent radiation in the two states of polarization take the forms

$$\left.\begin{aligned} I_l(0,\mu) = \tfrac{3}{8}F \Big\{ \mu + 0.705927 - (1-\mu^2)\Big[\frac{0.1402646}{1+2.718381\mu} \\ + \frac{0.06791696}{1+1.118216\mu}\Big] + 0.00718392(1-3.458589\mu) \\ + 0.01861255(1-1.327570\mu) - 0.0328664(1-1.046766\mu) \Big\} \end{aligned}\right\} \quad (122)$$

TABLE 1

THE LAW OF DARKENING IN THE EMERGENT RADIATION IN THE TWO STATES OF POLARIZATION GIVEN BY THE SECOND APPROXIMATION

μ	$I_l(0,\mu)/F$	$I_r(0,\mu)/F$	$I_l(0,\mu)/I_l(0,1)$	$I_r(0,\mu)/I_r(0,1)$
0.	0.1860	0.2302	0.2967	0.3673
0.1	.2331	.2735	0.3718	0.4363
0.2	.2789	.3150	0.4448	0.5025
0.3	.3238	.3554	0.5165	0.5670
0.4	.3681	.3951	0.5871	0.6303
0.5	.4119	.4344	0.6570	0.6929
0.6	.4553	.4733	0.7263	0.7549
0.7	.4985	.5119	0.7952	0.8166
0.8	.5415	.5504	0.8637	0.8779
0.9	.5843	.5887	0.9320	0.9391
1.0	0.6269	0.6269	1.0000	1.0000

and

$$\left.\begin{aligned} I_r(0,\mu) = \tfrac{3}{8}F \Big\{ \mu + 0.705927 - \frac{0.0787490}{1+3.458589\mu} - \frac{0.01419099}{1+1.327570\mu} \\ + \frac{0.00314593}{1+1.0467659\mu} \Big\}. \end{aligned}\right\} \quad (123)$$

Values of $I_l(0, \mu)$ and $I_r(0, \mu)$ obtained from the foregoing formulae are given in Table 2. Comparison with the values given in Table 1 indicates that the solution obtained in the third approximation is probably accurate to within 1 per cent over the entire range of the variables.

In Figure 2 we have illustrated the laws of darkening on the third approximation for the intensities $I_l(0, \mu)$ and $I_r(0, \mu)$. It is seen that, while they are equal at the center of the disk ($\mu = 1$), they differ by about 25 per cent at the limit ($\mu = 0$). The theory, therefore, predicts a polarization of the emergent radiation. And the degree of polarization $\delta(\mu)$, defined by

$$\delta(\mu) = \frac{I_r(0,\mu) - I_l(0,\mu)}{I_r(0,\mu) + I_l(0,\mu)}, \quad (124)$$

varies from 0 at $\mu = 1$ to 11 per cent at $\mu = 0$ (see Table 2). It is not impossible that this predicted polarization of the radiation of the early-type stars (in which scattering by free electrons is believed to play an important part in the transfer of radiation) could be detected under suitably favorable conditions.

One further comparison is of interest. In an earlier paper (paper III) we have worked out the theory of radiative transfer in an atmosphere in which radiation is scattered in accordance with Rayleigh's phase function. This is not a strictly correct procedure, inasmuch as no allowance is made for the polarization of the existing radiation field. However, a comparison (see Table 2) of the emergent intensity $I(0, \mu)/F$ derived on the theory of transfer incorporating Rayleigh's phase function with the total emergent in-

TABLE 2

THE LAWS OF DARKENING IN THE TWO STATES OF POLARIZATION GIVEN BY THE THIRD APPROXIMATION; THE DEGREE OF POLARIZATION OF THE EMERGENT RADIATION; COMPARISON OF THE TOTAL INTENSITIES GIVEN BY THE THEORY IGNORING THE POLARIZATION OF THE EXISTING FIELD OF RADIATION BUT INCORPORATING RAYLEIGH'S PHASE FUNCTION

μ	$\frac{I_l(0,\mu)}{F}$	$\frac{I_r(0,\mu)}{F}$	$\frac{I_l(0,\mu)}{I_l(0,1)}$	$\frac{I_r(0,\mu)}{I_r(0,1)}$	$\frac{I_r(0,\mu)}{I_l(0,\mu)}$	$\delta(\mu)$	$\frac{I_l(0,\mu)+I_r(0,\mu)}{F}$	$\frac{I(0,\mu)}{F}$ for Rayleigh's Phase Function
0	0.1840	0.2310	0.2914	0.3659	1.2557	0.1134	0.4151	0.4195
0.1	.2354	.2767	0.3728	0.4382	1.1753	.0806	0.5120	0.5175
0.2	.2832	.3190	0.4486	0.5053	1.1264	.0594	0.6023	0.6076
0.3	.3291	.3598	0.5213	0.5699	1.0932	.0445	0.6890	0.6937
0.4	.3738	.3997	0.5921	0.6330	1.0691	.0334	0.7735	0.7773
0.5	.4178	.4390	0.6616	0.6953	1.0508	.0248	0.8567	0.8593
0.6	.4611	.4779	0.7303	0.7569	1.0364	.0179	0.9390	0.9402
0.7	.5041	.5165	0.7983	0.8181	1.0247	.0122	1.0206	1.0203
0.8	.5467	.5549	0.8659	0.8789	1.0150	.0075	1.1017	1.0998
0.9	.5891	.5932	0.9331	0.9396	1.0069	0.0034	1.1824	1.1789
1.0	0.6314	0.6314	1.0000	1.0000	1.0000	0	1.2628	1.2576

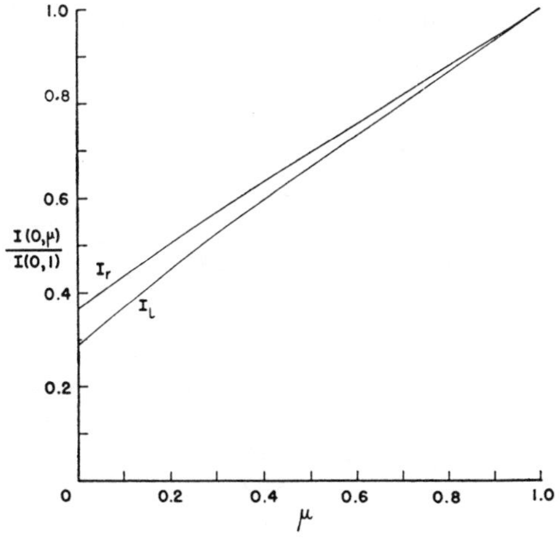

FIG. 2.—The laws of darkening in the two states of polarization. The symbol I_l refers to the component polarized with the electric vector in the meridian plane, while I_r refers to the component polarized with the electric vector at right angles to the meridian plane.

tensity $[I_l(0, \mu) + I_r(0, \mu)]/F$ given by our present more exact theory shows that the errors made in the *total intensities* by ignoring the polarization of the radiation field are small.

6. *The reduction of the laws of darkening in the two states of polarization to certain closed forms.*—In § 4 we derived expressions for the angular distribution of the emergent radiation in the two states of polarization (eqs. [111] and [112]). These expressions involve certain constants of integration ($2n$ of them in the nth approximation), and it would appear that these have to be evaluated before the solutions can be brought to their numerical forms. However, we shall now show how, for the purposes of characterizing the emergent radiation, we can avoid the necessity of solving explicitly for these constants by expressing the laws of darkening in forms in which they require only a knowledge of characteristic roots k_a and κ_β.

First, we may observe that equations (90) and (91), which determine the constants of integration, can be re-written as

$$S_l(\mu_i) = 0 \quad \text{and} \quad S_r(\mu_i) = 0 \qquad (i = 1, \ldots, n), \tag{125}$$

where

$$S_l(\mu) = (1 - \mu^2) \sum_{\beta=1}^{n-1} \frac{L_\beta}{1 - \mu\kappa_\beta} + \sum_{a=1}^{n} M_a (1 + k_a \mu) - \mu + Q \tag{126}$$

and

$$S_r(\mu) = -\mu + Q - \sum_{a=1}^{n} \frac{M_a (k_a^2 - 1)}{1 - \mu k_a}. \tag{127}$$

In terms of these same functions the angular distribution of the emergent radiation in the two states of polarization can also be expressed. For, according to equations (111) and (112), we can write

$$I_l(0, \mu) = \tfrac{3}{8} F S_l(-\mu) \quad \text{and} \quad I_r(0, \mu) = \tfrac{3}{8} F S_r(-\mu). \tag{128}$$

We shall now show how explicit expressions for the functions $S_l(\mu)$ and $S_r(\mu)$ can be obtained.

First, we shall define the functions $R(\mu)$, $R_a(\mu)$, $\rho(\mu)$ and $\rho_\beta(\mu)$ according to the formulae

$$R(\mu) = \prod_{a=1}^{n} (1 - k_a \mu); \qquad R_a(\mu) = \prod_{a \neq a} (1 - k_a \mu), \tag{129}$$

and

$$\rho(\mu) = \prod_{\beta=1}^{n-1} (1 - \kappa_\beta \mu); \qquad \rho_\beta(\mu) = \prod_{b \neq \beta} (1 - \kappa_b \mu). \tag{130}$$

Considering, now, the function $S_l(\mu)$, we see that $R(\mu)S_l(\mu)$ is a polynomial of degree n in μ which vanishes for $\mu = \mu_i$ $(i = 1, \ldots, n)$. We must, accordingly, have a relation of the form

$$S_l(\mu) = \sigma \frac{P(\mu)}{\rho(\mu)}, \tag{131}$$

where σ is a constant and

$$P(\mu) = \prod_{i=1}^{n} (\mu - \mu_i). \tag{132}$$

The function $S_l(\mu)$ is therefore determinate apart from a constant of proportionality. Apart from this same constant of proportionality, the constants L_β can also be determined. For, according to equations (126) and (131),

$$\left(1-\frac{1}{\kappa_\beta^2}\right)L_\beta=\sigma\frac{P(1/\kappa_\beta)}{\rho_\beta(1/\kappa_\beta)}\qquad(\beta=1,\ldots,n-1).\quad(133)$$

Moreover, setting $\mu=+1$, respectively, -1, in equations (126) and (131) we obtain

$$S_l(+1)=\sum_{a=1}^{n}M_a(1+k_a)-1+Q=\sigma\frac{P(1)}{\rho(1)},\qquad(134)$$

and

$$S_l(-1)=\sum_{a=1}^{n}M_a(1-k_a)+1+Q=\sigma\frac{P(-1)}{\rho(-1)}.\qquad(135)$$

Adding and subtracting these two equations, we have

$$\sum_{a=1}^{n}M_a=\alpha\sigma-Q\qquad(136)$$

and

$$\sum_{a=1}^{n}M_ak_a=\beta\sigma+1,\qquad(137)$$

where we have written

$$\alpha=\frac{1}{2}\left[\frac{P(1)}{\rho(1)}+\frac{P(-1)}{\rho(-1)}\right]\quad\text{and}\quad\beta=\frac{1}{2}\left[\frac{P(1)}{\rho(1)}-\frac{P(-1)}{\rho(-1)}\right].\qquad(138)$$

Considering, next, the function $S_r(\mu)$, we observe that we must have a proportionality of the form

$$R(\mu)S_r(\mu)\propto P(\mu)(\mu+c),\qquad(139)$$

where c is a constant, since the quantity on the right-hand side is a polynomial of degree $n+1$ in μ and has the zeros $\mu=\mu_i$ $(i=1,\ldots,n)$. The constant of proportionality in equation (139) can be found from a comparison of the coefficients of the highest powers of μ on either side. In this manner we find that

$$S_r(\mu)=(-1)^{n+1}k_1\ldots k_n\frac{P(\mu)}{R(\mu)}(\mu+c).\qquad(140)$$

From equations (127) and (140) we now conclude that

$$M_a=(-1)^n k_1\ldots k_n\frac{P(1/k_a)}{R_a(1/k_a)(k_a^2-1)}\left(\frac{1}{k_a}+c\right)\qquad(a=1,\ldots,n).\quad(141)$$

Also, setting $\mu=0$ in equations (127) and (140), we have

$$\sum_{a=1}^{n}M_a(k_a^2-1)-Q=k_1\ldots k_n\mu_1\ldots\mu_n c.\qquad(142)$$

On the other hand, from the characteristic equation (72) for k we infer that

$$k_1 \ldots . k_n \mu_1 \ldots . \mu_n = \frac{1}{\sqrt{2}}. \tag{143}$$ [12]

Hence,

$$\sum_{a=1}^{n} M_a (k_a^2 - 1) = Q + \frac{c}{\sqrt{2}}. \tag{144}$$

Adding equations (136) and (144), we have

$$\sum_{a=1}^{n} M_a k_a^2 = \alpha\sigma + \frac{c}{\sqrt{2}}. \tag{145}$$

Finally, substituting for M_a according to equation (141), we can re-write equations (136) and (137) in the forms

$$\xi_{-1} + c\,\xi_0 = \alpha\sigma - Q\,. \tag{146}$$

$$\xi_0 + c\,\xi_1 = \beta\sigma + 1\,, \tag{147}$$

and

$$\xi_1 + c\,\xi_2 = \alpha\sigma + \frac{c}{\sqrt{2}}, \tag{148}$$

where

$$\xi_m = (-1)^n k_1 \ldots . k_n \sum_{a=1}^{n} \frac{P(1/k_a)\, k_a^m}{R_a(1/k_a)(k_a^2 - 1)} \qquad (m = -1, 0, 1, 2). \tag{149}$$

To evaluate the sum occurring on the right-hand side of equation (149), we introduce the function

$$f_m(x) = \sum_{a=1}^{n} \frac{P(1/k_a)\, k_a^m}{R_a(1/k_a)(k_a^2 - 1)} R_a(x) \qquad (m = -1, 0, 1, 2), \tag{149a}$$

and express ξ_m in terms of it. Thus,

$$\xi_m = (-1)^n k_1 \ldots . k_n f_m(0)\,. \tag{150}$$

[12] This relation follows most readily from the characteristic equation written in the form

$$\sum_{j=0}^{n} p_{2j} \Delta_{2j} = 0\,,$$

where the p_{2j}'s are the coefficients of μ^{2j} in the Legendre polynomial $P_{2n}(\mu)$ and

$$\Delta_{2j} = \sum_{i=1}^{n} \frac{a_i(1 - \mu_i^2)\mu_i^{2j}}{1 - k^2\mu_i^2}.$$

The Δ's defined in this manner satisfy the recursion formula

$$\Delta_{2j} = \frac{1}{k^2}\left(\Delta_{2j-2} - \frac{2}{4j^2 - 1}\right).$$

For the characteristic equation (72), $\Delta_0 = \frac{4}{3}$ and $\Delta_2 = 2/3k^2$. From the recursion formula we therefore conclude that Δ_{2n} starts with $2/3k^{2n}$. The equation for k must accordingly have the form

$$\tfrac{4}{3} p_0 k^{2n} + \ldots . + \tfrac{2}{3} p_{2n} = 0\,.$$

Hence,

$$k_1^2 \ldots . k_n^2 = \frac{(-1)^n p_{2n}}{2p_0} = \frac{1}{2\mu_1^2 \ldots . \mu_n^2}.$$

Now, $f_m(x)$ defined as in equation (149) is a polynomial of degree $(n-1)$ in x, which takes the values

$$\frac{P(1/k_a)\,k_a^m}{k_a^2-1} \tag{151}$$

for $x=1/k_a$, $(a=1,\ldots\ldots,n)$. In other words,

$$(1-x^2)\,f_m(x)-x^{2-m}P(x)=0 \quad \text{for} \quad x=\frac{1}{k_a} \quad \text{and} \quad a=1,\ldots\ldots n. \tag{152}$$

The polynomial on the right-hand side of equation (152) must therefore divide $R(x)$. There must, accordingly, exist a relation of the form

$$(1-x^2)\,f_m(x)-x^{2-m}P(x)=R(x)\Psi(x), \tag{153}$$

where $\Psi(x)$ is a polynomial of degree 3, 2, 1, or 1 in x for $m=-1,0,1$, or 2, respectively. To determine $\Psi(x)$ more explicitly we must consider each case separately. We shall illustrate the procedure for the case $m=-1$.

For $m=-1$, equation (153) becomes

$$(1-x^2)\,f_{-1}(x)-x^3P(x)=R(x)(A_{-1}x^3+B_{-1}x^2+C_{-1}x+D_{-1}), \tag{154}$$

where A_{-1}, B_{-1}, C_{-1}, and D_{-1} are certain constants to be determined. The constants A_{-1} and B_{-1} readily follow from a comparison of the coefficients of x^{n+3} and x^{n+2} on either side of equation (154). In this manner we find

$$A_{-1}=\frac{(-1)^{n+1}}{k_1\ldots\ldots k_n} \quad \text{and} \quad B_{-1}=\frac{(-1)^n}{k_1\ldots\ldots k_n}\left(\sum_{j=1}^{n}\mu_j-\sum_{a=1}^{n}\frac{1}{k_a}\right). \tag{155}$$

Next, putting $x=+1$, respectively -1, in equation (154), we have

$$A_{-1}+B_{-1}+C_{-1}+D_{-1}=-\frac{P(1)}{R(1)}, \tag{156}$$

and

$$-A_{-1}+B_{-1}-C_{-1}+D_{-1}=+\frac{P(-1)}{R(-1)}. \tag{157}$$

These equations determine the remaining constants C_{-1} and D_{-1}. In particular,

$$D_{-1}=f_{-1}(0)=-b+\frac{(-1)^{n+1}}{k_1\ldots\ldots k_n}\left(\sum_{j=1}^{n}\mu_j-\sum_{a=1}^{n}\frac{1}{k_a}\right), \tag{158}$$

where (cf. eq. [138])

$$b=\frac{1}{2}\left[\frac{P(1)}{R(1)}-\frac{P(-1)}{R(-1)}\right]. \tag{159}$$

From equation (150) we now conclude that

$$\xi_{-1}=(-1)^{n+1}k_1\ldots\ldots k_n b-\left(\sum_{j=1}^{n}\mu_j-\sum_{a=1}^{n}\frac{1}{k_a}\right). \tag{160}$$

The evaluation of ξ_0, ξ_1, and ξ_2 proceeds along similar lines. We find

$$\xi_0=(-1)^{n+1}k_1\ldots\ldots k_n a+1, \tag{161}$$

$$\xi_1=(-1)^{n+1}k_1\ldots\ldots k_n b, \tag{162}$$

and

$$\xi_2=(-1)^{n+1}k_1\ldots\ldots k_n a+\frac{1}{\sqrt{2}}. \tag{163}$$

In equations (161) and (163) a stands for (cf. eq. [138])

$$a=\frac{1}{2}\left[\frac{P(1)}{R(1)}+\frac{P(-1)}{R(-1)}\right]. \tag{164}$$

With the foregoing expressions for ξ's, equations (147) and (148) become

$$a+bc=\frac{(-1)^{n+1}}{k_1\ldots\ldots k_n}\beta\sigma, \tag{165}$$

and

$$b+ac=\frac{(-1)^{n+1}}{k_1\ldots\ldots k_n}\alpha\sigma. \tag{166}$$

These equations determine c and σ. We find

$$c=-\frac{\alpha a-\beta b}{\alpha b-\beta a}, \tag{167}$$

and

$$\sigma=(-1)^n k_1\ldots\ldots k_n\frac{a^2-b^2}{\alpha b-\beta a}. \tag{168}$$

Equation (146) now determines Q. After some minor reductions we find that

$$Q=-c+\sum_{j=1}^{n}\mu_j-\sum_{\alpha=1}^{n}\frac{1}{k_\alpha}. \tag{169}$$

Finally, substituting for c and σ according to equations (167) and (168) in equations (131) and (140), we obtain

$$S_l(\mu)=\frac{1}{\sqrt{2}}\frac{a^2-b^2}{\alpha b-a\beta}\frac{(-1)^n}{\mu_1\ldots\ldots\mu_n}\frac{P(\mu)}{\rho(\mu)}, \tag{170}$$

and

$$S_r(\mu)=\frac{1}{\sqrt{2}}\frac{(-1)^{n+1}P(\mu)}{\mu_1\ldots\ldots\mu_n R(\mu)}\left(\mu-\frac{\alpha a-\beta b}{\alpha b-\beta a}\right). \tag{171}$$

The laws of darkening now follow according to equation (128).

We may note here that in the third approximation

$$\sigma=-3.3351 \quad \text{and} \quad c=-0.87134 .$$

7. *Concluding remarks.*—The successful solution of a specific problem in theory of radiative transfer distinguishing the different states of polarization justifies the hope that it will be possible to solve other astrophysical problems in which polarization plays a significant role. Thus it may be expected that a theory of diffuse reflection along the lines of an earlier paper of this series[13] but incorporating the polarization of the existing field of diffuse radiation will account, in a general way, for the remarkable observations of Lyot[14] on the polarization of the reflected light from Venus. We hope to return to these and similar essentially more difficult problems in the theory of radiative transfer in the near future.

I am indebted to Dr. G. Herzberg for helpful discussions on some of the physical aspects of the problem considered in this paper.

Note added May 6: The problem of diffuse reflection from a semi-infinite plane-parallel atmosphere, allowing for the partial polarization of the diffuse radiation has now been solved. It is hoped to publish the results of this investigation in the near future.

[13] *Ap. J.*, **103,** 165, 1946.

[14] *Ann. Obs. Meudon* (Paris), **8,** 66, 1929.

8

ON THE RADIATIVE EQUILIBRIUM OF A STELLAR ATMOSPHERE. XI

S. Chandrasekhar
Yerkes Observatory
Received May 13, 1946

ABSTRACT

In this paper the general equations of transfer allowing for the polarization of the scattered radiation in accordance with Rayleigh's law are formulated. It is shown that, for a general, partially plane-polarized radiation field, equations of transfer are required not only for the intensities I_l and I_r in two directions at right angles to each other in the plane of the electric and the magnetic vectors but also for a third quantity, $U = (I_l - I_r) \tan 2\chi$, where χ denotes the inclination of the plane of polarization to the direction to which I_l refers. This last quantity allows for the turning of the plane of polarization on scattering. It is further shown that, when partially plane-polarized light is scattered by an element of gas, the scattered radiation in any given direction can be analyzed as a mixture of ***independent*** plane-polarized streams: and the law of composition of such streams is formulated, following an early investigation of Stokes.

The equations of transfer appropriate to the problem of diffuse reflection of a parallel unpolarized beam by a semi-infinite plane-parallel atmosphere are solved in a general nth approximation, and explicit formulae are derived for the characterization of the state of polarization and the angular distribution of the reflected radiation. Tables of the necessary functions (four of them) are provided in the second and the third approximations.

1. *Introduction.*—In Paper X[1] of this series the theory of radiative transfer in a semi-infinite plane-parallel atmosphere, allowing for the polarization of the scattered radiation in accordance with Rayleigh's law,[2] has been worked out. However, under the circumstances in which the problem was formulated, there was an essential simplification arising from the axial symmetry of the radiation field at each point about the normal to the plane of stratification. For, under conditions of axial symmetry, the radiation field can be characterized completely by the specification of the intensities I_l and I_r referring, respectively, to the two states of polarization in which the electric vector vibrates in the meridian plane and at right angles to it. There is, however, a large class of problems in the theory of radiative transfer, as, for example, in the theory of diffuse reflection,[3] in which the axial symmetry of the radiation field is not present. The intensities I_l and I_r will not then completely suffice to characterize the radiation field. The question, therefore, arises as to how best we can characterize the radiation field under these more general conditions, in order that the relevant equations of transfer may be most conveniently formulated. This is a fundamental question; for on its answer will depend the solution of a variety of problems, including the important one of the illumination and the polarization of the sunlit sky. Indeed, it is somewhat surprising to find that even the basic equations of the problem have not so far been written down, though their necessity has been recognized since Lord Rayleigh[4] accounted in a general way for the color and polarization of the light from the sky.[5] In this paper we shall accordingly formulate the basic principles

[1] *Ap. J.*, **103**, 351, 1946. This paper will be referred to as "Paper X."

[2] Here and in the sequel we shall include under "Rayleigh's law" only that part of it which pertains to the state of polarization and the angular distribution of the scattered radiation. Therefore, the scattering by free electrons (in which context the problem was specifically formulated in paper X) also comes under the general scope of Rayleigh's law in this sense.

[3] *Ap. J.*, **103**, 165, 1946. This Paper will be referred to as "Paper IX."

[4] *Phil. Mag.*, **41**, 107, 274, 447, 1871; also *Scientific Papers of Lord Rayleigh*, **1**, 87, 104, 518, Cambridge, England, 1899. In this connection see also C. V. Raman, *Molecular Diffraction of Light*, chap. iii, University of Calcutta, 1922.

[5] E.g., L. V. King in his classic work, "On the Scattering and Absorption of Light in Gaseous Media with Applications to the Sky Radiation" (*Phil. Trans. Roy. Soc.*, A, **212**, 375, 1913), remarks:

"That portion of the sky radiation due to self-illumination is largely unpolarized and may to a large extent account for this deficiency from complete polarization: this point is mentioned by Rayleigh in his

on which the theory will have to be developed and then establish the equations of transfer which are valid for an atmosphere scattering radiation in accordance with Rayleigh's law. These equations will then be solved appropriately for the problem of diffuse reflection of a parallel beam of unpolarized radiation by a semi-infinite plane-parallel atmosphere. It may be stated here that for this latter problem simple closed expressions for the solution can be found in a general nth approximation. To reduce these solutions to their numerical forms it is only necessary to solve certain algebraic equations for characteristic roots.[6]

Applications of the theory developed in this paper to the interpretation of Lyot's observations[7] on the polarization of the light reflected from the various planets, particularly Venus, will be found in a forthcoming paper.

2. *The parametric representation of partially plane-polarized light and the composition of plane-polarized streams with no mutual phase relationships.*—The composition and resolution of streams of polarized light with no mutual phase relationships and the most convenient characterization of an arbitrarily polarized light was first considered by Sir George Stokes in a paper published in 1852.[8] This paper of Stokes is basic for our further considerations. However, the extreme generality of Stokes's considerations[9] are not essential for our purposes. It may, therefore, be useful to have them presented under the simplified conditions of our problem in which, as we shall see, only partially plane-polarized light need be considered.

Consider, then, a partially plane-polarized beam with the plane of polarization inclined at an angle χ to a certain fixed direction[10] l (say) and with intensities I_χ and $I_{\chi+\pi/2}$ in the directions[10] of maximum and minimum intensity. The intensity $I(\psi)$ in a direction inclined at an angle ψ to the direction of l is clearly

$$I(\psi) = I_\chi \cos^2(\psi - \chi) + I_{\chi+\pi/2} \sin^2(\psi - \chi) . \tag{1}$$

After some elementary transformations, the foregoing equation can be reduced to the form

$$\left.\begin{aligned} I(\psi) = \tfrac{1}{2}(I_\chi + I_{\chi+\pi/2}) + \tfrac{1}{2}(I_\chi - I_{\chi+\pi/2}) \cos 2\chi \cos 2\psi \\ + \tfrac{1}{2}(I_\chi - I_{\chi+\pi/2}) \sin 2\chi \sin 2\psi . \end{aligned}\right\} \tag{2}$$

On the other hand, if I_l and I_r denote the intensities in the direction l and in the direction r at right angles to it, we have

$$I_l = I_\chi \cos^2 \chi + I_{\chi+\pi/2} \sin^2 \chi \tag{3}$$

and

$$I_r = I_\chi \sin^2 \chi + I_{\chi+\pi/2} \cos^2 \chi . \tag{4}$$

1871 paper. . . . . The complete solution of the problem from this aspect would require us to split up the incident radiation into two components, one of which is polarized in the principal plane, the other at right angles to it: the effect of self-illumination would lead to two simultaneous integral equations in three variables, the solutions of which would be much too complicated to be useful."

Actually, the problem requires the solution of three simultaneous integral equations (and not two, as King mentions); but, even so, there is no difficulty in solving these equations in a manner appropriate to any problem and to any desired degree of accuracy.

[6] The degree of these equations depends on the order of the approximation in which the solutions are sought.

[7] *Ann. Obs. Meudon* (Paris), Vol. **8**, 1926.

[8] *Trans. Cambridge Phil. Soc.*, **9**, 399, 1852; or *Mathematical and Physical Papers of Sir George Stokes*, **1**, 233, Cambridge, England, 1901.

[9] It is surprising that these fundamental considerations of Stokes are not to be found in any of the standard books on physical optics. The only book which I have been able to discover in which the matter is considered is J. Walker, *The Analytical Theory of Light*, §§ 20–22, pp. 28–32, Cambridge, England, 1904.

[10] These directions are referred in the plane containing the electric and the magnetic vectors.

From these equations we conclude that

$$I_l + I_r = I_\chi + I_{\chi+\pi/2} \tag{5}$$

and

$$(I_l - I_r) = (I_\chi - I_{\chi+\pi/2}) \cos 2\chi \,. \tag{6}$$

Using these relations, we can re-write equation (2) in the form

$$I(\psi) = \tfrac{1}{2}(I_l + I_r) + \tfrac{1}{2}(I_l - I_r)\cos 2\psi + \tfrac{1}{2}(I_l - I_r)\tan 2\chi \sin 2\psi \,. \tag{7}$$

From equation (7) it follows that *optically equivalent*,[11] partially plane-polarized beams can be described completely in terms of the intensities I_l and I_r in two directions at right angles to each other and the further quantity,

$$U = (I_l - I_r)\tan 2\chi \,. \tag{8}$$

Conversely, whenever the intensity in a direction ψ of a polarized beam can be expressed in the form

$$I(\psi) = A + B\cos 2\psi + \tfrac{1}{2}U \sin 2\psi \,, \tag{9}$$

it represents a partially plane-polarized beam with intensities $A + B$ and $A - B$ in the directions l and r and with the plane of polarization inclined at an angle

$$\chi = \tfrac{1}{2}\tan^{-1}\frac{U}{2B} \tag{10}$$

to the direction of l.

We shall next examine what the result of composition of a number of streams of *plane*-polarized light with no mutual phase relationships is. Let $I^{(n)}$ denote the intensity of a typical stream and χ_n the inclination of its plane of polarization to the direction l. Further, let $I(\psi)$ denote the intensity of the mixture in the direction ψ. Since the different streams are assumed to have no phase relationships, it is evident that[12]

$$I(\psi) = \Sigma I^{(n)}(\psi) \,, \tag{11}$$

where $I^{(n)}(\psi)$ denotes the intensity of the stream $I^{(n)}$ in the direction ψ. Accordingly,

$$I(\psi) = \Sigma I^{(n)} \cos^2(\psi - \chi_n) \,, \tag{12}$$

or, after some minor reductions,

$$I(\psi) = \tfrac{1}{2}\Sigma I^{(n)} + \tfrac{1}{2}\Sigma I^{(n)}(\cos^2\chi_n - \sin^2\chi_n)\cos 2\psi + \tfrac{1}{2}\Sigma I^{(n)} \sin 2\chi_n \sin 2\psi \,, \tag{13}$$

Comparing this with equations (7) and (9), we conclude that the mixture of a number of independent plane-polarized streams is a partially plane-polarized beam. Further,

$$I_l = \Sigma I^{(n)} \cos^2\chi_n \,; \qquad I_r = \Sigma I^{(n)} \sin^2\chi_n \,; \tag{14}$$

and

$$U = (I_l - I_r)\tan 2\chi = \Sigma I^{(n)} \sin 2\chi_n \,. \tag{15}$$

In other words, the intensities I_l and I_r of the mixture are simply the sum of the intensities of the component streams in the directions l and r, while the plane of polarization of the resultant is determined by U, according to equation (15).

3. *The laws of scattering.*—We shall suppose that the laws of scattering as they pertain to the angular distribution and the state of polarization of the scattered radiation

[11] For a discussion of the concept of optical equivalence see Stokes (*op. cit.*).

[12] However, for an explicit demonstration see *ibid.*

are those of Rayleigh. For our purposes we may formulate them in the following manner (cf. Paper X, eqs. [1]–[3]):

If I denotes the intensity of a plane-polarized beam incident on an element of gas, then the amount of radiation scattered in a direction inclined at an angle Θ to the direction of incidence and confined to an element of solid angle $d\omega$ and per unit mass of the scattering material will be given by

$$\tfrac{3}{2}\sigma I \frac{d\omega}{4\pi} \quad \text{or} \quad \tfrac{3}{2}\sigma I \cos^2\Theta \frac{d\omega}{4\pi}, \tag{16}$$

depending on whether the incident radiation is polarized with the electric vector perpendicular or parallel to the plane of scattering[13] and where σ denotes the mass-scattering coefficient. More generally, if I_α denotes the intensity of an incident plane-polarized beam in which the electric vector is inclined at an angle α to the plane of scattering, we may resolve the incident amplitude[14] into its components

$$\xi_\alpha \sin\alpha \quad \text{and} \quad \xi_\alpha \cos\alpha \tag{17}$$

perpendicular and, respectively, parallel to the plane of scattering. According to equation (16), these incident amplitudes will give rise to scattered amplitudes proportional to $\xi_\alpha \sin\alpha$ and $\xi_\alpha \cos\alpha \cos\Theta$, which are perpendicular and parallel, respectively, to the plane of scattering. However, since the incident beam is plane-polarized, the scattered amplitudes in the two directions will be in phase. The resultant will therefore represent a plane-polarized beam with the electric vector inclined to the plane of scattering by an angle

$$\chi = \tan^{-1}(\tan\alpha \sec\Theta). \tag{18}$$

Further, the scattered intensity will be given by

$$\tfrac{3}{2}\sigma I_\alpha(\sin^2\alpha + \cos^2\alpha \cos^2\Theta)\frac{d\omega}{4\pi}. \tag{19}$$

4. *The source functions* $\mathfrak{J}_l$, $\mathfrak{J}_r$, *and* $\mathfrak{J}_U$.—We shall now show how, with the laws of scattering formulated in § 3 and with the rule of composition of plane-polarized beams given in § 2, we can obtain the equations of transfer valid under conditions when the radiation field is characterized by no special symmetry properties.

First, we may observe that, since on Rayleigh's laws of scattering a plane-polarized beam remains plane-polarized also after being scattered, it follows that the radiation scattered by an element of gas exposed to a partially plane-polarized radiation field will again be partially plane-polarized. For, under the circumstances, the radiation scattered in any given direction can be regarded as a mixture of several independent plane-polarized components, and this, according to our remarks in §2, can lead only to a partially plane-polarized beam. Hence, under conditions in which the diffuse reflection of an initially unpolarized radiation is the most that is contemplated, we need not consider a state of polarization more general than partial plane-polarization. Accordingly, we may characterize the radiation field at each point by the three quantities $I_l(\vartheta, \varphi)$, $I_r(\vartheta, \varphi)$, and $U(\vartheta, \varphi)$, where ϑ and φ are the polar angles referred to an appropriately chosen coordinate system through the point under consideration (see Fig. 1 in Paper X) and I_l and I_r are the intensities in the beam in the direction of the meridian plane through (ϑ, φ) and at right angles to it, respectively. And, finally, the quantity U is related to I_l and I_r according to equation (8), χ now denoting the inclination of the plane of maximum amplitude of the electric vector with the meridian plane.

[13] This is the plane which contains the direction of the incident and the scattered light.

[14] As in Paper X, we shall mean by "amplitude" a quantity whose square is the intensity.

It is now apparent that equations of transfer for I_l, I_r, and U must be expressible in the forms

$$\frac{dI_l(s, \vartheta, \varphi)}{\rho\sigma ds} = -I_l(s, \vartheta, \varphi) + \mathfrak{J}_l(s, \vartheta, \varphi), \tag{20}$$

$$\frac{dI_r(s, \vartheta, \varphi)}{\rho\sigma ds} = -I_r(s, \vartheta, \varphi) + \mathfrak{J}_r(s, \vartheta, \varphi), \tag{21}$$

$$\frac{dU(s, \vartheta, \varphi)}{\rho\sigma ds} = -U(s, \vartheta, \varphi) + \mathfrak{J}_U(s, \vartheta, \varphi), \tag{22}$$

where s measures the linear distance in the direction of (ϑ, φ), and $\mathfrak{J}_l$, $\mathfrak{J}_r$, and $\mathfrak{J}_U$ are the source functions for the respective quantities.

To evaluate $\mathfrak{J}_l(\vartheta, \varphi)$, $\mathfrak{J}_r(\vartheta, \varphi)$, and $\mathfrak{J}_U(\vartheta, \varphi)$ appropriate to any point of the atmosphere, we shall first consider the contributions to these source functions for the radiation in the direction (ϑ, φ) arising from the scattering of the radiation in the direction (ϑ', φ').

The radiation in the direction (ϑ', φ') is characterized by the intensities $I_l(\vartheta', \varphi')$, $I_r(\vartheta', \varphi')$, and $U(\vartheta', \varphi')$. Equivalently, we may, of course, also characterize it by the intensities $I_\chi(\vartheta', \varphi')$ and $I_{\chi+\pi/2}(\vartheta', \varphi')$ in the directions of maximum and minimum intensity, and the inclination χ of the direction of maximum amplitude of the electric vector to the meridian plane through (ϑ', φ'). The intensities I_χ and $I_{\chi+\pi/2}$ are related to I_l and I_r according to the equations

$$I_\chi + I_{\chi+\pi/2} = I_l + I_r \qquad \text{and} \qquad I_\chi - I_{\chi+\pi/2} = (I_l - I_r)\sec 2\chi\,. \tag{23}$$

Now a partially plane-polarized beam characterized by the maximum and minimum intensities I_χ and $I_{\chi+\pi/2}$ is optically equivalent to two plane-polarized beams of intensities, I_χ and $I_{\chi+\pi/2}$, with no correlation in their phases. We may, accordingly, consider the scattering of the radiation in the direction (ϑ', φ') as the result of scattering of the two plane-polarized components. According to the laws of scattering formulated in § 3, each of these components will give rise to plane-polarized scattered beams, which will again have no correlation in phases. In other words, the radiation scattered in the direction (ϑ, φ) from other directions can be considered as a mixture of a very large number of independent plane-polarized streams and can therefore be combined according to the laws of composition given in § 2.

Consider, then, the scattering of a plane-polarized beam of intensity $I_\chi(\vartheta', \varphi')$ in the direction (ϑ', φ') and confined to an element of solid angle, $d\omega'$, into the direction (ϑ, φ). Let ξ_χ denote the corresponding amplitude. The components of this amplitude parallel, respectively, perpendicular to the meridian plane through (ϑ', φ') are

$$\xi_\chi \cos\chi \qquad \text{and} \qquad \xi_\chi \sin\chi\,. \tag{24}$$

The amplitude $\xi_\chi \cos\chi$ has components parallel, respectively, perpendicular, to the plane of scattering, which are

$$\xi_\chi \cos\chi \cos i_1 \qquad \text{and} \qquad \xi_\chi \cos\chi \sin i_1 \tag{25}$$

where i_1 denotes the angle between the meridian plane through (ϑ', φ') and the plane of scattering (see Fig. 1 of Paper X). When this radiation is scattered in the direction (ϑ, φ), the components of the scattered amplitude that are parallel, respectively, perpendicular to the plane of scattering are proportional to

$$\xi_\chi \cos\chi \cos i_1 \cos\Theta \qquad \text{and} \qquad \xi_\chi \cos\chi \sin i_1\,. \tag{26}$$

The corresponding amplitudes in the meridian plane through (ϑ, φ) and at right angles to it are proportional (cf. Paper X, eqs. [12] and [13]) to

$$\left.\begin{aligned} &\xi_\chi \cos\chi\,(\sin i_1 \sin i_2 - \cos i_1 \cos i_2 \cos\Theta) \\ &\qquad = \xi_\chi \cos\chi\,(\sin\vartheta \sin\vartheta' + \cos\vartheta \cos\vartheta' \cos[\varphi' - \varphi]) , \end{aligned}\right\} \quad (27)$$

and

$$-\xi_\chi \cos\chi\,(\sin i_1 \cos i_2 + \sin i_2 \cos i_1 \cos\Theta) = -\xi_\chi \cos\chi \cos\vartheta' \sin(\varphi' - \varphi) , \quad (28)$$

where i_2 denotes the angle between the meridian plane through (ϑ, φ) and the scattering plane.

Similarly, the amplitude $\xi_\chi \sin\chi$, when scattered in the direction (ϑ, φ), leads to amplitudes in the meridian plane through (ϑ, φ) and at right angles to it which are proportional, respectively, to (cf. Paper X, eqs. [27] and [28]),

$$\xi_\chi \sin\chi\,(\sin i_1 \cos i_2 \cos\Theta + \cos i_1 \sin i_2) = \xi_\chi \sin\chi \cos\vartheta \sin(\varphi' - \varphi) \quad (29)$$

and

$$\xi_\chi \sin\chi\,(\sin i_1 \sin i_2 \cos\Theta - \cos i_1 \cos i_2) = \xi_\chi \sin\chi \cos(\varphi' - \varphi) . \quad (30)$$

Since the original amplitudes, $\xi_\chi \cos\chi$ and $\xi_\chi \sin\chi$, were derived from a plane-polarized beam, the scattered amplitudes represented by equations (27)–(30) are all in phase and should therefore be added as amplitudes, vectorially. Thus, the plane-polarized component $I_\chi(\vartheta', \varphi')$ of the radiation in the direction (ϑ', φ'), when scattered in the direction (ϑ, φ), leads to a plane-polarized beam, the amplitudes of which in the meridian plane and at right angles to it are proportional, respectively, to

$$\xi_\chi [(l, l) \cos\chi + (r, l) \sin\chi] , \quad (31)$$

and

$$\xi_\chi [(l, r) \cos\chi + (r, r) \sin\chi] , \quad (32)$$

where, for the sake of brevity, we have written

$$\left.\begin{aligned} (l, l) &= \sin\vartheta \sin\vartheta' + \cos\vartheta \cos\vartheta' \cos(\varphi' - \varphi) , \\ (r, l) &= +\cos\vartheta \sin(\varphi' - \varphi) , \\ (l, r) &= -\cos\vartheta' \sin(\varphi' - \varphi) , \\ (r, r) &= \cos(\varphi' - \varphi) . \end{aligned}\right\} \quad (33)$$

Writing the amplitudes (31) and (32) as

$$A_\chi \xi_\chi \quad \text{and} \quad B_\chi \xi_\chi , \quad (34)$$

we observe that the intensity of scattered plane-polarized beam is proportional to

$$I_\chi(\vartheta', \varphi')\,(A_\chi^2 + B_\chi^2) , \quad (35)$$

while its plane of vibration of the electric vector is inclined to the meridian plane through (ϑ, φ) by an angle

$$\tan^{-1} \frac{B_\chi}{A_\chi} . \quad (36)$$

Hence, the contributions $d\mathfrak{J}_{l,\chi}(\vartheta, \varphi; \vartheta', \varphi')$, $d\mathfrak{J}_{r,\chi}(\vartheta, \varphi; \vartheta,' \varphi')$, and $d\mathfrak{J}_{U,\chi}(\vartheta, \varphi; \vartheta', \varphi')$ to the source functions $\mathfrak{J}_l(\vartheta, \varphi)$, $\mathfrak{J}_r(\vartheta, \varphi)$, and $\mathfrak{J}_U(\vartheta, \varphi)$, arising from the scattering of the plane-polarized component $I_\chi(\vartheta', \varphi')$ of the radiation in the direction (ϑ', φ') and confined to an element of solid angle $d\omega'$, are

$$d\mathfrak{J}_{l,\chi}(\vartheta, \varphi; \vartheta', \varphi') = \frac{3}{8\pi} A_\chi^2 I_\chi(\vartheta', \varphi')\, d\omega', \tag{37}$$

$$d\mathfrak{J}_{r,\chi}(\vartheta, \varphi; \vartheta', \varphi') = \frac{3}{8\pi} B_\chi^2 I_\chi(\vartheta', \varphi')\, d\omega', \tag{38}$$

and

$$\left.\begin{aligned} d\mathfrak{J}_{U,\chi}(\vartheta, \varphi; \vartheta', \varphi') &= \frac{3}{8\pi} I_\chi(\vartheta', \varphi')(A_\chi^2+B_\chi^2) \sin 2\left(\tan^{-1}\frac{B_\chi}{A_\chi}\right) d\omega' \\ &= \frac{3}{4\pi} A_\chi B_\chi I_\chi(\vartheta', \varphi')\, d\omega' . \end{aligned}\right\} \tag{39}$$

Substituting for A_χ and B_χ, we have

$$d\mathfrak{J}_{l,\chi}(\vartheta, \varphi; \vartheta', \varphi') = \frac{3}{8\pi} I_\chi(\vartheta', \varphi')[(l, l)\cos\chi + (r, l)\sin\chi]^2 d\omega', \tag{40}$$

$$d\mathfrak{J}_{r,\chi}(\vartheta, \varphi; \vartheta', \varphi') = \frac{3}{8\pi} I_\chi(\vartheta', \varphi')[(l, r)\cos\chi + (r, r)\sin\chi]^2 d\omega', \tag{41}$$

and

$$\left.\begin{aligned} d\mathfrak{J}_{U,\chi}(\vartheta, \varphi; \vartheta' \varphi') = \frac{3}{4\pi} I_\chi(\vartheta', \varphi')&[(l, l)\cos\chi + (r, l)\sin\chi] \\ &\times [(l, r)\cos\chi + (r, r)\sin\chi]\, d\omega' . \end{aligned}\right\} \tag{42}$$

The contributions to the source functions for the radiation in the direction (ϑ, φ) arising from the scattering of the other plane-polarized component $I_{\chi+\pi/2}(\vartheta', \varphi')$ in the direction (ϑ', φ') can be readily obtained from equations (40)–(42) by simply replacing χ by $\chi+\pi/2$. In this manner we obtain

$$d\mathfrak{J}_{l,\chi+\pi/2}(\vartheta, \varphi; \vartheta' \varphi') = \frac{3}{8\pi} I_{\chi+\pi/2}(\vartheta', \varphi')[-(l, l)\sin\chi + (r, l)\cos\chi]^2 d\omega', \tag{43}$$

$$d\mathfrak{J}_{r,\chi+\pi/2}(\vartheta, \varphi; \vartheta', \varphi') = \frac{3}{8\pi} I_{\chi+\pi/2}(\vartheta', \varphi')[-(l, r)\sin\chi + (r, r)\cos\chi]^2 d\omega', \tag{44}$$

and

$$\left.\begin{aligned} d\mathfrak{J}_{U,\chi+\pi/2}(\vartheta, \varphi; \vartheta', \varphi') = \frac{3}{4\pi} I_{\chi+\pi/2}(\vartheta', \varphi')&[-(l, l)\sin\chi + (r, l)\cos\chi] \\ &\times [-(l, r)\sin\chi + (r, r)\cos\chi]\, d\omega' . \end{aligned}\right\} \tag{45}$$

Adding the respective equations of the two foregoing sets of equations and making use of the relations (23), we obtain for the contributions $d\mathfrak{J}_l(\vartheta, \varphi; \vartheta', \varphi')$, $d\mathfrak{J}_r(\vartheta, \varphi; \vartheta', \varphi')$, and $d\mathfrak{J}_U(\vartheta, \varphi; \vartheta', \varphi')$ to the source functions $\mathfrak{J}_l(\vartheta, \varphi)$, $\mathfrak{J}_r(\vartheta, \varphi)$, and $\mathfrak{J}_U(\vartheta, \varphi)$ arising from the scattering of the radiation in the direction (ϑ, φ),

$$\left.\begin{aligned} d\mathfrak{J}_l(\vartheta, \varphi; \vartheta', \varphi') = \frac{3}{8\pi}\{&(l, l)^2 I_l(\vartheta', \varphi') + (r, l)^2 I_r(\vartheta', \varphi') \\ &+ (l, l)(r, l)\, U(\vartheta', \varphi')\}\, d\omega', \end{aligned}\right\} \tag{46}$$

$$\left.\begin{aligned} d\mathfrak{J}_r(\vartheta, \varphi; \vartheta', \varphi') = \frac{3}{8\pi}\{&(l, r)^2 I_l(\vartheta', \varphi') + (r, r)^2 I_r(\vartheta', \varphi') \\ &+ (l, r)(r, r)\, U(\vartheta', \varphi')\}\, d\omega', \end{aligned}\right\} \tag{47}$$

and

$$d\mathfrak{J}_U(\vartheta,\varphi;\vartheta',\varphi')=\frac{3}{8\pi}\{2(l,l)(l,r)I_l(\vartheta',\varphi')+2(r,l)(r,r)I_r(\vartheta',\varphi')+[(l,l)(r,r)+(r,l)(l,r)]U(\vartheta',\varphi')\}d\omega'. \quad (48)$$

It is to be particularly noted that the foregoing equations involve only the intensities $I_l(\vartheta',\varphi')$, $I_r(\vartheta',\varphi')$, and $U(\vartheta',\varphi')$, in terms of which we agreed to characterize the radiation field.

Finally, integrating equations (46), (47), and (48) over the unit sphere, we obtain the required source functions. We have

$$\mathfrak{J}_l(\mu,\varphi)=\frac{3}{8\pi}\int_{-1}^{+1}\int_0^{2\pi}\{(l,l)^2I_l(\mu',\varphi')+(r,l)^2I_r(\mu',\varphi')+(l,l)(r,l)U(\mu',\varphi')\}d\mu'd\varphi', \quad (49)$$

$$\mathfrak{J}_r(\mu,\varphi)=\frac{3}{8\pi}\int_{-1}^{+1}\int_0^{2\pi}\{(l,r)^2I_l(\mu',\varphi')+(r,r)^2I_r(\mu',\varphi')+(l,r)(r,r)U(\mu',\varphi')\}d\mu'd\varphi', \quad (50)$$

and

$$\mathfrak{J}_U(\mu,\varphi)=\frac{3}{8\pi}\int_{-1}^{+1}\int_0^{2\pi}\{2(l,l)(l,r)I_l(\mu',\varphi')+2(r,l)(r,r)I_r(\mu',\varphi')+[(l,l)(r,r)+(r,l)(l,r)]U(\mu',\varphi')\}d\mu'd\varphi', \quad (51)$$

where the direction cosines μ and μ' have been used in place of cos ϑ and cos ϑ'. We may further note that, according to equation (33), we have

$$\left.\begin{aligned}
(l,l)^2&=\tfrac{1}{2}[2(1-\mu'^2)+\mu^2(3\mu'^2-2)]\\
&\quad+2\mu\mu'(1-\mu^2)^{\frac{1}{2}}(1-\mu'^2)^{\frac{1}{2}}\cos(\varphi'-\varphi)+\tfrac{1}{2}\mu^2\mu'^2\cos 2(\varphi'-\varphi),\\
(r,l)^2&=\tfrac{1}{2}\mu^2[1-\cos 2(\varphi'-\varphi)]; \qquad (l,r)^2=\tfrac{1}{2}\mu'^2[1-\cos 2(\varphi'-\varphi)],\\
(r,r)^2&=\tfrac{1}{2}[1+\cos 2(\varphi'-\varphi)],\\
(l,l)(r,l)&=\mu(1-\mu^2)^{\frac{1}{2}}(1-\mu'^2)^{\frac{1}{2}}\sin(\varphi'-\varphi)+\tfrac{1}{2}\mu^2\mu'\sin 2(\varphi'-\varphi),\\
(l,r)(r,r)&=-\tfrac{1}{2}\mu'\sin 2(\varphi'-\varphi); \qquad (r,l)(r,r)=\tfrac{1}{2}\mu\sin 2(\varphi'-\varphi),\\
(l,l)(l,r)&=-\mu'(1-\mu^2)^{\frac{1}{2}}(1-\mu'^2)^{\frac{1}{2}}\sin(\varphi'-\varphi)-\tfrac{1}{2}\mu\mu'^2\sin 2(\varphi'-\varphi),\\
(l,l)(r,r)+(r,l)(l,r)&=(1-\mu^2)^{\frac{1}{2}}(1-\mu'^2)^{\frac{1}{2}}\cos(\varphi'-\varphi)\\
&\quad+\mu\mu'\cos 2(\varphi'-\varphi).
\end{aligned}\right\} \quad (52)$$

5. *The reduction of the equations of transfer for the problem of diffuse reflection of a parallel beam of unpolarized radiation by a plane-parallel atmosphere.*—For the problem of diffuse reflection, it is convenient to distinguish between the incident radiation which penetrates to various depths, unaffected by any scattering or absorbing mechanisms, and the diffuse scattered radiation. Thus, if a parallel beam of unpolarized radiation of flux πF per unit area normal to itself is incident on a plane-parallel atmosphere at an angle β normal to the boundary, then at a level which is at an optical depth τ (measured in terms of the mass-scattering coefficient σ) below the surface, a fraction $e^{-\tau \sec \beta}$ of the incident radiation would not have suffered any scattering process and will, accordingly, be unpolarized: it is this unpolarized part of the radiation field which we wish to distinguish from the rest. With this understanding we may also distinguish

between the contributions to the various source functions arising from the unpolarized incident radiation field and the partially polarized diffuse radiation field. As in § 4, we shall now characterize the diffuse radiation field by the intensities $I_l(\tau, \vartheta, \varphi)$, $I_r(\tau, \vartheta, \varphi)$, and $U(\tau, \vartheta, \varphi)$, where ϑ and φ may now be chosen as the polar angles in a co-ordinate system in which the Z-axis is normal to the plane of stratification and the direction of the incident flux is $(\pi - \beta, 0)$.

The equations of transfer may now be written in the forms

$$\mu \frac{dI_l(\tau, \mu, \varphi)}{d\tau} = I_l(\tau, \mu, \varphi) - \mathfrak{J}_l(\tau, \mu, \varphi) - \mathfrak{J}_l^{(i)}(\tau, \mu, \varphi), \tag{53}$$

$$\mu \frac{dI_r(\tau, \mu, \varphi)}{d\tau} = I_r(\tau, \mu, \varphi) - \mathfrak{J}_r(\tau, \mu, \varphi) - \mathfrak{J}_r^{(i)}(\tau, \mu, \varphi), \tag{54}$$

and

$$\mu \frac{dU(\tau, \mu, \varphi)}{d\tau} = U(\tau, \mu, \varphi) - \mathfrak{J}_U(\tau, \mu, \varphi) - \mathfrak{J}_U^{(i)}(\tau, \mu, \varphi), \tag{55}$$

where $\mathfrak{J}_l(\tau, \mu, \varphi)$, $\mathfrak{J}_r(\tau, \mu, \varphi)$, and $\mathfrak{J}_U(\tau, \mu, \varphi)$ are the contributions to the respective source functions for the radiation at τ and in the direction (μ, φ), arising from the scattering of the diffuse radiation from all other directions, while $\mathfrak{J}_l^{(i)}(\tau, \mu, \varphi)$, $\mathfrak{J}_r^{(i)}(\tau, \mu, \varphi)$, and $\mathfrak{J}_U^{(i)}(\tau, \mu, \varphi)$ are those arising from the scattering of the unpolarized radiation of flux $\pi F e^{-\tau \sec \beta}$, which prevails at τ. By arguments similar to those used to establish formulae (46)–(48) it can readily be shown that

$$\left.\begin{aligned}\mathfrak{J}_l^{(i)}(\tau, \mu, \varphi) = \tfrac{3}{16} F e^{-\tau \sec\beta} \{ \tfrac{1}{2}[\, 2\sin^2\beta + \mu^2(3\cos^2\beta - 1)\,] \\ - 2\mu(1-\mu^2)^{\frac{1}{2}} \cos\beta \sin\beta \cos\varphi - \tfrac{1}{2}\mu^2 \sin^2\beta \cos 2\varphi \},\end{aligned}\right\} \tag{56}$$

$$\mathfrak{J}_r^{(i)}(\tau, \mu, \varphi) = \tfrac{3}{32} F e^{-\tau \sec\beta} \{ (1 + \cos^2\beta) + \sin^2\beta \cos 2\varphi \}, \tag{57}$$

and

$$\mathfrak{J}_U^{(i)}(\tau, \mu, \varphi) = -\tfrac{3}{16} F e^{-\tau \sec\beta} \{ 2(1-\mu^2)^{\frac{1}{2}} \sin\beta \cos\beta \sin\varphi + \mu \sin^2\beta \sin 2\varphi \}. \tag{58}$$

Combining equations (49)–(58), we have the required equations of transfer. An examination of these equations indicates that the solutions for $I_l(\tau, \mu\, \varphi)$, $I_r(\tau, \mu, \varphi)$, and $U(\tau, \mu, \varphi)$ must be expressible in the forms

$$I_l(\tau, \mu, \varphi) = I_l^{(0)}(\tau, \mu) + I_l^{(1)}(\tau, \mu)\cos\varphi + I_l^{(2)}(\tau, \mu)\cos 2\varphi, \tag{59}$$

$$I_r(\tau, \mu, \varphi) = I_r^{(0)}(\tau, \mu) + I_r^{(2)}(\tau, \mu)\cos 2\varphi, \tag{60}$$

and

$$U(\tau, \mu, \varphi) = U^{(1)}(\tau, \mu)\sin\varphi + U^{(2)}(\tau.\, \mu)\sin 2\varphi. \tag{61}$$

As the notation indicates, $I_l^{(0)}$, $I_l^{(1)}$, etc., are all functions of τ and μ only. Substituting the foregoing forms for the solution in the equations of transfer, we find that they break up into three sets of equations, which are

$$\left.\begin{aligned}\mu \frac{dI_l^{(0)}}{d\tau} &= I_l^{(0)} - \frac{3}{8}\Big[\int_{-1}^{+1} I_l^{(0)}(\tau, \mu') [\![2(1-\mu'^2) + \mu^2(3\mu'^2 - 2)]\!]\, d\mu' \\ &\quad + \mu^2 \int_{-1}^{+1} I_r^{(0)}(\tau, \mu')\, d\mu' \Big] - \tfrac{3}{32} F e^{-\tau \sec\beta}[\, 2\sin^2\beta + \mu^2(3\cos^2\beta - 1)\,], \\ \mu \frac{dI_r^{(0)}}{d\tau} &= I_r^{(0)} - \frac{3}{8}\Big[\int_{-1}^{+1} I_l^{(0)}(\tau, \mu')\,\mu'^2 d\mu' + \int_{-1}^{+1} I_r^{(0)}(\tau, \mu')\, d\mu' \Big] \\ &\quad - \tfrac{3}{32} F e^{-\tau \sec\beta}(1 + \cos^2\beta);\end{aligned}\right\} \tag{I}$$

$$
\left.
\begin{aligned}
\mu \frac{dI_l^{(1)}}{d\tau} &= I_l^{(1)} - \tfrac{3}{4}\mu(1-\mu^2)^{\frac{1}{2}} \int_{-1}^{+1} I_l^{(1)}(\tau,\mu')\,\mu'(1-\mu'^2)^{\frac{1}{2}} d\mu' \\
&- \tfrac{3}{8}\mu(1-\mu^2)^{\frac{1}{2}} \int_{-1}^{+1} U^{(1)}(\tau,\mu')(1-\mu'^2)^{\frac{1}{2}} d\mu' + \tfrac{3}{8} F e^{-\tau \sec\beta} \mu (1-\mu^2)^{\frac{1}{2}} \cos\beta \sin\beta\,, \\
\mu \frac{dU^{(1)}}{d\tau} &= U^{(1)} - \tfrac{3}{4}(1-\mu^2)^{\frac{1}{2}} \int_{-1}^{+1} I_l^{(1)}(\tau,\mu')\,\mu'(1-\mu'^2)^{\frac{1}{2}} d\mu' \\
&- \tfrac{3}{8}(1-\mu^2)^{\frac{1}{2}} \int_{-1}^{+1} U^{(1)}(\tau,\mu')(1-\mu'^2)^{\frac{1}{2}} d\mu' + \tfrac{3}{8} F e^{-\tau \sec\beta} (1-\mu^2)^{\frac{1}{2}} \cos\beta \sin\beta\,;
\end{aligned}
\right\} \quad \text{(II)}
$$

$$
\left.
\begin{aligned}
\mu \frac{dI_l^{(2)}}{d\tau} &= I_l^{(2)} - \tfrac{3}{16}\mu^2 \int_{-1}^{+1} I_l^{(2)}(\tau,\mu')\,\mu'^2 d\mu' + \tfrac{3}{16}\mu^2 \int_{-1}^{+1} I_r^{(2)}(\tau,\mu')\, d\mu' \\
&\qquad - \tfrac{3}{16}\mu^2 \int_{-1}^{+1} U^{(2)}(\tau,\mu')\,\mu' d\mu' + \tfrac{3}{32} F e^{-\tau\sec\beta} \mu^2 \sin^2\beta\,, \\
\mu \frac{dI_r^{(2)}}{d\tau} &= I_r^{(2)} + \frac{3}{16} \int_{-1}^{+1} I_l^{(2)}(\tau,\mu')\,\mu'^2 d\mu' - \tfrac{3}{16} \int_{-1}^{+1} I_r^{(2)}(\tau,\mu')\, d\mu' \\
&\qquad + \tfrac{3}{16} \int_{-1}^{+1} U^{(2)}(\tau,\mu')\,\mu' d\mu' - \tfrac{3}{32} F e^{-\tau\sec\beta} \sin^2\beta\,, \\
\mu \frac{dU^{(2)}}{d\tau} &= U^{(2)} - \tfrac{3}{8}\mu \int_{-1}^{+1} I_l^{(2)}(\tau,\mu')\,\mu'^2 d\mu' + \tfrac{3}{8}\mu \int_{-1}^{+1} I_r^{(2)}(\tau,\mu')\, d\mu' \\
&\qquad - \tfrac{3}{8}\mu \int_{-1}^{+1} U^{(2)}(\tau,\mu')\,\mu' d\mu' + \tfrac{3}{16} F e^{-\tau\sec\beta} \mu \sin^2\beta\,.
\end{aligned}
\right\} \quad \text{(III)}
$$

We now proceed to the solution of these equations appropriately for the problem of diffuse reflection from a semi-infinite plane-parallel atmosphere.

6. *The solutions of the equations for* $I_l^{(0)}$ *and* $I_r^{(0)}$.—Considering the equations of the first of the three systems to which we reduced our problem in § 5, we replace it in the nth approximation by the following system of $4n$ linear equations:

$$
\left.
\begin{aligned}
\mu_i \frac{dI_{l,i}^{(0)}}{d\tau} &= I_{l,i}^{(0)} - \tfrac{3}{8}\left[2\Sigma a_j I_{l,j}^{(0)} (1-\mu_j^2) + \mu_i^2 \{ \Sigma a_j I_{l,j}^{(0)} (3\mu_j^2 - 2) + \Sigma a_j I_{r,j}^{(0)} \} \right] \\
&- \tfrac{3}{32} F e^{-\tau\sec\beta} \left[2\sin^2\beta + \mu_i^2 (3\cos^2\beta - 1) \right] \qquad (i = \pm 1, \ldots, \pm n)\,,
\end{aligned}
\right\} \quad (62)
$$

and

$$
\left.
\begin{aligned}
\mu_i \frac{dI_{r,i}^{(0)}}{d\tau} &= I_{r,i}^{(0)} - \tfrac{3}{8}\left[\Sigma a_j I_{l,j}^{(0)} \mu_j^2 + \Sigma a_j I_{r,j}^{(0)} \right] - \tfrac{3}{32} F e^{-\tau\sec\beta} (1+\cos^2\beta) \\
&\qquad\qquad (i = \pm 1, \ldots, \pm n)\,,
\end{aligned}
\right\} \quad (63)
$$

where the various symbols have their usual meanings.

It is now seen that the homogeneous system associated with equations (62) and (63) is the same as that considered in Paper X, §§ 3–6. The complementary functions for $I_{l,i}^{(0)}$ and $I_{r,i}^{(0)}$ are therefore the same as those derived in the general solution (Paper X, eqs. [85] and [86]) of the homogeneous system. To complete the solution we therefore need only certain particular integrals for $I_{l,}^{(0)}$ and $I_{r,i}^{(0)}$; and these can be found in the following manner:

Setting

$$
I_{l,i}^{(0)} = \tfrac{3}{32} F g_i e^{-\tau\sec\beta} \quad \text{and} \quad I_{r,i}^{(0)} = \tfrac{3}{32} F h_i e^{-\tau\sec\beta} \qquad (i = \pm 1, \ldots, \pm n) \quad (64)
$$

in equations (62) and (63), we readily verify that the constants g_i and h_i must be expressible in the forms

$$g_i = \frac{\alpha\mu_i^2 + \delta}{1 + \mu_i \sec\beta} \quad \text{and} \quad h_i = \frac{\gamma}{1 + \mu_i \sec\beta} \qquad (i = \pm 1, \ldots, \pm n), \tag{65}$$

where α, δ, and γ are certain constants which are determined in accordance with the relations

$$\left.\begin{aligned} \alpha\mu_i^2 + \delta = \tfrac{3}{8}\,[2\alpha(E_2 - E_4) + 2\delta(E_0 - E_2) + \mu_i^2\{\alpha(3E_4 - 2E_2) \\ + \delta(3E_2 - 2E_0) + \gamma E_0\}] + 2\sin^2\beta + \mu_i^2(3\cos^2\beta - 1), \end{aligned}\right\} \tag{66}$$

and

$$\gamma = \tfrac{3}{8}\,[\alpha E_4 + \delta E_2 + \gamma E_0] + 1 + \cos^2\beta, \tag{67}$$

where we have used E_m to denote

$$E_m = \Sigma \frac{a_j \mu_j^m}{1 + \mu_j \sec\beta}. \tag{68}$$

Since equations (66) and (67) must be valid for all i, we must require that

$$\tfrac{8}{3}\alpha = \alpha(3E_4 - 2E_2) + \delta(3E_2 - 2E_0) + \gamma E_0 + \tfrac{8}{3}(3\cos^2\beta - 1), \tag{69}$$

$$\tfrac{4}{3}\delta = \alpha(E_2 - E_4) + \delta(E_0 - E_2) + \tfrac{8}{3}\sin^2\beta, \tag{70}$$

and

$$\tfrac{8}{3}\gamma = \alpha E_4 + \delta E_2 + \gamma E_0 + \tfrac{8}{3}(1 + \cos^2\beta). \tag{71}$$

These equations determine α, δ, and γ. We find

$$\alpha = + \frac{1}{1 - \frac{3}{8}(E_0 - E_2)} - \frac{2}{1 - \frac{3}{4}(E_0 - E_2)}, \tag{72}$$

$$\delta = - \frac{\cos^2\beta}{1 - \frac{3}{8}(E_0 - E_2)} + \frac{2}{1 - \frac{3}{4}(E_0 - E_2)}, \tag{73}$$

and

$$\gamma = + \frac{\sin^2\beta}{1 - \frac{3}{8}(E_0 - E_2)}. \tag{74}$$

In reducing the solutions for α, δ, and γ to the foregoing forms, repeated use has been made of the recursion formula (Paper IX, eq. [42]),

$$E_m = \cos\beta\left(\frac{2}{m}\epsilon_{m,\,\text{odd}} - E_{m-1}\right) \qquad (m \leqslant 4n), \tag{75}$$

which the E's satisfy. In terms of the quantities

$$C = \frac{1}{1 - \frac{3}{8}(E_0 - E_2)} = \frac{1}{1 - \dfrac{3}{4}\displaystyle\sum_{j=1}^{n} \frac{a_j(1 - \mu_j^2)}{1 - \mu_j^2 \sec^2\beta}} \tag{76}$$

and

$$\Gamma = \frac{1}{1 - \frac{3}{4}(E_0 - E_2)} = \frac{1}{1 - \dfrac{3}{2}\displaystyle\sum_{j=1}^{n} \frac{a_j(1 - \mu_j^2)}{1 - \mu_j^2 \sec^2\beta}}, \tag{77}$$

we can express α, δ, and γ more conveniently in the forms

$$\alpha = C - 2\Gamma \; ; \qquad \delta = -C\cos^2\beta + 2\Gamma \; ; \qquad \gamma = C\sin^2\beta \, . \tag{78}$$

We now observe that C and Γ, as defined in equations (76) and (77), bear the same relations to the characteristic equations (Paper X, eqs. [72] and [73])

$$1 = \frac{3}{4}\sum_{j=1}^{n} \frac{a_j(1-\mu_j^2)}{1-\mu_j^2 k^2} \quad \text{and} \quad 1 = \frac{3}{2}\sum_{j=1}^{n} \frac{a_j(1-\mu_j^2)}{1-\mu_j^2 k^2}\,, \tag{79}$$

as γ, defined in Paper VIII, equation (40), bears to the corresponding characteristic equation (Paper VIII, eq. [10]). We can therefore express C and Γ in the forms (cf. also Paper IX, eqs. [52], [123], and [166])

$$C = \frac{1}{\mu_1^2 \ldots . \mu_n^2} \frac{P(\cos\beta)P(-\cos\beta)}{R(\cos\beta)R(-\cos\beta)}\,, \tag{80}$$

and

$$\Gamma = \frac{1}{\mu_1^2 \ldots . \mu_n^2} \frac{P(\cos\beta)P(-\cos\beta)}{\rho(\cos\beta)\,\rho(-\cos\beta)}\,, \tag{81}$$

where the functions R, ρ, and P have the same meanings as in paper X, equations (129), (130), and (132).

Returning to equation (64), we can now express the particular integrals for $I_{l,i}^{(0)}$ and $I_{r,i}^{(0)}$ in the forms

$$I_{l,i}^{(0)} = \tfrac{3}{32}F e^{-\tau\sec\beta}\left[\frac{2(1-\mu_i^2)\Gamma}{1+\mu_i\sec\beta} - C(1-\mu_i\sec\beta)\cos^2\beta\right] \quad (i = \pm 1\,, \ldots . , \pm n) \tag{82}$$

and

$$I_{r,i}^{(0)} = \tfrac{3}{32}F e^{-\tau\sec\beta}\frac{C\sin^2\beta}{1+\mu_i\sec\beta} \qquad (i = \pm 1\,, \ldots . , \pm n)\,. \tag{83}$$

Adding these particular integrals for $I_{l,i}^{(0)}$ and $I_{r,i}^{(0)}$ to the corresponding solutions (Paper X, eqs. [87] and [88]) derived from the homogeneous equations and which are further compatible with our present requirement of the boundedness of the solutions for $\tau \to \infty$, we have

$$I_{l,i}^{(0)} = \tfrac{3}{32}F\Bigg[(1-\mu_i^2)\sum_{\beta=1}^{n-1}\frac{L_\beta e^{-\kappa_\beta\tau}}{1+\mu_i\kappa_\beta} + \sum_{\alpha=1}^{n} M_\alpha(1-k_\alpha\mu_i)\,e^{-k_\alpha\tau} + Q + e^{-\tau\sec\beta}\left\{\frac{2(1-\mu_i^2)\Gamma}{1+\mu_i\sec\beta} - C(1-\mu_i\sec\beta)\cos^2\beta\right\}\Bigg] \quad (i = \pm 1\,, \ldots . , \pm n) \tag{84}$$

and

$$I_{r,i}^{(0)} = \tfrac{3}{32}F\left[Q - \sum_{\alpha=1}^{n}\frac{M_\alpha(k_\alpha^2-1)\,e^{-k_\alpha\tau}}{1+\mu_i k_\alpha} + \frac{C\sin^2\beta}{1+\mu_i\sec\beta}\,e^{-\tau\sec\beta}\right] \quad (i = \pm 1\,, \ldots . , \pm n)\,, \tag{85}$$

where $L_\beta(\beta = 1, \ldots . , n-1)$, $M_\alpha(\alpha = 1, \ldots . , n)$, and Q are $2n$ constants of integration to be determined by the boundary conditions

$$I_{l,i}^{(0)} = I_{r,i}^{(0)} = 0 \qquad (i = -1\,, \ldots . , -n)\,. \tag{86}$$

In terms of the functions

$$\left.\begin{aligned} S_l(\mu) = (1-\mu^2)\sum_{\beta=1}^{n-1}\frac{L_\beta}{1-\mu\kappa_\beta}+\sum_{a=1}^{n}M_a(1+k_a\mu)+Q \\ +(1-\mu^2)\frac{2\Gamma}{1-\mu\sec\beta}-C(1+\mu\sec\beta)\cos^2\beta \end{aligned}\right\} \quad (87)$$

and

$$S_r(\mu) = Q-\sum_{a=1}^{n}\frac{M_a(k_a^2-1)}{1-\mu k_a}+\frac{C\sin^2\beta}{1-\mu\sec\beta}, \quad (88)$$

the boundary conditions are

$$S_l(\mu_i) = S_r(\mu_i) = 0 \qquad (=1, \ldots, n). \quad (89)$$

The angular distribution of the reflected radiation, corresponding to the parts $I_l^{(0)}$ and $I_r^{(0)}$ of the emergent intensities are also expressible in terms of the functions $S_l(\mu)$ and $S_r(\mu)$. We have

$$I_l^{(0)}(0,\mu) = \tfrac{3}{32}FS_l(-\mu) \quad \text{and} \quad I_r^{(0)}(0,\mu) = \tfrac{3}{32}FS_r(-\mu). \quad (90)$$

We shall now show how explicit formulae for $S_l(\mu)$ and $S_r(\mu)$ can be found without having the necessity of solving for the constants L_β, M_a, and Q.

Consider the function

$$(1-\mu\sec\beta)\,p(\mu)S_l(\mu) = (1-\mu\sec\beta)\prod_{\beta=1}^{n-1}(1-\kappa_\beta\mu)S_l(\mu). \quad (91)$$

This is a polynomial of degree $(n+1)$ in μ, which vanishes for $\mu=\mu_i$ $(i=1,\ldots,n)$. There must, therefore, exist a relation of the form

$$S_l(\mu) = X\frac{P(\mu)(\mu+a)}{(1-\mu\sec\beta)\,p(\mu)}, \quad (92)$$

where X and a are certain constants. Of these two constants, X is readily determined; for, from equations (87) and (92), it follows that

$$\lim_{\mu\to\cos\beta}(1-\mu\sec\beta)S_l(\mu) = X\frac{P(\cos\beta)}{p(\cos\beta)}(\cos\beta+a) = 2\Gamma\sin^2\beta, \quad (93)$$

and, substituting for Γ from equation (81), we obtain

$$X = \frac{2\sin^2\beta}{\mu_1^2\ldots\mu_n^2}\frac{P(-\cos\beta)}{p(-\cos\beta)(\cos\beta+a)}. \quad (94)$$

Hence,

$$S_l(\mu) = \frac{2\sin^2\beta}{\mu_1^2\ldots\mu_n^2}\frac{P(-\cos\beta)P(\mu)}{p(-\cos\beta)\,p(\mu)}\frac{\mu+a}{(\cos\beta+a)(1-\mu\sec\beta)}. \quad (95)$$

Similarly, it can be shown that

$$S_r(\mu) = \frac{\sin^2\beta}{\mu_1^2\ldots\mu_n^2}\frac{P(-\cos\beta)P(\mu)}{R(-\cos\beta)R(\mu)}\frac{\mu+b}{(\cos\beta+b)(1-\mu\sec\beta)}, \quad (96)$$

where b is another constant. It now remains to determine the constants a and b. For this purpose we proceed as follows:

Setting $\mu = +1$, respectively -1, in equation (87), we have

$$\sum_{a=1}^{n} M_a (1 + k_a) + Q - C (\cos\beta + 1) \cos\beta = S_l(+1) \tag{97}$$

and

$$\sum_{a=1}^{n} M_a (1 - k_a) + Q - C (\cos\beta - 1) \cos\beta = S_l(-1) . \tag{98}$$

Adding and subtracting these two equations, we obtain

$$\sum_{a=1}^{n} M_a + Q - C \cos^2\beta = \tfrac{1}{2} [S_l(+1) + S_l(-1)] \tag{99}$$

and

$$\sum_{a=1}^{n} M_a k_a - C \cos\beta = \tfrac{1}{2} [S_l(+1) - S_l(-1)] . \tag{100}$$

Next, setting $\mu = 0$ in equation (88), we have

$$Q - \sum_{a=1}^{n} M_a (k_a^2 - 1) + C \sin^2\beta = S_r(0) , \tag{101}$$

and, subtracting this equation from equation (99), we obtain

$$\sum_{a=1}^{n} M_a k_a^2 - C = \tfrac{1}{2} [S_l(+1) + S_l(-1)] - S_r(0) . \tag{102}$$

We shall now show how the constants a and b can be determined from equations (100) and (102).

First, substituting for $S_l(+1)$, $S_l(-1)$, and $S_r(0)$ according to equations (95) and (96) in equations (100) and (102), we find, after some minor reductions, that

$$\left.\begin{aligned} \sum_{a=1}^{n} M_a k_a - C \cos\beta = &- \frac{\cos\beta \; P(-\cos\beta)}{\mu_1^2 \ldots \mu_n^2 \rho(-\cos\beta)(a + \cos\beta)} \\ &\times \left[(a+1)(1+\cos\beta)\frac{P(1)}{\rho(1)} + (a-1)(1-\cos\beta)\frac{P(-1)}{\rho(-1)} \right] \end{aligned}\right\} \tag{103}$$

and

$$\left.\begin{aligned} \sum_{a=1}^{n} M_a k_a^2 - C + \frac{\sin^2\beta}{\mu_1^2 \ldots \mu_n^2} \frac{P(-\cos\beta)}{R(-\cos\beta)} \frac{(-1)^n \mu_1 \ldots \mu_n b}{b + \cos\beta} \\ = \frac{\cos\beta \; P(-\cos\beta)}{\mu_1^2 \ldots \mu_n^2 \rho(-\cos\beta)(a+\cos\beta)} \left[-(a+1)(1+\cos\beta)\frac{P(1)}{\rho(1)} \right. \\ \left. + (a-1)(1-\cos\beta)\frac{P(-1)}{\rho(-1)} \right] . \end{aligned}\right\} \tag{104}$$

On the other hand, from equations (88) and (96) it follows that

$$\left.\begin{aligned} M_a &= -\frac{1}{(k_a^2-1)}\lim_{\mu\to k_a^{-1}}(1-\mu k_a)\,S_r(\mu)\\ &= \frac{\sin^2\beta\cos\beta}{\mu_1^2\ldots\mu_n^2}\,\frac{P(-\cos\beta)}{R(-\cos\beta)(b+\cos\beta)}\,\frac{(1+bk_a)P(1/k_a)}{(k_a^2-1)(1-k_a\cos\beta)R_a(1/k_a)}, \end{aligned}\right\}\tag{105}$$

where $R_a(x)$ has the same meaning as in Paper X, equation (129). Substituting the foregoing expression for M_a in equations (103) and (104), we find that they can be reduced to the forms

$$\left.\begin{aligned} &\frac{(\zeta_1+b\zeta_2)\sin^2\beta}{R(-\cos\beta)(b+\cos\beta)}-\frac{P(\cos\beta)}{R(\cos\beta)R(-\cos\beta)}\\ &=\frac{-1}{\rho(-\cos\beta)(a+\cos\beta)}\left[(a+1)(1+\cos\beta)\frac{P(1)}{\rho(1)}+(a-1)(1-\cos\beta)\frac{P(-1)}{\rho(-1)}\right], \end{aligned}\right\}\tag{106}$$

$$\left.\begin{aligned} &\frac{\sin^2\beta}{R(-\cos\beta)(b+\cos\beta)}\left[(\zeta_2+b\zeta_3)\cos\beta+(-1)^n\mu_1\ldots\mu_n b\right]\\ &-\frac{P(\cos\beta)}{R(\cos\beta)R(-\cos\beta)}=\frac{\cos\beta}{\rho(-\cos\beta)(a+\cos\beta)}\left[-(a+1)(1+\cos\beta)\frac{P(1)}{\rho(1)}\right.\\ &\qquad\left.+(a-1)(1-\cos\beta)\frac{P(-1)}{\rho(-1)}\right], \end{aligned}\right\}\tag{107}$$

where

$$\zeta_m=\sum_{a=1}^{n}\frac{k_a^m}{(k_a^2-1)(1-k_a\cos\beta)}\,\frac{P(1/k_a)}{R_a(1/k_a)}\qquad(m=1,2,3).\tag{108}$$

To evaluate the sum occurring on the right side of equation (108), we introduce the function

$$f_m(x)=\sum_{a=1}^{n}\frac{P(1/k_a)\,k_a^m}{(k_a^2-1)R_a(1/k_a)}R_a(x)\tag{109}$$

and express ζ_m in terms of it. Thus

$$\zeta_m=\frac{f_m(\cos\beta)}{R(\cos\beta)}.\tag{110}$$

Now $f_m(x)$, defined as in equation (109), is a polynomial of degree $(n-1)$ in x, which takes the values

$$\frac{P(1/k_a)\,k_a^m}{(k_a^2-1)}\tag{111}$$

for $x=1/k_a$, $(a=1,\ldots,n)$. In other words,

$$(1-x^2)\,f_m(x)-x^{2-m}P(x)\tag{112}$$

vanishes for $x = 1/k_\alpha$, ($\alpha = 1, \ldots, n$). It must accordingly divide $R(x)$. This fact enables us to determine $f_m(x)$. To illustrate this, we shall consider the case $m = 3$. In this case there must exist a relation of the form

$$x(1-x^2) f_3(x) - P(x) = R(x)(Ax^2 + Bx + D), \tag{113}$$

since the polynomial on the right-hand side is of degree $n + 2$ in x and vanishes for $x = 1/k_\alpha$, $\alpha = 1, \ldots, n$. To determine the constants A, B, and D, we first set $x = 0$ and find at once that

$$D = (-1)^{n+1}\mu_1 \ldots \mu_n . \tag{114}$$

Next, setting $x = +1$, respectively -1, we have

$$A + B + D = -\frac{P(1)}{R(1)} \tag{115}$$

and

$$A - B + D = -\frac{P(-1)}{R(-1)} . \tag{116}$$

These equations determine A and B and make $f_3(x)$ determinate; ζ_3 then follows according to equation (110). In this manner we find that

$$\left.\begin{aligned}\zeta_3 = \frac{1}{\cos\beta\sin^2\beta}\Bigg\{&\frac{P(\cos\beta)}{R(\cos\beta)} - \frac{1}{2}\left[\frac{P(1)}{R(1)} + \frac{P(-1)}{R(-1)}\right]\cos^2\beta \\ &- \frac{1}{2}\left[\frac{P(1)}{R(1)} - \frac{P(-1)}{R(-1)}\right]\cos\beta + (-1)^{n+1}\mu_1 \ldots \mu_n \sin^2\beta\Bigg\} .\end{aligned}\right\} \tag{117}$$

Similarly,

$$\left.\begin{aligned}\zeta_1 = \frac{1}{\sin^2\beta}\Bigg\{&\frac{P(\cos\beta)}{R(\cos\beta)}\cos\beta - \frac{1}{2}\left[\frac{P(1)}{R(1)} + \frac{P(-1)}{R(-1)}\right]\cos\beta \\ &- \frac{1}{2}\left[\frac{P(1)}{R(1)} - \frac{P(-1)}{R(-1)}\right]\Bigg\} ,\end{aligned}\right\} \tag{118}$$

and

$$\left.\begin{aligned}\zeta_2 = \frac{1}{\sin^2\beta}\Bigg\{&\frac{P(\cos\beta)}{R(\cos\beta)} - \frac{1}{2}\left[\frac{P(1)}{R(1)} - \frac{P(-1)}{R(-1)}\right]\cos\beta \\ &- \frac{1}{2}\left[\frac{P(1)}{R(1)} + \frac{P(-1)}{R(-1)}\right]\Bigg\} .\end{aligned}\right\} \tag{119}$$

Substituting these expressions for ζ_1, ζ_2, and ζ_3 in equations (106) and (107) we obtain, after some further reductions, the equations

$$\frac{1}{R(1)R(-\cos\beta)}\,\frac{b+1}{b+\cos\beta} = \frac{2}{\rho(1)\,\rho(-\cos\beta)}\,\frac{a+1}{a+\cos\beta} \tag{120}$$

and

$$\frac{1}{R(-1)R(-\cos\beta)}\,\frac{b-1}{b+\cos\beta} = \frac{2}{\rho(-1)\,\rho(-\cos\beta)}\,\frac{a-1}{a+\cos\beta} . \tag{121}$$

Finally, solving these equations for a and b, we find

$$a=\frac{xy\cos\beta-(x\eta+y\xi)\cos\beta+(x\eta-y\xi)}{(x\eta+y\xi)-xy-(x\eta-y\xi)\cos\beta} \tag{122}$$

and

$$b=\frac{4\xi\eta\cos\beta-(x\eta+y\xi)\cos\beta-(x\eta-y\xi)}{(x\eta+y\xi)-4\xi\eta+(x\eta-y\xi)\cos\beta}, \tag{123}$$

where we have used the abbreviations

$$x=\frac{1}{R(1)R(-\cos\beta)}; \qquad y=\frac{1}{R(-1)R(-\cos\beta)}, \tag{124}$$

and

$$\xi=\frac{1}{\rho(1)\,\rho(-\cos\beta)}; \qquad \eta=\frac{1}{\rho(-1)\,\rho(-\cos\beta)}. \tag{125}$$

With the solutions (122) and (123) for a and b, we find that (cf. eqs. [95] and [96])

$$\frac{a+\mu}{a+\cos\beta}\sin^2\beta=1-\mu\cos\beta+\frac{x\eta+y\xi-xy}{x\eta-y\xi}(\mu-\cos\beta) \tag{126}$$

and

$$\frac{b+\mu}{b+\cos\beta}\sin^2\beta=1-\mu\cos\beta-\frac{x\eta+y\xi-4\xi\eta}{x\eta-y\xi}(\mu-\cos\beta). \tag{127}$$

Using these formulae in equations (95) and (96), we have

$$S_l(\mu)=\frac{2}{\mu_1^2\cdots\mu_n^2}\frac{P(-\cos\beta)P(\mu)}{\rho(-\cos\beta)\,\rho(\mu)}\left[1-\mu\cos\beta+\frac{x\eta+y\xi-xy}{x\eta-y\xi}(\mu-\cos\beta)\right]\frac{1}{1-\mu\sec\beta} \tag{128}$$

and

$$S_r(\mu)=\frac{1}{\mu_1^2\cdots\mu_n^2}\frac{P(-\cos\beta)P(\mu)}{R(-\cos\beta)R(\mu)}\left[1-\mu\cos\beta-\frac{x\eta+y\xi-4\xi\eta}{x\eta-y\xi}(\mu-\cos\beta)\right]\frac{1}{1-\mu\sec\beta}. \tag{129}$$

Now, using our definitions of x, y, ξ, and η, we readily verify that

$$\frac{x\eta+y\xi}{x\eta-y\xi}=\frac{R(-1)\,\rho(1)+R(1)\,\rho(-1)}{R(-1)\,\rho(1)-R(1)\,\rho(-1)}=c\text{ (say)}, \tag{130}$$ [15]

$$\frac{xy}{x\eta-y\xi}=\frac{\rho(-\cos\beta)}{R(-\cos\beta)}\,\frac{\rho(1)\,\rho(-1)}{R(-1)\,\rho(1)-R(1)\,\rho(-1)}, \tag{131}$$

and

$$\frac{4\xi\eta}{x\eta-y\xi}=\frac{R(-\cos\beta)}{\rho(-\cos\beta)}\,\frac{4R(1)R(-1)}{R(-1)\,\rho(1)-R(1)\,\rho(-1)}. \tag{132}$$

[15] It may be further verified that this is the same as the quantity $(\alpha a-\beta b)/(\alpha b-\beta a)$, which has already occurred in our analysis in Paper X (eq. [167]).

But it can be shown that

$$R(1)R(-1) = -\tfrac{1}{2}\rho(1)\,\rho(-1). \tag{133}$$ [16]

Hence, in terms of the quantity

$$q = \frac{2\rho(1)\,\rho(-1)}{R(-1)\,\rho(1) - R(1)\,\rho(-1)}, \tag{134}$$ [17]

we can write

$$\frac{xy}{x\eta - y\xi} = \tfrac{1}{2} q \frac{\rho(-\cos\beta)}{R(-\cos\beta)} \tag{135}$$

and

$$\frac{4\xi\eta}{x\eta - y\xi} = -q \frac{R(-\cos\beta)}{\rho(-\cos\beta)}. \tag{136}$$

In terms of c and q we can re-write equations (128) and (129) in the forms

$$\left.\begin{aligned} S_l(\mu) = \frac{1}{\mu_1^2 \ldots \mu_n^2} \frac{P(\mu)}{\rho(\mu)} \Bigg\{ & 2\frac{P(-\cos\beta)}{\rho(-\cos\beta)}[1 - \mu\cos\beta + c(\mu - \cos\beta)] \\ & - q\frac{P(-\cos\beta)}{R(-\cos\beta)}(\mu - \cos\beta) \Bigg\} \frac{1}{1 - \mu\sec\beta} \end{aligned}\right\} \tag{137}$$

[16] This relation follows, for example, from the identities

$$1 - \tfrac{3}{4}\sum_{j=1}^{n} \frac{a_j(1-\mu_j^2)}{1-\mu_j^2 x} \equiv (-1)^n \mu_1^2 \ldots \mu_n^2 \frac{\prod_{\alpha=1}^{n}(x - k_\alpha^2)}{\prod_{j=1}^{n}(1 - \mu_j^2 x)}$$

and

$$1 - \tfrac{3}{2}\sum_{j=1}^{n} \frac{a_j(1-\mu_j^2)}{1-\mu_j^2 x} \equiv (-1)^n \mu_1^2 \ldots \mu_n^2 \frac{x\prod_{\beta=1}^{n-1}(x - \kappa_\beta^2)}{\prod_{j=1}^{n}(1 - \mu_j^2 x)},$$

which can be readily established (cf. Paper VIII, eqs. [41] — [46]). Thus putting $x = 1$ in these identities, we find that

$$(-1)^n \mu_1^2 \ldots \mu_n^2 \frac{R(1)\,R(-1)}{\prod_{j=1}^{n}(1-\mu_j^2)} = \tfrac{1}{4}$$

and

$$(-1)^n \mu_1^2 \ldots \mu_n^2 \frac{\rho(1)\,\rho(-1)}{\prod_{j=1}^{n}(1-\mu_j^2)} = -\tfrac{1}{2},$$

from which the relation in question follows immediately.

[17] Again, it may be verified that this is the same as the quantity $(a^2 - b^2)/(ab - \beta a)$, which occurred in Paper X (eq. [168]).

and

$$S_r(\mu) = \frac{1}{\mu_1^2 \cdots \mu_n^2}\frac{P(\mu)}{R(\mu)}\left\{\frac{P(-\cos\beta)}{R(-\cos\beta)}[1-\mu\cos\beta - c(\mu-\cos\beta)] - q\frac{P(-\cos\beta)}{\rho(-\cos\beta)}(\mu-\cos\beta)\right\}\frac{1}{1-\mu\sec\beta}, \tag{138}$$

which are the required expressions for $S_l(\mu)$ and $S_r(\mu)$.

According to equation (90), we can now express the angular distributions $I_l^{(0)}(0, \mu)$ and $I_r^{(0)}(0, \mu)$ in the forms

$$I_l^{(0)}(0, \mu) = \tfrac{3}{32}F\{2H_l(\cos\beta)H_l(\mu)[1+\mu\cos\beta - c(\mu+\cos\beta)] + qH_r(\cos\beta)H_l(\mu)(\mu+\cos\beta)\}\frac{\cos\beta}{\cos\beta+\mu} \tag{139}$$

and

$$I_r^{(0)}(0, \mu) = \tfrac{3}{32}F\{H_r(\cos\beta)H_r(\mu)[1+\mu\cos\beta + c(\mu+\cos\beta)] + qH_l(\cos\beta)H_r(\mu)(\mu+\cos\beta)\}\frac{\cos\beta}{\cos\beta+\mu}, \tag{140}$$

where we have introduced the functions $H_l(\mu)$ and $H_r(\mu)$, defined according to

$$H_l(\mu) = \frac{(-1)^n}{\mu_1 \cdots \mu_n}\frac{P(-\mu)}{\rho(-\mu)} = \frac{1}{\mu_1 \cdots \mu_n}\frac{\prod_{i=1}^{n}(\mu+\mu_i)}{\prod_{\beta=1}^{n-1}(1+\kappa_\beta\mu)} \tag{141}$$

and

$$H_r(\mu) = \frac{(-1)^n}{\mu_1 \cdots \mu_n}\frac{P(-\mu)}{R(-\mu)} = \frac{1}{\mu_1 \cdots \mu_n}\frac{\prod_{i=1}^{n}(\mu+\mu_i)}{\prod_{\alpha=1}^{n}(1+k_\alpha\mu)}. \tag{142}$$

This completes the solution of system (I).

7. *The solution of the equations for* $I_l^{(1)}$ *and* $U^{(1)}$.—Considering, next, the equations of the second system (p. 119), we first observe that there must exist a simple proportionality between the functions $I_l^{(1)}(\tau, \mu)$ and $U^{(1)}(\tau, \mu)$. For, from the two equations belonging to this system, it immediately follows that

$$\mu\frac{d}{d\tau}[I_l^{(1)} - \mu U^{(1)}] = I_l^{(1)} - \mu U^{(1)}. \tag{143}$$

The boundedness of the solutions for $\tau \to \infty$ now requires that

$$I_l^{(1)}(\tau, \mu) \equiv \mu U^{(1)}(\tau, \mu). \tag{144}$$

Using this identity in the equation for $U^{(1)}$, we find

$$\left.\begin{aligned}\mu \frac{dU^{(1)}}{d\tau} = U^{(1)} - \tfrac{3}{8}(1-\mu^2)^{\frac{1}{2}} \int_{-1}^{+1} U^{(1)}(\tau,\mu')(1-\mu'^2)^{\frac{1}{2}}(1+2\mu'^2)\, a\mu \\ + \tfrac{3}{8} F e^{-\tau \sec\beta}(1-\mu^2)^{\frac{1}{2}} \sin\beta \cos\beta \, .\end{aligned}\right\} \qquad (145)$$

Now, writing

$$U^{(1)}(\tau,\mu) = \tfrac{3}{8} F (1-\mu^2)^{\frac{1}{2}} \sin\beta \cos\beta \; \phi(\tau,\mu)\,, \qquad (146)$$

we obtain for ϕ the integrodifferential equation

$$\mu \frac{d\phi}{d\tau} = \phi - \frac{3}{8} \int_{-1}^{+1} \phi(\tau,\mu')(1-\mu'^2)(1+2\mu'^2)\, d\mu' + e^{-\tau \sec\beta}\,. \qquad (147)$$

Equation (147) can be solved in a general nth approximation by following a procedure exactly similar to that described for the solution of a class of integrodifferential equations in Paper IX, § 4. In our present context, we are, however, most interested only in the angular distribution of ϕ at $\tau = 0$. For this we have (cf. Paper IX, eq. [136])

$$\phi(0,\mu) = -H^{(1)}(\cos\beta)\, H^{(1)}(\mu) \frac{\cos\beta}{\cos\beta + \mu}\,, \qquad (148)$$

where, in the nth approximation,

$$H^{(1)}(\mu) = \frac{1}{\mu_1 \ldots \mu_n} \frac{\prod_{i=1}^{n} (\mu + \mu_i)}{\prod_{a=1}^{n} (1 + k_a^{(1)} \mu)}\,, \qquad (149)$$

the $k_a^{(1)}$'s being the n positive roots of the characteristic equation (cf. Paper X, eq. [117])

$$1 = \frac{3}{4} \sum_{j=1}^{n} \frac{a_j (1-\mu_j^2)(1+2\mu_j^2)}{1-\mu_j^2 k^2}\,. \qquad (150)$$

For the angular distributions $I_l^{(1)}(0,\mu)$ and $U^{(1)}(0,\mu)$ we therefore have

$$I_l^{(1)}(0,\mu) = -\tfrac{3}{8} F \mu (1-\mu^2)^{\frac{1}{2}} \sin\beta \cos\beta \; H^{(1)}(\cos\beta)\, H^{(1)}(\mu) \frac{\cos\beta}{\cos\beta + \mu} \qquad (151)$$

and

$$U^{(1)}(0,\mu) = -\tfrac{3}{8} F (1-\mu^2)^{\frac{1}{2}} \sin\beta \cos\beta \; H^{(1)}(\cos\beta)\, H^{(1)}(\mu) \frac{\cos\beta}{\cos\beta + \mu}\,. \qquad (152)$$

8. *The solution of the equations for* $I_l^{(2)}$, $I_l^{(2)}$, *and* $U^{(2)}$.—As in the case of system (II), a simple inspection of the three equations belonging to the third system (p. 119) shows that we must identically require

$$I_l^{(2)}(\tau,\mu) \equiv -\mu^2 I_r^{(2)}(\tau,\mu) \qquad (153)$$

and

$$U^{(2)}(\tau,\mu) \equiv -2\mu I_r^{(2)}(\tau,\mu)\,. \qquad (154)$$

Using these identities in the equation for $I_r^{(2)}$, we obtain

$$\mu \frac{dI_r^{(2)}}{d\tau} = I_r^{(2)} - \frac{3}{16}\int_{-1}^{+1} I_r^{(2)}(\tau, \mu')(1+\mu'^2)^2 d\mu' - \tfrac{3}{32} F e^{-\tau \sec\beta} \sin^2\beta\,. \quad (155)$$

Equation (155) can again be solved quite readily by our standard procedures. In the nth approximation the characteristic equation is

$$1 = \frac{3}{8}\sum_{j=1}^{n} \frac{a_j(1+\mu_j^2)^2}{1-\mu_j^2 k^2}\,; \quad (156)$$

and the solution for the angular distributions of the various functions at $\tau = 0$ can all be expressed in terms of the function

$$H^{(2)}(\mu) = \frac{1}{\mu_1 \ldots \mu_n} \frac{\prod_{i=1}^{n}(\mu+\mu_i)}{\prod_{a-1}^{n}(1+k_a^{(2)}\mu)}\,, \quad (157)$$

where the $k_a^{(2)}$'s are the n positive characteristic roots. Thus,

$$I_l^{(2)}(0, \mu) = -\tfrac{3}{32} F\mu^2 \sin^2\beta\, H^{(2)}(\cos\beta)\, H^{(2)}(\mu) \frac{\cos\beta}{\cos\beta+\mu} \quad (158)$$

and

$$I_r^{(2)}(0, \mu) = +\tfrac{3}{32} F \sin^2\beta\, H^{(2)}(\cos\beta)\, H^{(2)}(\mu) \frac{\cos\beta}{\cos\beta+\mu} \quad (159)$$

and

$$U^{(2)}(0, \mu) = -\tfrac{3}{16} F\mu \sin^2\beta\, H^{(2)}(\cos\beta)\, H^{(2)}(\mu) \frac{\cos\beta}{\cos\beta+\mu}\,. \quad (160)$$

9. *The polarization and the angular distribution of the reflected radiation: tables of the necessary functions.*—Combining the solutions (139), (140), (151), (152), and (158)–(160) in accordance with equations (59)–(61), we obtain the equations which determine the polarization and the angular distribution of the radiation diffusely reflected by a semi-infinite plane-parallel atmosphere. We have

$$\left.\begin{aligned} I_l(0, \mu) = \tfrac{3}{32} F\{ & 2H_l(\mu) H_l(\mu')[1+\mu\mu' - c(\mu+\mu')] \\ & + qH_l(\mu)H_r(\mu')(\mu+\mu') - 4\mu\mu'(1-\mu^2)^{\frac{1}{2}}(1-\mu'^2)^{\frac{1}{2}} H^{(1)}(\mu)\, H^{(1)}(\mu')\cos\varphi \\ & - \mu^2(1-\mu'^2)\, H^{(2)}(\mu)\, H^{(2)}(\mu')\cos 2\varphi\} \frac{\mu'}{\mu+\mu'}\,, \end{aligned}\right\} \quad (161)$$

$$\left.\begin{aligned} I_r(0, \mu) = \tfrac{3}{32} F\{ & H_r(\mu) H_r(\mu')[1+\mu\mu' + c(\mu+\mu')] \\ & + qH_r(\mu) H_l(\mu')(\mu+\mu') + (1-\mu'^2)\, H^{(2)}(\mu)\, H^{(2)}(\mu')\cos 2\varphi\} \frac{\mu'}{\mu+\mu'}\,, \end{aligned}\right\} \quad (162)$$

and

$$\left.\begin{aligned} U(0, \mu) = -\tfrac{3}{16} F\{ & 2(1-\mu^2)^{\frac{1}{2}}(1-\mu'^2)^{\frac{1}{2}}\mu' H^{(1)}(\mu)\, H^{(1)}(\mu')\sin\varphi \\ & + \mu(1-\mu'^2)\, H^{(2)}(\mu)\, H^{(2)}(\mu')\sin 2\varphi\} \frac{\mu'}{\mu+\mu'}\,, \end{aligned}\right\} \quad (163)$$

where we have written μ' for $\cos\beta$.

In Table 1 we have collected the various constants which occur in the foregoing solution in the second and the third approximations. In Tables 2 and 3 the functions $H_l(\mu)$,

TABLE 1

THE CONSTANTS OCCURRING IN THE SOLUTION FOR THE DIFFUSELY REFLECTED RADIATION

Second Approximation	Third Approximation
$k_1=2.23607$	$k_1=3.45859$
$k_2=1.08012$	$k_2=1.32757$
	$k_3=1.04677$
$\kappa_1=1.52753$	$\kappa_1=2.71838$
	$\kappa_2=1.11822$
$c=0.86830$	$c=0.87134$
$q=0.70151$	$q=0.69392$
$k_1^{(1)}=2.08921$	$k_1^{(1)}=3.39756$
$k_2^{(1)}=0.89547$	$k_2^{(1)}=1.20271$
	$k_3^{(1)}=0.91108$
$k_1^{(2)}=2.50512$	$k_1^{(2)}=3.78851$
$k_2^{(2)}=0.74680$	$k_2^{(2)}=1.32582$
	$k_3^{(2)}=0.74119$

TABLE 2

THE FUNCTIONS $H_l(\mu)$, $H_r(\mu)$, $H^{(1)}(\mu)$, AND $H^{(2)}(\mu)$ IN THE SECOND APPROXIMATION

μ	SECOND APPROXIMATION			
	H_l	H_r	$H^{(1)}$	$H^{(2)}$
0	1.000	1.000	1.000	1.000
0.10	1.253	1.065	1.097	1.075
0.20	1.499	1.112	1.171	1.134
0.30	1.741	1.147	1.230	1.184
0.40	1.979	1.175	1.278	1.226
0.50	2.214	1.197	1.319	1.262
0.60	2.448	1.216	1.354	1.294
0.70	2.680	1.231	1.384	1.323
0.80	2.911	1.244	1.411	1.348
0.90	3.141	1.256	1.434	1.371
1.00	3.370	1.265	1.455	1.391

$H_r(\mu)$, $H^{(1)}(\mu)$, and $H^{(2)}(\mu)$ are all tabulated in the second and the third approximations.[18]

An inspection of these tables reveals that the accuracy of the third approximation is probably within 1 per cent over the entire range in which the functions are defined.

TABLE 3

THE FUNCTIONS $H_l(\mu)$, $H_r(\mu)$, $H^{(1)}(\mu)$, AND $H^{(2)}(\mu)$ IN THE THIRD APPROXIMATION

μ	THIRD APPROXIMATION			
	H_l	H_r	$H^{(1)}$	$H^{(2)}$
0	1.000	1.000	1.000	1.000
0.05	1.143	1.041	1.057	1.042
0.10	1.279	1.074	1.105	1.078
0.15	1.411	1.101	1.145	1.110
0.20	1.539	1.123	1.180	1.139
0.25	1.665	1.142	1.212	1.165
0.30	1.789	1.158	1.240	1.188
0.35	1.911	1.173	1.265	1.210
0.40	2.032	1.186	1.288	1.230
0.45	2.152	1.197	1.308	1.249
0.50	2.270	1.207	1.328	1.266
0.55	2.389	1.216	1.345	1.282
0.60	2.506	1.225	1.362	1.298
0.65	2.623	1.233	1.377	1.312
0.70	2.739	1.240	1.391	1.326
0.75	2.856	1.246	1.404	1.338
0.80	2.971	1.252	1.417	1.351
0.85	3.087	1.258	1.429	1.362
0.90	3.202	1.263	1.440	1.373
0.95	3.317	1.268	1.450	1.384
1.00	3.431	1.272	1.460	1.394

As stated in the introduction, applications of the theory developed in this paper to an interpretation of Lyot's polarization measurements on the planets will be found in a forthcoming paper.

[18] I am indebted to Mrs. Frances H. Breen for assistance with these calculations.

9

ON THE RADIATIVE EQUILIBRIUM OF A STELLAR ATMOSPHERE. XIII

S. CHANDRASEKHAR

Yerkes Observatory

Received October 28, 1946

ABSTRACT

In this paper the problem of diffuse reflection, which was considered in Paper XI, is extended to include the case of reflection of a partially polarized beam. It is shown that this extension leads us to consider that a *scattering matrix* S in terms of which Helmholtz' principle of reciprocity admits a simple mathematical formulation.

1. *Introduction.*—In Paper XI[1] of this series the general equations of transfer for an atmosphere scattering radiation in accordance with Rayleigh's law[2] and allowing for the imperfect polarization of the scattered radiation have been formulated and solved for the problem of diffuse reflection of a parallel beam of unpolarized radiation by a semi-infinite plane-parallel atmosphere. In this paper we shall extend this discussion to include the case of reflection of an incident partially polarized beam, principally with the object of clarifying and reformulating Helmholtz' principle of reciprocity[3] in terms of the symmetry properties of a certain *scattering matrix*, S, which the theory naturally leads us to consider. Moreover, this notion of a scattering matrix is basic for the further developments contained in the paper following this one.

2. *The reduction of the equations of transfer for the problem of diffuse reflection of a parallel beam of partially plane-polarized radiation by a plane-parallel atmosphere.*—According to the ideas outlined in Paper XI, § 2, we can characterize a pencil of partially plane-polarized light by the quantities I_l, I_r, and U, where I_l and I_r are the intensities in two directions at right angles to each other in the plane of vibration of the electric vector and $U = (I_l - I_r) \tan 2\chi$, where χ denotes the inclination of the plane of polarization to the direction to which I_l refers. An incident parallel beam of partially plane-polarized light can therefore be defined in terms of the three fluxes, πF_l, πF_r, and πU_0 per unit area normal to the beam, the indices l and r now referring to directions in the meridian plane and at right angles to it.[4]

Let the beam be incident at an angle β normal to the boundary. Then, as in Paper XI, § 5, we can distinguish at any level between a field of diffuse radiation (which has arisen as a result of multiple scattering and which we shall characterize by I_l, I_r, and U) and the incident radiation reduced by the factor $e^{-\tau \sec \beta}$, where τ denotes the optical depth of the level under consideration. Similarly, we can also distinguish at any level between the contributions to the various source functions arising from the scattering of the reduced incident radiation and the diffuse radiation. The equations of transfer can then be written as in Paper XI, equations (53)–(55), $\mathfrak{J}_l^{(i)}$, $\mathfrak{J}_r^{(i)}$, and $\mathfrak{J}_U^{(i)}$ now representing the contributions to the respective source functions arising from the scattering of the re-

[1] *Ap. J.*, **104**, 110, 1946. This paper will be referred to as "Paper XI." The earlier paper (*Ap. J.*, **103**, 351, 1946) will be referred to as "Paper X."

[2] As in Paper XI, we include under "Rayleigh's law" only that part of it which pertains to the state of polarization and the angular distribution of the scattered radiation.

[3] Cf. M. Minneart, *Ap. J.*, **93**, 403, 1941.

[4] Cf. Fig. 1 in Paper X.

duced incident fluxes $\pi F_l e^{-\tau \sec \beta}$, $\pi F_r e^{-\tau \sec \beta}$, and $\pi U_0 e^{-\tau \sec \beta}$ prevailing at τ. Accordingly, in place of Paper XI, equations (56)–(58), we now have

$$\left.\begin{aligned}\mathfrak{J}_l^{(i)}(\tau,\mu,\varphi) &= \tfrac{3}{16}F_l e^{-\tau\sec\beta}\{2\sin^2\beta+\mu^2(3\cos^2\beta-2)-4\mu(1-\mu^2)^{\frac{1}{2}}\\ &\quad\times\cos\beta\sin\beta\cos\varphi+\mu^2\cos^2\beta\cos 2\varphi\}+\tfrac{3}{16}F_r e^{-\tau\sec\beta}\mu^2(1-\cos 2\varphi)\\ &\quad+\tfrac{3}{16}U_0 e^{-\tau\sec\beta}\{-2\mu(1-\mu^2)^{\frac{1}{2}}\sin\beta\sin\varphi+\mu^2\cos\beta\sin 2\varphi\},\end{aligned}\right\}\quad(1)$$

$$\left.\begin{aligned}\mathfrak{J}_r^{(i)}(\tau,\mu,\varphi) &= \tfrac{3}{16}F_l e^{-\tau\sec\beta}\cos^2\beta(1-\cos 2\varphi)+\tfrac{3}{16}F_r e^{-\tau\sec\beta}(1+\cos 2\varphi)\\ &\quad-\tfrac{3}{16}U_0 e^{-\tau\sec\beta}\cos\beta\sin 2\varphi,\end{aligned}\right\}\quad(2)$$

and

$$\left.\begin{aligned}\mathfrak{J}_U^{(i)}(\tau,\mu,\varphi) &= \tfrac{3}{8}F_l e^{-\tau\sec\beta}\{-2(1-\mu^2)^{\frac{1}{2}}\cos\beta\sin\beta\sin\varphi+\mu\cos^2\beta\sin 2\varphi\}\\ &-\tfrac{3}{8}F_r e^{-\tau\sec\beta}\mu\sin 2\varphi+\tfrac{3}{8}U_0 e^{-\tau\sec\beta}\{(1-\mu^2)^{\frac{1}{2}}\sin\beta\cos\varphi-\mu\cos\beta\cos 2\varphi\}.\end{aligned}\right\}\quad(3)$$

Combining Paper XI, equations (49)–(55), and the foregoing equations (1)–(3), we have the required equations of transfer. An examination of these equations indicates that the solutions for $I_l(\tau,\mu,\varphi)$, $I_r(\tau,\mu,\varphi)$, and $U(\tau,\mu,\varphi)$ must be expressible in the forms

$$\left.\begin{aligned}I_l(\tau,\mu,\varphi) &= I_l^{(0)}(\tau,\mu)+I_l^{(1)}(\tau,\mu)\cos\varphi+I_l^{(-1)}(\tau,\mu)\sin\varphi\\ &\quad+I_l^{(2)}(\tau,\mu)\cos 2\varphi+I_l^{(-2)}(\tau,\mu)\sin 2\varphi,\end{aligned}\right\}\quad(4)$$

$$I_r(\tau,\mu,\varphi) = I_r^{(0)}(\tau,\mu)+I_r^{(2)}(\tau,\mu)\cos 2\varphi+I_r^{(-2)}(\tau,\mu)\sin 2\varphi,\quad(5)$$

and

$$\begin{aligned}U(\tau,\mu,\varphi) &= U^{(1)}(\tau,\mu)\sin\varphi+U^{(-1)}(\tau,\mu)\cos\varphi+U^{(2)}(\tau,\mu)\sin 2\varphi\\ &\quad+U^{(-2)}(\tau,\mu)\sin 2\varphi.\end{aligned}\quad(6)$$

As the notation indicates, $I_l^{(0)}$, $I_l^{(1)}$, etc., are all functions of τ and μ only. Substituting these forms for the solution in the equations of transfer, we find that they break up into the following five sets of equations:

$$\left.\begin{aligned}
\mu\frac{dI_l^{(0)}}{d\tau} &= I_l^{(0)} - \frac{3}{8}\Big[\int_{-1}^{+1} I_l^{(0)}(\tau,\mu')\,[\![2(1-\mu'^2)+\mu^2(3\mu'^2-2)]\!]\,d\mu' \\
&\quad + \mu^2\int_{-1}^{+1} I_r^{(0)}(\tau,\mu')\,d\mu'\Big] - \tfrac{3}{16}F_l e^{-\tau\sec\beta}\,[2\sin^2\beta+\mu^2(3\cos^2\beta-2)] \\
&\quad - \tfrac{3}{16}F_r e^{-\tau\sec\beta}\,\mu^2, \\
\mu\frac{dI_r^{(0)}}{d\tau} &= I_r^{(0)} - \frac{3}{8}\Big[\int_{-1}^{+1} I_l^{(0)}(\tau,\mu')\,\mu'^2 d\mu' + \int_{-1}^{+1} I_r^{(0)}(\tau,\mu')\,d\mu'\Big] \\
&\quad - \tfrac{3}{16}F_l e^{-\tau\sec\beta}\cos^2\beta - \tfrac{3}{16}F_r e^{-\tau\sec\beta};
\end{aligned}\right\}\quad \text{(I)}$$

$$\left.\begin{aligned}
\mu\frac{dI_l^{(1)}}{d\tau} &= I_l^{(1)} - \tfrac{3}{4}\mu(1-\mu^2)^{\frac{1}{2}}\int_{-1}^{+1} I_l^{(1)}(\tau,\mu')\,\mu'(1-\mu'^2)^{\frac{1}{2}}\,d\mu' - \tfrac{3}{8}\,\mu(1-\mu^2)^{\frac{1}{2}} \\
&\quad \times\int_{-1}^{+1} U^{(1)}(\tau,\mu')(1-\mu'^2)^{\frac{1}{2}}d\mu' + \tfrac{3}{4}F_l e^{-\tau\sec\beta}\mu(1-\mu^2)^{\frac{1}{2}}\cos\beta\sin\beta, \\
\mu\frac{dU^{(1)}}{d\tau} &= U^{(1)} - \tfrac{3}{4}(1-\mu^2)^{\frac{1}{2}}\int_{-1}^{+1} I_l^{(1)}(\tau,\mu')\,\mu'(1-\mu'^2)^{\frac{1}{2}}d\mu' - \tfrac{3}{8}(1-\mu^2)^{\frac{1}{2}} \\
&\quad \times\int_{-1}^{+1} U^{(1)}(\tau,\mu')(1-\mu'^2)^{\frac{1}{2}}d\mu' + \tfrac{3}{4}F_l e^{-\tau\sec\beta}(1-\mu^2)^{\frac{1}{2}}\cos\beta\sin\beta;
\end{aligned}\right\}\quad \text{(II)}$$

$$\left.\begin{aligned}
\mu\frac{dI_l^{(-1)}}{d\tau} &= I_l^{(-1)} - \tfrac{3}{4}\mu(1-\mu^2)^{\frac{1}{2}}\int_{-1}^{+1} I_l^{(-1)}(\tau,\mu')\,\mu'(1-\mu'^2)^{\frac{1}{2}}\,d\mu' \\
&\quad + \tfrac{3}{8}\mu(1-\mu^2)^{\frac{1}{2}}\int_{-1}^{+1} U^{(-1)}(\tau,\mu')(1-\mu'^2)^{\frac{1}{2}}d\mu' + \tfrac{3}{8}U_0 e^{-\tau\sec\beta}\mu(1-\mu^2)^{\frac{1}{2}}\sin\beta, \\
\mu\frac{dU^{(-1)}}{d\tau} &= U^{(-1)} + \tfrac{3}{4}(1-\mu^2)^{\frac{1}{2}}\int_{-1}^{+1} I_l^{(-1)}(\tau,\mu')\,\mu'(1-\mu'^2)^{\frac{1}{2}}d\mu' \\
&\quad - \tfrac{3}{8}(1-\mu^2)^{\frac{1}{2}}\int_{-1}^{+1} U^{(-1)}(\tau,\mu')(1-\mu'^2)^{\frac{1}{2}}d\mu' - \tfrac{3}{8}U_0 e^{-\tau\sec\beta}(1-\mu^2)^{\frac{1}{2}}\sin\beta;
\end{aligned}\right\}\quad \text{(III)}$$

$$\left.\begin{aligned}
\mu\frac{dI_l^{(2)}}{d\tau} &= I_l^{(2)} - \tfrac{3}{16}\mu^2\int_{-1}^{+1} I_l^{(2)}(\tau,\mu')\,\mu'^2 d\mu' + \tfrac{3}{16}\mu^2\int_{-1}^{+1} I_r^{(2)}(\tau,\mu')\,d\mu' \\
&\quad - \tfrac{3}{16}\mu^2\int_{-1}^{+1} U^{(2)}(\tau,\mu')\,\mu' d\mu' - \tfrac{3}{16}F_l e^{-\tau\sec\beta}\mu^2\cos^2\beta + \tfrac{3}{16}F_r e^{-\tau\sec\beta}\mu^2, \\
\mu\frac{dI_r^{(2)}}{d\tau} &= I_r^{(2)} + \tfrac{3}{16}\int_{-1}^{+1} I_l^{(2)}(\tau,\mu')\,\mu'^2 d\mu' - \tfrac{3}{16}\int_{-1}^{+1} I_r^{(2)}(\tau,\mu')\,d\mu' \\
&\quad + \tfrac{3}{16}\int_{-1}^{+1} U^{(2)}(\tau,\mu')\,\mu' d\mu' + \tfrac{3}{16}F_l e^{-\tau\sec\beta}\cos^2\beta - \tfrac{3}{16}F_r e^{-\tau\sec\beta}, \\
\mu\frac{dU^{(2)}}{d\tau} &= U^{(2)} - \tfrac{3}{8}\mu\int_{-1}^{+1} I_l^{(2)}(\tau,\mu')\,\mu'^2 d\mu' + \tfrac{3}{8}\mu\int_{-1}^{+1} I_r^{(2)}(\tau,\mu')\,d\mu' \\
&\quad - \tfrac{3}{8}\mu\int_{-1}^{+1} U^{(2)}(\tau,\mu')\,\mu' d\mu' - \tfrac{3}{8}F_l e^{-\tau\sec\beta}\mu\cos^2\beta + \tfrac{3}{8}F_r e^{-\tau\sec\beta}\mu;
\end{aligned}\right\}\quad \text{(IV)}$$

and

$$\left.\begin{aligned}
\mu \frac{dI_l^{(-2)}}{d\tau} &= I_l^{(-2)} - \tfrac{3}{16}\mu^2 \int_{-1}^{+1} I_l^{(-2)}(\tau, \mu')\, \mu'^2 d\mu' + \tfrac{3}{16}\mu^2 \int_{-1}^{+1} I_r^{(-2)}(\tau, \mu')\, d\mu' \\
&\quad + \tfrac{3}{16}\mu^2 \int_{-1}^{+1} U^{(-2)}(\tau, \mu')\, \mu' d\mu' - \tfrac{3}{16} U_0 e^{-\tau \sec\beta} \mu^2 \cos\beta\,, \\
\mu \frac{dI_r^{(-2)}}{d\tau} &= I_r^{(-2)} + \tfrac{3}{16} \int_{-1}^{+1} I_l^{(-2)}(\tau, \mu')\, \mu'^2 d\mu' - \tfrac{3}{16} \int_{-1}^{+1} I_r^{(-2)}(\tau, \mu')\, d\mu' \\
&\quad - \tfrac{3}{16} \int_{-1}^{+1} U^{(-2)}(\tau, \mu')\, \mu' d\mu' + \tfrac{3}{16} U_0 e^{-\tau \sec\beta} \cos\beta\,, \\
\mu \frac{dU^{(-2)}}{d\tau} &= U^{(-2)} + \tfrac{3}{8}\mu \int_{-1}^{+1} I_l^{(-2)}(\tau, \mu')\, \mu'^2 d\mu' - \tfrac{3}{8}\mu \int_{-1}^{+1} I_r^{(-2)}(\tau, \mu')\, d\mu' \\
&\quad - \tfrac{3}{8}\mu \int_{-1}^{+1} U^{(-2)}(\tau, \mu')\, \mu' d\mu' + \tfrac{3}{8} U_0 e^{-\tau \sec\beta} \mu \cos\beta\,.
\end{aligned}\right\} \quad \text{(V)}$$

We now proceed to the solution of these equations appropriately for the problem of diffuse reflection by a semi-infinite plane-parallel atmosphere.

3. *The solution of the equations of transfer.*—Of the five systems to which we reduced the equations of transfer in § 2, the only one which requires any detailed consideration is the first; the others are equivalent to the systems already considered in Paper XI, §§ 7 and 8.

Considering, then, System (I), we observe that, as the inhomogeneous parts consist of two terms proportional, respectively, to F_l and F_r, we can express $I_l^{(0)}$ and $I_r^{(0)}$ as the sum of the appropriate solutions of

$$\left.\begin{aligned}
\mu \frac{dI_l^{(0)}}{d\tau} &= I_l^{(0)} - \frac{3}{8}\Big[\int_{-1}^{+1} I_l^{(0)}(\tau, \mu') \,[\![2(1-\mu'^2) + \mu^2 (3\mu'^2 - 2)]\!]\, d\mu' \\
&\quad + \mu^2 \int_{-1}^{+1} I_r^{(0)}(\tau, \mu')\, d\mu' \Big] - \tfrac{3}{16} F_l e^{-\tau \sec\beta} [2 \sin^2\beta + \mu^2 (3\cos^2\beta - 2)]\,, \\
\mu \frac{dI_r^{(0)}}{d\tau} &= I_r^{(0)} - \frac{3}{8}\Big[\int_{-1}^{+1} I_l^{(0)}(\tau, \mu')\, \mu'^2 d\mu' + \int_{-1}^{+1} I_r^{(0)}(\tau, \mu')\, d\mu' \Big] \\
&\quad - \tfrac{3}{16} F_l e^{-\tau \sec\beta} \cos^2\beta\,;
\end{aligned}\right\} \quad (\text{I}_a)$$

and

$$\left.\begin{aligned}
\mu \frac{dI_l^{(0)}}{d\tau} &= I_l^{(0)} - \frac{3}{8}\Big[\int_{-1}^{+1} I_l^{(0)}(\tau, \mu') \,[\![2(1-\mu'^2) + \mu^2 (3\mu'^2 - 2)]\!]\, d\mu' \\
&\quad + \mu^2 \int_{-1}^{+1} I_r^{(0)}(\tau, \mu')\, d\mu' \Big] - \tfrac{3}{16} F_r e^{-\tau \sec\beta} \mu^2\,, \\
\mu \frac{dI_r^{(0)}}{d\tau} &= I_r^{(0)} - \frac{3}{8}\Big[\int_{-1}^{+1} I_l^{(0)}(\tau, \mu')\, \mu'^2 d\mu' + \int_{-1}^{+1} I_r^{(0)}(\tau, \mu')\, d\mu' \Big] \\
&\quad - \tfrac{3}{16} F_r e^{-\tau \sec\beta}\,.
\end{aligned}\right\} \quad (\text{I}_b)$$

i) *Solution of System* (I_o).—Replacing the equations of System (I_a) by an equivalent system of $4n$ linear equations in the usual manner, we find that the equations admit particular integrals of the form

$$I_{l,i}^{(0)} = \tfrac{3}{16} F_l e^{-\tau \sec\beta} \frac{2(1-\mu_i^2)\,\Gamma}{1+\mu_i \sec\beta} \qquad (i = \pm 1, \ldots, \pm n), \quad (7)$$

and

$$I_{r,i}^{(0)} \equiv 0 \qquad (i = \pm 1, \ldots, \pm n), \quad (8)$$

where Γ has the same meaning as in Paper XI, equation (81). Combining these particular integrals for $I_{l,i}^{(0)}$ and $I_{r,i}^{(0)}$ with the corresponding solutions (Paper X, eqs. [87] and [88]) of the homogeneous equations and which are further compatible with our present requirement of the boundedness of the solution for $\tau \to \infty$, we have

$$\left.\begin{aligned} I_{l,i}^{(0)} = \tfrac{3}{16} F_l \Big[(1-\mu_i^2) \sum_{\beta=1}^{n-1} \frac{L_\beta e^{-\kappa_\beta \tau}}{1+\mu_i \kappa_\beta} + \sum_{a=1}^{n} M_a (1-k_a \mu) e^{-k_a \tau} + Q \\ + e^{-\tau \sec\beta} \frac{2(1-\mu_i^2)\,\Gamma}{1+\mu_i \sec\beta} \Big] \qquad (i = \pm 1, \ldots, \pm n) \end{aligned}\right\} \quad (9)$$

and

$$I_{r,i}^{(0)} = \tfrac{3}{16} F_l \left[Q - \sum_{a=1}^{n} \frac{M_a (k_a^2 - 1)}{1+\mu_i k_a} e^{-k_a \tau} \right] \quad (i = \pm 1, \ldots, \pm n), \quad (10)$$

where $L_\beta (\beta = 1, \ldots, n-1)$, $M_a (a = 1, \ldots, n)$, and Q are $2n$ constants of integration to be determined from the boundary conditions

$$I_{l,i}^{(0)} = I_{r,i}^{(0)} = 0 \quad \text{at} \quad \tau = 0 \quad (i = -1, \ldots, -n). \quad (11)$$

In terms of the functions

$$\left.\begin{aligned} S_l(\mu) = (1-\mu^2) \sum_{\beta=1}^{n-1} \frac{L_\beta}{1-\mu \kappa_\beta} + \sum_{a=1}^{n} M_a (1+k_a \mu) + Q \\ + (1-\mu^2) \frac{2\Gamma}{1-\mu \sec\beta} \end{aligned}\right\} \quad (12)$$

and

$$S_r(\mu) = Q - \sum_{a=1}^{n} \frac{M_a (k_a^2 - 1)}{1 - \mu k_a}, \quad (13)$$

the boundary conditions are

$$S_l(\mu_i) = S_r(\mu_i) = 0 \qquad (i = 1, \ldots, n), \quad (14)$$

while the angular distributions of the corresponding emergent intensities at $\tau = 0$ are expressible in the forms

$$I_l^{(0)}(0, \mu) = \tfrac{3}{16} F_l S_l(-\mu) \quad \text{and} \quad I_r^{(0)}(0, \mu) = \tfrac{3}{16} F_l S_r(-\mu). \quad (15)$$

To obtain explicit formulae for the functions $S_l(\mu)$ and $S_r(\mu)$, we proceed as in Paper XI in a similar evaluation.

We first show that the functions $S_l(\mu)$ and $S_r(\mu)$ must have the forms

$$S_l(\mu) = \frac{2\sin^2\beta}{\mu_1^2\ldots\mu_n^2}\frac{P(-\cos\beta)}{\rho(-\cos\beta)}\frac{P(\mu)}{\rho(\mu)}\frac{\mu+a}{(\cos\beta+a)(1-\mu\sec\beta)} \tag{16}$$

and

$$S_r(\mu) = \frac{X\cos\beta}{\mu_1^2\ldots\mu_n^2}\frac{P(-\cos\beta)}{R(-\cos\beta)}\frac{P(\mu)}{R(\mu)}, \tag{17}$$

where a and X are certain constants to be determined and the functions R, ρ, and P have the same meanings as in Paper X, equations (129), (132), and (133). From equations (13) and (17) it now follows that

$$M_\alpha = -\frac{X\cos\beta}{\mu_1^2\ldots\mu_n^2}\frac{P(-\cos\beta)}{R(-\cos\beta)}\frac{P(1/k_\alpha)}{R_\alpha(1/k_\alpha)(k_\alpha^2-1)} \quad (\alpha=1,\ldots,n), \tag{18}$$

where $R_\alpha(x)$ is defined as in Paper X, equation (129).

To determine the constants X and a, we make use of the relations (Paper XI, eqs. [100] and [102])

$$\sum_{\alpha=1}^{n} M_\alpha k_\alpha = \tfrac{1}{2}\left[S_l(+1) - S_l(-1)\right] \tag{19}$$

and

$$\sum_{\alpha=1}^{n} M_\alpha k_\alpha^2 = \tfrac{1}{2}\left[S_l(+1) + S_l(-1)\right] - S_r(0), \tag{20}$$

which follow from the definitions of the functions S_l and S_r.

Substituting for $S_l(+1)$, $S_l(-1)$, $S_r(0)$, and M_α according to equations (16)–(18) in equations (19) and (20), we find after some minor reductions that

$$\left.\begin{aligned} X\sum_{\alpha=1}^{n}\frac{P(1/k_\alpha)\,k_\alpha}{R_\alpha(1/k_\alpha)(k_\alpha^2-1)} &= \frac{R(-\cos\beta)}{\rho(-\cos\beta)(a+\cos\beta)}\Big[(a+1)(1+\cos\beta)\frac{P(1)}{\rho(1)} \\ &\quad + (a-1)(1-\cos\beta)\frac{P(-1)}{\rho(-1)}\Big] \end{aligned}\right\} \tag{21}$$

and

$$\left.\begin{aligned} X\sum_{\alpha=1}^{n}\frac{P(1/k_\alpha)\,k_\alpha^2}{R_\alpha(1/k_\alpha)(k_\alpha^2-1)} - X(-1)^n\mu_1\ldots\mu_n &= \frac{R(-\cos\beta)}{\rho(-\cos\beta)(a+\cos\beta)} \\ \times\Big[(a+1)(1+\cos\beta)\frac{P(1)}{\rho(1)} &- (a-1)(1-\cos\beta)\frac{P(-1)}{\rho(-1)}\Big]. \end{aligned}\right\} \tag{22}$$

The summations which occur on the left-hand side of these equations have been evaluated in Paper X. We have (cf. Paper X, eqs. [149], [162], and [163])

$$\sum_{\alpha=1}^{n}\frac{P(1/k_\alpha)\,k_\alpha}{R_\alpha(1/k_\alpha)(k_\alpha^2-1)} = -\frac{1}{2}\left[\frac{P(1)}{R(1)} - \frac{P(-1)}{R(-1)}\right] \tag{23}$$

and

$$\sum_{\alpha=1}^{n}\frac{P(1/k_\alpha)\,k_\alpha^2}{R_\alpha(1/k_\alpha)(k_\alpha^2-1)} = -\frac{1}{2}\left[\frac{P(1)}{R(1)} + \frac{P(-1)}{R(-1)}\right] + (-1)^n\mu_1\ldots\mu_n. \tag{24}$$

Using these results in equations (21) and (22), we obtain, after some further reductions,

$$\frac{X}{R(1)R(-\cos\beta)} = -\frac{2}{\rho(1)\,\rho(-\cos\beta)}\,\frac{(a+1)(1+\cos\beta)}{a+\cos\beta} \tag{25}$$

and

$$\frac{X}{R(-1)R(-\cos\beta)} = +\frac{2}{\rho(-1)\,\rho(-\cos\beta)}\,\frac{(a-1)(1-\cos\beta)}{a+\cos\beta}\,. \tag{26}$$

Solving these equations for X and a, we find that

$$X = q\,\frac{R(-\cos\beta)}{\rho(-\cos\beta)} \tag{27}$$

and

$$\frac{a+\mu}{a+\cos\beta}\sin^2\beta = 1-\mu\cos\beta - c(\cos\beta-\mu)\,, \tag{28}$$

where q and c are two constants having the same values as in Paper XI, equations (130) and (134).

Combining equations (16), (17), (27), and (28), we finally have

$$S_l(\mu) = \frac{2}{\mu_1^2\ldots\mu_n^2}\,\frac{P(-\cos\beta)}{\rho(-\cos\beta)}\,\frac{P(\mu)}{\rho(\mu)}\,\frac{1-\mu\cos\beta-c(\cos\beta-\mu)}{1-\mu\sec\beta} \tag{29}$$

and

$$S_r(\mu) = \frac{q\cos\beta}{\mu_1^2\ldots\mu_n^2}\,\frac{P(-\cos\beta)}{\rho(-\cos\beta)}\,\frac{P(\mu)}{R(\mu)}\,. \tag{30}$$

The corresponding laws of the emergent distributions are

$$I_l^{(0)}(0,\mu) = \tfrac{3}{16}F_l\{2H_l(\mu)\,H_l(\mu')\,[1+\mu\mu'-c(\mu+\mu')]\}\frac{\mu'}{\mu+\mu'} \tag{31}$$

and

$$I_r^{(0)}(0,\mu) = \tfrac{3}{16}F_l\,q\,H_r(\mu)\,H_l(\mu')\,\mu'\,, \tag{32}$$

where the functions $H_l(\mu)$ and $H_r(\mu)$ are defined as in Paper XI, equations (141) and (142). Further, in equations (31) and (32) we have written $\mu' = \cos\beta$.

ii) *Solution of System* (I_b).—The discussion of this system proceeds quite analogously. Thus we find that, for this system also, the boundary conditions and the angular distribution of the emergent radiation can be expressed in terms of the positive zeros of certain functions and their values for the negative arguments, respectively. The functions we have to consider in this case are (cf. Paper XI, eqs. [87] and [88])

$$\left.\begin{aligned} S_l(\mu) = (1-\mu^2)\sum_{\beta=1}^{n-1}\frac{L_\beta}{1-\mu\kappa_\beta} + \sum_{\alpha=1}^{n} M_\alpha(1+k_\alpha\mu) + Q \\ -C(1+\mu\sec\beta)\cos^2\beta \end{aligned}\right\} \tag{33}$$

and

$$S_r(\mu) = Q - \sum_{\alpha=1}^{n}\frac{M_\alpha(k_\alpha^2-1)}{1-\mu k_\alpha} + \frac{C\sin^2\beta}{1-\mu\sec\beta}\,, \tag{34}$$

where C has the same meaning as in Paper XI, equation (80).

To obtain explicit formulae for the functions $S_l(\mu)$ and $S_r(\mu)$ defined as in equations (33) and (34), we first observe that they must have the forms

$$S_l(\mu) = \frac{2Y}{\mu_1^2 \cdots \mu_n^2} \frac{P(-\cos\beta)}{\rho(-\cos\beta)} \frac{P(\mu)}{\rho(\mu)} \cos\beta \tag{35}$$

and

$$S_r(\mu) = \frac{\sin^2\beta}{\mu_1^2 \cdots \mu_n^2} \frac{P(-\cos\beta)}{R(-\cos\beta)} \frac{P(\mu)}{R(\mu)} \frac{\mu + b}{(\cos\beta + b)(1 - \mu\sec\beta)}, \tag{36}$$

where Y and b are certain constants to be determined. The reductions now required to determine these constants are very similar to those used in Paper XI, pp. 122–27. Following this procedure, we find that the equations which determine Y and b can be brought to the forms (cf. eqs. [25] and [26])

$$\frac{2Y}{\rho(1)\,\rho(-\cos\beta)} = -\frac{1}{R(1)\,R(-\cos\beta)} \frac{(b+1)(1+\cos\beta)}{b+\cos\beta} \tag{37}$$

and

$$\frac{2Y}{\rho(-1)\,\rho(-\cos\beta)} = +\frac{1}{R(-1)\,R(-\cos\beta)} \frac{(b-1)(1-\cos\beta)}{b+\cos\beta}. \tag{38}$$

Solving these equations, we find that

$$Y = \tfrac{1}{2} q \frac{\rho(-\cos\beta)}{R(-\cos\beta)} \tag{39}$$

and

$$\frac{b+\mu}{b+\cos\beta} \sin^2\beta = 1 - \mu\cos\beta + c(\cos\beta - \mu). \tag{40}$$

Substituting these equations in our expressions for $S_l(\mu)$ and $S_r(\mu)$, we have

$$S_l(\mu) = \frac{q\cos\beta}{\mu_1^2 \cdots \mu_n^2} \frac{P(-\cos\beta)}{R(-\cos\beta)} \frac{P(\mu)}{\rho(\mu)} \tag{41}$$

and

$$S_r(\mu) = \frac{1}{\mu_1^2 \cdots \mu_n^2} \frac{P(-\cos\beta)}{R(-\cos\beta)} \frac{P(\mu)}{R(\mu)} \frac{1 - \mu\cos\beta + c(\cos\beta - \mu)}{1 - \mu\sec\beta}. \tag{42}$$

The angular distributions of the emergent radiation now take the forms

$$I_l^{(0)}(0, \mu) = \tfrac{3}{16} F_r q H_l(\mu) H_r(\mu') \mu' \tag{43}$$

and

$$I_r^{(0)}(0, \mu) = \tfrac{3}{16} F_r \{ H_r(\mu) H_r(\mu') [1 + \mu\mu' + c(\mu + \mu')] \} \frac{\mu'}{\mu + \mu'}. \tag{44}$$

This completes our discussion of Systems (I_a) and (I_b).

Combining solutions (31), (32), (43), and (44) of Systems (I_a) and (I_b), we obtain the required solution of System (I). We have,

$$\left.\begin{aligned} I_l^{(0)}(0, \mu) = \tfrac{3}{16} \{ & 2H_l(\mu) H_l(\mu') [1 + \mu\mu' - c(\mu + \mu')] F_l \\ & + qH_l(\mu) H_r(\mu')(\mu + \mu') F_r \} \frac{\mu'}{\mu + \mu'} \end{aligned}\right\} \tag{45}$$

and

$$\left.\begin{aligned} I_r^{(0)}(0, \mu) &= \tfrac{3}{16}\{qH_r(\mu)H_l(\mu')(\mu+\mu')F_l \\ &\quad + H_r(\mu)H_r(\mu')[1+\mu\mu'+c(\mu+\mu')F_r]\}\frac{\mu'}{\mu+\mu'}. \end{aligned}\right\} \tag{46}$$

Turning our attention, next, to the remaining Systems (II)–(V), we observe that they are essentially equivalent to the systems considered in Paper XI, §§ 7 and 8; and the solutions for the angular distribution of the emergent intensities can be written down after a simple inspection. We have

$$\left.\begin{aligned} I_l^{(1)}(0, \mu) &= \mu U^{(1)}(0, \mu) \\ &= -\tfrac{3}{4}F_l\mu\mu'(1-\mu^2)^{\frac{1}{2}}(1-\mu'^2)^{\frac{1}{2}}H^{(1)}(\mu)H^{(1)}(\mu')\frac{\mu'}{\mu+\mu'}; \end{aligned}\right\} \tag{47}$$

$$\left.\begin{aligned} I_l^{(-1)}(0, \mu) &= -\mu U^{(-1)}(0, \mu) \\ &= -\tfrac{3}{8}U_0\mu(1-\mu^2)^{\frac{1}{2}}(1-\mu'^2)^{\frac{1}{2}}H^{(1)}(\mu)H^{(1)}(\mu')\frac{\mu'}{\mu+\mu'}; \end{aligned}\right\} \tag{48}$$

$$\left.\begin{aligned} I_l^{(2)}(0, \mu) &= -\mu^2 I_r^{(2)}(0, \mu) = \tfrac{1}{2}\mu U^{(2)}(0, \mu) \\ &= \tfrac{3}{16}(F_l\mu'^2 - F_r)\mu^2 H^{(2)}(\mu)H^{(2)}(\mu')\frac{\mu'}{\mu+\mu'}; \end{aligned}\right\} \tag{49}$$

and

$$\left.\begin{aligned} I_l^{(-2)}(0, \mu) &= -\mu^2 I_r^{(-2)}(0, \mu) = -\tfrac{1}{2}\mu U^{(-2)}(0, \mu) \\ &= \tfrac{3}{16}U_0\mu^2\mu' H^{(2)}(\mu)H^{(2)}(\mu')\frac{\mu'}{\mu+\mu'}. \end{aligned}\right\} \tag{50}$$

In the foregoing equations the functions $H^{(1)}(\mu)$ and $H^{(2)}(\mu)$ are defined in the same way as in Paper XI, equations (149) and (157).

Finally, combining solutions (45)–(50) in accordance with equations (4)–(6), we obtain the complete set of equations which determine the polarization and the angular distribution of the radiation diffusely reflected by a semi-infinite plane-parallel atmosphere. We have

$$\left.\begin{aligned} I_l(0, \mu, \varphi) &= \tfrac{3}{16}[F_l\{2H_l(\mu)H_l(\mu')[\![1+\mu\mu'-c(\mu+\mu')]\!] - 4\mu\mu' \\ &\times(1-\mu^2)^{\frac{1}{2}}(1-\mu'^2)^{\frac{1}{2}}H^{(1)}(\mu)H^{(1)}(\mu')\cos\varphi + \mu^2\mu'^2H^{(2)}(\mu)H^{(2)}(\mu')\cos 2\varphi\} \\ &+F_r\{qH_l(\mu)H_r(\mu')(\mu+\mu') - \mu^2H^{(2)}(\mu)H^{(2)}(\mu')\cos 2\varphi\} \\ &+U_0\{-2\mu(1-\mu^2)^{\frac{1}{2}}(1-\mu'^2)^{\frac{1}{2}}H^{(1)}(\mu)H^{(1)}(\mu')\sin\varphi \\ &\qquad + \mu^2\mu'H^{(2)}(\mu)H^{(2)}(\mu')\sin 2\varphi\}]\frac{\mu'}{\mu+\mu'}; \end{aligned}\right\} \tag{51}$$

$$\left.\begin{aligned} I_r(0, \mu, \varphi) &= \tfrac{3}{16}[F_l\{qH_r(\mu)H_l(\mu')(\mu+\mu') - \mu'^2H^{(2)}(\mu)H^{(2)}(\mu')\cos 2\varphi\} \\ &+F_r\{H_r(\mu)H_r(\mu')[\![1+\mu\mu'+c(\mu+\mu')]\!] + H^{(2)}(\mu)H^{(2)}(\mu')\cos 2\varphi\} \\ &\qquad - U_0\mu'H^{(2)}(\mu)H^{(2)}(\mu')\sin 2\varphi]\frac{\mu'}{\mu+\mu'}; \end{aligned}\right\} \tag{52}$$

and

$$\left.\begin{aligned} U(0,\mu,\varphi) &= \tfrac{3}{8}\,[F_l\{-2\mu'(1-\mu'^2)^{\frac{1}{2}}(1-\mu^2)^{\frac{1}{2}}H^{(1)}(\mu)\,H^{(1)}(\mu')\sin\varphi+\mu\mu'^2 \\ &\times H^{(2)}(\mu)\,H^{(2)}(\mu')\sin 2\varphi\}-F_r\mu H^{(2)}(\mu)\,H^{(2)}(\mu')\sin 2\varphi+U_0\{(1-\mu^2)^{\frac{1}{2}} \\ &\times(1-\mu'^2)^{\frac{1}{2}}H^{(1)}(\mu)\,H^{(1)}(\mu')\cos\varphi-\mu\mu'H^{(2)}(\mu)\,H^{(2)}(\mu')\cos 2\varphi\}]\frac{\mu'}{\mu+\mu'}\,. \end{aligned}\right\} \quad (53)$$

For the case of an incident beam of unpolarized light,

$$F_l = F_r = \tfrac{1}{2}F, \qquad \text{and} \qquad U_0 = 0\,, \qquad (54)$$

and the solutions reduce to the ones given in Paper XI, § 9.

4. *The principle of reciprocity; the scattering matrix.*—The solution for the problem of diffuse reflection given in the preceding section allows us to verify Helmholtz' principle of reciprocity when polarization accompanying scattering is taken into account. Stated for our problem in its conventional form, the principle asserts that the intensity diffusely reflected in a direction (μ,φ) and in a direction in the plane of vibration of the electric vector, which is inclined at an angle ψ with the meridian plane from a parallel beam incident in the direction $(-\mu',\varphi')$ and which is plane-polarized in a direction inclined at an angle χ with the meridian plane, must be expressible in the form

$$I(0;\psi,\mu,\varphi;\chi,\mu',\varphi') = \mu' H(\psi,\mu;\chi,\mu';\varphi-\varphi')\,, \qquad (55)$$

where H is a symmetrical function in the variable pair (ψ,μ) and (χ,μ'). To verify this, we first observe that, for an incident plane-polarized beam in which the plane of polarization is inclined at an angle χ with the meridian plane,

$$F_l = F\cos^2\chi; \qquad F_r = F\sin^2\chi; \qquad \text{and} \qquad U_0 = F\sin 2\chi\,, \qquad (56)$$

where πF denotes the total net flux in the beam per unit area normal to itself. Further, to obtain the intensity scattered in a direction (μ,φ) and inclined at an angle ψ with the meridian plane, we must combine the solutions given in § 3 according to (cf. Paper XI, eq. [7])

$$I(0;\psi,\mu,\varphi) = I_l(0,\mu,\varphi)\cos^2\psi + I_r(0,\mu,\varphi)\sin^2\psi + \tfrac{1}{2}U(0,\mu,\varphi)\sin 2\psi\,. \quad (57)$$

In this manner we obtain

$$\left.\begin{aligned} I(0;\psi,\mu,\varphi;\chi,\mu',\varphi') &= \tfrac{3}{16}F\,[H_l(\mu;\mu')\cos^2\chi\cos^2\psi+H_r(\mu;\mu')\sin^2\chi\sin^2\psi \\ &+q(\mu+\mu')\{H_l(\mu)\,H_r(\mu')\cos^2\psi\sin^2\chi+H_r(\mu)\,H_l(\mu')\cos^2\chi\sin^2\psi\} \\ &+H^{(1)}(\mu)H^{(1)}(\mu')(1-\mu^2)^{\frac{1}{2}}(1-\mu'^2)^{\frac{1}{2}}\{(-4\mu\mu'\cos^2\chi\cos^2\psi+\sin 2\psi\sin 2\chi) \\ &\times\cos(\varphi-\varphi')-2(\mu\cos^2\psi\sin 2\chi+\mu'\cos^2\chi\sin 2\psi)\sin(\varphi-\varphi')\} \\ &+H^{(2)}(\mu)\,H^{(2)}(\mu')\{(\mu^2\mu'^2\cos^2\chi\cos^2\psi+\sin^2\chi\sin^2\psi-\mu^2\sin^2\chi\cos^2\psi \\ &-\mu'^2\cos^2\chi\sin^2\psi-\mu\mu'\sin 2\chi\sin 2\psi)\cos 2(\varphi-\varphi')+(\mu^2\mu'\cos^2\psi\sin 2\chi \\ &+\mu\mu'^2\cos^2\chi\sin 2\psi-\mu'\sin^2\psi\sin 2\chi-\mu\sin^2\chi\sin 2\psi)\sin 2(\varphi-\varphi')\}] \\ &\times\frac{\mu'}{\mu+\mu'} \end{aligned}\right\} \quad (58)$$

where we have written

$$H_l(\mu;\mu') = 2H_l(\mu)\,H_l(\mu')\,[1+\mu\mu'-c\,(\mu+\mu')] \tag{59}$$

and

$$H_r(\mu;\mu') = H_r(\mu)\,H_r(\mu')\,[1+\mu\mu'+c\,(\mu+\mu')]\,. \tag{60}$$

It is seen that this expression has indeed the symmetry property required by the reciprocity principle. It may be further pointed out in this connection that the *total* reflected intensity,

$$I(0,\mu,\varphi) = I_l(0,\mu,\varphi) + I_r(0,\mu,\varphi)\,, \tag{61}$$

has also the symmetry required by the reciprocity principle only for the case of an incident unpolarized beam. In no other case can the principle be applied for total intensities.

The demonstration of the reciprocity principle that we have just given makes it apparent that a precise mathematical formulation of the principle as generally stated will be involved and will lack simplicity. However, we shall show that there is an alternative formulation of the principle which has a simple mathematical expression. For this purpose it is convenient to replace the subscripts l, r, and U by the indices 1, 2, and 3 and let

$$I_l = I_1\,;\qquad I_r = I_2\,;\qquad \text{and}\qquad U = I_3\,. \tag{62}$$

An incident partially polarized beam will then be characterized by the fluxes $\pi F_k (k = 1, 2, 3)$, and the solutions for the emergent intensities given in § 3 can be abbreviated in the form

$$I_i(0,\mu,\varphi) = \frac{3}{16\mu}\,q_i \sum_{k=1}^{3} S_{ik}(\mu,\varphi;\mu',\varphi')\,F_k\,, \tag{63}$$

where

$$q_1 = q_2 = 1 \qquad \text{and} \qquad q_3 = 2\,, \tag{64}$$

and S_{ik} are certain coefficients which we can directly read from equations (51)–(53). Moreover, these coefficients may be regarded as the elements of matrix $\mathbf{S}$:

$$
\left(\frac{1}{\mu}+\frac{1}{\mu'}\right)\mathbf{S}(\mu, \varphi; \mu', \varphi') =
$$

$$
\left|
\begin{array}{lll}
\begin{array}{l}
2H_l(\mu)\,H_l(\mu')\,[1+\mu\mu'-c(\mu+\mu')] \\
-4\mu\mu'(1-\mu^2)^{\frac{1}{2}}(1-\mu'^2)^{\frac{1}{2}}H^{(1)}(\mu)\,H^{(1)}(\mu') \\
\qquad\qquad \times\cos(\varphi-\varphi') \\
+\mu^2\mu'^2H^{(2)}(\mu)\,H^{(2)}(\mu')\cos 2(\varphi-\varphi')
\end{array}
&
\begin{array}{l}
qH_l(\mu)\,H_r(\mu')\,(\mu+\mu') \\
\\
\\
-\mu^2H^{(2)}(\mu)\,H^{(2)}(\mu') \\
\qquad \times\cos 2(\varphi-\varphi')
\end{array}
&
\begin{array}{l}
\\
-2\mu(1-\mu^2)^{\frac{1}{2}}(1-\mu'^2)^{\frac{1}{2}}H^{(1)}(\mu)\,H^{(1)}(\mu') \\
\qquad\qquad \times\sin(\varphi-\varphi') \\
+\mu^2\mu'H^{(2)}(\mu)\,H^{(2)}(\mu')\sin 2(\varphi-\varphi')
\end{array}
\\[2em]
\begin{array}{l}
qH_r(\mu)\,H_l(\mu')\,(\mu+\mu') \\
\\
-\mu'^2H^{(2)}(\mu)\,H^{(2)}(\mu')\cos 2(\varphi-\varphi')
\end{array}
&
\begin{array}{l}
H_r(\mu)\,H_r(\mu')\,[1+\mu\mu' \\
\qquad +c(\mu+\mu')] \\
+H^{(2)}(\mu)\,H^{(2)}(\mu') \\
\qquad \times\cos 2(\varphi-\varphi')
\end{array}
&
\begin{array}{l}
\\
\\
-\mu'H^{(2)}(\mu)\,H^{(2)}(\mu')\sin 2(\varphi-\varphi')
\end{array}
\\[2em]
\begin{array}{l}
-2\mu'(1-\mu'^2)^{\frac{1}{2}}(1-\mu^2)^{\frac{1}{2}}H^{(1)}(\mu)\,H^{(1)}(\mu') \\
\qquad\qquad \times\sin(\varphi-\varphi') \\
+\mu\mu'^2H^{(2)}(\mu)\,H^{(2)}(\mu')\sin 2(\varphi-\varphi')
\end{array}
&
\begin{array}{l}
\\
\\
-\mu H^{(2)}(\mu)\,H^{(2)}(\mu') \\
\qquad \times\sin 2(\varphi-\varphi')
\end{array}
&
\begin{array}{l}
(1-\mu^2)^{\frac{1}{2}}(1-\mu'^2)^{\frac{1}{2}}H^{(1)}(\mu)\,H^{(1)}(\mu') \\
\qquad\qquad \times\cos(\varphi-\varphi') \\
-\mu\mu'H^{(2)}(\mu)\,H^{(2)}(\mu')\cos 2(\varphi-\varphi')
\end{array}
\end{array}
\right|
\qquad (65)
$$

We shall call this the *scattering matrix;* for in terms of it the angular distribution and the state of polarization of the reflected radiation can always be specified.

We now see that among the elements of the matrix $\boldsymbol{S}$ we have the relations

$$S_{ik}\,(\mu,\ \varphi\cdot\ \mu',\ \varphi') = S_{ki}\,(\mu',\ \varphi;\ \mu,\ \varphi')\ . \tag{66}$$

In other words, if $\tilde{\boldsymbol{S}}$ denotes the transposed matrix of $\boldsymbol{S}$, then

$$\tilde{\boldsymbol{S}}\,(\mu,\ \varphi;\ \mu',\ \varphi') = \boldsymbol{S}\,(\mu',\ \varphi;\ \mu,\ \varphi')\ . \tag{67}$$

It is on account of this symmetry property of $\boldsymbol{S}$ that the reciprocity principle arises. Conversely, we may regard this symmetry of the matrix $\boldsymbol{S}$ as the expression of the reciprocity principle.

In conclusion I wish to record that this investigation was stimulated by a series of lectures which Dr. M. Minneart gave at the Yerkes Observatory during the summer of 1946.

10

ON THE RADIATIVE EQUILIBRIUM OF A STELLAR ATMOSPHERE. XV

S. CHANDRASEKHAR
Yerkes Observatory

Received February 17, 1947

ABSTRACT

In this paper the general equations of transfer for an atmosphere scattering radiation in accordance with Rayleigh's law and allowing for a partial elliptic polarization of the radiation field are formulated. It is shown that, under these conditions, we must consider, in addition to the intensities I_l and I_r in two directions at right angles to each other in the plane of the electric and the magnetic vectors, the two further quantities

$$U = (I_l - I_r) \tan 2\chi \qquad \text{and} \qquad V = (I_l - I_r) \tan 2\beta \sec 2\chi ,$$

where χ denotes the inclination of the plane of polarization to the direction to which l refers and $-\pi/2 \leqslant \beta \leqslant +\pi/2$ is an angle, the tangent of which is equal to the ratio of axes of the ellipse characterizing the state of polarization. (The sign of β depends on whether the polarization is right handed or left handed.)

It is found that the equations of transfer for I_l, I_r, and U are of exactly the same forms as in cases in which only partial plane-polarization is contemplated and that the equation for V is independent of others. On Rayleigh's law, V is scattered in accordance with a phase function, $\frac{3}{2}\cos\Theta$.

The solution of the equation for V appropriate for the problem of diffuse reflection by a semi-infinite atmosphere is also given.

1. *Introduction.*—In earlier papers[1] of this series the general equations of transfer valid for an atmosphere scattering radiation in accordance with Rayleigh's law[2] and allowing for a partial plane-polarization of the radiation field have been formulated and solved for the case of an electron-scattering atmosphere and for the problem of diffuse reflection of a partially plane-polarized beam by a plane-parallel atmosphere. In this paper we shall extend this discussion to include the case of partial elliptic polarization of the radiation field.

2. *The parametric representation of partially elliptically polarized light and its resolution into two oppositely polarized streams. The composition of elliptically polarized streams with no mutual phase relationships.*—The most convenient characterization of an arbitrarily polarized light is due to Sir George Stokes,[3] and his representation with slight modifications will be used in this paper. However, in view of the general inaccessibility of Stokes's considerations, it may be useful to have them presented in a form suitable for our purposes.

Consider, first, an elliptically polarized beam with the plane of polarization[4] inclined at an angle χ to a certain fixed direction,[5] l (say). Let β denote the angle whose tangent is equal to the ratio of the axes of the ellipse traced by the end-point of the electric vector, the numerical value of β being supposed to lie between the limits 0 and $\pi/2$ and

[1] *Ap. J.*, **103**, 351, 1946; **104**, 110, 1946; and **105**, 151, 164, 1947. These papers will be referred to as "Papers X, XI, XIII, and XIV," respectively. We shall also have occasion to refer to Paper IX (*Ap. J.* **103**, 165, 1946).

[2] As in Papers XI, XIII, and XIV, we shall include under "Rayleigh's law" only that part of it which pertains to the state of polarization and the angular distribution of the scattered radiation.

[3] *Trans. Cambridge Phil. Soc.*, **9**, 399, 1852; or *Mathematical and Physical Papers of Sir George Stokes* (Cambridge, England, 1901), **3**, 233–51.

[4] We shall take this to coincide with the plane of vibration of the electric vector.

[5] These directions are referred in the plane containing the electric and the magnetic vectors.

the sign of β being positive or negative according to whether the polarization is right handed or left handed. Finally, let $\xi^{(0)}$ denote a quantity proportional to the mean amplitude of the electric vector, whose square is equal to the intensity of the beam:

$$I = [\xi^{(0)}]^2 . \tag{1}$$

Then, for the elliptically polarized beam with the specified characteristics, the *amplitudes* ξ_χ and $\xi_{\chi+\pi/2}$ in the directions χ and $\chi + \pi/2$ can be represented in the forms

$$\xi_\chi = \xi^{(0)} \cos\beta \sin\omega t \quad \text{and} \quad \xi_{\chi+\pi/2} = \xi^{(0)} \sin\beta \cos\omega t , \tag{2}$$

where ω denotes the circular frequency of the light considered.

From equations (2) it follows that the amplitudes ξ_l and ξ_r in the direction l and in the direction r at right angles to l are

$$\xi_l = \xi^{(0)} (\cos\beta \cos\chi \sin\omega t - \sin\beta \sin\chi \cos\omega t) \tag{3}$$

and

$$\xi_r = \xi^{(0)} (\cos\beta \sin\chi \sin\omega t + \sin\beta \cos\chi \cos\omega t) . \tag{4}$$

We can re-write the foregoing expressions for ξ_l and ξ_r in the forms

$$\xi_l = \xi_l^{(0)} \sin(\omega t - \epsilon_1) \quad \text{and} \quad \xi_r = \xi_r^{(0)} \sin(\omega t - \epsilon_2) , \tag{5}$$

where

$$\left.\begin{aligned} \xi_l^{(0)} &= \xi^{(0)} (\cos^2\beta \cos^2\chi + \sin^2\beta \sin^2\chi)^{1/2} , \\ \xi_r^{(0)} &= \xi^{(0)} (\cos^2\beta \sin^2\chi + \sin^2\beta \cos^2\chi)^{1/2} , \end{aligned}\right\} \tag{6}$$

and

$$\tan\epsilon_1 = \tan\chi \tan\beta ; \qquad \tan\epsilon_2 = -\cot\chi \tan\beta . \tag{7}$$

The intensities I_l and I_r in the directions l and r are therefore given by

$$I_l = [\xi_l^{(0)}]^2 = I (\cos^2\beta \cos^2\chi + \sin^2\beta \sin^2\chi) \tag{8}$$

and

$$I_r = [\xi_r^{(0)}]^2 = I (\cos^2\beta \sin^2\chi + \sin^2\beta \cos^2\chi) . \tag{9}$$

Furthermore, from equations (6) and (7) it follows that

$$2\,\xi_l^{(0)} \xi_r^{(0)} \cos(\epsilon_1 - \epsilon_2) = I \cos 2\beta \sin 2\chi \tag{10}$$

and

$$2\,\xi_l^{(0)} \xi_r^{(0)} \sin(\epsilon_1 - \epsilon_2) = I \sin 2\beta . \tag{11}$$

Thus, whenever the amplitudes of an elliptically polarized beam in two directions at right angles to each other can be expressed in the form (5), we can at once write the relations

$$I = I_l + I_r = [\xi_l^{(0)}]^2 + [\xi_r^{(0)}]^2 , \tag{12}$$

$$Q = I_l - I_r = I \cos 2\beta \cos 2\chi = [\xi_l^{(0)}]^2 - [\xi_r^{(0)}]^2 , \tag{13}$$

$$U = (I_l - I_r) \tan 2\chi = I \cos 2\beta \sin 2\chi = 2\,\xi_l^{(0)} \xi_r^{(0)} \cos\delta , \tag{14}$$

and

$$V = (I_l - I_r) \tan 2\beta \sec 2\chi = I \sin 2\beta = 2\,\xi_l^{(0)} \xi_r^{(0)} \sin\delta , \tag{15}$$

where

$$\delta = \epsilon_1 - \epsilon_2 \tag{16}$$

denotes the difference in phase with which the components of the electric vector vibrate in the directions l and r.

It will be observed that, according to the definitions of the quantities Q, U, and V for an elliptically polarized beam,

$$I = (Q^2 + U^2 + V^2)^{1/2} . \tag{17}$$

Now Stokes has shown that any elliptically polarized beam (and therefore, as it will appear, *any* partially polarized beam) can be resolved into two other elliptically polarized beams of specified states of polarization. Thus, a beam of intensity I polarized in the direction χ and of an ellipticity corresponding to β can be expressed as the result of superposition of two beams polarized in the directions χ_1 and χ_2 and with ellipticities corresponding to β_1 and β_2, and of intensities

$$I_1 = \frac{\sin^2(\beta_2 - \beta)\cos^2(\chi_2 - \chi) + \cos^2(\beta_2 + \beta)\sin^2(\chi_2 - \chi)}{\sin^2(\beta_2 - \beta_1)\cos^2(\chi_2 - \chi_1) + \cos^2(\beta_2 + \beta_1)\sin^2(\chi_2 - \chi_1)} I \tag{18}$$

and

$$I_2 = \frac{\sin^2(\beta_1 - \beta)\cos^2(\chi_1 - \chi) + \cos^2(\beta_1 + \beta)\sin^2(\chi_1 - \chi)}{\sin^2(\beta_2 - \beta_1)\cos^2(\chi_2 - \chi_1) + \cos^2(\beta_2 + \beta_1)\sin^2(\chi_2 - \chi_1)} I , \tag{19}$$

respectively. Of these various modes of resolution, the one of greatest interest is the resolution into ***oppositely polarized beams*** when

$$\beta_2 = -\beta_1 \qquad \text{and} \qquad \chi_2 = \chi_1 + \frac{\pi}{2} , \tag{20}$$

i.e., when the ellipses described are similar, their major axes perpendicular to each other, and the direction of revolution of one contrary to that in the other. The importance of this concept of opposite polarization arises from the fact that ***oppositely polarized beams cannot interfere with one another.***

For the case of resolution into oppositely polarized beams, equations (18) and (19) become

$$I_1 = I\,[\sin^2(\beta_1 + \beta)\sin^2(\chi_1 - \chi) + \cos^2(\beta_1 - \beta)\cos^2(\chi_1 - \chi)] \tag{21}$$

and

$$I_2 = I\,[\sin^2(\beta_1 - \beta)\cos^2(\chi_1 - \chi) + \cos^2(\beta_1 + \beta)\sin^2(\chi_1 - \chi)] . \tag{22}$$

When $\beta_1 = 0$, we have the resolution of the beam into plane-polarized components at right angles to each other, and equations (21) and (22) become equivalent to equations (8) and (9).

The expression (21) for I_1 can be expanded into the form

$$\left.\begin{aligned} I_1 = \tfrac{1}{2}\,[I + I\cos 2\beta\cos 2\chi\cos 2\beta_1\cos 2\chi_1 + I\cos 2\beta\sin 2\chi\cos 2\beta_1\sin 2\chi_1 \\ + I\sin 2\beta\sin 2\beta_1] , \end{aligned}\right\} \tag{23}$$

or, in terms of the quantities Q, U, and V defined as in equations (13)–(15), we have

$$I_1 = \tfrac{1}{2}\,[I + Q\cos 2\beta_1\cos 2\chi_1 + U\cos 2\beta_1\sin 2\chi_1 + V\sin 2\beta_1) . \tag{24}$$

The corresponding expression for I_2 can be obtained by simply changing the sign of β_1 and replacing χ_1 by $\chi_1 + \pi/2$. We have

$$I_2 = \tfrac{1}{2}\,[I - Q\cos 2\beta_1\cos 2\chi_1 - U\cos 2\beta_1\sin 2\chi_1 - V\sin 2\beta_1] . \tag{25}$$

Turning, next, to the consideration of the result of the superposition of a number of *independent* streams of elliptically polarized light, i.e., of polarized streams which have no phase relationships, we start with the following principle enunciated by Stokes: "When any number of polarized streams from different sources mix together, after having been variously modified by reflection, refraction, transmission through doubly refracting media, tourmalines, etc., the intensity of the mixture is equal to the sum of the intensities due to the separate streams."

From this principle it follows, in particular, that, if each of the separate streams is resolved into two states of opposite polarization, the resultant mixture will have intensities in the two states which are the sum of respective intensities of the separate streams. Thus, if $I^{(n)}$, $\chi^{(n)}$, and $\beta^{(n)}$ define the intensity, the plane of polarization, and the ellipticity of a typical stream and if $I_1^{(n)}$ and $I_2^{(n)}$ are its intensities in the states of polarization (β_1, χ_1) and $(-\beta_1, \chi_1 + \pi/2)$, respectively, then

$$I_1 = \Sigma I_1^{(n)} \quad \text{and} \quad I_2 = \Sigma I_2^{(n)} \tag{26}$$

where I_1 and I_2 refer to the mixture.

Using equation (24) to express the intensities of the various streams in the component (β_1, χ_1), we have

$$I_1 = \tfrac{1}{2}\left[\Sigma I^{(n)} + \Sigma Q^{(n)} \cos 2\beta_1 \cos 2\chi_1 + \Sigma U^{(n)} \cos 2\beta_1 \sin 2\chi_1 + \Sigma V^{(n)} \sin 2\beta_1\right]. \tag{27}$$

The intensity of the resultant mixture in the state of polarization (χ_1, β_1) can therefore be expressed in the form

$$I_1 = \tfrac{1}{2}\left[I + Q \cos 2\beta_1 \cos 2\chi_1 + U \cos 2\beta_1 \sin 2\chi_1 + V \sin 2\beta_1\right], \tag{28}$$

where

$$I = \Sigma I^{(n)}; \quad Q = \Sigma Q^{(n)}; \quad U = \Sigma U^{(n)} \cdot \quad V = \Sigma V^{(n)}. \tag{29}$$

We have, of course, a similar expression for I_2.

From equations (28) and (29) it follows that a beam resulting from the superposition of a number of independent streams of elliptically polarized light can again be characterized by a set of parameters, I, Q, U, and V, which are the sums of the respective parameters characterizing the individual streams. Moreover, any two polarized beams characterized by the same set of parameters, I, Q, U, and V, will be *optically equivalent* in the sense that "they will present exactly the same appearance on being viewed through a crystal followed by a Nicol's prism or other analyzer" (Stokes). On the other hand, it should be noted that, for a polarized beam obtained in this manner, the relation (17) (derived for an elliptically polarized beam) will not, in general, be valid, indicating the fact that the mixture of a number of independent streams of elliptically polarized light will, in general, lead to a beam which is only partially polarized. However, it is clear that, under the circumstances of partial polarization, we can regard the light as a mixture of an unpolarized beam of natural light, of intensity

$$I^{(u)} = I - (Q^2 + U^2 + V^2)^{1/2}, \tag{30}$$

and a polarized beam, of intensity

$$I^{(p)} = (Q^2 + U^2 + V^2)^{1/2}, \tag{31}$$

the plane of polarization and the ratio of the axes of the ellipse of this polarized part being given by

$$\tan 2\chi = \frac{U}{Q} \quad \text{and} \quad \sin 2\beta = \frac{V}{(Q^2 + U^2 + V^2)^{1/2}}. \tag{32}$$

An alternative way of regarding a partially polarized beam, defined in terms of the parameters I, Q, U, and V, is to express it as the resultant of two streams in the states of opposite polarization (β, χ) and $(-\beta, \chi + \pi/2)$, where χ and β are given by equations (32). The intensities of the component streams in the two states can be readily written down when it is remembered that unpolarized, or natural, light is equivalent to any two independent oppositely polarized streams of half the intensity. The unpolarized part (30) of the beam is, therefore, equivalent to a mixture of two independent polarized beams, each of intensity $\frac{1}{2}I^{(u)}$, in the states of polarization (β, χ) and $(-\beta, \chi + \pi/2)$. Combining the former with the polarized part $I^{(p)}$ (eq. [31]) in the same state of polarization, we conclude that a beam characterized by the parameters I, Q, U, and V is equivalent to two independent beams of elliptically polarized light, of intensities

$$I^{(+)} = \tfrac{1}{2}\left[I + (Q^2 + U^2 + V^2)^{1/2}\right] \tag{33}$$

and

$$I^{(-)} = \tfrac{1}{2}\left[I - (Q^2 + U^2 + V^2)^{1/2}\right], \tag{34}$$

in the states of polarization

$$(\beta, \chi) \quad \text{and} \quad \left(-\beta, \chi + \frac{\pi}{2}\right), \tag{35}$$

respectively, where

$$\chi = \tfrac{1}{2}\tan^{-1}\frac{U}{Q} \quad \text{and} \quad \beta = \tfrac{1}{2}\sin^{-1}\frac{V}{(Q^2 + U^2 + V^2)^{1/2}}. \tag{36}$$

The particular importance of this resolution arises from the fact that we may regard the two streams $I^{(+)}$ and $I^{(-)}$ as *independent.*

3. *The source functions* $\mathfrak{J}_l$, $\mathfrak{J}_r$, $\mathfrak{J}_U$, *and* $\mathfrak{J}_V$.—From the discussion of Stokes's representation of partially polarized light in the preceding section, it follows that, for the purposes of formulating the equation of transfer, we may uniquely characterize the radiation field at any given point by the four intensities $I_l(\vartheta, \varphi)$, $I_r(\vartheta, \varphi)$, $U(\vartheta, \varphi)$, and $V(\vartheta, \varphi)$, where ϑ and φ are the polar angles, referred to an appropriately chosen co-ordinate system through the point under consideration (see Fig. 1 in Paper X), and l and r refer to the directions in the meridian plane and at right angles to it, respectively.

To evaluate the source functions, $\mathfrak{J}_l(\vartheta, \varphi)$, $\mathfrak{J}_r(\vartheta, \varphi)$, $\mathfrak{J}_U(\vartheta, \varphi)$, and $\mathfrak{J}_V(\vartheta, \varphi)$, appropriate to any point in the atmosphere, we shall first consider the contributions to these source functions arising from the scattering of the radiation in the direction (ϑ', φ').

The radiation in the direction (ϑ', φ') will be characterized by the intensities $I_l(\vartheta', \varphi)'$, $I_r(\vartheta', \varphi')$, $U(\vartheta', \varphi')$, and $V(\vartheta', \varphi')$. Equivalently, we may also characterize it by the intensities

$$I^{(+)}(\vartheta', \varphi') = \tfrac{1}{2}\left\{I_l + I_r + \left[(I_l - I_r)^2 + U^2 + V^2\right]^{1/2}\right\}_{\vartheta', \varphi'} \tag{37}$$

and

$$I^{(-)}(\vartheta', \varphi') = \tfrac{1}{2}\left\{I_l + I_r - \left[(I_l - I_r)^2 + U^2 + V^2\right]^{1/2}\right\}_{\vartheta', \varphi'}, \tag{38}$$

in the states of opposite polarization,

$$(\beta, \chi) \quad \text{and} \quad \left(-\beta, \chi + \frac{\pi}{2}\right), \tag{39}$$

respectively, where

$$\left.\begin{aligned} \tan 2\chi &= \frac{U(\vartheta', \varphi')}{I_l(\vartheta', \varphi') - I_r(\vartheta', \varphi')} \\ \text{and} \quad \sin 2\beta &= \frac{V(\vartheta', \varphi')}{\sqrt{[I_l(\vartheta', \varphi') - I_r(\vartheta', \varphi')]^2 + U^2(\vartheta', \varphi') + V^2(\vartheta', \varphi')}}. \end{aligned}\right\} \tag{40}$$

And, as we have already explained in § 2, in this type of resolution the polarized components $I^{(+)}(\vartheta', \varphi')$ and $I^{(-)}(\vartheta', \varphi')$ may be regarded as uncorrelated in their phases.

We may accordingly consider the scattering of the radiation in the direction (ϑ', φ') as the result of scattering of the two independent, oppositely polarized components, $I^{(+)}(\vartheta', \varphi')$ and $I^{(-)}(\vartheta', \varphi')$. Each of these polarized components will give rise to polarized scattered beams, which will again have no correlation in phases. In other words, the radiation scattered in the direction (ϑ, φ) from other directions can be considered as a mixture of a very large number of independent, elliptically polarized streams and can, therefore, be combined according to the additive law of composition of the parameters I_l, I_r, U, and V.

Consider, then, the scattering of the elliptically polarized component $I^{(+)}(\vartheta', \varphi')$ in the direction (ϑ', φ') and confined to an element of solid angle, $d\omega'$, into the direction (ϑ, φ). Introducing the qualities, ξ's, as in § 2, we may express the amplitudes $\xi_\chi^{(+)}$ and $\xi_{\chi+\pi/2}^{(+)}$ in the forms

$$\xi_\chi^{(+)} = \xi^{(+,\,0)} \cos\beta \sin\omega t \quad \text{and} \quad \xi_{\chi+\pi/2}^{(+)} = \xi^{(+,\,0)} \sin\beta \cos\omega t\,, \tag{41}$$

where $\xi^{(+,\,0)}$ is so defined that (cf. eq. [1])

$$I^{(+)}(\vartheta', \varphi') = [\xi^{(+,\,0)}]^2\,. \tag{42}$$

The amplitude, $\xi_\chi^{(+)}$, when scattered in the direction (ϑ, φ) will lead to amplitudes in the meridian plane through (ϑ, φ) and at right angles to it which are proportional, respectively, to (cf. Paper XI, eqs. [31] and [32])

$$\xi_\chi^{(+)}[(l, l)\cos\chi + (r, l)\sin\chi] \tag{43}$$

and

$$\xi_\chi^{(+)}[(l, r)\cos\chi + (r, r)\sin\chi]\,, \tag{44}$$

where (cf. Paper XI, eq. [33]),

$$\left.\begin{aligned}(l, l) &= \sin\vartheta \sin\vartheta' + \cos\vartheta \cos\vartheta' \cos(\varphi' - \varphi)\,,\\ (r, l) &= +\cos\vartheta \sin(\varphi' - \varphi)\,,\\ (l, r) &= -\cos\vartheta' \sin(\varphi' - \varphi)\,,\\ (r, r) &= \cos(\varphi' - \varphi)\,.\end{aligned}\right\} \tag{45}$$

Similarly, the amplitude $\xi_{\chi+\pi/2}^{(+)}$ will lead to scattered amplitudes in the directions, parallel, respectively, perpendicular to the meridian plane through (ϑ, φ) of amounts proportional to

$$\xi_{\chi+\pi/2}^{(+)}[-(l, l)\sin\chi + (r, l)\cos\chi] \tag{46}$$

and

$$\xi_{\chi+\pi/2}^{(+)}[-(l, r)\sin\chi + (r, r)\cos\chi]\,. \tag{47}$$

The phase relationship between $\xi_\chi^{(+)}$ and $\xi_{\chi+\pi/2}^{(+)}$ will be maintained in these scattered amplitudes and must, therefore, be added as amplitudes with the correct phase differences. Therefore, the elliptically polarized component, $I^{(+)}(\vartheta', \varphi')$, of the radiation in the direction (ϑ', φ'), when scattered in the direction (ϑ, φ), will give rise to an elliptically polarized beam, the amplitudes of which in the meridian plane and at right angles to it will be proportional, respectively, to

$$\xi_l^{(s)} = \xi^{(+,\,0)}(A_\chi \cos\beta \sin\omega t + A_{\chi+\pi/2} \sin\beta \cos\omega t) \tag{48}$$

and

$$\xi_r^{(s)} = \xi^{(+,\,0)}(B_\chi \cos\beta \sin\omega t + B_{\chi+\pi/2} \sin\beta \cos\omega t)\,, \tag{49}$$

where, for the sake of brevity, we have written

$$\left.\begin{aligned} A_\chi &= +(l,l)\cos\chi + (r,l)\sin\chi\,,\\ B_\chi &= +(l,r)\cos\chi + (r,r)\sin\chi\,,\\ A_{\chi+\pi/2} &= -(l,l)\sin\chi + (r,l)\cos\chi\,,\\ B_{\chi+\pi/2} &= -(l,r)\sin\chi + (r,r)\cos\chi\,. \end{aligned}\right\} \qquad (50)$$

We can now re-write equations (48) and (49) in the standard form (cf. eq. [5]),

$$\left.\begin{aligned} \xi_l^{(s)} &= \xi_l^{(s,\,0)}\sin(\omega t-\epsilon_1)\,,\\ \xi_r^{(s)} &= \xi_r^{(s,\,0)}\sin(\omega t-\epsilon_2)\,, \end{aligned}\right\} \qquad (51)$$

where

$$\left.\begin{aligned} \xi_l^{(s,\,0)} &= \xi^{(+,\,0)}(A_\chi^2\cos^2\beta + A_{\chi+\pi/2}^2\sin^2\beta)^{1/2}\,,\\ \xi_r^{(s,\,0)} &= \xi^{(+,\,0)}(B_\chi^2\cos^2\beta + B_{\chi+\pi/2}^2\sin^2\beta)^{1/2}\,, \end{aligned}\right\} \qquad (52)$$

and

$$\tan\epsilon_1 = -\frac{A_{\chi+\pi/2}}{A_\chi}\tan\beta\,;\qquad \tan\epsilon_2 = -\frac{B_{\chi+\pi/2}}{B_\chi}\tan\beta\,. \qquad (53)$$

With the amplitudes of the scattered radiation expressed in the form (51), we can write down the contributions, $d\mathfrak{J}_l^{(+)}(\vartheta,\varphi;\vartheta',\varphi')$, $d\mathfrak{J}_r^{(+)}(\vartheta,\varphi;\vartheta'\,\varphi')$, $d\mathfrak{J}_U^{(+)}(\vartheta,\varphi;\vartheta'\,\varphi')$, and $d\mathfrak{J}_V^{(+)}(\vartheta,\varphi;\vartheta',\varphi')$, to the source functions, $\mathfrak{J}_l(\vartheta,\varphi)$, $\mathfrak{J}_r(\vartheta,\varphi)$, $\mathfrak{J}_U(\vartheta,\varphi)$, and $\mathfrak{J}_V(\vartheta,\varphi)$, arising from the scattering of the elliptically polarized component, $I^{(+)}(\vartheta',\varphi')$, of the radiation in the direction (ϑ',φ') and confined to an element of solid angle, $d\omega'$, in the following forms (cf. eqs. [8]–[15]):

$$\left.\begin{aligned} d\mathfrak{J}_l^{(+)}(\vartheta,\varphi;\vartheta',\varphi') &= \frac{3}{8\pi}[\xi_l^{(s,\,0)}]^2 d\omega'\,,\\ d\mathfrak{J}_r^{(+)}(\vartheta,\varphi;\vartheta',\varphi') &= \frac{3}{8\pi}[\xi_r^{(s,\,0)}]^2 d\omega'\,,\\ d\mathfrak{J}_U^{(+)}(\vartheta,\varphi;\vartheta',\varphi') &= \frac{3}{8\pi}[2\xi_l^{(s,\,0)}\xi_r^{(s,\,0)}\cos(\epsilon_1-\epsilon_2)]\,d\omega'\,,\\ d\mathfrak{J}_V^{(+)}(\vartheta,\varphi;\vartheta',\varphi') &= \frac{3}{8\pi}[2\xi_l^{(s,\,0)}\xi_r^{(s,\,0)}\sin(\epsilon_1-\epsilon_2)]\,d\omega'\,. \end{aligned}\right\} \qquad (54)$$

The various quantities which occur on the right-hand sides of the foregoing expressions can be evaluated according to equations (50)–(53). Thus

$$\left.\begin{aligned} [\xi_l^{(s,\,0)}]^2 &= [\xi^{(+,\,0)}]^2(A_\chi^2\cos^2\beta + A_{\chi+\pi/2}^2\sin^2\beta)\\ &= I^{(+)}(\vartheta',\varphi')\{(l,l)^2[\cos^2\chi\cos^2\beta+\sin^2\chi\sin^2\beta]\\ &\quad+(r,l)^2[\sin^2\chi\cos^2\beta+\cos^2\chi\sin^2\beta]+(l,l)(r,l)\sin 2\chi\cos 2\beta\}\,; \end{aligned}\right\} \qquad (55)$$

or, using equations (8), (9), and (14), we have

$$[\xi_l^{(s,\,0)}]^2 = (l,l)^2 I_l^{(+)}(\vartheta',\varphi') + (r,l)^2 I_r^{(+)}(\vartheta',\varphi') + (l,l)(r,l)\,U^{(+)}(\vartheta',\varphi')\,. \qquad (56)$$

Similarly,

$$[\xi_r^{(s,\,0)}]^2 = (l,r)^2 I_l^{(+)}(\vartheta',\varphi') + (r,r)^2 I_r^{(+)}(\vartheta',\varphi') + (l,r)(r,r)\,U^{(+)}(\vartheta',\varphi')\,. \qquad (57)$$

Again, according to equations (50), (52), and (53),

$$\left.\begin{aligned} 2\,\xi_l^{(s,\,0)}\xi_r^{(s,\,0)}\cos(\epsilon_1-\epsilon_2) &= 2I^{(+)}(\vartheta',\varphi')\,(A_\chi B_\chi\cos^2\beta + A_{\chi+\pi/2}B_{\chi+\pi/2}\sin^2\beta) \\ &= 2I^{(+)}(\vartheta',\varphi')\,\{(l,l)(l,r)\,[\cos^2\chi\cos^2\beta+\sin^2\chi\sin^2\beta] \\ &\quad + (r,l)(r,r)\,[\sin^2\chi\cos^2\beta+\cos^2\chi\sin^2\beta] \\ &\quad + [(l,l)(r,r)+(r,l)(l,r)]\cos\chi\sin\chi\cos 2\beta\} \end{aligned}\right\} \quad (58)$$

or

$$\left.\begin{aligned} 2\,\xi_l^{(s,\,0)}\xi_r^{(s,\,0)}\cos(\epsilon_1-\epsilon_2) &= 2(l,l)(l,r)\,I_l^{(+)}(\vartheta',\varphi')+2(r,l)(r,r)\,I_r^{(+)}(\vartheta',\varphi') \\ &\quad +[(l,l)(r,r)+(r,l)(l,r)]\,U^{(+)}(\vartheta',\varphi')\,. \end{aligned}\right\} \quad (59)$$

And, finally,

$$\left.\begin{aligned} 2\,\xi_l^{(s,\,0)}\xi_r^{(s,\,0)}\sin(\epsilon_1-\epsilon_2) &= I^{(+)}(\vartheta',\varphi')\,(A_\chi B_{\chi+\pi/2}-B_\chi A_{\chi+\pi/2})\sin 2\beta \\ &= I^{(+)}(\vartheta',\varphi')\,[(l,l)(r,r)-(r,l)(l,r)]\sin 2\beta\,, \end{aligned}\right\} \quad (60)$$

or

$$2\,\xi_l^{(s,\,0)}\xi_r^{(s,\,0)}\sin(\epsilon_1-\epsilon_2) = V^{(+)}(\vartheta',\varphi')\,[(l,l)(r,r)-(r,l)(l,r)]\,. \quad (61)$$

Equations (54) now become

$$\left.\begin{aligned} d\mathfrak{J}_l^{(+)}(\vartheta,\varphi;\vartheta',\varphi') &= \frac{3}{8\pi}\{(l,l)^2 I_l^{(+)}(\vartheta',\varphi') + (r,l)^2 I_r^{(+)}(\vartheta',\varphi') \\ &\quad + (l,l)(r,l)\,U^{(+)}(\vartheta',\varphi')\}\,d\omega'\,, \end{aligned}\right\} \quad (62)$$

$$\left.\begin{aligned} d\mathfrak{J}_r^{(+)}(\vartheta,\varphi;\vartheta',\varphi') &= \frac{3}{8\pi}\{(l,r)^2 I_l^{(+)}(\vartheta',\varphi') + (r,r)^2 I_r^{(+)}(\vartheta',\varphi') \\ &\quad + (l,r)(r,r)\,U^{(+)}(\vartheta',\varphi')\}\,d\omega'\,, \end{aligned}\right\} \quad (63)$$

$$\left.\begin{aligned} d\mathfrak{J}_U^{(+)}(\vartheta,\varphi;\vartheta',\varphi') &= \frac{3}{8\pi}\{2(l,l)(l,r)\,I_l^{(+)}(\vartheta',\varphi')+2(r,l)(r,r)\,I_r^{(+)}(\vartheta',\varphi') \\ &\quad +[(l,l)(r,r)+(r,l)(l,r)]\,U^{(+)}(\vartheta',\varphi')\}\,d\omega'\,, \end{aligned}\right\} \quad (64)$$

and

$$d\mathfrak{J}_V^{(+)}(\vartheta,\varphi;\vartheta',\varphi') = \frac{3}{8\pi}[(l,l)(r,r)-(r,l)(l,r)]\,V^{(+)}(\vartheta',\varphi')\,d\omega'. \quad (65)$$

The corresponding contributions to the various source functions arising from the scattering of the other polarized component, $I^{(-)}(\vartheta',\varphi')$, in the opposite state of polarization, $(-\beta,\ \chi+\pi/2)$, can be obtained by simply writing $(-)$ in place of $(+)$ in the foregoing equations. And, since the intensities I_l, I_r, U, and V are simply additive when streams of polarized light with no correlation in their phases are mixed, it is clear that the contributions to the various source functions arising from the scattering of the radiation $[I_l(\vartheta',\varphi'),\ I_r(\vartheta',\varphi'),\ U(\vartheta',\varphi'),\ V(\vartheta',\varphi')]$ in the direction (ϑ',φ') and confined to an element of solid angle, $d\omega'$, are

$$\left.\begin{aligned} d\mathfrak{J}_l(\vartheta,\varphi;\vartheta',\varphi') &= \frac{3}{8\pi}\{(l,l)^2 I_l(\vartheta',\varphi') + (r,l)^2 I_r(\vartheta',\vartheta') \\ &\quad + (l,l)(r,l)\,U(\vartheta',\varphi')\}\,d\omega'\,, \end{aligned}\right\} \quad (66)$$

$$\left.\begin{aligned} d\mathfrak{J}_r(\vartheta,\varphi;\vartheta',\varphi') &= \frac{3}{8\pi}\{(l,r)^2 I_l(\vartheta',\varphi') + (r,r)^2 I_r(\vartheta',\varphi') \\ &\quad + (l,r)(r,r)\,U(\vartheta',\varphi')\}\,d\omega'\,, \end{aligned}\right\} \quad (67)$$

$$\left.\begin{aligned} d\mathfrak{J}_U(\vartheta,\varphi;\vartheta',\varphi') &= \frac{3}{8\pi}\{2(l,l)(l,r)\,I_l(\vartheta',\varphi')+2(r,l)(r,r)\,I_r(\vartheta',\varphi') \\ &\quad +[(l,l)(r,r)+(r,l)(l,r)]\,U(\vartheta',\varphi')\}\,d\omega'\,, \end{aligned}\right\} \quad (68)$$

and

$$d\mathfrak{J}_V(\vartheta, \varphi; \vartheta', \varphi') = \frac{3}{8\pi}[(l, l)(r, r) - (r, l)(l, r)]\, V(\vartheta', \varphi')\, d\omega'. \tag{69}$$

Integrating these equations over the unit sphere, we obtain the required source functions. We have

$$\left.\begin{aligned}\mathfrak{J}_l(\mu, \varphi) = \frac{3}{8\pi}\int_{-1}^{+1}\int_0^{2\pi} \{ &(l, l)^2 I_l(\mu', \varphi') + (r, l)^2 I_r(\mu', \varphi') \\ &+ (l, l)(r, l)\, U(\mu', \varphi')\}\, d\mu' d\varphi' ,\end{aligned}\right\} \tag{70}$$

$$\left.\begin{aligned}\mathfrak{J}_r(\mu, \varphi) = \frac{3}{8\pi}\int_{-1}^{+1}\int_0^{2\pi} \{ &(l, r)^2 I_l(\mu', \varphi') + (r, r)^2 I_r(\mu', \varphi') \\ &+ (l, r)(r, r)\, U(\mu', \varphi')\}\, d\mu' d\varphi' ,\end{aligned}\right\} \tag{71}$$

$$\left.\begin{aligned}\mathfrak{J}_U(\mu, \varphi) = \frac{3}{8\pi}\int_{-1}^{+1}\int_0^{2\pi} \{ &2(l, l)(l, r) I_l(\mu', \varphi') + 2(r, l)(r, r) I_r(\mu', \varphi') \\ &+ [(l, l)(r, r) + (r, l)(l, r)]\, U(\mu', \varphi')\}\, d\mu' d\varphi' ,\end{aligned}\right\} \tag{72}$$

and

$$\mathfrak{J}_V(\mu, \varphi) = \frac{3}{8\pi}\int_{-1}^{+1}\int_0^{2\pi} [(l, l)(r, r) - (r, l)(l, r)]\, V(\mu', \varphi')\, d\mu' d\varphi' , \tag{73}$$

where the direction cosines, μ and μ', have been used in place of $\cos \vartheta$ and $\cos \vartheta'$.

Comparing equations (70)–(72) with the corresponding equations obtained in Paper XI (eqs. [49]–[51]) for a partially plane-polarized radiation field, we observe that they are of identical forms. The intensities I_l, I_r, and U, therefore, satisfy the same equations of transfer as in cases in which only partial plane-polarization is contemplated. The equations of transfer for I_l, I_r, and U can therefore be expressed quite generally in vector form as in Paper XIV, § 2. However, when elliptic polarization is contemplated,[6] an additional equation for V must be considered. This equation is, however, independent of the others and is given by (cf. eq. [73])

$$\left.\begin{aligned}\frac{dV(s, \mu, \varphi)}{\rho\sigma\, ds} = -V(s, \mu, \varphi) + \frac{3}{8\pi}\int_{-1}^{+1}\int_0^{2\pi} &[(l, l)(r, r) - (r, l)(l, r)] \\ &\times V(s, \mu', \varphi')\, d\mu' d\varphi' ,\end{aligned}\right\} \tag{74}$$

where s measures the linear distance in the direction (ϑ, φ) and σ denotes the mass-scattering coefficient.

Finally, we may notice that, according to equations (45),

$$(l, l)(r, r) - (r, l)(l, r) = \mu\mu' + (1 - \mu^2)(1 - \mu'^2)^{1/2} \cos(\varphi - \varphi'). \tag{75}$$

The intensity V *is therefore scattered in accordance with a phase function* $\frac{3}{2}\cos\Theta$.

4. *The solution of the equation of transfer for* V *for the problem of diffuse reflection by a semi-infinite plane-parallel atmosphere.*—In the problem of diffuse reflection we distinguish, as usual, between the part of the incident radiation which penetrates to various depths and the diffuse scattered radiation. Similarly, we also distinguish between the contributions to the source function arising from the scattering of the reduced incident radiation prevailing at any level and from the scattering of the diffuse radiation.

If πV_0 denotes the flux in V incident as a parallel beam on a plane-parallel atmosphere

[6] It is, of course, evident—and it is, indeed, required by the equations of transfer—that, on Rayleigh scattering, partial elliptic polarization in the radiation field can be induced only by the direct incidence of such radiation.

at an angle $\cos^{-1}\mu_0$ normal to the boundary, the equation of transfer (74) can be rewritten in the following form:

$$\left.\begin{aligned}\mu\frac{dV(\tau,\mu,\varphi)}{d\tau} &= V(\tau,\mu,\varphi)\\ -\frac{3}{8\pi}\int_{-1}^{+1}\int_0^{2\pi}&[\mu\mu'+(1-\mu^2)^{1/2}(1-\mu'^2)^{1/2}\cos(\varphi-\varphi')]\,V(\mu',\varphi')\,d\mu'd\varphi'\\ &-\tfrac{3}{8}V_0[-\mu\mu_0+(1-\mu^2)^{1/2}(1-\mu_0^2)^{1/2}\cos\varphi]\,e^{-\tau/\mu_0}\,.\end{aligned}\right\}\tag{76}$$

From equation (76) it follows that the solution $V(\tau,\mu,\varphi)$ must be expressible in the form

$$V(\tau,\mu,\varphi) = V^{(0)}(\tau,\mu)+V^{(1)}(\tau,\mu)\cos\varphi\,,\tag{77}$$

where, as the notation implies, $V^{(0)}$ and $V^{(1)}$ are functions of τ and μ only. Equation (76) now breaks up into the following two equations:

$$\mu\frac{dV^{(0)}}{d\tau} = V^{(0)}-\tfrac{3}{4}\mu\int_{-1}^{+1}V(\tau,\mu')\,\mu'd\mu'+\tfrac{3}{8}V_0\mu\mu_0e^{-\tau/\mu_0}\tag{78}$$

and

$$\left.\begin{aligned}\mu\frac{dV^{(1)}}{d\tau} = V^{(1)}-\tfrac{3}{8}(1-\mu^2)^{1/2}\int_{-1}^{+1}&V^{(1)}(\tau,\mu')(1-\mu'^2)^{1/2}d\mu'\\ &-\tfrac{3}{8}V_0(1-\mu^2)^{1/2}(1-\mu'^2)^{1/2}e^{-\tau/\mu_0}\,.\end{aligned}\right\}\tag{79}$$

Equations (78) and (79) are of the standard form considered in Paper IX, § 4. We can therefore write down at once the angular distribution of the radiation reflected from a semi-infinite atmosphere. We have

$$V^{(0)}(0,\mu) = -\tfrac{3}{8}V_0\mu\mu_0H_v(\mu)\,H_v(\mu_0)\frac{\mu_0}{\mu+\mu_0}\tag{80}$$

and

$$V^{(1)}(0,\mu) = \tfrac{3}{8}V_0(1-\mu^2)^{1/2}(1-\mu'^2)^{1/2}H_v^{(1)}(\mu)\,H_v^{(1)}(\mu_0)\frac{\mu_0}{\mu+\mu_0}\,,\tag{81}$$

where, in the nth approximation of the method of solution of the earlier papers, H_v and $H_v^{(1)}$ are H-functions (cf. Paper XIV) defined in terms of the roots of the characteristic equations,

$$1 = \frac{3}{2}\sum_{j=1}^{n}\frac{a_j\mu_j^2}{1-k^2\mu_j^2}\tag{82}$$

and

$$1 = \frac{3}{4}\sum_{j=1}^{n}\frac{a_j(1-\mu_j^2)}{1-k^2\mu_j^2}\,,\tag{83}$$

respectively.

It will be noticed that the characteristic equation defining $H_v^{(1)}$ is the same as the one defining $H_r(\mu)$ in Papers X and XI. Hence $H_v^{(1)}(\mu)$ is identical with $H_r(\mu)$.

Combining solutions (80) and (81) in accordance with equation (77), we have the law of reflection:

$$\left.\begin{aligned}V(\mu,\varphi;\mu_0,\varphi_0) &= \tfrac{3}{8}V_0[-\mu\mu_0H_v(\mu)\,H_v(\mu_0)\\ &+(1-\mu^2)^{1/2}(1-\mu_0^2)^{1/2}H_r(\mu)\,H_r(\mu_0)\cos(\varphi-\varphi_0)]\frac{\mu_0}{\mu+\mu_0}\,.\end{aligned}\right\}\tag{84}$$

A table of the function $H_r(\mu)$ in the third approximation will be found in Paper XI (Table 3). A similar tabulation of the function H_v is now provided (Table 1).

TABLE 1

THE FUNCTION $H_v(\mu)$ IN THE THIRD APPROXIMATION

μ	$H_v(\mu)$	μ	$H_v(\mu)$	μ	$H_v(\mu)$
0	1.000	0.35	1.107	0.70	1.168
0.05	1.021	.40	1.118	0.75	1.175
.10	1 039	45	1.128	0.80	1.182
.15	1.055	.50	1.137	0.85	1.188
.20	1.070	.55	1.146	0.90	1.193
.25	1.084	.60	1.154	0.95	1.199
0.30	1.096	0.65	1.161	1.00	1.204

$k_1 = 0.8540755$; $k_2 = 1.3690329$; $k_3 = 4.1105114$.

Finally, we may remark that, according to the ideas developed in Paper XIV, the exact solution for $V(\mu, \varphi; \mu_0, \varphi_0)$ can be obtained by simply redefining the functions H_v and H_r, which occur in equation (84) as solutions of the functional equations,

$$H_v(\mu) = 1 + \tfrac{3}{4}\mu H_v(\mu) \int_0^1 \frac{H_v(\mu')}{\mu + \mu'} \mu'^2 d\mu' \tag{85}$$

and

$$H_r(\mu) = 1 + \tfrac{3}{8}\mu H_r(\mu) \int_0^1 \frac{H_r(\mu')}{\mu + \mu'} (1 - \mu'^2)\, d\mu' . \tag{86}$$

Tables of solutions of these and other functional equations will be found in Paper XVI.

11

THE ILLUMINATION AND POLARIZATION OF THE SUNLIT SKY ON RAYLEIGH SCATTERING

S. CHANDRASEKHAR AND DONNA D. ELBERT

CONTENTS

1. INTRODUCTION

Since 1871 when Lord Rayleigh first accounted for the principal features of the brightness and polarization of the sunlit sky in terms of the laws of scattering now associated with his name, it has been generally recognized that a problem of fundamental importance both for meteorological optics and for theories of planetary illumination is the following:

A parallel beam of radiation in a given state of polarization is incident on a plane-parallel atmosphere of optical thickness τ_1 in some specified direction. Each element of the atmosphere scatters radiation in accordance with Rayleigh's laws. It is required to find the distribution of intensity and polarization of the light diffusely transmitted by the atmosphere below $\tau = \tau_1$ and of the light diffusely reflected by the atmosphere above $\tau = 0$.

In the theory of planetary illumination one is principally interested in the reflected light while in the theory of sky illumination one is similarly interested in the transmitted light. In this paper we shall be concerned only with the latter.

It is clear that an exact treatment of the foregoing problem in the theory of diffuse reflection and transmission will require the formulation and solution of the appropriate equations of radiative transfer. This was accomplished six years ago[1] and the theory is described and briefly illustrated in the book *Radiative Transfer* (Oxford, 1950) by one of us. A general account of the theory, together with a comparison of its predictions with observations particularly those relating to the polarization of the sunlit sky, was published in 1951 in a brief article.[2]

In this paper we shall present the calculations which we have made (at intervals) during the past five years with the object of giving the theory a concrete form. At one time it was our hope to present our calculations based on the exact mathematical solution of the problem with detailed comparisons not only with the available observational data but also with the calculations of the earlier investigators based on approximations of various kinds. But pressure of time and circumstance have forced us to abandon this plan: this paper will be restricted to giving the results of our calculations with only such comparisons with observations as seemed to us of particular interest.

[1] S. Chandrasekhar, On the radiative equilibrium of a stellar atmosphere. XXII. (V. Rayleigh scattering), *Astrophys. Jour.* **107**: 199, 1947.

[2] S. Chandrasekhar and Donna Elbert, Polarization of the sunlit sky, *Nature* **167**: 51, 1951.

2. THE SOLUTION OF THE FUNDAMENTAL PROBLEM

As we have stated, the problem in the theory of diffuse reflection and transmission formulated in § 1 has been exactly solved; the solution is given in *Radiative Transfer* (§§ 69–73). We shall not describe in any detail how the solution was obtained. But a few explanatory remarks on the parameters in terms of which the solution was obtained and on the structure of the solution itself may be useful in the present connection.

Since on Rayleigh's laws light gets partially plane-polarized whenever it is scattered, it is clear that in formulating the equations of radiative transfer we must allow for the partial plane-polarization of the radiation field. Now to describe a radiation field which is partially plane-polarized we need three parameters to specify the intensity, the degree of polarization, and the plane of polarization. It would scarcely be expected that one could include such diverse quantities as an intensity, a ratio, and an angle in any satisfactory way in formulating the basic equations of the problem. It appears that for these latter purposes the most convenient representation of polarized light is a set of parameters first introduced by Stokes[3] in 1852. The meaning of these parameters for a partially plane-polarized beam is simple: Let l and r refer to two arbitrarily chosen directions at right-angles to one another in the plane transverse to the direction of propagation of the beam. The intensity $I(\psi)$ in a direction making an angle ψ (measured clock-wise) to the direction of l can be expressed in the form

$$I(\psi) = I_l \cos^2 \psi + I_r \sin^2 \psi + \tfrac{1}{2} U \sin 2\psi. \quad (1)$$

The coefficients I_l, I_r and U in this representation are the Stokes parameters. In terms of these parameters the angle χ which the plane of polarization makes with the direction l and the degree of polarization, δ, are given by

$$\tan 2\chi = U/(I_l - I_r) \quad (2)$$

[3] G. G. Stokes, On the composition and resolution of streams of polarized light from different sources, *Trans. Camb. Phil. Soc.* **9**: 399, 1852.

and

$$\delta = (I_l - I_r) \sec 2\chi / (I_l + I_r). \tag{3}$$

The additive property of the Stokes parameters which makes them so convenient for treating problems of radiative transfer is evident from the representation (1): If two independent streams of polarized light are mixed, then the Stokes parameter characterizing the mixture is the sum of the Stokes parameters of the individual streams.

In terms of the Stokes parameters a law of scattering is specified by a matrix, since an elementary act of scattering results in a linear transformation of the parameters. Consequently, by considering the intensity as a vector $\boldsymbol{I}$ with the components I_l, I_r and U (where l and r from now on refer to directions parallel and perpendicular, respectively, to the meridian through the point under consideration and in the plane containing the directions of the beam and of the normal to the plane of stratification of the atmosphere) and by replacing the "phase function" commonly introduced to describe the angular distribution of the scattered radiation by a *phase matrix*, $\boldsymbol{P}$, we can formulate the basic equation of transfer without any difficulty of principle. In this manner we find that the equation we have to solve is[4]

$$\mu \frac{d\boldsymbol{I}(\tau, \mu, \varphi)}{d\tau} = \boldsymbol{I}(\tau, \mu, \varphi) - \frac{1}{4\pi} \int_{-1}^{+1} \int_0^{2\pi} \boldsymbol{P}(\mu, \varphi; \mu', \varphi') \boldsymbol{I}(\tau, \mu', \varphi') \, d\mu' \, d\varphi' - \tfrac{1}{4} e^{-\tau/\mu_0} \boldsymbol{P}(\mu, \varphi; -\mu_0, \varphi_0) \boldsymbol{F}. \tag{4}$$

where

$$\boldsymbol{F} = (F_l, F_r, F_U) \tag{5}$$

is the (Stokes) vector which represents the parallel beam of radiation incident on the atmosphere in the direction $(-\mu_0, \varphi_0)$: πF_l, πF_r and πF_U denote the net fluxes per unit area normal to the beam in the three Stokes parameters. Further, in equation (4) μ denotes the cosine of the angle to the outward normal and the azimuthal angle. And, finally, for the case of Rayleigh scattering the phase matrix $\boldsymbol{P}(\mu, \varphi; \mu', \varphi')$ has the explicit form:

$$\boldsymbol{P}(\mu, \varphi; \mu', \varphi') = \boldsymbol{Q}[\boldsymbol{P}^{(0)}(\mu, \mu') + (1 - \mu^2)^{\frac{1}{2}}(1 - \mu'^2)^{\frac{1}{2}} \boldsymbol{P}^{(1)}(\mu, \varphi; \mu', \varphi') + \boldsymbol{P}^{(2)}(\mu, \varphi; \mu', \varphi')], \tag{6}$$

where

$$\boldsymbol{Q} = \begin{pmatrix} 1 & 0 & 0 \\ 0 & 1 & 0 \\ 0 & 0 & 2 \end{pmatrix}, \tag{7}$$

$$\boldsymbol{P}^{(0)}(\mu, \mu') = \frac{3}{4} \begin{pmatrix} 2(1 - \mu^2)(1 - \mu'^2) + \mu^2 \mu'^2 & \mu^2 & 0 \\ \mu'^2 & 1 & 0 \\ 0 & 0 & 0 \end{pmatrix}, \tag{8}$$

$$\boldsymbol{P}^{(1)}(\mu, \varphi; \mu', \varphi') = \frac{3}{4} \begin{pmatrix} 4\mu\mu' \cos(\varphi' - \varphi) & 0 & 2\mu \sin(\varphi' - \varphi) \\ 0 & 0 & 0 \\ -2\mu' \sin(\varphi' - \varphi) & 0 & \cos(\varphi' - \varphi) \end{pmatrix} \tag{9}$$

and

$$\boldsymbol{P}^{(2)}(\mu, \varphi; \mu', \varphi') = \frac{3}{4} \begin{pmatrix} \mu^2\mu'^2 \cos 2(\varphi' - \varphi) & -\mu^2 \cos 2(\varphi' - \varphi) & \mu^2\mu' \sin 2(\varphi' - \varphi) \\ -\mu'^2 \cos 2(\varphi' - \varphi) & \cos 2(\varphi' - \varphi) & -\mu' \sin 2(\varphi' - \varphi) \\ -\mu\mu'^2 \sin 2(\varphi' - \varphi) & \mu \sin 2(\varphi' - \varphi) & \mu\mu' \cos 2(\varphi' - \varphi) \end{pmatrix}. \tag{10}$$

The solution of equation (4) appropriate to the problem on hand must satisfy the boundary conditions

$$\left.\begin{aligned} &\boldsymbol{I}(0, -\mu, \varphi) \equiv 0 \quad (0 < \mu \leqslant 1, \ 0 \leqslant \varphi \leqslant 2\pi) \\ &\text{and} \\ &\boldsymbol{I}(\tau_1, +\mu, \varphi) \equiv 0 \quad (0 < \mu \leqslant 1, \ 0 \leqslant \varphi \leqslant 2\pi), \end{aligned}\right\} \tag{11}$$

since there is no diffuse radiation in any inward direction at $\tau = 0$ and in any outward direction at $\tau = \tau_1$. And the solution to the problem of diffuse reflection and transmission will be completed when we specify the angular distribution and the state of polarization of the diffuse light which emerges from $\tau = 0$ and $\tau = \tau_1$.

The laws of diffuse reflection and transmission by a plane-parallel atmosphere are generally expressed (*cf. Radiative Transfer*, 44) in terms of a *scattering matrix*, $\boldsymbol{S}(\mu, \varphi; \mu_0, \varphi_0)$ and a *transmission matrix*, $\boldsymbol{T}(\mu, \varphi; \mu_0, \varphi_0)$ such that the reflected and the transmitted intensities are given by

$$\boldsymbol{I}(0; \mu, \varphi; \mu_0, \varphi_0) = \frac{1}{4\mu} \boldsymbol{S}(\mu, \varphi; \mu_0, \varphi_0) \boldsymbol{F}$$

and

$$\boldsymbol{I}(\tau_1; -\mu, \varphi; \mu_0, \varphi_0) = \frac{1}{4\mu} \boldsymbol{T}(\mu, \varphi; \mu_0, \varphi_0) \boldsymbol{F}. \tag{12}$$

Our problem, then, is to specify $\boldsymbol{S}$ and $\boldsymbol{T}$ for an atmosphere scattering radiation in accordance with Rayleigh's laws.

[4] The derivation of the equations which follow will be found in *Radiative Transfer* § 16: 35–45.

The elements of $\boldsymbol{S}$ and $\boldsymbol{T}$ are clearly functions of the four variables μ, φ, μ_0 and φ_0 in addition, of course, to the optical thickness τ_1 which may, however, be treated as a parameter. If the variables μ, φ, μ_0 and φ_0 were not separable the problem of tabulating $\boldsymbol{S}$ and $\boldsymbol{T}$ may indeed be considered as impracticable. But the essential feature of the solution for $\boldsymbol{S}$ and $\boldsymbol{T}$ (which we shall presently write down) which makes the problem a practicable one is that $\boldsymbol{S}$ and $\boldsymbol{T}$ involve only four pairs of functions $X_l(\mu)$, $Y_l(\mu)$; $X_r(\mu)$, $Y_r(\mu)$; $X^{(1)}(\mu)$, $Y^{(1)}(\mu)$; and $X^{(2)}(\mu)$, $Y^{(2)}(\mu)$ all of the single variable μ. Further, these four pairs of functions belong to a general class (the X- and Y-functions) which satisfy a simultaneous pair of integral equations of the form[5]

$$X(\mu) = 1 + \mu\int_0^1 \frac{\Psi(\mu')}{\mu+\mu'} \times [X(\mu)X(\mu') - Y(\mu)Y(\mu')]\,d\mu' \quad (13)$$

and

$$Y(\mu) = e^{-\tau_1/\mu} + \mu\int_0^1 \frac{\Psi(\mu')}{\mu-\mu'} \times [Y(\mu)X(\mu') - X(\mu)Y(\mu')]\,d\mu', \quad (14)$$

where the *characteristic function* $\Psi(\mu)$ is in problems of radiative transfer, an even polynomial in μ satisfying the condition

$$\int_0^1 \Psi(\mu)\,d\mu \leqslant \tfrac{1}{2}. \quad (15)$$

The case when equality occurs in (15) (the so-called *conservative case*) is special: The solutions of equations (13) and (14) are, then, no longer unique; they form instead a one-parameter family. In conservative cases one therefore defines what are called *standard solutions* which have the property:

$$\int_0^1 X(\mu)\Psi(\mu)\,d\mu = 1$$

and

$$\int_0^1 Y(\mu)\Psi(\mu)\,d\mu = 0. \quad (16)$$

As we have already stated, the solutions for $\boldsymbol{S}$ and $\boldsymbol{T}$ (for an atmosphere scattering radiation in accordance with Rayleigh's laws) involve only four pairs of X- and Y-functions: X_l, Y_l; X_r, Y_r; $X^{(1)}$, $Y^{(1)}$; $X^{(2)}$, $Y^{(2)}$; and the characteristic functions in terms of which these are defined are:

$$\Psi_l(\mu) = \tfrac{3}{4}(1-\mu^2);$$
$$\Psi_r(\mu) = \tfrac{3}{8}(1-\mu^2),$$
$$\Psi^{(1)}(\mu) = \tfrac{3}{8}(1-\mu^2)(1+2\mu^2)$$

and

$$\Psi^{(2)}(\mu) = \tfrac{3}{16}(1+\mu^2)^2, \quad (17)$$

respectively. The function $\Psi_l(\mu)$ belongs to the conservative class; accordingly, in this case we define $X_l(\mu)$, $Y_l(\mu)$ as the standard solutions having the property

$$\tfrac{3}{4}\int_0^1 X_l(\mu)(1-\mu^2)\,d\mu = 1$$

and

$$\int_0^1 Y_l(\mu)(1-\mu^2)\,d\mu = 0. \quad (18)$$

After these explanatory remarks we shall now write down the solutions for $\boldsymbol{S}$ and $\boldsymbol{T}$ given in *Radiative Transfer:*

The scattering and the transmission matrices allow a decomposition into azimuth independent and azimuth dependent terms in the same manner as the phase matrix [equation (6)] and have the forms:

$$\boldsymbol{S}(\mu,\varphi;\mu_0,\varphi_0) = \boldsymbol{Q}[\tfrac{3}{4}\boldsymbol{S}^{(0)}(\mu;\mu_0) + (1-\mu^2)^{\frac{1}{2}}(1-\mu_0^2)^{\frac{1}{2}}\boldsymbol{S}^{(1)}(\mu,\varphi;\mu_0,\varphi_0) + \boldsymbol{S}^{(2)}(\mu,\varphi;\mu_0,\varphi_0)] \quad (19)$$

and

$$\boldsymbol{T}(\mu,\varphi;\mu_0,\varphi_0) = \boldsymbol{Q}[\tfrac{3}{4}\boldsymbol{T}^{(0)}(\mu;\mu_0) + (1-\mu^2)^{\frac{1}{2}}(1-\mu_0^2)^{\frac{1}{2}}\boldsymbol{T}^{(1)}(\mu,\varphi;\mu_0,\varphi_0) + \boldsymbol{T}^{(2)}(\mu,\varphi;\mu_0,\varphi_0)]. \quad (20)$$

The dependence of the azimuth dependent terms $(\boldsymbol{S}^{(1)}, \boldsymbol{T}^{(1)})$ and $(\boldsymbol{S}^{(2)}, \boldsymbol{T}^{(2)})$ on $\varphi_0 - \varphi$ are essentially the same as $\boldsymbol{P}^{(1)}$ and $\boldsymbol{P}^{(2)}$; indeed, we have

$$\left(\frac{1}{\mu_0}+\frac{1}{\mu}\right)\boldsymbol{S}^{(i)} = [X^{(i)}(\mu)X^{(i)}(\mu_0) - Y^{(i)}(\mu)Y^{(i)}(\mu_0)]\boldsymbol{P}^{(i)}(\mu,\varphi;-\mu_0,\varphi_0)$$

and

$$\left(\frac{1}{\mu_0}-\frac{1}{\mu}\right)\boldsymbol{T}^{(i)} = [Y^{(i)}(\mu)X^{(i)}(\mu_0) - X^{(i)}(\mu)Y^{(i)}(\mu_0)]\boldsymbol{P}^{(i)}(-\mu,\varphi;-\mu_0,\varphi_0) \quad (i=1,2). \quad (21)$$

In contrast, the solutions for the azimuth independent terms $\boldsymbol{S}^{(0)}$ and $\boldsymbol{T}^{(0)}$ are very complicated. They are given by

$$\left(\frac{1}{\mu_0}+\frac{1}{\mu}\right)\boldsymbol{S}^{(0)}(\mu;\mu_0) = \begin{pmatrix}\psi(\mu) & 2^{\frac{1}{2}}\phi(\mu) & 0\\ \chi(\mu) & 2^{\frac{1}{2}}\zeta(\mu) & 0\\ 0 & 0 & 0\end{pmatrix}\begin{pmatrix}\psi(\mu_0) & \chi(\mu_0) & 0\\ 2^{\frac{1}{2}}\phi(\mu_0) & 2^{\frac{1}{2}}\zeta(\mu_0) & 0\\ 0 & 0 & 0\end{pmatrix} - \begin{pmatrix}\xi(\mu) & 2^{\frac{1}{2}}\eta(\mu) & 0\\ \sigma(\mu) & 2^{\frac{1}{2}}\theta(\mu) & 0\\ 0 & 0 & 0\end{pmatrix}\begin{pmatrix}\xi(\mu_0) & \sigma(\mu_0) & 0\\ 2^{\frac{1}{2}}\eta(\mu_0) & 2^{\frac{1}{2}}\theta(\mu_0) & 0\\ 0 & 0 & 0\end{pmatrix} \quad (22)$$

[5] For the theory of the X- and Y-functions see *Radiative Transfer*, chap. VIII.

and

$$\left(\frac{1}{\mu_0}-\frac{1}{\mu}\right) \boldsymbol{T}^{(0)}(\mu;\mu_0) = \begin{pmatrix} \xi(\mu) & 2^{\frac{1}{2}}\eta(\mu) & 0 \\ \sigma(\mu) & 2^{\frac{1}{2}}\theta(\mu) & 0 \\ 0 & 0 & 0 \end{pmatrix} \begin{pmatrix} \psi(\mu_0) & \chi(\mu_0) & 0 \\ 2^{\frac{1}{2}}\phi(\mu_0) & 2^{\frac{1}{2}}\zeta(\mu_0) & 0 \\ 0 & 0 & 0 \end{pmatrix}$$

$$- \begin{pmatrix} \psi(\mu) & 2^{\frac{1}{2}}\phi(\mu) & 0 \\ \chi(\mu) & 2^{\frac{1}{2}}\zeta(\mu) & 0 \\ 0 & 0 & 0 \end{pmatrix} \begin{pmatrix} \xi(\mu_0) & \sigma(\mu_0) & 0 \\ 2^{\frac{1}{2}}\eta(\mu_0) & 2^{\frac{1}{2}}\theta(\mu_0) & 0 \\ 0 & 0 & 0 \end{pmatrix}, \qquad (23)$$

where ψ, ϕ, χ, etc., are eight functions expressible in terms of the two pairs of X- and Y-functions, X_l, Y_l and X_r, Y_r in the forms

$$\begin{aligned}
\psi(\mu) &= \mu[\nu_1 Y_l(\mu) - \nu_2 X_l(\mu)], \\
\xi(\mu) &= \mu[\nu_2 Y_l(\mu) - \nu_1 X_l(\mu)], \\
\phi(\mu) &= (1+\nu_4\mu)X_l(\mu) - \nu_3\mu Y_l(\mu), \\
\eta(\mu) &= (1-\nu_4\mu)Y_l(\mu) + \nu_3\mu X_l(\mu), \\
\chi(\mu) &= (1-u_4\mu)X_r(\mu) + u_3\mu Y_r(\mu) + Q(u_4-u_3)\mu^2[X_r(\mu)-Y_r(\mu)], \\
\sigma(\mu) &= (1+u_4\mu)Y_r(\mu) - u_3\mu X_r(\mu) - Q(u_4-u_3)\mu^2[X_r(\mu)-Y_r(\mu)], \\
\zeta(\mu) &= \tfrac{1}{2}\mu[\nu_1 Y_r(\mu) - \nu_2 X_r(\mu)] + \tfrac{1}{2}Q(\nu_2-\nu_1)\mu^2[X_r(\mu)-Y_r(\mu)], \\
\theta(\mu) &= \tfrac{1}{2}\mu[\nu_2 Y_r(\mu) - \nu_1 X_r(\mu)] - \tfrac{1}{2}Q(\nu_2-\nu_1)\mu^2[X_r(\mu)-Y_r(\mu)],
\end{aligned} \qquad (24)$$

where the constants ν_1, ν_2, ν_3, ν_4, u_3, u_4 and Q are to be determined by the following formulae:

$$\begin{gathered}
\nu_2+\nu_1 = 2\Delta_1(\kappa_1\delta_1-\kappa_2\delta_2); \qquad \nu_2-\nu_1 = 2\Delta_2(\kappa_1\delta_1-\kappa_2\delta_2), \\
\nu_4+\nu_3 = \Delta_1(d_1\kappa_1-d_0\kappa_2); \qquad \nu_4-\nu_3 = \Delta_2[c_1\delta_1-c_0\delta_2-2Q(d_0\delta_1-d_1\delta_2)], \\
u_4+u_3 = \Delta_1(c_1\delta_1-c_0\delta_2); \qquad u_4-u_3 = \Delta_2(d_1\kappa_1-d_0\kappa_2), \\
\Delta_1 = (d_0\delta_1-d_1\delta_2)^{-1}; \quad \Delta_2 = [c_0\kappa_1-c_1\kappa_2-2Q(d_1\kappa_1-d_0\kappa_2)]^{-1}, \\
Q = (c_0-c_2)[(d_0-d_2)\tau_1+2(d_1-d_3)]^{-1}, \\
c_0 = A_0+B_0-\frac{8}{3}; \quad d_0 = A_0-B_0-\frac{8}{3}, \\
c_n = A_n+B_n; \quad d_n = A_n-B_n; \quad \kappa_n = \alpha_n+\beta_n; \quad \delta_n = \alpha_n-\beta_n \qquad (n=1,2,3,\cdots),
\end{gathered} \qquad (25)$$

α_n, β_n, A_n and B_n are the moments of order n of X_l, Y_l, X_r and Y_r, respectively. (It may be recalled here that X_l and Y_l are the standard solutions for the case.)

In the theory of the illumination of the sky we are interested in the transmitted light in the case of incident natural light. In this latter case $F_l = F_r = \frac{1}{2}F$ (where πF denotes the net flux of the incident natural light) and $F_U = 0$. The equations governing the intensity and polarization of the sky as witnessed by an observer at $\tau = \tau_1$ in these circumstances readily follow from the solutions already given; thus by setting $\boldsymbol{F} = \frac{1}{2}(F_l, F_r, 0)$ in equation (12) and combining equations (20), (21), and (23) appropriately, we find:

$$\begin{aligned}
I_l(\tau_1;-\mu,\varphi;\mu_0,\varphi_0) = \tfrac{3}{32}\big[&\{\psi(\mu_0)+\chi(\mu_0)\}\xi(\mu) + 2\{\phi(\mu_0)+\zeta(\mu_0)\}\eta(\mu) - \{\xi(\mu_0)+\sigma(\mu_0)\}\psi(\mu) \\
&- 2\{\theta(\mu_0)+\eta(\mu_0)\}\phi(\mu) + 4\mu\mu_0(1-\mu^2)^{\frac{1}{2}}(1-\mu_0^2)^{\frac{1}{2}}\{X^{(1)}(\mu_0)Y^{(1)}(\mu) - Y^{(1)}(\mu_0)X^{(1)}(\mu)\}\cos(\varphi_0-\varphi) \\
&- \mu^2(1-\mu_0^2)\{X^{(2)}(\mu_0)Y^{(2)}(\mu) - Y^{(2)}(\mu_0)X^{(2)}(\mu)\}\cos 2(\varphi_0-\varphi)\big]\frac{F\mu_0}{\mu-\mu_0}, \\
I_r(\tau_1;-\mu,\varphi;\mu_0,\varphi_0) = \tfrac{3}{32}\big[&\{\psi(\mu_0)+\chi(\mu_0)\}\sigma(\mu) + 2\{\phi(\mu_0)+\zeta(\mu_0)\}\theta(\mu) - \{\xi(\mu_0)+\sigma(\mu_0)\}\chi(\mu) \\
&- 2\{\theta(\mu_0)+\eta(\mu_0)\}\zeta(\mu) + (1-\mu_0^2)\{X^{(2)}(\mu_0)Y^{(2)}(\mu) - Y^{(2)}(\mu_0)X^{(2)}(\mu)\}\cos 2(\varphi_0-\varphi)\big]\frac{F\mu_0}{\mu-\mu_0}
\end{aligned}$$

and

$$\begin{aligned}
U(\tau_1;-\mu,\varphi;\mu_0,\varphi_0) = \tfrac{3}{16}\big[&2(1-\mu^2)^{\frac{1}{2}}(1-\mu_0^2)^{\frac{1}{2}}\mu_0\{X^{(1)}(\mu_0)Y^{(1)}(\mu) - Y^{(1)}(\mu_0)X^{(1)}(\mu)\}\sin(\varphi_0-\varphi) \\
&- \mu(1-\mu_0^2)\{X^{(2)}(\mu_0)Y^{(2)}(\mu) - Y^{(2)}(\mu_0)X^{(2)}(\mu)\}\sin 2(\varphi_0-\varphi)\big]\frac{F\mu_0}{\mu-\mu_0}.
\end{aligned} \qquad (26)$$

3. THE EFFECT OF REFLECTION BY THE GROUND

Before we can apply the solution for $\boldsymbol{S}$ and $\boldsymbol{T}$ given in § 2 to the problem of the illumination of the sky, we must consider the effect of the ground at $\tau = \tau_1$. The solution for $\boldsymbol{S}$ and $\boldsymbol{T}$ given in § 2 was derived on the assumption that at $\tau = \tau_1$ there is no diffuse radiation in the outward direction [*cf.* equation (11)]. The presence of the ground will alter this. However, if the law of reflection by the ground is specified then it is not a difficult matter to relate the solution of the problem when there is a ground to the solution of the problem when there is no ground. This reduction is particularly simple if the ground reflects according to Lambert's law with a certain albedo λ_0; that is, if the light reflected by the ground is unpolarized and uniform in the outward hemisphere independently of the state of polarization and the angular distribution of the incident light, and if, further, the outward flux of the reflected light is always a certain fixed fraction, λ_0, of the inward flux of the radiation incident on the surface. Under these latter circumstances it can be shown (*Radiative Transfer*, § 73) that the effect of the ground is to increase the diffuse intensities emergent at $\tau = 0$ and directed inward at $\tau = \tau_1$ by amounts $\boldsymbol{I}^*(0; \mu, \varphi)$ and $\boldsymbol{I}^*(\tau_1; -\mu, \varphi)$ given by

$$\boldsymbol{I}^*(0; \mu, \varphi) = \frac{\lambda_0\mu_0}{2(1-\lambda_0\bar{s})}\boldsymbol{\Gamma}(\mu; \mu_0)\boldsymbol{F} \tag{27}$$

and

$$\boldsymbol{I}^*(\tau_1; -\mu, \varphi) = \frac{\lambda_0\mu_0}{2(1-\lambda_0\bar{s})}\boldsymbol{\Lambda}(\mu; \mu_0)\boldsymbol{F}, \tag{28}$$

where $\bar{s}$ is a constant (to be defined presently) and

$$\boldsymbol{\Gamma}(\mu; \mu_0) = \begin{pmatrix} \gamma_l(\mu)\gamma_l(\mu_0) & \gamma_l(\mu)\gamma_r(\mu_0) & 0 \\ \gamma_r(\mu)\gamma_l(\mu_0) & \gamma_r(\mu)\gamma_r(\mu_0) & 0 \\ 0 & 0 & 0 \end{pmatrix} \tag{29}$$

and

$$\boldsymbol{\Lambda}(\mu; \mu_0) = \begin{pmatrix} \{1-\gamma_l(\mu)\}\gamma_l(\mu_0) & \{1-\gamma_l(\mu)\}\gamma_r(\mu_0) & 0 \\ \{1-\gamma_r(\mu)\}\gamma_l(\mu_0) & \{1-\gamma_r(\mu)\}\gamma_r(\mu_0) & 0 \\ 0 & 0 & 0 \end{pmatrix}. \tag{30}$$

In equations (29) and (30) $\gamma_l(\mu)$ and $\gamma_r(\mu)$ are two functions which are related to X_l, Y_l, X_r and Y_r by

$$\gamma_l(\mu) = \tfrac{3}{8}Q(\nu_2-\nu_1)(d_0-d_2)[X_l(\mu)+Y_l(\mu)], \tag{31}$$

and

$$\gamma_r(\mu) = \tfrac{3}{8}Q(d_0-d_2) \times [(u_4-u_3)\{X_r(\mu)+Y_r(\mu)\} - u_5\mu\{X_r(\mu)+Y_r(\mu)\}], \tag{32}$$

where

$$u_5 = \Delta_2(c_0\kappa_1 - c_1\kappa_2), \tag{33}$$

and the remaining constants have the same meanings as in equations (25). Finally, the constant $\bar{s}$ in equations (27) and (28) is given by

$$\bar{s} = 1 - \tfrac{3}{8}Q(d_0-d_2) \times [(\nu_2-\nu_1)\kappa_1 + (u_4-u_3)c_1 - u_5d_2]. \tag{34}$$

Again, when the incident light is natural the corrections which have to be made to the intensities given by equations (26) to allow for a ground surface at $\tau = \tau_1$ which reflects according to Lambert's law with an albedo λ_0 are given by

$$I_l^*(\tau_1; -\mu, \varphi; \mu_0, \varphi_0) = \frac{\lambda_0}{4(1-\lambda_0\bar{s})}\{\gamma_l(\mu_0)+\gamma_r(\mu_0)\} \times \{1-\gamma_l(\mu)\}\mu_0F,$$

$$I_r^*(\tau_1; -\mu, \varphi; \mu_0, \varphi_0) = \frac{\lambda_0}{4(1-\lambda_0\bar{s})}\{\gamma_l(\mu_0)+\gamma_r(\mu_0)\} \times \{1-\gamma_r(\mu)\}\mu_0F,$$

$$U^*(\tau_1; -\mu, \varphi; \mu_0, \varphi_0) = 0. \tag{35}$$

4. DESCRIPTION OF THE TABLES*

The solution of the fundamental problem in the theory of the illumination of the sky given in the two preceding sections was obtained some six years ago. A detailed examination of its predictions had to await the tabulation of the basic eight functions X_l, Y_l, X_r, Y_r, $X^{(1)}$, $Y^{(1)}$, $X^{(2)}$, and $Y^{(2)}$ of the variable μ for various values of τ_1. This tabulation has now been completed for $\tau_1 = 0.05$, 0.10, 0.15, 0.20, 0.25, 0.50, and 1.00 with the cooperation of the Watson Scientific Computing Laboratory (New York).[6] The solutions were obtained by a direct process of iteration applied to the governing integral equations. The

*The tables are omitted from this volume. See, instead, Kinsell L. Coulson, Jitendra V. Dave, and Zdenek Sekera, *Tables Related to Radiation Emerging from a Planetary Atmosphere with Rayleigh Scattering* (Berkeley: University of California Press, 1960).

[6] While these calculations were in progress, Dr. Z. Sekera initiated a similar program at the Department of Meteorology of the University of California at Los Angeles in cooperation with the Institute for Numerical Analysis of the National Bureau of Standards at Los Angeles. Their Report No. 3 (prepared by the Air Material Command, Air Force Cambridge Research Center) provides some calculations for $\tau_1 = 0.15$, 0.25, and 1.0. Their calculations, while they are much less extensive than ours, do provide a valuable check.

iterations were started with the solutions in the corrected second approximation described in *Radiative Transfer*, Chapter VIII [§ 60, see particularly equations (117), (118), and (120)]. The corrected second approximations were computed by one of us (D. D. E.) at the Yerkes Observatory. The iterations were carried out at the Watson Scientific Laboratory with IBM pluggable sequence relay calculators by Miss Ann Franklin to whom and to Dr. Wallace Eckert we are very greatly indebted. The functions obtained after the iterations showed, however, a certain "raggedness" between $\mu = 0.9$ and 1.0. The solutions have therefore been "smoothed" by plotting the deviations of the iterated solutions from the corrected second approximations. Table 1 presents these smoothed solutions. While the solutions have been tabulated to five decimals the last place is definitely not reliable. But it is expected that if the tabulated solutions are rounded to one less place the solution may be trusted to two or three units in the surviving place. The solutions for $\tau_1 \leqslant 0.20$ are very probably more accurate than this while for $\tau_1 = 0.5$ and 1.0 they may be less accurate. Nevertheless, the solutions are given to five places since the functions as tabulated do have smooth differences and as such they can be used for further iterations to improve their accuracy if the need for it should arise.

The moments α_n, β_n, A_n and B_n of order n of X_l, Y_l, X_r and Y_r, respectively, which are needed in the evaluation of the various terms of $\boldsymbol{S}$ and $\boldsymbol{T}$ are given in table 2. The theory of the X- and Y-functions leads to a number of identical relations which must exist between their moments; these relations among the moments listed in table 2 have been verified within the accuracy of the tabulated values. Table 2 also includes the values of the constants ν_1, ν_2, ν_3, ν_4, u_3, u_4 and Q which occur in the definitions of the functions ψ, ϕ, etc. [equations (24) and (25)]; the value of the constant $\bar{s}$ [equation (34)] which occurs in the expressions for the ground corrections [equations (27) and (28)] is also listed in this table.

According to equations (22) and (23) the calculation of $\boldsymbol{S}$ and $\boldsymbol{T}$ in a given case can be most easily carried out in terms of the auxiliary functions ψ, ϕ, χ, ζ, ξ, η, σ and θ. These functions computed with the aid of tables 1 and 2 are given in table 3. Similarly, table 4 gives the functions $\gamma_l(\mu)$ and $\gamma_r(\mu)$ which are needed to allow for reflection by the ground [*cf.* equations (31) and (32)].

With the basic functions tabulated we can calculate the theoretical illumination and polarization of the sky on Rayleigh scattering for plane-parallel atmospheres. To illustrate the use of tables 1–4 we have made some model calculations to which the remaining tables are devoted.

The most extensive calculations were made for $\tau_1 = 0.15$; this is approximately the value of the optical thickness at λ4500A. For $\tau_1 = 0.15$, angles of incidence (or, zenith distances) $\theta_0 = 90°$, 85.4°, 76.1°, 58.7°, 50.1°, 43.9°, 36.9°, 19.95°, and 0° (corresponding to $\mu_0 = 0$, 0.08, 0.24, 0.52, 0.64, 0.72, 0.80, 0.94, and 1.00) and azimuthal differences $\varphi_0 - \varphi$ (denoted, simply, by φ in the tables) $= 0°(10°)90°$ were considered. The results of the calculations are summarized in table 5; it gives the total intensity, $I_l + I_r$, in units of F, the inclination, χ, of the plane of polarization with the meridian through the direction of observation and the degree of polarization, δ. Less extensive calculations were made for $\tau_1 = 0.10$ and 0.20. For these two values of the optical thicknesses the total intensity $(I_l + I_r)$ and the degree of polarization, δ, were found only in the principal meridian $(\varphi_0 - \varphi = 0°)$ containing the sun and in the plane at right-angles $(\varphi_0 - \varphi = 90°)$ for zenith distances $\theta_0 = 90°$, 80.8°, 60°, 30.7°, and 0° (corresponding to $\mu_0 = 0$, 0.16, 0.50, 0.86, and 1.00). The results of the calculations are given in tables 6 and 7.

In the calculations presented in tables 5–7 no allowance has been made for the reflection by the ground surface. This is taken into account in tables 8–10 in accordance with equations (35). Again, the most extensive calculations were made for $\tau_1 = 0.15$. For angles of incidence corresponding to $\mu_0 = 0$, 0.24, 0.52, 0.64, 0.72, 0.80, and 1.00 the effect of a ground reflecting according to Lambert's law for two values of the albedo, λ_0, on the intensity and the degree of polarization in the principal meridian $(\varphi_0 - \varphi = 0°)$ was determined. For $\mu_0 = 0.64$ the effect on the intensity for $\varphi_0 - \varphi = 0°(10°)90°$ and $\lambda_0 = 0.10$ was determined; and for the same angle of incidence the effect on the intensity for $\varphi_0 - \varphi = 90°$ was also found for $\lambda_0 = 0.25$. The results of all these calculations are given in table 8. For $\tau_1 = 0.10$ and 0.20 the effect of ground reflection on the intensity and polarization in the principal meridian is illustrated for $\lambda_0 = 0.10$ and 0.20 and for angles of incidence corresponding to $\mu_0 = 0$, 0.16, 0.50, 0.86, and 1.00. The results of these calculations are given in tables 9 and 10.

Table 11 gives the positions of the neutral points. The results given in this table will be described and discussed in the following section.

Finally the supplementary table 12 gives the functions ψ, ϕ, χ, ζ, ξ, η, σ, θ, $X^{(1)}$, $Y^{(1)}$, $X^{(2)}$, $Y^{(2)}$, γ_l, and γ_r for $\tau_1 = 0.01(0.01)0.20(0.05)0.50(0.10)1.0$ and for representative values of μ. At the head of the table for each value τ_1, the value of $\bar{s}$ (needed for the evaluation of the ground correction) is also given. The values of the functions listed in this table are *not* based on the solutions of the basic X- and Y-functions derived from the integral equations they satisfy; they are based, instead, on the corrected second approximations (*cf. Radiative Transfer*, § 60) for the relevant X- and Y-functions. However, for $\tau_1 \leqslant 0.25$ the table should suffice to calculate $\boldsymbol{S}$ and $\boldsymbol{T}$ to well within a fraction of a per cent; for the larger values of τ_1 accuracy within a few per cent may be expected. But one could, if one wished, obtain from these tables values of considerably higher precision by differencing the values of the functions for $\tau_1 = 0.05$, 0.10, 0.15,

0.20, 0.25, 0.50, and 1.0 given in tables 3 and 12 and interpolating among these differences to estimate the corrections (to the values given by the corrected second approximation) for any other intermediate value of τ_1. With these supplementary tables, then, the theory developed and described in *Radiative Transfer* has been, finally, brought to a point where it is capable of giving numerical values for any of the desired quantities under most conditions in which they are likely to be of interest.

5. THE POLARIZATION OF THE SUNLIT SKY: THE THEORY OF THE NEUTRAL POINTS AND LINES

As we have already stated in the introductory section we shall not attempt in this paper any detailed comparison between the calculations presented here and those of earlier investigators based on approximations of various kinds. But an exception might be made with regard to the quantitative explanation which the present theory affords for the phenomena associated with the *neutral points* of Arago, Babinet, and Brewster. The phenomena in question are these:

The neutral points are the points of zero polarization; from symmetry we should, of course, expect them to occur in the principal meridian. And for a long time it has been known that there are in general two such neutral points. For angles of incidence not exceeding 70° these neutral points occur between 0° and 20° above and below the sun; these are the neutral points of Babinet and Brewster, respectively. But when the sun is low, the neutral point occurs about 20° above the anti-solar point in the opposite sky: this is the Arago point. These facts concerning the

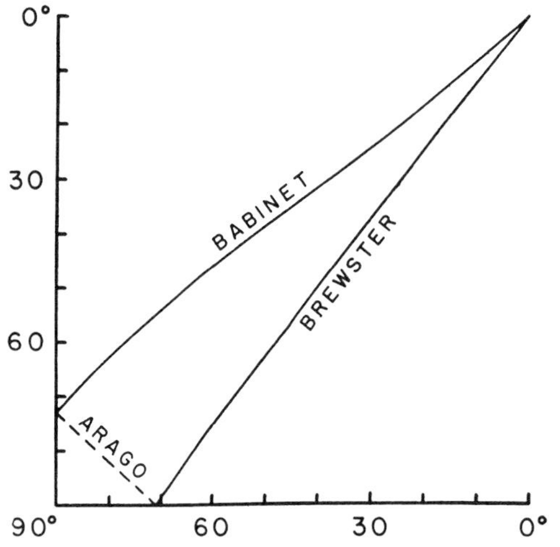

FIG. 1. Calculated positions of the neutral points for various angles of incidence for an atmosphere of optical thickness $\tau_1 = 0.10$. The abscissa gives the zenith distance of the sun and the ordinate gives the corresponding positions of the neutral points. The Arago point occurs on the side of the horizon opposite the sun; to emphasize this its position on the sky is indicated by the dashed curve.

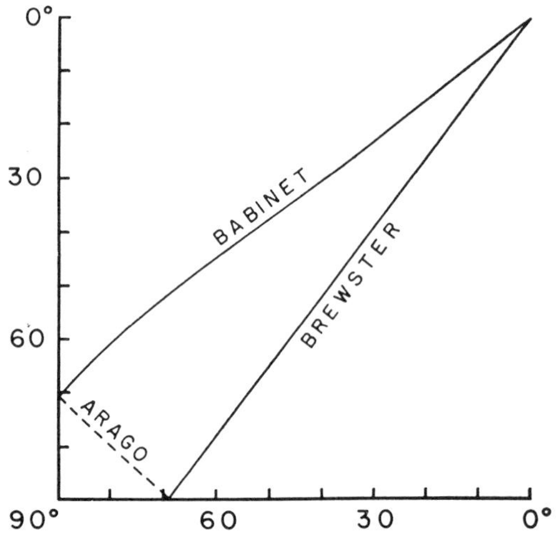

FIG. 2. Calculated positions of the neutral points for various angles of incidence for an atmosphere of optical thickness $\tau_1 = 0.15$. The abscissa and the ordinate have otherwise the same meanings as in fig. 1.

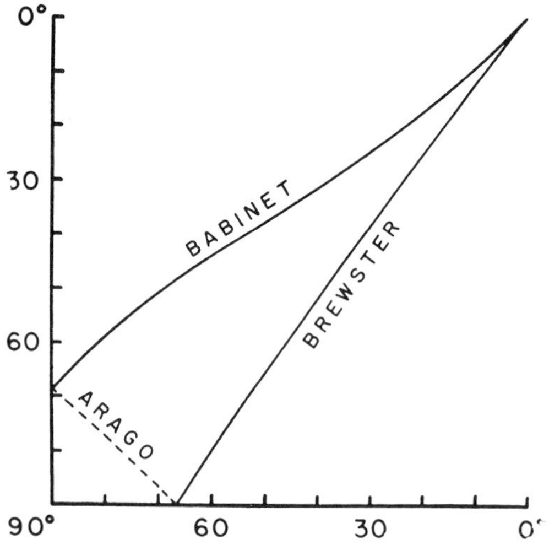

FIG. 3. Calculated positions of the neutral points for various angles of incidence for an atmosphere of optical thickness $\tau_1 = 0.20$. The abscissa and the ordinate have otherwise the same meanings as in fig. 1.

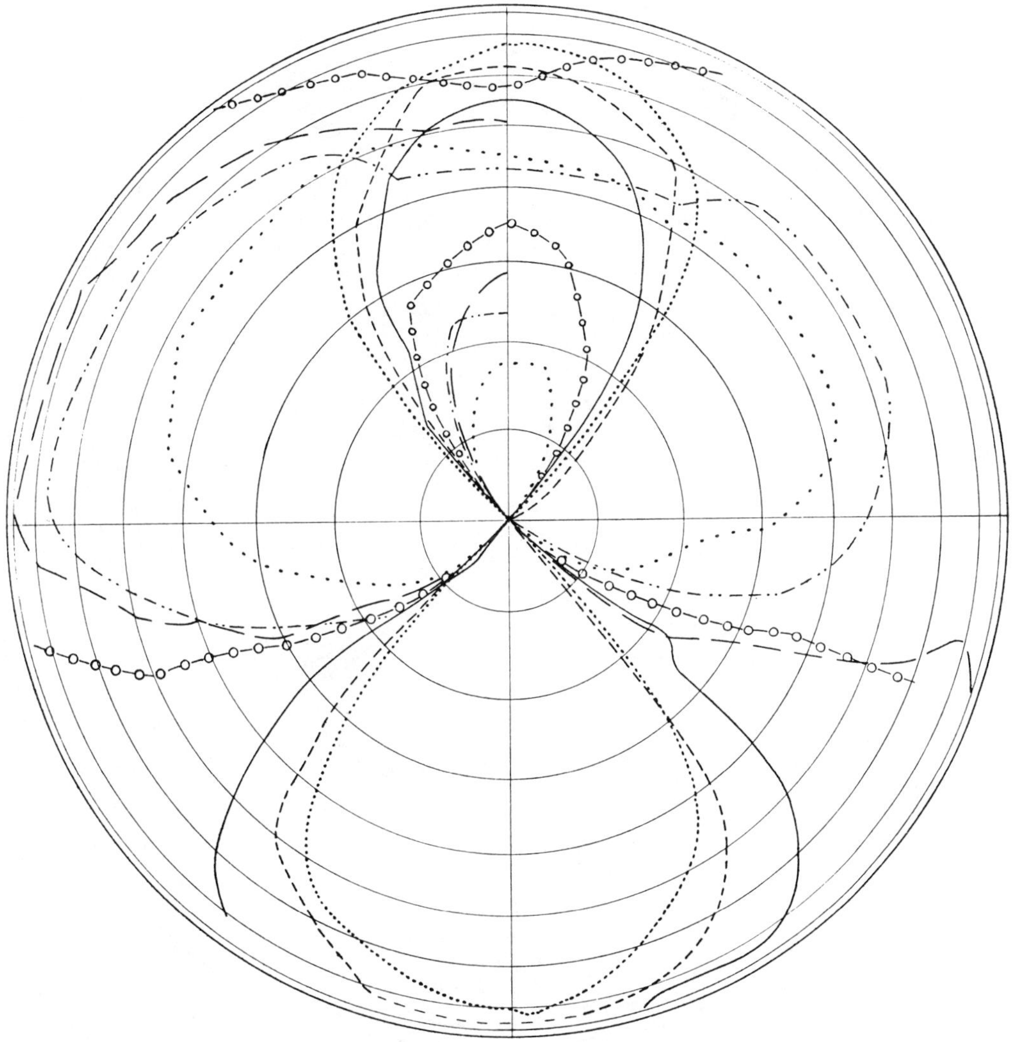

FIG. 4. Dorno's observations of the neutral lines on May 17, 1917, at Davos. The various curves were determined during the following times when the zenith distance of the sun varied by the amounts given:

....	$7^{p}\,19$– $7^{p}\,28$	$\theta_0 = 87°\ 10'$– $0°$
- - - - - - - - -	$6^{p}\,13$– $6^{p}\,33$	$\theta_0 = 78°\ 14'$–$81°\ 30'$
————	$5^{p}\,14$– $5^{p}\,34$	$\theta_0 = 68°\ 20'$–$71°\ 44'$
-o-o-o-o-	$3^{p}\,15$– $3^{p}\,42$	$\theta_0 = 48°\ 12'$–$52°\ 40'$
— — — —	$9^{a}\ \ 9$– $9^{a}\,34$	$\theta_0 = 44°\ 21'$–$40°\ 33'$
— · · — — · ·	$9^{a}\,44$–$10^{a}\,11$	$\theta_0 = 39°\ \ 5'$–$35°\ 26'$
. . . .	$12^{p}\,56$– $1^{p}\,14$	$\theta_0 = 29°\ 50'$–$31°\ 26'$

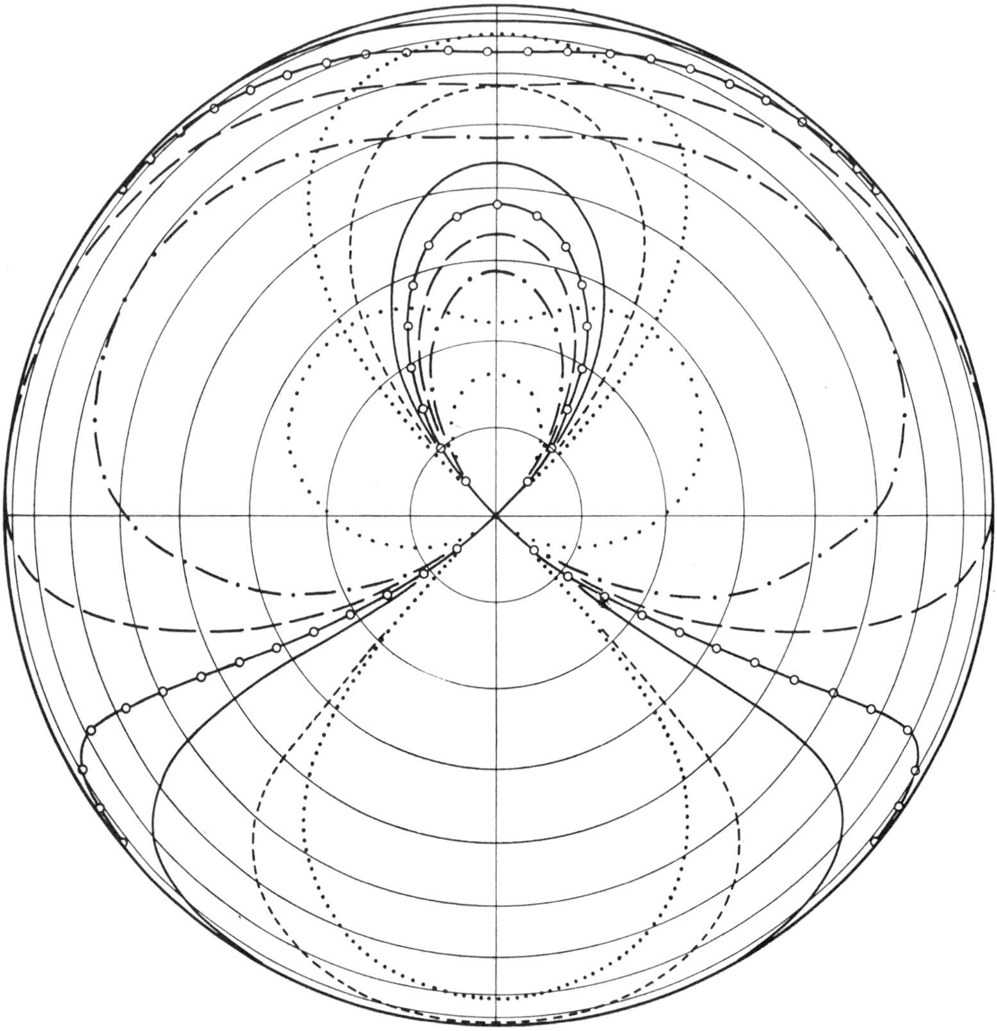

FIG. 5. The neutral lines as predicted by the theory. The various curves refer to the following zenith distances of the sun:

··········· = 90°;	———— = 58.7°;	— — — — = 43.9°;
- - - - - - - - - = 76.1°;	-o-o-o-o- = 50.2°;	—·—·— = 36.9°;
	· · · · · · · · · = 19.9°.	

The curves in this figure which roughly correspond to Dorno's observations are marked similarly.

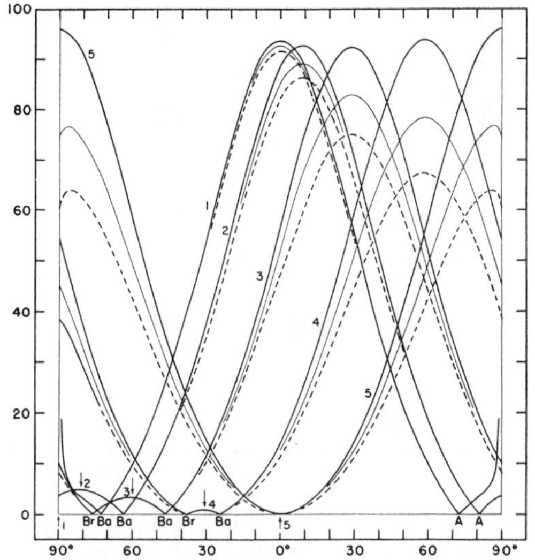

FIG. 6. Variation of the degree of polarization in the principal meridian for various angles of incidence for an atmosphere of optical thickness $\tau_1 = 0.10$. The abscissa gives the zenith distance and the ordinate gives the degree of polarization in per cent. The curves marked 1, 2, 3, 4, and 5 represent the variation for the angles of incidence $\theta_0 = 90°$, 80.8°, 60.0°, 30.7° and 0°, respectively. The thick solid curves are obtained before any ground corrections have been applied. The dashed curves are obtained if we allow for a ground reflecting according to Lambert's law with an albedo $\lambda_0 = 0.20$. The thin intermediate curves are obtained if $\lambda_0 = 0.10$. The positions of the neutral points are also indicated.

points of zero polarization should be contrasted with what should be expected on the laws of single scattering, namely that the polarization should tend to zero as we approach the direction towards the sun.

During the nineteenth century the existence of these neutral points and their behavior with the direction of the sun were regarded as among the most remarkable phenomena in meteorological optics. As such they were studied with great care and attention and by none more than Carl Dorno whose monumental work on the subject[7] contains a wealth of information painstakingly gathered. Dorno not only observed the neutral points on the principal meridian, but he also investigated in detail the continuation of these neutral points over the entire hemisphere along what he called the neutral lines. These lines separate the regions of positive from the regions of negative polarization;[8] they show a remarkable dependence on the direction of the sun. A diagram representing Dorno's principal results is reproduced in figure 4. It will be seen from this figure that when the sun is nearly on the horizon the neutral line connects the Babinet and the Arago points by a closed symmetrical curve of the shape of a lemniscate: this is the so-called lemniscate of Busch. As the sun rises the lemniscate becomes more and more asymmetrical and when the angle of incidence exceeds about 70° the lemniscate opens out and a part of the locus appears on the horizon below the sun and passes through the Brewster point which has now risen. The neutral line consists of two such separated curves until the angle of incidence becomes about 45° when the opposite ends join together to form a closed re-entrant curve. For still smaller angles of incidence the neutral line collapses towards the center and finally reduces to a point when the sun is at the zenith.

We shall now see how this entire range of phenomena associated with the neutral points and lines are faithfully reproduced by our calculations. In

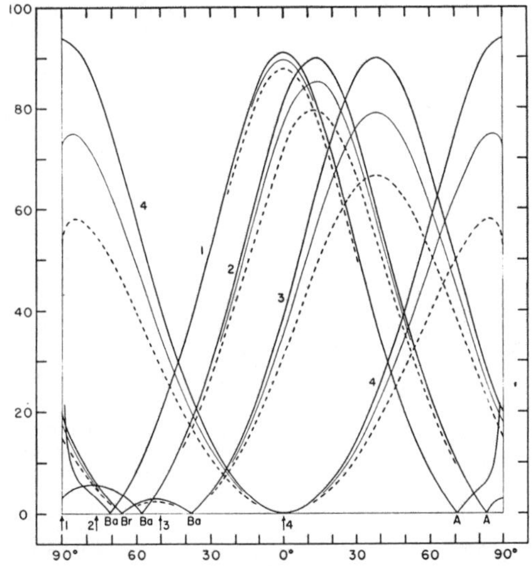

FIG. 7. Variation of the degree of polarization in the principal meridian for various angles of incidence for an atmosphere of optical thickness $\tau_1 = 0.15$. The abscissa gives the zenith distance and the ordinate gives the degree of polarization in per cent. The curves marked 1, 2, 3 and 4 represent the variation for the angles of incidence $\theta_0 = 90°$, 76.1°, 50.2° and 0°, respectively. The thick solid curves are obtained before any ground corrections have been applied. The dashed curves are obtained if we allow for a ground reflecting according to Lambert's law with an albedo $\lambda_0 = 20.5$. The thin intermediate curves are obtained if $\lambda_0 = 0.10$. The positions of the neutral points are also indicated.

[7] C. Dorno, Himmelshelligkeit, Himmelspolarisation und Sonnenintensität in Davos 1911 bis 1918, *Veröffentl. Preuss. Met. Inst.*, No. 303, Berlin, 1919.

[8] The polarization is assumed positive if I_l is less than I_r and negative if the reverse is true. On this convention the polarization is negative on the principal meridian between the neutral points of Babinet and Brewster; these points, therefore, separate the regions of positive from the regions of negative polarization on this meridian. The neutral lines of Dorno do the same for the entire hemisphere.

table 11 we have collected all the information contained in tables 5–10 regarding the points at which the polarization changes sign for various zenith distances of the sun; they are further illustrated in figures 1, 2, 3, and 5. Considering first figures 1–3 (which refer to the principal meridian), we observe that the calculations predict the occurrence of the neutral points as observed. In particular it will be noticed that the Brewster point sets when the angle of incidence is about 70°: its dependence on the values of the optical thickness in the range of interest is not pronounced. Also as the Brewster point sets the Arago point rises in the opposite sky. And as the sun sinks lower, the Arago point continues to rise until, when the sun sets, the Babinet and the Arago points are both at an equal

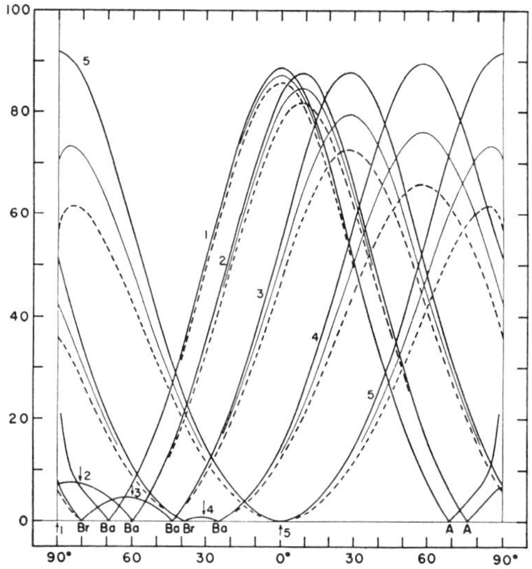

FIG. 8. Variation of the degree of polarization in the principal meridian for various angles of incidence for an atmosphere of optical thickness $\tau_1 = 0.20$. The abscissa gives the zenith distance and the ordinate gives the degree of polarization in per cent. The curves marked 1, 2, 3, 4, and 5 represent the variation for the angles of incidence $\theta_0 = 90°$, 80.8°, 60.0°, 30.7°, and 0°, respectively. The thick solid curves are obtained before any ground corrections have been applied. The dashed curves are obtained if we allow for a ground reflecting according to Lambert's law with an albedo $\lambda_0 = 0.20$. The thin intermediate curves are obtained if $\lambda_0 = 0.10$. The positions of the neutral points are also indicated.

elevation of about 20° (it varies from 17° to 21° for τ_1 in the range $0.10 \leqslant \tau_1 \leqslant 0.20$) from the horizon.

With the calculations for the different values of $\varphi_0 - \varphi$ for $\tau_1 = 0.15$ given in table 5, we can draw an entire system of calculated neutral lines. This has been done in figure 5. Comparing it with the results of Dorno's observations (fig. 4) we observe how well the two sets of curves match.

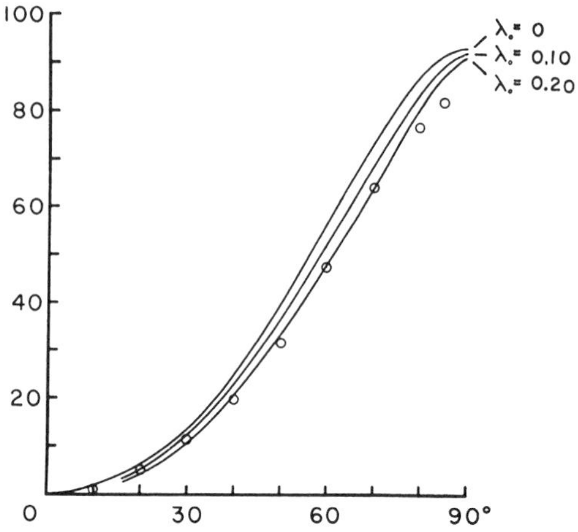

FIG. 9. Variation of the degree of polarization at the zenith for various angles of incidence and for an atmosphere of optical thickness $\tau_1 = 0.10$. The curve $\lambda_0 = 0$ allows for no ground reflection. The other two curves allow for reflection by a ground with the albedos indicated. The circles represent the observations of Tousey and Hulburt for values of τ_1 and λ_0 estimated at 0.10 and 0.20, respectively.

Turning next to the calculated degrees of polarization, we have illustrated its variation on the principal meridian for various angles of incidence and for the three values of the optical thickness for which calculations have been made in figures 6, 7, and 8. It will be noticed from these figures that the effect of a ground surface with an albedo even as high as 0.25 does not

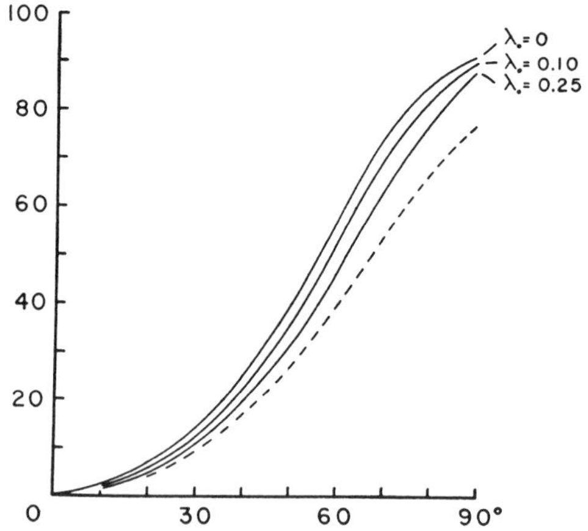

FIG. 10. Variation of the degree of polarization at the zenith for various angles of incidence and for an atmosphere of optical thickness $\tau_1 = 0.15$. The curve $\lambda_0 = 0$ allows for no ground reflection. The other two curves allow for reflection by a ground with the albedos indicated. The dashed curve represents the observations of Tichanowsky.

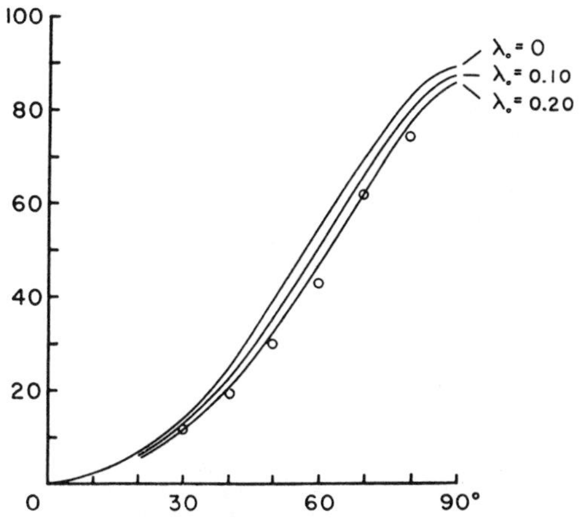

FIG. 11. Variation of the degree of polarization at the zenith for various angles of incidence and for an atmosphere of optical thickness $\tau_1 = 0.20$. The curve $\lambda_0 = 0$ allows for no ground reflection. The other two curves allow for reflection by a ground with the albedos indicated. The circles represent the observations of Richardson and Hulburt for a value of τ_1 estimated at 0.3.

make any essential difference to the predicted positions of the neutral points. This independence of the neutral points (and lines) on ground reflection is not difficult to understand. As is well known, the laws of Rayleigh scattering give the maximum polarization for the scattered light; all other laws give much less polarization. A ground reflecting according to Lambert's law can, therefore, hardly compete with Rayleigh scattering for producing polarization. The effect of reflection by the ground is therefore essentially one of adding a component of natural light to the polarized light already present. This last statement is, of course, not strictly true. The difference between γ_l and γ_r is precisely a measure of the polarization of the ground contribution to the sky brightness; but as is apparent from the tables the difference between γ_l and γ_r is generally very small. For these same reasons we should not expect that the direction of maximum polarization will be influenced by ground reflection. This is, indeed, the case: the direction of maximum polarization always occurs in a direction which is very nearly at right-angles to the direction of the sun. In contrast, ground reflection has a very pronounced effect on the *degree* of maximum polarization: In the absence of ground corrections, the maximum polarization varies between 94 and 89 per cent depending on τ_1 but is very nearly independent of the altitude of the sun (though there is a slight increase for higher altitudes). However, when the effect of ground reflection is taken into account the maximum polarization shows a very marked decrease with altitude. Observations do show such a behavior and we conclude that this is probably due to the effect of ground reflection.

Finally, in figures 9, 10, and 11 we have compared the degree of polarization at the zenith for various angles of incidence with the observations of Tichanowsky,[9] Tousey and Hulburt,[10] and Richardson and Hulburt.[11] It will be seen that the agreement between the theory and the observations is as good as one might expect.

In concluding this paper we should again like to record our thanks to Dr. Wallace Eckert and Miss Ann Franklin of the Watson Scientific Computing Laboratory for their generous co-operation in obtaining the solutions for the basic *X*- and *Y*-functions.

[9] J. J. Tichanowsky, Resultate der Messungen der Himmelspolarisation in verschiedenen Spektrumabschnitten, *Meteor. Ztschr.* **43**: 288, 1926.

[10] R. Tousey and E. O. Hulburt, Brightness and polarization of the daylight sky at various altitudes above sea level, *Jour. Opt. Soc. Amer.* **37**: 78, 1947.

[11] R. A. Richardson and E. O. Hulburt, Sky-brightness measurements near Bocaiuva, Brazil, *Jour. Geophys. Research* **54**: 215, 1949.

3. *Principles of Invariance: The H- and the X- and Y-functions*

ON THE RADIATIVE EQUILIBRIUM OF A STELLAR ATMOSPHERE. XIV

S. CHANDRASEKHAR
Yerkes Observatory

Received October 28, 1946

ABSTRACT

The main object of this paper is to show how, by incorporating into the methods developed in the earlier papers of this series a basic idea of Ambarzumian, we are able to solve exactly a large class of problems in the theory of radiative transfer. More specifically, we show that, when the solutions for the emergent (or the reflected) radiation (obtained in the nth approximation of the earlier papers) involve H-functions of the form

$$H(\mu) = \frac{1}{\mu_1 \ldots \mu_n} \frac{\prod_i (\mu + \mu_i)}{\prod_a (1 + k_a \mu)}, \tag{i}$$

where the μ_i's are the positive zeros of the Legendre polynomial $P_{2n}(\mu)$ and the k_a's are the roots of a characteristic equation

$$1 = 2\sum_{j=1}^{n} \frac{a_j \Psi(\mu_j)}{1 - k^2\mu_j^2} \qquad \left(\Psi \text{ an even polynomial in } \mu \text{ and } \int_0^1 \Psi(\mu)d\mu \leqslant \tfrac{1}{2}\right), \tag{ii}$$

then, in the limit of infinite approximation, the H-functions become solutions of functional equations of a certain standard form, namely,

$$H(\mu) = 1 + \mu H(\mu) \int_0^1 \frac{H(\mu')\Psi(\mu')}{\mu + \mu'} d\mu' . \tag{iii}$$

Further, when the solution involves, in addition to the H-functions, certain constants, then in the exact solution these constants are related in a determinate way with the moments of the H-functions themselves.

The paper is divided into five main sections. In Section I we derive (following Ambarzumian's general ideas) the functional equations for the problem of diffuse reflection when allowance is also made for the polarization of the scattered radiation in accordance with Rayleigh's law. This problem requires us to find a functional equation for the scattering matrix $\boldsymbol{S}$ introduced in Paper XIII. In Section II the functional equation for $\boldsymbol{S}$ is solved and the solution expressed in terms of four functional equations, all of type (iii). The exact solutions for the laws of darkening in the two states of polarization for an electron-scattering atmosphere are also given in this section. Section III is devoted to a general study of functional equations of the form (iii). The basic theorem relating the H-function as defined in terms of the Gaussian division and the characteristic roots and as defined as the solution of the associated functional equation is proved in this section. In Section IV the problem of diffuse reflection in accordance with the phase function $\lambda(1 + x \cos \Theta)$ is treated. And, finally, in Section V we give the exact solutions for transfer problems involving Rayleigh's phase function.

1. *Introduction.*—The problem of diffuse reflection by a semi-infinite plane-parallel atmosphere has been considered under a variety of circumstances in earlier papers of this series.[1] And it has been shown for many of the problems considered that the angular distribution (and in some cases, also, the state of polarization) of the reflected radiation can be given in closed forms in a general nth approximation. The essential feature of

[1] *Ap. J.*, **101**, 348, 1945; **103**, 165, 1946; **104**, 110, 191, 1946; and **105**, 151, 1947. These papers will be referred to as Papers VIII, IX, XI, XII, and XIII, respectively. Other papers of this series to which we shall have occasion to refer are Papers II (*Ap. J.*, **100**, 76, 1944), III (*ibid.*, **100**, 117, 1944), and X (*ibid.*, **103**, 351, 1946).

these solutions is that, in addition to certain constants, they involve only certain "H-functions" of the form

$$H(\mu) = \frac{1}{\mu_1 \cdots \mu_n} \frac{\prod_i (\mu + \mu_i)}{\prod_\alpha (1 + k_\alpha \mu)}, \tag{1}$$

where the μ_i's are the positive zeros of the Legendre polynomial $P_{2n}(\mu)$ and the k_α's are the positive roots of a characteristic equation of the form

$$1 = 2 \sum_{j=1}^{n} \frac{a_j \Psi(\mu_j)}{1 - k^2 \mu_j^2}, \tag{2}$$

where the a_j's are the Gaussian weights and $\Psi(\mu)$ is an even polynomial in μ, satisfying the condition[2]

$$\int_0^1 \Psi(\mu)\, d\mu \leqslant \tfrac{1}{2}. \tag{3}$$

Now the solutions we have found for the various problems of diffuse reflection were obtained by directly solving the appropriate equations of transfer, in contrast to a method recently devised by V. A. Ambarzumian[3] in which an explicit solution of the transfer problem is avoided by deriving functional equations for the reflection functions $\sigma(\mu, \varphi; \mu_0, \varphi_0)$, which give the intensity reflected in a given direction (μ, φ) when a parallel beam of unit flux per unit area normal to itself is incident in the direction $(-\mu_0, \varphi_0)$. These functional equations of Ambarzumian are of rather complicated forms, even in cases in which our method gives closed solutions of relatively simple forms.[4] This is somewhat surprising, and a close examination of the correspondence between the two methods undertaken in this paper reveals that a knowledge of the *forms* of the solutions obtained by our method enables us, in most cases, to reduce Ambarzumian's type of functional equations to the following standard form:

$$H(\mu) = 1 + \mu H(\mu) \int_0^1 \frac{H(\mu') \Psi(\mu')}{\mu + \mu'} d\mu'. \tag{4}$$

Indeed, we shall show that in a certain nth approximation the solution of the functional equation (4) is, in fact, our H-function. More precisely, it will appear that $H(\mu)$, as we have defined it in equations (1) and (2), is an exact solution of the associated *difference equation*

$$H(\mu) = 1 + \mu H(\mu) \sum_{j=1}^{n} \frac{a_j H(\mu_j) \Psi(\mu_j)}{\mu + \mu_j}. \tag{5}$$

It would therefore appear that, with the solutions in the forms we have given them, we can pass directly to the limit of infinite approximation by letting the H-functions become solutions of functional equations of the form (4) rather than of the associated difference equations. And as for the constants which appear in our solutions, they can be shown to be related in some way with the moments of the H-functions themselves. The conclusions we have just stated are of practical importance as well; for, on the basis of

[2] This condition is necessary for $H(\mu)$ to be real (see theorem 3, corollary 1, § 12).

[3] *J. Phys. Acad. Sci. U.S.S.R.*, **8**, 64, 1944.

[4] Thus compare his pair of functional equations on p. 72, eq. (30) (*op. cit.*), with our solution (108) in Paper IX. Both refer to the same physical problem.

tests that we have made, it appears that, starting with a solution for $H(\mu)$ in the third approximation, we can derive by a rapid method of iteration the exact function numerically. Thus, by incorporating Ambarzumian's basic idea in the methods developed in this series of papers, we are able to solve a very large class of transfer problems *exactly*.

The plan of this paper is as follows:

In Section I we derive, following Ambarzumian's general ideas, the functional equations for the problem of diffuse reflection when allowance is also made for the polarization of the scattered radiation in accordance with Rayleigh's law. As we shall see, this problem requires us to find a functional equation for the scattering matrix **S**, which we introduced in the preceding paper. The problem is, accordingly, more advanced than the ones considered by Ambarzumian; but it serves to illustrate the power of his fundamental idea. In Section II we show that the solution of the functional equations (which, for the most important set of functions, is a simultaneous system of order 4) derived in Section I can all be expressed in terms of four independent functional equations of the standard type (4) and a constant, which is expressible in terms of the moments of two of these functions. Section III is devoted to a general study of functional equations of the form (4). In particular, we prove in this section the basic theorem relating the H-function as defined by equations (1) and (2) and the solution of the functional equation (4). The analysis of these sections throws some interesting sidelights on integrals of the type of Hopf-Bronstein. Several other theorems relating to this type of functional equation are also proved. In Section IV we treat a problem considered by Ambarzumian and show how the solution of a pair of simultaneous functional equations which he had derived can also be reduced to a single functional equation of the form (4). Finally, in Section V we solve the functional equations for the problem of diffuse reflection with Rayleigh's phase function (without allowing for polarization).

I. FUNCTIONAL EQUATION FOR THE SCATTERING MATRIX DESCRIBING THE DIFFUSE REFLECTION FROM A SEMI-INFINITE ATMOSPHERE, SCATTERING RADIATION ACCORDING TO RAYLEIGH'S LAW

2. *The vector form of the equations of transfer.*—For the purpose of deriving the functional equations for the problem of diffuse reflection according to Rayleigh's law and allowing for the polarization of the scattered radiation, we must first re-write the equations of transfer given in Papers XI and XIII in more suitable forms. First, it is convenient to introduce subscripts 1, 2, and 3 instead of the letters l, r, and U and to let

$$I_l = I_1 , \qquad I_r = I_2 , \qquad \text{and} \qquad U = I_3 . \tag{6}$$

Now suppose that a parallel beam of partially polarized light, characterized by the fluxes

$$\pi F_k \qquad (k = 1, 2, 3) , \tag{7}$$

is incident on the atmosphere in the direction $(-\mu_0, \varphi_0)$. The equation of transfer for I_i can be written in the form

$$\mu \frac{dI_i}{d\tau} = I_i - B_i(\tau, \mu, \varphi) \qquad (i = 1, 2, 3) , \tag{8}$$

where the source-function, B_i, can be regarded as the sum of two source-functions, one arising from the scattering of the diffuse radiation at τ and the other arising from the scattering of the reduced incident radiation, $\pi F_k e^{-\tau/\mu_0}$, prevailing at τ. According to Paper XIII, equations (1)–(3), we can write

$$B_i(\tau, \mu, \varphi) = \mathfrak{J}_i(\tau, \mu, \varphi) + \tfrac{3}{16} q_i \sum_{k=1}^{3} j_{ik}(\mu, \varphi; \mu_0, \varphi_0) F_k e^{-\tau/\mu_0} \qquad (i = 1, 2, 3) , \tag{9}$$

where $\mathfrak{J}_i(\tau, \mu, \varphi)$ denotes the contribution to the source-function arising from the diffuse radiation field at τ and $j_{ik}(\mu, \varphi; \mu_0, \varphi_0)$ are certain coefficients which can be regarded as the elements of the matrix

$$J(\mu, \varphi; \mu_0, \varphi_0) = \left\| \begin{array}{lll} 2(1-\mu^2)(1-\mu_0^2) + \mu^2\mu_0^2 & & \\ - 4\mu\mu_0(1-\mu^2)^{\frac{1}{2}}(1-\mu_0^2)^{\frac{1}{2}}\cos(\varphi-\varphi_0) & & - 2\mu(1-\mu^2)^{\frac{1}{2}}(1-\mu_0^2)^{\frac{1}{2}}\sin(\varphi-\varphi_0) \\ + \mu^2\mu_0^2\cos 2(\varphi-\varphi_0) & \mu^2 - \mu^2\cos 2(\varphi-\varphi_0) & + \mu^2\mu_0\sin 2(\varphi-\varphi_0) \\ \mu_0^2 - \mu_0^2\cos 2(\varphi-\varphi_0) & 1 + \cos 2(\varphi-\varphi_0) & - \mu_0\sin 2(\varphi-\varphi_0) \\ - 2\mu_0(1-\mu_0^2)^{\frac{1}{2}}(1-\mu^2)^{\frac{1}{2}}\sin(\varphi-\varphi_0) & & (1-\mu^2)^{\frac{1}{2}}(1-\mu_0^2)^{\frac{1}{2}}\cos(\varphi-\varphi_0) \\ + \mu\mu_0^2\sin 2(\varphi-\varphi_0) & - \mu\sin 2(\varphi-\varphi_0) & - \mu\mu_0\cos 2(\varphi-\varphi_0) \end{array} \right\| \quad (10)$$

Further, in equation (9)

$$q_1 = q_2 = 1 \,, \qquad \text{and} \qquad q_3 = 2 \,. \tag{11}$$

It will be observed that the matrix J (which is appropriate for *single scattering*) has the same symmetry that we have noted for the scattering matrix S in Paper XIII (eq. [67]). Thus,

$$\tilde{J}(\mu, \varphi; \mu_0, \varphi_0) = J(\mu_0, \varphi; \mu, \varphi_0) \,. \tag{12}$$

Returning to the source-functions $\mathfrak{J}_i(\tau, \mu, \varphi)$, we verify that the expressions given for these functions in Paper XI (eqs. [49]–[52]) can be abbreviated in the form

$$\left.\begin{aligned}\mathfrak{J}_i(\tau, \mu, \varphi) = \frac{3}{16\pi} q_i \int_{-1}^{+1}\int_0^{2\pi} \sum_{k=1}^{3} j_{ik}(\mu, \varphi; -\mu', \varphi') I_k(\tau, \mu', \varphi') \, d\mu' d\varphi' \\ (i = 1, 2, 3) \,.\end{aligned}\right\} \tag{13}$$

Combining equations (8), (9), and (13), we have

$$\left.\begin{aligned}\mu \frac{dI_i}{d\tau} = I_i - \frac{3}{16\pi} q_i \int_{-1}^{+1}\int_0^{2\pi} \sum_{k=1}^{3} j_{ik}(\mu, \varphi; -\mu', \varphi') I_k(\tau, \mu', \varphi') \, d\mu' d\varphi' \\ - \tfrac{3}{16} q_i \sum_{k=1}^{3} j_{ik}(\mu, \varphi; \mu_0, \varphi_0) F_k e^{-\tau/\mu_0} \qquad (i = 1, 2, 3) \,.\end{aligned}\right\} \tag{14}$$

If we now regard the intensities $I_i (i = 1, 2, 3)$ as the components of a vector I, we can re-write the three equations of transfer which equation (14) represents as a single vector equation of the form

$$\left.\begin{aligned}\mu \frac{dI}{d\tau} = I - \frac{3}{16\pi} Q \int_{-1}^{+1}\int_0^{2\pi} J(\mu, \varphi; -\mu', \varphi') I(\tau, \mu', \varphi') \, d\mu' d\varphi' \\ - \tfrac{3}{16} Q J(\mu, \varphi; \mu_0, \varphi_0) F e^{-\tau/\mu_0} \,,\end{aligned}\right\} \tag{15}$$

where

$$F = (F_1, F_2, F_3) \tag{16}$$

and

$$Q = \begin{vmatrix} 1 & 0 & 0 \\ 0 & 1 & 0 \\ 0 & 0 & 2 \end{vmatrix} \,. \tag{17}$$

Since the equation of transfer is linear in F, it is clear that the intensity $I_i(\mu, \varphi; \mu_0, \varphi_0)$, diffusely reflected by the atmosphere in the direction (μ, φ), must be expressible in the form

$$I_i(\mu, \varphi; \mu_0, \varphi_0) = \tfrac{3}{16} q_i \sum_{k=1}^{3} \sigma_{ik}(\mu, \varphi; \mu_0, \varphi_0) F_k \qquad (i = 1, 2, 3) \tag{18}$$

or as a vector equation,

$$I(\mu, \varphi; \mu_0, \varphi_0) = \tfrac{3}{16} Q \Sigma(\mu, \varphi; \mu_0, \varphi_0) F \,. \tag{19}$$

3. *The functional equation for the scattering matrix.*—To obtain the functional equations for the quantities $\sigma_{ik}(\mu, \varphi; \mu_0, \varphi_0)$, introduced in § 2, the procedure is as follows:

Consider the case when the incident beam has only one nonvanishing component in F, say F_k. Denote the corresponding intensities by $I_i^{(k)} (i = 1, 2, 3)$. The equations of transfer now reduce to

$$\mu \frac{dI_i^{(k)}}{d\tau} = I_i^{(k)} - B_i^{(k)}(\tau, \mu, \varphi) \qquad (i = 1, 2, 3) \,, \tag{20}$$

where

$$B_i^{(k)}(\tau, \mu, \varphi) = \mathfrak{J}_i^{(k)}(\tau, \mu, \varphi) + \tfrac{3}{16} q_i j_{ik}(\mu, \varphi; \mu_0, \varphi_0) F_k e^{-\tau/\mu_0} \qquad (i = 1, 2, 3) . \quad (21)$$

However, in general, there will be nonvanishing intensities in all three components in the reflected light. In fact, according to equation (18), we must have

$$I_i^{(k)}(\mu, \varphi; \mu_0, \varphi_0) = \tfrac{3}{16} q_i \sigma_{ik}(\mu, \varphi; \mu_0, \varphi_0) F_k \qquad (i = 1, 2, 3) . \quad (22)$$

Now consider a level at depth $d\tau$ below the boundary of the atmosphere at $\tau = 0$. At this level the radiation field present can be decomposed into two parts: first, there is the reduced incident flux of amount

$$\pi F_k \left(1 - \frac{d\tau}{\mu_0}\right), \quad (23)$$

and, second, there is a diffuse radiation field. The amount of this diffuse radiation field which is directed inward can be inferred from the equation of transfer: for, since at $\tau = 0$ there is no inward intensity, at the level $d\tau$ we must have an inward intensity (cf. eq. [20]),

$$I_i^{(k)}(d\tau, -\mu', \varphi') = \frac{1}{\mu'} B_i^{(k)}(0, -\mu', \varphi')\, d\tau \qquad (i = 1, 2, 3) , \quad (24)$$

in the direction $(-\mu', \varphi')$. Both the radiation fields (23) and (24) will be reflected by the atmosphere below $d\tau$ by the *same laws* as those by which the atmosphere below $\tau = 0$ reflects. The reason for this invariance is simply that the removal of a layer of arbitrary optical thickness from an infinite atmosphere cannot alter its reflecting-power. This is Ambarzumian's basic idea. Accordingly, the reflection of radiations (23) and (24) by the atmosphere below $d\tau$ will contribute to an *outward* intensity in the direction (μ, φ) the amount

$$\left.\begin{aligned} I_i^{(k)}(d\tau, \mu, \varphi) &= \tfrac{3}{16} q_i \sigma_{ik}(\mu, \varphi; \mu_0, \varphi_0)\left(1 - \frac{d\tau}{\mu_0}\right) F_k \\ &+ \frac{3}{16\pi} q_i d\tau \int_0^1 \int_0^{2\pi} \sum_{l=1}^{3} \sigma_{il}(\mu, \varphi; \mu', \varphi') B_l^{(k)}(0, -\mu', \varphi') \frac{d\mu'}{\mu'} d\varphi' . \end{aligned}\right\} \quad (25)$$

On the other hand, from the equation of transfer, we can directly conclude that

$$\left.\begin{aligned} I_i^{(k)}(d\tau, \mu, \varphi) &= I_i^{(k)}(0, \mu, \varphi) + \frac{d\tau}{\mu}[I_i^{(k)}(0, \mu, \varphi) - B_i^{(k)}(0, \mu, \varphi)] \\ &= \tfrac{3}{16} q_i \left(1 + \frac{d\tau}{\mu}\right) \sigma_{ik}(\mu, \varphi; \mu_0, \varphi_0) F_k - \frac{d\tau}{\mu} B_i^{(k)}(0, \mu, \varphi) . \end{aligned}\right\} \quad (26)$$

Combining equations (25) and (26) and passing to the limit $d\tau = 0$, we have

$$\left.\begin{aligned} \tfrac{3}{16} q_i \left(\frac{1}{\mu} + \frac{1}{\mu_0}\right) \sigma_{ik}(\mu, \varphi; \mu_0, \varphi_3) F_k &= \frac{1}{\mu} B_i^{(k)}(0, \mu, \varphi) \\ &+ \frac{3}{16\pi} q_i \int_0^1 \int_0^{2\pi} \sum_{l=1}^{3} \sigma_{il}(\mu, \varphi; \mu', \varphi') B_l^{(k)}(0, -\mu', \varphi') \frac{d\mu'}{\mu'} d\varphi' . \end{aligned}\right\} \quad (27)$$

But, according to equations (13) and (21),

$$\left.\begin{aligned} B_i^{(k)}(0, \mu, \varphi) &= \mathfrak{J}_i^{(k)}(0, \mu, \varphi) + \tfrac{3}{16} q_i j_{ik}(\mu, \varphi; \mu_0, \varphi_0) F_k \\ &= \tfrac{3}{16} q_i j_{ik}(\mu, \varphi; \mu_0, \varphi_0) F_k + \frac{3}{16\pi} q_i \int_0^1 \int_0^{2\pi} \sum_{m=1}^{3} j_{im}(\mu, \varphi; -\mu', \varphi') \\ &\times I_m^{(k)}(0, \mu', \varphi')\, d\mu' d\varphi' = \tfrac{3}{16} q_i F_k \Big[j_{ik}(\mu, \varphi; \mu_0, \varphi_0) \\ &+ \frac{3}{16\pi} \int_0^1 \int_0^{2\pi} \sum_{m=1}^{3} j_{im}(\mu, \varphi; -\mu', \varphi')\, q_m \sigma_{mk}(\mu', \varphi'; \mu_0, \varphi_0)\, d\mu' d\varphi' \Big] . \end{aligned}\right\} \quad (28)$$

Using equation (28) in equation (27) and writing

$$\mu\sigma_{ik}(\mu,\varphi;\mu_0,\varphi_0)=S_{ik}(\mu,\varphi;\mu_0,\varphi_0)\,, \tag{29}$$

we obtain, after some further reductions,

$$\left.\begin{aligned}\left(\frac{1}{\mu}+\frac{1}{\mu_0}\right)S_{ik}(\mu,\varphi;\mu_0,\varphi_0)&=j_{ik}(\mu,\varphi;\mu_0,\varphi_0)\\&+\frac{3}{16\pi}\int_0^1\int_0^{2\pi}\sum_{m=1}^{3}j_{im}(\mu,\varphi;-\mu'',\varphi'')\,q_mS_{mk}(\mu'',\varphi'';\mu_0,\varphi_0)\frac{d\mu''}{\mu''}d\varphi''\\&+\frac{3}{16\pi}\int_0^1\int_0^{2\pi}\sum_{l=1}^{3}S_{il}(\mu,\varphi;\mu',\varphi')\,q_lj_{lk}(-\mu',\varphi';\mu_0,\varphi_0)\frac{d\mu'}{\mu'}d\varphi'\\&+\frac{9}{256\pi^2}\int_0^1\int_0^{2\pi}\int_0^1\int_0^{2\pi}\sum_{l=1}^{3}\sum_{m=1}^{3}S_{il}(\mu,\varphi;\mu',\varphi')\,q_lj_{lm}(-\mu',\varphi';-\mu'',\varphi'')\\&\qquad\times q_mS_{mk}(\mu'',\varphi'';\mu_0,\varphi_0)\frac{d\mu'}{\mu'}d\varphi'\frac{d\mu''}{\mu''}d\varphi''\,.\end{aligned}\right\} \tag{30}$$

Equation (30) can be written as a matrix equation in the form

$$\left.\begin{aligned}\left(\frac{1}{\mu}+\frac{1}{\mu_0}\right)\boldsymbol{S}(\mu,\varphi;\mu_0,\varphi_0)&=\boldsymbol{J}(\mu,\varphi;\mu_0,\varphi_0)\\&+\frac{3}{16\pi}\int_0^1\int_0^{2\pi}\boldsymbol{J}(\mu,\varphi;-\mu'',\varphi'')\boldsymbol{QS}(\mu'',\varphi'';\mu_0,\varphi_0)\frac{d\mu''}{\mu''}d\varphi''\\&+\frac{3}{16\pi}\int_0^1\int_0^{2\pi}\boldsymbol{S}(\mu,\varphi;\mu',\varphi')\boldsymbol{QJ}(-\mu',\varphi';\mu_0,\varphi_0)\frac{d\mu'}{\mu'}d\varphi'\\&+\frac{9}{256\pi^2}\int_0^1\int_0^{2\pi}\int_0^1\int_0^{2\pi}\boldsymbol{S}(\mu,\varphi;\mu',\varphi')\boldsymbol{QJ}(-\mu',\varphi';-\mu'',\varphi'')\boldsymbol{Q}\\&\qquad\times\boldsymbol{S}(\mu'',\varphi'';\mu_0,\varphi_0)\frac{d\mu'}{\mu'}d\varphi'\frac{d\mu''}{\mu''}d\varphi''\,.\end{aligned}\right\} \tag{31}$$

This is the required functional equation for the scattering matrix.

From equation (31) it readily follows that the scattering matrix $\boldsymbol{S}$ must have the same properties for transposition as the matrix $\boldsymbol{J}$. Hence,

$$\tilde{\boldsymbol{S}}(\mu,\varphi;\mu_0,\varphi_0)=\boldsymbol{S}(\mu_0,\varphi;\mu,\varphi_0)\,. \tag{32}$$

In other words, we are able to deduce very simply the reciprocity principle in the form in which we reformulated it in Paper XIII (§ 4).

II. THE SOLUTION OF THE FUNCTIONAL EQUATION FOR THE SCATTERING MATRIX

4. *The reduction of the functional equation for* **S**.—From the form of the functional equation for $\boldsymbol{S}$ and the manner of its relation to $\boldsymbol{J}$, it is evident that in a Fourier analysis of the elements of $\boldsymbol{S}$ in $(\varphi-\varphi_0)$ we must have the same nonvanishing components as in the corresponding elements of $\boldsymbol{J}$. We may accordingly assume without loss of generality that $\boldsymbol{S}$ has the form

$$\mathbf{S}(\mu, \varphi; \mu_0, \varphi_0) = \left| \begin{array}{lll} S_{11}^{(0)}(\mu, \mu_0) & S_{12}^{(0)}(\mu, \mu_0) & \\ -4\mu\mu_0(1-\mu^2)^{\frac{1}{2}}(1-\mu_0^2)^{\frac{1}{2}}S_{11}^{(1)}(\mu, \mu_0)\cos(\varphi-\varphi_0) & & -2\mu(1-\mu^2)^{\frac{1}{2}}(1-\mu_0^2)^{\frac{1}{2}}S_{13}^{(1)}(\mu, \mu_0)\sin(\varphi-\varphi_0) \\ +\mu^2\mu_0^2S_{11}^{(2)}(\mu, \mu_0)\cos 2(\varphi-\varphi_0) & -\mu^2S_{21}^{(2)}(\mu, \mu_0)\cos 2(\varphi-\varphi_0) & +\mu^2\mu_0S_{13}^{(2)}(\mu, \mu_0)\sin 2(\varphi-\varphi_0) \\ & & \\ S_{21}^{(0)}(\mu, \mu_0) & S_{22}^{(0)}(\mu, \mu_0) & \\ -\mu_0^2S_{21}^{(2)}(\mu, \mu_0)\cos 2(\varphi-\varphi_0) & +S_{22}^{(2)}(\mu, \mu_0)\cos 2(\varphi-\varphi_0) & -\mu_0S_{23}^{(2)}(\mu, \mu_0)\sin 2(\varphi-\varphi_0) \\ & & \\ -2\mu_0(1-\mu_0^2)^{\frac{1}{2}}(1-\mu^2)^{\frac{1}{2}}S_{31}^{(1)}(\mu, \mu_0)\sin(\varphi-\varphi_0) & & (1-\mu^2)^{\frac{1}{2}}(1-\mu_0^2)^{\frac{1}{2}}S_{33}^{(1)}(\mu, \mu_0)\cos(\varphi-\varphi_0) \\ +\mu_0^2\mu S_{31}^{(2)}(\mu, \mu_0)\sin 2(\varphi-\varphi_0) & -\mu S_{32}^{(2)}(\mu, \mu_0)\sin 2(\varphi-\varphi_0) & -\mu\mu_0S_{33}^{(2)}(\mu, \mu_0)\cos 2(\varphi-\varphi_0), \end{array} \right. \tag{3.3}$$

where, as the notation indicates, $S_{11}^{(0)}$, etc., are all functions of μ and μ_0 only. From the property (32) of **S** for transposition, we now conclude that

$$S_{jk}^{(i)}(\mu, \mu_0) = S_{kj}^{(i)}(\mu_0, \mu) . \tag{34}$$

Substituting in equation (31) the form (33) for **S** and equating the different Fourier components of the various elements, we find, after a long but straightforward series of reductions, that we are led to the following three systems of equations:

$$
\left.
\begin{aligned}
\left(\frac{1}{\mu}+\frac{1}{\mu_0}\right) S_{11}^{(0)}(\mu,\mu_0) &= j_{11}^{(0)}(\mu,\mu_0) + \frac{3}{8}\int_0^1 \frac{d\mu''}{\mu''}\left\{ j_{11}^{(0)}(\mu,\mu'') S_{11}^{(0)}(\mu'',\mu_0) \right. \\
&\quad \left. + \mu^2 S_{21}^{(0)}(\mu'',\mu_0)\right\} + \frac{3}{8}\int_0^1 \frac{d\mu'}{\mu'}\left\{ j_{11}^{(0)}(\mu',\mu_0) S_{11}^{(0)}(\mu,\mu') + \mu_0^2 S_{12}^{(0)}(\mu,\mu')\right\} \\
&\quad + \frac{9}{64}\int_0^1\int_0^1 \frac{d\mu''}{\mu''}\frac{d\mu'}{\mu'}\{ S_{11}^{(0)}(\mu,\mu')[j_{11}^{(0)}(\mu',\mu'') S_{11}^{(0)}(\mu'',\mu_0) + \mu'^2 S_{21}^{(0)}(\mu'',\mu_0)] \\
&\qquad + S_{12}^{(0)}(\mu,\mu')[\mu''^2 S_{11}^{(0)}(\mu'',\mu_0) + S_{21}^{(0)}(\mu'',\mu_0)]\}, \\
\left(\frac{1}{\mu}+\frac{1}{\mu_0}\right) S_{12}^{(0)}(\mu,\mu_0) &= \mu^2 + \frac{3}{8}\int_0^1 \frac{d\mu''}{\mu''}\left\{ j_{11}^{(0)}(\mu,\mu'') S_{12}^{(0)}(\mu'',\mu_0) + \mu^2 S_{22}^{(0)}(\mu'',\mu_0)\right\} \\
&\quad + \frac{3}{8}\int_0^1 \frac{d\mu'}{\mu'}\{ S_{11}^{(0)}(\mu,\mu')\mu'^2 + S_{12}^{(0)}(\mu,\mu')\} \\
&\quad + \frac{9}{64}\int_0^1\int_0^1 \frac{d\mu''}{\mu''}\frac{d\mu'}{\mu'}\{ S_{12}^{(0)}(\mu,\mu')\,[\mu''^2 S_{12}^{(0)}(\mu'',\mu_0) + S_{22}^{(0)}(\mu'',\mu_0)] \\
&\qquad + S_{11}^{(0)}(\mu,\mu')[j_{11}^{(0)}(\mu',\mu'') S_{12}^{(0)}(\mu'',\mu_0) + \mu'^2 S_{22}^{(0)}(\mu'',\mu_0)]\}, \\
\left(\frac{1}{\mu}+\frac{1}{\mu_0}\right) S_{21}^{(0)}(\mu,\mu_0) &= \mu_0^2 + \frac{3}{8}\int_0^1 \frac{d\mu''}{\mu''}\{\mu''^2 S_{11}^{(0)}(\mu'',\mu_0) + S_{21}^{(0)}(\mu'',\mu_0)\} \\
&\quad + \frac{3}{8}\int_0^1 \frac{d\mu'}{\mu'}\{ S_{21}^{(0)}(\mu,\mu')\, j_{11}^{(0)}(\mu',\mu_0) + S_{22}^{(0)}(\mu,\mu')\,\mu_0^2\} \\
&\quad + \frac{9}{64}\int_0^1\int_0^1 \frac{d\mu''}{\mu''}\frac{d\mu'}{\mu'}\{ S_{21}^{(0)}(\mu'',\mu_0)[S_{21}^{(0)}(\mu,\mu')\mu'^2 + S_{22}^{(0)}(\mu,\mu')] \\
&\qquad + S_{11}^{(0)}(\mu'',\mu_0)[S_{21}^{(0)}(\mu,\mu')\, j_{11}^{(0)}(\mu',\mu'') + S_{22}^{(0)}(\mu,\mu')\mu''^2]\}, \\
\left(\frac{1}{\mu}+\frac{1}{\mu_0}\right) S_{22}^{(0)}(\mu,\mu_0) &= 1 + \frac{3}{8}\int_0^1 \frac{d\mu''}{\mu''}\{\mu''^2 S_{12}^{(0)}(\mu'',\mu_0) + S_{22}^{(0)}(\mu'',\mu_0)\} \\
&\quad + \frac{3}{8}\int_0^1 \frac{d\mu'}{\mu'}\{ S_{21}^{(0)}(\mu,\mu')\mu'^2 + S_{22}^{(0)}(\mu,\mu')\} \\
&\quad + \frac{9}{64}\int_0^1\int_0^1 \frac{d\mu''}{\mu''}\frac{d\mu'}{\mu'}\{ S_{22}^{(0)}(\mu,\mu')\,[\mu''^2 S_{12}^{(0)}(\mu'',\mu_0) + S_{22}^{(0)}(\mu'',\mu_0)] \\
&\qquad + S_{21}^{(0)}(\mu,\mu')[j_{11}^{(0)}(\mu',\mu'') S_{12}^{(0)}(\mu'',\mu_0) + \mu'^2 S_{22}^{(0)}(\mu'',\mu_0)]\},
\end{aligned}
\right\} \quad \text{(I)}
$$

where we have used the abbreviation

$$
j_{11}^{(0)}(\mu,\mu') = 2(1-\mu^2)(1-\mu'^2) + \mu^2\mu'^2\,; \tag{35}
$$

$$
\left.
\begin{aligned}
\left(\frac{1}{\mu}+\frac{1}{\mu_0}\right) S_{11}^{(1)}(\mu,\mu_0) &= \left\{1 + \frac{3}{8}\int_0^1 \frac{d\mu''}{\mu''}(1-\mu''^2)[S_{31}^{(1)}(\mu'',\mu_0) + 2\mu''^2 S_{11}^{(1)}(\mu'',\mu_0)]\right\} \\
&\quad \times\left\{1 + \frac{3}{8}\int_0^1 \frac{d\mu'}{\mu'}(1-\mu'^2)[S_{13}^{(1)}(\mu,\mu') + 2\mu'^2 S_{11}^{(1)}(\mu,\mu')]\right\}, \\
\left(\frac{1}{\mu}+\frac{1}{\mu_0}\right) S_{13}^{(1)}(\mu,\mu_0) &= \left\{1 + \frac{3}{8}\int_0^1 \frac{d\mu''}{\mu''}(1-\mu''^2)[S_{33}^{(1)}(\mu'',\mu_0) + 2\mu''^2 S_{13}^{(1)}(\mu'',\mu_0)]\right\} \\
&\quad \times\left\{1 + \frac{3}{8}\int_0^1 \frac{d\mu'}{\mu'}(1-\mu'^2)[S_{13}^{(1)}(\mu,\mu') + 2\mu'^2 S_{11}^{(1)}(\mu,\mu')]\right\}, \\
\left(\frac{1}{\mu}+\frac{1}{\mu_0}\right) S_{31}^{(1)}(\mu,\mu_0) &= \left\{1 + \frac{3}{8}\int_0^1 \frac{d\mu''}{\mu''}(1-\mu''^2)[S_{31}^{(1)}(\mu'',\mu_0) + 2\mu''^2 S_{11}^{(1)}(\mu'',\mu_0)]\right\} \\
&\quad \times\left\{1 + \frac{3}{8}\int_0^1 \frac{d\mu'}{\mu'}(1-\mu'^2)[S_{33}^{(1)}(\mu,\mu') + 2\mu'^2 S_{31}^{(1)}(\mu,\mu')]\right\}, \\
\left(\frac{1}{\mu}+\frac{1}{\mu_0}\right) S_{33}^{(1)}(\mu,\mu_0) &= \left\{1 + \frac{3}{8}\int_0^1 \frac{d\mu''}{\mu''}(1-\mu''^2)[S_{33}^{(1)}(\mu'',\mu_0) + 2\mu''^2 S_{13}^{(1)}(\mu'',\mu_0)]\right\} \\
&\quad \times\left\{1 + \frac{3}{8}\int_0^1 \frac{d\mu'}{\mu'}(1-\mu'^2)[S_{33}^{(1)}(\mu,\mu') + 2\mu'^2 S_{31}^{(1)}(\mu,\mu')]\right\};
\end{aligned}
\right\} \quad \text{(II)}
$$

and

$$\left.\begin{aligned}
\left(\frac{1}{\mu}+\frac{1}{\mu_0}\right) S_{11}^{(2)}(\mu, \mu_0) &= \left\{1+\frac{3}{16}\int_0^1 \frac{d\mu''}{\mu''}\left[\mu''^4 S_{11}^{(2)}(\mu'', \mu_0)+2\mu''^2 S_{31}^{(2)}(\mu'', \mu_0)\right.\right. \\
&\quad \left.\left.+S_{21}^{(2)}(\mu'', \mu_0)\right]\right\}\left\{1+\frac{3}{16}\int_0^1 \frac{d\mu'}{\mu'}\left[\mu'^4 S_{11}^{(2)}(\mu, \mu')+2\mu'^2 S_{13}^{(2)}(\mu, \mu')+S_{12}^{(2)}(\mu, \mu')\right]\right\}, \\
\left(\frac{1}{\mu}+\frac{1}{\mu_0}\right) S_{22}^{(2)}(\mu, \mu_0) &= \left\{1+\frac{3}{16}\int_0^1 \frac{d\mu''}{\mu''}\left[\mu''^4 S_{12}^{(2)}(\mu'', \mu_0)+2\mu''^2 S_{32}^{(2)}(\mu'', \mu_0)\right.\right. \\
&\quad \left.\left.+S_{22}^{(2)}(\mu'', \mu_0)\right]\right\}\left\{1+\frac{3}{16}\int_0^1 \frac{d\mu'}{\mu'}\left[\mu'^4 S_{21}^{(2)}(\mu, \mu')+2\mu'^2 S_{23}^{(2)}(\mu, \mu')+S_{22}^{(2)}(\mu, \mu')\right]\right\}, \\
\left(\frac{1}{\mu}+\frac{1}{\mu_0}\right) S_{33}^{(2)}(\mu, \mu_0) &= \left\{1+\frac{3}{16}\int_0^1 \frac{d\mu''}{\mu''}\left[\mu''^4 S_{13}^{(2)}(\mu'', \mu_0)+2\mu''^2 S_{33}^{(2)}(\mu'', \mu_0)\right.\right. \\
&\quad \left.\left.+S_{23}^{(2)}(\mu'', \mu_0)\right]\right\}\left\{1+\frac{3}{16}\int_0^1 \frac{d\mu'}{\mu'}\left[\mu'^4 S_{31}^{(2)}(\mu, \mu')+2\mu'^2 S_{33}^{(2)}(\mu, \mu')+S_{32}^{(2)}(\mu, \mu')\right]\right\}, \\
\left(\frac{1}{\mu}+\frac{1}{\mu_0}\right) S_{12}^{(2)}(\mu, \mu_0) &= \left\{1+\frac{3}{16}\int_0^1 \frac{d\mu''}{\mu''}\left[\mu''^4 S_{12}^{(2)}(\mu'', \mu_0)+2\mu''^2 S_{32}^{(2)}(\mu'', \mu_0)\right.\right. \\
&\quad \left.\left.+S_{22}^{(2)}(\mu'', \mu_0)\right]\right\}\left\{1+\frac{3}{16}\int_0^1 \frac{d\mu'}{\mu'}\left[\mu'^4 S_{11}^{(2)}(\mu, \mu')+2\mu'^2 S_{13}^{(2)}(\mu, \mu')+S_{12}^{(2)}(\mu, \mu')\right]\right\}, \\
\left(\frac{1}{\mu}+\frac{1}{\mu_0}\right) S_{21}^{(2)}(\mu, \mu_0) &= \left\{1+\frac{3}{16}\int_0^1 \frac{d\mu''}{\mu''}\left[\mu''^4 S_{11}^{(2)}(\mu'', \mu_0)+2\mu''^2 S_{31}^{(2)}(\mu'', \mu_0)\right.\right. \\
&\quad \left.\left.+S_{21}^{(2)}(\mu'', \mu_0)\right]\right\}\left\{1+\frac{3}{16}\int_0^1 \frac{d\mu'}{\mu'}\left[\mu'^4 S_{21}^{(2)}(\mu, \mu')+2\mu'^2 S_{23}^{(2)}(\mu, \mu')+S_{22}^{(2)}(\mu, \mu')\right]\right\}, \\
\left(\frac{1}{\mu}+\frac{1}{\mu_0}\right) S_{23}^{(2)}(\mu, \mu_0) &= \left\{1+\frac{3}{16}\int_0^1 \frac{d\mu''}{\mu''}\left[\mu''^4 S_{13}^{(2)}(\mu'', \mu_0)+2\mu''^2 S_{33}^{(2)}(\mu'', \mu_0)\right.\right. \\
&\quad \left.\left.+S_{23}^{(2)}(\mu'', \mu_0)\right]\right\}\left\{1+\frac{3}{16}\int_0^1 \frac{d\mu'}{\mu'}\left[\mu'^4 S_{21}^{(2)}(\mu, \mu')+2\mu'^2 S_{23}^{(2)}(\mu, \mu')+S_{22}^{(2)}(\mu, \mu')\right]\right\}, \\
\left(\frac{1}{\mu}+\frac{1}{\mu_0}\right) S_{32}^{(2)}(\mu, \mu_0) &= \left\{1+\frac{3}{16}\int_0^1 \frac{d\mu''}{\mu''}\left[\mu''^4 S_{12}^{(2)}(\mu'', \mu_0)+2\mu''^2 S_{32}^{(2)}(\mu'', \mu_0)\right.\right. \\
&\quad \left.\left.+S_{22}^{(2)}(\mu'', \mu_0)\right]\right\}\left\{1+\frac{3}{16}\int_0^1 \frac{d\mu'}{\mu'}\left[\mu'^4 S_{31}^{(2)}(\mu, \mu')+2\mu'^2 S_{33}^{(2)}(\mu, \mu')+S_{32}^{(2)}(\mu, \mu')\right]\right\}, \\
\left(\frac{1}{\mu}+\frac{1}{\mu_0}\right) S_{31}^{(2)}(\mu, \mu_0) &= \left\{1+\frac{3}{16}\int_0^1 \frac{d\mu''}{\mu''}\left[\mu''^4 S_{11}^{(2)}(\mu'', \mu_0)+2\mu''^2 S_{31}^{(2)}(\mu'', \mu_0)\right.\right. \\
&\quad \left.\left.+S_{21}^{(2)}(\mu'', \mu_0)\right]\right\}\left\{1+\frac{3}{16}\int_0^1 \frac{d\mu'}{\mu'}\left[\mu'^4 S_{31}^{(2)}(\mu, \mu')+2\mu'^2 S_{33}^{(2)}(\mu, \mu')+S_{32}^{(2)}(\mu, \mu')\right]\right\}, \\
\left(\frac{1}{\mu}+\frac{1}{\mu_0}\right) S_{13}^{(2)}(\mu, \mu_0) &= \left\{1+\frac{3}{16}\int_0^1 \frac{d\mu''}{\mu''}\left[\mu''^4 S_{13}^{(2)}(\mu'', \mu_0)+2\mu''^2 S_{33}^{(2)}(\mu'', \mu_0)\right.\right. \\
&\quad \left.\left.+S_{23}^{(2)}(\mu'', \mu_0)\right]\right\}\left\{1+\frac{3}{16}\int_0^1 \frac{d\mu'}{\mu'}\left[\mu'^4 S_{11}^{(2)}(\mu, \mu')+2\mu'^2 S_{13}^{(2)}(\mu, \mu')+S_{12}^{(2)}(\mu, \mu')\right]\right\}.
\end{aligned}\right\} \quad \text{(III)}$$

5. *The functional equations for* $S_{11}^{(0)}$, $S_{12}^{(0)}$, $S_{21}^{(0)}$, *and* $S_{22}^{(0)}$.—The first of the three systems of equations to which we have reduced the functional equation for **S** is the most important and, at the same time, the most difficult. We shall accordingly consider this system in some detail.

Substituting for $j_{11}^{(0)}$, according to equation (35) in the equations of System (I), we find on examination that they can be brought to the forms

$$\left.\begin{aligned}\left(\frac{1}{\mu}+\frac{1}{\mu_0}\right)S_{11}^{(0)}(\mu,\mu_0) &= \left\{\mu^2+\frac{3}{8}\int_0^1\frac{d\mu'}{\mu'}[\mu'^2S_{11}^{(0)}(\mu,\mu')+S_{12}^{(0)}(\mu,\mu')]\right\}\\ &\quad\times\left\{\mu_0^2+\frac{3}{8}\int_0^1\frac{d\mu''}{\mu''}[\mu''^2S_{11}^{(0)}(\mu'',\mu_0)+S_{21}^{(0)}(\mu'',\mu_0)]\right\}\\ &\quad+2\left\{1-\mu^2+\frac{3}{8}\int_0^1\frac{d\mu'}{\mu'}(1-\mu'^2)S_{11}^{(0)}(\mu,\mu')\right\}\\ &\quad\times\left\{1-\mu_0^2+\frac{3}{8}\int_0^1\frac{d\mu''}{\mu''}(1-\mu''^2)S_{11}^{(0)}(\mu'',\mu_0)\right\};\end{aligned}\right\}\quad(36)$$

$$\left.\begin{aligned}\left(\frac{1}{\mu}+\frac{1}{\mu_0}\right)S_{12}^{(0)}(\mu,\mu_0) &= \left\{\mu^2+\frac{3}{8}\int_0^1\frac{d\mu'}{\mu'}[\mu'^2S_{11}^{(0)}(\mu,\mu')+S_{12}^{(0)}(\mu,\mu')]\right\}\\ &\quad\times\left\{1+\frac{3}{8}\int_0^1\frac{d\mu''}{\mu''}[\mu''^2S_{12}^{(0)}(\mu'',\mu_0)+S_{22}^{(0)}(\mu'',\mu_0)]\right\}\\ &\quad+2\left\{\frac{3}{8}\int_0^1\frac{d\mu''}{\mu''}(1-\mu''^2)S_{12}^{(0)}(\mu'',\mu_0)\right\}\\ &\quad\times\left\{1-\mu^2+\frac{3}{8}\int_0^1\frac{d\mu'}{\mu'}(1-\mu'^2)S_{11}^{(0)}(\mu,\mu')\right\};\end{aligned}\right\}\quad(37)$$

$$\left.\begin{aligned}\left(\frac{1}{\mu}+\frac{1}{\mu_0}\right)S_{21}^{(0)}(\mu,\mu_0) &= \left\{\mu_0^2+\frac{3}{8}\int_0^1\frac{d\mu''}{\mu''}[\mu''^2S_{11}^{(0)}(\mu'',\mu_0)+S_{21}^{(0)}(\mu'',\mu_0)]\right\}\\ &\quad\times\left\{1+\frac{3}{8}\int_0^1\frac{d\mu'}{\mu'}[\mu'^2S_{21}^{(0)}(\mu,\mu')+S_{22}^{(0)}(\mu,\mu')]\right\}\\ &\quad+2\left\{\frac{3}{8}\int_0^1\frac{d\mu'}{\mu'}(1-\mu'^2)S_{21}^{(0)}(\mu,\mu')\right\}\\ &\quad\times\left\{1-\mu_0^2+\frac{3}{8}\int_0^1\frac{d\mu''}{\mu''}(1-\mu''^2)S_{11}^{(0)}(\mu'',\mu_0)\right\};\end{aligned}\right\}\quad(38)$$

and

$$\left.\begin{aligned}\left(\frac{1}{\mu}+\frac{1}{\mu_0}\right)S_{22}^{(0)}(\mu,\mu_0) &= \left\{1+\frac{3}{8}\int_0^1\frac{d\mu''}{\mu''}[\mu''^2S_{12}^{(0)}(\mu'',\mu_0)+S_{22}^{(0)}(\mu'',\mu_0)]\right\}\\ \times\left\{1+\frac{3}{8}\int_0^1\frac{d\mu'}{\mu'}[\mu'^2S_{21}^{(0)}(\mu,\mu')+S_{22}^{(0)}(\mu,\mu')]\right\}&\\ +\frac{9}{32}\int_0^1\frac{d\mu'}{\mu'}(1-\mu'^2)S_{21}^{(0)}(\mu,\mu')\int_0^1\frac{d\mu''}{\mu''}(1-\mu''^2)S_{12}^{(0)}(\mu'',\mu_0)&\,.\end{aligned}\right\}\quad(39)$$

An inspection of these equations shows what we have already seen directly from the functional equation for **S**, that, among these functions of "zero order," the relation (34) must hold in particular. In other words,

$$S_{11}^{(0)}(\mu,\mu_0)=S_{11}^{(0)}(\mu_0,\mu)\,;\;S_{12}^{(0)}(\mu,\mu_0)=S_{21}^{(0)}(\mu_0,\mu)\,;\;S_{22}^{(0)}(\mu,\mu_0)=S_{22}^{(0)}(\mu_0,\mu)\,.\quad(40)$$

In view of these relations, it follows from equations (36)–(39) that we can express the functions $S_{11}^{(0)}$, $S_{12}^{(0)}$, $S_{21}^{(0)}$ and $S_{22}^{(0)}$ in the forms

$$\left(\frac{1}{\mu}+\frac{1}{\mu_0}\right) S_{11}^{(0)}(\mu, \mu_0) = \psi(\mu)\psi(\mu_0) + 2\phi(\mu)\phi(\mu_0), \tag{41}$$

$$\left(\frac{1}{\mu}+\frac{1}{\mu_0}\right) S_{12}^{(0)}(\mu, \mu_0) = \psi(\mu)\chi(\mu_0) + 2\phi(\mu)\zeta(\mu_0), \tag{42}$$

$$\left(\frac{1}{\mu}+\frac{1}{\mu_0}\right) S_{21}^{(0)}(\mu, \mu_0) = \psi(\mu_0)\chi(\mu) + 2\phi(\mu_0)\zeta(\mu), \tag{43}$$

and

$$\left(\frac{1}{\mu}+\frac{1}{\mu_0}\right) S_{22}^{(0)}(\mu, \mu_0) = \chi(\mu)\chi(\mu_0) + 2\zeta(\mu)\zeta(\mu_0), \tag{44}$$

where

$$\psi(\mu) = \mu^2 + \frac{3}{8}\int_0^1 \frac{d\mu'}{\mu'}[\mu'^2 S_{11}^{(0)}(\mu, \mu') + S_{12}^{(0)}(\mu, \mu')], \tag{45}$$

$$\phi(\mu) = 1 - \mu^2 + \frac{3}{8}\int_0^1 \frac{d\mu'}{\mu'}(1-\mu'^2) S_{11}^{(0)}(\mu, \mu'), \tag{46}$$

$$\chi(\mu) = 1 + \frac{3}{8}\int_0^1 \frac{d\mu'}{\mu'}[\mu'^2 S_{21}^{(0)}(\mu, \mu') + S_{22}^{(0)}(\mu, \mu')], \tag{47}$$

and

$$\zeta(\mu) = \frac{3}{8}\int_0^1 \frac{d\mu'}{\mu'}(1-\mu'^2) S_{21}^{(0)}(\mu, \mu'). \tag{48}$$

Substituting for $S_{11}^{(0)}$, etc., from equations (41)–(44) back into equations (45)–(48), we obtain the functional equations for our problem in their normal forms. We have

$$\left.\begin{aligned} \psi(\mu) &= \mu^2 + \tfrac{3}{8}\mu\psi(\mu)\int_0^1 \frac{d\mu'}{\mu+\mu'}[\mu'^2\psi(\mu') + \chi(\mu')] \\ &\qquad + \tfrac{3}{4}\mu\phi(\mu)\int_0^1 \frac{d\mu'}{\mu+\mu'}[\mu'^2\phi(\mu') + \zeta(\mu')], \end{aligned}\right\} \tag{49}$$

$$\left.\begin{aligned} \phi(\mu) &= 1 - \mu^2 + \tfrac{3}{8}\mu\psi(\mu)\int_0^1 \frac{d\mu'}{\mu+\mu'}(1-\mu'^2)\psi(\mu') \\ &\qquad + \tfrac{3}{4}\mu\phi(\mu)\int_0^1 \frac{d\mu'}{\mu+\mu'}(1-\mu'^2)\phi(\mu'), \end{aligned}\right\} \tag{50}$$

$$\left.\begin{aligned} \chi(\mu) &= 1 + \tfrac{3}{8}\mu\chi(\mu)\int_0^1 \frac{d\mu'}{\mu+\mu'}[\mu'^2\psi(\mu') + \chi(\mu')] \\ &\qquad + \tfrac{3}{4}\mu\zeta(\mu)\int_0^1 \frac{d\mu'}{\mu+\mu'}[\mu'^2\phi(\mu') + \zeta(\mu')], \end{aligned}\right\} \tag{51}$$

and

$$\zeta(\mu) = \tfrac{3}{8}\mu\chi(\mu)\int_0^1 \frac{d\mu'}{\mu+\mu'}(1-\mu'^2)\psi(\mu') + \tfrac{3}{4}\mu\zeta(\mu)\int_0^1 \frac{d\mu'}{\mu+\mu'}(1-\mu'^2)\phi(\mu'). \tag{52}$$

Equations (49)–(52) represent a nonlinear, nonhomogeneous system of four simultaneous functional equations. The solution of such a system would seem, on first sight, to be an impossible task. But we shall show how a knowledge of the forms of the solutions gained from the analysis in Papers XI and XIII enables us to reduce the solution of equations (49)–(52) to two simple functional equations, each of form (4). And, as we have already indicated in § 1, such functional equations can be solved quite readily by a rapidly converging method of iteration.

6. *The forms of the solutions of the functional equations* (49)–(52).—In solving the functional equations (49)–(52), we shall be guided by the forms of the solutions which we have obtained by directly solving the relevant equations of transfer. We shall, therefore, first re-write the solutions given in Papers XI and XIII in forms which can be compared with equations (41)–(44).

From the expression for the scattering matrix given in Paper XIII (eq. [65]) we see, for example, that

$$\left(\frac{1}{\mu}+\frac{1}{\mu_0}\right) S_{11}^{(0)}(\mu, \mu_0) = 2H_l(\mu) H_l(\mu_0)[1+\mu\mu_0 - c(\mu+\mu_0)], \tag{53}$$

$$\left(\frac{1}{\mu}+\frac{1}{\mu_0}\right) S_{12}^{(0)}(\mu, \mu_0) = qH_l(\mu) H_r(\mu_0)(\mu+\mu_0), \tag{54}$$

$$\left(\frac{1}{\mu}+\frac{1}{\mu_0}\right) S_{21}^{(0)}(\mu, \mu_0) = qH_r(\mu) H_l(\mu_0)(\mu+\mu_0), \tag{55}$$

and

$$\left(\frac{1}{\mu}+\frac{1}{\mu_0}\right) S_{22}^{(0)}(\mu, \mu_0) = H_r(\mu) H_r(\mu_0)[1+\mu\mu_0 + c(\mu+\mu_0)]. \tag{56}$$

Here q and c are two constants which are defined in Paper XI, equations (130) and (134); and $H_l(\mu)$ and $H_r(\mu)$ are H-functions defined in terms of the roots of the characteristic equations

$$1 = \frac{3}{2}\sum_{j=1}^{n} \frac{a_j(1-\mu_j^2)}{1-k^2\mu_j^2} \tag{57}$$

and

$$1 = \frac{3}{4}\sum_{j=1}^{n} \frac{a_j(1-\mu_j^2)}{1-k^2\mu_j^2}, \tag{58}$$

respectively.

Now, between the constants q and c there exists the relation

$$q^2 = 2(1-c^2). \tag{59}$$ [5]

Hence

$$1+\mu\mu_0 \pm c(\mu+\mu_0) = (1\pm c\mu)(1\pm c\mu_0) + \tfrac{1}{2}q^2\mu\mu_0. \tag{60}$$

[5] This follows from the definitions of the constants q and c. Thus, from Paper XI, eq. (130), we find that

$$2(1-c^2) = -8\frac{R(1)R(-1)\rho(1)\rho(-1)}{[R(-1)\rho(1) - R(1)\rho(-1)]^2};$$

or, using Paper XI, eqs. (133) and (134), we have

$$2(1-c^2) = \left\{\frac{2\rho(1)\rho(-1)}{R(-1)\rho(1) - R(1)\rho(-1)}\right\}^2 = q^2.$$

We may accordingly re-write equations (53)–(56) in the forms

$$\left(\frac{1}{\mu}+\frac{1}{\mu_0}\right)S_{11}^{(0)}(\mu,\mu_0) = [q\mu H_l(\mu)][q\mu_0 H_l(\mu_0)] + 2[H_l(\mu)(1-c\mu)][H_l(\mu_0)(1-c\mu_0)]\,, \tag{61}$$

$$\left(\frac{1}{\mu}+\frac{1}{\mu_0}\right)S_{12}^{(0)}(\mu,\mu_0) = [q\mu H_l(\mu)][H_r(\mu_0)(1+c\mu_0)] + 2[H_l(\mu)(1-c\mu)][\tfrac{1}{2}q\mu_0 H_r(\mu_0)]\,, \tag{62}$$

$$\left(\frac{1}{\mu}+\frac{1}{\mu_0}\right)S_{21}^{(0)}(\mu,\mu_0) = [q\mu_0 H_l(\mu_0)][H_r(\mu)(1+c\mu)] + 2[H_l(\mu_0)(1-c\mu_0)][\tfrac{1}{2}q\mu H_r(\mu)]\,, \tag{63}$$

and

$$\left(\frac{1}{\mu}+\frac{1}{\mu_0}\right)S_{22}^{(0)}(\mu,\mu_0) = [H_r(\mu)(1+c\mu)][H_r(\mu_0)(1+c\mu_0)] + 2[\tfrac{1}{2}q\mu H_r(\mu)][\tfrac{1}{2}q\mu_0 H_r(\mu_0)]\,. \tag{64}$$

The similarity of equations (61)–(64) to (41)–(44) is apparent. While the solutions (53)–(56) can be brought into such a correspondence with equations (41)–(44) in a variety of ways, it appeared after some examination that the particular choice we have made is the correct one. Indeed, we shall prove in § 8 that the solutions of the functional equations (49)–(52) are of the forms

$$\psi(\mu) = q\mu H_l(\mu)\,, \tag{65}$$

$$\phi(\mu) = H_l(\mu)(1-c\mu)\,, \tag{66}$$

$$\chi(\mu) = H_r(\mu)(1+c\mu) \tag{67}$$

and

$$\zeta(\mu) = \tfrac{1}{2}q\mu H_r(\mu)\,, \tag{68}$$

where $H_l(\mu)$ and $H_r(\mu)$ are defined as the solutions of the functional equations (cf. our remarks in § 1)

$$H_l(\mu) = 1+\tfrac{3}{4}\mu H_l(\mu)\int_0^1 \frac{H_l(\mu')(1-\mu'^2)}{\mu+\mu'}\,d\mu'\,, \tag{69}$$

and

$$H_r(\mu) = 1+\tfrac{3}{8}\mu H_r(\mu)\int_0^1 \frac{H_r(\mu')(1-\mu'^2)}{\mu+\mu'}\,d\mu'\,, \tag{70}$$

and q and c are two constants satisfying the relation (59) and related in a way that we shall later specify with the moments of $H_l(\mu)$ and $H_r(\mu)$.

7. *Some integral properties of the functions* $H_l(\mu)$ *and* $H_r(\mu)$.—Before we can prove that the solutions of the functional equations (49)–(52) are as we have stated them in § 6, it is necessary to establish certain integral properties of the functions $H_l(\mu)$ and $H_r(\mu)$ which satisfy the functional equations (69) and (70).

We shall state and prove the required properties in the form of three lemmas.

Lemma 1.—

$$\int_0^1 H_l(\mu)(1-\mu^2)\,d\mu = \tfrac{4}{3}\,, \tag{71}$$

and

$$\int_0^1 H_r(\mu)(1-\mu^2)\,d\mu = \tfrac{4}{3}(2-\sqrt{2})\,. \tag{72}$$

To prove equation (71), for example, we multiply the functional equation satisfied by $H_l(\mu)$ by $(1-\mu^2)$ and integrate over the range of μ. We obtain

$$\left.\begin{aligned}\int_0^1 H_l(\mu)(1-\mu^2)\,d\mu = \tfrac{2}{3}+\frac{3}{4}\int_0^1\int_0^1 \frac{\mu}{\mu+\mu'} H_l(\mu)(1-\mu^2)\\ \times H_l(\mu')(1-\mu'^2)\,d\mu d\mu'\,.\end{aligned}\right\} \quad (73)$$

Interchanging μ and μ' in the double integral in equation (73) and taking the average of the two equations, we have

$$\int_0^1 H_l(\mu)(1-\mu^2)\,d\mu = \tfrac{2}{3}+\frac{3}{8}\int_0^1\int_0^1 H_l(\mu)(1-\mu^2)\,H_l(\mu')(1-\mu'^2)\,d\mu d\mu' \quad (74)$$

or, alternatively,

$$\frac{3}{8}\left[\int_0^1 H_l(\mu)(1-\mu^2)\,d\mu\right]^2-\int_0^1 H_l(\mu)(1-\mu^2)\,d\mu+\tfrac{2}{3}=0\,. \quad (75)$$

From equation (75) the relation in question follows.

In a similar way, from the functional equation satisfied by $H_r(\mu)$ we obtain (cf. eq. [75])

$$\frac{3}{16}\left[\int_0^1 H_r(\mu)(1-\mu^2)\,d\mu\right]^2-\int_0^1 H_r(\mu)(1-\mu^2)\,d\mu+\tfrac{2}{3}=0\,. \quad (76)$$

Solving this equation, we find

$$\int_0^1 H_r(\mu)(1-\mu^2)\,d\mu = \tfrac{4}{3}\,(2\pm\sqrt{2})\,. \quad (77)$$

The ambiguity in sign in the foregoing equation can be removed by considerations which we shall not go into here. But it may be stated that the positive sign in the quantity in parenthesis on the right-hand side of equation (77) is excluded on the ground that

$$H_r(\mu) < H_l(\mu) \qquad (\mu<1)\,. \quad (78)$$

Corollary.—The functional equations for $H_l(\mu)$ and $H_r(\mu)$ have the alternative forms

$$1=\tfrac{3}{4}H_l(\mu)\int_0^1 H_l(\mu')(1-\mu'^2)\frac{\mu'}{\mu+\mu'}\,d\mu' \quad (79)$$

and

$$1=H_r(\mu)\left[\frac{1}{\sqrt{2}}+\frac{3}{8}\int_0^1 H_r(\mu')(1-\mu'^2)\frac{\mu'}{\mu+\mu'}\,d\mu'\right]. \quad (80)$$

Equation (80) can be proved in the following manner:

$$\left.\begin{aligned}&\tfrac{3}{8}H_r(\mu)\int_0^1 H_r(\mu')(1-\mu'^2)\frac{\mu'}{\mu+\mu'}\,d\mu'\\ &\qquad=\tfrac{3}{8}H_r(\mu)\int_0^1 H_r(\mu')(1-\mu'^2)\left[1-\frac{\mu}{\mu+\mu'}\right]d\mu\\ &\qquad=H_r(\mu)\left(1-\frac{1}{\sqrt{2}}\right)-\tfrac{3}{8}\mu H_r(\mu)\int_0^1\frac{H_r(\mu')(1-\mu'^2)}{\mu+\mu'}\,d\mu'\\ &\qquad=H_r(\mu)\left(1-\frac{1}{\sqrt{2}}\right)-H_r(\mu)+1\\ &\qquad=1-\frac{1}{\sqrt{2}}H_r(\mu)\,.\end{aligned}\right\} \quad (81)$$

Equation (79) can be established in a similar way.

Lemma 2.—If α_0 and α_1 are the moments of orders zero and one, respectively, of the function $H_l(\mu)$, then

$$\alpha_0 = 1 + \tfrac{3}{8}\,(\alpha_0^2 - \alpha_1^2)\,. \tag{82}$$

Similarly, if A_0 and A_1 are the zero and first moments of $H_r(\mu)$, then

$$A_0 = 1 + \tfrac{3}{16}\,(A_0^2 - A_1^2)\,. \tag{83}$$

To prove equation (82) we simply integrate over μ the functional equation defining $H_l(\mu)$. We find

$$\left.\begin{aligned}\int_0^1 H_l(\mu)\,d\mu &= 1 + \frac{3}{4}\int_0^1\!\!\int_0^1 \frac{H_l(\mu)\,H_l(\mu')}{\mu+\mu'}\,\mu\,(1-\mu'^2)\,d\mu\,d\mu' \\ &= 1 + \frac{3}{8}\int_0^1\!\!\int_0^1 \frac{H_l(\mu)\,H_l(\mu')}{\mu+\mu'}\,[\mu(1-\mu'^2)+\mu'(1-\mu^2)]\,d\mu\,d\mu' \\ &= 1 + \frac{3}{8}\int_0^1\!\!\int_0^1 H_l(\mu)\,H_l(\mu')(1-\mu\mu')\,d\mu\,d\mu' \\ &= 1 + \frac{3}{8}\left[\int_0^1 H_l(\mu)\,d\mu\right]^2 - \frac{3}{8}\left[\int_0^1 H_l(\mu)\,\mu\,d\mu\right]^2,\end{aligned}\right\} \tag{84}$$

from which the relation in question follows.

Relation (83) between the moments of $H_r(\mu)$ can be similarly proved.

Lemma 3.—

$$(1-\mu^2)\int_0^1 \frac{H_l(\mu')}{\mu+\mu'}\,d\mu' = \frac{H_l(\mu)-1}{\frac{3}{4}\mu H_l(\mu)} + (\alpha_1 - \mu\alpha_0)\,, \tag{85}$$

and

$$(1-\mu^2)\int_0^1 \frac{H_r(\mu')}{\mu+\mu'}\,d\mu' = \frac{H_r(\mu)-1}{\frac{3}{8}\mu H_r(\mu)} + (A_1 - \mu A_0)\,, \tag{86}$$

where α_0, α_1, A_0, and A_1 have the same meanings as in lemma 2.

The integral relation (85) can be proved in the following manner:

$$\left.\begin{aligned}\int_0^1 \frac{H_l(\mu')}{\mu+\mu'}\,d\mu' &= \int_0^1 \frac{H_l(\mu')(1-\mu'^2)}{\mu+\mu'}\,d\mu' + \int_0^1 \frac{H_l(\mu')\,\mu'^2}{\mu+\mu'}\,d\mu' \\ &= \frac{H_l(\mu)-1}{\frac{3}{4}\mu H_l(\mu)} + \int_0^1 H_l(\mu')\left[\mu' - \mu + \frac{\mu^2}{\mu+\mu'}\right]d\mu' \\ &= \frac{H_l(\mu)-1}{\frac{3}{4}\mu H_l(\mu)} + (\alpha_1 - \mu\alpha_0) + \mu^2\int_0^1 \frac{H_l(\mu')}{\mu+\mu'}\,d\mu'\,.\end{aligned}\right\} \tag{87}$$

Hence, the result. Equation (86) follows quite similarly.

8. *Verification of the solution.*—We shall now show that the solutions of the functional equations (49)–(52) are as stated in § 6. First, we may observe that if the constants q and c are related as in equation (59), then equations (41)–(44) and (65)–(68) will together lead to expressions for $S_{11}^{(0)}$, etc., which will be formally identical with equations (53)–(56); however, $H_l(\mu)$ and $H_r(\mu)$ will have to be understood (here as elsewhere) as satisfying the functional equations (69) and (70).[6] The proof that equations (65)–(68)

[6] We shall not repeat this every time. The expressions for the functions as products involving the Gaussian division of the interval (0, 1) and the relevant characteristic roots is only an approximation. As our discussion in Sec. III will emphasize, the exact H-functions always satisfy functional equations of the form (4), and in this paper we are seeking only exact solutions.

do represent the solution of equations (49)–(52) will therefore consist in showing that, with a suitable choice of q and c (compatible with eq. [59]), the substitution of $S_{11}^{(0)}$, etc., according to equations (53)–(56), in equations (45)–(48) will lead to ψ, ϕ, etc., which will be in agreement with equations (65)–(68).

Considering, first, $\phi(\mu)$, we have (cf. eqs. [46] and [53])

$$\left.\begin{aligned}\phi(\mu) &= 1-\mu^2+\tfrac{3}{4}\mu H_l(\mu)\int_0^1 \frac{d\mu'}{\mu+\mu'}(1-\mu'^2)\,H_l(\mu')\,[1-c\,(\mu+\mu')+\mu\mu'] \\ &= 1-\mu^2+\tfrac{3}{4}\mu H_l(\mu)\int_0^1 d\mu'\,(1-\mu'^2)\,H_l(\mu')\left[\mu-c+\frac{1-\mu^2}{\mu+\mu'}\right] \\ &= 1-\mu^2+\tfrac{3}{4}\mu\,(\mu-c)\,H_l(\mu)\int_0^1 H_l(\mu')\,(1-\mu'^2)\,d\mu' \\ &\qquad +\tfrac{3}{4}\mu\,(1-\mu^2)\,H_l(\mu)\int_0^1 \frac{H_l(\mu')\,(1-\mu'^2)}{\mu+\mu'}\,d\mu'.\end{aligned}\right\} \tag{88}$$

Using lemma 1 (eq. [71]) and recalling the functional equation satisfied by $H_l(\mu)$, we find

$$\phi(\mu) = 1-\mu^2+\mu\,(\mu-c)\,H_l(\mu)+(1-\mu^2)\,[H_l(\mu)-1] \tag{89}$$

or

$$\phi(\mu) = H_l(\mu)\,(1-c\mu)\,. \tag{90}$$

This is in agreement with equation (66).

Considering, next, $\phi+\psi$, we have (cf. eqs. [45] and [46])

$$\left.\begin{aligned}\phi(\mu)+\psi(\mu) &= 1+\frac{3}{8}\int_0^1 \frac{d\mu'}{\mu'}\,[S_{11}^{(0)}(\mu,\mu')+S_{12}^{(0)}(\mu,\mu')] \\ &= 1+\tfrac{3}{8}\mu H_l(\mu)\int_0^1 \frac{d\mu'}{\mu+\mu'}\Big[2H_l(\mu')[\![1-c\,(\mu+\mu')+\mu\mu']\!] \\ &\qquad\qquad +qH_r(\mu')\,(\mu+\mu')\Big] \\ &= 1+\tfrac{3}{8}q\mu H_l(\mu)\,A_0+\tfrac{3}{4}\mu H_l(\mu)\int_0^1 d\mu' H_l(\mu') \\ &\qquad\qquad \times\left[\mu-c+\frac{1-\mu^2}{\mu+\mu'}\right] \\ &= 1+\tfrac{3}{8}q\mu H_l(\mu)\,A_0+\tfrac{3}{4}\mu\,(\mu-c)\,H_l(\mu)\,\alpha_0 \\ &\qquad +\tfrac{3}{4}\mu\,(1-\mu^2)\,H_l(\mu)\int_0^1 \frac{H_l(\mu')\,d\mu'}{\mu+\mu'},\end{aligned}\right\} \tag{91}$$

where, as in lemmas 2 and 3 (§ 7), α_0 and A_0 are the zero-order moments of $H_l(\mu)$ and $H_r(\mu)$. Using lemma 3 (eq. [85]) in the last step of the preceding reduction, we find

$$\phi(\mu)+\psi(\mu) = H_l(\mu)+\mu H_l(\mu)\,[\tfrac{3}{8}qA_0+\tfrac{3}{4}(\alpha_1-\alpha_0 c)]\,. \tag{92}$$

From equations (90) and (92) we now obtain

$$\psi(\mu) = \mu H_l(\mu)\,[\tfrac{3}{8}qA_0+\tfrac{3}{4}(\alpha_1-\alpha_0 c)+c]\,. \tag{93}$$

In order that this last equation may be in agreement with equation (65), we must require that

$$q = \tfrac{3}{8}qA_0+\tfrac{3}{4}(\alpha_1-\alpha_0 c)+c\,. \tag{94}$$

Turning our attention next to $\zeta(\mu)$, we have (cf. eqs. [48] and [55])

$$\zeta(\mu) = \tfrac{3}{8} q \mu H_r(\mu) \int_0^1 H_l(\mu')(1-\mu'^2)\, d\mu', \tag{95}$$

or, using lemma 1 (eq. [71]), we obtain

$$\zeta(\mu) = \tfrac{1}{2} q \mu H_r(\mu), \tag{96}$$

in agreement with equation (68).

Finally, considering $\zeta(\mu) + \chi(\mu)$, we have (cf. eqs. [47] and [48])

$$\left.\begin{aligned}\zeta(\mu)+\chi(\mu) &= 1+\frac{3}{8}\int_0^1 \frac{d\mu'}{\mu'}\left[S_{21}^{(0)}(\mu,\mu')+S_{22}^{(0)}(\mu,\mu')\right]\\ &= 1+\tfrac{3}{8}\mu H_r(\mu)\int_0^1 \frac{d\mu'}{\mu+\mu'}\left[qH_l(\mu')(\mu+\mu')+H_r(\mu')\right.\\ &\qquad\qquad \left.\times [\![1+c(\mu+\mu')+\mu\mu']\!]\right]\\ &= 1+\tfrac{3}{8} q\mu H_r(\mu)\,\alpha_0+\tfrac{3}{8}\mu(\mu+c)H_r(\mu)A_0+\tfrac{3}{8}\mu(1-\mu^2)\\ &\qquad\qquad \times H_r(\mu)\int_0^1 \frac{H_r(\mu')}{\mu+\mu'}\,d\mu'.\end{aligned}\right\} \tag{97}$$

Using lemma 3 (eq. [86]) and equation (96), we find, after some minor reductions, that

$$\chi(\mu) = H_r(\mu) + \mu H_r(\mu)\left[\tfrac{3}{8} q\alpha_0 + \tfrac{3}{8}(A_1 + A_0 c) - \tfrac{1}{2} q\right]. \tag{98}$$

In order, now, that equation (98) may agree with equation (67), we must require that

$$c = \tfrac{3}{8} q\alpha_0 + \tfrac{3}{8}(A_1 + A_0 c) - \tfrac{1}{2} q. \tag{99}$$

Equations (94) and (99), for the determination of the constants q and c, can be rewritten in the forms

$$(3A_0 - 8)\, q - 2(3\alpha_0 - 4)\, c + 6\alpha_1 = 0 \tag{100}$$

and

$$(3\alpha_0 - 4)\, q + (3A_0 - 8)\, c + 3A_1 = 0. \tag{101}$$

Solving these equations, we find

$$q = -6\,\frac{\alpha_1(3A_0-8)+A_1(3\alpha_0-4)}{(3A_0-8)^2+2(3\alpha_0-4)^2} \tag{102}$$

and

$$c = +3\,\frac{2\alpha_1(3\alpha_0-4)-A_1(3A_0-8)}{(3A_0-8)^2+2(3\alpha_0-4)^2}. \tag{103}$$

It remains to prove that q and c, as given by equations (102) and (103), are compatible with the condition (eq. [59])

$$2(1-c^2) = q^2, \tag{104}$$

which we have assumed. To show that this is the case, we evaluate $2(1-c^2)-q^2$ according to equations (102) and (103). We find

$$2(1-c^2)-q^2 = 2\,\frac{[(3A_0-8)^2-9A_1^2]-2[9\alpha_1^2-(3\alpha_0-4)^2]}{(3A_0-8)^2+2(3\alpha_0-4)^2}. \tag{105}$$

On the other hand, using lemma 2 (eq. [82]), we find

$$(3A_0-8)^2-9A_1^2=9(A_0^2-A_1^2)-48A_0+64=16\,. \tag{106}$$

Similarly, using equation (83), we find

$$9\alpha_1^2-(3\alpha_0-4)^2=9(\alpha_1^2-\alpha_0^2)+24\alpha_0-16=8\,. \tag{107}$$

Hence, the right-hand side of equation (105) vanishes, and q and c are indeed related, as required.

This completes the verification.

It may be noted that, since we have incidentally shown that

$$(3A_0-8)^2+2(3\alpha_0-4)^2=9(A_1^2+2\alpha_1^2)\,, \tag{108}$$

the expressions (102) and (103) for q and c have the alternative forms

$$q=2\,\frac{4(A_1+2\alpha_1)-3(A_0\alpha_1+\alpha_0A_1)}{3(A_1^2+2\alpha_1^2)} \tag{109}$$

and

$$c=\frac{8(A_1-\alpha_1)+3(2\alpha_1\alpha_0-A_0A_1)}{3(A_1^2+2\alpha_1^2)}\,. \tag{110}$$

9. *The solutions of Systems (II) and (III).*—In contrast to System (I), the solutions of Systems (II) and (III) can be found without any difficulty, since an inspection of the equations of these two systems shows that all the functions $S^{(1)}$ are equal to each other, and similarly all the functions $S^{(2)}$ are equal to each other. Therefore, writing

$$S_{ij}^{(1)}(\mu,\mu_0)=S^{(1)}(\mu,\mu_0)\qquad (i,j=1,3) \tag{111}$$

and

$$S_{ij}^{(2)}(\mu,\mu_0)=S^{(2)}(\mu,\mu_0)\qquad (i,j=1,2,3)\,, \tag{112}$$

we find that the equations of Systems (II) and (III) reduce to a single equation each for the two functions $S^{(1)}(\mu,\mu_0)$ and $S^{(2)}(\mu,\mu_0)$. We have

$$\left.\begin{aligned}\left(\frac{1}{\mu}+\frac{1}{\mu_0}\right)S^{(1)}(\mu,\mu_0)&=\left\{1+\frac{3}{8}\int_0^1\frac{d\mu''}{\mu''}(1-\mu''^2)(1+2\mu''^2)S^{(1)}(\mu'',\mu_0)\right\}\\&\times\left\{1+\frac{3}{8}\int_0^1\frac{d\mu'}{\mu'}(1-\mu'^2)(1+2\mu'^2)S^{(1)}(\mu,\mu')\right\}\end{aligned}\right\} \tag{113}$$

and

$$\left.\begin{aligned}\left(\frac{1}{\mu}+\frac{1}{\mu_0}\right)S^{(2)}(\mu,\mu_0)&=\left\{1+\frac{3}{16}\int_0^1\frac{d\mu''}{\mu''}(1+\mu''^2)^2S^{(2)}(\mu'',\mu_0)\right\}\\&\times\left\{1+\frac{3}{16}\int_0^1\frac{d\mu'}{\mu'}(1+\mu'^2)^2S^{(2)}(\mu,\mu')\right\}.\end{aligned}\right\} \tag{114}$$

From equations (113) and (114) it follows that the functions $S^{(1)}(\mu,\mu_0)$ and $S^{(2)}(\mu,\mu_0)$ are symmetrical in the variables μ and μ_0 and that they are expressible in the forms

$$\left(\frac{1}{\mu}+\frac{1}{\mu_0}\right)S^{(1)}(\mu,\mu_0)=H^{(1)}(\mu)H^{(1)}(\mu_0) \tag{115}$$

and

$$\left(\frac{1}{\mu}+\frac{1}{\mu_0}\right) S^{(2)}(\mu, \mu_0) = H^{(2)}(\mu)\, H^{(2)}(\mu_0)\,, \tag{116}$$

where $H^{(1)}(\mu)$ and $H^{(2)}(\mu)$ are solutions of the functional equations

$$H^{(1)}(\mu) = 1 + \tfrac{3}{8}\mu H^{(1)}(\mu) \int_0^1 \frac{H^{(1)}(\mu')}{\mu+\mu'}(1-\mu'^2)(1+2\mu'^2)\, d\mu \tag{117}$$

and

$$H^{(2)}(\mu) = 1 + \tfrac{3}{16}\mu H^{(2)}(\mu) \int_0^1 \frac{H^{(2)}(\mu')}{\mu+\mu'}(1+\mu'^2)^2 d\mu'\,. \tag{118}$$

It will be noticed that these functional equations for $H^{(1)}(\mu)$ and $H^{(2)}(\mu)$ again illustrate the correspondence between the functional equations which define them exactly and the approximate forms in terms of the Gaussian division and characteristic roots in which they appear in the method of solution of the earlier papers of this series.

We have now completed the solution of the functional equation for the scattering matrix **S**. It will be observed that the solution for **S** which we have now obtained is of exactly the same form as that given in Paper XIII (eq. [65]), with the only difference that the four H-functions which appear in the solution are now defined in terms of the exact functional equations which they satisfy; further, the two constants, q and c, are shown to be related in a definite way with the moments of $H_l(\mu)$ and $H_r(\mu)$.

10. *The exact laws of darkening in the two states of polarization in an electron-scattering atmosphere.*—In Paper X we worked out a detailed theory of the radiative equilibrium of an atmosphere in which the Thomson scattering by free electrons governs the transfer of radiation. It was predicted that, under the circumstances considered, the emergent radiation should be polarized, the intensities in the two states of polarization showing different laws of darkening. We shall now see how an exact solution for this latter problem can be written down, using the results of the preceding sections.

In Paper X we showed that the emergent intensities $I_l(0, \mu)$ and $I_r(0, \mu)$ in the two states of polarization can be expressed in the forms (cf. Paper X, eqs. [128], [170], and [171] and nn. 15 and 17 in Paper XI)

$$I_l(0, \mu) = \tfrac{3}{8}F\frac{q}{\sqrt{2}}H_l(\mu) \tag{119}$$

and

$$I_r(0, \mu) = \tfrac{3}{8}F\frac{1}{\sqrt{2}}H_r(\mu)(\mu+c)\,, \tag{120}$$

where the constants q and c have the same meanings as in the problem of diffuse reflection which we have just considered. It therefore follows that the exact solutions for the emergent intensities $I_l(0, \mu)$ and $I_r(0, \mu)$ can be obtained by simply redefining the functions $H_l(\mu)$ and $H_r(\mu)$ as satisfying the functional equations (69) and (70) and determining the constants q and c according to equations (109) and (110). That we are justified in doing this is further confirmed by the fact that the total emergent flux given by equations (119) and (120) is exactly πF, as it should be; for, in order for this to be the case, it is clearly necessary that

$$\left.\begin{aligned} F = F_l + F_r &= 2\int_0^1 [I_l(0, \mu) + I_r(0, \mu)]\,\mu d\mu \\ &= \frac{3}{4\sqrt{2}}F(q\alpha_1 + A_2 + cA_1) \end{aligned}\right\} \tag{121}$$

or that

$$q\alpha_1 + A_2 + cA_1 = \tfrac{4}{3}\sqrt{2}\,. \tag{122}$$

The validity of equation (122) can be established as follows:

Evaluating $q\alpha_1 + cA_1$ according to equations (109) and (110), we find that

$$q\alpha_1 + cA_1 = \tfrac{8}{3} - A_0 . \tag{123}$$

Hence

$$q\alpha_1 + A_2 + cA_1 = \tfrac{8}{3} + (A_2 - A_0) . \tag{124}$$

But lemma 1, equation (72), is equivalent to

$$A_0 - A_2 = \tfrac{4}{3}(2 - \sqrt{2}) ; \tag{125}$$

using this result in equation (124), we prove that equation (122) is, in fact, true.

III. ON A CLASS OF FUNCTIONAL EQUATIONS WHICH OCCURS IN THE THEORY OF RADIATIVE TRANSFER

***11.** The basic theorems.*—The discussion of the problem of diffuse reflection from the point of view of the functional equation satisfied by the scattering matrix has disclosed two things: first, that there exists a class of functional equations of the form

$$H(\mu) = 1 + \mu H(\mu) \int_0^1 \frac{H(\mu')}{\mu + \mu'} \Psi(\mu')\, d\mu' , \tag{126}$$

which plays an important role in the theory of radiative transfer, and, second, that in a certain nth approximation the solution of equation (126) is given by equations (1) and (2). In Section III, we propose to examine this correspondence and establish some general theorems relating to functional equations of the type (126).

We shall begin our discussion by proving the following theorem, which appears to be the basic one in this subject.

Theorem 1.—Let $\Psi(\mu)$ be an even polynomial of degree $2m$ in μ, such that

$$\int_0^1 \Psi(\mu)\, d\mu \leqslant \tfrac{1}{2} . \tag{127}$$

Let $\mu_j (j = \pm 1, \ldots, \pm n)$ denote the division of the interval $(-1, +1)$ according to the zeros of the Legendre polynomial of order $2n (> m)$; further, let $a_j (= a_{-j})$ denote the corresponding Gaussian weights. Finally, let $k_\alpha (\alpha = 1, \ldots, n)$ denote the distinct positive (or zero) roots of the characteristic equation

$$1 = \sum_j \frac{a_j \Psi(\mu_j)}{1 + k\mu_j} = 2 \sum_{j=1}^{n} \frac{a_j \Psi(\mu_j)}{1 - k^2 \mu_j^2} . \tag{128}$$

Then the function

$$H(\mu) = \frac{1}{\mu_1 \ldots \mu_n} \frac{\prod_i (\mu + \mu_i)}{\prod_\alpha (1 + k_\alpha \mu)} \tag{129}$$

satisfies the equation

$$H(\mu) \equiv 1 + \mu H(\mu) \sum_{j=1}^{n} \frac{a_j H(\mu_j) \Psi(\mu_j)}{\mu + \mu_j} . \tag{130}$$

Proof.—Let us first consider the case

$$\int_0^1 \Psi(\mu)\, d\mu < \tfrac{1}{2} . \tag{131}$$

In this case the characteristic equation admits of n distinct nonvanishing positive roots, and we consider the function

$$S(\mu) = \sum_{\alpha=1}^{n} \frac{L_\alpha}{1 - k_\alpha \mu} + 1\,, \tag{132}$$

where $L_\alpha(\alpha = 1, \ldots, n)$ are certain constants to be determined from the equations

$$\sum_{\alpha=1}^{n} \frac{L_\alpha}{1 - k_\alpha \mu_i} + 1 = 0 \qquad (i = 1, \ldots, n)\,, \tag{133}$$

or, equivalently,

$$S(\mu_i) = 0 \qquad (i = 1, \ldots, n)\,. \tag{134}$$

By considerations of a type that we are now familiar with, it can be shown that

$$S(\mu) = k_1 \ldots k_n \mu_1 \ldots \mu_n \frac{(-1)^n}{\mu_1 \ldots \mu_n} \frac{\prod_i (\mu - \mu_i)}{\prod_\alpha (1 - k_\alpha \mu)}\,. \tag{135}$$

In other words,

$$S(\mu) = k_1 \ldots k_n \mu_1 \ldots \mu_n H(-\mu)\,. \tag{136}$$

Since $H(0) = 1$, we can re-write equation (136) alternatively in the form

$$S(\mu) = S(0)\, H(-\mu)\,, \tag{137}$$

where

$$S(0) = \sum_{\alpha=1}^{n} L_\alpha + 1 = k_1 \ldots k_n \mu_1 \ldots \mu_n\,. \tag{138}$$

Now, since $\Psi(\mu)$ is even in μ, we can write the equation which a characteristic root satisfies in either of the forms

$$1 = \sum_j \frac{a_j \Psi(\mu_j)}{1 + k\mu_j} \tag{139}$$

or

$$1 = \sum_j \frac{a_j \Psi(\mu_j)}{1 - k\mu_j}\,. \tag{140}$$

Let k_α denote a particular characteristic root. Then, on account of equations (139) and (140), which are satisfied by any of the characteristic roots, we can clearly write:

$$\left.\begin{aligned} S(0) &= \sum_{\beta=1}^{n} L_\beta + 1 \\ &= \sum_{\beta=1}^{n} \frac{L_\beta}{k_\alpha + k_\beta} \left[\sum_j a_j \Psi(\mu_j) \left\{ \frac{k_\alpha}{1 + k_\alpha \mu_j} + \frac{k_\beta}{1 - k_\beta \mu_j} \right\} \right] \\ &\qquad + \sum_j \frac{a_j \Psi(\mu_j)}{1 + k_\alpha \mu_j}\,. \end{aligned}\right\} \tag{141}$$

Simplifying the quantity in brackets in the foregoing equation, we have

$$S(0) = \sum_{\beta=1}^{n} L_\beta \Big[\sum_j \frac{a_j \Psi(\mu_j)}{(1+k_\alpha \mu_j)(1-k_\beta \mu_j)} \Big] + \sum_j \frac{a_j \Psi(\mu_j)}{1+k_\alpha \mu_j}, \tag{142}$$

or, inverting the order of the summation,

$$S(0) = \sum_j \frac{a_j \Psi(\mu_j)}{1+k_\alpha \mu_j} \Big[\sum_{\beta=1}^{n} \frac{L_\beta}{1-k_\beta \mu_j} + 1 \Big]. \tag{143}$$

But the quantity in brackets in equation (143) is $S(\mu_j)$ (cf. eq. [132]). Hence

$$S(0) = \sum_j \frac{a_j S(\mu_j) \Psi(\mu_j)}{1+k_\alpha \mu_j}. \tag{144}$$

In equation (144) (as in eqs. [139]–[143]) the summation is, of course, extended over all values of j, positive and negative. However, since $S(\mu_j) = 0$ (eq. [134]), in equation (144), only the terms with negative j make a nonzero contribution. We can therefore write

$$S(0) = \sum_{j=1}^{n} \frac{a_j S(-\mu_j) \Psi(\mu_j)}{1-k_\alpha \mu_j} \quad (\alpha = 1, \ldots, n), \tag{145}$$

or, in view of equation (137),

$$1 = \sum_{j=1}^{n} \frac{a_j H(\mu_j) \Psi(\mu_j)}{1-k_\alpha \mu_j} \quad (\alpha = 1, \ldots, n). \tag{146}$$

Now consider the function

$$1 - \mu \sum_{j=1}^{n} \frac{a_j H(\mu_j) \Psi(\mu_j)}{\mu + \mu_j}. \tag{147}$$

According to equation (146), this vanishes for $\mu = -1/k_\alpha$ $(\alpha = 1, \ldots, n)$; for

$$1 + \frac{1}{k_\alpha} \sum_{j=1}^{n} \frac{a_j H(\mu_j) \Psi(\mu_j)}{(-1/k_\alpha) + \mu_j} = 1 - \sum_{j=1}^{n} \frac{a_j H(\mu_j) \Psi(\mu_j)}{1-k_\alpha \mu_j} = 0. \tag{148}$$

Hence

$$\prod_{j=1}^{n} (\mu + \mu_j) - \mu \sum_{j=1}^{n} a_j H(\mu_j) \Psi(\mu_j) \prod_{i \neq j} (\mu + \mu_i) \tag{149}$$

also vanishes for

$$\mu = -\frac{1}{k_\alpha} \quad (\alpha = 1, \ldots, n). \tag{150}$$

But the expression (149) is a polynomial of degree n in μ. It cannot, therefore, differ from

$$\prod_{\alpha=1}^{n} (1 + k_\alpha \mu) \tag{151}$$

by more than a constant factor; and the constant of proportionality is seen to be

$$\mu_1 \ldots \mu_n \tag{152}$$

from a comparison of the two functions at $\mu = 0$. It therefore follows that

$$1 - \mu \sum_{j=1}^{n} \frac{a_j H(\mu_j) \Psi(\mu_j)}{\mu + \mu_j} = \mu_1 \ldots \mu_n \frac{\prod_{\alpha=1}^{n} (1 + k_\alpha \mu)}{\prod_{j=1}^{n} (\mu + \mu_j)} = \frac{1}{H(\mu)} . \quad (153)$$

Hence

$$H(\mu) = 1 + \mu H(\mu) \sum_{j=1}^{n} \frac{a_j H(\mu_j) \Psi(\mu_j)}{\mu + \mu_j} , \quad (154)$$

which proves the theorem for the case (131).

Turning, now, to the case

$$\int_0^1 \Psi(\mu)\, d\mu = \tfrac{1}{2} , \quad (155)$$

we observe that, in this case, $k = 0$ is a root of the characteristic equation, and we are left with only $(n - 1)$ positive roots. We therefore consider in this case the function

$$S(\mu) = \sum_{\alpha=1}^{n-1} \frac{L_\alpha}{1 - k_\alpha \mu} + L_0 - \mu \quad (156)$$

in place of equation (132). However, the constants L_0 and L_α $(\alpha = 1, \ldots, n - 1)$ are again to be determined by the conditions

$$S(\mu_i) = 0 \qquad (i = 1, \ldots, n) . \quad (157)$$

With this definition of $S(\mu)$, equation (137) continues to be valid, and the rest of the proof follows on similar lines. The only essential point of departure that needs to be noted is that, at the stage of the proof corresponding to equation (142) and before inverting the order of the summation, we must add the extra term

$$- \sum_j \frac{a_j \Psi(\mu_j)\, \mu_j}{1 + k_\alpha \mu_j} \quad (158)$$

to the right-hand side of the equation. We can do this without altering anything, since the quantity we thus add is zero; for

$$\left.\begin{aligned} \sum_j \frac{a_j \Psi(\mu_j)\, \mu_j}{1 + k_\alpha \mu_j} &= \frac{1}{k_\alpha} \left[\sum_j a_j \Psi(\mu_j) \left\{ 1 - \frac{1}{1 + k_\alpha \mu_j} \right\} \right] \\ &= \frac{1}{k_\alpha} \left[1 - \sum_j \frac{a_j \Psi(\mu_j)}{1 + k_\alpha \mu_j} \right] = 0 . \end{aligned}\right\} \quad (159)^{7}$$

For the rest, the proof is exactly the same as in case (127).

This completes the proof of the theorem.

[7] Note that we are permitted to set $\sum_j a_j \Psi(\mu_j) = 1$, since the Gauss sum in the "nth approximation" evaluates the integrals exactly for all polynomials of degree less than or equal to $4n - 1$; and we have assumed that $2n > m$.

Theorem 2.—The solution of the functional equation

$$H(\mu) = 1 + \mu H(\mu) \int_0^1 \frac{H(\mu')\Psi(\mu')}{\mu + \mu'} d\mu', \tag{160}$$

where $\Psi(\mu)$ is an even polynomial satisfying the condition

$$\int_0^1 \Psi(\mu)\, d\mu \leqslant \tfrac{1}{2}, \tag{161}$$

is the limit function

$$\lim_{n \to \infty} \frac{1}{\mu_1 \ldots \mu_n} \frac{\prod_{i=1}^{n} (\mu + \mu_i)}{\prod_{\alpha} (1 + k_\alpha \mu)}, \tag{162}$$

where the μ_i's and the k_α's have the same meanings as in theorem 1.

Proof.—The theorem arises in the following way: It is known that the integral of a bounded function over the interval (0, 1) can be approximated by a Gauss sum with any desired degree of accuracy by choosing a division of the interval according to the zeros of a Legendre polynomial of a sufficiently high degree. The integral which occurs on the right-hand side of equation (160) can therefore be replaced by the Gauss sum

$$\sum_{j=1}^{n} \frac{a_j H(\mu_j)\Psi(\mu_j)}{\mu + \mu_j} \tag{163}$$

to any desired accuracy by choosing a sufficiently large n. But by theorem 1, for a finite n, no matter how large, the solution of the equation

$$H(\mu) = 1 + \mu H(\mu) \sum_{j=1}^{n} \frac{a_j H(\mu_j)\Psi(\mu_j)}{\mu + \mu_j} \tag{164}$$

is always given by equation (129). Hence, if we now let $n \to \infty$, equation (164) becomes equation (160), and the solution for $H(\mu)$ is recovered as the limit function (162).

It is realized that the proof of theorem 2 which we have given is only a sketch and that it does not meet the full demands of a rigorous mathematical demonstration. But there can be hardly any doubt that a proof meeting such demands could be constructed, if one so desired, along the lines indicated.[8]

12. *Some integral properties of the functions* H(μ).—There are a number of integral theorems (of an essentially elementary kind) which can be proved for functions satisfying equations of the form (126). These theorems have applications in the further development of the theory, but they have also some physical interest.

Theorem 3.—

$$\int_0^1 H(\mu)\Psi(\mu)\, d\mu = 1 - \left[1 - 2\int_0^1 \Psi(\mu)\, d\mu\right]^{1/2}. \tag{165}$$

[8] It is probable that questions of "uniqueness" and "existence" will cause the principal difficulties in the construction of a rigorous mathematical proof. In our discussion we have implicitly assumed both, since the equations arise in a physical context and the physical situations are such as to leave no room for ambiguity. In any event, to go into mathematical subtleties here will be to go beyond the scope of a physical investigation.

Proof.—Multiplying the equation satisfied by $H(\mu)$ by $\Psi(\mu)$ and integrating over the range of μ, we have

$$\left.\begin{aligned}&\int_0^1 H(\mu)\Psi(\mu)\,d\mu\\&\quad=\int_0^1\Psi(\mu)\,d\mu+\int_0^1\int_0^1\frac{\mu}{\mu+\mu'}H(\mu)\Psi(\mu)H(\mu')\Psi(\mu')\,d\mu\,d\mu'.\end{aligned}\right\}\qquad(166)$$

Interchanging μ and μ' in the double integral in equation (166) and taking the average of the two equations, we obtain

$$\left.\begin{aligned}&\int_0^1 H(\mu)\Psi(\mu)\,d\mu\\&\quad=\int_0^1\Psi(\mu)\,d\mu+\frac{1}{2}\int_0^1\int_0^1 H(\mu)\Psi(\mu)H(\mu')\Psi(\mu')\,d\mu\,d\mu'\end{aligned}\right\}\qquad(167)$$

or, alternatively,

$$\frac{1}{2}\left[\int_0^1 H(\mu)\Psi(\mu)\,d\mu\right]^2-\int_0^1 H(\mu)\Psi(\mu)\,d\mu+\int_0^1\Psi(\mu)\,d\mu=0\,.\qquad(168)$$

Solving this equation for the integral in question, we have

$$\int_0^1 H(\mu)\Psi(\mu)\,d\mu=1\pm\left[1-2\int_0^1\Psi(\mu)\,d\mu\right]^{1/2}.\qquad(169)$$

The ambiguity in the sign in equation (169) can be removed by the consideration that the integral on the left-hand side must uniformly converge to zero when $\Psi(\mu)$ tends to zero uniformly in the interval (0, 1). This requires us to choose the negative sign in equation (169), and the result stated follows.

Corollary 1.—A necessary and sufficient condition that $H(\mu)$ be real is

$$\int_0^1\Psi(\mu)\,d\mu\leqslant\tfrac{1}{2}\,.\qquad(170)$$

This is, of course, an immediate consequence of the theorem.

The physical meaning of the limitation (170) is interesting; for it is really equivalent to the condition that, on each scattering, more radiation should not be emitted than was incident. Further, it may be noted that we have the equality sign in (170) only in the *conservative case*, i.e., only when there is no absorption and the efficiency of scattering is unity.

Corollary 2.—An alternative form of the functional equation satisfied by $H(\mu)$ is

$$1=H(\mu)\left\{\left[1-2\int_0^1\Psi(\mu)\,d\mu\right]^{1/2}+\int_0^1\frac{\mu'}{\mu+\mu'}H(\mu')\Psi(\mu')\,d\mu'\right\}.\qquad(171)$$

Proof.—This can be proved as follows: Using the result of theorem 3 and recalling the functional equation satisfied by $H(\mu)$, we have

$$\left.\begin{aligned}H(\mu)\int_0^1\frac{\mu'}{\mu+\mu'}H(\mu')\Psi(\mu')\,d\mu'&=H(\mu)\int_0^1 H(\mu')\Psi(\mu')\left[1-\frac{\mu}{\mu+\mu'}\right]d\mu'\\&=H(\mu)\left\{1-\left[1-2\int_0^1\Psi(\mu)\,d\mu\right]^{1/2}\right\}-H(\mu)+1\\&=1-H(\mu)\left[1-2\int_0^1\Psi(\mu)\,d\mu\right]^{1/2},\end{aligned}\right\}\qquad(172)$$

which is equivalent to equation (171).

We may note in this connection that for the practical purposes of iterating known approximate solutions of $H(\mu)$, the equation in the form (171) is more efficient than the original.

Theorem 4.—

$$\left[1-2\int_0^1 \Psi(\mu)\,d\mu\right]^{1/2}\int_0^1 H(\mu)\Psi(\mu)\,\mu^2 d\mu + \frac{1}{2}\left[\int_0^1 H(\mu)\Psi(\mu)\,\mu\,d\mu\right]^2 = \int_0^1 \Psi(\mu)\,\mu^2 d\mu\,. \tag{173}$$

Proof.—To prove equation (173) we multiply the equation defining $H(\mu)$ by $\Psi(\mu)\mu^2$ and integrate over the range of μ. We find

$$\begin{aligned}
&\int_0^1 H(\mu)\Psi(\mu)\,\mu^2 d\mu \\
&=\int_0^1 \Psi(\mu)\,\mu^2 d\mu + \int_0^1\int_0^1 \frac{H(\mu)H(\mu')\Psi(\mu)\Psi(\mu')}{\mu+\mu'}\,\mu^3 d\mu\,d\mu' \\
&=\int_0^1 \Psi(\mu)\,\mu^2 d\mu + \frac{1}{2}\int_0^1\int_0^1 \frac{H(\mu)H(\mu')\Psi(\mu)\Psi(\mu')}{\mu+\mu'}(\mu^3+\mu'^3)\,d\mu\,d\mu' \\
&=\int_0^1 \Psi(\mu)\,\mu^2 d\mu + \frac{1}{2}\int_0^1\int_0^1 H(\mu)H(\mu')\Psi(\mu)\Psi(\mu')(\mu^2-\mu\mu'+\mu'^2)\,d\mu\,d\mu' \\
&=\int_0^1 \Psi(\mu)\,\mu^2 d\mu + \left[\int_0^1 H(\mu)\Psi(\mu)\,\mu^2 d\mu\right]\left[\int_0^1 H(\mu)\Psi(\mu)\,d\mu\right] \\
&\qquad -\frac{1}{2}\left[\int_0^1 H(\mu)\Psi(\mu)\,\mu\,d\mu\right]^2 .
\end{aligned} \tag{174}$$

Using theorem 3, we obtain, after some minor reductions,

$$\left[1-2\int_0^1 \Psi(\mu)\,d\mu\right]^{1/2}\int_0^1 H(\mu)\Psi(\mu)\,\mu^2 d\mu + \frac{1}{2}\left[\int_0^1 H(\mu)\Psi(\mu)\,\mu\,d\mu\right]^2 = \int_0^1 \Psi(\mu)\,\mu^2 d\mu\,, \tag{175}$$

which is the required result.

Corollary 1.—For the conservative case

$$\int_0^1 \Psi(\mu)\,d\mu = \tfrac{1}{2}\,, \tag{176}$$

we have the further integral,

$$\int_0^1 H(\mu)\Psi(\mu)\,\mu\,d\mu = \left[2\int_0^1 \Psi(\mu)\,\mu^2 d\mu\right]^{1/2}. \tag{177}$$

This last relation (177) throws some interesting light on the Hopf-Bronstein relation (Paper II, eq. [71]) between the boundary and the effective temperatures in a gray atmosphere in radiative equilibrium; for, in this case, the appropriate functional equation is

$$H(\mu) = 1 + \tfrac{1}{2}\mu H(\mu)\int_0^1 \frac{H(\mu')}{\mu+\mu'}\,d\mu'\,. \tag{178}$$

Consequently,

$$\Psi(\mu) = \text{constant} = \tfrac{1}{2}\,, \tag{179}$$

and the case is a conservative one. We have, accordingly, the two integrals,[9]

$$\int_0^1 H(\mu)\, d\mu = 2 \tag{180}$$

and

$$\int_0^1 H(\mu)\, \mu d\mu = \frac{2}{\sqrt{3}}\,. \tag{181}$$

On the other hand, since for this problem (cf. Paper II)

$$I(0, \mu) \propto H(\mu)\,, \tag{182}$$

we have

$$\frac{J(0)}{F} = \frac{\int_0^1 H(\mu)\, d\mu}{4\int_0^1 H(\mu)\, \mu d\mu} = \frac{\sqrt{3}}{4}\,, \tag{183}$$

which is the Hopf-Bronstein integral. We therefore conclude that for all conservative cases of radiative transfer we have an integral of the Hopf-Bronstein type, which is essentially that given by equation (177). We may give two further examples of such integrals.

i) Considering, first, the case of a semi-infinite atmosphere in radiative equilibrium (with no incident radiation) and scattering radiation in accordance with Rayleigh's phase function, we have a functional equation of our standard form with (cf. Sec. V)

$$\Psi(\mu) = \tfrac{3}{16}(3 - \mu^2)\,. \tag{184}$$

The integral of this function over μ is $\frac{1}{2}$, and we have a conservative case. The two integrals are

$$\int_0^1 H(\mu)(3 - \mu^2)\, d\mu = \tfrac{16}{3} \tag{185}$$

and

$$\int_0^1 H(\mu)(3 - \mu^2)\, \mu d\mu = \tfrac{16}{3}\sqrt{0.3}\,. \tag{186}$$

These equations imply that

$$\frac{\int_0^1 I(0, \mu)(3 - \mu^2)\, d\mu}{\int_0^1 I(0, \mu)(3 - \mu^2)\, \mu d\mu} = \frac{1}{\sqrt{0.3}}\,. \tag{187}$$

It can be shown that equation (187) is equivalent to a curious relation which was established in Paper III (eq. [67]).

ii) The equation for $H_l(\mu)$, which we have discussed in Section II, also corresponds to a conservative case, since this equation is characterized by (cf. eq. [69])

$$\Psi(\mu) = \tfrac{3}{4}(1 - \mu^2)\,, \tag{188}$$

and the integral of this over μ is $\frac{1}{2}$. The two integrals now are

$$\int_0^1 H_l(\mu)(1 - \mu^2)\, d\mu = \tfrac{4}{3} \tag{189}$$

and

$$\int_0^1 H_l(\mu)(1 - \mu^2)\, \mu d\mu = \tfrac{4}{3}\sqrt{0.2}\,. \tag{189'}$$

[9] The first of these is contained in Ambarzumian's paper (*op. cit.*), but the second is not.

Hence, for the problem of an electron-scattering atmosphere,

$$\frac{\int_0^1 I_l(0,\mu)(1-\mu^2)\,d\mu}{\int_0^1 I_l(0,\mu)(1-\mu^2)\,\mu\,d\mu}=\sqrt{5}\,. \tag{190}$$

Corollary 2.—For the conservative case, an alternative form of the functional equation satisfied by $H(\mu)$ is

$$H(\mu)\left\{\left[2\int_0^1 \Psi(\mu)\,\mu^2 d\mu\right]^{1/2}-\int_0^1 \frac{\mu'^2}{\mu+\mu'}H(\mu')\Psi(\mu')\,d\mu'\right\}=\mu\,. \tag{191}$$

This readily follows from equations (165) and (177).

Theorem 5.—

$$\left.\begin{aligned}\int_0^1 H(\mu)\Psi(\mu)\,\mu^{2n}d\mu &= \int_0^1 \Psi(\mu)\,\mu^{2n}d\mu+\tfrac{1}{2}(-1)^n\left[\int_0^1 H(\mu)\Psi(\mu)\,\mu^n d\mu\right]^2 \\ &+\sum_{m=0}^{n-1}(-1)^m\left[\int_0^1 H(\mu)\Psi(\mu)\,\mu^{2n-m}d\mu\right]\left[\int_0^1 H(\mu)\Psi(\mu)\,\mu^m d\mu\right].\end{aligned}\right\} \tag{192}$$

This is a generalization of theorem 4, and the proof follows on similar lines.

Theorem 6.—For a function $H(\mu)$ satisfying a functional equation of the form

$$H(\mu)=1+\mu H(\mu)\int_0^1 \frac{H(\mu')(a+b\mu'^2)}{\mu+\mu'}\,d\mu'\,, \tag{193}$$

where a and b are two constants, we have the relations

$$\alpha_0=1+\tfrac{1}{2}(a\alpha_0^2+b\alpha_1^2) \tag{194}$$

and

$$(a+b\mu^2)\int_0^1 \frac{H(\mu')}{\mu+\mu'}\,d\mu'=\frac{H(\mu)-1}{\mu H(\mu)}-b(\alpha_1-\mu\alpha_0)\,, \tag{195}$$

where α_0 and α_1 are the moments of order zero and one of $H(\mu)$.

The proof of this theorem follows on exactly the same lines as those of lemmas 2 and 3 (§ 7).

IV. THE SOLUTION OF THE FUNCTIONAL EQUATIONS FOR THE PROBLEM OF DIFFUSE REFLECTION IN ACCORDANCE WITH THE PHASE FUNCTION $\lambda(1+x\cos\Theta)$

13. *The functional equations of the problem.*—The problem of diffuse reflection in accordance with the phase function $\lambda(1+x\cos\Theta)$ ($\lambda\leqslant 1$ and $-1\leqslant x\leqslant 1$) was treated in Paper IX, and Ambarzumian (*op. cit.*) derived the relevant functional equations of the problem by his method. However, our object in considering this problem again is principally to show how a knowledge of the form of the solution gained from our analysis in Paper IX enables us to reduce a pair of functional equations, which Ambarzumian has derived for the azimuth independent term in the *scattering function*, to a single functional equation of the standard form.

Now, if we express the intensity $I(\mu,\varphi;\mu_0,\varphi_0)$ reflected in the direction (μ,φ) when a parallel beam of radiation of net flux πF per unit area normal to itself is incident in the direction $(-\mu_0,\varphi_0)$ in the form

$$I(\mu,\varphi;\mu_0,\varphi_0)=\frac{\lambda}{4\mu}S(\mu,\varphi;\mu_0,\varphi_0)F\,, \tag{196}$$

the scattering function $S(\mu, \varphi; \mu_0, \varphi_0)$ satisfies the functional equation (cf. Ambarzumian, *op. cit.*, eq. [10])

$$\left.\begin{aligned}\left(\frac{1}{\mu}+\frac{1}{\mu_0}\right)S(\mu,\varphi;\mu_0,\varphi_0) &= j(\mu,\varphi;\mu_0,\varphi_0)+\frac{\lambda}{4\pi}\int_0^1\int_0^{2\pi} j(\mu,\varphi;-\mu'',\varphi'')\\ &\times S(\mu'',\varphi'';\mu_0,\varphi_0)\frac{d\mu''}{\mu''}\,d\varphi''+\frac{\lambda}{4\pi}\int_0^1\int_0^{2\pi} S(\mu,\varphi;\mu',\varphi')\\ &\times j(-\mu',\varphi';\mu_0,\varphi_0)\frac{d\mu'}{\mu'}\,d\varphi'+\frac{\lambda^2}{16\pi^2}\int_0^1\int_0^{2\pi}\int_0^1\int_0^{2\pi} S(\mu,\varphi;\mu',\varphi')\\ &\times j(-\mu',\varphi';-\mu'',\varphi'')\,S(\mu'',\varphi'';\mu_0,\varphi_0)\frac{d\mu'}{\mu'}\,d\varphi'\,\frac{d\mu''}{\mu''}\,d\varphi'',\end{aligned}\right\}\quad(197)$$

where, to exhibit the analogy with the equation for the scattering matrix derived in Section I (eq. [31]), we have written

$$j(\mu,\varphi;\mu_0,\varphi_0) = 1 - x\mu\mu_0 + x(1-\mu^2)^{\frac{1}{2}}(1-\mu_0^2)^{\frac{1}{2}}\cos(\varphi-\varphi_0). \quad (198)$$

An examination of equation (197) shows that we can express the scattering function in the form

$$S(\mu,\varphi;\mu_0,\varphi_0) = S^{(0)}(\mu,\mu_0) + x(1-\mu^2)^{\frac{1}{2}}(1-\mu_0^2)^{\frac{1}{2}}S^{(1)}(\mu,\mu_0)\cos(\varphi-\varphi_0), \quad (199)$$

where, as the notation indicates, $S^{(0)}$ and $S^{(1)}$ are functions of μ and μ_0 only. Substituting this form for S in equation (197), we obtain the two equations

$$\left.\begin{aligned}\left(\frac{1}{\mu}+\frac{1}{\mu_0}\right)S^{(0)}(\mu,\mu_0) &= 1 - x\mu\mu_0+\tfrac{1}{2}\lambda\int_0^1(1+x\mu\mu'')\,S^{(0)}(\mu'',\mu_0)\frac{d\mu''}{\mu''}\\ +\tfrac{1}{2}\lambda\int_0^1 S^{(0)}(\mu,\mu')(1+x\mu'\mu_0)\frac{d\mu'}{\mu'}&+\tfrac{1}{4}\lambda^2\int_0^1\int_0^1 S^{(0)}(\mu,\mu')(1-x\mu'\mu'')\\ &\times S^{(0)}(\mu'',\mu_0)\frac{d\mu'}{\mu'}\frac{d\mu''}{\mu''}\end{aligned}\right\}\quad(200)$$

and

$$\left.\begin{aligned}\left(\frac{1}{\mu}+\frac{1}{\mu_0}\right)S^{(1)}(\mu,\mu_0) &= \left\{1+\tfrac{1}{4}x\lambda\int_0^1\frac{d\mu''}{\mu''}(1-\mu''^2)\,S^{(1)}(\mu'',\mu_0)\right\}\\ &\times\left\{1+\tfrac{1}{4}x\lambda\int_0^1\frac{d\mu'}{\mu'}(1-\mu'^2)\,S^{(1)}(\mu,\mu')\right\}.\end{aligned}\right\}\quad(201)$$

Of these two functional equations, only equation (200) requires any detailed consideration, since it is at once apparent that the solution of equation (201) must be of the form

$$\left(\frac{1}{\mu}+\frac{1}{\mu_0}\right)S^{(1)}(\mu,\mu_0) = H^{(1)}(\mu)\,H^{(1)}(\mu_0), \quad (202)$$

where $H^{(1)}(\mu)$ satisfies the equation

$$H^{(1)}(\mu) = 1+\tfrac{1}{4}x\lambda\mu H^{(1)}(\mu)\int_0^1\frac{H^{(1)}(\mu')}{\mu+\mu'}(1-\mu'^2)\,d\mu'. \quad (203)$$

Our subsequent discussions in this section will therefore be concerned with only equation (200).

14. *The reduction of the functional equation for* $S^{(0)}(\mu, \mu_0)$ *to their normal forms.*—On examination it is seen that equation (200) can be written in the form

$$\left.\begin{aligned}\left(\frac{1}{\mu}+\frac{1}{\mu_0}\right) S^{(0)}(\mu, \mu_0) &= \left\{1+\tfrac{1}{2}\lambda\int_0^1 S^{(0)}(\mu'', \mu_0)\frac{d\mu''}{\mu''}\right\} \\ &\times\left\{1+\tfrac{1}{2}\lambda\int_0^1 S^{(0)}(\mu, \mu')\frac{d\mu'}{\mu'}\right\} \\ &- x\left\{\mu_0-\tfrac{1}{2}\lambda\int_0^1 S^{(0)}(\mu'', \mu_0)\,d\mu''\right\}\left\{\mu-\tfrac{1}{2}\lambda\int_0^1 S^{(0)}(\mu, \mu')\,d\mu'\right\}.\end{aligned}\right\} \tag{204}$$

From this equation it follows that $S^{(0)}(\mu, \mu_0)$ is symmetrical in μ and μ_0 and, further, that it can be expressed in the form

$$\left(\frac{1}{\mu}+\frac{1}{\mu_0}\right) S^{(0)}(\mu, \mu_0) = \psi(\mu)\,\psi(\mu_0) - x\phi(\mu)\,\phi(\mu_0)\,, \tag{205}$$

where

$$\psi(\mu) = 1+\tfrac{1}{2}\lambda\int_0^1 S^{(0)}(\mu, \mu')\frac{d\mu'}{\mu'} \tag{206}$$

and

$$\phi(\mu) = \mu-\tfrac{1}{2}\lambda\int_0^1 S^{(0)}(\mu, \mu')\,d\mu'\,. \tag{207}$$

Now, substituting for $S^{(0)}$ from equation (205) back into equations (206) and (207), we obtain the functional equations in their normal forms. We have

$$\psi(\mu) = 1+\tfrac{1}{2}\lambda\mu\psi(\mu)\int_0^1\frac{\psi(\mu')}{\mu+\mu'}\,d\mu' - \tfrac{1}{2}x\lambda\mu\phi(\mu)\int_0^1\frac{\phi(\mu')}{\mu+\mu'}\,d\mu' \tag{208}$$

and

$$\phi(\mu) = \mu-\tfrac{1}{2}\lambda\mu\psi(\mu)\int_0^1\frac{\psi(\mu')}{\mu+\mu'}\,\mu' d\mu' + \tfrac{1}{2}x\lambda\mu\phi(\mu)\int_0^1\frac{\phi(\mu')}{\mu+\mu'}\,\mu' d\mu'\,. \tag{209}$$

These equations agree with those given by Ambarzumian (*op. cit.*).

15. *The solutions of the functional equations (208) and (209).*—As in our discussion of the functional equations (49)–(52) in Section II, we shall be guided in the present instance also by the form of the solutions obtained by directly solving the relevant equation of transfer. We shall, therefore, first re-write the solution given in Paper IX in a form in which it can be compared with equation (205).

Now the solution for $I^{(0)}(0, \mu)$ given in Paper IX (eq. [108]) is equivalent to

$$\left(\frac{1}{\mu}+\frac{1}{\mu_0}\right) S^{(0)}(\mu, \mu_0) = H^{(0)}(\mu)\,H^{(0)}(\mu_0)\,[1-c(\mu+\mu_0)-x(1-\lambda)\,\mu\mu_0]\,, \tag{210}$$

where $H^{(0)}(\mu)$ is defined in terms of the roots of the characteristic equation (Paper IX, eq. [26])

$$1 = \lambda\sum_{j=1}^{n}\frac{a_j\,[1+x(1-\lambda)\,\mu_j^2]}{1-k^2\mu_j^2}\,. \tag{211}$$

Further, in equation (210), we have written c in place of $x(1-\lambda)\sigma/\rho$, which occurs in Paper IX, equation (108).

We now re-write solution (210) in the form

$$\left.\begin{aligned}\left(\frac{1}{\mu}+\frac{1}{\mu_0}\right) S^{(0)}(\mu,\mu_0) &= [H^{(0)}(\mu)(1-c\mu)][H^{(0)}(\mu_0)(1-c\mu_0)] \\ &\quad - x[q\mu H^{(0)}(\mu)][q\mu_0 H^{(0)}(\mu_0)],\end{aligned}\right\} \tag{212}$$

where

$$x q^2 = c^2 + x(1-\lambda). \tag{213}$$

The similarity of equations (212) and (205) is evident. Again, though solution (208) can be brought into such a correspondence with equation (205) in other ways, it will appear that we have made the correct choice in writing it as we have done. Indeed, we shall presently prove that the solution of the functional equations (208) and (209) are of the forms

$$\psi(\mu) = H^{(0)}(\mu)(1-c\mu) \tag{214}$$

and

$$\phi(\mu) = q\mu H^{(0)}(\mu), \tag{215}$$

where $H^{(0)}(\mu)$ is now defined as the solution of the functional equation

$$H^{(0)}(\mu) = 1 + \tfrac{1}{2}\lambda\mu H^{(0)}(\mu)\int_0^1 \frac{H^{(0)}(\mu')}{\mu+\mu'}[1+x(1-\lambda)\mu'^2]\,d\mu', \tag{216}$$

and q and c are two constants related in the manner of equation (213) and expressible in terms of the moments of $H^{(0)}(\mu)$.

16. *The verification of the solution.*—Before we proceed to prove that the solution of equations (208) and (209) is as we have stated it in § 15, we may note that, since the functional equation (216) is of the form considered in theorem 6 (§ 12), we have the relations

$$\alpha_0 = 1 + \tfrac{1}{4}\lambda[\alpha_0^2 + x(1-\lambda)\alpha_1^2] \tag{217}$$

and

$$[1+x(1-\lambda)\mu^2]\int_0^1 \frac{H^{(0)}(\mu')}{\mu+\mu'}d\mu' = \frac{H^{(0)}(\mu)-1}{\frac{1}{2}\lambda\mu H^{(0)}(\mu)} - x(1-\lambda)(\alpha_1-\mu\alpha_0), \tag{218}$$

where α_0 and α_1 are the moments of order zero and one of $H^{(0)}(\mu)$.

Our proof that equations (214) and (215) do represent the solution of equations (208) and (209) will consist in showing that, with a suitable choice of q and c (compatible with eq. [213]), the substitution of $S^{(0)}$ according to equation (210) (with $H^{(0)}$ now defined as the solution of eq. [216]) in equations (206) and (207) leads to ψ and ϕ, which are in agreement with equations (214) and (215).

Considering, first, $\psi(\mu)$, we have

$$\left.\begin{aligned}\psi(\mu) &= 1+\tfrac{1}{2}\lambda\mu H^{(0)}(\mu)\int_0^1 \frac{d\mu'}{\mu+\mu'}H^{(0)}(\mu')[1-c(\mu+\mu') \\ &\qquad - x(1-\lambda)\mu\mu'] \\ &= 1+\tfrac{1}{2}\lambda\mu H^{(0)}(\mu)\int_0^1 d\mu' H^{(0)}(\mu')\Big[\frac{1+x(1-\lambda)\mu^2}{\mu+\mu'} \\ &\qquad - c - x(1-\lambda)\mu\Big].\end{aligned}\right\} \tag{219}$$

The right-hand side of equation (219) can be simplified by using equation (218). We find in this manner that

$$\psi(\mu) = H^{(0)}(\mu)[1-\tfrac{1}{2}\lambda\mu\{c\alpha_0 + x(1-\lambda)\alpha_1\}]. \tag{220}$$

In order that equation (220) may be in agreement with equation (214), we must require that

$$c = \tfrac{1}{2}\lambda\,[c\,\alpha_0 + x\,(1-\lambda)\,\alpha_1] \tag{221}$$

or that

$$c = x\lambda\,(1-\lambda)\,\frac{\alpha_1}{2-\lambda\alpha_0}. \tag{222}$$

Considering, next, $\phi(\mu)$, we find in a similar way from equations (207) and (210) that

$$\phi(\mu) = [1 + \tfrac{1}{2}\lambda\,(c\,\alpha_1 - \alpha_0)]\,\mu H^{(0)}(\mu). \tag{223}$$

The agreement of equations (215) and (223) now requires that

$$q = 1 + \tfrac{1}{2}\lambda\,(c\,\alpha_1 - \alpha_0). \tag{224}$$

Substituting for c from equation (22) in equation (224), we find, after some minor rearranging of the terms, that

$$q = \frac{4 - 4\alpha_0\lambda + \lambda^2\,[\alpha_0^2 + x\,(1-\lambda)\,\alpha_1^2]}{2\,(2-\lambda\alpha_0)}. \tag{225}$$

The term in brackets in the numerator of equation (225) can be simplified by using relation (217) between the moments of $H^{(0)}(\mu)$. We find that

$$q = \frac{2\,(1-\lambda)}{2-\lambda\alpha_0}. \tag{226}$$

It remains to verify that q and c, as given by equations (222) and (226), are consistent with relation (213), which we have assumed. To show that this is the case, we evaluate

$$q^2 - \frac{c^2}{x} \tag{227}$$

according to equations (222) and (226). We find that

$$\left.\begin{aligned} q^2 - \frac{c^2}{x} &= \frac{1}{(2-\lambda\alpha_0)^2}\Big[4\,(1-\lambda)^2 - x\lambda^2\,(1-\lambda^2)\,\alpha_1^2\Big] \\ &= \frac{1-\lambda}{(2-\lambda\alpha_0)^2}\Big[4\,(1-\lambda) - x\lambda^2\,(1-\lambda)\,\alpha_1^2\Big]. \end{aligned}\right\} \tag{228}$$

But (cf. eq. [217])

$$-x\lambda^2\,(1-\lambda)\,\alpha_1^2 = 4\lambda\,(1 - \alpha_0 + \tfrac{1}{4}\lambda\alpha_0^2); \tag{229}$$

using this relation in equation (228), we find that

$$q^2 - \frac{c^2}{x} = 1 - \lambda, \tag{230}$$

which shows that q and c, as defined by equations (222) and (226), are indeed related, as required.

This completes the verification.

V. THE SOLUTION OF THE FUNCTIONAL EQUATIONS FOR THE PROBLEM OF DIFFUSE REFLECTION IN ACCORDANCE WITH RAYLEIGH'S PHASE FUNCTION

17. *The functional equations of the problem.*—For this problem we write the reflected intensity $I(\mu, \varphi; \mu_0, \varphi_0)$ in the form

$$\left.\begin{aligned}\mu I(\mu, \varphi; \mu_0, \varphi_0) = \tfrac{3}{32}F\,[S^{(0)}(\mu, \mu_0) - 4\mu\mu_0(1-\mu^2)^{\frac{1}{2}}(1-\mu_0^2)^{\frac{1}{2}} \\ \times S^{(1)}(\mu, \mu_0)\cos(\varphi-\varphi_0) + (1-\mu^2)(1-\mu_0^2)\,S^{(2)}(\mu, \mu_0)\cos 2(\varphi-\varphi_0)],\end{aligned}\right\} \quad (231)$$

and find for $S^{(0)}(\mu, \mu_0)$, $S^{(1)}(\mu, \mu_0)$, and $S^{(2)}(\mu, \mu_0)$ the functional equations

$$\left.\begin{aligned}\left(\frac{1}{\mu}+\frac{1}{\mu_0}\right)S^{(0)}(\mu, \mu_0) &= j^{(0)}(\mu, \mu_0) + \tfrac{3}{16}\int_0^1 j^{(0)}(\mu, \mu'')\,S^{(0)}(\mu'', \mu_0)\frac{d\mu''}{\mu''} \\ &+ \tfrac{3}{16}\int_0^1 S^{(0)}(\mu, \mu')\,j^{(0)}(\mu', \mu_0)\frac{d\mu'}{\mu'} + \frac{9}{256}\int_0^1\int_0^1 S^{(0)}(\mu, \mu')\,j^{(0)}(\mu', \mu'') \\ &\qquad \times S^{(0)}(\mu'', \mu_0)\frac{d\mu'}{\mu'}\frac{d\mu''}{\mu''},\end{aligned}\right\} \quad (232)$$

$$\left.\begin{aligned}\left(\frac{1}{\mu}+\frac{1}{\mu_0}\right)S^{(1)}(\mu, \mu_0) &= \left\{1+\frac{3}{8}\int_0^1 \mu''^2(1-\mu''^2)\,S^{(1)}(\mu'', \mu_0)\frac{d\mu''}{\mu''}\right\} \\ &\times\left\{1+\frac{3}{8}\int_0^1 \mu'^2(1-\mu'^2)\,S^{(1)}(\mu, \mu')\frac{d\mu'}{\mu'}\right\},\end{aligned}\right\} \quad (233)$$

and

$$\left.\begin{aligned}\left(\frac{1}{\mu}+\frac{1}{\mu_0}\right)S^{(2)}(\mu, \mu_0) &= \left\{1+\frac{3}{32}\int_0^1 (1-\mu''^2)^2\,S^{(2)}(\mu'', \mu_0)\frac{d\mu''}{\mu''}\right\} \\ &\times\left\{1+\frac{3}{32}\int_0^1 (1-\mu'^2)\,S^{(2)}(\mu, \mu')\frac{d\mu'}{\mu'}\right\},\end{aligned}\right\} \quad (234)$$

where in equation (231) we have written

$$\left.\begin{aligned}j^{(0)}(\mu, \mu_0) &= 3-\mu^2-\mu_0^2+3\mu^2\mu_0^2 \\ &= \tfrac{1}{3}(3-\mu^2)(3-\mu_0^2)+\tfrac{8}{3}\mu^2\mu_0^2.\end{aligned}\right\} \quad (235)$$

Of the three functional equations, (231)–(233), only the first need be considered in any detail, since it is at once apparent that the solutions for the other two must be of the forms

$$\left(\frac{1}{\mu}+\frac{1}{\mu_0}\right)S^{(1)}(\mu, \mu_0) = H^{(1)}(\mu)\,H^{(1)}(\mu_0) \quad (236)$$

and

$$\left(\frac{1}{\mu}+\frac{1}{\mu_0}\right)S^{(2)}(\mu, \mu_0) = H^{(2)}(\mu)\,H^{(2)}(\mu_0), \quad (237)$$

where $H^{(1)}(\mu)$ and $H^{(2)}(\mu)$ satisfy the functional equations

$$H^{(1)}(\mu) = 1+\tfrac{3}{8}\mu H^{(1)}(\mu)\int_0^1 \frac{H^{(1)}(\mu')}{\mu+\mu'}\,\mu'^2(1-\mu'^2)\,d\mu' \quad (238)$$

and

$$H^{(2)}(\mu) = 1+\tfrac{3}{32}\mu H^{(2)}(\mu)\int_0^1 \frac{H^{(2)}(\mu')}{\mu+\mu'}(1-\mu'^2)^2 d\mu'. \quad (239)$$

Returning to equation (231), we find, on examination, that this equation can be written in the form

$$\left(\frac{1}{\mu}+\frac{1}{\mu_0}\right) S^{(0)}(\mu,\mu_0) = \frac{1}{3}\left\{3-\mu^2+\frac{3}{16}\int_0^1 (3-\mu'^2)\, S^{(0)}(\mu,\mu')\frac{d\mu'}{\mu'}\right\} \times\left\{3-\mu_0^2+\frac{3}{16}\int_0^1 (3-\mu''^2)\, S^{(0)}(\mu'',\mu_0)\frac{d\mu''}{\mu''}\right\} + \frac{8}{3}\left\{\mu^2+\frac{3}{16}\int_0^1 S^{(0)}(\mu,\mu')\,\mu' d\mu'\right\}\left\{\mu_0^2+\frac{3}{16}\int_0^1 S^{(0)}(\mu'',\mu_0)\,\mu'' d\mu''\right\}. \tag{240}$$

From equation (240) it follows that $S^{(0)}(\mu, \mu_0)$ is symmetrical in μ, μ_0 and, further, that it can be expressed in the form

$$\left(\frac{1}{\mu}+\frac{1}{\mu_0}\right) S^{(0)}(\mu,\mu_0) = \tfrac{1}{3}\psi(\mu)\,\psi(\mu_0) + \tfrac{8}{3}\phi(\mu)\,\phi(\mu_0)\,, \tag{241}$$

where

$$\psi(\mu) = 3-\mu^2+\frac{3}{16}\int_0^1 (3-\mu'^2)\, S^{(0)}(\mu,\mu')\frac{d\mu'}{\mu'} \tag{242}$$

and

$$\phi(\mu) = \mu^2+\frac{3}{16}\int_0^1 \mu'^2 S^{(0)}(\mu,\mu')\frac{d\mu'}{\mu'}. \tag{243}$$

Now, substituting for $S^{(0)}$ according to equation (241) in equations (242) and (243), we obtain the functional equations

$$\psi(\mu) = 3-\mu^2+\tfrac{1}{16}\mu\psi(\mu)\int_0^1 \frac{\psi(\mu')}{\mu+\mu'}(3-\mu'^2)\,d\mu' + \tfrac{1}{2}\mu\phi(\mu)\int_0^1 \frac{\phi(\mu')}{\mu+\mu'}(3-\mu'^2)\,d\mu' \tag{244}$$

and

$$\phi(\mu) = \mu^2+\tfrac{1}{16}\mu\psi(\mu)\int_0^1 \frac{\psi(\mu')}{\mu+\mu'}\mu'^2 d\mu' + \tfrac{1}{2}\mu\phi(\mu)\int_0^1 \frac{\phi(\mu')}{\mu+\mu'}\mu'^2 d\mu'\,, \tag{245}$$

which are now in their normal forms.

18. *The solution of the functional equations (244) and (245).*—As in our earlier discussion of similar systems of functional equations, we shall again be guided by the form of the solution obtained by the direct solution of the equation of transfer.

Now, the solution for $I^{(0)}(0, \mu)$ given in Paper IX, equation (189), is readily seen to be equivalent to

$$\left(\frac{1}{\mu}+\frac{1}{\mu_0}\right) S^{(0)}(\mu,\mu_0) = H^{(0)}(\mu)\,H^{(0)}(\mu_0)\,[3-c(\mu+\mu_0)+\mu\mu_0]\,, \tag{246}$$

where $H^{(0)}(\mu)$ is defined in terms of the roots of the characteristic equation

$$1 = \frac{3}{8}\sum_{j=1}^{n}\frac{a_j(3-\mu_j^2)}{1-k^2\mu_j^2}. \tag{247}$$

Further, in equation (246) we have written

$$c = (1-2\xi)\sqrt{3}\,, \tag{248}$$

ξ being defined as in Paper IX, equations (184) and (188).

We now re-write solution (246) in the form

$$\left.\begin{aligned}\left(\frac{1}{\mu}+\frac{1}{\mu_0}\right) S^{(0)}(\mu, \mu_0) &= \tfrac{1}{3}\,[H^{(0)}(\mu)(3-c\mu)]\,[H^{(0)}(\mu_0)(3-c\mu_0)] \\ &\quad + \tfrac{8}{3}\,[q\mu H^{(0)}(\mu)]\,[q\mu_0 H^{(0)}(\mu^0)]\,,\end{aligned}\right\} \tag{249}$$

where

$$q^2 = \tfrac{1}{8}(3-c^2)\,. \tag{250}$$

The similarity of equations (241) and (249) is apparent, and, though solution (246) can be brought into this correspondence with equation (241) in other ways, it is nevertheless true that

$$\psi(\mu) = H^{(0)}(\mu)(3-c\mu) \tag{251}$$

and

$$\phi(\mu) = q\mu H^{(0)}(\mu) \tag{252}$$

represent the solution of the functional equations (244) and (245) if we define $H^{(0)}(\mu)$ as satisfying the equation

$$H^{(0)}(\mu) = 1 + \tfrac{3}{16}\mu H^{(0)}(\mu)\int_0^1 \frac{H^{(0)}(\mu')}{\mu+\mu'}(3-\mu'^2)\,d\mu' \tag{253}$$

and choose q and c consistent with equation (250) and in a manner we shall specify.

The proof that the solutions of equations (244) and (245) are as we have just stated proceeds as in the earlier cases, use now being made of the following relations:

$$\int_0^1 H^{(0)}(\mu)(3-\mu^2)\,d\mu = \tfrac{16}{3}\,, \tag{254}$$

$$\alpha_0 = 1 + \tfrac{3}{32}(3\alpha_0^2 - \alpha_1^2)\,, \tag{255}$$

and

$$(3-\mu^2)\int_0^1 \frac{H^{(0)}(\mu')}{\mu+\mu'}\,d\mu' = \frac{H^{(0)}(\mu)-1}{\tfrac{3}{16}\mu H^{(0)}(\mu)} + (\alpha_1 - \mu\alpha_0)\,, \tag{256}$$

where α_0 and α_1 are the moments of order zero and one of $H^{(0)}(\mu)$. Relations (254)-(256) follow from theorems 3 and 6 (§ 12) for the particular case of equation (253).

First, considering $\psi(\mu)$ as defined in equation (242), we find that the substitution of $S^{(0)}$ according to equation (246) leads to an expression for ψ which is in agreement with equation (251). On the other hand, if we evaluate $\phi(\mu)$ in a similar fashion, we find that

$$\phi(\mu) = \mu H^{(0)}(\mu)\,[\tfrac{9}{16}\alpha_1 - c(\tfrac{9}{16}\alpha_0 - 1)]\,. \tag{257}$$

To bring this into agreement with equation (252), we must set

$$q = \tfrac{9}{16}\alpha_1 - c(\tfrac{9}{16}\alpha_0 - 1)\,. \tag{258}$$

But, as we have already assumed that q and c are related as in equation (250), we must require that

$$\left(\frac{3-c^2}{8}\right)^{\frac{1}{2}} = \tfrac{9}{16}\alpha_1 - c(\tfrac{9}{16}\alpha_0 - 1) \tag{259}$$

or that

$$32(3-c^2) = [9\alpha_1 - c(9\alpha_0 - 16)]^2\,. \tag{260}$$

Expanding this last equation, we find

$$(288 - 288\alpha_0 + 81\alpha_0^2)\,c^2 - 18(9\alpha_0 - 16)\,\alpha_1 c + 81\alpha_1^2 - 96 = 0\,. \tag{261}$$

The coefficient of c^2 in equation (261) can be simplified by using equation (255). We find

$$9\alpha_1^2 c^2 - 6(9\alpha_0 - 16)\alpha_1 c + 27\alpha_1^2 - 32 = 0 . \tag{262}$$

The discriminant of this equation is

$$36\alpha_1^2[(9\alpha_0 - 16)^2 - 27\alpha_1^2 + 32] = 36\alpha_1^2[(288 - 288\alpha_0 + 81\alpha_0^2) - 27\alpha_1^2] ; \tag{263}$$

but this vanishes because of equation (255). Hence the quantity on the left-hand side of equation (262) is a perfect square, and we conclude that

$$c = \left(3 - \frac{32}{9\alpha_1^2}\right)^{\frac{1}{2}} \quad \text{and} \quad q = \frac{2}{3\alpha_1} . \tag{264}$$ [10]

Thus, with q and c as defined in equation (264), equations (251) and (252) do represent the solution of the functional equations (244) and (245).

19. *The exact law of darkening for an atmosphere with a constant net flux and scattering radiation in accordance with Rayleigh's phase function.*—In Paper III we worked out the theory of the radiative equilibrium of an atmosphere (with no incident radiation) in which scattering in accordance with Rayleigh's phase function governs the transfer of radiation. And in the appendix to Paper IX we gave the law of darkening for such an atmosphere in closed form. We have (Paper IX, appendix, eq. [9])

$$I(0, \mu) = \lambda F H^{(0)}(\mu) , \tag{265}$$

where $H^{(0)}(\mu)$ has the same meaning as in the problem of diffuse reflection which we have just considered. But the constant λ, defined in the following manner (Paper IX, appendix, eq. [10]):

$$\lambda = (-1)^n \mu_1 \ldots . \mu_n \frac{3\sqrt{3}}{2(B-A)} , \tag{266}$$

appears to bear no relation to the constant c of § 18 (eq. [248]). However, it can be shown that, in the framework of the theory of Papers III and IX, there is, in fact, a simple relation between λ and c. To show this, we must make use of the relation

$$\frac{P(\sqrt{3})P(-\sqrt{3})}{R(\sqrt{3})R(-\sqrt{3})} = -8\mu_1^2 \ldots . \mu_n^2 , \tag{267}$$

which follows from the identity[11]

$$1 - \frac{3}{8}\sum_{j=1}^{n} \frac{a_j(3-\mu_j^2)}{1-\mu_j^2 x} = (-1)^n \mu_1^2 \ldots . \mu_n^2 \frac{x \prod_{\alpha=1}^{n-1}(x - k_\alpha^2)}{\prod_{j=1}^{n}(1-\mu_j^2 x)} \tag{268}$$

when we set $x = \frac{1}{3}$. Now, from the definition of ξ, we have (cf. Paper IX, eqs. [184] and [188])

$$\xi(1-\xi) = -\frac{AB}{(A-B)^2} = -\frac{1}{(A-B)^2}\frac{P(\sqrt{3})P(-\sqrt{3})}{R(\sqrt{3})R(-\sqrt{3})} \tag{269}$$

or

$$\xi(1-\xi) = \frac{8\mu_1^2 \ldots . \mu_n^2}{(A-B)^2} . \tag{270}$$

[10] In eq. (264) we have chosen the positive sign for c, since it is found that $9\alpha_0 > 16$.

[11] Cf. Papers VIII, eqs. (41) and (46), and XI, n. 16.

Using this last relation in equation (266), we have

$$\lambda = \left[\tfrac{27}{32}\xi(1-\xi)\right]^{\frac{1}{2}}; \tag{271}$$

or, expressing ξ in terms of c in accordance with equation (248), we find

$$\lambda = \left[\tfrac{9}{128}(3-c^2)\right]^{\frac{1}{2}} = \tfrac{3}{4}q\,. \tag{272}$$

With this value of λ, the law of darkening (265) becomes

$$I(0,\mu) = \tfrac{3}{4}qFH^{(0)}(\mu)\,. \tag{273}$$

In the form (273) the law of darkening involves only constants and functions which also appear in the associated problem of diffuse reflection and for which exact expressions have been given. The exact law of darkening for our present problem is therefore obtained by simply redefining $H^{(0)}(\mu)$ in terms of the functional equation (253) and determining q as in equation (264). Thus, substituting for q from equation (264), we have

$$I(0,\mu) = \frac{1}{2\alpha_1}FH^{(0)}(\mu)\,, \tag{274}$$

in which form it is evident that solution (274) provides for the emergent radiation an outward flux which is exactly πF. Finally, we may note that for this problem there is an equivalent of the Hopf-Bronstein relation which is given by equation (187).

20. *Remarks on future work.*—The analyses of the various problems presented in this paper have served to illustrate and amplify the statement made in the introduction, that exact solutions for a large class of problems in the theory of radiative transfer can be obtained by letting the H-functions (defined in terms of the Gaussian division and characteristic roots) which appear in their solution become solutions of associated functional equations. And, as we have already indicated, the task of determining these exact H-functions by a process of iteration is considerably lightened by starting with as good an approximation as our "third approximation" provides. Mrs. Frances H. Breen and the writer have started on this work of evaluating the exact forms of the various H-functions which have been tabulated in their second or third approximations in earlier papers. We hope to have these calculations completed in the near future. But, apart from this matter of getting the exact numerical forms of the solutions of problems which have already been solved in principle, there are a number of other directions in which the theory developed in these papers can be extended. We may briefly indicate the nature of some of these.

First, there is the problem of scattering on laws more general than Rayleigh's. It would appear that in these more general cases the equations of transfer could still be written in vector form, like equation (15), the physical problem being one of determining the analogue of the matrix $\boldsymbol{QJ}$ which will describe the result of a single scattering. Second, there is the related question of what the equations of transfer are in cases in which the radiation field is one of general elliptic polarization. Under these circumstances it would follow from Stokes's considerations[12] that four intensities will be required to characterize the field (instead of three, as in the case of partial plane polarization) and that the equations of transfer could be expressed as a single vector equation in four-dimensional space. Scattering and diffuse reflection will then be described by matrices of order 4. These generalizations of the theory developed in Papers XI and XIII are likely to be fairly straightforward ones; but it would be of interest to have at least these formal developments completed.

[12] *Mathematical and Physical Papers of Sir George Stokes* (Cambridge, England, 1901), **3**, 233.

Third, there is the problem of transfer in plane-parallel atmospheres of finite thicknesses. It is not difficult to write down the formal solutions for such problems (e.g., Paper XII, eqs. [45] and [74]); the main difficulty (if such it is!) is that, in the forms in which the solutions have been written down, they involve constants of integration explicitly. And the problem of the elimination of the constants is not far advanced. But, once we can accomplish the elimination of the constants, we shall also be in a position to reduce the Ambarzumian type of functional equations to manageable proportions. However, it is not impossible that some deep generalizations of the methods of the earlier papers will be necessary before significant advances can be made in these directions.

Finally, there is the whole field of time-dependent transfer problems which requires to be explored. So far there has been only one investigation in this subject, namely, that by E. A. Milne.[13] In view of our present enormously increased resources for solving transfer problems, it would clearly be profitable to reconsider time-dependent situations and formulate problems of interest.

It is our intention to examine these and other problems in the near future.

[13] *J. London Math. Soc.*, **1**, 40, 1926.

ON THE RADIATIVE EQUILIBRIUM OF A STELLAR ATMOSPHERE XVII

S. CHANDRASEKHAR
Yerkes Observatory
Received February 25, 1947

ABSTRACT

In this paper a systematic study is made of the various functional equations which characterize the transfer problems in plane-parallel atmospheres.

First, a functional equation relating the angular distribution of the emergent radiation from a semi-infinite atmosphere and the law of diffuse reflection by the same atmosphere is derived. This equation arises in consequence of the invariance of the emergent radiation from a semi-infinite atmosphere to the addition (or removal) of layers of arbitrary optical thickness to (or from) the atmosphere. Next, four functional equations governing the problem of transmission and diffuse reflection by an atmosphere of finite optical thickness are formulated. These equations have been derived under very general conditions; and they have been further reduced to a basic system of functional equations for the case in which scattering takes place in accordance with a phase function expressible as a series in Legendre polynomials. And, finally, further functional equations are formulated which determine the radiation field in the interior in terms of the scattering and transmission functions of atmospheres of finite optical thicknesses.

1. *Introduction.*—In the study of the transfer of radiation in stellar atmospheres the two basic problems are, first, the radiative equilibrium of a semi-infinite atmosphere with a constant net flux and, second, the law of diffuse reflection by the same atmosphere. It is, of course, apparent that the first of these problems is significant only in conservative cases; for in all other cases the equation of transfer will not admit the integral which insures the constancy of the net flux through the atmosphere.

In the problem of a semi-infinite atmosphere with a constant net flux, the radiation field will be axially symmetrical at each point about the direction normal to the plane of stratification, and greatest interest is attached to the *law of darkening*,

$$I(0, \mu) \qquad (0 \leqslant \mu \leqslant 1)\,, \quad (1)$$

which expresses the angular distribution of the emergent radiation. On the other hand, in the problem of diffuse reflection, our principal interest is in the intensity,

$$I(\mu, \varphi; \mu_0, \varphi_0)\,, \quad (2)$$

diffusely reflected in the direction (μ, φ) when a parallel beam of radiation of net flux πF per unit area normal to itself is incident on the atmosphere in the direction $(-\mu_0, \varphi_0)$. This *reflected intensity* is generally expressed in terms of a *scattering function* $S(\mu, \varphi; \mu_0, \varphi_0)$ in the form

$$I(\mu, \varphi; \mu_0, \varphi_0) = \frac{1}{4\mu} S(\mu, \varphi; \mu_0, \varphi_0) F\,. \quad (3)$$

When the polarization of the scattered radiation has also to be allowed for, then, in the problem with a constant net flux, we must distinguish between the intensities I_l and I_r in the two states of polarization in which the electric vector vibrates in the meridian plane and at right angles to it, respectively. Correspondingly, in the problem of diffuse reflection we must consider a *scattering matrix*, $\mathbf{S}(\mu, \varphi; \mu_0, \varphi_0)$, defined suitably (cf. Papers XIII and XIV[1]).

Explicit solutions for the two basic problems have been found under a variety of con-

[1] *Ap. J.*, **105**, 151 (eq. [63]), and 164 (eq. [19]), 1947.

ditions in the earlier papers of this series.[2] And an examination of these solutions reveals certain remarkable relationships between the angular distribution of the emergent radiation in the problem with a constant net flux and the law of diffuse reflection. The relationship is naturally the simplest in the case of isotropic scattering, when[3]

$$I(0, \mu) = \frac{\sqrt{3}}{4} F H(\mu) \tag{4}$$

and

$$\left(\frac{1}{\mu}+\frac{1}{\mu_0}\right) S(\mu, \mu_0) = H(\mu) H(\mu_0), \tag{5}$$

where $H(\mu)$ satisfies the functional equation,

$$H(\mu) = 1 + \tfrac{1}{2}\mu H(\mu) \int_0^1 \frac{H(\mu')}{\mu+\mu'} d\mu'. \tag{6}$$

In the other cases the relationship is of a much more complex nature and has to be sought between the darkening function and the *azimuth independent* term,

$$S^{(0)}(\mu, \mu_0) = \frac{1}{2\pi} \int_0^{2\pi} S(\mu, \varphi; \mu_0, \varphi_0) \, d\varphi, \tag{7}$$

in the law of diffuse reflection.[4] The question now arises as to the origin and meaning of this relationship. While it should, in principle, be possible to go back to the original equations of transfer and derive the relationships in question as integrals of the problem, it would seem, in view of the complex nature of the relationships to be established, that such a procedure would not disclose the *physical* origin of the relationships. However, in this paper we shall show that there is an alternative way of looking at the problem which enables us to obtain in a simple way a functional equation relating $I(0, \mu)$ and $S^{(0)}(\mu, \mu_0)$. As we shall see, this functional equation arises essentially from the invariance of $I(0, \mu)$ to the addition (or removal) of layers of arbitrary thickness to (or from) the atmosphere. And it is to this invariance that we should attribute the relationship between the two basic problems.

The ideas leading to the establishment of the functional equation that we have just mentioned enable us to formulate further functional equations, which together appear to characterize the problems of radiative transfer in plane-parallel atmospheres (both finite and semi-infinite) in their entirety. In §§ 5–8 of this paper these other functional equations of the problem are derived.

I. THE FUNCTIONAL EQUATION RELATING THE LAW OF DARKENING AND THE SCATTERING FUNCTION FOR SEMI-INFINITE PLANE-PARALLEL ATMOSPHERES

2. *The functional equation relating* $I(0, \mu)$ *and* $S^{(0)}(\mu, \mu_0)$.—Considering, first, the case in which the scattering of radiation is described simply in terms of a phase function $p(\cos\Theta)$, we shall suppose that

$$p(\cos\Theta) = \Sigma \varpi_l P_l(\cos\Theta) \tag{8}$$

where ϖ_l's are constants and

$$\varpi_0 = 1, \tag{9}$$

[2] See esp. Paper XVI (*Ap. J.*, **105**, 435, 1947), where the solutions for the important cases of Rayleigh phase function and Rayleigh scattering are tabulated.

[3] Cf. Papers VIII (*Ap. J.*, **101**, 348, 1945) and XIV (*ibid.*, **105**, 164, eqs. [178]–[183]).

[4] When polarization is taken into account, we must consider the corresponding matrix, $\mathbf{S}^{(0)}(\mu, \mu_0)$ (see eq. [51], below).

to insure the constancy of the net flux through the atmosphere. For this case the equation of transfer for the problem with a constant net flux (and no incident radiation) can be expressed in the form

$$\mu \frac{dI(\tau, \mu)}{d\tau} = I(\tau, \mu) - \frac{1}{2}\int_{-1}^{+1} j(\mu, \mu') I(\tau, \mu') d\mu', \tag{10}$$

where

$$j(\mu, \mu') = \Sigma \varpi_l P_l(\mu) P_l(\mu') \tag{11}$$

is a symmetrical function in the variables μ and μ'.

We shall find it convenient to re-write equation (10) in the form

$$\mu \frac{dI(\tau, \mu)}{d\tau} = I(\tau, \mu) - B(\tau, \mu), \tag{12}$$

where the source function

$$B(\tau, \mu) = \frac{1}{2}\int_{-1}^{+1} j(\mu, \mu') I(\tau, \mu') d\mu'. \tag{13}$$

Now consider the radiation at a depth $d\tau$ below the boundary of the atmosphere at $\tau = 0$. The radiation field at this level, which is directed *inward*, can be inferred from the equation of transfer (12); for, since at $\tau = 0$ there is no incident intensity, at the level $d\tau$ we must have an inward intensity,

$$I(d\tau, -\mu') = \frac{1}{\mu'} B(0, -\mu') d\tau, \tag{14}$$

in the direction $-\mu'$. This inward-directed radiation will be reflected by the atmosphere below $d\tau$ by the law of diffuse reflection of a semi-infinite plane-parallel atmosphere. If this law is expressed in the form (3), the reflection of the radiation (14) will contribute to the *outward* intensity at the level $d\tau$, the amount

$$\frac{d\tau}{4\pi\mu}\int_0^1\int_0^{2\pi} S(\mu, \varphi; \mu', \varphi') B(0, -\mu') \frac{d\mu'}{\mu'} d\varphi', \tag{15}$$

or, in view of the axial symmetry of $B(0, -\mu')$,

$$\frac{d\tau}{2\mu}\int_0^1 S^{(0)}(\mu, \mu') B(0, -\mu') \frac{d\mu'}{\mu'}, \tag{16}$$

where $S^{(0)}(\mu, \mu')$ is defined as in equation (7).

Now *the outward intensity* $I(d\tau, \mu)$ *at the level* $d\tau$ *can differ from* $I(0, \mu)$ *only by the amount (16)*; for the intensity and the angular distribution of the emergent radiation from a semi-infinite plane-parallel atmosphere cannot be altered by the removal (or addition) of layers of arbitrary optical thickness from (or to) the atmosphere. Consequently, the removal of the layers above the level $d\tau$ must restore $I(d\tau, \mu)$ to $I(0, \mu)$. We must therefore have

$$I(d\tau, \mu) = I(0, \mu) + \frac{d\tau}{2\mu}\int_0^1 S^{(0)}(\mu, \mu') B(0, -\mu') \frac{d\mu'}{\mu'}. \tag{17}$$

On the other hand, from the equation of transfer (12), we can directly conclude that

$$\left.\begin{aligned} I(d\tau, \mu) &= I(0, \mu) + d\tau\left(\frac{dI}{d\tau}\right)_{\tau=0} \\ &= I(0, \mu) + \frac{d\tau}{\mu}[I(0, \mu) - B(0, \mu)]. \end{aligned}\right\} \tag{18}$$

Combining equations (17) and (18) and passing to the limit $d\tau = 0$, we obtain

$$I(0, \mu) = B(0, \mu) + \frac{1}{2}\int_0^1 S^{(0)}(\mu, \mu') B(0, -\mu') \frac{d\mu'}{\mu'} . \tag{19}$$

But, according to equation (13),

$$B(0, \mu) = \frac{1}{2}\int_0^1 j(\mu, \mu'') I(0, \mu'') d\mu'' . \tag{20}$$

Using this expression for $B(0, \mu)$ in equation (19), we have

$$\left.\begin{aligned} I(0, \mu) &= \frac{1}{2}\int_0^1 j(\mu, \mu'') I(0, \mu'') d\mu'' \\ &\quad + \frac{1}{4}\int_0^1\int_0^1 S^{(0)}(\mu, \mu') j(-\mu', \mu'') I(0, \mu'') \frac{d\mu'}{\mu'} d\mu'' . \end{aligned}\right\} \tag{21}$$

Finally, substituting for $j(\mu, \mu')$ according to equation (11), we find

$$\left.\begin{aligned} I(0, \mu) &= \frac{1}{2}\sum_l \varpi_l \left[\int_0^1 P_l(\mu') I(0, \mu') d\mu'\right] \\ &\quad \times \left[P_l(\mu) + (-1)^l \frac{1}{2}\int_0^1 S^{(0)}(\mu, \mu') P_l(\mu') \frac{d\mu'}{\mu'}\right] . \end{aligned}\right\} \tag{22}$$

Equation (22) is a functional equation relating $I(0, \mu)$ and $S^{(0)}(\mu, \mu')$; and it is apparent how this equation will determine $I(0, \mu)$ uniquely in terms of $S^{(0)}(\mu, \mu')$, and conversely

3. *Illustrations of the functional equation (22).*—We shall illustrate equation (22) by considering, first, the case of isotropic scattering. In this case

$$\varpi_0 = 1 , \qquad \text{and} \qquad \varpi_l = 0 \qquad (l \neq 0) , \tag{23}$$

and (cf. eq. [5])

$$\left(\frac{1}{\mu} + \frac{1}{\mu'}\right) S^{(0)}(\mu, \mu') = H(\mu) H(\mu') , \tag{24}$$

and equation (22) reduces to

$$I(0, \mu) = \left[\frac{1}{2}\int_0^1 I(0, \mu') d\mu'\right]\left[1 + \tfrac{1}{2}\mu H(\mu) \int_0^1 \frac{H(\mu')}{\mu + \mu'} d\mu'\right] . \tag{25}$$

Using equation (6), satisfied by $H(\mu)$, we find the foregoing equation to be

$$I(0, \mu) = \left[\frac{1}{2}\int_0^1 I(0, \mu') d\mu'\right] H(\mu) . \tag{26}$$

It is seen that equation (26) is consistent with itself, since, according to Paper XIV, equation (180),

$$\int_0^1 H(\mu) d\mu = 2 . \tag{27}$$

We may therefore write

$$I(0, \mu) = \text{constant } H(\mu) \tag{28}$$

and determine the constant of proportionality from the condition

$$2\int_0^1 I(0, \mu) \mu d\mu = F . \tag{29}$$

We find in this manner that (cf. Paper XIV, eqs. [182]–[183])

$$I(0, \mu) = \frac{\sqrt{3}}{4} F H(\mu), \tag{30}$$

in agreement with equation (4).

As a second illustration of equation (22), we shall consider the case of scattering in accordance with Rayleigh's phase function. In this case

$$\varpi_0 = 1, \quad \varpi_2 = \tfrac{1}{2}, \quad \text{and} \quad \varpi_l = 0 \quad (l \neq 0, 2), \tag{31}$$

and equation (22) reduces to

$$\left.\begin{aligned} I(0, \mu) = \frac{1}{2}\left[\int_0^1 I(0, \mu')\, d\mu'\right]\left[1 + \frac{1}{2}\int_0^1 S^{(0)}(\mu, \mu')\frac{d\mu'}{\mu'}\right] \\ + \frac{1}{16}\left[\int_0^1 I(0, \mu')(3\mu'^2 - 1)\, d\mu'\right] \\ \times\left[3\mu^2 - 1 + \frac{1}{2}\int_0^1 S^{(0)}(\mu, \mu')(3\mu'^2 - 1)\frac{d\mu'}{\mu'}\right]. \end{aligned}\right\} \tag{32}$$

Also, for this case (Paper XIV, eqs. [231] and [246] or Paper XVI, Sec. II),

$$\left(\frac{1}{\mu} + \frac{1}{\mu'}\right) S^{(0)}(\mu, \mu') = \tfrac{3}{8} H^{(0)}(\mu) H^{(0)}(\mu')\left[3 - c(\mu + \mu') + \mu\mu'\right], \tag{33}$$

where $H^{(0)}(\mu)$ is defined in terms of the functional equation (cf. Paper XIV, eq. [253]),

$$H^{(0)}(\mu) = 1 + \tfrac{3}{16}\mu H^{(0)}(\mu)\int_0^1 \frac{H^{(0)}(\mu')}{\mu + \mu'}(3 - \mu'^2)\, d\mu', \tag{34}$$

and

$$c = \frac{\alpha_2}{\alpha_1}, \tag{35}$$ [5]

α_2 and α_1 being the moments of order 1 and 2 of $H^{(0)}(\mu)$.

Inserting the reflection function (33) in equation (32), we find, after some reductions in which repeated use is made of Paper XIV, equations (254)–(256), that

$$\left.\begin{aligned} I(0, \mu) = \tfrac{3}{16}(3A_0 - A_2) H^{(0)}(\mu) \\ + \tfrac{3}{32}\left[-2A_0\alpha_2 + 3A_2\alpha_1^2 + 3A_2\alpha_2(2 - \alpha_0)\right]\frac{1}{\alpha_1}\mu H^{(0)}(\mu), \end{aligned}\right\} \tag{36}$$

where A_0 and A_2 are the (unknown!) moments of order 0 and 2 of $I(0, \mu)$.

The constants A_0 and A_2 in equation (36) can be determined by taking the zero and second moments of this equation. We find that

$$A_0 = \tfrac{3}{16}(3A_0 - A_2)\alpha_0 + \tfrac{3}{32}\left[-2A_0\alpha_2 + 3A_2\alpha_1^2 + 3A_2\alpha_2(2 - \alpha_0)\right] \tag{37}$$

and

$$A_2 = \tfrac{3}{16}(3A_0 - A_2)\alpha_2 + \tfrac{3}{32}\left[-2A_0\alpha_2 + 3A_2\alpha_1^2 + 3A_2\alpha_2(2 - \alpha_0)\right]\frac{\alpha_3}{\alpha_1}. \tag{38}$$

[5] In Paper XIV the constant c was defined somewhat differently (cf. eq. [264]). But, since the discriminant of eq. (262) vanishes, we could equally well have written

$$c = \frac{9\alpha_0 - 16}{3\alpha_1}.$$

Using the relation $9\alpha_0 - 3\alpha_2 = 16$ (Paper XIV, eq. [254]) in this equation for c, we obtain formula (35).

These equations determine A_0 and A_2. However, a simple inspection shows that

$$\frac{A_0}{a_0} = \frac{A_2}{a_2} = \text{constant} \tag{39}$$

satisfies equations (37) and (38); for (cf. Paper XIV, eqs. [254] and [255])

$$3a_0 - a_2 = \tfrac{16}{3}, \tag{40}$$

and

$$\left.\begin{aligned} -2a_0a_2 + 3a_2a_1^2 + 3a_2^2(2-a_0) &= a_2[-2a_0 + 3a_1^2 + (9a_0-16)(2-a_0)] \\ &= a_2[3a_1^2 - 9a_0^2 + 32(a_0-1)] = 0 . \end{aligned}\right\} \tag{41}$$

Hence,

$$I(0, \mu) = \text{constant } H^{(0)}(\mu); \tag{42}$$

and the constant of proportionality can again be determined from the flux condition (29). In this manner we find that

$$I(0, \mu) = \frac{1}{2a_1} FH^{(0)}(\mu), \tag{43}$$

in agreement with Paper XIV, equation (274).

4. *The functional equations relating the laws of darkening and diffuse reflection in cases in which the polarization of the scattered radiation is taken into account.*—As sufficiently illustrative of the general problem, we shall consider the case of Rayleigh scattering for which the general equations of transfer have already been formulated in earlier papers.[6] However, in our present context we require the equations of transfer for the case of axial symmetry when it is sufficient to distinguish only between the intensities I_l and I_r in the two states of polarization in which the electric vector vibrates in the meridian plane and at right angles to it, respectively.[7] And the equations of transfer relevant to this problem have been derived in Paper X (eqs. [37] and [38]). For our present purposes it is convenient to combine these equations into a single vector equation, considering $I_l(\tau, \mu)$ and $I_r(\tau, \mu)$ as the components of a *two*-dimensional vector,

$$\boldsymbol{I} = (I_l, I_r). \tag{44}$$

The equation of transfer (Paper X, eqs. [37] and [38]) for I_l and I_r can then be expressed in the form (cf. Paper XIV, eqs. [10] and [15])

$$\mu \frac{d\boldsymbol{I}}{d\tau} = \boldsymbol{I}(\tau, \mu) - \frac{3}{8}\int_{-1}^{+1} J(\mu, \mu')\boldsymbol{I}(\tau, \mu')\, d\mu', \tag{45}$$

where $J(\mu, \mu')$ denotes the matrix,

$$J(\mu, \mu') = \begin{pmatrix} 2(1-\mu^2)(1-\mu'^2) + \mu^2\mu'^2 & \mu^2 \\ \mu'^2 & 1 \end{pmatrix}. \tag{46}$$

Introducing the source function,

$$\boldsymbol{B}(\tau, \mu) = \frac{3}{8}\int_{-1}^{+1} J(\mu, \mu')\boldsymbol{I}(\tau, \mu')\, d\mu', \tag{47}$$

[6] Papers X (*Ap. J.*, **103**, 351, 1946), XI (*ibid.*, **104**, 110, 1946), XIII (*ibid.*, **105**, 151, 1947), XIV (*ibid.*, p. 164), and XV (*ibid.*, p. 424).

[7] The other Stokes parameters, $U = (I_l - I_r)\tan 2\chi$ and $V = (I_l - I_r)\sec 2\chi \tan 2\beta$ are identically zero for the problem with a constant net flux.

we can re-write equation (45) in the form

$$\mu \frac{dI}{d\tau} = I(\tau, \mu) - B(\tau, \mu). \tag{48}$$

To derive the functional equation for $I(0, \mu)$, we proceed exactly as in § 2 by considering the radiation field at a level $d\tau$ below the boundary of the atmosphere at $\tau = 0$. At this level there will be an inward intensity, of amount

$$\frac{d\tau}{\mu'} B(0, \mu'), \tag{49}$$

in the direction $-\mu'$. On account of the axial symmetry of this field and the consequent vanishing of the component U, the radiation (49) will be reflected by the atmosphere below $d\tau$ in accordance with the two-dimensional scattering matrix (cf. Paper XVI, § 3),

$$\frac{3}{16\mu} S^{(0)}(\mu, \mu'), \tag{50}$$

where

$$\left(\frac{1}{\mu}+\frac{1}{\mu'}\right) S^{(0)}(\mu, \mu') = \begin{pmatrix} 2H_l(\mu) H_l(\mu')[1 - c(\mu+\mu') + \mu\mu'] & qH_l(\mu) H_r(\mu')(\mu+\mu') \\ qH_r(\mu) H_l(\mu')(\mu+\mu') & H_r(\mu) H_r(\mu')[1 + c(\mu+\mu') + \mu\mu'] \end{pmatrix}. \tag{51}$$

(The definitions of the various quantities occurring on the right-hand side of the foregoing equation will be found in Paper XIV, Part II).

The reflection of the radiation (49) by the atmosphere below $d\tau$ will, accordingly, contribute to the outward intensity, $I(d\tau, \mu)$, at the level $d\tau$, the amount

$$\frac{3}{8\mu} d\tau \int_0^1 S^{(0)}(\mu, \mu') B(0, -\mu') \frac{d\mu'}{\mu'}. \tag{52}$$

Again, from the invariance of $I(0, \mu)$ to the removal (or addition) of layers of arbitrary thickness from (or to) the atmosphere, we conclude that $I(d\tau, \mu)$ can differ from $I(0, \mu)$ only by the amount (52). Hence,

$$I(d\tau, \mu) = I(0, \mu) + \frac{3}{8\mu} d\tau \int_0^1 S^{(0)}(\mu, \mu') B(0, -\mu') \frac{d\mu'}{\mu'}. \tag{53}$$

On the other hand, from the equation of transfer we directly find that

$$I(d\tau, \mu) = I(0, \mu) + \frac{d\tau}{\mu} [I(0, \mu) - B(0, \mu)]. \tag{54}$$

Combining equations (53) and (54) and passing to the limit $d\tau = 0$, we obtain

$$I(0, \mu) = B(0, \mu) + \frac{3}{8} \int_0^1 S^{(0)}(\mu, \mu') B(0, -\mu') \frac{d\mu'}{\mu'}. \tag{55}$$

Finally, substituting for $B(0, \mu)$ according to equation (47) in equation (55), we have

$$\left.\begin{aligned} I(0, \mu) = \frac{3}{8}\Big[& \int_0^1 J(\mu, \mu') I(0, \mu') \, d\mu' \\ & + \frac{3}{8} \int_0^1 \int_0^1 S^{(0)}(\mu, \mu') J(-\mu', \mu'') I(0, \mu'') \frac{d\mu'}{\mu'} d\mu'' \Big]. \end{aligned}\right\} \tag{56}$$

This is the required functional equation relating $I(0, \mu)$ and $S^{(0)}(\mu, \mu')$.

It should, of course, be possible to deduce the solution for $I(0, \mu)$ from the known form of $S^{(0)}(\mu, \mu')$. The necessary calculations are likely to be somewhat tedious, since it is not too easy even to verify that the solution (Paper XIV, eqs. [119] and [120])

$$I(0, \mu) = \frac{3}{8\sqrt{2}} F \begin{pmatrix} qH_l(\mu) \\ H_r(\mu)(\mu + c) \end{pmatrix} \tag{57}$$

actually satisfies equation (56). Thus, for $I(0, \mu)$ given by equation (57),

$$B(0, \mu) = \frac{3}{8\sqrt{2}} F \begin{pmatrix} q(1-\mu^2) + c\mu^2 \\ c \end{pmatrix}, \tag{58}$$

and the validity of equation (56) requires that the equation

$$\left.\begin{aligned} \begin{pmatrix} qH_l(\mu) \\ H_r(\mu)(\mu + c) \end{pmatrix} = \begin{pmatrix} q(1-\mu^2) + c\mu^2 \\ c \end{pmatrix} + \tfrac{3}{8}\mu \int_0^1 \frac{d\mu'}{\mu + \mu'} \\ \times \begin{pmatrix} 2H_l(\mu)H_l(\mu')[1 - c(\mu + \mu') + \mu\mu'] & qH_l(\mu)H_r(\mu')(\mu + \mu') \\ qH_r(\mu)H_l(\mu')(\mu + \mu') & H_r(\mu)H_r(\mu')[1 - c(\mu + \mu') + \mu\mu'] \end{pmatrix} \\ \times \begin{pmatrix} q(1-\mu'^2) + c\mu'^2 \\ c \end{pmatrix} \end{aligned}\right\} \tag{59}$$

be true. By a series of reductions, in which the various integral properties[8] of the functions $H_l(\mu)$ and $H_r(\mu)$ are used repeatedly it can be shown that this is indeed the case. But we shall not go into the details of these reductions here.

II. FUNCTIONAL EQUATIONS FOR TRANSFER PROBLEMS IN PLANE-PARALLEL ATMOSPHERES

5. *Functional equations for the problem of transmission and diffuse reflection by a plane-parallel atmosphere of finite optical thickness.*—In Paper XIV we formulated the functional equation satisfied by the scattering matrix in the problem of diffuse reflection, on Rayleigh's law, by a semi-infinite plane-parallel atmosphere. The corresponding functional equation for the case of scattering according to a general phase function (but not allowing for the polarization of the scattered radiation) had been derived earlier by V. A. Ambarzumian.[9] The principle underlying the derivation of these functional equations is the invariance, first pointed out byAmbarzumian, of the intensity and the angular distribution of the reflected light to the addition (or removal) of layers of arbitrary optical thickness to (or from) the atmosphere. It is apparent that similar functional equations must govern the problem of transmission and reflection by a plane-parallel atmosphere of finite optical thickness, τ_1 (say). This possibility has, indeed, been envisaged by Ambarzumian,[10] who stated that the required functional equations can be obtained from the invariance of the reflected and the transmitted light to the addition of a layer of a certain optical thickness to the atmosphere above $\tau = 0$ and the *simultaneous* removal of a layer of equal optical thickness from the atmosphere, below, at $\tau = \tau_1$. This invariance will lead to two functional equations governing the intensities and the angular distributions of the reflected and the transmitted radiations. Ambarzumian has, in fact, written down these two equations for the case of isotropic scattering.[11] However, on consideration it

[8] Cf. Paper XIV, §§ 7 and 8.

[9] *J. Phys. Acad. Sc. U.S.S.R.*, **8**, 64, 1944.

[10] *C.R.* (Doklady) *Acad. d. Sc. U.R.S.S.*, **38**, 229, 1943.

[11] *Ibid.*, eqs. (8) and (9). The derivation of these equations is not given in this short note. However, from the details of the treatment of the semi-infinite case, which are given, it is not difficult to reconstruct the proof that the author probably had in mind. The present writer has been unable to trace any later publication in which Ambarzumian has returned to these matters.

appears that the two functional equations which can be derived from the invariance referred to by Ambarzumian are not sufficient to characterize the problem completely. Actually, it would seem that four equations are necessary to make the problem determinate, and we shall show how these four equations arise by considering in detail the case of scattering according to a general phase function. The modifications required to include the polarization of the scattered radiation will be indicated later.

The equation of transfer for the problem of diffuse reflection by a plane-parallel atmosphere can be written in the form[12]

$$\mu \frac{dI(\tau, \mu, \varphi)}{d\tau} = I(\tau, \mu, \varphi) - B(\tau, \mu, \varphi), \tag{60}$$

where the source function $B(\tau, \mu, \varphi)$ is given by

$$\left.\begin{aligned} B(\tau, \mu, \varphi) = \frac{1}{4\pi}\int_{-1}^{+1}\int_0^{2\pi} & p(\mu, \varphi; \mu', \varphi')\, I(\tau, \mu', \varphi')\, d\mu' d\varphi' \\ & + \tfrac{1}{4}F e^{-\tau/\mu_0} p(\mu, \varphi; -\mu_0, \varphi_0), \end{aligned}\right\} \tag{61}$$

when a parallel beam of radiation of flux πF per unit area normal to itself is incident on the atmosphere in the direction $(-\mu_0, \varphi_0)$. In equation (61), $p(\mu, \varphi; \mu', \varphi')$ is the phase function governing the probability that the radiation in the direction (μ', φ') will be scattered in the direction (μ, φ).

We shall suppose that the optical thickness of the atmosphere is τ_1. We shall then be interested not only in the intensity $I(0, \mu, \varphi)$, $(0 \leqslant \mu \leqslant 1)$, reflected by the atmosphere in the direction (μ, φ), but also in the intensity $I(\tau_1, \mu, \varphi)$ transmitted in the direction $(-\mu, \varphi)$.[13] We shall express these intensities in the forms

$$I(0, \mu, \varphi) = \frac{1}{4\mu} FS(\tau_1; \mu, \varphi; \mu_0, \varphi_0) \tag{62}$$

and

$$I(\tau_1, \mu, \varphi) = \frac{1}{4\mu} FT(\tau_1; \mu, \varphi; \mu_0, \varphi_0), \tag{63}$$

to emphasize the fact that we are considering the reflection and transmission by an atmosphere of optical thickness τ_1.

Now consider the radiation field present at a depth $d\tau$ below the boundary of the atmosphere at $\tau = 0$. At this level there will be the incident flux, reduced to the amount

$$\pi F\left(1 - \frac{d\tau}{\mu_0}\right), \tag{64}$$

and a diffuse radiation field. The intensity in this latter diffuse field, which is directed inward, can be inferred from the equation of transfer: for, since at $\tau = 0$ there is no inward intensity, at the level $d\tau$ we must have the intensity

$$\frac{d\tau}{\mu'} B(0, -\mu', \varphi') \tag{65}$$

in the direction $(-\mu', \varphi')$. Both the radiation fields (64) and (65) will be reflected by the atmosphere of optical thickness $\tau_1 - d\tau$ below $d\tau$ and will contribute

$$\left.\begin{aligned} & \frac{F}{4\mu}\left(1 - \frac{d\tau}{\mu_0}\right) S(\tau_1 - d\tau; \mu, \varphi; \mu_0, \varphi_0) \\ & \quad + \frac{d\tau}{4\pi\mu}\int_0^1\int_0^{2\pi} S(\tau_1 - d\tau; \mu, \varphi; \mu', \varphi')\, B(0, -\mu', \varphi') \frac{d\mu'}{\mu'} d\varphi' \end{aligned}\right\} \tag{66}$$

[12] Cf. Paper XII (*Ap. J.*, **104**, 191, 1946), eqs. (1)–(3).

[13] Note the minus sign here.

to the intensity in the outward direction (μ, φ) at $d\tau$. Hence

$$\left.\begin{aligned} I(d\tau, \mu, \varphi) = \frac{F}{4\mu}\Big[\Big(1-\frac{d\tau}{\mu_0}\Big) S(\tau_1;\ \mu, \varphi;\ \mu_0, \varphi_0) - \frac{\partial S(\tau_1;\ \mu, \varphi;\ \mu_0, \varphi_0)}{\partial \tau_1}\, d\tau\Big] \\ + \frac{d\tau}{4\pi\mu}\int_0^1\int_0^{2\pi} S(\tau_1;\ \mu, \varphi;\ \mu', \varphi') B(0, -\mu', \varphi') \frac{d\mu'}{\mu'}\, d\varphi', \end{aligned}\right\} \tag{67}$$

where we have neglected all quantities of order higher than the first in $d\tau$. On the other hand, from the equation of transfer it directly follows that

$$I(d\tau, \mu, \varphi) = \frac{F}{4\mu}\Big(1+\frac{d\tau}{\mu}\Big) S(\tau_1;\ \mu, \varphi;\ \mu_0, \varphi_0) - \frac{d\tau}{\mu} B(0, \mu, \varphi). \tag{68}$$

Combining equations (67) and (68), we have

$$\left.\begin{aligned} \tfrac{1}{4}F\Big[\Big(\frac{1}{\mu}+\frac{1}{\mu_0}\Big) S(\tau_1;\ \mu, \varphi;\ \mu_0, \varphi_0) + \frac{\partial S(\tau_1;\ \mu, \varphi;\ \mu_0, \varphi_0)}{\partial \tau_1}\Big] \\ = B(0, \mu, \varphi) + \frac{1}{4\pi}\int_0^1\int_0^{2\pi} S(\tau_1;\ \mu, \varphi;\ \mu', \varphi') B(0, -\mu', \varphi') \frac{d\mu'}{\mu'}\, d\varphi'. \end{aligned}\right\} \tag{69}$$

In addition to equation (69), there is a further relation which can be derived from considerations relating to the radiation field present at the depth $d\tau$. The relation arises from the fact that the transmission of the flux πF incident on $\tau = 0$ by the entire atmosphere of optical thickness τ_1 must be the same as the transmission of the radiations (64) and (65) by the atmosphere of optical thickness $\tau_1 - d\tau$ below the level $d\tau$. Of the flux (64), the amount transmitted in the direction $(-\mu, \varphi)$ by the atmosphere below $d\tau$ is given by

$$\frac{F}{4\mu}\Big(1-\frac{d\tau}{\mu_0}\Big) T(\tau_1 - d\tau;\ \mu, \varphi;\ \mu_0, \varphi_0), \tag{70}$$

or, to the first order in $d\tau$,

$$\frac{F}{4\mu}\Big[\Big(1-\frac{d\tau}{\mu_0}\Big) T(\tau_1;\ \mu, \varphi;\ \mu_0, \varphi_0) - \frac{\partial T(\tau_1;\ \mu, \varphi;\ \mu_0, \varphi_0)}{\partial \tau_1}\, d\tau\Big]. \tag{71}$$

On the other hand, the transmission of the diffuse field (65) will contribute to the radiation in the direction $(-\mu, \varphi)$ the additional intensity

$$\left.\begin{aligned} \frac{d\tau}{4\pi\mu}\int_0^1\int_0^{2\pi} T(\tau_1;\ \mu, \varphi;\ \mu', \varphi') B(0, -\mu', \varphi') \frac{d\mu'}{\mu'}\, d\varphi' \\ + \frac{d\tau}{\mu} B(0, -\mu, \varphi)\, e^{-\tau_1/\mu}, \end{aligned}\right\} \tag{72}$$

where the second term arises from the intensity in the *diffuse* field (65), which is already in the direction $(-\mu, \varphi)$.

Adding contributions (71) and (72) and remembering that this must equal

$$\frac{F}{4\mu} T(\tau_1;\ \mu, \varphi;\ \mu_0, \varphi_0), \tag{73}$$

we obtain

$$\left.\begin{aligned} \tfrac{1}{4}F\Big[\frac{1}{\mu_0} T(\tau_1;\ \mu, \varphi;\ \mu_0, \varphi_0) + \frac{\partial T(\tau_1;\ \mu, \varphi;\ \mu_0, \varphi_0)}{\partial \tau_1}\Big] \\ = B(0, -\mu, \varphi)\, e^{-\tau_1/\mu} + \frac{1}{4\pi}\int_0^1\int_0^{2\pi} T(\tau_1;\ \mu, \varphi;\ \mu', \varphi') B(0, -\mu', \varphi') \frac{d\mu'}{\mu'}\, d\varphi'. \end{aligned}\right\} \tag{74}$$

Consider, next, the radiation field present at the level $\tau_1 - d\tau$. Since there is no outward intensity at $\tau = \tau_1$, at the level $\tau_1 - d\tau$ there must be the intensity

$$\frac{d\tau}{\mu'} B(\tau_1, \mu', \varphi') , \tag{75}$$

in the direction $(+\mu', \varphi')$. The reflection of this radiation by the atmosphere *above* $\tau_1 - d\tau$ will contribute to the intensity in the direction $(-\mu, \varphi)$ the amount

$$\frac{d\tau}{4\pi\mu} \int_0^1 \int_0^{2\pi} S(\tau_1; \mu, \varphi; \mu', \varphi') B(\tau_1, \mu', \varphi') \frac{d\mu'}{\mu'} d\varphi' . \tag{76}$$

In addition, there will be the radiation directly transmitted in the direction $(-\mu, \varphi)$ by the atmosphere above the level $\tau_1 - d\tau$. The intensity arising from this account is given by

$$\frac{F}{4\mu}\left[T(\tau_1; \mu; \varphi; \mu_0, \varphi_0) - \frac{\partial T(\tau_1; \mu, \varphi; \mu_0, \varphi_0)}{\partial \tau_1} d\tau\right] . \tag{77}$$

Hence

$$\left.\begin{aligned} I(\tau_1 - d\tau, -\mu, \varphi) = {} & \frac{F}{4\mu}\left[T(\tau_1; \mu, \varphi; \mu_0, \varphi_0) - \frac{\partial T(\tau_1; \mu, \varphi; \mu_0, \varphi_0)}{\partial \tau_1} d\tau\right] \\ & + \frac{d\tau}{4\pi\mu} \int_0^1 \int_0^{2\pi} S(\tau_1; \mu, \varphi; \mu', \varphi') B(\tau_1, \mu', \varphi') \frac{d\mu'}{\mu'} d\varphi' . \end{aligned}\right\} \tag{78}$$

But this must equal

$$I(\tau_1 - d\tau, -\mu, \varphi) = \frac{F}{4\mu}\left(1 + \frac{d\tau}{\mu}\right) T(\tau_1; \mu, \varphi; \mu_0, \varphi_0) - \frac{d\tau}{\mu} B(\tau_1, -\mu, \varphi) , \tag{79}$$

which follows from the equation of transfer. From equations (78) and (79) we now obtain

$$\left.\begin{aligned} & \tfrac{1}{4}F\left[\frac{1}{\mu} T(\tau_1; \mu, \varphi; \mu_0, \varphi_0) + \frac{\partial T(\tau_1; \mu, \varphi; \mu_0, \varphi_0)}{\partial \tau_1}\right] \\ & \quad = B(\tau_1, -\mu, \varphi) + \frac{1}{4\pi} \int_0^1 \int_0^{2\pi} S(\tau_1; \mu, \varphi; \mu', \varphi') B(\tau_1, \mu', \varphi') \frac{d\mu'}{\mu'} d\varphi' . \end{aligned}\right\} \tag{80}$$

Again, in addition to equation (80), there is a further relation which can be derived from considerations relating to the field present at the level $\tau_1 - d\tau$. The relation arises from the fact that the reflection of the flux πF incident on $\tau = 0$ by the entire atmosphere of optical thickness τ_1 must be the same as the reflection of the same flux by the atmosphere of optical thickness $\tau_1 - d\tau$ and the transmission of the radiation (75) incident on the level $\tau_1 - d\tau$ from below, by the atmosphere above it. In other words, we must have

$$\left.\begin{aligned} \frac{F}{4\mu} S(\tau_1, \mu, \varphi; \mu_0, \varphi_0) = {} & \frac{F}{4\mu}\left[S(\tau_1, \mu, \varphi; \mu_0, \varphi_0) - \frac{\partial S(\tau_1; \mu, \varphi; \mu_0, \varphi_0)}{\partial \tau_1} d\tau\right] \\ & + \frac{d\tau}{4\pi\mu} \int_0^1 \int_0^{2\pi} T(\tau_1; \mu, \varphi; \mu', \varphi') B(\tau_1, \mu', \varphi') \frac{d\mu'}{\mu'} d\varphi' \\ & \qquad + \frac{d\tau}{\mu} B(\tau_1, \mu, \varphi) e^{-\tau_1/\mu} , \end{aligned}\right\} \tag{81}$$

where the three terms on the right-hand side arise, respectively, from the reflection of the flux πF by the atmosphere of optical thickness $\tau_1 - d\tau$, from the *diffuse* transmission of the radiation (75) incident on the level $\tau_1 - d\tau$, and, finally, from the intensity in the field (75) already in the direction (μ, φ). From equation (81) it now follows that

$$\left.\begin{aligned}\tfrac{1}{4}F\frac{\partial S(\tau_1;\mu,\varphi;\mu_0,\varphi_0)}{\partial\tau_1} &= B(\tau_1,\mu,\varphi)\,e^{-\tau_1/\mu}\\ &\quad+\frac{1}{4\pi}\int_0^1\int_0^{2\pi}T(\tau_1;\mu,\varphi;\mu',\varphi')B(\tau_1,\mu',\varphi')\frac{d\mu'}{\mu'}\,d\varphi'.\end{aligned}\right\}\qquad(82)$$

Now, according to equations (61)–(63)

$$\left.\begin{aligned}B(0,\mu,\varphi) &= \tfrac{1}{4}F\Big[p(\mu,\varphi;-\mu_0,\varphi_0)\\ &\quad+\frac{1}{4\pi}\int_0^1\int_0^{2\pi}p(\mu,\varphi;\mu'',\varphi'')\,S(\tau_1;\mu'',\varphi'';\mu_0,\varphi_0)\frac{d\mu''}{\mu''}\,d\varphi''\Big],\end{aligned}\right\}\qquad(83)$$

and

$$\left.\begin{aligned}B(\tau_1,\mu,\varphi) &= \tfrac{1}{4}F\Big[p(\mu,\varphi;-\mu_0,\varphi_0)\,e^{-\tau_1/\mu_0}\\ &\quad+\frac{1}{4\pi}\int_0^1\int_0^{2\pi}p(\mu,\varphi;-\mu'',\varphi'')\,T(\tau_1;\mu'',\varphi'';\mu_0,\varphi_0)\frac{d\mu''}{\mu''}\,d\varphi''\Big].\end{aligned}\right\}\qquad(84)$$

Using equations (83) and (84) in equations (69), (74), (80), and (82), we obtain

$$
\left. \begin{aligned}
\left(\frac{1}{\mu}+\frac{1}{\mu_0}\right) S(\tau_1;\mu,\varphi;\mu_0,\varphi_0) + \frac{\partial S(\tau_1;\mu,\varphi;\mu_0,\varphi_0)}{\partial \tau_1} &= p(\mu,\varphi;-\mu_0,\varphi_0) \\
&+ \frac{1}{4\pi}\int_0^1\int_0^{2\pi} p(\mu,\varphi;\mu'',\varphi'')\, S(\tau_1;\mu'',\varphi'';\mu_0,\varphi_0)\frac{d\mu''}{\mu''}\,d\varphi'' \\
&+ \frac{1}{4\pi}\int_0^1\int_0^{2\pi} S(\tau_1;\mu,\varphi;\mu',\varphi')\, p(\mu',\varphi';\mu_0,\varphi_0)\frac{d\mu'}{\mu'}\,d\varphi' \\
+ \frac{1}{16\pi^2}\int_0^1\int_0^{2\pi}\int_0^1\int_0^{2\pi} S(\tau_1;\mu,\varphi;\mu',\varphi')\, p(-\mu'\ \varphi';\mu'',\varphi'')\, S(\tau_1;\mu'',\varphi'';\mu_0,\varphi_0) &\frac{d\mu'}{\mu'}\,d\varphi'\,\frac{d\mu''}{\mu''}\,d\varphi'',
\end{aligned} \right\} \quad (85)
$$

$$
\left. \begin{aligned}
\frac{\partial S(\tau_1;\mu,\varphi;\mu_0,\varphi_0)}{\partial \tau_1} &= p(\mu,\varphi;-\mu_0,\varphi_0)\exp\left\{-\tau_1\left(\frac{1}{\mu}+\frac{1}{\mu_0}\right)\right\} \\
&+ \frac{e^{-\tau_1/\mu}}{4\pi}\int_0^1\int_0^{2\pi} p(\mu,\varphi;-\mu'',\varphi'')\, T(\tau_1;\mu'',\varphi'';\mu_0,\varphi_0)\frac{d\mu''}{\mu''}\,d\varphi'' \\
&+ \frac{e^{-\tau_1/\mu_0}}{4\pi}\int_0^1\int_0^{2\pi} T(\tau_1;\mu,\varphi;\mu',\varphi')\, p(\mu',\varphi';-\mu_0,\varphi_0)\frac{d\mu'}{\mu'}\,d\varphi' \\
+ \frac{1}{16\pi^2}\int_0^1\int_0^{2\pi}\int_0^1\int_0^{2\pi} T(\tau_1;\mu,\varphi;\mu',\varphi')\, p(\mu',\varphi';-\mu'',\varphi'')\, T(\tau_1;\mu'',\varphi'';\mu_0,\varphi_0) &\frac{d\mu'}{\mu'}\,d\varphi'\,\frac{d\mu''}{\mu''}\,d\varphi'',
\end{aligned} \right\} \quad (86)
$$

$$
\left.\begin{aligned}
\frac{1}{\mu_0} T(\tau_1; \mu, \varphi; \mu_0, \varphi_0) + \frac{\partial T(\tau_1; \mu, \varphi; \mu_0, \varphi_0)}{\partial \tau_1} &= e^{-\tau_1/\mu} p(\mu, \varphi; \mu_0, \varphi_0) \\
&+ \frac{e^{-\tau_1/\mu}}{4\pi} \int_0^1 \int_0^{2\pi} p(-\mu, \varphi; \mu'', \varphi'') S(\tau_1; \mu'', \varphi''; \mu_0, \varphi_0) \frac{d\mu''}{\mu''} d\varphi'' \\
&+ \frac{1}{4\pi} \int_0^1 \int_0^{2\pi} T(\tau_1; \mu, \varphi; \mu', \varphi') p(\mu', \varphi'; \mu_0, \varphi_0) \frac{d\mu'}{\mu'} d\varphi' \\
+ \frac{1}{16\pi^2} \int_0^1 \int_0^{2\pi} \int_0^1 \int_0^{2\pi} T(\tau_1; \mu, \varphi; \mu', \varphi') p(-\mu', \varphi'; \mu'', \varphi'') & S(\tau_1; \mu'', \varphi''; \mu_0, \varphi_0) \frac{d\mu'}{\mu'} d\varphi' \frac{d\mu''}{\mu''} d\varphi'',
\end{aligned}\right\} \quad (87)
$$

and

$$
\left.\begin{aligned}
\frac{1}{\mu} T(\tau_1; \mu, \varphi; \mu_0, \varphi_0) + \frac{\partial T(\tau_1; \mu, \varphi; \mu_0, \varphi_0)}{\partial \tau_1} &= e^{-\tau_1/\mu_0} p(\mu, \varphi; \mu_0, \varphi_0) \\
&+ \frac{1}{4\pi} \int_0^1 \int_0^{2\pi} p(\mu, \varphi; \mu'', \varphi'') T(\tau_1; \mu'', \varphi''; \mu_0, \varphi_0) \frac{d\mu''}{\mu''} d\varphi'' \\
&+ \frac{e^{-\tau_1/\mu_0}}{4\pi} \int_0^1 \int_0^{2\pi} S(\tau_1; \mu, \varphi; \mu', \varphi') p(\mu', \varphi'; -\mu_0, \varphi_0) \frac{d\mu'}{\mu'} d\varphi' \\
+ \frac{1}{16\pi^2} \int_0^1 \int_0^{2\pi} \int_0^1 \int_0^{2\pi} S(\tau_1; \mu, \varphi; \mu', \varphi') p(\mu', \varphi'; -\mu'', \varphi'') & T(\tau_1; \mu'', \varphi''; \mu_0, \varphi_0) \frac{d\mu'}{\mu'} d\varphi' \frac{d\mu''}{\mu''} d\varphi''.
\end{aligned}\right\} \quad (88)
$$

Equations (85)–(88) represent the four functional equations governing the problem of transmission and reflection by a plane-parallel atmosphere of finite optical thickness. A simple inspection of these equations shows that

$$S(\tau_1;\ \mu,\ \varphi;\ \mu_0,\ \varphi_0) = S(\tau_1;\ \mu_0,\ \varphi_0;\ \mu,\ \varphi) \tag{89}$$

and

$$T(\tau_1;\ \mu,\ \varphi;\ \mu_0,\ \varphi_0) = T(\tau_1;\ \mu_0,\ \varphi_0;\ \mu,\ \varphi)\,. \tag{90}$$

Equations (89) and (90) are the expression of the reciprocity principle for the problem on hand.

Finally, we may remark that, when we allow for the polarization of the scattered radiation, we must consider a scattering matrix, $\boldsymbol{S}(\tau_1;\ \mu,\ \varphi;\ \mu_0,\ \varphi_0)$, and a *transmission matrix*, $\boldsymbol{T}(\tau_1;\ \mu,\ \varphi;\ \mu_0,\ \varphi_0)$, in place of the functions $S(\tau_1;\ \mu,\ \varphi;\ \mu_0,\ \varphi_0)$ and $T(\tau_1;\ \mu,\ \varphi;\ \mu_0,\ \varphi_0)$. It is, however, clear that the functional equations satisfied by $\boldsymbol{S}(\tau_1;\ \mu,\ \varphi;\ \mu_0,\ \varphi_0)$ and $\boldsymbol{T}(\tau_1;\ \mu,\ \varphi;\ \mu_0,\ \varphi_0)$ will be of the same forms as equations (85)–(88), only they will be matrix equations in which a *phase matrix* $\boldsymbol{P}(\mu,\ \varphi;\ \mu',\ \varphi')$ will play the same role as the phase function $p(\mu,\ \varphi;\ \mu',\ \varphi')$. Thus, in the case of Rayleigh scattering, the matrix $-\frac{3}{4}\boldsymbol{Q}\,\boldsymbol{J}\,(\mu,\ \varphi;\ -\mu',\ \varphi')$, defined as in Paper XIV, equations (10) and (17), will replace $p(\mu,\ \varphi;\ \mu',\ \varphi')$.

6. *The reduction of the functional equations (85)–(88) for the case in which the phase function is expressible as a series in Legendre polynomials.*—For the case in which the phase function p (cos Θ) is expressible as a series in Legendre polynomials in the form

$$p(\cos\Theta) = \sum_l \varpi_l P_l(\cos\Theta)\,, \tag{91}$$ [14]

the functional equations (85)–(88) can be reduced in the following manner:

First, we may observe that, for a phase function of the form (91),

$$\left.\begin{aligned} p(\mu,\ \varphi;\ \mu',\ \varphi') = \sum_l \varpi_l \Big[& P_l(\mu)\,P_l(\mu') \\ & + 2\sum_{m=1}^{l} \frac{(l-m)!}{(l+m)!} P_l^m(\mu)\,P_l^m(\mu')\cos m(\varphi-\varphi') \Big]\,. \end{aligned}\right\} \tag{92}$$

Rearranging the series on the right-hand side of this equation, we can write

$$p(\mu,\ \varphi;\ \mu',\ \varphi') = \sum_m \Big[\sum_{l=m} \varpi_l^m P_l^m(\mu)\,P_l^m(\mu') \Big] \cos m(\varphi-\varphi')\,, \tag{93}$$

where

$$\varpi_l^m = (2-\delta_{0,m})\,\varpi_l \frac{(l-m)!}{(l+m)!} \qquad (l = m,\ m+1\ ,\ \ldots.) \tag{94}$$

and

$$\left.\begin{aligned} \delta_{0,m} &= 1 \quad \text{if} \quad m = 0 \\ &= 0 \quad \text{otherwise}\,. \end{aligned}\right\} \tag{95}$$

From the expansion (93) for the phase function, it follows that the scattering and the transmission functions, $S(\tau_1;\ \mu,\ \varphi;\ \mu_0,\ \varphi_0)$ and $T(\tau_1;\ \mu,\ \varphi;\ \mu_0,\ \varphi_0)$, must be expressible in the forms

$$S(\tau_1;\ \mu,\ \varphi;\ \mu_0,\ \varphi_0) = \sum_m S^{(m)}(\tau_1;\ \mu,\ \mu_0)\cos m(\varphi-\varphi_0) \tag{96}$$

[14] In all practical cases the series on the right-hand side will be a terminating one; but it is not necessary to make this restriction at this point.

and

$$T(\tau_1;\mu,\varphi;\mu_0,\varphi_0)=\sum_m T^{(m)}(\tau_1;\mu,\mu_0)\cos m(\varphi-\varphi_0)\,, \tag{97}$$

where, as the notation implies, $S^{(m)}$ and $T^{(m)}$ are functions of τ_1, μ, and μ_0 only.

Substituting the forms (96) and (97) for S and T in equations (85)–(88), we find that the equations for the various Fourier components separate and that they can be further reduced to the following forms:

$$\left.\begin{aligned}&\left(\frac{1}{\mu}+\frac{1}{\mu_0}\right)S^{(m)}(\tau_1;\mu,\mu_0)+\frac{\partial S^{(m)}(\tau_1;\mu,\mu_0)}{\partial\tau_1}\\&=\sum_{l=m}\varpi_l^m(-1)^{m+l}\left[P_l^m(\mu)+\frac{(-1)^{m+l}}{2(2-\delta_{0,m})}\int_0^1 S^{(m)}(\tau_1;\mu,\mu')P_l^m(\mu')\frac{d\mu'}{\mu'}\right]\\&\qquad\times\left[P_l^{(m)}(\mu_0)+\frac{(-1)^{m+l}}{2(2-\delta_{0,m})}\int_0^1 P_l^m(\mu'')S^{(m)}(\tau_1;\mu'',\mu_0)\frac{d\mu''}{\mu''}\right],\end{aligned}\right\}\tag{98}$$

$$\left.\begin{aligned}&\frac{\partial S^{(m)}(\tau_1;\mu,\mu_0)}{\partial\tau_1}=\sum_{l=m}\varpi_l^m(-1)^{m+l}\\&\qquad\times\left[e^{-\tau_1/\mu}P_l^m(\mu)+\frac{1}{2(2-\delta_{0,m})}\int_0^1 T^{(m)}(\tau_1;\mu,\mu')P_l^m(\mu')\frac{d\mu'}{\mu'}\right]\\&\qquad\times\left[e^{-\tau_1/\mu_0}P_l^m(\mu_0)+\frac{1}{2(2-\delta_{0,m})}\int_0^1 P_l^m(\mu'')T^{(m)}(\tau_1;\mu'',\mu_0)\frac{d\mu''}{\mu''}\right],\end{aligned}\right\}\tag{99}$$

$$\left.\begin{aligned}&\frac{1}{\mu_0}T^{(m)}(\tau_1;\mu,\mu_0)+\frac{\partial T^{(m)}(\tau_1;\mu,\mu_0)}{\partial\tau_1}\\&=\sum_{l=m}\varpi_l^m\left[P_l^m(\mu_0)+\frac{(-1)^{m+l}}{2(2-\delta_{0,m})}\int_0^1 P_l^m(\mu'')S^{(m)}(\tau_1;\mu'',\mu_0)\frac{d\mu''}{\mu''}\right]\\&\qquad\times\left[e^{-\tau_1/\mu}P_l^m(\mu)+\frac{1}{2(2-\delta_{0,m})}\int_0^1 T^{(m)}(\tau_1;\mu,\mu')P_l^m(\mu')\frac{d\mu'}{\mu'}\right],\end{aligned}\right\}\tag{100}$$

and

$$\left.\begin{aligned}&\frac{1}{\mu}T^{(m)}(\tau_1;\mu,\mu_0)+\frac{\partial T^{(m)}(\tau_1;\mu,\mu_0)}{\partial\tau_1}\\&=\sum_{l=m}\varpi_l^m\left[P_l^m(\mu)+\frac{(-1)^{m+l}}{2(2-\delta_{0,m})}\int_0^1 S^{(m)}(\tau_1;\mu,\mu')P_l^m(\mu')\frac{d\mu'}{\mu'}\right]\\&\qquad\times\left[e^{-\tau_1/\mu_0}P_l^m(\mu_0)+\frac{1}{2(2-\delta_{0,m})}\int_0^1 P_l^m(\mu'')T^{(m)}(\tau_1;\mu'',\mu_0)\frac{d\mu''}{\mu''}\right].\end{aligned}\right\}\tag{101}$$

If we now let

$$\psi_l^m(\tau_1,\mu)=P_l^m(\mu)+\frac{(-1)^{m+l}}{2(2-\delta_{0,m})}\int_0^1 S^{(m)}(\tau_1;\mu,\mu')P_l^m(\mu')\frac{d\mu'}{\mu'} \tag{102}$$

and

$$\phi_l^m(\tau_1,\mu)=e^{-\tau_1/\mu}P_l^m(\mu)+\frac{1}{2(2-\delta_{0,m})}\int_0^1 T^{(m)}(\tau_1;\mu,\mu')P_l^m(\mu')\frac{d\mu'}{\mu'}, \tag{103}$$

then, in view of equations (89) and (90), we can re-write equations (98)–(101) in the forms

$$\left.\begin{aligned}&\left(\frac{1}{\mu}+\frac{1}{\mu_0}\right) S^{(m)}(\tau_1;\mu,\mu_0)\\&\qquad+\frac{\partial S^{(m)}(\tau_1;\mu,\mu_0)}{\partial\tau_1}=\sum_{l=m}\varpi_l^m(-1)^{m+l}\psi_l^m(\tau_1,\mu)\psi_l^m(\tau_1,\mu_0),\end{aligned}\right\}\tag{104}$$

$$\frac{\partial S^{(m)}(\tau_1;\mu,\mu_0)}{\partial\tau_1}=\sum_{l=m}\varpi_l^m(-1)^{m+l}\phi_l^m(\tau_1,\mu)\phi_l^m(\tau_1,\mu_0),\tag{105}$$

$$\frac{1}{\mu}T^{(m)}(\tau_1;\mu,\mu_0)+\frac{\partial T^{(m)}(\tau_1;\mu,\mu_0)}{\partial\tau_1}=\sum_{l=m}\varpi_l^m\psi_l^m(\tau_1,\mu)\phi_l^m(\tau_1,\mu_0)\tag{106}$$

and

$$\frac{1}{\mu_0}T^{(m)}(\tau_1;\mu,\mu_0)+\frac{\partial T^{(m)}(\tau_1;\mu,\mu_0)}{\partial\tau_1}=\sum_{l=m}\varpi_l^m\psi_l^m(\tau_1,\mu_0)\phi_l^m(\tau_1,\mu).\tag{107}$$

Alternative forms of the foregoing equations are

$$\left.\begin{aligned}&\left(\frac{1}{\mu}+\frac{1}{\mu_0}\right) S^{(m)}(\tau_1;\mu,\mu_0)\\&\quad=\sum_{l=m}\varpi_l^m(-1)^{m+l}\left[\psi_l^m(\tau_1,\mu)\psi_l^m(\tau_1,\mu_0)-\phi_l^m(\tau_1,\mu)\phi_l^m(\tau_1,\mu_0)\right],\end{aligned}\right\}\tag{108}$$

$$\left.\begin{aligned}&\left(\frac{1}{\mu}-\frac{1}{\mu_0}\right) T^{(m)}(\tau_1;\mu,\mu_0)\\&\quad=\sum_{l=m}\varpi_l^m\left[\psi_l^m(\tau_1,\mu)\phi_l^m(\tau_1,\mu_0)-\psi_l^m(\tau_1,\mu_0)\phi_l^m(\tau_1,\mu)\right],\end{aligned}\right\}\tag{109}$$

$$\frac{\partial S^{(m)}(\tau_1;\mu,\mu_0)}{\partial\tau_1}=\sum_{l=m}\varpi_l^m(-1)^{m+l}\phi_l^m(\tau_1,\mu)\phi_l^m(\tau_1,\mu_0),\tag{110}$$

and

$$\left.\begin{aligned}&\left(\frac{1}{\mu}-\frac{1}{\mu_0}\right)\frac{\partial T^{(m)}(\tau_1;\mu,\mu_0)}{\partial\tau_1}\\&\quad=\sum_{l=m}\varpi_l^m\left[\frac{1}{\mu}\psi_l^m(\tau_1,\mu_0)\phi_l^m(\tau_1,\mu)-\frac{1}{\mu_0}\psi_l^m(\tau_1,\mu)\phi_l^m(\tau_1,\mu_0)\right].\end{aligned}\right\}\tag{111}$$

Finally, substituting for $S^{(m)}(\tau_1;\mu,\mu')$ and $T^{(m)}(\tau_1;\mu,\mu')$ from equations (108) and (109) in equations (102) and (103), we obtain

$$\left.\begin{aligned}\psi_l^m(\tau_1,\mu)&=P_l^m(\mu)+\frac{1}{2}\sum_{k=m}(-1)^{k+l}\varpi_k\frac{(k-m)!}{(k+m)!}\mu\\&\times\int_0^1\frac{d\mu'}{\mu+\mu'}\left[\psi_k^m(\tau_1,\mu)\psi_k^m(\tau_1,\mu')-\phi_k^m(\tau_1,\mu)\phi_k^m(\tau_1,\mu')\right]P_l^m(\mu')\end{aligned}\right\}\tag{112}$$

and

$$\left.\begin{aligned}\phi_l^m(\tau_1,\mu) &= P_l^m(\mu)\,e^{-\tau_1/\mu}+\frac{1}{2}\sum_{k=m}\varpi_k\frac{(k-m)!}{(k+m)!}\,\mu\\ &\times\int_0^1\frac{d\mu'}{\mu-\mu'}\left[\psi_k^m(\tau_1,\mu')\,\phi_k^m(\tau_1,\mu)-\psi_k^m(\tau_1,\mu)\,\phi_k^m(\tau_1,\mu')\right]P_l^m(\mu')\,,\end{aligned}\right\}\quad(113)$$

where we have further substituted for ϖ_k^m in accordance with equation (94).

7. *The functional equations for the case of isotropic scattering.*—In the case of isotropic scattering,

$$\varpi_l = 0 \qquad (l\neq 0)\,, \quad (114)$$

and equations (108)–(113) reduce to

$$\left(\frac{1}{\mu}+\frac{1}{\mu_0}\right)S(\tau_1;\mu,\mu_0) = \varpi_0\left[\psi(\tau_1,\mu)\,\psi(\tau_1,\mu_0)-\phi(\tau_1,\mu)\,\phi(\tau_1,\mu_0)\right], \quad (115)$$

$$\left(\frac{1}{\mu}-\frac{1}{\mu_0}\right)T(\tau_1;\mu,\mu_0) = \varpi_0\left[\psi(\tau_1,\mu)\,\phi(\tau_1,\mu_0)-\psi(\tau_1,\mu_0)\,\phi(\tau_1,\mu)\right], \quad (116)$$

$$\frac{\partial S(\tau_1;\mu,\mu_0)}{\partial\tau_1} = \varpi_0\phi(\tau_1,\mu)\,\phi(\tau_1,\mu_0)\,, \quad (117)$$

$$\left(\frac{1}{\mu}-\frac{1}{\mu_0}\right)\frac{\partial T(\tau_1;\mu,\mu_0)}{\partial\tau_1} = \varpi_0\left[\frac{1}{\mu}\psi(\tau_1,\mu_0)\,\phi(\tau_1,\mu)-\frac{1}{\mu_0}\psi(\tau_1,\mu)\,\phi(\tau_1,\mu_0)\right], \quad (118)$$

$$\psi(\tau_1,\mu) = 1+\tfrac{1}{2}\varpi_0\mu\int_0^1\frac{d\mu'}{\mu+\mu'}\left[\psi(\tau_1,\mu)\,\psi(\tau_1,\mu')-\phi(\tau_1,\mu)\,\phi(\tau_1,\mu')\right], \quad (119)$$

and

$$\phi(\tau_1,\mu) = e^{-\tau_1/\mu}+\tfrac{1}{2}\varpi_0\mu\int_0^1\frac{d\mu'}{\mu-\mu'}\left[\psi(\tau_1,\mu')\,\phi(\tau_1,\mu)-\psi(\tau_1,\mu)\,\phi(\tau_1,\mu')\right]. \quad (120)$$

Equations (115), (116), (119), and (120) agree with the equations given by Ambarzumian.[10] But equations (117) and (118) are new.

The question now arises as to whether the solution of the system of equations (115)–(120) can be obtained in closed forms when the integrals on the right-hand sides of equations (119) and (120) are replaced by Gauss sums in the nth approximation. This question is related to the elimination of the constants in the method of solution of the earlier papers of this series (Paper XII, for example) and to the still larger question of whether the systems of functional equations to which we were led in § 6 can be reduced to single (or, more possibly, a pair of) functional equations of standard forms. All these questions, as they arise in connection with transfer problems in semi-infinite atmospheres, were essentially answered in Paper XIV; but they remain to be investigated in the present more general context. (See note added at end of paper.)

8. *Functional equations governing the radiation field in the interior of plane-parallel atmospheres.*—We have concerned ourselves, so far, with only the radiations emerging from the boundaries of plane-parallel atmospheres. But it is clear that the ideas, particularly those leading to the functional equations derived in §§ 2, 4, and 5, can be applied equally to formulate functional equations for the radiation field in the interior. In this section we shall give some examples of such functional equations.

We shall consider, first, the radiation field in an atmosphere of optical thickness τ_1, on which is incident a parallel beam of radiation of flux πF, in the direction $(-\mu_0, \varphi_0)$. At a depth τ in this atmosphere, there will be the incident flux reduced to the amount

$$\pi F\,e^{-\tau/\mu_0}\,, \quad (121)$$

as well as a diffuse radiation field characterized by the intensity $I(\tau, \mu, \varphi)$. To distinguish between the radiation in the outward ($0 \leqslant \mu \leqslant 1$) and the inward ($0 > \mu \geqslant -1$) directions, we shall write

$$I(\tau, +\mu, \varphi) \qquad (0 \leqslant \mu \leqslant 1) \tag{122}$$

and

$$I(\tau, -\mu, \varphi) \qquad (0 < \mu \leqslant 1). \tag{123}$$

Now the atmosphere below τ will reflect the radiations (121) and (123) according to the reflective power of an atmosphere of optical thickness $(\tau_1 - \tau)$ and will contribute to an outward intensity in the direction $(+\mu, \varphi)$ which must equal $I(\tau, +\mu, \varphi)$. In other words,

$$\left.\begin{aligned} I(\tau, +\mu, \varphi) &= \frac{1}{4\mu} F e^{-\tau/\mu_0} S(\tau_1 - \tau; \mu, \varphi; \mu_0, \varphi_0) \\ &\quad + \frac{1}{4\pi\mu} \int_0^1 \int_0^{2\pi} S(\tau_1 - \tau; \mu, \varphi; \mu', \varphi') I(\tau; -\mu', \varphi') d\mu' d\varphi'. \end{aligned}\right\} \tag{124}$$

Similarly, we must have

$$\left.\begin{aligned} I(\tau, -\mu, \varphi) &= \frac{1}{4\mu} FT(\tau; \mu, \varphi; \mu_0, \varphi_0) \\ &\quad + \frac{1}{4\pi\mu} \int_0^1 \int_0^{2\pi} S(\tau; \mu, \varphi; \mu', \varphi') I(\tau, +\mu', \varphi') d\mu' d\varphi', \end{aligned}\right\} \tag{125}$$

expressing the fact that the intensity in the direction $(-\mu, \varphi)$ may be regarded as resulting from the transmission of the incident flux by the part of the atmosphere above τ and the reflection of the radiation (122) incident on τ, from below.

Functional equations of a different sort arise from considerations of the type which led to the functional equations (74) and (82) in § 5. Thus, from the equivalence of the transmission by the atmosphere of optical thickness τ_1 and the transmission of the radiations (121) and (123) by the atmosphere of optical thickness $(\tau_1 - \tau)$ below the level τ, we conclude that

$$\left.\begin{aligned} &\frac{1}{4\mu} FT(\tau_1; \mu, \varphi; \mu_0, \varphi_0) = \\ &\frac{1}{4\mu} F e^{-\tau/\mu_0} T(\tau_1 - \tau; \mu, \varphi; \mu_0, \varphi_0) + I(\tau, -\mu, \varphi) e^{-(\tau_1 - \tau)/\mu} \\ &\quad + \frac{1}{4\pi\mu} \int_0^1 \int_0^{2\pi} T(\tau_1 - \tau; \mu, \varphi; \mu', \varphi') I(\tau; -\mu', \varphi') d\mu' d\varphi'. \end{aligned}\right\} \tag{126}$$

The three terms on the right-hand side of equation (126) represent, respectively, the contributions from the transmission of the flux (121) by the atmosphere below τ, the intensity in the diffuse field (123) already in the direction $(-\mu, \varphi)$ and the diffuse transmission of the field (123) by the atmosphere of optical thickness $(\tau_1 - \tau)$.

Similarly, we must have

$$\left.\begin{aligned} \frac{1}{4\mu} FS(\tau_1; \mu, \varphi; \mu_0, \varphi_0) &= \frac{1}{4\mu} FS(\tau; \mu, \varphi; \mu_0, \varphi_0) + I(\tau + \mu, \varphi) e^{-\tau/\mu} \\ &\quad + \frac{1}{4\pi\mu} \int_0^1 \int_0^{2\pi} T(\tau; \mu, \varphi; \mu', \varphi') I(\tau, +\mu', \varphi') d\mu' d\varphi', \end{aligned}\right\} \tag{127}$$

which expresses the equivalence of the reflection by the atmosphere of optical thickness τ_1 and the reflection by the part of the atmosphere above τ and the transmission of the radiation (122) incident on τ from below.

Equations (126) and (127) can be re-written in the forms

$$\left.\begin{aligned} I(\tau, +\mu, \varphi)\, e^{-\tau/\mu} = \frac{F}{4\mu}\left[S(\tau_1; \mu, \varphi; \mu_0, \varphi_0) - S(\tau; \mu, \varphi; \mu_0, \varphi_0)\right] \\ - \frac{1}{4\pi\mu}\int_0^1\int_0^{2\pi} T(\tau; \mu, \varphi; \mu' \ \varphi')\, I(\tau, +\mu', \varphi')\, d\mu' d\varphi' \end{aligned}\right\} \quad (128)$$

and

$$\left.\begin{aligned} I(\tau, -\mu, \varphi)\, e^{-(\tau_1 - \tau)/\mu} = \\ \frac{F}{4\mu}\left[T(\tau_1; \mu, \varphi; \mu_0, \varphi_0) - e^{-\tau/\mu_0} T(\tau_1 - \tau; \mu, \varphi; \mu_0, \varphi_0)\right] \\ - \frac{1}{4\pi\mu}\int_0^1\int_0^{2\pi} T(\tau_1 - \tau; \mu, \varphi; \mu', \varphi')\, I(\tau, -\mu', \varphi')\, d\mu' d\varphi' . \end{aligned}\right\} \quad (129)$$

It is clear that equations (124) and (125) or (128) and (129) will suffice to determine the radiation field in the interior uniquely, in terms of the scattering and transmission functions of atmospheres of finite optical thicknesses.

The functional equations analogous to equations (124)–(129) for the axially symmetric radiation field in the interior of a semi-infinite atmosphere with a constant net flux are

$$I(\tau, +\mu) = I(0, +\mu) + \frac{1}{2\mu}\int_0^1 S^{(0)}(\infty; \mu, \mu')\, I(\tau, -\mu')\, d\mu' , \quad (130)$$

$$I(0, +\mu) = I(\tau, +\mu)\, e^{-\tau/\mu} + \frac{1}{2\mu}\int_0^1 T^{(0)}(\tau; \mu, \mu')\, I(\tau, +\mu')\, d\mu' , \quad (131)$$

and

$$I(\tau, -\mu) = \frac{1}{2\mu}\int_0^1 S^{(0)}(\tau; \mu, \mu')\, I(\tau, +\mu')\, d\mu' , \quad (132)$$

where $S^{(0)}$ and $T^{(0)}$ are the azimuth independent terms in the scattering and the transmission functions defined as in equation (7). Equations (130), (131), and (132) express, respectively, the invariance of the emergent intensity to the removal of layers of arbitrary optical thickness from the atmosphere, the consideration that the emergent intensity may be regarded as the transmission of the radiation $I(\tau, +\mu)$ by the atmosphere above the level τ, and the fact that the inward intensity prevailing at any level arises in consequence of the reflection of the outward radiation field by the atmosphere above τ.

It is evident that the functional equations (124)–(132), together with the equations (85)–(88), satisfied by the scattering and the transmission functions, are entirely equivalent to the transfer problems formulated in terms of the equations of transfer and boundary conditions.

In a later paper we propose to undertake a detailed study of the various functional equations which we have formulated in this paper; but it is of interest to recall, meantime, that the basic ideas underlying the formulation of these functional equations resemble those which were introduced by Sir George Stokes and Lord Rayleigh in their treatment of the reflection and transmission of light by piles of plates.[15]

Note added April 15.—The questions raised at the end of § 7 have now been answered. In particular, solutions in closed forms for equations (119) and (120) have been found when the integrals on the right-hand sides are replaced by the corresponding Gauss sums. It is hoped to publish the results of these investigations in the near future.

[15] *Mathematical and Physical Papers of Sir George Stokes*, IV (Cambridge, England, 1904), 145; and *Scientific Papers of Lord Rayleigh*, VI (Cambridge, England, 1920), 492. I am indebted to Sir K. S. Krishnan for drawing my attention to these early investigations.

14

ON THE RADIATIVE EQUILIBRIUM OF A STELLAR ATMOSPHERE. XX

S. Chandrasekhar
Yerkes Observatory
Received June 20, 1947

ABSTRACT

In this paper the exact solution for a standard problem in the theory of line formation in stellar atmospheres is obtained. The problem allows for the combined effects of monochromatic line scattering and general continuous absorption but assumes that the ratio of the line (σ_ν) to the continuous (κ_ν) absorption coefficient is constant through the atmosphere and that the Planck intensity, B_ν, is a linear function of the optical thickness in ($\sigma_\nu + \kappa_\nu$).

The exact solution involves the H-functions already tabulated in Paper XIX and their first and second moments. A table of the latter is provided in this paper.

1. *Introduction.*—As is well known, an important case in the theory of line formation in stellar atmospheres arises when the combined effects of monochromatic line scattering and general continuous absorption are considered. The equation of transfer appropriate to these conditions is

$$\mu \frac{dI_\nu}{dt_\nu} = I_\nu - \tfrac{1}{2}(1-\lambda_\nu)\int_{-1}^{+1} I_\nu(t_\nu, \mu')\,d\mu' - \lambda_\nu B_\nu , \tag{1}$$

where B_ν is the Planck intensity at the prevailing temperature, t_ν is the optical thickness in the combined line (σ_ν) and continuous (κ_ν) absorption coefficients, and

$$\lambda_\nu = \frac{\kappa_\nu}{\sigma_\nu + \kappa_\nu} . \tag{2}$$

An assumption which appears not unreasonable in the context of line formation is that B_ν is a linear function of the optical thickness, τ_ν, in the *continuum*. In other words, we may write

$$B_\nu = B_\nu^{(10)} + B_\nu^{(10)} \tau_\nu , \tag{3}$$

where $B_\nu^{(0)}$ and $B_\nu^{(1)}$ are certain appropriately chosen constants. (It may be remarked here that in $B_\nu^{(0)}$ and $B_\nu^{(1)}$ we may replace ν by the frequency at the center of the line, ν_0, without introducing any sensible error; this cannot, of course, be done for I_ν and λ_ν.) If we now further suppose (and this is not always an equally justifiable assumption) that σ_ν/κ_ν is constant through the atmosphere,[1] then

$$B_\nu = B_\nu^{(0)} + B_\nu^{(1)} \lambda_\nu t_\nu . \tag{4}$$

With B_ν given by equation (4), the equation of transfer becomes

$$\mu \frac{dI}{dt} = I - \tfrac{1}{2}(1-\lambda)\int_{-1}^{+1} I(t, \mu')\,d\mu' - \lambda\left[B^{(0)} + \lambda B^{(1)} t\right] , \tag{5}$$

where, for the sake of convenience, we have suppressed the subscript ν to the various quantities.

[1] For a theory along the lines of this series of papers, in which this assumption is not made, see M. Tuberg, *Ap. J.*, **103**, 145, 1946. The basic ideas in this context are, however, due to B. Strömgren, *Ap. J.*, **86**, 1, 1937.

Equation (5) has generally been regarded as one of the fundamental equations in the theory of stellar atmospheres.

The solution of equation (5) in the method of approximation of the earlier papers of this series was given in Paper II (§ 7),[2] and numerical results relating to the solution in the second and third approximations will be found in Papers IV and VI.[3] However, in this paper we shall show how the ideas developed in Paper XIV[4] enable us to pass to the limit of infinite approximation and obtain the exact solution for this problem in the theory of line formation.

2. *The solution for the emergent intensity in closed form in the* n*th approximation.*—The solution of equation (5) in the nth approximation can be written in the following form (Paper II, eq. [90]):

$$I_i = \lambda B^{(1)} \left[\sum_{a=1}^{n} \frac{L_a e^{-k_a \tau}}{1 + k_a \mu_i} + \mu_i + t + \frac{B^{(0)}}{\lambda B^{(1)}} \right] \qquad (i = \pm 1, \ldots, \pm n), \quad (6)$$

where the k_a's ($a = 1, \ldots, n$) are the positive roots of the characteristic equation

$$1 = (1 - \lambda) \sum_{j=1}^{n} \frac{a_j}{1 - \mu_j^2 k^2}; \quad (7)$$

the L_a's ($a = 1, \ldots, n$) are the n constants of integration; and the rest of the symbols have their usual meanings.

Now in transfer problems in semi-infinite plane-parallel atmospheres the boundary conditions which determine the constants of integration and the equation which gives the intensity distribution of the emergent radiation can both be expressed in terms of the same function. Thus, in the present case, the boundary conditions and the emergent intensity $I(0, \mu)$ can be expressed in the following forms:

$$S(\mu_i) = 0, \qquad (i = 1, \ldots, n) \quad (8)$$

and

$$I(0, \mu) = \lambda B^{(1)} S(-\mu), \quad (9)$$

where

$$S(\mu) = \sum_{a=1}^{n} \frac{L_a}{1 - k_a \mu} - \mu + \frac{B^{(0)}}{\lambda B^{(1)}}. \quad (10)$$

In view of the boundary conditions (eq. [8]) we can write

$$S(\mu) = (-1)^{n+1} k_1 \ldots k_n \frac{P(\mu)}{R(\mu)} (\mu - c), \quad (11)$$

where c is a constant and $P(\mu)$ and $R(\mu)$ have their standard meanings.

Moreover, according to equations (10) and (11)

$$L_a = (-1)^{n+1} k_1 \ldots k_n \frac{P(1/k_a)}{R_a(1/k_a)} \left(\frac{1}{k_a} - c \right) \qquad (a = 1, \ldots, n). \quad (12)$$

For the roots of the characteristic equation (7), it can be readily shown that

$$k_1 \ldots k_n \mu_1 \ldots \mu_n = \lambda^{\frac{1}{2}}. \quad (13)$$

[2] *Ap. J.*, **100**, 76, 1944. [3] *Ibid.*, p. 355, and **101**, 320, 1945. [4] *Ibid.*, **105**, 164, 1947.

Equation (11) therefore becomes

$$S(\mu) = -\lambda^{\frac{1}{2}} H(-\mu)(\mu - c). \tag{14}$$

To determine the constant c, we proceed as follows:
First, putting $\mu = 0$ in equations (10) and (11), we have

$$\sum_{\alpha=1}^{n} L_\alpha + \frac{B^{(0)}}{\lambda B^{(1)}} = c\lambda^{\frac{1}{2}}. \tag{15}$$

We next evaluate ΣL_α in accordance with equation (12). Thus

$$\left.\begin{aligned} \sum_{\alpha=1}^{n} L_\alpha &= (-1)^{n+1} k_1 \ldots\ldots k_n \sum_{\alpha=1}^{n} \frac{P(1/k_\alpha)}{R_\alpha(1/k_\alpha)} \left(\frac{1}{k_\alpha} - c\right) \\ &= (-1)^{n+1} k_1 \ldots\ldots k_n f(0), \end{aligned}\right\} \tag{16}$$

where

$$f(x) = \sum_{\alpha=1}^{n} \frac{P(1/k_\alpha)}{R_\alpha(1/k_\alpha)} \left(\frac{1}{k_\alpha} - c\right) R_\alpha(x). \tag{17}$$

Defined in this manner, $f(x)$ is a polynomial of degree $(n - 1)$ in x, which takes the values

$$\left(\frac{1}{k_\alpha} - c\right) P(1/k_\alpha) \tag{18}$$

for $x = 1/k_\alpha$ $(\alpha = 1, \ldots\ldots, n)$. There must, accordingly, exist a relation of the form

$$f(x) = (x - c)P(x) + R(x)(Ax + C), \tag{19}$$

where A and C are certain constants; and the constants can be determined from the condition that the coefficients of x^{n+1} and x^n must vanish on the right-hand side. We thus find that

$$A = \frac{(-1)^{n+1}}{k_1 \ldots\ldots k_n} \quad \text{and} \quad C = \frac{(-1)^n}{k_1 \ldots\ldots k_n} \left(\sum_{j=1}^{n} \mu_j - \sum_{\alpha=1}^{n} \frac{1}{k_\alpha} + c\right). \tag{20}$$

Hence

$$f(0) = (-1)^{n+1} \mu_1 \ldots\ldots \mu_n c + \frac{(-1)^n}{k_1 \ldots\ldots k_n} \left(\sum_{j=1}^{n} \mu_j - \sum_{\alpha=1}^{n} \frac{1}{k_\alpha} + c\right), \tag{21}$$

and (cf. eqs. [13] and [16])

$$\sum_{\alpha=1}^{n} L_\alpha = c\lambda^{\frac{1}{2}} + \left(\sum_{\alpha=1}^{n} \frac{1}{k_\alpha} - \sum_{j=1}^{n} \mu_j - c\right). \tag{22}$$

Inserting this value of ΣL_α in equation (15), we obtain

$$c = \frac{B^{(0)}}{\lambda B^{(1)}} + \sum_{\alpha=1}^{n} \frac{1}{k_\alpha} - \sum_{j=1}^{n} \mu_j. \tag{23}$$

Finally, substituting for c from equation (23) in equation (14), we have

$$S(\mu) = -\lambda^{\frac{1}{2}} H(-\mu)\left[\mu - \left(\frac{B^{(0)}}{\lambda B^{(1)}} + \sum_{a=1}^{n} \frac{1}{k_a} - \sum_{j=1}^{n} \mu_j\right)\right]. \tag{24}$$

Equation (9) governing the intensity distribution of the emergent radiation now becomes

$$I(0, \mu) = \lambda^{3/2} B^{(1)} H(\mu)\left(\mu + \frac{B^{(0)}}{\lambda B^{(1)}} + \sum_{a=1}^{n} \frac{1}{k_a} - \sum_{j=1}^{n} \mu_j\right). \tag{25}$$

This is the required solution in closed form.

3. *Passage to the limit of infinite approximation.*—According to theorem 2 of Paper XIV (§ 11), any H-function defined in the usual manner in terms of the points of the Gaussian division and the roots of a characteristic equation becomes in the limit of infinite approximation the solution of a functional equation of the standard form,

$$H(\mu) = 1 + \mu H(\mu) \int_0^1 \frac{H(\mu')}{\mu + \mu'} \Psi(\mu') \, d\mu', \tag{26}$$

considered in Paper XIV (Sec. III). Consequently, from equation (25) we conclude that the exact solution for the emergent intensity for the problem under consideration must be of the form

$$I(0, \mu) = \lambda^{3/2} B^{(1)} H(\mu)\left(\mu + \frac{B^{(0)}}{\lambda B^{(1)}} + q\right), \tag{27}$$

where $H(\mu)$ is now defined as a solution of a functional equation of the form (26) with the characteristic function

$$\Psi(\mu) = \tfrac{1}{2}(1 - \lambda) = \text{constant}, \tag{28}$$

and

$$q = \lim_{n \to \infty} \left(\sum_{a=1}^{n} \frac{1}{k_a} - \sum_{j=1}^{n} \mu_j\right). \tag{29}$$

The exact H-functions characterizing the problem of line formation are therefore the same as the ones which occur in the problem of diffuse reflection by a semi-infinite plane-parallel atmosphere with an albedo

$$\varpi_0 = 1 - \lambda. \tag{30}$$

The H-functions tabulated in the preceding paper[5] are therefore equally applicable to this problem. However, to complete the solution we must determine q as defined in equation (29).

4. *Evaluation of the limit* $\left(\sum_1^n k_a^{-1} - \sum_1^n \mu_j\right)$ *as* n $\to \infty$. *The exact solution.*—Consider the function (cf. Paper XIV, § 11)

$$s(\mu) = \sum_{a=1}^{n} \frac{l_a}{1 - k_a \mu} + 1, \tag{31}$$

where the k_a's ($a = 1, \ldots, n$) are the roots of the characteristic equation (7) and the l_a's ($a = 1, \ldots, n$) are n constants determined by the conditions

$$s(\mu_i) = 0 \qquad (i = 1, \ldots, n). \tag{32}$$

[5] Paper XIX, *Ap. J.*, **106**, 143, 1947.

It is known that, defined in this manner (Paper XIV, eqs. [132]–[138]),

$$s(\mu) = k_1 \ldots k_n \mu_1 \ldots \mu_n H(-\mu) . \tag{33}$$

Using relation (13), we have

$$s(\mu) = \lambda^{\frac{1}{2}} H(-\mu) . \tag{34}$$

Moreover, according to equations (31) and (32),

$$l_a = (-1)^n k_1 \ldots k_n \frac{P(1/k_a)}{R_a(1/k_a)} \qquad (a = 1, \ldots, n) . \tag{35}$$

Now consider

$$\sum_{a=1}^{n} \frac{l_a}{k_a} = (-1)^n k_1 \ldots k_n \sum_{a=1}^{n} \frac{P(1/k_a)}{R_a(1/k_a)} \frac{1}{k_a} . \tag{36}$$

It is seen that the summation which occurs on the right-hand side of equation (36) is a special case ($c = 0$) of the one considered in § 2 (eq. [16]). We therefore have (cf. eq. [21])

$$\sum_{a=1}^{n} \frac{l_a}{k_a} = \sum_{j=1}^{n} \mu_j - \sum_{a=1}^{n} \frac{1}{k_a} . \tag{37}$$

But $\Sigma l_a/k_a$ can also be expressed in terms of $H(\mu)$. Thus, consider

$$\sum_{i=1}^{n} a_i \mu_i s(-\mu_i) . \tag{38}$$

Since $s(+\mu_i) = 0$ $(i = 1, \ldots, n)$, we can extend the summation in (38) for negative values of i also. Hence

$$\sum_{i=1}^{n} a_i \mu_i s(-\mu_i) = \sum_{i=-n}^{+n} a_i \mu_i s(-\mu_i) . \tag{39}$$ [6]

Now substitute for $s(-\mu_i)$ according to equation (31) on the right-hand side of equation (39). We obtain

$$\left.\begin{aligned} \sum_{i=1}^{n} a_i \mu_i s(-\mu_i) &= \sum_{i=-n}^{+n} a_i \mu_i \left(\sum_{a=1}^{n} \frac{l_a}{1 + k_a \mu_i} + 1 \right) \\ &= \sum_{i=-n}^{+n} a_i \mu_i \left(\sum_{a=1}^{n} \frac{l_a}{1 + k_a \mu_i} \right) ; \end{aligned}\right\} \tag{40}$$

or, inverting the order of the summation, we have

$$\left.\begin{aligned} \sum_{i=1}^{n} a_i \mu_i s(-\mu_i) &= \sum_{a=1}^{n} l_a \left(\sum_{i=-n}^{+n} \frac{a_i \mu_i}{1 + k_a \mu_i} \right) \\ &= \sum_{a=1}^{n} \frac{l_a}{k_a} \sum_{i=-n}^{+n} a_i \left(1 - \frac{1}{1 + k_a \mu_i} \right) . \end{aligned}\right\} \tag{41}$$

[6] In the summation on the right-hand side there is no term with $i = 0$.

Using a well-known property of the Gaussian weights and also the equation defining the characteristic roots (eq. [7]), we have

$$\sum_{i=1}^{n} a_i \mu_i s(-\mu_i) = \left(2 - \frac{2}{1-\lambda}\right) \sum_{\alpha=1}^{n} \frac{l_\alpha}{k_\alpha} = -\frac{2\lambda}{1-\lambda} \sum_{\alpha=1}^{n} \frac{l_\alpha}{k_\alpha}. \tag{42}$$

Hence (cf. eqs. [34] and [37])

$$\sum_{i=1}^{n} a_i \mu_i H(\mu_i) = \frac{2\lambda^{\frac{1}{2}}}{1-\lambda} \left(\sum_{\alpha=1}^{n} \frac{1}{k_\alpha} - \sum_{j=1}^{n} \mu_j \right), \tag{43}$$

or

$$\sum_{\alpha=1}^{n} \frac{1}{k_\alpha} - \sum_{j=1}^{n} \mu_j = \frac{1-\lambda}{2\lambda^{\frac{1}{2}}} \sum_{i=1}^{n} a_i \mu_i H(\mu_i). \tag{44}$$

We now pass to the limit $n \to \infty$. Then $H(\mu)$ becomes the solution of the functional equation

$$H(\mu) = 1 + \tfrac{1}{2}(1-\lambda)\mu H(\mu) \int_0^1 \frac{H(\mu')}{\mu + \mu'} d\mu' \tag{45}$$

and

$$\lim_{n \to \infty} \left(\sum_{\alpha=1}^{n} \frac{1}{k_\alpha} - \sum_{j=1}^{n} \mu_j \right) = \frac{1-\lambda}{2\lambda^{\frac{1}{2}}} \alpha_1 \tag{46}$$

where α_1 is the first moment of $H(\mu)$.

The exact solution for $I(0, \mu)$ is therefore given by

$$I(0, \mu) = \lambda^{3/2} B^{(1)} H(\mu) \left(\mu + \frac{B^{(0)}}{\lambda B^{(1)}} + \frac{1-\lambda}{2\lambda^{\frac{1}{2}}} \alpha_1 \right). \tag{47}$$

5. *Exact formulae for the residual intensity. A table of the moments of* H(μ).—The intensity, $I^{(\text{cont})}(0, \mu)$, in the continuum is obtained by letting $\lambda \to 1$. In this limit

$$H(\mu) \equiv 1 \qquad (\lambda = 1), \tag{48}$$

and

$$I^{(\text{cont})}(0, \mu) = B^{(1)} \left(\mu + \frac{B^{(0)}}{B^{(1)}} \right). \tag{49}$$

(This solution can, of course, be obtained directly from eq. [5].) The residual intensity, r, in the line is therefore given by

$$r(\mu) = \frac{\lambda^{3/2}}{\mu + \dfrac{B^{(0)}}{B^{(1)}}} H(\mu) \left(\mu + \frac{B^{(0)}}{\lambda B^{(1)}} + \frac{1-\lambda}{2\lambda^{\frac{1}{2}}} \alpha_1 \right). \tag{50}$$

Again, according to equation (47), the emergent flux is given by

$$F = 2\lambda^{3/2} B^{(1)} \left(\alpha_2 + \frac{B^{(0)}}{\lambda B^{(1)}} \alpha_1 + \frac{1-\lambda}{2\lambda^{\frac{1}{2}}} \alpha_1^2 \right), \tag{51}$$

where α_2 is the second moment of $H(\mu)$. The residual intensity, R_ν, in the emergent flux is therefore given by

$$R(\nu) = \frac{\lambda^{3/2}}{\dfrac{1}{3} + \dfrac{1}{2} \dfrac{B^{(0)}}{B^{(1)}}} \left(\alpha_2 + \frac{B^{(0)}}{\lambda B^{(1)}} \alpha_1 + \frac{1-\lambda}{2\lambda^{\frac{1}{2}}} \alpha_1^2 \right). \tag{52}$$

As we have already remarked, the H-functions which occur in this problem are the same as those tabulated in Paper XIX. To facilitate the use of solutions (47), (50), and (52) we now provide a table of values of the first and the second moments of $H(\mu)$ for various values of λ (Table 1).

TABLE 1

THE FIRST AND THE SECOND MOMENTS OF $H(\mu)$

ϖ_0	λ	First Moment	Second Moment	ϖ_0	λ	First Moment	Second Moment
0	1.0	0.500000	0.333333	0.7	0.3	0.678674	0.461423
0.1	0.9	.515609	.344357	0.8	.2	0.735808	.503218
.2	0.8	.533154	.356787	0.9	.1	0.825318	.569449
.3	0.7	.553123	.370985	0.925	.075	0.858734	.594404
.4	0.6	.576210	.387466	0.950	.050	0.901864	.626785
.5	0.5	.603495	.407030	0.975	0.025	0.964471	.674134
0.6	0.4	0.636636	0.430922	1.000	0	1.154701*	0.820352*

* The exact values of these entries are 1.154701 and 0.820354, respectively.

It is a pleasure to record here my indebtedness to Mrs. Frances H. Breen, who has been of very valuable assistance to me in the numerical work of this entire series of papers.

15

ON THE RADIATIVE EQUILIBRIUM OF A STELLAR ATMOSPHERE. XXII

S. CHANDRASEKHAR
Yerkes Observatory
Received November 3, 1947

ABSTRACT

In the present paper exact solutions are found for the problems of diffuse reflection and transmission considered in Paper XXI in a general finite approximation. The method consists in starting with the functional equations of Paper XVII governing the laws of diffuse reflection and transmission; reducing them to pairs of functional equations of the standard form,

$$X(\mu) = 1 + \mu \int_0^1 \frac{\Psi(\mu')}{\mu + \mu'} [X(\mu)X(\mu') - Y(\mu)Y(\mu')]d\mu' \quad \text{(i)}$$

and

$$Y(\mu) = e^{-\tau_1/\mu} + \mu \int_0^1 \frac{\Psi(\mu')}{\mu - \mu'} [Y(\mu)X(\mu') - X(\mu)Y(\mu')]d\mu' , \quad \text{(ii)}$$

where $\Psi(\mu)$ is an even polynomial in μ satisfying the condition

$$\int_0^1 \Psi(\mu)d\mu \leqslant \tfrac{1}{2} ,$$

and τ_1 is the optical thickness of the atmosphere; and, finally, relating in a unique manner the various constants occurring in the solutions with the moments of the X- and Y-functions appropriate for the problem.

There is, however, one important difference between the present theory and the corresponding theory of transfer in semi-infinite atmospheres as developed in Paper XIV. It is that, in all conservative cases of perfect scattering, the solutions of the functional equations incorporating the invariances of the problem are not unique but form a one-parametric family. For the removal of the resulting arbitrariness in the solutions, appeal must be made to the flux and the K-integrals, which conservative problems of perfect scattering always admit.

The paper is divided into five main sections. Section I is devoted to a general study of functional equations of the form (i) and (ii) and to deriving various integral properties of these functions useful in the subsequent analysis. The one-parametric nature of the solution of these equations for the case

$$\int_0^1 \Psi(\mu)d\mu = \tfrac{1}{2}$$

is proved in this section; also the basic correspondence between the solutions of equations (i) and (ii) and the *rational functions* X and Y introduced in Paper XXI is established. The following sections deal with the problem of diffuse reflection and transmission under conditions of (i) isotropic scattering with an albedo $\varpi_0 \leqslant 1$; (ii) scattering in accordance with Rayleigh's phase function; (iii) scattering in accordance with the phase function $\lambda(1 + x \cos \Theta)$; and (iv) Rayleigh scattering with proper allowance for the polarization of the radiation field.

1. *Introduction.*—This paper is a continuation of Paper XXI[1] and completes the theory of diffuse reflection and transmission by plane-parallel atmospheres of finite optical thicknesses. By considering the functional equations for the laws of diffuse reflection and transmission derived in Paper XVII[2] we shall show how the exact solutions for the various problems can be found. Now these functional equations governing the angular distributions of the reflected and the transmitted radiations are simultaneous

[1] *Ap. J.*, **106**, 152, 1947. [2] *Ibid.*, **105**, 441, 1947.

nonlinear nonhomogeneous systems of such high order[3] that they might be considered impossible for practical solution if it were not for the guidance provided by the analysis of Paper XXI regarding the forms of the solutions to be sought. Indeed, it will appear that the solutions of the reflected and the transmitted radiations in the various cases have exactly the same forms as those found in Paper XXI, with, however, the X- and Y-functions occurring in them redefined as solutions of a simultaneous pair of functional equations of the form.

$$X(\mu) = 1 + \mu \int_0^1 \frac{\Psi(\mu')}{\mu + \mu'} [X(\mu) X(\mu') - Y(\mu) Y(\mu')] d\mu' \tag{1}$$

and

$$Y(\mu) = e^{-\tau_1/\mu} + \mu \int_0^1 \frac{\Psi(\mu')}{\mu - \mu'} [Y(\mu) X(\mu') - X(\mu) Y(\mu')] d\mu', \tag{2}$$

where τ_1 denotes the optical thickness of the atmosphere and $\Psi(\mu)$ is an even polynomial in μ, satisfying the condition

$$\int_0^1 \Psi(\mu) d\mu \leqslant \tfrac{1}{2}. \tag{3}$$

Equations (1) and (2) therefore play the same basic role in the theory of radiative transfer in atmospheres of finite optical thicknesses as the equation

$$H(\mu) = 1 + \mu H(\mu) \int_0^1 \frac{\Psi(\mu')}{\mu + \mu'} H(\mu') d\mu' \tag{4}$$

played in the theory of semi-infinite atmospheres.[4] It is, in fact, clear that

$$X(\mu) \rightarrow H(\mu) \quad \text{and} \quad Y(\mu) \rightarrow 0 \quad \text{as} \quad \tau_1 \rightarrow \infty . \tag{5}$$

There is, however, one important respect in which the present theory differs from the theory of radiative transfer in semi-infinite atmospheres, namely, that, in all conservative cases of perfect scattering, the functional equations governing the angular distributions of the emergent radiations derived from the invariances discussed in Paper XVII do not suffice to characterize the physical solutions uniquely; for, as we shall see, the general solutions of the relevant equations have a single arbitrary parameter in them. Thus, for the case

$$\int_0^1 \Psi(\mu) d\mu = \tfrac{1}{2}, \tag{6}$$

we shall show that if $X(\mu)$ and $Y(\mu)$ are solutions of equations (1) and (2), then so are

$$X(\mu) + Q\mu [X(\mu) + Y(\mu)] \tag{7}$$

and

$$Y(\mu) - Q\mu [X(\mu) + Y(\mu)], \tag{8}$$

where Q is an arbitrary constant. Similar ambiguities arise in the solutions of the more complicated systems representing general cases of perfect scattering. The physical origin of this nonuniqueness in the solution is not clear; but we shall see that in all cases the ambiguity can be removed by appealing to the "K-integral,"

$$K = \frac{1}{2} \int_{-1}^{+1} I(\tau, \mu) \mu^2 d\mu = \tfrac{1}{4} \mu_0 F (-\mu_0 e^{-\tau/\mu_0} + \gamma_1 \tau + \gamma_2), \tag{9}$$

[3] For example, in the case of Rayleigh scattering, the order of the system is eight.

[4] See Paper XIV (*Ap. J.*, **105**, 164, 1947); also the author's Josiah Willard Gibbs Lecture in the *Bull. Amer. Math. Soc.*, **53**, 641–711, 1947.

which all conservative problems admit.[5] (In eq. [9], μ_0 is the direction cosine of the angle of incidence of a parallel beam of radiation of net flux πF per unit area normal to itself, and γ_1 and γ_2 are two constants.)

The plan of this paper is as follows:

Section I is devoted to a general study of the functional equations (1) and (2) and to deriving certain relations useful in our subsequent analysis. The ambiguity in the solutions of equations (1) and (2) for the case (6) is proved in this section. The basic correspondence between the solutions of equations (1) and (2) and the *rational functions*, X and Y, introduced in Paper XXI (eqs. [125] and [126]) is also established in this section. Sections II, III, IV, and V deal with the problem of diffuse reflection and transmission under conditions, respectively, of (i) isotropic scattering with an albedo $\varpi_0 \leqslant 1$; (ii) scattering in accordance with Rayleigh's phase function; (iii) scattering in accordance with the phase function $\lambda(1 + x \cos \Theta)$; and, finally, (iv) Rayleigh scattering with proper allowance for the polarization of the radiation field.

I. ON THE FUNCTIONAL EQUATIONS SATISFIED BY X AND Y

2. *Definitions and alternative forms of the basic equations.*—In dealing with the solutions of equations (1) and (2) it is convenient to introduce the following abbreviations:

$$x_n = \int_0^1 X(\mu)\Psi(\mu)\mu^n d\mu\,, \qquad y_n = \int_0^1 Y(\mu)\Psi(\mu)\mu^n d\mu\,, \tag{10}$$

$$\alpha_n = \int_0^1 X(\mu)\mu^n d\mu\,, \quad \text{and} \quad \beta_n = \int_0^1 Y(\mu)\mu^n d\mu\,; \tag{11}$$

i.e., x_n and y_n are the moments of order n and $X(\mu)$ and $Y(\mu)$, weighted by the *characteristic function* $\Psi(\mu)$, while α_n and β_n are the simple moments themselves.

Certain alternative forms of the basic equations which we shall find useful may also be noted here. Writing

$$\frac{\mu}{\mu+\mu'} = 1 - \frac{\mu'}{\mu+\mu'}, \quad \text{respectively}\,, \quad \frac{\mu}{\mu-\mu'} = 1 + \frac{\mu'}{\mu-\mu'}, \tag{12}$$

in equations (1) and (2), we readily find that

$$\int_0^1 \frac{\mu'\Psi(\mu')}{\mu+\mu'}[X(\mu)X(\mu') - Y(\mu)Y(\mu')]d\mu' = 1 - [(1-x_0)X(\mu) + y_0 Y(\mu)] \tag{13}$$

and

$$\int_0^1 \frac{\mu'\Psi(\mu')}{\mu-\mu'}[Y(\mu)X(\mu') - X(\mu)Y(\mu')]d\mu' = -e^{-\tau_1/\mu} + [y_0 X(\mu) + (1-x_0)Y(\mu)]\,. \tag{14}$$

We also have

$$\int_0^1 \frac{\mu'^2\Psi(\mu')}{\mu+\mu'}[X(\mu)X(\mu') - Y(\mu)Y(\mu')]d\mu' = x_1 X(\mu) - y_1 Y(\mu) - \mu + \mu[(1-x_0)X(\mu) + y_0 Y(\mu)] \tag{15}$$

and

$$\int_0^1 \frac{\mu'^2\Psi(\mu')}{\mu-\mu'}[Y(\mu)X(\mu') - X(\mu)Y(\mu')]d\mu' = y_1 X(\mu) - x_1 Y(\mu) - \mu e^{-\tau_1/\mu} + \mu[y_0 X(\mu) + (1-x_0)Y(\mu)]\,. \tag{16}$$

[5] In the case of Rayleigh scattering there are two such integrals to be considered (cf. Sec. V).

The foregoing equations can be verified by writing

$$\frac{\mu'^2}{\mu+\mu'}=\mu'-\frac{\mu\mu'}{\mu+\mu'}, \qquad \text{respectively}, \qquad \frac{\mu'^2}{\mu-\mu'}=-\mu'+\frac{\mu\mu'}{\mu-\mu'}, \tag{17}$$

and using equations (13) and (14).

3. *Integrodifferential equations for* X(μ, τ_1) *and* Y(μ, τ_1).—In equations (1) and (2), $0<\tau_1<\infty$ is, of course, to be regarded as some assigned constant. Nevertheless, it is sometimes convenient to emphasize explicitly the dependence of the solutions X and Y on τ_1. We shall then write $X(\mu, \tau_1)$ and $Y(\mu, \tau_1)$ instead of simply as $X(\mu)$ and $Y(\mu)$. And, considered as functions of τ_1 also, X and Y satisfy certain integrodifferential equations which are of importance. We shall state them in the form of the following theorem:

Theorem 1.—If $X(\mu, \tau_1)$ and $Y(\mu, \tau_1)$ are solutions of equations (1) and (2) for a particular value of τ_1, then solutions for other values of τ_1 can be obtained from the integrodifferential equations

$$\begin{aligned}\frac{\partial X(\mu,\tau_1)}{\partial\tau_1}&=Y(\mu,\tau_1)\int_0^1\frac{d\mu'}{\mu'}\Psi(\mu')\,Y(\mu',\tau_1)\\&=y_{-1}(\tau_1)\,Y(\mu,\tau_1)\end{aligned} \tag{18}$$

and

$$\begin{aligned}\frac{\partial Y(\mu,\tau_1)}{\partial\tau_1}+\frac{Y(\mu,\tau_1)}{\mu}&=X(\mu,\tau_1)\int_0^1\frac{d\mu'}{\mu'}\Psi(\mu')\,Y(\mu',\tau_1)\\&=y_{-1}(\tau_1)\,X(\mu,\tau_1)\,.\end{aligned} \tag{19}$$

Proof.—According to equations (18) and (19),

$$\begin{aligned}&\mu\frac{\partial}{\partial\tau_1}\int_0^1\frac{\Psi(\mu')}{\mu+\mu'}[X(\mu)\,X(\mu')-Y(\mu)\,Y(\mu')]\,d\mu'\\&\qquad=\mu\int_0^1\frac{\Psi(\mu')}{\mu+\mu'}\Big\{y_{-1}X(\mu)\,Y(\mu')+y_{-1}X(\mu')\,Y(\mu)\\&\qquad-Y(\mu)\Big[-\frac{Y(\mu')}{\mu'}+y_{-1}X(\mu')\Big]-Y(\mu')\Big[-\frac{Y(\mu)}{\mu}+y_{-1}X(\mu)\Big]\Big\}\,d\mu'.\end{aligned} \tag{20}$$

Hence

$$\mu\frac{\partial}{\partial\tau_1}\int_0^1\frac{\Psi(\mu')}{\mu+\mu'}[X(\mu)\,X(\mu')-Y(\mu)\,Y(\mu')]\,d\mu'=y_{-1}Y(\mu)\,. \tag{21}$$

Similarly,

$$\begin{aligned}&\mu\frac{\partial}{\partial\tau_1}\int_0^1\frac{\Psi(\mu')}{\mu-\mu'}[Y(\mu)\,X(\mu')-X(\mu)\,Y(\mu')]\,d\mu'\\&\qquad=\int_0^1\frac{\Psi(\mu')}{\mu-\mu'}\Big[-Y(\mu)\,X(\mu')+\frac{\mu}{\mu'}X(\mu)\,Y(\mu')\Big]d\mu'.\end{aligned} \tag{22}$$

We therefore have

$$\begin{aligned}&\int_0^1\frac{\Psi(\mu')}{\mu-\mu'}[Y(\mu)\,X(\mu')-X(\mu)\,Y(\mu')]\,d\mu'\\&\qquad+\mu\frac{\partial}{\partial\tau_1}\int_0^1\frac{\Psi(\mu')}{\mu-\mu'}[Y(\mu)\,X(\mu')-X(\mu)\,Y(\mu')]\,d\mu'=y_{-1}X(\mu)\,.\end{aligned} \tag{23}$$

On the other hand, if X and Y are solutions of equations (1) and (2), we must have

$$\frac{\partial X}{\partial \tau_1} = \mu \frac{\partial}{\partial \tau_1} \int_0^1 \frac{\Psi(\mu')}{\mu + \mu'} [X(\mu) X(\mu') - Y(\mu) Y(\mu')] d\mu' \tag{24}$$

and

$$\begin{aligned} \frac{\partial Y}{\partial \tau_1} + \frac{Y}{\mu} = & \int_0^1 \frac{\Psi(\mu')}{\mu - \mu'} [Y(\mu) X(\mu') - X(\mu) Y(\mu')] d\mu' \\ & + \mu \frac{\partial}{\partial \tau_1} \int_0^1 \frac{\Psi(\mu')}{\mu - \mu'} [Y(\mu) X(\mu') - X(\mu) Y(\mu')] d\mu' . \end{aligned} \tag{25}$$

From equations (21), (23), (24), and (25) we now conclude that, if $X(\mu, \tau_1)$ and $Y(\mu, \tau_1)$ are solutions of equations (1) and (2) for a particular value of τ_1, then

$$X(\mu, \tau_1) + y_{-1} Y(\mu, \tau_1) d\tau_1 \tag{26}$$

and

$$Y(\mu, \tau_1) + \left[-\frac{Y(\mu, \tau_1)}{\mu} + y_{-1} X(\mu, \tau_1) \right] d\tau_1 \tag{27}$$

are solutions of the same equations for an infinitesimally larger value of τ_1, namely, $\tau_1 + d\tau_1$. This proves the theorem.

Corollary.—

$$X^2(\mu, \tau_1) - Y^2(\mu, \tau_1) = H^2(\mu) - \frac{2}{\mu} \int_{\tau_1}^{\infty} Y^2(\mu, t) dt . \tag{28}$$

Proof.—Eliminating y_{-1} between equations (18) and (19), we have

$$X \frac{\partial X}{\partial \tau_1} = Y \frac{\partial Y}{\partial \tau_1} + \frac{Y^2}{\mu} \tag{29}$$

or

$$\frac{\partial}{\partial \tau_1} (X^2 - Y^2) = \frac{2}{\mu} Y^2 . \tag{30}$$

Integrating equation (30) and remembering that

$$X(\mu, \tau_1) \to H(\mu) \quad \text{and} \quad Y(\mu, \tau_1) \to 0 \quad \text{as} \quad \tau_1 \to \infty \tag{31}$$

we obtain the result stated.

4. *Some integral properties of the functions* X *and* Y.—As in the case of the H-functions, there are a number of integral theorems (of an essentially elementary kind) which can be proved for functions satisfying equations of the form (1) and (2). The theorems which follow are the analogues for the X- and Y-functions, of the theorems proved for the H-functions in Paper XIV, § 12.

Theorem 2.—

$$\int_0^1 X(\mu) \Psi(\mu) d\mu = 1 - \left[1 - 2 \int_0^1 \Psi(\mu) d\mu + \left\{ \int_0^1 Y(\mu) \Psi(\mu) d\mu \right\}^2 \right]^{1/2} . \tag{32}$$

Proof.—Multiplying the equation satisfied by $X(\mu)$ by $\Psi(\mu)$ and integrating over the range of μ, we have (cf. eq. [10])

$$\begin{aligned} x_0 = & \int_0^1 \Psi(\mu) d\mu \\ & + \int_0^1 \int_0^1 \frac{\mu}{\mu + \mu'} \Psi(\mu) \Psi(\mu') [X(\mu) X(\mu') - Y(\mu) Y(\mu')] d\mu d\mu' . \end{aligned} \tag{33}$$

Interchanging μ and μ' in the double integral on the right-hand side and taking the average of the two equations, we have

$$\begin{aligned} x_0 &= \int_0^1 \Psi(\mu)\, d\mu + \frac{1}{2} \int_0^1 \int_0^1 \Psi(\mu) \Psi(\mu') [X(\mu) X(\mu') - Y(\mu) Y(\mu')]\, d\mu d\mu' \\ &= \int_0^1 \Psi(\mu)\, d\mu + \tfrac{1}{2}(x_0^2 - y_0^2). \end{aligned} \tag{34}$$

Solving this equation for x_0, we have

$$x_0 = 1 \pm \left[1 - 2\int_0^1 \Psi(\mu)\, d\mu + y_0^2\right]^{1/2}. \tag{35}$$

The ambiguity in the sign in equation (35) can be removed by the consideration that the quantity on the right-hand side must uniformly converge to zero when $\Psi(\mu) \to 0$ uniformly in the interval (0, 1). This requires us to choose the negative sign in equation (35), and the result stated follows.

Corollary.—In the conservative case

$$\int_0^1 \Psi(\mu)\, d\mu = \tfrac{1}{2}, \tag{36}$$

we have

$$\int_0^1 [X(\mu) + Y(\mu)] \Psi(\mu)\, d\mu = 1. \tag{37}$$

Theorem 3.—

$$(1 - x_0) x_2 + y_0 y_2 + \tfrac{1}{2}(x_1^2 - y_1^2) = \int_0^1 \Psi(\mu) \mu^2 d\mu. \tag{38}$$

Proof.—Multiplying equation (1) by $\Psi(\mu)\mu^2$ and integrating over the range of μ, we have

$$\begin{aligned} x_2 &= \int_0^1 \Psi(\mu) \mu^2 d\mu \\ &\quad + \int_0^1 \int_0^1 \frac{\mu^3}{\mu + \mu'} \Psi(\mu) \Psi(\mu') [X(\mu) X(\mu') - Y(\mu) Y(\mu')]\, d\mu d\mu' \\ &= \int_0^1 \Psi(\mu) \mu^2 d\mu \\ &\quad + \frac{1}{2} \int_0^1 \int_0^1 (\mu^2 - \mu\mu' + \mu'^2) \Psi(\mu) \Psi(\mu') [X(\mu) X(\mu') - Y(\mu) Y(\mu')]\, d\mu d\mu'. \end{aligned} \tag{39}$$

Hence

$$x_2 = \int_0^1 \Psi(\mu) \mu^2 d\mu + x_2 x_0 - y_2 y_0 - \tfrac{1}{2}(x_1^2 - y_1^2), \tag{40}$$

which is equivalent to equation (38).

Corollary.—In the conservative case,

$$y_0 (x_2 + y_2) + \tfrac{1}{2}(x_1^2 - y_1^2) = \int_0^1 \Psi(\mu) \mu^2 d\mu. \tag{41}$$

This follows from equation (38) and the corollary of theorem 2 (eq. [37]), according to which

$$x_0 + y_0 = 1. \tag{42}$$

Theorem 4.—When the characteristic function $\Psi(\mu)$ has the form

$$\Psi(\mu) = a + b\mu^2 , \tag{43}$$

where a and b are two constants,[6] we have the relations

$$\alpha_0 = 1 + \tfrac{1}{2}[a(\alpha_0^2 - \beta_0^2) + b(\alpha_1^2 - \beta_1^2)] , \tag{44}$$

$$\begin{aligned}(a + b\mu^2)\int_0^1 \frac{d\mu'}{\mu+\mu'}[X(\mu)X(\mu') - Y(\mu)Y(\mu')] \\ = \frac{1}{\mu}[X(\mu) - 1] - b[(\alpha_1 - \mu\alpha_0)X(\mu) - (\beta_1 - \mu\beta_0)Y(\mu)] ,\end{aligned} \tag{45}$$

and

$$\begin{aligned}(a + b\mu^2)\int_0^1 \frac{d\mu'}{\mu-\mu'}[Y(\mu)X(\mu') - X(\mu)Y(\mu')] \\ = \frac{1}{\mu}[Y(\mu) - e^{-\tau_1/\mu}] - b[(\beta_1 + \mu\beta_0)X(\mu) - (\alpha_1 + \mu\alpha_0)Y(\mu)] ,\end{aligned} \tag{46}$$

where α_0, β_0 and α_1, β_1 are moments of order zero and one of $X(\mu)$ and $Y(\mu)$, respectively.

To prove equation (44), we simply integrate the equation satisfied by $X(\mu)$. We find

$$\begin{aligned}\alpha_0 &= 1 + \int_0^1\int_0^1 \frac{(a + b\mu'^2)\mu}{\mu+\mu'}[X(\mu)X(\mu') - Y(\mu)Y(\mu')]\,d\mu\,d\mu' \\ &= 1 + \frac{1}{2}\int_0^1\int_0^1 (a + b\mu\mu')[X(\mu)X(\mu') - Y(\mu)Y(\mu')]\,d\mu\,d\mu' \\ &= 1 + \tfrac{1}{2}[a(\alpha_0^2 - \beta_0^2) + b(\alpha_1^2 - \beta_1^2)] .\end{aligned} \tag{47}$$

The relation (46) can be proved in the following manner:

$$\begin{aligned}a\int_0^1 \frac{d\mu'}{\mu-\mu'}[Y(\mu)X(\mu') - X(\mu)Y(\mu')] \\ = \int_0^1 \frac{a + b\mu'^2}{\mu-\mu'}[Y(\mu)X(\mu') - X(\mu)Y(\mu')]\,d\mu' \\ + b\int_0^1 \left(\mu + \mu' - \frac{\mu^2}{\mu-\mu'}\right)[Y(\mu)X(\mu') - X(\mu)Y(\mu')]\,d\mu' \\ = \frac{1}{\mu}[Y(\mu) - e^{-\tau_1/\mu}] + b[(\alpha_1 + \mu\alpha_0)Y(\mu) - (\beta_1 + \mu\beta_0)X(\mu)] \\ - b\mu^2\int_0^1 \frac{d\mu'}{\mu-\mu'}[Y(\mu)X(\mu') - X(\mu)Y(\mu')] .\end{aligned} \tag{48}$$

Hence the result. Equation (45) follows quite similarly.

5. *The nonuniqueness of the solution in the conservative case. The standard solution.*—We shall now prove the following theorem:

Theorem 5.—In the conservative case,

$$\int_0^1 \Psi(\mu)\,d\mu = \tfrac{1}{2} , \tag{49}$$

[6] The condition $\int_0^1 \Psi(\mu)d\mu \leqslant \frac{1}{2}$ requires that $a + \frac{1}{3}b \leqslant \frac{1}{2}$.

the solutions of equations (1) and (2) are not unique; more particularly, if $X(\mu)$ and $Y(\mu)$ are solutions, then so are

$$X(\mu)+Q\mu[X(\mu)+Y(\mu)] \tag{50}$$

and

$$Y(\mu)-Q\mu[X(\mu)+Y(\mu)], \tag{51}$$

where Q is an arbitrary constant.

Proof.—Writing

$$F(\mu)=X(\mu)+Q\mu[X(\mu)+Y(\mu)] \tag{52}$$

and

$$G(\mu)=Y(\mu)-Q\mu[X(\mu)+Y(\mu)], \tag{53}$$

we verify that

$$\begin{aligned}F(\mu)F(\mu')-G(\mu)G(\mu')&=X(\mu)X(\mu')-Y(\mu)Y(\mu')\\&\quad+Q(\mu+\mu')[X(\mu)+Y(\mu)][X(\mu')+Y(\mu')]\end{aligned} \tag{54}$$

and

$$\begin{aligned}G(\mu)F(\mu')-F(\mu)G(\mu')&=Y(\mu)X(\mu')-X(\mu)Y(\mu')\\&\quad-Q(\mu-\mu')[X(\mu)+Y(\mu)][X(\mu')+Y(\mu')].\end{aligned} \tag{55}$$

Hence

$$\begin{aligned}&\mu\int_0^1\frac{\Psi(\mu')}{\mu+\mu'}[F(\mu)F(\mu')-G(\mu)G(\mu')]\,d\mu'\\&\quad=\mu\int_0^1\frac{\Psi(\mu')}{\mu+\mu'}[X(\mu)X(\mu')-Y(\mu)Y(\mu')]\,d\mu'\\&\quad+Q\mu[X(\mu)+Y(\mu)]\int_0^1[X(\mu')+Y(\mu')]\Psi(\mu')\,d\mu'.\end{aligned} \tag{56}$$

Using equation (1) and the corollary of theorem 2 (eq. [37]), we have

$$\begin{aligned}&\mu\int_0^1\frac{\Psi(\mu')}{\mu+\mu'}[F(\mu)F(\mu')-G(\mu)G(\mu')]\,d\mu'\\&\quad=X(\mu)-1+Q\mu[X(\mu)+Y(\mu)]\\&\quad=F(\mu)-1.\end{aligned} \tag{57}$$

Similarly,

$$\begin{aligned}&\mu\int_0^1\frac{\Psi(\mu')}{\mu-\mu'}[G(\mu)F(\mu')-F(\mu)G(\mu')]\,d\mu'\\&\quad=Y(\mu)-e^{-\tau_1/\mu}-Q\mu[X(\mu)+Y(\mu)]\\&\quad=G(\mu)-e^{-\tau_1/\mu}.\end{aligned} \tag{58}$$

Hence $F(\mu)$ and $G(\mu)$ satisfy the same equations as $X(\mu)$ and $Y(\mu)$, and the theorem follows.

Corollary.—The solutions derivable from a given one according to equations (52) and (53) form a one-parametric family which can be generated by any of its members.

Proof.—Let

$$F_1(\mu)=F(\mu)+Q_1\mu[F(\mu)+G(\mu)] \tag{59}$$

and

$$G_1(\mu)=G(\mu)-Q_1\mu[F(\mu)+G(\mu)], \tag{60}$$

where Q_1 is an arbitrary constant. According to theorem 5, F_1 and G_1 are also solutions of equations (1) and (2). On the other hand, since (cf. eqs. [52] and [53])

$$F(\mu) + G(\mu) = X(\mu) + Y(\mu)\,, \tag{61}$$

we can express F_1 and G_1 alternatively in the forms

$$F_1(\mu) = X(\mu) + (Q+Q_1)\,\mu\,[X(\mu) + Y(\mu)] \tag{62}$$

and

$$G_1(\mu) = Y(\mu) - (Q+Q_1)\,\mu\,[X(\mu) + Y(\mu)]\,. \tag{63}$$

In other words, $F_1(\mu)$ and $G_1(\mu)$ can also be derived directly from $X(\mu)$ and $Y(\mu)$.

It would appear that, in a given conservative case, all the solutions are included in one and only one family. In nonconservative cases, on the other hand, it would seem that the solutions are unique.

In view of the ambiguity in the solutions of equations (1) and (2) in conservative cases, it would be convenient to select, in each case, a particular member of the one-parametric family of solutions as a *standard solution.*

Definition.—In a conservative case we shall define the solutions which have the property

$$x_0 = \int_0^1 X(\mu)\Psi(\mu)\,d\mu = 1 \tag{64}$$

and

$$y_0 = \int_0^1 Y(\mu)\Psi(\mu)\,d\mu = 0 \tag{65}$$ [7]

as the standard solutions of equations (1) and (2).

Such solutions can always be found; for, if a particular X and Y do not satisfy equations (64) and (65), we can always find a Q such that the solutions derived from X and Y in the manner of equations (52) and (53) have the required property. Standard solutions defined in this manner have several interesting properties. We shall state them in the form of the following theorems:

Theorem 6.—The standard solutions are invariant to increments of τ_1 according to the integrodifferential equations of theorem 1.

Multiplying equations (18) and (19) by $\Psi(\mu)$ and integrating over the range of μ, we have

$$\frac{dx_0}{d\tau_1} = y_0 y_{-1} = 0 \tag{66}$$

and

$$\frac{dy_0}{d\tau_1} = -(1-x_0)\,y_{-1} = 0\,. \tag{67}$$

Theorem 7.—Let $X(\mu, \tau_1)$ and $Y(\mu, \tau_1)$ denote the standard solutions of equations (1) and (2) in a conservative case for a particular value of τ_1. Consider the solutions

$$F(\mu, \tau_1) = X(\mu, \tau_1) + Q\mu\,[X(\mu, \tau_1) + Y(\mu, \tau_1)] \tag{68}$$

and

$$G(\mu, \tau_1) = Y(\mu, \tau_1) - Q\mu\,[X(\mu, \tau_1) + Y(\mu, \tau_1)] \tag{69}$$

of equations (1) and (2) derived from X and Y and continue them for other values of τ_1 according to the equations of theorem 1. These solutions for other values of τ_1 can, in turn, be derived from the standard solutions appropriate for these values of τ_1 with vary-

[7] Since $x_0 + y_0 = 1$, eq. (64) implies eq. (65) and vice versa.

ing values of Q. The quantity Q, considered as a function of τ_1 in this manner, satisfies the differential equation

$$\frac{d}{d\tau_1}\left(\frac{1}{Q}\right)-\frac{2y_{-1}}{Q}=-1\,. \tag{70}$$

Proof.—According to equations (18) and (19),

$$\frac{\partial F}{\partial \tau_1}=G\int_0^1 \frac{d\mu'}{\mu'}\Psi(\mu')\,G(\mu')\,, \tag{71}$$

and

$$\frac{\partial G}{\partial \tau_1}+\frac{G}{\mu}=F\int_0^1 \frac{d\mu'}{\mu'}\Psi(\mu')\,G(\mu')\,. \tag{72}$$

Now (cf. eq. [69])

$$\begin{aligned}\int_0^1 \frac{d\mu'}{\mu'}\Psi(\mu')\,G(\mu') &= \int_0^1 \frac{d\mu'}{\mu'}\Psi(\mu')\,Y(\mu')-Q\int_0^1 [X(\mu')+Y(\mu')]\Psi(\mu')\,d\mu' \\ &= y_{-1}-Q\,.\end{aligned} \tag{73}$$

Hence

$$\frac{\partial F}{\partial \tau_1}=(y_{-1}-Q)\,G\,, \tag{74}$$

and

$$\frac{\partial G}{\partial \tau_1}+\frac{G}{\mu}=(y_{-1}-Q)\,F\,. \tag{75}$$

On the other hand, since X and Y remain standard solutions when continued for other values of τ_1, we must have

$$\begin{aligned}\frac{\partial F}{\partial \tau_1}&=\frac{\partial X}{\partial \tau_1}+Q\mu\left(\frac{\partial X}{\partial \tau_1}+\frac{\partial Y}{\partial \tau_1}\right)+\mu(X+Y)\frac{dQ}{d\tau_1} \\ &=y_{-1}Y+Q\mu\left(y_{-1}Y-\frac{Y}{\mu}+y_{-1}X\right)+\mu(X+Y)\frac{dQ}{d\tau_1} \\ &=(y_{-1}-Q)\,Y+\mu(X+Y)\left(y_{-1}Q+\frac{dQ}{d\tau_1}\right).\end{aligned} \tag{76}$$

We can re-write the foregoing equation in the form

$$\frac{\partial F}{\partial \tau_1}=(y_{-1}-Q)\,[\,Y-Q\mu(X+Y)]+\mu(X+Y)\left[Q(y_{-1}-Q)+y_{-1}Q+\frac{dQ}{d\tau_1}\right], \tag{77}$$

or

$$\frac{\partial F}{\partial \tau_1}=(y_{-1}-Q)\,G+\mu(X+Y)\left(2y_{-1}Q-Q^2+\frac{dQ}{d\tau_1}\right). \tag{78}$$

Comparing equations (74) and (78), we must have

$$\frac{dQ}{d\tau_1}+2y_{-1}Q-Q^2=0\,. \tag{79}$$

A similar consideration of the equation for $\partial G/\partial\tau_1$ leads to the same equation for Q.

Equation (79) can be re-written in the form

$$\frac{1}{Q^2}\frac{dQ}{d\tau_1}+\frac{2y_{-1}}{Q}=1\,, \tag{80}$$

which is equivalent to equation (70).

The various relations (eqs. [13]–[16] and [41]) derived in the preceding sections for solutions of equations (1) and (2) in general take particularly simple forms for standard solutions of conservative cases. We shall collect these relations in the form of the following theorem:

Theorem 8.—For the standard solutions in a conservative case we have the relations

$$x_0=1\,,\qquad y_0=0\,, \tag{81}$$

$$x_1^2-y_1^2=2\int_0^1\Psi(\mu)\,\mu^2d\mu\,, \tag{82}$$

$$\int_0^1\frac{\mu'\Psi(\mu')}{\mu+\mu'}\left[X(\mu)\,X(\mu')-Y(\mu)\,Y(\mu')\right]d\mu'=1\,, \tag{83}$$

$$\int_0^1\frac{\mu'\Psi(\mu')}{\mu-\mu'}\left[Y(\mu)\,X(\mu')-X(\mu)\,Y(\mu')\right]d\mu'=-\,e^{-\tau_1/\mu}\,, \tag{84}$$

$$\int_0^1\frac{\mu'^2\Psi(\mu')}{\mu+\mu'}\left[X(\mu)\,X(\mu')-Y(\mu)\,Y(\mu')\right]d\mu'=x_1X(\mu)-y_1Y(\mu)-\mu\,, \tag{85}$$

and

$$\int_0^1\frac{\mu'^2\Psi(\mu')}{\mu-\mu'}\left[Y(\mu)\,X(\mu')-X(\mu)\,Y(\mu')\right]d\mu'=y_1X(\mu)-x_1Y(\mu)-\mu\,e^{-\tau_1/\mu}\,. \tag{86}$$

6. *The correspondence between the solutions of equations* (1) *and* (2) *and the functions* X *and* Y *introduced into the solution of the equations of transfer in a finite approximation.*—In solving the equations of transfer appropriately for the problem of diffuse reflection and transmission in Paper XXI, we found that we had to introduce certain functions, X and Y, involving the nonvanishing roots of a characteristic equation of the form

$$1=2\sum_{j=1}^{n}\frac{a_j\Psi(\mu_j)}{1-k^2\mu_j^2}\,, \tag{87}$$

where, as usual, the μ_j's are the zeros of $P_{2n}(\mu)$ and the a_j's are the corresponding Gaussian weights. In terms of these functions X and Y it was possible to express the solutions of the emergent radiations in closed forms in all cases considered. In analogy with the theory of the H-functions (Paper XIV, § 11), we may therefore expect that the functions X and Y appearing in the solutions in a finite approximation are rational approximations to the solutions of equations (1) and (2) when they are replaced by their "finite forms," namely,

$$X(\mu)=1+\mu\sum_{j=1}^{n}\frac{a_j\Psi(\mu_j)}{\mu+\mu_j}\left[X(\mu)\,X(\mu_j)-Y(\mu)\,Y(\mu_j)\right] \tag{88}$$

and

$$Y(\mu)=e^{-\tau_1/\mu}+\mu\sum_{j=1}^{n}\frac{a_j\Psi(\mu_j)}{\mu-\mu_j}\left[Y(\mu)\,X(\mu_j)-X(\mu)\,Y(\mu_j)\right]. \tag{89}$$

We shall now examine in what sense the functions X and Y introduced in Paper XXI, equations (125) and (126), are related to equations (88) and (89).

The definitions of the functions X and Y in Paper XXI (eqs. [125] and [126]) suggest that, in seeking solutions of the equations (88) and (89), we try the forms

$$X(\mu) = F(\mu) - e^{-\tau_1/\mu} G(-\mu) \tag{90}$$

and

$$Y(\mu) = e^{-\tau_1/\mu} F(-\mu) - G(\mu) , \tag{91}$$

where $F(\mu)$ and $G(\mu)$ are certain rational functions in μ, satisfying the conditions

$$F(-\mu_j) = G(-\mu_j) = 0 \qquad (j = 1, \ldots, n) . \tag{92}$$

For the forms (90) and (91)

$$X(\mu) = e^{-\tau_1/\mu} Y(-\mu) \quad \text{and} \quad Y(\mu) = e^{-\tau_1/\mu} X(-\mu) . \tag{93}$$

For X and Y related in this manner, it may be directly verified that equations (88) and (89) are equivalent to each other and that therefore it would suffice to consider only one of them.[8]

Now substituting for X and Y according to equations (90) and (91) in equations (88) and (89) and remembering the further conditions (eq. [92]) imposed on F and G, we find, after some minor reductions, that

$$\begin{aligned} F(\mu) - e^{-\tau_1/\mu} G(-\mu) = 1 + \mu \sum_{j=1}^{n} \frac{a_j \Psi(\mu_j)}{\mu + \mu_j} [F(\mu) F(\mu_j) - G(\mu) G(\mu_j)] \\ - \mu e^{-\tau_1/\mu} \sum_{j=1}^{n} \frac{a_j \Psi(\mu_j)}{\mu + \mu_j} [G(-\mu) F(\mu_j) - F(-\mu) G(\mu_j)] . \end{aligned} \tag{94}$$

Equating the terms with and without the exponential factor in this equation, we obtain

$$F(\mu) = 1 + \mu \sum_{j=1}^{n} \frac{a_j \Psi(\mu_j)}{\mu + \mu_j} [F(\mu) F(\mu_j) - G(\mu) G(\mu_j)] \tag{95}$$

and

$$G(-\mu) = \mu \sum_{j=1}^{n} \frac{a_j \Psi(\mu_j)}{\mu + \mu_j} [G(-\mu) F(\mu_j) - F(-\mu) G(\mu_j)] . \tag{96}$$

We can re-write these equations alternatively in the forms

$$F(\mu) \left[1 - \mu \sum_{j=1}^{n} \frac{a_j \Psi(\mu_j)}{\mu + \mu_j} F(\mu_j)\right] + G(\mu) \left[\mu \sum_{j=1}^{n} \frac{a_j \Psi(\mu_j)}{\mu + \mu_j} G(\mu_j)\right] = 1 \tag{97}$$

and

$$G(-\mu) \left[1 - \mu \sum_{j=1}^{n} \frac{a_j \Psi(\mu_j)}{\mu + \mu_j} F(\mu_j)\right] + F(-\mu) \left[\mu \sum_{j=1}^{n} \frac{a_j \Psi(\mu_j)}{\mu + \mu_j} G(\mu_j)\right] = 0 . \tag{98}$$

Solving for the quantities in brackets in equations (97) and (98), we find

$$F(-\mu) = [F(\mu) F(-\mu) - G(\mu) G(-\mu)] \left[1 - \mu \sum_{j=1}^{n} \frac{a_j \Psi(\mu_j)}{\mu + \mu_j} F(\mu_j)\right] \tag{99}$$

[8] It is of interest to note in this connection that the substitution (93) makes the functional equations (1) and (2) also equivalent to each other.

and

$$G(-\mu) = [F(\mu)F(-\mu) - G(\mu)G(-\mu)]\left[-\mu\sum_{j=1}^{n}\frac{a_j\Psi(\mu_j)}{\mu+\mu_j}G(\mu_j)\right]. \tag{100}$$

So far we have pursued only the consequences of assumptions (90), (91), and (92) regarding the form of the solutions of equations (88) and (89) adopted. We shall now write down explicitly the formulae for $F(\mu)$ and $G(\mu)$ suggested by the expressions for $X(\mu)$ and $Y(\mu)$ given in Paper XXI (eqs. [125] and [126]) and see how well they satisfy equations (99) and (100).

From a comparison of equations (90) and (91) and the equations (125) and (126) of Paper XXI, we conclude that

$$\begin{aligned} F(\mu) &= \frac{(-1)^n}{\mu_1\ldots\mu_n}\frac{P(-\mu)}{W(\mu)}\frac{C_0(-\mu)}{[C_0^2(0)-C_1^2(0)]^{1/2}} \\ &= \frac{1}{\mu_1\ldots\mu_n}\frac{\prod_{i=1}^{n}(\mu+\mu_i)}{\prod_{a}(1+k_a\mu)}\frac{C_0(-\mu)}{[C_0^2(0)-C_1^2(0)]^{1/2}} \end{aligned} \tag{101}$$

and

$$\begin{aligned} G(\mu) &= \frac{(-1)^n}{\mu_1\ldots\mu_n}\frac{P(-\mu)}{W(\mu)}\frac{C_1(-\mu)}{[C_0^2(0)-C_1^2(0)]^{1/2}} \\ &= \frac{1}{\mu_1\ldots\mu_n}\frac{\prod_{i=1}^{n}(\mu+\mu_i)}{\prod_{a}(1+k_a\mu)}\frac{C_1(-\mu)}{[C_0^2(0)-C_1^2(0)]^{1/2}}, \end{aligned} \tag{102}$$

where $C_0(\mu)$ and $C_1(\mu)$ are certain polynomials in μ, of degree n in nonconservative cases and $n-1$ in conservative cases, satisfying the conditions (cf. Paper XXI, eqs. [50], [108], and [109])

$$C_0(1/k_a) = \lambda_a C_1(-1/k_a) \tag{103}$$

and

$$\lambda_a = e^{k_a\tau_1}\frac{P(-1/k_a)}{P(+1/k_a)}. \tag{104}$$ [9]

According to the theorems proved in Paper XXI, § 4, the relations (103) and (104) are sufficient to determine $C_0(\mu)$ and $C_1(\mu)$ uniquely, apart from two arbitrary constants of proportionality in $C_0(\mu)+C_1(\mu)$ and $C_0(\mu)-C_1(\mu)$. For the particular "normalization" adopted in Paper XXI (eqs. [100] and [101])

$$C_0(\mu)\to\prod_{a>0}(1+k_a\mu) = R(-\mu) \qquad \text{as} \qquad \tau_1\to\infty, \tag{105}$$

and

$$C_1(\mu)\to 0 \qquad \text{as} \qquad \tau_1\to\infty. \tag{106}$$

[9] In equations (103) and (104) (as in eqs. [101] and [102]), a runs through positive and negative indices corresponding to all the nonvanishing roots k_a($a=\pm1,\ldots,\pm n$ or $\pm n\mp1$ and $k_a=-k_{-a}$) of the characteristic equation.

These further conditions, which we shall now also require, suffice to characterize the functions $C_0(\mu)$ and $C_1(\mu)$ without any arbitrariness.

Remembering that, in the approximation in which we are at present working,

$$H(\mu) = \frac{1}{\mu_1 \cdots \mu_n} \frac{\prod_{i=1}^{n} (\mu + \mu_i)}{\prod_{a>0} (1 + k_a \mu)}, \tag{107}$$

we can re-write equations (101) and (102) in the forms

$$F(\mu) = \frac{H(\mu)}{R(\mu)} \frac{C_0(-\mu)}{[C_0^2(0) - C_1^2(0)]^{1/2}} \quad \text{and} \quad G(\mu) = \frac{H(\mu)}{R(\mu)} \frac{C_1(-\mu)}{[C_0^2(0) - C_1^2(0)]^{1/2}}. \tag{108}$$

For $F(\mu)$ and $G(\mu)$, defined, in this manner (cf. Paper XXI, eq. [105]),

$$\begin{aligned} &F(\mu)F(-\mu) - G(\mu)G(-\mu) \\ &= \frac{H(\mu)H(-\mu)}{[C_0^2(0) - C_1^2(0)]W(\mu)} [C_0(\mu)C_0(-\mu) - C_1(\mu)C_1(-\mu)] = H(\mu)H(-\mu). \end{aligned} \tag{109}$$

Using this result and also equations (108) in equations (99) and (100), we find that our problem is reduced to examining the validity of the equations

$$\frac{C_0(\mu)}{[C_0^2(0) - C_1^2(0)]^{1/2}} = H(\mu)R(-\mu)\left[1 - \mu \sum_{j=1}^{n} \frac{a_j \Psi(\mu_j)}{\mu + \mu_j} F(\mu_j)\right] \tag{110}$$

and

$$\frac{C_1(\mu)}{[C_0^2(0) - C_1^2(0)]^{1/2}} = H(\mu)R(-\mu)\left[-\mu \sum_{j=1}^{n} \frac{a_j \Psi(\mu_j)}{\mu + \mu_j} G(\mu_j)\right], \tag{111}$$

or, equivalently,

$$C_0(\mu) = \frac{(-1)^n}{\mu_1 \cdots \mu_n} [C_0^2(0) - C_1^2(0)]^{1/2} P(-\mu)\left[1 - \mu \sum_{j=1}^{n} \frac{a_j \Psi(\mu_j)}{\mu + \mu_j} F(\mu_j)\right] \tag{112}$$

and

$$C_1(\mu) = \frac{(-1)^n}{\mu_1 \cdots \mu_n} [C_0^2(0) - C_1^2(0)]^{1/2} P(-\mu)\left[-\mu \sum_{j=1}^{n} \frac{a_j \Psi(\mu_j)}{\mu + \mu_j} G(\mu_j)\right]. \tag{113}$$

The validity (or otherwise) of equations (112) and (113) will depend essentially on whether the quantities on the right-hand sides of these equations are related in the manner required by equations (103). To examine this we have to evaluate the summations which occur in equations (112) and (113).

To carry out the summations in equations (112) and (113), we have first to break $F(\mu)$ and $G(\mu)$ into partial fractions. This requires us to treat the conservative and the nonconservative cases separately.

Considering, first, the nonconservative case, we have $2n$ distinct roots for the characteristic equation (87), which occur in pairs ($k_a = -k_{-a}$, $a = 1, \ldots, n$), and $F(\mu)$ and $G(\mu)$ can be expressed in the forms

$$F(\mu) = \sum_{a=-n}^{+n} \frac{L_a}{1+k_a\mu} + \frac{1}{k_1^2 \ldots k_n^2 \mu_1 \ldots \mu_n} \frac{c_0^{(n)}}{[C_0^2(0) - C_1^2(0)]^{1/2}} \tag{114}$$

and

$$G(\mu) = \sum_{a=-n}^{+n} \frac{L_a e^{-k_a\tau_1}}{1-k_a\mu} + \frac{1}{k_1^2 \ldots k_n^2 \mu_1 \ldots \mu_n} \frac{c_1^{(n)}}{[C_0^2(0) - C_1^2(0)]^{1/2}}, \tag{115}$$ [10]

where the $2n$ constants $L_a (a = \pm 1, \ldots, \pm n)$ are to be determined from the conditions (cf. eq. [92])

$$F(-\mu_j) = G(-\mu_j) = 0 \qquad (j = 1, \ldots, n), \tag{116}$$

and $c_0^{(n)}$ and $c_1^{(n)}$ are the coefficients of the highest power, μ^n, in $C_0(\mu)$ and $C_1(\mu)$.

To verify that $F(\mu)$ and $G(\mu)$, defined in the manner of the foregoing equations, agree with our earlier definitions (eqs. [108]), we first observe that conditions (116) enable us to express $F(\mu)$ and $G(\mu)$ in the forms

$$F(\mu) = \frac{(-1)^n}{\mu_1 \ldots \mu_n} \frac{P(-\mu)}{W(\mu)} f(-\mu) \tag{117}$$

and

$$G(\mu) = \frac{(-1)^n}{\mu_1 \ldots \mu_n} \frac{P(-\mu)}{W(\mu)} g(-\mu), \tag{118}$$

where $f(\mu)$ and $g(\mu)$ are polynomials of degree n in μ; and that, further,

$$f(1/k_a) = \lambda_a g(-1/k_a) \qquad (a = \pm 1, \ldots, \pm n), \tag{119}$$

where λ_a has the same meaning as in equation (104).[11] These latter conditions arise from a comparison of the values

$$L_a = \frac{(-1)^n}{\mu_1 \ldots \mu_n} \frac{P(+1/k_a)}{W_a(1/k_a)} f(1/k_a) \qquad (a = \pm 1, \ldots, \pm n) \tag{120}$$

and

$$L_a e^{-k_a\tau_1} = \frac{(-1)^n}{\mu_1 \ldots \mu_n} \frac{P(-1/k_a)}{W_a(1/k_a)} g(-1/k_a) \qquad (a = \pm 1, \ldots, \pm n), \tag{121}$$

which follow from equations (114), (115), (117), and (118). In accordance with the theorems of Paper XXI, § 4, we therefore conclude that $f(\mu)$ and $g(\mu)$ must be expressible in the forms

$$f(\mu) = q_0 C_0(\mu) + q_1 C_1(\mu) \quad \text{and} \quad g(\mu) = q_0 C_1(\mu) + q_1 C_0(\mu), \tag{122}$$

where q_0 and q_1 are constants. And, finally, from a comparison of the coefficients of the highest power of μ in f and g as deducible from equations (114) and (115) and equations (117) and (118), respectively, we readily verify that

$$q_0 = \frac{1}{[C_0^2(0) - C_1^2(0)]^{1/2}} \quad \text{and} \quad q_1 = 0, \tag{123}$$

as required.

[10] As in Paper XXI (cf. p. 153, n. 6), in all summations and products over a there is no term with $a = 0$.

[11] Negative values of a are permitted (cf. n. 9).

For convenience we shall re-write equations (114) and (115) in the forms

$$F(\mu) = \sum_{a=-n}^{+n} \frac{L_a}{1+k_a\mu} + a \tag{124}$$

and

$$G(\mu) = \sum_{a=-n}^{+n} \frac{L_a e^{-k_a\tau_1}}{1-k_a\mu} + b \,, \tag{125}$$

where

$$a = \frac{1}{k_1^2 \dots k_n^2 \mu_1 \dots \mu_n} \frac{c_0^{(n)}}{[C_0^2(0) - C_1^2(0)]^{1/2}}$$

and

$$b = \frac{1}{k_1^2 \dots k_n^2 \mu_1 \dots \mu_n} \frac{c_1^{(n)}}{[C_0^2(0) - C_1^2(0)]^{1/2}} \,. \tag{126}$$

Moreover (cf. eqs. [120] and [121]),

$$L_a = \frac{(-1)^n P(+1/k_a)}{\mu_1 \dots \mu_n W_a(1/k_a)} \frac{C_0(1/k_a)}{[C_0^2(0) - C_1^2(0)]^{1/2}} \,, \tag{127}$$

and

$$L_a e^{-k_a\tau_1} = \frac{(-1)^n P(-1/k_a)}{\mu_1 \dots \mu_n W_a(1/k_a)} \frac{C_1(-1/k_a)}{[C_0^2(0) - C_1^2(0)]^{1/2}} \,. \tag{128}$$

Returning, now, to the evaluation of the summations on the right-hand sides of equations (112) and (113), we consider, first,

$$\Sigma_1(\mu) = 1 - \mu \sum_{j=1}^{n} \frac{a_j \Psi(\mu_j)}{\mu+\mu_j} F(\mu_j) \,. \tag{129}$$

Since $F(-\mu_j) = 0$ ($j = 1, \dots, n$) we can, without altering anything, extend the summation also over negative values of j. We thus have

$$\Sigma_1(\mu) = 1 - \mu \sum_{j=-n}^{+n} \frac{a_j \Psi(\mu_j)}{\mu+\mu_j} \left(\sum_{\beta=-n}^{+n} \frac{L_\beta}{1+k_\beta \mu_j} + a \right), \tag{130}$$

where we have further substituted for $F(\mu_j)$ according to equation (124). Remembering that, according to the characteristic equation defining the roots k_a and k_β (cf. eq. [87]),

$$1 = \sum_{j=-n}^{+n} \frac{a_j \Psi(\mu_j)}{1+k_a\mu_j} = \sum_{j=-n}^{+n} \frac{a_j \Psi(\mu_j)}{1+k_\beta\mu_j} \tag{131}$$

for all a's and β's ($= \pm 1, \dots, \pm n$), we have for $\mu = 1/k_a$

$$\begin{aligned} \Sigma_1(1/k_a) &= 1 - \sum_{j=-n}^{+n} \frac{a_j \Psi(\mu_j)}{1+k_a\mu_j} \left(\sum_{\beta=-n}^{+n} \frac{L_\beta}{1+k_\beta\mu_j} + a \right) \\ &= 1 - a - \sum_{j=-n}^{+n} \sum_{\beta=-n}^{+n} L_\beta \frac{a_j \Psi(\mu_j)}{(1+k_a\mu_j)(1+k_\beta\mu_j)} \,. \end{aligned} \tag{132}$$

Alternatively, inverting the order of the summation, we can also write

$$\Sigma_1(1/k_a) = 1 - a - \sum_{\substack{\beta=-n\\ \beta\neq a}}^{+n} \frac{L_\beta}{k_a - k_\beta} \sum_{j=-n}^{+n} a_j\Psi(\mu_j)\left(\frac{k_a}{1+k_a\mu_j} - \frac{k_\beta}{1+k_\beta\mu_j}\right) - L_a \sum_{j=-n}^{+n} \frac{a_j\Psi(\mu_j)}{(1+k_a\mu_j)^2}. \tag{133}$$

Hence

$$\Sigma_1(1/k_a) = 1 - a - \sum_{\beta=-n}^{+n} L_\beta + L_a\left[1 - \sum_{j=-n}^{+n} \frac{a_j\Psi(\mu_j)}{(1+k_a\mu_j)^2}\right]. \tag{134}$$

But

$$1 - \sum_{j=-n}^{+n} \frac{a_j\Psi(\mu_j)}{(1+k_a\mu_j)^2} = \sum_{j=-n}^{+n} \frac{a_j\Psi(\mu_j)}{1+k_a\mu_j}\left(1 - \frac{1}{1+k_a\mu_j}\right) = k_a \sum_{j=-n}^{+n} \frac{a_j\mu_j\Psi(\mu_j)}{(1+k_a\mu_j)^2}. \tag{135}$$

We therefore have (cf. eq. [124])

$$\Sigma_1(1/k_a) = 1 - F(0) + L_a k_a \sum_{j=-n}^{+n} \frac{a_j\mu_j\Psi(\mu_j)}{(1+k_a\mu_j)^2} \qquad (a = \pm 1, \ldots, \pm n). \tag{136}$$

The summation

$$\Sigma_2(\mu) = -\mu \sum_{j=1}^{n} \frac{a_j\Psi(\mu_j)}{\mu+\mu_j} G(\mu_j) = -\mu \sum_{j=-n}^{+n} \frac{a_j\Psi(\mu_j)}{\mu+\mu_j} G(\mu_j) \tag{137}$$

can be similarly reduced. For $\mu = -1/k_a$ we find

$$\Sigma_2(-1/k_a) = -G(0) + L_a k_a e^{-k_a\tau_1} \sum_{j=-n}^{+n} \frac{a_j\mu_j\Psi(\mu_j)}{(1+k_a\mu_j)^2} \qquad (a = \pm 1, \ldots, \pm n). \tag{138}$$

An expression for the quantity

$$\sum_{j=-n}^{+n} \frac{a_j\mu_j\Psi(\mu_j)}{(1+k_a\mu_j)^2}, \tag{139}$$

which occurs in both equations (136) and (138), can be found by differentiating the identity[12]

$$1 - \sum_{j=-n}^{+n} \frac{a_j\Psi(\mu_j)}{1+\mu_j/z} = \frac{1}{H(z)H(-z)} = \frac{\mu_1^2 \ldots \mu_n^2}{P(z)P(-z)} \prod_{a=-n}^{+n} (1+k_a z), \tag{140}$$

[12] Cf. eq. (285) in the author's Gibbs Lecture (reference given in n. 4).

with respect to z and setting $z = 1/k_a$. In this manner we find

$$k_a \sum_{j=-n}^{+n} \frac{a_j \mu_j \Psi(\mu_j)}{(1 + k_a \mu_j)^2} = \mu_1^2 \ldots \mu_n^2 \frac{W_a(1/k_a)}{P(1/k_a) P(-1/k_a)} . \tag{141}$$

Using this result in equations (136) and (138) and substituting also for L_a and $L_a e^{-k_a \tau_1}$ according to equations (127) and (128), we have

$$\Sigma_1(1/k_a) = 1 - F(0) + (-1)^n \mu_1 \ldots \mu_n \frac{C_0(1/k_a)}{[C_0^2(0) - C_1^2(0)]^{1/2} P(-1/k_a)} \tag{142}$$

$$(a = \pm 1, \ldots, \pm n)$$

and

$$\Sigma_2(-1/k_a) = -G(0) + (-1)^n \mu_1 \ldots \mu_n \frac{C_1(-1/k_a)}{[C_0^2(0) - C_1^2(0)]^{1/2} P(+1/k_a)} \tag{143}$$

$$(a = \pm 1, \ldots, \pm n) .$$

The right-hand sides of equations (112) and (113) for $\mu = +1/k_a$, respectively $-1/k_a$, therefore become

$$\frac{(-1)^n}{\mu_1 \ldots \mu_n} [C_0^2(0) - C_1^2(0)]^{1/2} P(-1/k_a)[1 - F(0)] + C_0(1/k_a) \tag{144}$$

and

$$\frac{(-1)^{n+1}}{\mu_1 \ldots \mu_n} [C_0^2(0) - C_1^2(0)]^{1/2} P(+1/k_a) G(0) + C_1(-1/k_a) . \tag{145}$$

Now the validity of equations (112) and (113) requires that expressions (144) and (145) be *exactly* $C_0(1/k_a)$ and $C_1(-1/k_a)$. This will be the case only if $F(0) = 1$ and $G(0) = 0$. But, according to equations (108),

$$F(0) = \frac{C_0(0)}{[C_0^2(0) - C_1^2(0)]^{1/2}} \quad \text{and} \quad G(0) = \frac{C_1(0)}{[C_0^2(0) - C_1^2(0)]^{1/2}} ; \tag{146}$$

and it is *not* true that $C_1(0) = 0$ identically, in all approximations, and for all values of τ_1. However, according to equations (105) and (106), the conditions $F(0) = 1$ and $G(0) = 0$ will be met with increasing accuracy as $\tau_1 \to \infty$. Also, actual numerical calculations have shown that the errors with which the conditions $F(0) = 1$ and $G(0) = 0$ are met in the third or the fourth approximations (in our method of solution) are not large even for values of τ_1 of the order of 0.25 or less.

Turning, next, to the consideration of conservative cases, we have only $2n - 2$ non-vanishing roots for the characteristic equation. The expressions corresponding to (114) and (115) for $F(\mu)$ and $G(\mu)$ in partial fractions are, in consequence,

$$F(\mu) = \sum_{a=-n+1}^{n-1} \frac{L_a}{1 + k_a \mu} + L_0 - \frac{1}{k_1^2 \ldots k_{n-1}^2 \mu_1 \ldots \mu_n} \frac{c_0^{(n-1)}}{[C_0^2(0) - C_1^2(0)]^{1/2}} \mu \tag{147}$$

and

$$G(\mu) = \sum_{a=-n+1}^{n-1} \frac{L_a e^{-k_a \tau_1}}{1 - k_a \mu} + \mathfrak{L}_0 - \frac{1}{k_1^2 \ldots k_{n-1}^2 \mu_1 \ldots \mu_n} \frac{c_1^{(n-1)}}{[C_0^2(0) - C_1^2(0)]^{1/2}} \mu , \tag{148}$$

where the $2n$ constants, $L_{\pm a}(a = 1, \ldots, n-1)$, L_0, and $\mathfrak{L}_0$ are again to be determined by conditions of the form (116) and $c_0^{(n-1)}$ and $c_1^{(n-1)}$ are, respectively, the coefficients of the highest power, μ^{n-1}, in $C_0(\mu)$ and $C_1(\mu)$, defined in terms of the reduced number of characteristic roots appropriate for the conservative case.[13] With expressions (147) and (148) for $F(\mu)$ and $G(\mu)$, the rest of the analysis proceeds exactly as in the nonconservative case. The only difference is that use must also be made of the equation

$$\sum_{j=-n}^{+n} \frac{a_j \mu_j \Psi(\mu_j)}{1+k_\alpha \mu_j} = 0 \qquad (\alpha = \pm 1, \ldots, \pm n \mp 1), \tag{149}$$

which is valid only in conservative cases (cf. Paper XIV, eq. [159]).

One special characteristic of the solution (148) should be noted. We have

$$\begin{aligned} -\sum_{j=1}^{n} a_j \Psi(\mu_j)\, Y(\mu_j) &= \sum_{j=1}^{+n} a_j \Psi(\mu_j)\, G(\mu_j) = \sum_{j=-n}^{+n} a_j \Psi(\mu_j)\, G(\mu_j) \\ &= \sum_{\alpha=-n+1}^{n-1} L_\alpha e^{-k_\alpha \tau_1} \left(\sum_{j=-n}^{+n} \frac{a_j \Psi(\mu_j)}{1+k_\alpha \mu_j} \right) + \mathfrak{L}_0 \sum_{j=-n}^{+n} a_j \Psi(\mu_j). \end{aligned} \tag{150}$$

Since (conservative case!)

$$\sum_{j=-n}^{+n} a_j \Psi(\mu_j) = 1, \tag{151}$$

we have

$$-\sum_{j=1}^{n} a_j \Psi(\mu_j)\, Y(\mu_j) = \sum_{\alpha=-n+1}^{n-1} L_\alpha e^{-k_\alpha \tau_1} + \mathfrak{L}_0 = G(0). \tag{152}$$

We have already seen that the nonvanishing of $G(0)$ is a measure of the inaccuracy of our scheme of approximation. We therefore conclude that, for the exact solutions in the limit of infinite approximation,

$$\int_0^1 Y(\mu)\Psi(\mu)\, d\mu = 0, \tag{153}$$

and that the functions X and Y, defined in terms of the reduced number of the characteristic roots in conservative cases, must be associated with the *standard solutions* of the functional equations (1) and (2) as defined in § 5 (eqs. [64] and [65]).

We now summarize the conclusions that we have reached in the form of the following theorem:

Theorem 9.—The functions $X(\mu)$ and $Y(\mu)$ defined in terms of the nonvanishing roots of the characteristic equation

$$1 = 2\sum_{j=1}^{n} \frac{a_j \Psi(\mu_j)}{1-k^2\mu_j^2} \tag{154}$$

in the manner

$$\begin{aligned} X(\mu) = \frac{(-1)^n}{\mu_1 \ldots \mu_n} \frac{1}{[C_0^2(0) - C_1^2(0)]^{1/2}} \frac{1}{W(\mu)} \\ [P(-\mu)C_0(-\mu) - e^{-\tau_1/\mu} P(\mu) C_1(\mu)] \end{aligned} \tag{155}$$

[13] See particularly the remarks in Paper XXI, n. 12.

and

$$Y(\mu) = \frac{(-1)^n}{\mu_1 \ldots \mu_n} \frac{1}{[C_0^2(0) - C_1^2(0)]^{1/2}} \frac{1}{W(\mu)} \tag{156}$$

$$[e^{-\tau_1/\mu} P(\mu) C_0(\mu) - P(-\mu) C_1(-\mu)],$$

where

$$C_0(\mu) = \sum_{2^{n-1}\text{ terms}}^{l=n,n-2,\ldots} \epsilon_l^{(0)} \frac{\prod_{i=1}^{l} \prod_{m=1}^{n-l} (k_{r_i} + k_{s_m})}{\prod_{i=1}^{l} \prod_{m=1}^{n-l} (k_{r_i} - k_{s_m})} \prod_{i=1}^{l} (1 + k_{r_i}\mu) \prod_{m=1}^{n-l} \frac{1}{\lambda_{s_m}} (1 - k_{s_m}\mu) \tag{157}$$

and

$$C_1(\mu) = (-1)^{n-1} \sum_{2^{n-1}\text{ terms}}^{l=n-1,n-3,\ldots} \epsilon_l^{(1)} \frac{\prod_{i=1}^{l} \prod_{m=1}^{n-l} (k_{r_i} + k_{s_m})}{\prod_{i=1}^{l} \prod_{m=1}^{n-l} (k_{r_i} - k_{s_m})} \tag{158}$$

$$\times \prod_{i=1}^{l} (1 + k_{r_i}\mu) \prod_{m=1}^{n-l} \frac{1}{\lambda_{s_m}} (1 - k_{s_m}\mu),$$

where

$$\begin{aligned} \epsilon_l^{(0)} &= +1 \text{ for integers of the form } n - 4l \\ &= -1 \text{ for integers of the form } n - 4l - 2 \\ &= \quad 0 \text{ otherwise}, \end{aligned} \tag{159}$$

and

$$\begin{aligned} \epsilon_l^{(1)} &= +1 \text{ for integers of the form } n - 4l - 1 \\ &= -1 \text{ for integers of the form } n - 4l - 3 \\ &= \quad 0 \text{ otherwise}, \end{aligned} \tag{160}$$

and

$$\lambda_a = e^{k_a \tau_1} \frac{P(-1/k_a)}{P(+1/k_a)}, \tag{161}$$

are, in the limit of infinite approximation, to be associated with the solutions of the functional equations

$$X(\mu) = 1 + \mu \int_0^1 \frac{\Psi(\mu')}{\mu + \mu'} [X(\mu) X(\mu') - Y(\mu) Y(\mu')] \, d\mu' \tag{162}$$

and

$$Y(\mu) = e^{-\tau_1/\mu} + \mu \int_0^1 \frac{\Psi(\mu')}{\mu - \mu'} [Y(\mu) X(\mu') - X(\mu) Y(\mu')] \, d\mu'. \tag{163}$$

In conservative cases, $X(\mu)$ and $Y(\mu)$ (defined in terms of the reduced number of nonvanishing characteristic roots) are in the limit of infinite approximation to be associated with the standard solutions of equations (162) and (163), having the property

$$\int_0^1 X(\mu) \Psi(\mu) \, d\mu = 1 \quad \text{and} \quad \int_0^1 Y(\mu) \Psi(\mu) \, d\mu = 0. \tag{164}$$

II. ISOTROPIC SCATTERING WITH AN ALBEDO $\varpi_0 \leqslant 1$

7. *Equations of the problem.*—For the problem of diffuse reflection and transmission by an atmosphere scattering radiation isotropically with an albedo $\varpi_0 \leqslant 1$, the basic equations are (cf. Paper XVII, eqs. [115]–[120])

$$I(0, \mu) = \frac{\varpi_0}{4\mu} FS(\mu, \mu_0) ; \qquad I(\tau_1, -\mu) = \frac{\varpi_0}{4\mu} FT(\mu, \mu_0) , \tag{165}$$

$$\left(\frac{1}{\mu_0}+\frac{1}{\mu}\right) S(\mu, \mu_0) = X(\mu) X(\mu_0) - Y(\mu) Y(\mu_0) , \tag{166}$$

$$\left(\frac{1}{\mu_0}-\frac{1}{\mu}\right) T(\mu, \mu_0) = Y(\mu) X(\mu_0) - X(\mu) Y(\mu_0) , \tag{167}$$

$$\frac{\partial S}{\partial \tau_1} = Y(\mu) Y(\mu_0) , \tag{168}$$

and

$$\left(\frac{1}{\mu_0}-\frac{1}{\mu}\right) \frac{\partial T}{\partial \tau_1} = \frac{1}{\mu_0} X(\mu) Y(\mu_0) - \frac{1}{\mu} Y(\mu) X(\mu_0) . \tag{169}$$

Further, the definitions of $X(\mu)$ and $Y(\mu)$ in terms of $S(\mu, \mu_0)$ and $T(\mu, \mu_0)$ are

$$X(\mu) = 1 + \tfrac{1}{2}\varpi_0 \int_0^1 S(\mu, \mu') \frac{d\mu'}{\mu'} \tag{170}$$

and

$$Y(\mu) = e^{-\tau_1/\mu} + \tfrac{1}{2}\varpi_0 \int_0^1 T(\mu, \mu') \frac{d\mu'}{\mu'} . \tag{171}$$

In virtue of equations (166), (167), (170), and (171), we have the equations

$$X(\mu) = 1 + \tfrac{1}{2}\varpi_0 \mu \int_0^1 \frac{d\mu'}{\mu+\mu'} [X(\mu) X(\mu') - Y(\mu) Y(\mu')] \tag{172}$$

and

$$Y(\mu) = e^{-\tau_1/\mu} + \tfrac{1}{2}\varpi_0 \mu \int_0^1 \frac{d\mu'}{\mu-\mu'} [Y(\mu) X(\mu') - X(\mu) Y(\mu')] . \tag{173}$$

Thus X and Y satisfy functional equations of the form considered in Section I, with the characteristic function

$$\Psi(\mu) = \tfrac{1}{2}\varpi_0 = \text{constant} . \tag{174}$$

In considering the foregoing equations, it is of interest to establish, first, that equations (168) and (169) are really equivalent to the integrodifferential equations of theorem 1 (§ 3).

Thus, differentiating equation (170) with respect to τ_1 and using equation (168), we have

$$\frac{\partial X}{\partial \tau_1} = \tfrac{1}{2}\varpi_0 Y(\mu) \int_0^1 \frac{d\mu'}{\mu'} Y(\mu') . \tag{175}$$

Next, differentiating equation (171), we have

$$\frac{\partial Y}{\partial \tau_1} = -\frac{1}{\mu} e^{-\tau_1/\mu} + \tfrac{1}{2}\varpi_0 \int_0^1 \frac{d\mu'}{\mu-\mu'} \left[\frac{\mu}{\mu'} X(\mu) Y(\mu') - Y(\mu) X(\mu')\right] ; \tag{176}$$

and, combining this with equation (173), we obtain

$$\frac{\partial Y}{\partial \tau_1}+\frac{Y}{\mu}=\tfrac{1}{2}\varpi_0 X(\mu)\int_0^1 \frac{d\mu'}{\mu'}\,Y(\mu')\,. \tag{177}$$

It is seen that equations (175) and (177) are in agreement with equations (18) and (19) of theorem 1.

Finally, we may note that, according to equations (165)–(167) we can express the reflected and the transmitted intensities in the forms

$$I(0,\mu)=\tfrac{1}{4}\varpi_0 F\frac{\mu_0}{\mu+\mu_0}[X(\mu)X(\mu_0)-Y(\mu)Y(\mu_0)] \tag{178}$$

and

$$I(\tau_1,-\mu)=\tfrac{1}{4}\varpi_0 F\frac{\mu_0}{\mu-\mu_0}[Y(\mu)X(\mu_0)-X(\mu)Y(\mu_0)]\,. \tag{179}$$

8. *The case $\varpi_0<1$.*—Comparing the expressions for the emergent intensities given in the preceding section (eqs. [178] and [179]) with those given in Paper XXI (Sec. I, eqs. [127] and [128]), we observe that we have here a confirmation and an illustration of the correspondence enunciated in theorem 9 between the functions X and Y occurring in the solutions of the equations of transfer in a finite approximation, and the functions defined in terms of the functional equations in the exact theory.

9. *The ambiguity in the solution of the functional equations in the case $\varpi_0=1$ and its resolution by an appeal to the* K-*integral.*—When $\varpi_0=1$, the equations (172) and (173) belong to the conservative class discussed in § 5, and, according to theorem 5, the solutions of these equations, in this case, are not unique, the general solutions being, in fact, expressible in the forms

$$X(\mu)+Q\mu[X(\mu)+Y(\mu)] \tag{180}$$

and

$$Y(\mu)-Q\mu[X(\mu)+Y(\mu)]\,, \tag{181}$$

where Q is an arbitrary constant and $X(\mu)$ and $Y(\mu)$ are the standard solutions, having for the characteristic function $\frac{1}{2}$ the property

$$\alpha_0=\int_0^1 X(\mu)\,d\mu=2 \qquad \text{and} \qquad \beta_0=\int_0^1 Y(\mu)\,d\mu=0\,. \tag{182}$$

With solutions (180) and (181) of the equations

$$X(\mu)=1+\tfrac{1}{2}\mu\int_0^1\frac{d\mu'}{\mu+\mu'}[X(\mu)X(\mu')-Y(\mu)Y(\mu')] \tag{183}$$

and

$$Y(\mu)=e^{-\tau_1/\mu}+\tfrac{1}{2}\mu\int_0^1\frac{d\mu'}{\mu-\mu'}[Y(\mu)X(\mu')-X(\mu)Y(\mu')]\,, \tag{184}$$

the expressions (178) and (179) for the emergent intensities take the forms

$$I(0,\mu)=\tfrac{1}{4}\mu_0 F\left\{\frac{1}{\mu_0+\mu}[X(\mu_0)X(\mu)-Y(\mu_0)Y(\mu)] + Q[X(\mu_0)+Y(\mu_0)][X(\mu)+Y(\mu)]\right\} \tag{185}$$

and

$$I(\tau_1, -\mu) = \tfrac{1}{4}\mu_0 F \left\{ \frac{1}{\mu_0 - \mu} [Y(\mu_0) X(\mu) - X(\mu_0) Y(\mu)] \right. \\ \left. - Q [X(\mu_0) + Y(\mu_0)][X(\mu) + Y(\mu)] \right\}. \tag{186}$$

Solutions (185) and (186) for the emergent intensities involve the arbitrary constant Q, and there is nothing in the framework of the equations of § 7, for the case $\varpi_0 = 1$, which will remove this arbitrariness. We therefore conclude that the various invariances considered in Paper XVII are not sufficient to determine the physical solutions uniquely in conservative cases. We shall encounter further examples of this in Sections III and V. But it should be noted in the present context that a comparison of solutions (185) and (186) with those obtained in Paper XXI (Sec. II, eqs. [167]–[172]) provides a confirmation and an illustration of what is stated in theorem 9, namely, that the X- and Y-functions defined in terms of the reduced number of nonvanishing characteristic roots in conservative cases and which occur in the solutions of the equations of transfer in a finite approximation, are, in the framework of the exact theory, to be associated with the standard solutions of the corresponding functional equations.

We now turn to the matter of the arbitrariness in solutions (185) and (186) and the manner in which it is to be resolved.

The equation of transfer appropriate to the problem on hand is

$$\mu \frac{dI}{d\tau} = I(\tau, \mu) - \frac{1}{2}\int_{-1}^{+1} I(\tau, \mu')\, d\mu' - \tfrac{1}{4} F e^{-\tau/\mu_0}. \tag{187}$$

From this equation, two integrals which the problem admits can be derived. They are

$$F(\tau) = 2\int_{-1}^{+1} I(\tau, \mu)\, \mu\, d\mu = \mu_0 F (e^{-\tau/\mu_0} + \gamma_1) \tag{188}$$

and

$$K(\tau) = \frac{1}{2}\int_{-1}^{+1} I(\tau, \mu)\, \mu^2 d\mu = \tfrac{1}{4}\mu_0 F (-\mu_0 e^{-\tau/\mu_0} + \gamma_1 \tau + \gamma_2), \tag{189}$$

where γ_1 and γ_2 are two constants. The first of these represents the flux integral. We shall refer to the second as the "K-integral."

Applying the integrals (188) and (189) at $\tau = 0$ and $\tau = \tau_1$, we have

$$F(0) = 2\int_0^1 I(0, \mu)\, \mu\, d\mu = \mu_0 F (1 + \gamma_1), \tag{190}$$

$$F(\tau_1) = -2\int_0^1 I(\tau_1, -\mu)\, \mu\, d\mu = \mu_0 F (e^{-\tau_1/\mu_0} + \gamma_1), \tag{191}$$

$$K(0) = \frac{1}{2}\int_0^1 I(0, \mu)\, \mu^2 d\mu = \tfrac{1}{4}\mu_0 F (-\mu_0 + \gamma_2), \tag{192}$$

and

$$K(\tau_1) = \frac{1}{2}\int_0^1 I(\tau_1, -\mu)\, \mu^2 d\mu = \tfrac{1}{4}\mu_0 F (-\mu_0 e^{-\tau_1/\mu_0} + \gamma_1 \tau_1 + \gamma_2). \tag{193}$$

On the other hand, we can also evaluate $F(0)$, $F(\tau_1)$, $K(0)$, and $K(\tau_1)$ according to the solutions (185) and (186) for $I(0, \mu)$ and $I(\tau_1, -\mu)$. In this manner we shall obtain four relations between the three constants γ_1, γ_2, and Q. However, it will appear that two of these four relations are equivalent and that, in fact, they just suffice to determine all the constants uniquely.

The integrals defining $F(0)$, $F(\tau_1)$, $K(0)$, and $K(\tau_1)$ in terms of $I(0, \mu)$ and $I(\tau_1, -\mu)$, given by equations (185) and (186), can all be evaluated by using the various relations valid for standard solutions and collected under theorem 8 (eqs. [81]–[86]). We find

$$F(0) = \mu_0 F\{1+\tfrac{1}{2}Q(\alpha_1+\beta_1)[X(\mu_0)+Y(\mu_0)]\}, \tag{194}$$

$$F(\tau_1) = \mu_0 F\{e^{-\tau_1/\mu_0}+\tfrac{1}{2}Q(\alpha_1+\beta_1)[X(\mu_0)+Y(\mu_0)]\}, \tag{195}$$

$$K(0) = \tfrac{1}{4}\mu_0 F\{-\mu_0+\tfrac{1}{2}\alpha_1 X(\mu_0)-\tfrac{1}{2}\beta_1 Y(\mu_0)+\tfrac{1}{2}Q(\alpha_2+\beta_2)[X(\mu_0)+Y(\mu_0)]\}, \tag{196}$$

and

$$K(\tau_1) = \tfrac{1}{4}\mu_0 F\{-\mu_0 e^{-\tau_1/\mu_0}+\tfrac{1}{2}\beta_1 X(\mu_0)-\tfrac{1}{2}\alpha_1 Y(\mu_0) - \tfrac{1}{2}Q(\alpha_2+\beta_2)[X(\mu_0)+Y(\mu_0)]\}, \tag{197}$$

where α_n and β_n are the moments of order n of $X(\mu)$ and $Y(\mu)$, respectively (cf. eq. 11).

It is now seen that equations (190) and (194) and (191) and (195), in agreement with each other, determine

$$\gamma_1 = \tfrac{1}{2}Q(\alpha_1+\beta_1)[X(\mu_0)+Y(\mu_0)]. \tag{198}$$

From equations (192) and (196) we next find that

$$\gamma_2 = \tfrac{1}{2}\alpha_1 X(\mu_0)-\tfrac{1}{2}\beta_1 Y(\mu_0)+\tfrac{1}{2}Q(\alpha_2+\beta_2)[X(\mu_0)+Y(\mu_0)]. \tag{199}$$

Finally, from equations (193) and (197) we obtain

$$\gamma_1\tau_1+\gamma_2 = \tfrac{1}{2}\beta_1 X(\mu_0)-\tfrac{1}{2}\alpha_1 Y(\mu_0)-\tfrac{1}{2}Q(\alpha_2+\beta_2)[X(\mu_0)+Y(\mu_0)]. \tag{200}$$

Now, substituting for γ_1 and γ_2 in equation (200), according to equations (198) and (199), we find

$$\tfrac{1}{2}Q(\alpha_1+\beta_1)\tau_1 = -\tfrac{1}{2}(\alpha_1-\beta_1)-Q(\alpha_2+\beta_2). \tag{201}$$

Hence,

$$Q = -\frac{\alpha_1-\beta_1}{(\alpha_1+\beta_1)\tau_1+2(\alpha_2+\beta_2)}. \tag{202}$$

With this determination of Q in terms of the optical thickness, τ_1, of the atmosphere and the moments of the standard solutions of equations (183) and (184), we have removed the arbitrariness left by the functional equations in the solutions for the emergent intensities. It is in some ways remarkable that an explicit appeal to the K-integral is necessary to resolve the arbitrariness left by the functional equations. We shall see later that similar appeals to the K-integrals are necessary in the two other cases of perfect scattering that we shall consider (namely, Rayleigh scattering and scattering in accordance with Rayleigh's phase function) to resolve the ambiguities in the solutions of the functional equations incorporating the various invariances of the problem.

10. *The verification that Q satisfies the differential equation of theorem 7.*—It is apparent that the quantity Q as introduced in § 9 must satisfy the differential equation of theorem 7. In our present context we can write this equation (eq. [70]) in the form

$$\frac{d}{d\tau_1}\left(\frac{1}{Q}\right)-\frac{\beta_{-1}}{Q} = -1, \tag{203}$$

since

$$y_{-1} = \frac{1}{2}\int_0^1 \frac{d\mu'}{\mu'}Y(\mu') = \tfrac{1}{2}\beta_{-1}. \tag{204}$$

We shall now show that Q as defined by equation (202) satisfies equation (203).

Making use of the relation (cf. theorem 8, eq. [82])

$$\alpha_1^2 - \beta_1^2 = 4\int_0^1 \mu^2 d\mu = \frac{4}{3}, \tag{205}$$

we first re-write equation (202) in the form

$$\frac{1}{Q} = -\tfrac{3}{4}\left[(\alpha_1+\beta_1)^2\tau_1 + 2(\alpha_2+\beta_2)(\alpha_1+\beta_1)\right]. \tag{206}$$

From this equation we then obtain

$$\frac{d}{d\tau_1}\left(\frac{1}{Q}\right) = -\tfrac{3}{4}\Big[(\alpha_1+\beta_1)\left\{(\alpha_1+\beta_1) + 2\tau_1\frac{d}{d\tau_1}(\alpha_1+\beta_1)\right\} + 2(\alpha_2+\beta_2)\frac{d}{d\tau_1}(\alpha_1+\beta_1) + 2(\alpha_1+\beta_1)\frac{d}{d\tau_1}(\alpha_2+\beta_2)\Big]. \tag{207}$$

To simplify equation (207) further, we observe that, according to equations (175) and (177), we now have

$$\frac{d}{d\tau_1}(X+Y) = \tfrac{1}{2}\beta_{-1}(X+Y) - \frac{Y}{\mu}. \tag{208}$$

Multiplying this equation by μ^n and integrating over the range of μ, we obtain

$$\frac{d}{d\tau_1}(\alpha_n+\beta_n) = \tfrac{1}{2}\beta_{-1}(\alpha_n+\beta_n) - \beta_{n-1}. \tag{209}$$

In particular,

$$\frac{d}{d\tau_1}(\alpha_1+\beta_1) = \tfrac{1}{2}\beta_{-1}(\alpha_1+\beta_1) \tag{210}$$

(since, according to eq. [182], $\beta_0 = 0$), and

$$\frac{d}{d\tau_1}(\alpha_2+\beta_2) = \tfrac{1}{2}\beta_{-1}(\alpha_2+\beta_2) - \beta_1. \tag{211}$$

Using the foregoing relations in equation (207), we find, after some minor reductions, that

$$\frac{d}{d\tau_1}\left(\frac{1}{Q}\right) = -\tfrac{3}{4}\left[\beta_{-1}\left\{(\alpha_1+\beta_1)^2\tau_1 + 2(\alpha_2+\beta_2)(\alpha_1+\beta_1)\right\} + (\alpha_1+\beta_1)^2 - 2\beta_1(\alpha_1+\beta_1)\right]. \tag{212}$$

Hence (cf. eqs. [205] and [206])

$$\frac{d}{d\tau_1}\left(\frac{1}{Q}\right) = \frac{\beta_{-1}}{Q} - \tfrac{3}{4}(\alpha_1^2 - \beta_1^2) = \frac{\beta_{-1}}{Q} - 1. \tag{213}$$

This completes the verification.

(*To be continued*)

16

ON THE RADIATIVE EQUILIBRIUM OF A STELLAR ATMOSPHERE. XXII (*Concluded*)*

S. Chandrasekhar
Yerkes Observatory
Received November 3, 1947

III. SCATTERING IN ACCORDANCE WITH RAYLEIGH'S PHASE FUNCTION

11. *The equations of the problem.*—We have already indicated in Paper XVII, § 6, how the functional equations governing the angular distributions of the reflected and the transmitted radiations from an atmosphere scattering according to a general phase function, expressible as a series in Legendre polynomials, can be reduced to independent systems of functional equations.

In the case of scattering according to Rayleigh's phase function, we can express the reflected and the transmitted intensities in the forms (cf. Paper XIV, eq. [231])

$$I(0;\mu,\varphi;\mu_0,\varphi_0)=\frac{3}{32\mu}F\,[S^{(0)}(\mu,\mu_0)-4\mu\mu_0(1-\mu^2)^{\frac{1}{2}}(1-\mu_0^2)^{\frac{1}{2}}\times S^{(1)}(\mu,\mu_0)\cos(\varphi-\varphi_0)+(1-\mu^2)(1-\mu_0^2)\,S^{(2)}(\mu,\mu_0)\cos 2(\varphi-\varphi_0)] \quad (214)$$

and

$$I(\tau_1;-\mu,\varphi;\mu_0,\varphi_0)=\frac{3}{32\mu}F\,[T^{(0)}(\mu,\mu_0)+4\mu\mu_0(1-\mu^2)^{\frac{1}{2}}(1-\mu_0^2)^{\frac{1}{2}}\times T^{(1)}(\mu,\mu_0)\cos(\varphi-\varphi_0)+(1-\mu^2)(1-\mu_0^2)\,T^{(2)}(\mu,\mu_0)\cos 2(\varphi-\varphi_0)]\,, \quad (215)$$

and the functions of the different orders (distinguished by the superscripts) satisfy independent systems of equations. Of these systems, the two governing the functions of order one and two are directly reducible to the standard forms considered in Section I. And the terms in the reflected and the transmitted intensities proportional to $\cos(\varphi-\varphi_0)$ and $\cos 2(\varphi-\varphi_0)$ are of exactly the same forms as those given in Paper XXI, equations (223) and (224); only the functions $X^{(1)}$, $Y^{(1)}$, and $X^{(2)}$, $Y^{(2)}$, must now be redefined in terms of the functional equations which they satisfy. These terms require, therefore, no further consideration.

Turning to the functions $S^{(0)}(\mu,\mu_0)$ and $T^{(0)}(\mu,\mu_0)$ of zero order, we find that these functions must be expressible in the forms (cf. Paper XIV, eqs. [241]–[245])

$$\left(\frac{1}{\mu_0}+\frac{1}{\mu}\right)S^{(0)}(\mu,\mu_0)=\tfrac{1}{3}\,[\psi(\mu)\psi(\mu_0)-\chi(\mu)\chi(\mu_0)]+\tfrac{8}{3}\,[\phi(\mu)\phi(\mu_0)-\zeta(\mu)\zeta(\mu_0)] \quad (216)$$

and

$$\left(\frac{1}{\mu_0}-\frac{1}{\mu}\right)T^{(0)}(\mu,\mu_0)=\tfrac{1}{3}\,[\chi(\mu)\psi(\mu_0)-\psi(\mu)\chi(\mu_0)]+\tfrac{8}{3}\,[\zeta(\mu)\phi(\mu_0)-\phi(\mu)\zeta(\mu_0)]\,, \quad (217)$$

* Sections I and II of this paper have already appeared in *Ap. J.*, **107**, 48, 1948. The remaining Sections III–V of the paper are published here. The numbering of the sections and equations continue from those of the earlier part.

where

$$\psi(\mu) = 3 - \mu^2 + \tfrac{3}{16}\int_0^1 (3 - \mu'^2)\, S^{(0)}(\mu, \mu') \frac{d\mu'}{\mu'}, \tag{218}$$

$$\phi(\mu) = \mu^2 + \tfrac{3}{16}\int_0^1 \mu'^2 S^{(0)}(\mu, \mu') \frac{d\mu'}{\mu'}, \tag{219}$$

$$\chi(\mu) = (3 - \mu^2)\, e^{-\tau_1/\mu} + \tfrac{3}{16}\int_0^1 (3 - \mu'^2)\, T^{(0)}(\mu, \mu') \frac{d\mu'}{\mu'} \tag{220}$$

and

$$\zeta(\mu) = \mu^2 e^{-\tau_1/\mu} + \tfrac{3}{16}\int_0^1 \mu'^2 T^{(0)}(\mu, \mu') \frac{d\mu'}{\mu'}. \tag{221}$$

Further, we must also have

$$\frac{\partial S^{(0)}}{\partial \tau_1} = \tfrac{1}{3}\chi(\mu)\,\chi(\mu_0) + \tfrac{8}{3}\zeta(\mu)\,\zeta(\mu_0) \tag{222}$$

and

$$\left(\frac{1}{\mu_0} - \frac{1}{\mu}\right)\frac{\partial T^{(0)}}{\partial \tau_1} = \frac{1}{\mu_0}\left[\tfrac{1}{3}\psi(\mu)\,\chi(\mu_0) + \tfrac{8}{3}\phi(\mu)\,\zeta(\mu_0)\right] - \frac{1}{\mu}\left[\tfrac{1}{3}\chi(\mu)\,\psi(\mu_0) + \tfrac{8}{3}\zeta(\mu)\,\phi(\mu_0)\right]. \tag{223}$$

Substituting for $S^{(0)}$ and $T^{(0)}$ according to equations (216) and (217) in equations (218)–(221), we obtain the following system of functional equations of fourth order:

$$\psi(\mu) = 3 - \mu^2 + \tfrac{1}{16}\mu\int_0^1 \frac{3 - \mu'^2}{\mu + \mu'}\left[\psi(\mu)\,\psi(\mu') - \chi(\mu)\,\chi(\mu')\right] d\mu' + \tfrac{1}{2}\mu\int_0^1 \frac{3 - \mu'^2}{\mu + \mu'}\left[\phi(\mu)\,\phi(\mu') - \zeta(\mu)\,\zeta(\mu')\right] d\mu', \tag{224}$$

$$\phi(\mu) = \mu^2 + \tfrac{1}{16}\mu\int_0^1 \frac{\mu'^2}{\mu + \mu'}\left[\psi(\mu)\,\psi(\mu') - \chi(\mu)\,\chi(\mu')\right] d\mu' + \tfrac{1}{2}\mu\int_0^1 \frac{\mu'^2}{\mu + \mu'}\left[\phi(\mu)\,\phi(\mu') - \zeta(\mu)\,\zeta(\mu')\right] d\mu', \tag{225}$$

$$\chi(\mu) = (3 - \mu^2)\, e^{-\tau_1/\mu} + \tfrac{1}{16}\mu\int_0^1 \frac{3 - \mu'^2}{\mu - \mu'}\left[\chi(\mu)\,\psi(\mu') - \psi(\mu)\,\chi(\mu')\right] d\mu' + \tfrac{1}{2}\mu\int_0^1 \frac{3 - \mu'^2}{\mu - \mu'}\left[\zeta(\mu)\,\phi(\mu') - \phi(\mu)\,\zeta(\mu')\right] d\mu', \tag{226}$$

and

$$\zeta(\mu) = \mu^2 e^{-\tau_1/\mu} + \tfrac{1}{16}\mu\int_0^1 \frac{\mu'^2}{\mu - \mu'}\left[\chi(\mu)\,\psi(\mu') - \psi(\mu)\,\chi(\mu')\right] d\mu' + \tfrac{1}{2}\mu\int_0^1 \frac{\mu'^2}{\mu - \mu'}\left[\zeta(\mu)\,\phi(\mu') - \phi(\mu)\,\zeta(\mu')\right] d\mu'. \tag{227}$$

12. *The form of the solution.*—In solving systems of functional equations of the type of equations (224)–(227), we shall be guided by the forms of the solutions obtained in the direct solution of the equations of transfer in a general finite approximation and the correspondence enunciated in theorem 9 between the X- and Y-functions occurring in such approximate solutions and the exact functions defined in terms of functional equations they satisfy.

Accordingly, in the present instance, we shall assume that $S^{(0)}(\mu, \mu')$ and $T^{(0)}(\mu, \mu')$ are of the forms (cf. Paper XXI, eqs. [221] and [222])

$$\left(\frac{1}{\mu'}+\frac{1}{\mu}\right) S^{(0)}(\mu, \mu') = X(\mu) X(\mu') [3 + c_1(\mu+\mu') + \mu\mu'] - Y(\mu) Y(\mu') [3 - c_1(\mu+\mu') + \mu\mu'] + c_2(\mu+\mu') [X(\mu) Y(\mu') + Y(\mu) X(\mu')] \quad (228)$$

and

$$\left(\frac{1}{\mu'}-\frac{1}{\mu}\right) T^{(0)}(\mu, \mu') = Y(\mu) X(\mu') [3 - c_1(\mu-\mu') - \mu\mu'] - X(\mu) Y(\mu') [3 + c_1(\mu-\mu') - \mu\mu'] - c_2(\mu-\mu') [X(\mu) X(\mu') + Y(\mu) Y(\mu')] , \quad (229)$$

where c_1 and c_2 are certain constants unspecified for the present, and $X(\mu)$ and $Y(\mu)$ are the *standard solutions* of the equations

$$X(\mu) = 1 + \tfrac{3}{16}\mu \int_0^1 \frac{3-\mu'^2}{\mu+\mu'} [X(\mu) X(\mu') - Y(\mu) Y(\mu')] d\mu' \quad (230)$$

and

$$Y(\mu) = e^{-\tau_1/\mu} + \tfrac{3}{16}\mu \int_0^1 \frac{3-\mu'^2}{\mu-\mu'} [Y(\mu) X(\mu') - X(\mu) Y(\mu')] d\mu' , \quad (231)$$

having the property

$$\tfrac{3}{16}\int_0^1 (3-\mu^2) X(\mu) d\mu = \tfrac{3}{16}(3\alpha_0 - \alpha_2) = 1 \quad (232)$$

and

$$\int_0^1 (3-\mu^2) Y(\mu) d\mu = (3\beta_0 - \beta_2) = 0 \quad (233)$$

where α_n and β_n have their usual meanings (cf. eq. [11]).

An alternative form of equations (228) and (229) which we shall find useful may be noted here:

$$S^{(0)}(\mu, \mu') = \{ (3-\mu^2) [X(\mu) X(\mu') - Y(\mu) Y(\mu')] + (\mu+\mu') X(\mu) [(c_1+\mu) X(\mu') + c_2 Y(\mu')] + (\mu+\mu') Y(\mu) [c_2 X(\mu') + (c_1-\mu) Y(\mu')] \} \frac{\mu\mu'}{\mu+\mu'} \quad (234)$$

and

$$T^{(0)}(\mu, \mu') = \{ (3-\mu^2) [Y(\mu) X(\mu') - X(\mu) Y(\mu')] - (\mu-\mu') X(\mu) [(c_1+\mu) Y(\mu') + c_2 X(\mu')] - (\mu-\mu') Y(\mu) [c_2 Y(\mu') + (c_1-\mu) X(\mu')] \} \frac{\mu\mu'}{\mu-\mu'} . \quad (235)$$

13. *Verification of the solution and a relation between the constants* c_1 *and* c_2.—The verification that the solutions for $S^{(0)}$ and $T^{(0)}$ have the forms assumed in § 12 will consist in first evaluating ψ, ϕ, χ, and ζ according to equations (218)–(221) and then showing that when the resulting expressions for ψ, ϕ, χ, and ζ are substituted back into equations (216) and (217) we shall recover the form of the solutions assumed. In general, such a procedure will lead to certain conditions which the constants introduced into the solution (such as c_1 and c_2 in the present instance) must satisfy. We shall see that, in the particular case under discussion, the conditions derived in the manner indicated do not suffice to determine c_1 and c_2 without an ambiguity and an arbitrariness. This is a further example of the nonuniqueness of the solution, in conservative cases, of the functional equations incorporating the invariances of the problem. But, again, an appeal to the integrals of the problem resolves the ambiguity and the arbitrariness.

Our first step, then, is to evaluate ψ, ϕ, χ, and ζ according to equations (218)–(221), when $S^{(0)}(\mu, \mu')$ and $T^{(0)}(\mu, \mu')$ have the forms given by equations (234) and (235). The evaluation of the integrals defining ψ, ϕ, etc., is fairly straightforward if appropriate use is made of the various integral properties of the standard solutions of equations (230) and (231). It may be noted that, in addition to equations (232) and (233), use must also be made of the relations (cf. theorem 4, eqs. [44]–[46])

$$\alpha_0 = 1 + \tfrac{3}{32}\left[3\left(\alpha_0^2 - \beta_0^2\right) - \left(\alpha_1^2 - \beta_1^2\right)\right], \tag{236}$$

$$(3-\mu^2)\int_0^1 \frac{X(\mu)X(\mu') - Y(\mu)Y(\mu')}{\mu+\mu'}\,d\mu' = \frac{X(\mu)-1}{\tfrac{3}{16}\mu} + (\alpha_1 - \mu\alpha_0)X(\mu) - (\beta_1 - \mu\beta_0)Y(\mu) \tag{237}$$

and

$$(3-\mu^2)\int_0^1 \frac{Y(\mu)X(\mu') - X(\mu)Y(\mu')}{\mu-\mu'}\,d\mu' = \frac{Y(\mu) - e^{-\tau_1/\mu}}{\tfrac{3}{16}\mu} + (\beta_1 + \mu\beta_0)X(\mu) - (\alpha_1 + \mu\alpha_0)Y(\mu)\,. \tag{238}$$

Evaluating ψ, ϕ, χ, and ζ in the manner indicated, we find that

$$\psi(\mu) = (3 + c_1\mu)X(\mu) + c_2\mu Y(\mu)\,, \tag{239}$$

$$\chi(\mu) = (3 - c_1\mu)Y(\mu) - c_2\mu X(\mu) \tag{240}$$

$$\phi(\mu) = +\mu\left[q_1 X(\mu) + q_2 Y(\mu)\right], \tag{241}$$

and

$$\zeta(\mu) = -\mu\left[q_2 X(\mu) + q_1 Y(\mu)\right] \tag{242}$$

where

$$q_1 = \tfrac{3}{16}\left(c_1\alpha_2 + c_2\beta_2 + 3\alpha_1\right) \tag{243}$$

and

$$q_2 = \tfrac{3}{16}\left(c_1\beta_2 + c_2\alpha_2 - 3\beta_1\right). \tag{244}$$

Using the expressions (239)–(242) for ψ, χ, ϕ, and ζ, we next evaluate $S^{(0)}$ and $T^{(0)}$ according to equations (216) and (217). We find

$$\begin{aligned}\left(\frac{1}{\mu'}+\frac{1}{\mu}\right) S^{(0)}(\mu,\mu') &= X(\mu)X(\mu')\left[3 + c_1(\mu+\mu') + \tfrac{1}{3}\left\{c_1^2 - c_2^2 + 8\left(q_1^2 - q_2^2\right)\right\}\mu\mu'\right] \\ &\quad - Y(\mu)Y(\mu')\left[3 - c_1(\mu+\mu') + \tfrac{1}{3}\left\{c_1^2 - c_2^2 + 8\left(q_1^2 - q_2^2\right)\right\}\mu\mu'\right] \\ &\quad + c_2(\mu+\mu')\left[X(\mu)Y(\mu') + Y(\mu)X(\mu')\right]\end{aligned} \tag{245}$$

and

$$\left(\frac{1}{\mu'}-\frac{1}{\mu}\right)T^{(0)}(\mu,\mu')$$
$$= Y(\mu)\,X(\mu')\left[3-c_1(\mu-\mu')-\tfrac{1}{3}\{c_1^2-c_2^2+8(q_1^2-q_2^2)\}\mu\mu'\right]$$
$$-X(\mu)\,Y(\mu')\left[3+c_1(\mu-\mu')-\tfrac{1}{3}\{c_1^2-c_2^2+8(q_1^2-q_2^2)\}\mu\mu'\right] \tag{246}$$
$$-c_2(\mu-\mu')\left[X(\mu)\,X(\mu')+Y(\mu)\,Y(\mu')\right].$$

A comparison of equations (245) and (246) and (228) and (229) now shows that, among the constants c_1, c_2, q_1, and q_2, we must require that there exist the relation (cf. Paper XIV, eq. [250])

$$c_1^2-c_2^2+8(q_1^2-q_2^2)=3\,. \tag{247}$$

Substituting for q_1 and q_2 according to equations (243) and (244) in equation (247), we obtain

$$32(c_1^2-c_2^2)+9\left[(c_1+c_2)(\alpha_2+\beta_2)+3(\alpha_1-\beta_1)\right]$$
$$\times\left[(c_1-c_2)(\alpha_2-\beta_2)+3(\alpha_1+\beta_1)\right]-96=0\,. \tag{248}$$

After some minor rearranging of the terms, the foregoing equation can be reduced to the form

$$[32+9(\alpha_2^2-\beta_2^2)](c_1^2-c_2^2)+27(\alpha_1+\beta_1)(\alpha_2+\beta_2)(c_1+c_2)$$
$$+27(\alpha_1-\beta_1)(\alpha_2-\beta_2)(c_1-c_2)+81(\alpha_1^2-\beta_1^2)-96=0\,. \tag{249}$$

On the other hand, according to equations (232), (233), and (236),

$$32+9(\alpha_2^2-\beta_2^2)=32+(9\alpha_0-16)^2-81\beta_0^2$$
$$=288(1-\alpha_0)+81(\alpha_0^2-\beta_0^2) \tag{250}$$
$$=27(\alpha_1^2-\beta_1^2)\,.$$

Equation (249) therefore becomes

$$(\alpha_1^2-\beta_1^2)(c_1^2-c_2^2)+(\alpha_1+\beta_1)(\alpha_2+\beta_2)(c_1+c_2)$$
$$+(\alpha_1-\beta_1)(\alpha_2-\beta_2)(c_1-c_2)+(\alpha_2^2-\beta_2^2)=0\,. \tag{251}$$

Hence

$$[(\alpha_1+\beta_1)(c_1+c_2)+(\alpha_2-\beta_2)][(\alpha_1-\beta_1)(c_1-c_2)+(\alpha_2+\beta_2)]=0\,. \tag{252}$$

It is apparent that one of the two factors in equation (252) must vanish. But within the framework of equations (224)–(227) it is impossible to decide which of the two it must be; and in either case we shall have only one relation between the two constants c_1 and c_2. The problem is therefore characterized by an ambiguity and an arbitrariness. We shall show in the following section how this can be resolved.

14. *The resolution of the ambiguity and the arbitrariness in the solution.*—It can be readily verified that the problem of diffuse reflection and transmission in accordance with Rayleigh's phase function admits, as in the conservative isotropic case, the flux and the K-integrals. The emergent values of F and K must therefore be given by equations of the form (cf. eqs. [190]–[193])

$$F(0) = \mu_0 F(1+\gamma_1)\,; \qquad F(\tau_1) = \mu_0 F(e^{-\tau_1/\mu_0}+\gamma_1)\,, \tag{253}$$

$$K(0) = \tfrac{1}{4}\mu_0 F(-\mu_0+\gamma_2)\,, \tag{254}$$

and

$$K(\tau_1) = \tfrac{1}{4}\mu_0 F(-\mu_0\, e^{-\tau_1/\mu_0}+\gamma_1\tau_1+\gamma_2)\,, \tag{255}$$

where γ_1 and γ_2 are constants.

It is evident that only the azimuth independent terms in the intensity will contribute to F and K. We have, accordingly, to evaluate $F(0)$, $F(\tau_1)$, $K(0)$, and $K(\tau_1)$ for emergent intensities of the forms (cf. eqs. [214], [215], [234], and [235])

$$\begin{aligned} I(0,\mu) = \tfrac{1}{2}\mu_0 F \Big\{ \tfrac{3}{16}\frac{3-\mu^2}{\mu_0+\mu}[X(\mu_0)X(\mu) - Y(\mu_0)Y(\mu)] \\ + \tfrac{3}{16}X(\mu_0)[(c_1+\mu)X(\mu)+c_2 Y(\mu)] \\ + \tfrac{3}{16}Y(\mu_0)[c_2 X(\mu)+(c_1-\mu)Y(\mu)]\Big\} \end{aligned} \tag{256}$$

and

$$\begin{aligned} I(\tau_1,-\mu) = \tfrac{1}{2}\mu_0 F \Big\{ \tfrac{3}{16}\frac{3-\mu^2}{\mu_0-\mu}[Y(\mu_0)X(\mu) - X(\mu_0)Y(\mu)] \\ - \tfrac{3}{16}X(\mu_0)[c_2 X(\mu)+(c_1-\mu)Y(\mu)] \\ - \tfrac{3}{16}Y(\mu_0)[(c_1+\mu)X(\mu)+c_2 Y(\mu)]\Big\}\,. \end{aligned} \tag{257}$$

With $I(0,\mu)$ and $I(\tau_1,-\mu)$ given by equations (256) and (257), the integrals defining $F(0)$, $F(\tau_1)$, $K(0)$, and $K(\tau_1)$ can all be evaluated quite simply by using the various relations given in theorem 8 (eqs. [81]–[86]) and remembering that in the present case

$$x_1 = \tfrac{3}{16}(3\alpha_1-\alpha_3) \qquad \text{and} \qquad y_1 = \tfrac{3}{16}(3\beta_1-\beta_3)\,. \tag{258}$$

We thus find

$$F(0) = \mu_0 F\{1+\tfrac{3}{16}X(\mu_0)(c_1\alpha_1+c_2\beta_1+\alpha_2)+\tfrac{3}{16}Y(\mu_0)(c_1\beta_1+c_2\alpha_1-\beta_2)\}\,, \tag{259}$$

$$\begin{aligned} F(\tau_1) = \mu_0 F\{ e^{-\tau_1/\mu_0}+\tfrac{3}{16}X(\mu_0)(c_1\beta_1+c_2\alpha_1-\beta_2) \\ +\tfrac{3}{16}Y(\mu_0)(c_1\alpha_1+c_2\beta_1+\alpha_2)\}\,, \end{aligned} \tag{260}$$

$$\begin{aligned} K(0) = \tfrac{1}{4}\mu_0 F\{-\mu_0+\tfrac{3}{16}X(\mu_0)(c_1\alpha_2+c_2\beta_2+3\alpha_1) \\ +\tfrac{3}{16}Y(\mu_0)(c_1\beta_2+c_2\alpha_2-3\beta_1)\}\,, \end{aligned} \tag{261}$$

and

$$\begin{aligned} K(\tau_1) = \tfrac{1}{4}\mu_0 F\{-\mu_0\, e^{-\tau_1/\mu_0}-\tfrac{3}{16}X(\mu_0)(c_1\beta_2+c_2\alpha_2-3\beta_1) \\ -\tfrac{3}{16}Y(\mu_0)(c_1\alpha_2+c_2\beta_2+3\alpha_1)\}\,. \end{aligned} \tag{262}$$

Comparing the reflected and the transmitted fluxes given by equations (259) and (260) with those given by the flux integral (eq. [253]), we find that

$$\gamma_1 = \tfrac{3}{16} X(\mu_0)(c_1\alpha_1 + c_2\beta_1 + \alpha_2) + \tfrac{3}{16} Y(\mu_0)(c_1\beta_1 + c_2\alpha_1 - \beta_2) \qquad (263)$$

and also that

$$\gamma_1 = \tfrac{3}{16} X(\mu_0)(c_1\beta_1 + c_2\alpha_1 - \beta_2) + \tfrac{3}{16} Y(\mu_0)(c_1\alpha_1 + c_2\beta_1 + \alpha_2)\,. \qquad (264)$$

We must therefore require that

$$c_1\alpha_1 + c_2\beta_1 + \alpha_2 = c_1\beta_1 + c_2\alpha_1 - \beta_2\,, \qquad (265)$$

or

$$(c_1 - c_2)(\alpha_1 - \beta_1) + \alpha_2 + \beta_2 = 0\,; \qquad (266)$$

but this is one of the factors in equation (252). The appeal to the flux integral has therefore decided which of the two factors in equation (252) must be set equal to zero.

In view of equation (266) we can combine equations (263) and (264) to give

$$\gamma_1 = \tfrac{3}{32}[(c_1 + c_2)(\alpha_1 + \beta_1) + (\alpha_2 - \beta_2)][X(\mu_0) + Y(\mu_0)]\,. \qquad (267)$$

Next, from equations (254) and (261) we find that

$$\gamma_2 = \tfrac{3}{16} X(\mu_0)(c_1\alpha_2 + c_2\beta_2 + 3\alpha_1) + \tfrac{3}{16} Y(\mu_0)(c_1\beta_2 + c_2\alpha_2 - 3\beta_1)\,. \qquad (268)$$

And, finally, from equations (255) and (262) we obtain

$$\gamma_1\tau_1 + \gamma_2 = -\tfrac{3}{16} X(\mu_0)(c_1\beta_2 + c_2\alpha_2 - 3\beta_1) - \tfrac{3}{16} Y(\mu_0)(c_1\alpha_2 + c_2\beta_2 + 3\alpha_1)\,. \qquad (269)$$

Now, substituting for γ_1 and γ_2 according to equations (267) and (268) in equation (269), we find

$$[(\alpha_1 + \beta_1)(c_1 + c_2) + (\alpha_2 - \beta_2)]\,\tau_1 = -2\,[(\alpha_2 + \beta_2)(c_1 + c_2) + 3\,(\alpha_1 - \beta_1)]\,; \qquad (270)$$

or, solving for $(c_1 + c_2)$, we have

$$c_1 + c_2 = -\frac{(\alpha_2 - \beta_2)\,\tau_1 + 6\,(\alpha_1 - \beta_1)}{(\alpha_1 + \beta_1)\,\tau_1 + 2\,(\alpha_2 + \beta_2)}\,. \qquad (271)$$

Since we have already shown that (cf. eq. [266])

$$c_1 - c_2 = -\frac{\alpha_2 + \beta_2}{\alpha_1 - \beta_1}\,, \qquad (272)$$

the solution to the problem is completed.

IV. SCATTERING IN ACCORDANCE WITH THE PHASE FUNCTION $\lambda(1 + x\cos\Theta)$

15. *The equations of the problem.*—In the problem of diffuse reflection and transmission according to the phase function $\lambda(1 + x\cos\Theta)(\lambda < 1,\ 1 \geqslant x \geqslant -1)$ we can express the reflected and the transmitted intensities in the following forms (cf. Paper XIV, eqs. [196] and [199]):

$$I(0; \mu, \varphi; \mu_0, \varphi_0) = \frac{\lambda}{4\mu} F [S^{(0)}(\mu, \mu_0) + x(1-\mu^2)^{\frac{1}{2}}(1-\mu_0^2)^{\frac{1}{2}} \times S^{(1)}(\mu, \mu_0)\cos(\varphi - \varphi_0)] \tag{273}$$

and

$$I(\tau_1; -\mu, \varphi; \mu_0, \varphi_0) = \frac{\lambda}{4\mu} F [T^{(0)}(\mu, \mu_0) + x(1-\mu^2)^{\frac{1}{2}}(1-\mu_0^2)^{\frac{1}{2}} \times T^{(1)}(\mu, \mu_0)\cos(\varphi - \varphi_0)]\,. \tag{274}$$

The system of equations governing $S^{(1)}$ and $T^{(1)}$ are directly reducible to the standard forms considered in Section I. And the terms in the emergent intensities proportional to $\cos(\varphi - \varphi_0)$ are of exactly the same forms as those given in Paper XXI, equations (278) and (279); only the functions $X^{(1)}$ and $Y^{(1)}$ must now be redefined in terms of the functional equations which they satisfy.

Turning to the "zero-order" functions $S^{(0)}(\mu, \mu_0)$ and $T^{(0)}(\mu, \mu_0)$, we find that these functions must be expressible in the forms (cf. Paper XIV, eqs. [205]–[209])

$$\left(\frac{1}{\mu_0}+\frac{1}{\mu}\right) S^{(0)}(\mu, \mu_0) = \psi(\mu)\psi(\mu_0) - \chi(\mu)\chi(\mu_0) - x[\phi(\mu)\phi(\mu_0) - \zeta(\mu)\zeta(\mu_0)] \tag{275}$$

and

$$\left(\frac{1}{\mu_0}-\frac{1}{\mu}\right) T^{(0)}(\mu, \mu_0) = \chi(\mu)\psi(\mu_0) - \psi(\mu)\chi(\mu_0) + x[\zeta(\mu)\phi(\mu_0) - \phi(\mu)\zeta(\mu_0)]\,, \tag{276}$$

where ψ, ϕ, χ, and ζ are defined in terms of $S^{(0)}$ and $T^{(0)}$ in the following manner:

$$\psi(\mu) = 1 + \tfrac{1}{2}\lambda \int_0^1 S^{(0)}(\mu, \mu')\frac{d\mu'}{\mu'}\,, \tag{277}$$

$$\phi(\mu) = \mu - \tfrac{1}{2}\lambda \int_0^1 S^{(0)}(\mu, \mu')\, d\mu'\,, \tag{278}$$

$$\chi(\mu) = e^{-\tau_1/\mu} + \tfrac{1}{2}\lambda \int_0^1 T^{(0)}(\mu, \mu')\frac{d\mu'}{\mu'}\,, \tag{279}$$

and

$$\zeta(\mu) = \mu\, e^{-\tau_1/\mu} + \tfrac{1}{2}\lambda \int_0^1 T^{(0)}(\mu, \mu')\, d\mu'\,. \tag{280}$$

Further, we must also have

$$\frac{\partial S^{(0)}}{\partial \tau_1} = \chi(\mu)\chi(\mu_0) - x\zeta(\mu)\zeta(\mu_0) \tag{281}$$

and

$$\left(\frac{1}{\mu_0}-\frac{1}{\mu}\right)\frac{\partial T^{(0)}}{\partial \tau_1} = \frac{1}{\mu_0}[\psi(\mu)\chi(\mu_0) + x\phi(\mu)\zeta(\mu_0)] - \frac{1}{\mu}[\chi(\mu)\psi(\mu_0) + x\zeta(\mu)\phi(\mu_0)]\,. \tag{282}$$

Substituting for $S^{(0)}$ and $T^{(0)}$ according to equations (275) and (276) in equations (277)–(280), we obtain the following system of functional equations of fourth order:

$$\psi(\mu) = 1 + \tfrac{1}{2}\lambda\mu \int_0^1 \frac{d\mu'}{\mu+\mu'} [\psi(\mu)\psi(\mu') - \chi(\mu)\chi(\mu')] - \tfrac{1}{2}x\lambda\mu \int_0^1 \frac{d\mu'}{\mu+\mu'} [\phi(\mu)\phi(\mu') - \zeta(\mu)\zeta(\mu')] , \tag{283}$$

$$\phi(\mu) = \mu - \tfrac{1}{2}\lambda\mu \int_0^1 \frac{\mu' d\mu'}{\mu+\mu'} [\psi(\mu)\psi(\mu') - \chi(\mu)\chi(\mu')] + \tfrac{1}{2}x\lambda\mu \int_0^1 \frac{\mu' d\mu'}{\mu+\mu'} [\phi(\mu)\phi(\mu') - \zeta(\mu)\zeta(\mu')] , \tag{284}$$

$$\chi(\mu) = e^{-\tau_1/\mu} + \tfrac{1}{2}\lambda\mu \int_0^1 \frac{d\mu'}{\mu-\mu'} [\chi(\mu)\psi(\mu') - \psi(\mu)\chi(\mu')] + \tfrac{1}{2}x\lambda\mu \int_0^1 \frac{d\mu'}{\mu-\mu'} [\zeta(\mu)\phi(\mu') - \phi(\mu)\zeta(\mu')] , \tag{285}$$

and

$$\zeta(\mu) = \mu e^{-\tau_1/\mu} + \tfrac{1}{2}\lambda\mu \int_0^1 \frac{\mu' d\mu'}{\mu-\mu'} [\chi(\mu)\psi(\mu') - \psi(\mu)\chi(\mu')] + \tfrac{1}{2}x\lambda\mu \int_0^1 \frac{\mu' d\mu'}{\mu-\mu'} [\zeta(\mu)\phi(\mu') - \phi(\mu)\zeta(\mu')] . \tag{286}$$

16. *The form of the solution.*—The solutions for the reflected and the transmitted intensities in a general finite approximation have been found in Paper XXI (eqs. [276] and [277]). Applying to these solutions the correspondence enunciated in theorem 9, we are led to assume for $S^{(0)}(\mu, \mu_0)$ and $T^{(0)}(\mu, \mu_0)$ the following forms:

$$\begin{aligned} S^{(0)}(\mu, \mu') = \{ & X(\mu) X(\mu') [1 - x(1-\lambda) c_1 (\mu+\mu') - x(1-\lambda)\mu\mu'] \\ & - Y(\mu) Y(\mu') [1 + x(1-\lambda) c_1 (\mu+\mu') - x(1-\lambda)\mu\mu'] \\ & - x(1-\lambda) c_2 (\mu+\mu') [X(\mu) Y(\mu') + Y(\mu) X(\mu')] \} \frac{\mu\mu'}{\mu+\mu'} \end{aligned} \tag{287}$$

and

$$\begin{aligned} T^{(0)}(\mu, \mu') = \{ & Y(\mu) X(\mu') [1 + x(1-\lambda) c_1 (\mu-\mu') + x(1-\lambda)\mu\mu'] \\ & - X(\mu) Y(\mu') [1 - x(1-\lambda) c_1 (\mu-\mu') + x(1-\lambda)\mu\mu'] \\ & + x(1-\lambda) c_2 (\mu-\mu') [X(\mu) X(\mu') + Y(\mu) Y(\mu')] \} \frac{\mu\mu'}{\mu-\mu'} , \end{aligned} \tag{288}$$

where c_1 and c_2 are certain constants unspecified for the present and $X(\mu)$ and $Y(\mu)$ are solutions of the equations

$$X(\mu) = 1 + \tfrac{1}{2}\lambda\mu \int_0^1 \frac{1 + x(1-\lambda)\mu'^2}{\mu+\mu'} [X(\mu) X(\mu') - Y(\mu) Y(\mu')] d\mu' \tag{289}$$

and

$$Y(\mu) = e^{-\tau_1/\mu} + \tfrac{1}{2}\lambda\mu \int_0^1 \frac{1 + x(1-\lambda)\mu'^2}{\mu-\mu'} [Y(\mu) X(\mu') - X(\mu) Y(\mu')] d\mu' . \tag{290}$$

17. *Verification of the solution and the evaluation of the constants in the solution in terms of the moments of* X(μ) *and* Y(μ).—The verification that the solutions for $S^{(0)}(\mu, \mu')$

and $T^{(0)}(\mu, \mu')$ have the forms assumed in § 16, will consist in first evaluating ψ, ϕ, χ, and ζ according to equations (277)–(280); then requiring that, when the resulting expressions for ψ, ϕ, χ, and ζ are substituted back into equations (275) and (276), we shall recover the form of the solutions assumed; and, finally, showing that the various requirements can be met. In the present case it will appear that the procedure outlined makes the solution determinate.

The evaluation of ψ, ϕ, χ, and ζ according to equations (277)–(280) for $S^{(0)}(\mu, \mu')$ and $T^{(0)}(\mu, \mu')$ given by equations (287) and (288) is straightforward if proper use is made of the integral properties of the functions $X(\mu)$ and $Y(\mu)$. Since these functions are defined in terms of the characteristic function

$$\Psi(\mu) = \tfrac{1}{2}\lambda\,[1 + x(1-\lambda)\,\mu^2]\,, \tag{291}$$

we have (cf. theorem 4, eqs. [44]–[46])

$$\alpha_0 = 1 + \tfrac{1}{4}\lambda\,[\alpha_0^2 - \beta_0^2 + x(1-\lambda)(\alpha_1^2 - \beta_1^2)]\,, \tag{292}$$

$$\begin{aligned}[1 + x(1-\lambda)\,\mu^2] \int_0^1 \frac{X(\mu)\,X(\mu') - Y(\mu)\,Y(\mu')}{\mu + \mu'}\,d\mu' &= \frac{X(\mu) - 1}{\tfrac{1}{2}\lambda\mu} \\ &\quad - x(1-\lambda)\,[(\alpha_1 - \mu\alpha_0)\,X(\mu) - (\beta_1 - \mu\beta_0)\,Y(\mu)]\,,\end{aligned} \tag{293}$$

and

$$\begin{aligned}[1 + x(1-\lambda)\,\mu^2] \int_0^1 \frac{Y(\mu)\,X(\mu') - X(\mu)\,Y(\mu')}{\mu - \mu'}\,d\mu' &= \frac{Y(\mu) - e^{-\tau_1/\mu}}{\tfrac{1}{2}\lambda\mu} \\ &\quad - x(1-\lambda)\,[(\beta_1 + \mu\beta_0)\,X(\mu) - (\alpha_1 + \mu\alpha_0)\,Y(\mu)]\,.\end{aligned} \tag{294}$$

Evaluating ψ, ϕ, χ, and ζ in the manner indicated, we find

$$\psi(\mu) = (1 - q_0\mu)\,X(\mu) - p_0\mu\,Y(\mu)\,, \tag{295}$$

$$\chi(\mu) = (1 + q_0\mu)\,Y(\mu) + p_0\mu\,X(\mu)\,, \tag{296}$$

$$\phi(\mu) = \mu\,[q_1 X(\mu) + p_1 Y(\mu)]\,, \tag{297}$$

and

$$\zeta(\mu) = \mu\,[p_1 X(\mu) + q_1 Y(\mu)]\,, \tag{298}$$

where

$$q_0 = \tfrac{1}{2}x\lambda(1-\lambda)(c_1\alpha_0 + c_2\beta_0 + \alpha_1)\,, \tag{299}$$

$$p_0 = \tfrac{1}{2}x\lambda(1-\lambda)(c_1\beta_0 + c_2\alpha_0 - \beta_1)\,, \tag{300}$$

$$q_1 = 1 + \tfrac{1}{2}\lambda\,[x(1-\lambda)(c_1\alpha_1 + c_2\beta_1) - \alpha_0]\,, \tag{301}$$

and

$$p_1 = \tfrac{1}{2}\lambda\,[x(1-\lambda)(c_1\beta_1 + c_2\alpha_1) + \beta_0]\,. \tag{302}$$

Using the expressions (295)–(298) for ψ, χ, ϕ, and ζ in equations (275) and (276) for $S^{(0)}$ and $T^{(0)}$, we obtain

$$\begin{aligned}\left(\frac{1}{\mu'} + \frac{1}{\mu}\right) S^{(0)}(\mu, \mu') &\\ = X(\mu)\,X(\mu')\,[1 - q_0(\mu + \mu') + \{q_0^2 - p_0^2 - x(q_1^2 - p_1^2)\}\,\mu\mu'] &\\ - Y(\mu)\,Y(\mu')\,[1 + q_0(\mu + \mu') + \{q_0^2 - p_0^2 - x(q_1^2 - p_1^2)\}\,\mu\mu'] &\\ - p_0(\mu + \mu')\,[X(\mu)\,Y(\mu') + Y(\mu)\,X(\mu')] &\end{aligned} \tag{303}$$

and

$$\left(\frac{1}{\mu'}-\frac{1}{\mu}\right)T^{(0)}(\mu,\mu')$$
$$= Y(\mu)X(\mu')[1+q_0(\mu-\mu')-\{q_0^2-p_0^2-x(q_1^2-p_1^2)\}\mu\mu']$$
$$-X(\mu)Y(\mu')[1-q_0(\mu-\mu')-\{q_0^2-p_0^2-x(q_1^2-p_1^2)\}\mu\mu'] \tag{304}$$
$$+p_0(\mu-\mu')[X(\mu)X(\mu')+Y(\mu)Y(\mu')].$$

Now, comparing equations (303) and (304) and (287) and (288), we observe that we must have

$$q_0 = x(1-\lambda)c_1 \quad \text{and} \quad p_0 = x(1-\lambda)c_2; \tag{305}$$

further, we must also require that

$$q_0^2-p_0^2-x(q_1^2-p_1^2) = -x(1-\lambda). \tag{306}$$

According to equations (299), (300), and (305), we have

$$c_1 = \tfrac{1}{2}\lambda(c_1\alpha_0+c_2\beta_0+\alpha_1) \tag{307}$$

and

$$c_2 = \tfrac{1}{2}\lambda(c_1\beta_0+c_2\alpha_0-\beta_1). \tag{308}$$

Solving these equations for c_1 and c_2, we obtain

$$c_1 = \frac{q_0}{x(1-\lambda)} = \lambda\frac{(2-\lambda\alpha_0)\alpha_1-\lambda\beta_0\beta_1}{(2-\lambda\alpha_0)^2-\lambda^2\beta_0^2} \tag{309}$$

and

$$c_2 = \frac{p_0}{x(1-\lambda)} = \lambda\frac{-(2-\lambda\alpha_0)\beta_1+\lambda\beta_0\alpha_1}{(2-\lambda\alpha_0)^2-\lambda^2\beta_0^2}. \tag{310}$$

Inserting for c_1 and c_2 from equations (309) and (310) in equation (301) for q_1, we find

$$q_1 = \tfrac{1}{2}(2-\lambda\alpha_0)\left[1+\lambda\frac{x\lambda(1-\lambda)(\alpha_1^2-\beta_1^2)}{(2-\lambda\alpha_0)^2-\lambda^2\beta_0^2}\right]. \tag{311}$$

Using the relation (cf. eq. [292])

$$4(\alpha_0-1)-\lambda(\alpha_0^2-\beta_0^2) = x\lambda(1-\lambda)(\alpha_1^2-\beta_1^2), \tag{312}$$

we can now reduce equation (311) to the form

$$q_1 = \frac{2(1-\lambda)(2-\lambda\alpha_0)}{(2-\lambda\alpha_0)^2-\lambda^2\beta_0^2}. \tag{313}$$

Similarly,

$$p_1 = \frac{2(1-\lambda)\lambda\beta_0}{(2-\lambda\alpha_0)^2-\lambda^2\beta_0^2}. \tag{314}$$

It remains to verify that equation (306) is valid for q_0, p_0, q_1, and p_1 given by equations (309), (310), (313), and (314). To show that this is the case, we first observe that, according to equations (309) and (310),

$$q_0^2-p_0^2 = x^2\lambda^2(1-\lambda)^2\frac{\alpha_1^2-\beta_1^2}{(2-\lambda\alpha_0)^2-\lambda^2\beta_0^2}, \tag{315}$$

while, according to equations (313) and (314),

$$q_1^2 - p_1^2 = \frac{4(1-\lambda)^2}{(2-\lambda\alpha_0)^2 - \lambda^2\beta_0^2}. \tag{316}$$

Hence

$$q_0^2 - p_0^2 - x(q_1^2 - p_1^2) = \frac{x(1-\lambda)}{(2-\lambda\alpha_0)^2 - \lambda^2\beta_0^2}[x\lambda^2(1-\lambda)(\alpha_1^2 - \beta_1^2) - 4(1-\lambda)]; \tag{317}$$

or, using equation (312), we have

$$\begin{aligned} q_0^2 - p_0^2 - x(q_1^2 - p_1^2) \\ &= \frac{x(1-\lambda)}{(2-\lambda\alpha_0)^2 - \lambda^2\beta_0^2}[4\lambda(\alpha_0 - 1) - \lambda^2(\alpha_0^2 - \beta_0^2) - 4(1-\lambda)] \\ &= \frac{x(1-\lambda)}{(2-\lambda\alpha_0)^2 - \lambda^2\beta_0^2}[\lambda^2\beta_0^2 - 4 + 4\lambda\alpha_0 - \lambda^2\alpha_0^2] \\ &= -x(1-\lambda). \end{aligned} \tag{318}$$

The constants q_0, p_0, q_1, and p_1, as defined by equations (309), (310), (313), and (314), are therefore related, as required.

This completes the verification.

V. RAYLEIGH SCATTERING

18. *The equations of the problem.*—When proper allowance is made for the polarization characteristics of the radiation field, the laws of diffuse reflection and transmission are best formulated in terms of a scattering matrix $\boldsymbol{S}$ and a transmission matrix $\boldsymbol{T}$ (cf. Paper XXI, eqs. [533] and [534]). And, as we have already indicated in Paper XVII (the last paragraph of § 5 on p. 455), the equations governing $\boldsymbol{S}$ and $\boldsymbol{T}$ are of the same forms as equations (85)–(88) of Paper XVII, provided that these equations are interpreted as matrix equations in which a *phase matrix* plays the same role as the *phase function* in the more conventional problems. For the particular case of Rayleigh scattering the phase matrix is explicitly known (cf. Paper XIV, eq. [10]), and the required equations for the field quantities I_l, I_r, U, and V[14] can be written down. However, in view of the form of the solutions for U, V, and the azimuth dependent terms in I_l and I_r found in Paper XXI (eqs. [290]–[303] and [548]–[551]) it is evident that the exact solutions for these terms in the scattering and the transmission matrices must be of identically the same forms: only the various X- and Y-functions occurring in the solutions must be redefined in terms of the exact functional equations which they satisfy. Consequently, it is sufficient to confine our detailed considerations to the azimuth independent terms in I_l and I_r which we shall now regard as the components of a *two-dimensional* vector

$$\boldsymbol{I} = (I_l, I_r). \tag{319}$$ [15]

Let

$$\boldsymbol{F} = (F_l, F_r), \tag{320}$$

[14] For a definition of these quantities see Paper XXI, § 15, and the references there given.

[15] Strictly, superscripts (0) should be attached to these and similar azimuth independent quantities describing the diffuse-radiation field. We have suppressed them for the sake of convenience. They should, however, be restored when writing down the complete solution (cf. Paper XXI, § 19, eqs. [540]–[546]).

where πF_l and πF_r are the incident fluxes in the intensities in the meridian plane and at right angles to it,[16] respectively. The reflected and the transmitted intensities can then be expressed in terms of a scattering and a transmission matrix (with two rows and columns) in the forms

$$I(0, \mu) = \frac{3}{16\mu} S(\mu, \mu_0) F \qquad \text{and} \qquad I(\tau_1, -\mu) = \frac{3}{16\mu} T(\mu, \mu_0) F . \tag{321}$$ [15]

The equations governing S and T can be written down in analogy with equations (85)–(88) of Paper XVII by replacing $p(\mu, \varphi; \mu', \varphi')$ by the matrix

$$\tfrac{3}{4} J(\mu, \mu') = \tfrac{3}{4} \begin{pmatrix} 2(1-\mu^2)(1-\mu'^2) + \mu^2\mu'^2 & \mu^2 \\ \mu'^2 & 1 \end{pmatrix}. \tag{322}$$

The resulting equations can be written most compactly by adopting the following notation:

Let the "product" $[A, B]_{\mu,\mu'}$ of two matrices, $A(\mu, \mu')$ and $B(\mu, \mu')$, be defined by the formula

$$[A, B]_{\mu,\mu'} = \frac{3}{8} \int_0^1 A(\mu, \mu'') B(\mu'', \mu') \frac{d\mu''}{\mu''}, \tag{323}$$

where, under the integral sign, the ordinary matrix product is intended. With this product notation, the equations satisfied by S and T take the forms

$$\left(\frac{1}{\mu_0} + \frac{1}{\mu}\right) S + \frac{\partial S}{\partial \tau_1} = J + [J, S] + [S, J] + [[S, J], S], \tag{324}$$

$$\frac{\partial S}{\partial \tau_1} = \exp\left\{-\tau_1\left(\frac{1}{\mu} + \frac{1}{\mu_0}\right)\right\} J + e^{-\tau_1/\mu}[J, T] + e^{-\tau_1/\mu_0}[T, J] + [[T, J], T], \tag{325}$$

$$\frac{1}{\mu_0} T + \frac{\partial T}{\partial \tau_1} = e^{-\tau_1/\mu} J + e^{-\tau_1/\mu}[J, S] + [T, J] + [[T, J], S], \tag{326}$$

and

$$\frac{1}{\mu} T + \frac{\partial T}{\partial \tau_1} = e^{-\tau_1/\mu_0} J + [J, T] + e^{-\tau_1/\mu_0}[S, J] + [[S, J], T]. \tag{327}$$

A discussion of equations (324)–(327) shows that S and T must be expressible in the forms

$$\begin{aligned} \left(\frac{1}{\mu'} + \frac{1}{\mu}\right) S(\mu, \mu') &= \begin{pmatrix} \psi(\mu) & 2^{\frac{1}{2}}\phi(\mu) \\ \chi(\mu) & 2^{\frac{1}{2}}\zeta(\mu) \end{pmatrix} \begin{pmatrix} \psi(\mu') & \chi(\mu') \\ 2^{\frac{1}{2}}\phi(\mu') & 2^{\frac{1}{2}}\zeta(\mu') \end{pmatrix} \\ &\quad - \begin{pmatrix} \xi(\mu) & 2^{\frac{1}{2}}\eta(\mu) \\ \sigma(\mu) & 2^{\frac{1}{2}}\theta(\mu) \end{pmatrix} \begin{pmatrix} \xi(\mu') & \sigma(\mu') \\ 2^{\frac{1}{2}}\eta(\mu') & 2^{\frac{1}{2}}\theta(\mu') \end{pmatrix} \end{aligned} \tag{328}$$

and

$$\begin{aligned} \left(\frac{1}{\mu'} - \frac{1}{\mu}\right) T(\mu, \mu') &= \begin{pmatrix} \xi(\mu) & 2^{\frac{1}{2}}\eta(\mu) \\ \sigma(\mu) & 2^{\frac{1}{2}}\theta(\mu) \end{pmatrix} \begin{pmatrix} \psi(\mu') & \chi(\mu') \\ 2^{\frac{1}{2}}\phi(\mu') & 2^{\frac{1}{2}}\zeta(\mu') \end{pmatrix} \\ &\quad - \begin{pmatrix} \psi(\mu) & 2^{\frac{1}{2}}\phi(\mu) \\ \chi(\mu) & 2^{\frac{1}{2}}\zeta(\mu) \end{pmatrix} \begin{pmatrix} \xi(\mu') & \sigma(\mu') \\ 2^{\frac{1}{2}}\eta(\mu') & 2^{\frac{1}{2}}\theta(\mu') \end{pmatrix}, \end{aligned} \tag{329}$$

[16] These directions are referred in the transverse plane containing the electric and the magnetic vectors.

where

$$\psi(\mu) = \mu^2 + \frac{3}{8}\int_0^1 \frac{d\mu'}{\mu'}[\mu'^2 S_{11}(\mu, \mu') + S_{12}(\mu, \mu')], \tag{330}$$

$$\phi(\mu) = 1 - \mu^2 + \frac{3}{8}\int_0^1 \frac{d\mu'}{\mu'}(1 - \mu'^2) S_{11}(\mu, \mu'), \tag{331}$$

$$\chi(\mu) = 1 + \frac{3}{8}\int_0^1 \frac{d\mu'}{\mu'}[\mu'^2 S_{21}(\mu, \mu') + S_{22}(\mu, \mu')], \tag{332}$$

$$\zeta(\mu) = \frac{3}{8}\int_0^1 \frac{d\mu'}{\mu'}(1 - \mu'^2) S_{21}(\mu, \mu'), \tag{333}$$

$$\xi(\mu) = \mu^2 e^{-\tau_1/\mu} + \frac{3}{8}\int_0^1 \frac{d\mu'}{\mu'}[\mu'^2 T_{11}(\mu, \mu') + T_{12}(\mu, \mu')], \tag{334}$$

$$\eta(\mu) = (1 - \mu^2) e^{-\tau_1/\mu} + \frac{3}{8}\int_0^1 \frac{d\mu'}{\mu'}(1 - \mu'^2) T_{11}(\mu, \mu'). \tag{335}$$

$$\sigma(\mu) = e^{-\tau_1/\mu} + \frac{3}{8}\int_0^1 \frac{d\mu'}{\mu'}[\mu'^2 T_{21}(\mu, \mu') + T_{22}(\mu, \mu')], \tag{336}$$

and

$$\theta(\mu) = \frac{3}{8}\int_0^1 \frac{d\mu'}{\mu'}(1 - \mu'^2) T_{21}(\mu, \mu'). \tag{337}$$

Substituting for S_{11}, etc., according to equations (328) and (329) in equations (330)–(337), we shall obtain a simultaneous system of functional equations of order *eight*. It is, however, not necessary to write down these equations explicitly.

19. *The form of the solution.*—The solutions for $\boldsymbol{S}$ and $\boldsymbol{T}$ in a general finite approximation have already been found in Paper XXI (eqs. [539]–[546]). Applying to these solutions the correspondence enunciated in theorem 9, we are led to assume the following forms for $\boldsymbol{S}$ and $\boldsymbol{T}$:

$$\begin{aligned}\left(\frac{1}{\mu'}+\frac{1}{\mu}\right) S_{11}(\mu, \mu') = 2\{ & X_l(\mu) X_l(\mu')[1 + \nu_4(\mu + \mu') + \mu\mu'] \\ & - Y_l(\mu) Y_l(\mu')[1 - \nu_4(\mu + \mu') + \mu\mu'] \\ & - \nu_3(\mu + \mu')[X_l(\mu) Y_l(\mu') + Y_l(\mu) X_l(\mu')]\},\end{aligned} \tag{338}$$

$$\begin{aligned}\left(\frac{1}{\mu'}+\frac{1}{\mu}\right) S_{12}(\mu, \mu') = (\mu + \mu')\{ & \nu_1[Y_l(\mu) X_r(\mu') + X_l(\mu) Y_r(\mu')] \\ & - \nu_2[X_l(\mu) X_r(\mu') + Y_l(\mu) Y_r(\mu')] \\ & + Q(\nu_2 - \nu_1)\mu'[X_l(\mu) + Y_l(\mu)][X_r(\mu') - Y_r(\mu')]\},\end{aligned} \tag{339}$$

$$\begin{aligned}\left(\frac{1}{\mu'}+\frac{1}{\mu}\right) S_{21}(\mu, \mu') = (\mu + \mu')\{ & \nu_1[X_r(\mu) Y_l(\mu') + Y_r(\mu) X_l(\mu')] \\ & - \nu_2[X_r(\mu) X_l(\mu') + Y_r(\mu) Y_l(\mu')] \\ & + Q(\nu_2 - \nu_1)\mu[X_r(\mu) - Y_r(\mu)][X_l(\mu') + Y_l(\mu')]\},\end{aligned} \tag{340}$$

$$\left(\frac{1}{\mu'}+\frac{1}{\mu}\right) S_{22}(\mu; \mu') = X_r(\mu) X_r(\mu')[1 - u_4(\mu+\mu') + u_5\mu\mu'] \\ - Y_r(\mu) Y_r(\mu')[1 + u_4(\mu+\mu') + u_5\mu\mu'] \\ + u_3(\mu+\mu')[X_r(\mu) Y_r(\mu') + Y_r(\mu) X_r(\mu')] \\ - Qu_5\mu\mu'(\mu+\mu')[X_r(\mu) - Y_r(\mu)][X_r(\mu') - Y_r(\mu')] \\ + Q(u_4 - u_3)\{\mu^2[X_r(\mu) - Y_r(\mu)][X_r(\mu') + Y_r(\mu')] \\ + \mu'^2[X_r(\mu) + Y_r(\mu)][X_r(\mu') - Y_r(\mu')]\}, \tag{341}$$

$$\left(\frac{1}{\mu'}-\frac{1}{\mu}\right) T_{11}(\mu, \mu') = 2\{Y_l(\mu) X_l(\mu')[1 - \nu_4(\mu-\mu') - \mu\mu'] \\ - X_l(\mu) Y_l(\mu')[1 + \nu_4(\mu-\mu') - \mu\mu'] \\ + \nu_3(\mu-\mu')[X_l(\mu) X_l(\mu') + Y_l(\mu) Y_l(\mu')]\}, \tag{342}$$

$$\left(\frac{1}{\mu'}-\frac{1}{\mu}\right) T_{12}(\mu, \mu') = (\mu-\mu')\{\nu_2[X_l(\mu) Y_r(\mu') + Y_l(\mu) X_r(\mu')] \\ - \nu_1[X_l(\mu) X_r(\mu') + Y_l(\mu) Y_r(\mu')] \\ - Q(\nu_2-\nu_1)\mu'[X_l(\mu) + Y_l(\mu)][X_r(\mu') - Y_r(\mu')]\}, \tag{343}$$

$$\left(\frac{1}{\mu'}-\frac{1}{\mu}\right) T_{21}(\mu, \mu') = (\mu-\mu')\{\nu_2[X_r(\mu) Y_l(\mu') + Y_r(\mu) X_l(\mu')] \\ - \nu_1[X_r(\mu) X_l(\mu') + Y_r(\mu) Y_l(\mu')] \\ - Q(\nu_2-\nu_1)\mu[X_r(\mu) - Y_r(\mu)][X_l(\mu') + Y_l(\mu')]\}, \tag{344}$$

and

$$\left(\frac{1}{u'}-\frac{1}{\mu}\right) T_{22}(\mu, \mu') = Y_r(\mu) X_r(\mu')[1 + u_4(\mu-\mu') - u_5\mu\mu'] \\ - X_r(\mu) Y_r(\mu')[1 - u_4(\mu-\mu') - u_5\mu\mu'] \\ - u_3(\mu-\mu')[X_r(\mu) X_r(\mu') + Y_r(\mu) Y_r(\mu')] \\ + Qu_5\mu\mu'(\mu-\mu')[X_r(\mu) - Y_r(\mu)][X_r(\mu') - Y_r(\mu')] \\ - Q(u_4 - u_3)\{\mu^2[X_r(\mu) - Y_r(\mu)][X_r(\mu') + Y_r(\mu')] \\ - \mu'^2[X_r(\mu) + Y_r(\mu)][X_r(\mu') - Y_r(\mu')]\}, \tag{345}$$

where ν_1, ν_2, ν_3, ν_4, u_3, u_4, and Q are certain constants, unspecified for the present;

$$u_5 = 1 + 2Q(u_4 - u_3); \tag{346}$$[17]

$X_r(\mu)$ and $Y_r(\mu)$ are the solutions of the equations

$$X_r(\mu) = 1 + \tfrac{3}{8}\mu\int_0^1 \frac{1-\mu'^2}{\mu+\mu'}[X_r(\mu) X_r(\mu') - Y_r(\mu) Y_r(\mu')]d\mu' \tag{347}$$

and

$$Y_r(\mu) = e^{-\tau_1/\mu} + \tfrac{3}{8}\mu\int_0^1 \frac{1-\mu'^2}{\mu-\mu'}[Y_r(\mu) X_r(\mu') - X_r(\mu) Y_r(\mu')]d\mu'; \tag{348}$$

[17] Cf. Paper XXI, eqs. (512)–(514).

and $X_l(\mu)$ and $Y_l(\mu)$ are the *standard solutions* of the equations

$$X_l(\mu) = 1 + \tfrac{3}{4}\mu \int_0^1 \frac{1-\mu'^2}{\mu+\mu'} [X_l(\mu) X_l(\mu') - Y_l(\mu) Y_l(\mu')] d\mu' \tag{349}$$

and

$$Y_l(\mu) = e^{-\tau_1/\mu} + \tfrac{3}{4}\mu \int_0^1 \frac{1-\mu'^2}{\mu-\mu'} [Y_l(\mu) X_l(\mu') - X_l(\mu) Y_l(\mu')] d\mu', \tag{350}$$

having the property

$$\tfrac{3}{4} \int_0^1 (1-\mu^2) X_l(\mu) d\mu = \tfrac{3}{4}(\alpha_0 - \alpha_2) = 1 \tag{351}$$

and

$$\int_0^1 (1-\mu^2) Y_l(\mu) d\mu = (\beta_0 - \beta_2) = 0. \tag{352}$$

For the purposes of the various evaluations in §§ 20 and 21, it is convenient to have equations (338)–(345) re-written in the following forms:

$$\begin{aligned} S_{11}(\mu,\mu') = 2\mu\mu' \Big\{ &\frac{1-\mu^2}{\mu+\mu'} [X_l(\mu) X_l(\mu') - Y_l(\mu) Y_l(\mu')] \\ &+ X_l(\mu)[(\nu_4+\mu) X_l(\mu') - \nu_3 Y_l(\mu')] \\ &+ Y_l(\mu)[-\nu_3 X_l(\mu') + (\nu_4-\mu) Y_l(\mu')] \Big\}, \end{aligned} \tag{353}$$

$$\begin{aligned} S_{12}(\mu,\mu') = \mu\mu' \{ &\nu_1 [X_l(\mu) Y_r(\mu') + Y_l(\mu) X_r(\mu')] \\ &- \nu_2 [X_l(\mu) X_r(\mu') + Y_l(\mu) Y_r(\mu')] \\ &+ Q(\nu_2-\nu_1)\mu' [X_l(\mu) + Y_l(\mu)][X_r(\mu') - Y_r(\mu')] \}, \end{aligned} \tag{354}$$

$$\begin{aligned} S_{21}(\mu,\mu') = \mu\mu' \{ &\nu_1 [X_r(\mu) Y_l(\mu') + Y_r(\mu) X_l(\mu')] \\ &- \nu_2 [X_r(\mu) X_l(\mu') + Y_r(\mu) Y_l(\mu')] \\ &+ Q(\nu_2-\nu_1)\mu [X_r(\mu) - Y_r(\mu)][X_l(\mu') + Y_l(\mu')] \} \end{aligned} \tag{355}$$

$$\begin{aligned} S_{22}(\mu,\mu') = \mu\mu' \Big\{ &\frac{1-\mu^2}{\mu+\mu'} [X_r(\mu) X_r(\mu') - Y_r(\mu) Y_r(\mu')] \\ &+ X_r(\mu)[(-u_4+\mu) X_r(\mu') + u_3 Y_r(\mu')] \\ &+ Y_r(\mu)[u_3 X_r(\mu') - (u_4+\mu) Y_r(\mu')] \\ &- Q u_5 \mu\mu' [X_r(\mu) - Y_r(\mu)][X_r(\mu') - Y_r(\mu')] \\ &+ Q(u_4-u_3)(\mu+\mu')[X_r(\mu) X_r(\mu') - Y_r(\mu) Y_r(\mu')] \\ &+ Q(u_4-u_3)(\mu-\mu')[X_r(\mu) Y_r(\mu') - Y_r(\mu) X_r(\mu')] \Big\}, \end{aligned} \tag{356}$$[18]

$$\begin{aligned} T_{11}(\mu,\mu') = 2\mu\mu' \Big\{ &\frac{1-\mu^2}{\mu-\mu'} [Y_l(\mu) X_l(\mu') - X_l(\mu) Y_l(\mu')] \\ &- X_l(\mu)[-\nu_3 X_l(\mu') + (\nu_4+\mu) Y_l(\mu')] \\ &- Y_l(\mu)[(\nu_4-\mu) X_l(\mu') - \nu_3 Y_l(\mu')] \Big\}, \end{aligned} \tag{357}$$

[18] The reduction of eqs. (341) and (345) to the forms (356) and (360) requires the use of eq. (346).

$$T_{12}(\mu, \mu') = \mu\mu'\{\nu_2[X_l(\mu)\,Y_r(\mu') + Y_l(\mu)\,X_r(\mu')] - \nu_1[X_l(\mu)\,X_r(\mu') + Y_l(\mu)\,Y_r(\mu')] - Q(\nu_2-\nu_1)\,\mu'[X_l(\mu) + Y_l(\mu)][X_r(\mu') - Y_r(\mu')]\}, \tag{358}$$

$$T_{21}(\mu, \mu') = \mu\mu'\{\nu_2[X_r(\mu)\,Y_l(\mu') + Y_r(\mu)\,X_l(\mu')] - \nu_1[X_r(\mu)\,X_l(\mu') + Y_r(\mu)\,Y_l(\mu')] - Q(\nu_2-\nu_1)\,\mu[X_r(\mu) - Y_r(\mu)][X_l(\mu') + Y_l(\mu')]\}, \tag{359}$$

and

$$\begin{aligned} T_{22}(\mu, \mu') = \mu\mu'\Big\{\frac{1-\mu^2}{\mu-\mu'}&[Y_r(\mu)\,X_r(\mu') - X_r(\mu)\,Y_r(\mu')] \\ &- X_r(\mu)[u_3X_r(\mu') + (-u_4+\mu)\,Y_r(\mu')] \\ &- Y_r(\mu)[-(u_4+\mu)\,X_r(\mu') + u_3Y_r(\mu')] \\ &+ Qu_5\mu\mu'[X_r(\mu) - Y_r(\mu)][X_r(\mu') - Y_r(\mu')] \\ &- Q(u_4-u_3)(\mu-\mu')[X_r(\mu)\,Y_r(\mu') - Y_r(\mu)\,X_r(\mu')] \\ &- Q(u_4-u_3)(\mu+\mu')[X_r(\mu)\,X_r(\mu') - Y_r(\mu)\,Y_r(\mu')]\Big\}. \end{aligned} \tag{360}$$ [18]

20. *The verification of the solution and the expression of the constants* ν_1, ν_2, ν_3, ν_4, u_3, *and* u_4 *in terms of the moments of* $X_l(\mu)$, $Y_l(\mu)$, $X_r(\mu)$, *and* $Y_r(\mu)$ *and a single arbitrary constant* Q.—We shall first evaluate ψ, ϕ, χ, ζ, ξ, η, σ, and θ according to equations (330)–(337) for $\boldsymbol{S}$ and $\boldsymbol{T}$ given by equations (353)–(360); then require that, when the resulting expressions for ψ, ϕ, etc., are substituted back into equations (328) and (329), we shall recover the form of the solutions assumed. As we should expect, this procedure will lead to several conditions[19] among the constants ν_1, ν_2, ν_3, ν_4, u_3, u_4, and Q introduced into the solution. We shall show that all these conditions can be met and that six of the constants (ν_1, ν_2, ν_3, ν_4, u_3, and u_4) can be expressed in terms of Q and the various moments of X_l, Y_l, X_r, and Y_r. The constant Q itself will be found to be left arbitrary. This is a further example of the one-parametric nature of the solutions of the functional equations incorporating the invariances of the problem in conservative cases. In § 21 we shall then finally show how this last arbitrariness in the solutions can be removed by appealing to the K-integrals of the problem.

The evaluation of ψ, ϕ, etc., according to equations (330)–(337) for $\boldsymbol{S}$ and $\boldsymbol{T}$ given by equations (353)–(360) is straightforward if proper use is made of the various integral properties of the functions X_l, Y_l, X_r, and Y_r. In addition to equations (351) and (352), use must also be made of the following relations (cf. theorem 4, eqs. [44]–[46]):

$$\alpha_0 = 1 + \tfrac{3}{8}[(\alpha_0^2 - \beta_0^2) - (\alpha_1^2 - \beta_1^2)], \tag{361}$$

$$(1-\mu^2)\int_0^1 \frac{X_l(\mu)\,X_l(\mu') - Y_l(\mu)\,Y_l(\mu')}{\mu+\mu'}\,d\mu' = \frac{X_l(\mu)-1}{\tfrac{3}{4}\mu} + (\alpha_1 - \mu\alpha_0)\,X_l(\mu) - (\beta_1 - \mu\beta_0)\,Y_l(\mu), \tag{362}$$

[19] Actually, we shall see that there are twelve of them.

$$(1-\mu^2)\int_0^1 \frac{Y_l(\mu)\,X_l(\mu')-X_l(\mu)\,Y_l(\mu')}{\mu-\mu'}\,d\mu' = \frac{Y_l(\mu)-e^{-\tau_1/\mu}}{\frac{3}{4}\mu} \tag{363}$$

$$+(\beta_1+\mu\beta_0)\,X_l(\mu)-(\alpha_1+\mu\alpha_0)\,Y_l(\mu)\,,$$

$$A_0 = 1+\tfrac{3}{16}\left[(A_0^2-B_0^2)-(A_1^2-B_1^2)\right]\,, \tag{364}$$

$$(1-\mu^2)\int_0^1 \frac{X_r(\mu)\,X_r(\mu')-Y_r(\mu)\,Y_r(\mu')}{\mu+\mu'}\,d\mu' = \frac{X_r(\mu)-1}{\frac{3}{8}\mu} \tag{365}$$

$$+(A_1-\mu A_0)\,X_r(\mu)-(B_1-\mu B_0)\,Y_r(\mu)\,,$$

and

$$(1-\mu^2)\int_0^1 \frac{Y_r(\mu)\,X_r(\mu')-X_r(\mu)\,Y_r(\mu')}{\mu-\mu'}\,d\mu' = \frac{Y_r(\mu)-e^{-\tau_1/\mu}}{\frac{3}{8}\mu} \tag{366}$$

$$+(B_1+\mu B_0)\,X_r(\mu)-(A_1+\mu A_0)\,Y_r(\mu)\,,$$

where (cf. eq. [11])

$$\alpha_n=\int_0^1 X_l(\mu)\,\mu^n d\mu\,,\qquad \beta_n=\int_0^1 Y_l(\mu)\,\mu^n d\mu\,,$$
$$A_n=\int_0^1 X_r(\mu)\,\mu^n d\mu\,,\quad\text{and}\quad B_n=\int_0^1 Y_r(\mu)\,\mu^n d\mu\,. \tag{367}$$

Evaluating ψ, ϕ, etc., in the manner indicated, we find that

$$\psi(\mu) = +\mu\left[q_1X_l(\mu)+q_2Y_l(\mu)\right]\,, \tag{368}$$

$$\xi(\mu) = -\mu\left[q_2X_l(\mu)+q_1Y_l(\mu)\right]\,, \tag{369}$$

$$\phi(\mu) = (1+\nu_4\mu)\,X_l(\mu)-\nu_3\mu\,Y_l(\mu)\,, \tag{370}$$

$$\eta(\mu) = (1-\nu_4\mu)\,Y_l(\mu)+\nu_3\mu X_l(\mu)\,, \tag{371}$$

$$\chi(\mu) = (1+p_1\mu)\,X_r(\mu)+p_2\mu\,Y_r(\mu)-t\mu^2\left[X_r(\mu)-Y_r(\mu)\right]\,, \tag{372}$$

$$\sigma(\mu) = (1-p_1\mu)\,Y_r(\mu)-p_2\mu X_r(\mu)+t\mu^2\left[X_r(\mu)-Y_r(\mu)\right]\,, \tag{373}$$

$$\zeta(\mu) = -\tfrac{1}{2}\mu\left[\nu_2X_r(\mu)-\nu_1Y_r(\mu)\right]+\tfrac{1}{2}Q(\nu_2-\nu_1)\,\mu^2\left[X_r(\mu)-Y_r(\mu)\right]\,, \tag{374}$$

and

$$\theta(\mu) = +\tfrac{1}{2}\mu\left[-\nu_1X_r(\mu)+\nu_2Y_r(\mu)\right]-\tfrac{1}{2}Q(\nu_2-\nu_1)\,\mu^2\left[X_r(\mu)-Y_r(\mu)\right]\,, \tag{375}$$

where

$$q_1=\tfrac{3}{4}\left[\alpha_2\nu_4-\beta_0\nu_3+\alpha_1+\tfrac{1}{2}B_0\nu_1-\tfrac{1}{2}A_0\nu_2+\tfrac{1}{2}Q(\nu_2-\nu_1)(A_1-B_1)\right]\,, \tag{376}$$

$$q_2=\tfrac{3}{4}\left[\beta_0\nu_4-\alpha_2\nu_3-\beta_1+\tfrac{1}{2}A_0\nu_1-\tfrac{1}{2}B_0\nu_2+\tfrac{1}{2}Q(\nu_2-\nu_1)(A_1-B_1)\right]\,, \tag{377}$$

$$p_1=\tfrac{3}{8}\left[+A_1-A_0u_4+B_0u_3+Q(u_4-u_3)(A_1-B_1)+\beta_0\nu_1-\alpha_2\nu_2\right]\,, \tag{378}$$

$$p_2=\tfrac{3}{8}\left[-B_1-B_0u_4+A_0u_3+Q(u_4-u_3)(A_1-B_1)+\alpha_2\nu_1-\beta_0\nu_2\right]\,, \tag{379}$$

and

$$t=\tfrac{3}{8}Q\left[-(\alpha_2+\beta_0)(\nu_2-\nu_1)-(A_0+B_0)(u_4-u_3)+u_5(A_1-B_1)\right]\,. \tag{380}$$

We now substitute for ψ, ϕ, etc., according to equations (368)–(375) in equations (328) and (329) and compare them with the solutions (353)–(360), which were originally assumed in the evaluation of ψ, ϕ, etc.

Considering, first, T_{11}, we have

$$\left(\frac{1}{\mu'}-\frac{1}{\mu}\right)T_{11}(\mu,\mu') = \xi(\mu)\psi(\mu') - \psi(\mu)\xi(\mu') + 2[\eta(\mu)\phi(\mu') - \phi(\mu)\eta(\mu')]$$

$$\begin{aligned} = 2\{ & Y_l(\mu)X_l(\mu')[1-\nu_4(\mu-\mu') - \{\nu_4^2-\nu_3^2+\tfrac{1}{2}(q_1^2-q_2^2)\}\mu\mu'] \\ & - X_l(\mu)Y_l(\mu')[1+\nu_4(\mu-\mu') - \{\nu_4^2-\nu_3^2+\tfrac{1}{2}(q_1^2-q_2^2)\}\mu\mu'] \\ & +\nu_3(\mu-\mu')[X_l(\mu)X_l(\mu') + Y_l(\mu)Y_l(\mu')]\}. \end{aligned} \quad (381)$$

Comparing this with equation (342), we conclude that we must have

$$\nu_4^2-\nu_3^2+\tfrac{1}{2}(q_1^2-q_2^2) = 1\,. \quad (382)$$

The consideration of S_{11} leads to the same condition, (382).

Considering, next, $S_{21}(\mu, \mu')$, we have

$$\left(\frac{1}{\mu'}+\frac{1}{\mu}\right)S_{21}(\mu,\mu') = \chi(\mu)\psi(\mu') - \sigma(\mu)\xi(\mu') + 2[\zeta(\mu)\phi(\mu') - \theta(\mu)\eta(\mu')]$$

$$\begin{aligned} = & X_r(\mu)Y_l(\mu')[+\nu_1\mu+q_2\mu'+\{p_1q_2-p_2q_1+\nu_2\nu_3-\nu_1\nu_4-Q(\nu_2-\nu_1)\}\mu\mu'] \\ & + Y_r(\mu)X_l(\mu')[+\nu_1\mu+q_2\mu'-\{p_1q_2-p_2q_1+\nu_2\nu_3-\nu_1\nu_4-Q(\nu_2-\nu_1)\}\mu\mu'] \\ & + X_r(\mu)X_l(\mu')[-\nu_2\mu+q_1\mu'+\{p_1q_1-p_2q_2+\nu_1\nu_3-\nu_2\nu_4-Q(\nu_2-\nu_1)\}\mu\mu'] \\ & + Y_r(\mu)Y_l(\mu')[-\nu_2\mu+q_1\mu'-\{p_1q_1-p_2q_2+\nu_1\nu_3-\nu_2\nu_4-Q(\nu_2-\nu_1)\}\mu\mu'] \\ & +Q(\nu_2-\nu_1)\mu(\mu+\mu')[X_r(\mu)-Y_r(\mu)][X_l(\mu')+Y_l(\mu')] \\ & +\mu^2\mu'[Q(\nu_2-\nu_1)(\nu_4+\nu_3)+t(q_2-q_1)][X_r(\mu)-Y_r(\mu)][X_l(\mu')-Y_l(\mu')]\,. \end{aligned} \quad (383)$$

Comparing equations (340) and (383), we find that we must have

$$q_2=\nu_1\,; \qquad q_1=-\nu_2\,, \quad (384)$$

$$Q(\nu_2-\nu_1)(\nu_4+\nu_3)+t(q_2-q_1)=0\,, \quad (385)$$

$$p_1q_2-p_2q_1+\nu_2\nu_3-\nu_1\nu_4=Q(\nu_2-\nu_1)\,, \quad (386)$$

and

$$p_1q_1-p_2q_2+\nu_1\nu_3-\nu_2\nu_4=Q(\nu_2-\nu_1)\,. \quad (387)$$

The consideration of the other cross-terms, S_{12}, T_{12}, and T_{21}, leads to the same conditions as do equations (384)–(387).

Finally, considering $S_{22}(\mu, \mu')$, we have

$$\left(\frac{1}{\mu'}+\frac{1}{\mu}\right)S_{22}(\mu,\mu') = \chi(\mu)\chi(\mu') - \sigma(\mu)\sigma(\mu') + 2[\zeta(\mu)\zeta(\mu') - \theta(\mu)\theta(\mu')]$$

$$\begin{aligned} = & X_r(\mu)X_r(\mu')[1+p_1(\mu+\mu')+\{p_1^2-p_2^2+\tfrac{1}{2}(\nu_2^2-\nu_1^2)\}\mu\mu'] \\ & - Y_r(\mu)Y_r(\mu')[1-p_1(\mu+\mu')+\{p_1^2-p_2^2+\tfrac{1}{2}(\nu_2^2-\nu_1^2)\}\mu\mu'] \\ & +p_2(\mu+\mu')[X_r(\mu)Y_r(\mu')+Y_r(\mu)X_r(\mu')] \\ & -\mu\mu'(\mu+\mu')[(p_1-p_2)t+\tfrac{1}{2}Q(\nu_2^2-\nu_1^2)][X_r(\mu)-Y_r(\mu)][X_r(\mu')-Y_r(\mu')] \\ & -t\{\mu^2[X_r(\mu)-Y_r(\mu)][X_r(\mu')+Y_r(\mu')] \\ & +\mu'^2[X_r(\mu)+Y_r(\mu)][X_r(\mu')-Y_r(\mu')]\}. \end{aligned} \quad (388)$$

From equations (341) and (388) we now obtain the further conditions

$$p_1 = -u_4 ; \qquad p_2 = u_3 , \tag{389}$$

$$p_1^2 - p_2^2 + \tfrac{1}{2}(\nu_2^2 - \nu_1^2) = u_5 = 1 + 2Q(u_4 - u_3) , \tag{390}$$

$$(p_1 - p_2)\, t + \tfrac{1}{2}Q(\nu_2^2 - \nu_1^2) = Qu_5 , \tag{391}$$

and

$$t = -Q(u_4 - u_3) . \tag{392}$$

The consideration of $T_{22}(\mu, \mu')$ leads to the same set of conditions as the foregoing.

Collecting all the conditions among the constants that we have found and combining them with equations (376)–(380), we have

$$\nu_1 = q_2 = \tfrac{3}{4}[-\beta_1 + \beta_0\nu_4 - \alpha_2\nu_3 + \tfrac{1}{2}A_0\nu_1 - \tfrac{1}{2}B_0\nu_2 + \tfrac{1}{2}Q(\nu_2 - \nu_1)(A_1 - B_1)] \tag{393}$$

$$-\nu_2 = q_1 = \tfrac{3}{4}[+\alpha_1 + \alpha_2\nu_4 - \beta_0\nu_3 + \tfrac{1}{2}B_0\nu_1 - \tfrac{1}{2}A_0\nu_2 + \tfrac{1}{2}Q(\nu_2 - \nu_1)(A_1 - B_1)] , \tag{394}$$

$$u_3 = p_2 = \tfrac{3}{8}[-B_1 - B_0u_4 + A_0u_3 + \alpha_2\nu_1 - \beta_0\nu_2 + Q(u_4 - u_3)(A_1 - B_1)] , \tag{395}$$

$$-u_4 = p_1 = \tfrac{3}{8}[+A_1 - A_0u_4 + B_0u_3 + \beta_0\nu_1 - \alpha_2\nu_2 + Q(u_4 - u_3)(A_1 - B_1)] , \tag{396}$$

$$\nu_4^2 - \nu_3^2 + \tfrac{1}{2}(\nu_2^2 - \nu_1^2) = 1 , \tag{397}$$

$$\nu_2(u_3 + \nu_3) - \nu_1(u_4 + \nu_4) = Q(\nu_2 - \nu_1) , \tag{398}$$

$$-\nu_1(u_3 - \nu_3) + \nu_2(u_4 - \nu_4) = Q(\nu_2 - \nu_1) , \tag{399}$$

$$Q(\nu_2 - \nu_1)(\nu_4 + \nu_3) + t(\nu_2 + \nu_1) = 0 , \tag{400}$$

$$u_4^2 - u_3^2 + \tfrac{1}{2}(\nu_2^2 - \nu_1^2) = u_5 = 1 + 2Q(u_4 - u_3) , \tag{401}$$

$$-t(u_4 + u_3) + \tfrac{1}{2}Q(\nu_2^2 - \nu_1^2) = Qu_5 , \tag{402}$$

$$t = -Q(u_4 - u_3) , \tag{403}$$

and

$$t = \tfrac{3}{8}Q[-(\nu_2 - \nu_1)(\alpha_2 + \beta_0) - (u_4 - u_3)(A_0 + B_0) + u_5(A_1 - B_1)] . \tag{404}$$

In considering the foregoing set of twelve equations, we first observe that, according to equation (403), equations (401) and (402) are equivalent. Further (cf. eqs. [400] and [403]),

$$(\nu_2 - \nu_1)(\nu_4 + \nu_3) = (u_4 - u_3)(\nu_2 + \nu_1) . \tag{405}$$

Next, adding and subtracting equations (398) and (399), we obtain

$$\nu_2(u_4 + u_3 - \nu_4 + \nu_3) - \nu_1(u_4 + u_3 + \nu_4 - \nu_3) = 2Q(\nu_2 - \nu_1) \tag{406}$$

and

$$\nu_2(\nu_4 + \nu_3 - u_4 + u_3) - \nu_1(\nu_4 + \nu_3 + u_4 - u_3) = 0 , \tag{407}$$

or

$$(\nu_2 - \nu_1)(u_4 + u_3 - 2Q) = (\nu_2 + \nu_1)(\nu_4 - \nu_3) \tag{408}$$

and

$$(\nu_2 - \nu_1)(\nu_4 + \nu_3) = (\nu_2 + \nu_1)(u_4 - u_3) . \tag{409}$$

Equations (405), (408), and (409) can be combined in the form

$$\frac{\nu_2 + \nu_1}{\nu_2 - \nu_1} = \frac{\nu_4 + \nu_3}{u_4 - u_3} = \frac{u_4 + u_3 - 2Q}{\nu_4 - \nu_3} = \frac{1}{\lambda}\ (\text{say}) . \tag{410}$$

It is now seen that equations (397) and (401) are equivalent; for, according to equation (410),

$$\nu_4^2 - \nu_3^2 = u_4^2 - u_3^2 - 2Q(u_4 - u_3)\,; \tag{411}$$

or, using equation (397), we have

$$1 - \tfrac{1}{2}(\nu_2^2 - \nu_1^2) = u_4^2 - u_3^2 - 2Q(u_4 - u_3)\,; \tag{412}$$

but this is the same as equation (401).

Now turning to equations (393)–(396) and rearranging the terms, we can re-write them in the forms

$$(3A_0 - 8)\nu_2 - 3B_0\nu_1 = 6(a_1 + a_2\nu_4 - \beta_0\nu_3) + 3Q(\nu_2 - \nu_1)(A_1 - B_1)\,, \tag{413}$$

$$3B_0\nu_2 - (3A_0 - 8)\nu_1 = 6(-\beta_1 + \beta_0\nu_4 - a_2\nu_3) + 3Q(\nu_2 - \nu_1)(A_1 - B_1)\,, \tag{414}$$

$$(3A_0 - 8)u_4 - 3B_0u_3 = 3(A_1 + \beta_0\nu_1 - a_2\nu_2) + 3Q(u_4 - u_3)(A_1 - B_1)\,, \tag{415}$$

and

$$3B_0u_4 - (3A_0 - 8)u_3 = 3(-B_1 + a_2\nu_1 - \beta_0\nu_2) + 3Q(u_4 - u_3)(A_1 - B_1)\,. \tag{416}$$

From these equations the following set can be derived:

$$P_1(\nu_2 - \nu_1) - 2\varpi_1(\nu_4 - \nu_3) = a_2\,. \tag{417}$$

$$P_2(\nu_2 + \nu_1) - 2\varpi_2(\nu_4 + \nu_3) = a_1\,, \tag{418}$$

$$P_1(u_4 - u_3) + \varpi_1(\nu_2 - \nu_1) = b_2\,, \tag{419}$$

$$P_2(u_4 + u_3) + \varpi_2(\nu_2 + \nu_1) = b_1\,, \tag{420}$$

where we have used the abbreviations

$$\begin{aligned} 6(a_1 + \beta_1) &= a_1\,; & 3(A_1 + B_1) &= b_1\,; & 3(a_2 + \beta_0) &= \varpi_1\,, \\ 6(a_1 - \beta_1) &= a_2\,; & 3(A_1 - B_1) &= b_2\,; & 3(a_2 - \beta_0) &= \varpi_2\,, \end{aligned} \tag{421}$$

$$P_1 = 3(A_0 + B_0) - 8 - 2Qb_2\,, \quad \text{and} \quad P_2 = 3(A_0 - B_0) - 8\,. \tag{422}$$

Using equation (410), we can reduce equations (417)–(420) to the forms

$$P_1(\nu_2 + \nu_1) - 2\varpi_1(u_4 + u_3) = \frac{a_2}{\lambda} - 4\varpi_1 Q\,, \tag{423}$$

$$\varpi_2(\nu_2 + \nu_1) + P_2(u_4 + u_3) = b_1\,, \tag{424}$$

$$P_1(\nu_4 + \nu_3) + \varpi_1(\nu_2 + \nu_1) = \frac{b_2}{\lambda}\,, \tag{425}$$

and

$$-2\varpi_2(\nu_4 + \nu_3) + P_2(\nu_2 + \nu_1) = a_1\,. \tag{426}$$

Solving equations (423) and (424) for $(\nu_2 + \nu_1)$ and $(u_4 + u_3)$, we have

$$\Delta(\nu_2 + \nu_1) = P_2\left(\frac{a_2}{\lambda} - 4\varpi_1 Q\right) + 2\varpi_1 b_1 \tag{427}$$

and

$$\Delta(u_4 + u_3) = -\varpi_2\left(\frac{a_2}{\lambda} - 4\varpi_1 Q\right) + b_1 P_1\,, \tag{428}$$

where

$$\Delta = P_1 P_2 + 2\varpi_1\varpi_2\,. \tag{429}$$

Similarly, from equations (424) and (425) we find

$$\Delta\,(\nu_4+\nu_3) = P_2\frac{b_2}{\lambda} - a_1\varpi_1 \tag{430}$$

and

$$\Delta\,(\nu_2+\nu_1) = 2\varpi_2\frac{b_2}{\lambda} + a_1P_1\,. \tag{431}$$

Equations (427) and (431) now determine λ; for, according to these equations, we must have

$$P_2\left(\frac{a_2}{\lambda} - 4\varpi_1 Q\right) + 2\varpi_1 b_1 = 2\varpi_2\frac{b_2}{\lambda} + a_1P_1\,, \tag{432}$$

or

$$\lambda = \frac{P_2a_2 - 2b_2\varpi_2}{a_1P_1 - 2b_1\varpi_1 + 4\varpi_1P_2Q}\,. \tag{433}$$

It is now seen that equations (410), (427), (or [431]), (428), (430), and (433) determine the six constants ν_1, ν_2, ν_3, ν_4, u_3, and u_4 uniquely in terms of the various moments of $X_l(\mu)$, $Y_l(\mu)$, $X_r(\mu)$, and $Y_r(\mu)$ *and* the constant Q. It remains to verify that, with the constants determined in this fashion, equations (397) and (404) (which we have not used so far) are also satisfied.

Considering, first, condition (397), we observe that, according to equation (410), this is equivalent to

$$(\nu_4+\nu_3)\,\lambda\,(u_4+u_3-2Q) + \tfrac{1}{2}\,(\nu_2+\nu_1)\,\lambda\,(\nu_2+\nu_1) = 1\,. \tag{434}$$

Substituting for $(\nu_4+\nu_3)$, (u_4+u_3), and $(\nu_2+\nu_1)$ from equations (427), (428), (430), and (431) in equation (434), we have

$$\begin{aligned}(b_2P_2 - a_1\varpi_1\lambda)\,(-a_2\varpi_2 + b_1P_1\lambda + 4\varpi_1\varpi_2\lambda Q - 2\lambda Q\Delta)&\\ + \tfrac{1}{2}\,(a_2P_2 + 2\varpi_1b_1\lambda - 4\varpi_1\lambda P_2Q)\,(a_1\lambda P_1 + 2\varpi_2b_2) &= \lambda\Delta^2\,.\end{aligned} \tag{435}$$

After some straightforward reductions, equation (435) becomes

$$\tfrac{1}{2}\lambda\,(a_1a_2 + 2b_1b_2)\,(P_1P_2 + 2\varpi_1\varpi_2) - 2b_2\lambda P_2Q\Delta = \lambda\Delta^2\,, \tag{436}$$

or (cf. eq. [429])

$$\tfrac{1}{2}\,(a_1a_2 + 2b_1b_2) = \Delta + 2b_2P_2Q\,. \tag{437}$$

Hence we have only to verify the truth of equation (437).

Now (cf. eqs. [351], [361], and [364])

$$\begin{aligned}[3\,(A_0+B_0) - 8]\,[3\,(A_0-B_0) - 8] &+ 18\,(\alpha_2^2 - \beta_0^2)\\ &= 9\,(A_0^2 - B_0^2) - 48A_0 + 64 + 18\,(\alpha_2^2 - \beta_0^2)\\ &= 9\,(A_1^2 - B_1^2) + 16 + 18\,[\,(\tfrac{4}{3} - \alpha_0)^2 - \beta_0^2]\\ &= 9\,(A_1^2 - B_1^2) + 48 + 18\,(\alpha_0^2 - \beta_0^2) - 48\alpha_0\\ &= 9\,(A_1^2 - B_1^2) + 18\,(\alpha_1^2 - \beta_1^2) = \tfrac{1}{2}\,(a_1a_2 + 2b_1b_2)\,.\end{aligned} \tag{438}$$

Hence

$$\begin{aligned}\tfrac{1}{2}\,(a_1a_2 + 2b_1b_2) &= P_2\,(P_1 + 2Qb_2) + 2\varpi_1\varpi_2\\ &= \Delta + 2b_2P_2Q\,,\end{aligned} \tag{439}$$

as required.

Finally, considering equation (404), we can re-write this in the form (cf. eq. [403])

$$8\,(u_4-u_3) = 3\,[\,(\nu_2-\nu_1)(\alpha_2+\beta_0) + (u_4-u_3)(A_0+B_0) - (A_1-B_1)\{1+2Q\,(u_4-u_3)\}\,] \tag{440}$$

or

$$(u_4-u_3)\,[3\,(A_0+B_0) - 8 - 6Q\,(A_1-B_1)] + 3\,(\nu_2-\nu_1)(\alpha_2+\beta_0) = 3\,(A_1-B_1)\,. \tag{441}$$

With the abbreviations (421) and (422), the foregoing equation is equivalent to

$$P_1\,(u_4-u_3) + \varpi_1\,(\nu_2-\nu_1) = b_2\,\cdot \tag{442}$$

but this is the same as equation (419), which we have already satisfied. With this we have satisfied all the equations (393)–(404).

Substituting for λ according to equation (433) in equations (428), (430), and (431), we find that the solutions for the constants can be expressed in the following forms:

$$\nu_2+\nu_1 = \frac{1}{\lambda}(\nu_2-\nu_1) = \frac{a_1a_2-4\varpi_1\varpi_2}{P_2a_2-2b_2\varpi_2}, \tag{443}$$

$$\nu_4+\nu_3 = \frac{1}{\lambda}(u_4-u_3) = \frac{a_1b_2-2\varpi_1P_2}{P_2a_2-2b_2\varpi_2}, \tag{444}$$

$$u_4+u_3 = \frac{a_2b_1-2\varpi_2P_1-4\varpi_2b_2Q}{P_2a_2-2b_2\varpi_2}, \tag{445}$$

$$u_4+u_3-2Q = \frac{1}{\lambda}(\nu_4-\nu_3) = \frac{a_2b_1-2\varpi_2P_1-2a_2P_2Q}{P_2a_2-2b_2\varpi_2}, \tag{446}$$

$$u_5 = 1+2Q\,(u_4-u_3) = \lambda\,\frac{a_1P_1-2\varpi_1b_1+2a_1b_2Q}{P_2a_2-2b_2\varpi_2}, \tag{447}$$

and

$$\frac{1}{\lambda} = \frac{a_1P_1-2\varpi_1b_1+4\varpi_1P_2Q}{P_2a_2-2b_2\varpi_2}, \tag{448}$$

where it may be recalled that a_1, a_2, b_1, b_2, ϖ_1, ϖ_2, P_1, and P_2 are defined in equations (421) and (422).

The constant Q is, however, left entirely arbitrary.

21. *The removal of the arbitrariness in the solution and the determination of the constant* Q.—In the preceding section we have verified that the solutions for the scattering and the transmission matrices are of the forms given by equations (338)–(345) (or, equivalently, [353]–[360]) and have further shown that the constants, ν_1, ν_2, ν_3, ν_4, u_3, u_4, and u_5, occurring in the solutions can all be expressed in a unique manner in terms of the constant Q (which is left arbitrary) and the moments of the functions $X_r(\mu)$, $Y_r(\mu)$, $X_l(\mu)$, and $Y_l(\mu)$—the latter two functions being the *standard solutions* of equations (349) and (350). The functional equations governing $\boldsymbol{S}$ and $\boldsymbol{T}$ therefore admit a one-parametric family of solutions. As in the two other cases of conservative scattering that we have considered (Secs. II and III), this arbitrariness in the solution of the equations incorporating the invariances of the problem can be removed by appealing to the flux and the K-integrals. However, in the present instance, there are (formally) two such sets of integrals corresponding to the fact that F_l and F_r can be specified independently of each other. Indeed, starting from the equations of transfer (Paper XIV, System I, p.

153) appropriate to the problem on hand, we can show that the problem admits the integrals

$$F_l(\tau) = 2\int_{-1}^{+1} [I_{ll}(\tau, \mu) + I_{rl}(\tau, \mu)]\, \mu d\mu = \mu_0 F_l [e^{-\tau/\mu_0} + \gamma_l^{(1)}], \tag{449}$$

$$K_l(\tau) = \frac{1}{2}\int_{-1}^{+1} [I_{ll}(\tau, \mu) + I_{rl}(\tau, \mu)]\, \mu^2 d\mu = \tfrac{1}{4}\mu_0 F_l [-\mu_0 e^{-\tau/\mu_0} + \gamma_l^{(1)}\tau + \gamma_l^{(2)}], \tag{450}$$

$$F_r(\tau) = 2\int_{-1}^{+1} [I_{lr}(\tau, \mu) + I_{rr}(\tau, \mu)]\, \mu d\mu = \mu_0 F_r [e^{-\tau/\mu_0} + \gamma_r^{(1)}], \tag{451}$$

and

$$K_r(\tau) = \frac{1}{2}\int_{-1}^{+1} [I_{lr}(\tau, \mu) + I_{rr}(\tau, \mu)]\, \mu^2 d\mu = \tfrac{1}{4}\mu_0 F_r [-\mu_0 e^{-\tau/\mu_0} + \gamma_r^{(1)}\tau + \gamma_r^{(2)}], \tag{452}$$

where $(I_{ll} + I_{rl})$ and $(I_{lr} + I_{rr})$ are the *total* intensities in the diffuse radiation field which are proportional, respectively, to F_l and F_r and where $\gamma_l^{(1)}$, $\gamma_l^{(2)}$, $\gamma_r^{(1)}$, and $\gamma_r^{(2)}$ are constants.

We shall now show how the integrals (449)–(452) enable us to eliminate the arbitrariness in the solution found in § 20 and determine Q explicitly in terms of the moments of $X_r(\mu)$ and $Y_r(\mu)$.

First, we may observe that, according to equations (321),

$$I_{ll}(0, \mu) + I_{rl}(0, \mu) = \frac{3}{16\mu}[S_{11}(\mu, \mu_0) + S_{21}(\mu, \mu_0)] F_l \tag{453}$$

$$I_{ll}(\tau_1, -\mu) + I_{rl}(\tau_1, -\mu) = \frac{3}{16\mu}[T_{11}(\mu, \mu_0) + T_{21}(\mu, \mu_0)] F_l, \tag{454}$$

$$I_{lr}(0, \mu) + I_{rr}(0, \mu) = \frac{3}{16\mu}[S_{12}(\mu, \mu_0) + S_{22}(\mu, \mu_0)] F_r, \tag{455}$$

and

$$I_{lr}(\tau_1, -\mu) + I_{rr}(\tau_1, -\mu) = \frac{3}{16\mu}[T_{12}(\mu, \mu_0) + T_{22}(\mu, \mu_0)] F_r. \tag{456}$$

Considering the part of the emergent intensities proportional to F_l and substituting for the relevant matrix elements of $\boldsymbol{S}$ and $\boldsymbol{T}$ from equations (353)–(360), we have

$$\begin{aligned} I_{ll}(0, \mu) + I_{rl}(0, \mu) = \tfrac{1}{2}\mu_0 \Big\{ & \frac{3}{4}\frac{1-\mu^2}{\mu_0+\mu}[X_l(\mu_0) X_l(\mu) - Y_l(\mu_0) Y_l(\mu)] \\ & + \tfrac{3}{4} X_l(\mu_0)[(\nu_4+\mu) X_l(\mu) - \nu_3 Y_l(\mu) + \tfrac{1}{2}\nu_1 Y_r(\mu) - \tfrac{1}{2}\nu_2 X_r(\mu) \\ & \qquad + \tfrac{1}{2}Q(\nu_2-\nu_1)\mu\{X_r(\mu) - Y_r(\mu)\}] \\ & + \tfrac{3}{4} Y_l(\mu_0)[(\nu_4-\mu) Y_l(\mu) - \nu_3 X_l(\mu) + \tfrac{1}{2}\nu_1 X_r(\mu) - \tfrac{1}{2}\nu_2 Y_r(\mu) \\ & \qquad + \tfrac{1}{2}Q(\nu_2-\nu_1)\mu\{X_r(\mu) - Y_r(\mu)\}]\Big\} F_l \end{aligned} \tag{457}$$

and

$$I_{ll}(\tau_1,-\mu)+I_{rl}(\tau_1,-\mu)=\tfrac{1}{2}\mu_0\Big\{\frac{3}{4}\frac{1-\mu^2}{\mu_0-\mu}[Y_l(\mu_0)X_l(\mu)-X_l(\mu_0)Y_l(\mu)]$$
$$-\tfrac{3}{4}X_l(\mu_0)[(\nu_4-\mu)Y_l(\mu)-\nu_3X_l(\mu)+\tfrac{1}{2}\nu_1X_r(\mu)-\tfrac{1}{2}\nu_2Y_r(\mu)$$
$$+\tfrac{1}{2}Q(\nu_2-\nu_1)\mu\{X_r(\mu)-Y_r(\mu)\}] \quad (458)$$
$$-\tfrac{3}{4}Y_l(\mu_0)[(\nu_4+\mu)X_l(\mu)-\nu_3Y_l(\mu)+\tfrac{1}{2}\nu_1Y_r(\mu)-\tfrac{1}{2}\nu_2X_r(\mu)$$
$$+\tfrac{1}{2}Q(\nu_2-\nu_1)\mu\{X_r(\mu)-Y_r(\mu)\}]\Big\}F_l.$$

Using equations (457) and (458), we can determine the emergent fluxes and the K's by evaluating the various integrals defining these quantities. The evaluations can all be carried out explicitly if proper use is made of the integral properties of the X- and Y-functions.[20] Comparing the resulting expressions for $F_l(0)$, $F_l(\tau_1)$, $K_l(0)$, and $K_l(\tau_1)$ with those given by equations (449) and (450) for $\tau=0$ and $\tau=\tau_1$, we find, respectively,

$$\gamma_l^{(1)}=\tfrac{3}{4}X_l(\mu_0)[\nu_4\alpha_1-\nu_3\beta_1+\alpha_2+\tfrac{1}{2}\nu_1B_1-\tfrac{1}{2}\nu_2A_1+\tfrac{1}{2}Q(\nu_2-\nu_1)(A_2-B_2)] \quad (459)$$
$$+\tfrac{3}{4}Y_l(\mu_0)[\nu_4\beta_1-\nu_3\alpha_1-\beta_2+\tfrac{1}{2}\nu_1A_1-\tfrac{1}{2}\nu_2B_1+\tfrac{1}{2}Q(\nu_2-\nu_1)(A_2-B_2)],$$

$$\gamma_l^{(1)}=\tfrac{3}{4}X_l(\mu_0)[\nu_4\beta_1-\nu_3\alpha_1-\beta_2+\tfrac{1}{2}\nu_1A_1-\tfrac{1}{2}\nu_2B_1+\tfrac{1}{2}Q(\nu_2-\nu_1)(A_2-B_2)] \quad (460)$$
$$+\tfrac{3}{4}Y_l(\mu_0)[\nu_4\alpha_1-\nu_3\beta_1+\alpha_2+\tfrac{1}{2}\nu_1B_1-\tfrac{1}{2}\nu_2A_1+\tfrac{1}{2}Q(\nu_2-\nu_1)(A_2-B_2)],$$

$$\gamma_l^{(2)}=\tfrac{3}{4}X_l(\mu_0)[\nu_4\alpha_2-\nu_3\beta_2+\alpha_1+\tfrac{1}{2}\nu_1B_2-\tfrac{1}{2}\nu_2A_2+\tfrac{1}{2}Q(\nu_2-\nu_1)(A_3-B_3)] \quad (461)$$
$$+\tfrac{3}{4}Y_l(\mu_0)[\nu_4\beta_2-\nu_3\alpha_2-\beta_1+\tfrac{1}{2}\nu_1A_2-\tfrac{1}{2}\nu_2B_2+\tfrac{1}{2}Q(\nu_2-\nu_1)(A_3-B_3)],$$

and

$$\gamma_l^{(1)}\tau_1+\gamma_l^{(2)}$$
$$=-\tfrac{3}{4}X_l(\mu_0)[\nu_4\beta_2-\nu_3\alpha_2-\beta_1+\tfrac{1}{2}\nu_1A_2-\tfrac{1}{2}\nu_2B_2+\tfrac{1}{2}Q(\nu_2-\nu_1)(A_3-B_3)] \quad (462)$$
$$-\tfrac{3}{4}Y_l(\mu_0)[\nu_4\alpha_2-\nu_3\beta_2+\alpha_1+\tfrac{1}{2}\nu_1B_2-\tfrac{1}{2}\nu_2A_2+\tfrac{1}{2}Q(\nu_2-\nu_1)(A_3-B_3)].$$

It should, first, be observed that equations (459) and (460) are consistent with each other; for, to be consistent,

$$(\nu_4+\nu_3)(\alpha_1-\beta_1)+\alpha_2+\beta_2-\tfrac{1}{2}(\nu_2+\nu_1)(A_1-B_1)=0 \quad (463)$$

must be true. With the abbreviations (421), equation (463) is equivalent to

$$a_2(\nu_4+\nu_3)+2\varpi_1-b_2(\nu_2+\nu_1)=0. \quad (464)$$

With the solutions (443) and (444) for $(\nu_2+\nu_1)$ and $(\nu_4+\nu_3)$, it is readily verified that equation (464) is indeed satisfied. We can accordingly combine equations (459) and (460) to give

$$\gamma_l^{(1)}=\tfrac{3}{8}[X_l(\mu_0)+Y_l(\mu_0)][(\nu_4-\nu_3)(\alpha_1+\beta_1)-\tfrac{1}{2}(\nu_2-\nu_1)(A_1+B_1) \quad (465)$$
$$+\alpha_2-\beta_2+Q(\nu_2-\nu_1)(A_2-B_2)].$$

[20] In the present context the relations of theorem 8 have to be used.

The factor of $X_l(\mu_0) + Y_l(\mu_0)$ on the right-hand side can be simplified considerably by using equations (443), (446), and (448). Thus

$$
\begin{aligned}
6(\nu_4-\nu_3)(\alpha_1+\beta_1) &- 3(\nu_2-\nu_1)(A_1+B_1) + 6(\alpha_2-\beta_2) \\
&= \lambda\left[a_1(u_4+u_3-2Q) - b_1(\nu_2+\nu_1) + \frac{2\varpi_2}{\lambda}\right] \\
&= \frac{\lambda}{P_2 a_2 - 2b_2\varpi_2}[a_1(a_2 b_1 - 2\varpi_2 P_1 - 2a_2 P_2 Q) \\
&\quad - b_1(a_1 a_2 - 4\varpi_1\varpi_2) + 2\varpi_2(a_1 P_1 - 2\varpi_1 b_1 + 4\varpi_1 P_2 Q] \\
&= \frac{2\lambda P_2 Q}{P_2 a_2 - 2b_2\varpi_2}(-a_1 a_2 + 4\varpi_1\varpi_2) \\
&= -2P_2 Q(\nu_2-\nu_1) = -2Q(\nu_2-\nu_1)[3(A_0-B_0)-8]\,.
\end{aligned}
\tag{466}
$$

Inserting this result in equation (465), we have

$$
\gamma_l^{(1)} = -\tfrac{1}{8}Q(\nu_2-\nu_1)[3(A_0-B_0)-8-3(A_2-B_2)][X_l(\mu_0)+Y_l(\mu_0)]\,. \tag{467}
$$

Next, from equations (461), and (462) we obtain

$$
\begin{aligned}
\gamma_l^{(1)}\tau_1 = -\tfrac{3}{4}[X_l(\mu_0)+Y_l(\mu_0)][(\nu_4-\nu_3)(\alpha_2+\beta_2) - \tfrac{1}{2}(\nu_2-\nu_1)(A_2+B_2) \\
+(\alpha_1-\beta_1) + Q(\nu_2-\nu_1)(A_3-B_3)]\,.
\end{aligned}
\tag{468}
$$

Again, the factor of $X_l(\mu_0) + Y_l(\mu_0)$ can be simplified by using equations (443) and (446); we find

$$
\begin{aligned}
\gamma_l^{(1)}\tau_1 = -\tfrac{1}{8}(\nu_2-\nu_1)[3(A_0-A_2)+3(B_0-B_2)-8 \\
-6Q\{(A_1-A_3)-(B_1-B_3)\}][X_l(\mu_0)+Y_l(\mu_0)]\,.
\end{aligned}
\tag{469}
$$

From equations (467) and (469) we now obtain

$$
\tau_1 = \frac{3(A_0-A_2)+3(B_0-B_2)-8-6Q[(A_1-A_3)-(B_1-B_3)]}{Q[3(A_0-A_2)-3(B_0-B_2)-8]}\,. \tag{470}
$$

Solving this equation for Q, we have

$$
Q = \frac{3(A_0-A_2)+3(B_0-B_2)-8}{[3(A_0-A_2)-3(B_0-B_2)-8]\tau_1+6(A_1-A_3)-6(B_1-B_3)}\,. \tag{471}
$$

Introducing the notation (cf. eq. [10])

$$
x_n^{(r)} = \frac{3}{8}\int_0^1 (1-\mu^2)\,X_r(\mu)\,\mu^n\,d\mu \tag{472}
$$

and

$$
y_n^{(r)} = \frac{3}{8}\int_0^1 (1-\mu^2)\,Y_r(\mu)\,\mu^n\,d\mu\,, \tag{473}
$$

we can re-write equation (471) in the form

$$
Q = \frac{x_0^{(r)}+y_0^{(r)}-1}{[x_0^{(r)}-y_0^{(r)}-1]\tau_1+2[x_1^{(r)}-y_1^{(r)}]}\,. \tag{474}
$$

In this form we recognize the similarity of the present expression for Q with equation (202).

A similar consideration of the integrals (451) and (452) leads to the same value of Q, though the details of the calculation are somewhat more complicated.[21] However, it may be of interest to note that the constant $\gamma_r^{(1)}$ in equations (451) and (452) has the value (cf. eq. [467])

$$\gamma_r^{(1)} = -\tfrac{1}{8}Q[3(A_0 - A_2) - 3(B_0 - B_2) - 8][(u_4 - u_3)\{X_r(\mu_0) + Y_r(\mu_0)\} - u_5\mu_0\{X_r(\mu_0) - Y_r(\mu_0)\}]. \tag{475}$$

With the foregoing determination of Q in terms of the moments of $X_r(\mu)$ and $Y_r(\mu)$, we have completed the solution of the problem.

22. *Concluding remarks.*—The analysis of the various problems of diffuse reflection and transmission presented in this paper has shown how problems of radiative transfer in plane-parallel atmospheres of finite optical thicknesses can be solved exactly; for, in every case considered, it was possible to reduce the complicated systems of functional equations representing the problem to pairs of equations of the standard form

$$X(\mu) = 1 + \mu\int_0^1 \frac{\Psi(\mu')}{\mu + \mu'}[X(\mu)X(\mu') - Y(\mu)Y(\mu')]\,d\mu' \tag{476}$$

and

$$Y(\mu) = e^{-\tau_1/\mu} + \mu\int_0^1 \frac{\Psi(\mu')}{\mu - \mu'}[Y(\mu)X(\mu') - X(\mu)Y(\mu')]\,d\mu'. \tag{477}$$

And, moreover, the expressions (155) and (156) for $X(\mu)$ and $Y(\mu)$, as rational functions involving the points of the Gaussian division and the roots of the characteristic equation

$$1 = \sum_{j=1}^{n} \frac{a_j\Psi(\mu_j)}{1 - k^2\mu_j^2}, \tag{478}$$

provide approximate solutions of equations (476) and (477). Starting with these approximate solutions,[22] we can solve equations (476) and (477) by a process of iteration. Since the iteration will have to be performed for every required value of τ_1, the problem of tabulating the X- and Y-functions is much more elaborate than in the case of the H-functions. However, the existence of the *differential equations* (eqs. [18] and [19]),

$$\frac{\partial X(\mu, \tau_1)}{\partial \tau_1} = Y(\mu, \tau_1)\int_0^1 \frac{d\mu'}{\mu'}\Psi(\mu')Y(\mu', \tau_1) \tag{479}$$

and

$$\frac{\partial Y(\mu, \tau_1)}{\partial \tau_1} = -\frac{Y(\mu, \tau_1)}{\mu} + X(\mu, \tau_1)\int_0^1 \frac{d\mu'}{\mu'}\Psi(\mu')Y(\mu', \tau_1), \tag{480}$$

simplifies the tabulation problem considerably, since corrections for small changes in τ_1 can always be found with the aid of these equations.[23]

[21] In the reductions, use must be made of eqs. (13)–(16).

[22] For $\tau_1 \to 0$, a generalization of the method described by van de Hulst (*Ap. J.*, **107**, 220, 1948) in the context of the simpler equations (172) and (173) can also be used with considerable advantage. While the necessary generalizations of van de Hulst's method will be considered in a later paper of this series, it may be remarked here that the method is essentially one of solving equations (476) and (477) by an iteration scheme which is started with the "trial solutions" $X(\mu) = 1$ and $Y(\mu) = e^{-\tau_1/\mu}$.

[23] Expressions for the second- and higher-order derivatives can be easily derived from eqs. (479) and (480), so that Taylor expansions with as many terms as may be desired could be used.

In view of the importance of the problem and its long-standing nature, it is worthy of comment here that the basic problem underlying the theories relating to the illumination and polarization of the sunlit sky has now been solved exactly. The solution presented in Section V assumes that beyond $\tau = \tau_1$ there is a vacuum (or, equivalently, that there is a perfect absorber at $\tau = \tau_1$). However, the solution for the case in which there is a "ground" can be reduced to the "standard problem" considered in this paper.[24]

Again, while attention was concentrated in this paper on problems of diffuse reflection and transmission, it is evident that solutions for other problems in which there is a distribution of external sources through the medium can also be reduced to the X- and Y-functions of this paper.[25] We shall consider such problems in later papers of this series.

Finally, it should be remarked that, while the present paper solves the mathematical problem of the transfer equations, the practical use of the solutions must await the construction of tables of the X- and Y-functions appropriate for the various problems. The preparation of these tables is now being undertaken by Mrs. Frances Breen and the writer.

[24] Cf. van de Hulst, *op. cit.*

[25] E.g., a case in the theory of formation of stellar absorption lines leads to an equation of transfer with an external source function which increases linearly with the optical depth (cf. Paper XX, *Ap. J.*, **106**, 145, eq. [5], 1947). The exact solution for this problem can, nevertheless, be reduced to an H-function and its moments (Paper XX, eq. [47]).

THE ANGULAR DISTRIBUTION OF THE RADIATION AT THE INTERFACE OF TWO ADJOINING MEDIA[1]

By S. Chandrasekhar

Abstract

In this paper the results contained in a recent paper by B. Davison entitled "Angular Distribution of Neutrons at the Interface of Two Adjoining Media" are derived by certain elementary methods based on certain simple invariances characterizing the problem.

1. In a recent paper published in this Journal, B. Davison (2) has considered a problem in the theory of radiative transfer which can be formulated as follows:

The half-space ($\tau < 0$) above a semi-infinite plane-parallel isotropically scattering atmosphere with an albedo ϖ_1 for single scattering is filled with an isotropically scattering atmosphere, similarly stratified, and with an albedo ϖ_2 different from ϖ_1. The equations of transfer valid for $0 < \tau < \infty$ and $0 > \tau > -\infty$ are, respectively,

$$\mu \frac{dI(\tau, \mu)}{d\tau} = I(\tau, \mu) - \tfrac{1}{2}\varpi_1 \int_{-1}^{+1} I(\tau, \mu')d\mu' \quad (\tau > 0) \tag{1}$$

and

$$\mu \frac{dI(\tau, \mu)}{d\tau} = I(\tau, \mu) - \tfrac{1}{2}\varpi_2 \int_{-1}^{+1} I(\tau, \mu')d\mu' \quad (\tau < 0), \tag{2}$$

where τ is the optical thickness measured from the interface of the two media inwards and μ is the cosine of the angle to the outward normal.

It is known that Equation (1) admits an integral of the form

$$I(\tau, \mu) = L_0 \frac{e^{k_1\tau}}{1 - k_1\mu}, \tag{3}$$

where L_0 is a constant and k_1 is the positive root less than 1 of the characteristic equation

$$\varpi_1 \log [(1 + k_1)/(1 - k_1)] = 2k_1. \tag{4}$$

Consider the solution of Equations (1) and (2) which has the asymptotic behavior (3) for $\tau \to \infty$ and which vanishes for $\tau \to -\infty$. The problem is: What is the angular distribution of the radiation at the interface, $\tau = 0$, of the two adjoining media? In the paper we have referred to, Davison has solved this problem by making use of the integral equations of the Schwarzschild–Milne type which govern the problem. However, we shall show that all of Davison's results can be obtained by entirely elementary methods by appealing to certain invariances of the problem in the manner developed by the writer in recent

[1] *Manuscript received August 31, 1950.*
Contribution from Yerkes Observatory, Williams Bay, Wisconsin, U.S.A.

years and summarized in his book (1) (hereafter referred to as *Radiative Transfer*).

2. Let the law of diffuse reflection by a semi-infinite plane-parallel atmosphere be expressed in terms of a scattering function S in the usual fashion (*Radiative Transfer*, p. 90). For an isotropically scattering atmosphere with an albedo ϖ the scattering function is given by (*Radiative Transfer*, pp. 97-98)

$$S(\mu, \mu') = \varpi \frac{\mu\mu'}{\mu + \mu'} H(\mu)H(\mu'), \tag{5}$$

where $-\mu'$ refers to the incident direction and μ to the scattered direction and

$$H(\mu) = 1 + \tfrac{1}{2}\int_0^1 S(\mu, \mu') \frac{d\mu'}{\mu'}, \tag{6}$$

satisfies the integral equation

$$H(\mu) = 1 + \tfrac{1}{2}\varpi\mu H(\mu)\int_0^1 \frac{H(\mu')}{\mu + \mu'} d\mu'. \tag{7}$$

An alternative form of Equation (7) which we shall find useful is

$$\frac{1}{H(\mu)} = 1 - \tfrac{1}{2}\varpi\mu\int_0^1 \frac{H(\mu')}{\mu + \mu'} d\mu'. \tag{8}$$

3. Let $I(\tau, +\mu)$ and $I(\tau, -\mu)$ $(0 < \mu \leqslant 1)$ refer to the outward and the inward directed radiations at the level τ. Now the inward directed radiation $I(\tau, -\mu)$ at a level $\tau > 0$ must be reflected by the semi-infinite atmosphere below τ according to the scattering function $S_1(\mu, \mu')$, appropriate for an isotropically scattering atmosphere with an albedo ϖ_1. (We shall use a subscript 1 to all functions referring to an atmosphere with an albedo ϖ_1 and similarly a subscript 2 to all functions referring to an atmosphere with an albedo ϖ_2.) It is evident that this reflection of the radiation $I(\tau, -\mu)$ by the atmosphere below τ must account for the difference between $I(\tau, +\mu)$ and $I(0, +\mu)$. Indeed, a little consideration shows that (cf. *Radiative Transfer*, p. 81, for the analogous formulation in the conservative case)

$$I(\tau, +\mu) = e^{k_1\tau} I(0, +\mu) + \frac{e^{-k_1\tau}}{2\mu}\int_0^1 S_1(\mu, \mu') I(\tau, -\mu')d\mu' \quad (\tau > 0). \tag{9}$$

The reason for the appearance of the factors $e^{k_1\tau}$ and $e^{-k_1\tau}$ in Equation (9) is that we are considering a situation in which the asymptotic behavior at $\tau = \infty$ is

$$I(\tau, \mu) \to L_0 \frac{e^{k_1\tau}}{1 - k_1\mu}. \tag{10}$$

Accordingly, there is an infinite source of radiation at great depths and the transformation $\tau \to \tau + t$, where t is a constant, is equivalent to multiplying all outward intensities by $e^{k_1 t}$ and all the inward intensities by $e^{-k_1 t}$.

Similarly, considering the inward directed radiation $I(\tau, -\mu)$ at a level $\tau < 0$, we must have

$$I(\tau, -\mu) = e^{k_1\tau}\, I(0, -\mu) + \frac{e^{-k_1\tau}}{2\mu}\int_0^1 S_2(\mu, \mu')\, I(\tau, +\mu')d\mu' \quad (\tau < 0). \qquad (11)$$

We can obtain a pair of integral equations connecting $I(0, +\mu)$ and $I(0, -\mu)$ by differentiating Equations (9) and (11) and passing to the limits $\tau = +0$ and $\tau = -0$, respectively. Thus, from Equation (9) we obtain

$$\left[\frac{dI(\tau, +\mu)}{d\tau}\right]_{\tau=+0} = k_1 I(0, +\mu) - \frac{k_1}{2\mu}\int_0^1 S_1(\mu, \mu') I(0, -\mu')\, d\mu' + \frac{1}{2\mu}\int_0^1 S_1(\mu, \mu')\left[\frac{dI(\tau, -\mu')}{d\tau}\right]_{\tau=+0} d\mu'. \qquad (12)$$

On the other hand from the equation of transfer (1) (which is appropriate for $\tau > 0$) we conclude that

$$\left[\frac{dI(\tau, +\mu)}{d\tau}\right]_{\tau=+0} = \frac{1}{\mu}\,[I(0, +\mu) - \varpi_1 J(0)]$$

and

$$\left[\frac{dI(\tau, -\mu)}{d\tau}\right]_{\tau=+0} = \frac{1}{\mu}\,[\varpi_1 J(0) - I(0, -\mu)], \qquad (13)$$

where we have written

$$J(0) = \tfrac{1}{2}\int_{-1}^{+1} I(0, \mu)\, d\mu. \qquad (14)$$

Substituting from Equations (13) in Equation (11) we obtain after some minor rearranging of the terms:

$$(1 - k_1\mu)\, I(0, +\mu) = \varpi_1 J(0)\left[1 + \tfrac{1}{2}\int_0^1 S_1(\mu, \mu')\frac{d\mu'}{\mu'}\right] - \tfrac{1}{2}\int_0^1 S_1(\mu, \mu')\,(1 + k_1\mu') I(0, -\mu')\frac{d\mu'}{\mu'}\,. \qquad (15)$$

Similarly, from Equation (11) we obtain

$$(1 + k_1\mu)\, I(0, -\mu) = \varpi_2 J(0)\left[1 + \tfrac{1}{2}\int_0^1 S_2(\mu, \mu')\frac{d\mu'}{\mu'}\right] - \tfrac{1}{2}\int_0^1 S_2(\mu, \mu')\,(1 - k_1\mu')\, I(0, +\mu')\frac{d\mu'}{\mu'}\,. \qquad (16)$$

Equations (15) *and* (16) *are the mathematical expressions of the invariance of the angular distribution of the radiation at the interface adjoining the two media for arbitrary normal displacements of the interface.*

Now substituting for S_1 and S_2 according to Equations (5) and (6), we obtain

$$(1 - k_1\mu)\, I(0, +\mu) = \varpi_1 J(0) H_1(\mu) - \tfrac{1}{2}\varpi_1 \mu H_1(\mu) \int_0^1 \frac{H_1(\mu')}{\mu + \mu'} (1 + k_1\mu') I(0, -\mu') d\mu' \tag{17}$$

and

$$(1 + k_1 \mu) I(0, -\mu) = \varpi_2 J(0) H_2(\mu) - \tfrac{1}{2}\varpi_2 \mu H_2(\mu) \int_0^1 \frac{H_2(\mu')}{\mu + \mu'} (1 - k_1\mu')\, I(0, +\mu') d\mu'. \tag{18}$$

Equations (17) and (18) represent a pair of integral equations governing $I(0, +\mu)$ and $I\,(0, -\mu)$.

Writing

$$(1 - k_1\mu) I(0, +\mu) = \varpi_1 J(0) F_1(\mu)$$

and

$$(1 + k_1\mu) I(0, -\mu) = \varpi_2 J(0) F_2(\mu), \tag{19}$$

we observe that Equations (17) and (18) become

$$F_1(\mu) = H_1(\mu)\left[1 - \tfrac{1}{2}\varpi_2 \mu \int_0^1 \frac{H_1(\mu')}{\mu + \mu'} F_2(\mu') d\mu'\right]$$

and

$$F_2(\mu) = H_2(\mu)\left[1 - \tfrac{1}{2}\varpi_1 \mu \int_0^1 \frac{H_2(\mu')}{\mu + \mu'} F_1(\mu') d\mu'\right]. \tag{20}$$

From these equations it is at once evident that

$$F_1(\mu) = \frac{H_1(\mu)}{H_2(\mu)} \text{ and } F_2(\mu) = \frac{H_2(\mu)}{H_1(\mu)}. \tag{21}$$

The angular distribution of the radiation at $\tau = 0$ is, therefore, given by

$$I(0, +\mu) = \varpi_1 J(0) \frac{H_1(\mu)}{H_2(\mu)(1 - k_1\mu)}$$

and

$$I(0, -\mu) = \varpi_2 J(0) \frac{H_2(\mu)}{H_1(\mu)(1 + k_1\mu)}. \tag{22}$$

It can be verified that the foregoing solution is equivalent to that given by Davison (*loc. cit.* Equations [21] and [30]).

4. For the solution (22) to be self-consistent it is necessary that

$$J(0) = \tfrac{1}{2}\int_0^1 [I(0, +\mu) + I(0, -\mu)]\, d\mu$$
$$= J(0)\left\{\tfrac{1}{2}\varpi_1 \int_0^1 \frac{H_1(\mu)}{H_2(\mu)(1 - k_1\mu)} d\mu + \tfrac{1}{2}\varpi_2 \int_0^1 \frac{H_2(\mu)}{H_1(\mu)(1 + k_1\mu)} d\mu\right\}. \tag{23}$$

In other words, we must have

$$\tfrac{1}{2}\varpi_1 \int_0^1 \frac{H_1(\mu)}{H_2(\mu)(1 - k_1\mu)} d\mu + \tfrac{1}{2}\varpi_2 \int_0^1 \frac{H_2(\mu)}{H_1(\mu)(1 + k_1\mu)} d\mu = 1. \tag{24}$$

The truth of this identity can be verified as follows:

Substituting for $1/H_2(\mu)$ according to Equation (8) in the first integral in Equation (24), we have

$$\begin{aligned}\tfrac{1}{2}\varpi_1\int_0^1 \frac{H_1(\mu)}{H_2(\mu)(1-k_1\mu)}\,d\mu \\ = \tfrac{1}{2}\varpi_1\int_0^1 d\mu\,\frac{H_1(\mu)}{1-k_1\mu}\left\{1-\tfrac{1}{2}\varpi_2\,\mu\int_0^1\frac{H_2(\mu')}{\mu+\mu'}\,d\mu'\right\}.\end{aligned} \tag{25}$$

On the other hand, since $-1/k_1$ is a pole of $H_1(\mu)$ (*Radiative Transfer*, pp. 123 and 350),

$$\tfrac{1}{2}\varpi_1\int_0^1\frac{H_1(\mu)}{1-k_1\mu}\,d\mu = 1. \tag{26}$$

The right-hand side of Equation (25) therefore becomes

$$1-\tfrac{1}{4}\varpi_1\varpi_2\int_0^1\int_0^1\frac{\mu H_1(\mu)H_2(\mu')}{(1-k_1\mu)(\mu+\mu')}\,d\mu\,d\mu'. \tag{27}$$

Now it can be readily verified* that

$$\tfrac{1}{2}\varpi_1\int_0^1\frac{\mu H_1(\mu)}{(1+b\mu)(\mu+\mu')}\,d\mu = \frac{1}{1-b\mu'}\left[\frac{1}{H_1(\mu')}-\frac{1}{H_1(1/b)}\right], \tag{28}$$

where b is a constant. For $b=-1/k_1$, this relation reduces to (cf. the remarks preceding Equation 26)

$$\tfrac{1}{2}\varpi_1\int_0^1\frac{\mu H_1(\mu)}{(1-k_1\mu)(\mu+\mu')}\,d\mu = \frac{1}{(1+k_1\mu')\,H_1(\mu')}. \tag{29}$$

Accordingly the integration over μ in (27) can be performed and we are left with

$$\tfrac{1}{2}\varpi_1\int_0^1\frac{H_1(\mu)}{H_2(\mu)(1-k_1\mu)}\,d\mu = 1-\tfrac{1}{2}\varpi_2\int_0^1\frac{H_2(\mu')}{(1+k_1\mu')H_1(\mu')}\,d\mu'. \tag{30}$$

This completes the verification of the identity (24).

5. The net flux $\pi F(0)$ of the radiation at $\tau=0$ can also be directly evaluated. Thus

$$\begin{aligned}2\int_0^1 I(0,+\mu)\,\mu d\mu &= 2\,\varpi_1 J(0)\int_0^1\frac{H_1(\mu)\mu}{H_2(\mu)(1-k_1\mu)}\,d\mu \\ &= \frac{2\varpi_1}{k_1}J(0)\left[\int_0^1 d\mu H_1(\mu)\left\{\frac{1}{1-k_1\mu}-1\right\}\left\{1-\tfrac{1}{2}\varpi_2\mu\int_0^1 d\mu'\frac{H_2(\mu')}{\mu+\mu'}\right\}\right] \\ &= \frac{4}{k_1}J(0)\left[1-\{1-(1-\varpi_1)^{1/2}\}-\tfrac{1}{4}\varpi_1\varpi_2\int_0^1 d\mu' H_2(\mu')\right.\end{aligned}$$

* *By expressing* $\mu/[(1+b\mu)(\mu+\mu')]$ *in partial fractions and using the equation satisfied by* $H_1(\mu)$.

$$\times\left\{\int_0^1 d\mu \frac{H_1(\mu)\mu}{(\mu+\mu')(1-k_1\mu)} - \int_0^1 d\mu \frac{H_1(\mu)\mu}{\mu+\mu'}\right\}\Bigg]$$

$$= \frac{4}{k_1} J(0)\left[(1-\varpi_1)^{1/2} - \tfrac{1}{2}\varpi_2 \int_0^1 d\mu' H_2(\mu')\left\{\frac{1}{(1+k_1\mu')H_1(\mu')} - \frac{1}{H_1(\mu')} + (1-\varpi_1)^{1/2}\right\}\right] \tag{31}$$

$$= \frac{4}{k_1} J(0)\left[(1-\varpi_1)^{1/2} - (1-\varpi_1)^{1/2}\{1-(1-\varpi_2)^{1/2}\} + \tfrac{1}{2}\varpi_2 k_1 \int_0^1 \frac{H_2(\mu')\mu'}{(1+k_1\mu')H_1(\mu')} d\mu'\right]$$

$$= \frac{4}{k_1}(1-\varpi_1)^{1/2}(1-\varpi_2)^{1/2} J(0) + 2\int_0^1 I(0, -\mu)\mu d\mu.$$

In the foregoing reductions we have made use of the various integral properties of the H-functions (*Radiative Transfer*, Chap. V.). We have thus proved that

$$\frac{F(0)}{J(0)} = \frac{4}{k_1}(1-\varpi_1)^{1/2}(1-\varpi_2)^{1/2}. \tag{32}$$

6. Finally, we may draw attention to a further invariance characterizing the problem. Considering the radiation at the interface $\tau = 0$, we observe that the inward directed radiation $I(0, -\mu)$ at $\tau = 0$ can be regarded as resulting from the reflection of the outward directed radiation $I(0, +\mu)$ by the semi-infinite atmosphere overlying $\tau = 0$. Thus:

$$I(0, -\mu) = \frac{1}{2\mu}\int_0^1 S_2(\mu, \mu') I(0, +\mu') d\mu',$$

or

$$I(0, -\mu) = \tfrac{1}{2}\varpi_2 H_2(\mu) \int_0^1 \frac{\mu' H_2(\mu')}{\mu+\mu'} I(0, +\mu')\, d\mu'. \tag{33}$$

On the other hand the reflection of $I(0, -\mu)$ by the atmosphere below $\tau = 0$ cannot account for all of $I(0, +\mu)$; it can account for only that part of the outward directed direction at $\tau = 0$ which is in *excess* of what will be present in the absence of an atmosphere above $\tau = 0$. Now we know that when there is no atmosphere above $\tau = 0$ the angular distribution of the emergent radiation is given by (*Radiative Transfer*, p. 346)

$$I(0, +\mu) = \frac{L_0}{H_1(1/k_1)} \frac{H_1(\mu)}{1-k_1\mu}. \tag{34}$$

Accordingly

$$I(0, +\mu) - \frac{L_0}{H_1(1/k_1)} \frac{H_1(\mu)}{1-k_1\mu} = \frac{1}{2\mu}\int_0^1 S_1(\mu, \mu') I(0, -\mu')\, d\mu'$$

$$= \tfrac{1}{2}\varpi_1 H_1(\mu) \int_0^1 \frac{H_1(\mu')\mu'}{\mu+\mu'} I(0, -\mu')\, d\mu'. \tag{35}$$

It can be directly verified that with $I(0, +\mu)$ and $I(0, -\mu)$ given by Equations (22) the invariance (33) is identically satisfied. Moreover, the evaluation of the right-hand side of Equation (35) for $I(0, -\mu)$ given by (22) leads to a determination of the constant L_0 in the asymptotic behavior of the solution at $\tau = \infty$. We find

$$L_0 = \varpi_1 J(0) \frac{H_1(1/k_1)}{H_2(1/k_1)}. \tag{36}$$

7. In conclusion it may be stated that there is no difficulty in extending the methods of this paper to more general laws of scattering (cf. *Radiative Transfer*, Chaps. VI, IX, X, and Appendix III).

References

1. CHANDRASEKHAR, S. Radiative transfer. Oxford University Press. 1950.
2. DAVISON, B. Can. J. Research, A, 28: 303. 1950.

ON THE DIFFUSE REFLECTION OF A PENCIL OF RADIATION BY A PLANE-PARALLEL ATMOSPHERE

By S. Chandrasekhar

UNIVERSITY OF CHICAGO

Communicated June 19, 1958

1. *Introduction.*—The problem of the diffuse reflection and transmission of an incident parallel beam of radiation by a plane-parallel atmosphere has been considered extensively in the literature.[1] Several basic problems in this subject have received exact solutions (*R.T.*, chaps. iv, ix, and x). In obtaining these exact solutions, certain general principles of invariance (*R.T.*, chaps. iv and vii) have played a central role.

Problems of a different order of complexity arise when we consider the incidence of a narrow pencil of radiation (such as a searchlight beam) instead of a parallel beam. In many ways these are fundamental problems: on their solution depends the solutions of others. Thus the solution for the problem of the diffuse reflection and transmission of radiation from a point source of light can be expressed as an integral over the solution for the *searchlight problem*. Nevertheless, exact solutions for any of the basic problems in this general category do not exist, though approximate methods for the solution of some of them have been developed.[2] It would clearly be of value if the effectiveness of these latter approximate methods could be tested against an exact solution of some specific problem which has the same inherent difficulties. In this paper a method based on invariances will be described for the solution of the simplest problem in this subject, namely, the problem of the diffuse reflection of an infinitesimal pencil of radiation of finite flux incident in some direction and at some point on an isotropically scattering, uniform, semi-infinite atmosphere.

2. *The Equation of Transfer.*—Let a pencil of radiation of infinitesimal cross-section, but of finite net flux πF normal to itself, be incident at some point on a plane-parallel atmosphere with the properties described in section 1. Let θ_0 be the angle of incidence.

Choose a Cartesian system of co-ordinates with the z-axis passing through the point of incidence and normal to the atmosphere; z is measured inward from the boundary of the atmosphere. Let the x-axis be so chosen that the (x, z)-plane contains the incident pencil. Further, let z, ϖ, and ψ define a system of cylindrical co-ordinates with the same z-axis: ϖ denotes the distance from the axis, and ψ is the polar angle in the (x, y)-plane (see Figs. 1 and 2).

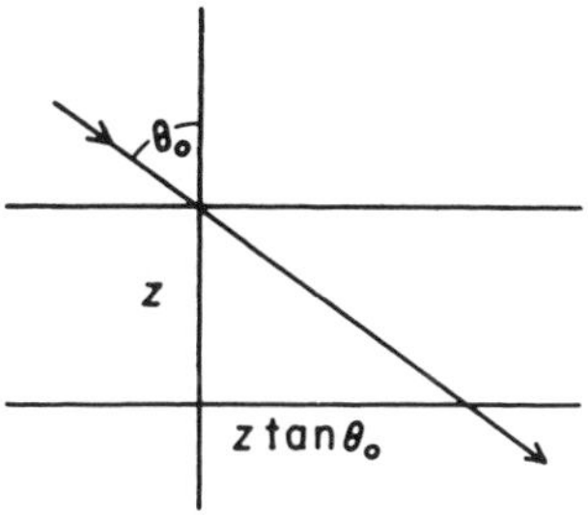

Fig. 1.

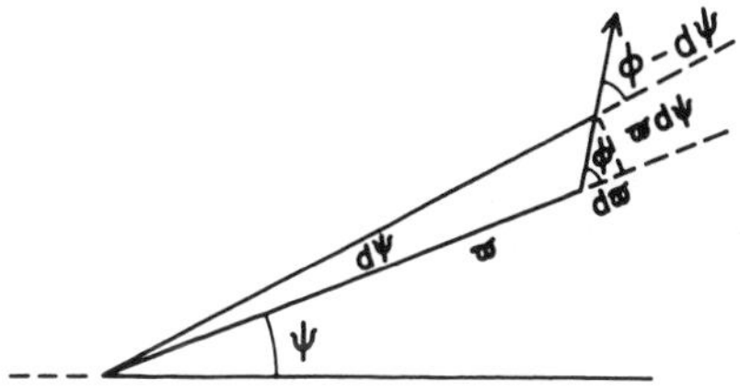

Fig. 2.

In the problem under consideration the specific intensity, I, at any point in the medium depends not only on its position, (z, ϖ, ψ), but also on the direction through the point. We shall specify the latter by the polar angles θ and ϕ (see Figs. 1 and 2). Instead of θ, it will sometimes by convenient to use $\mu = \cos\theta$ as the variable.

We may now write

$$I \equiv I(z, \varpi, \psi;\ \mu, \phi), \tag{1}$$

where we have separately grouped the variables referring to the position and the direction.

Since the atmosphere has been assumed to be uniform, there will be no loss of generality in taking $1/\kappa\rho$ (where κ denotes the mass-scattering coefficient and ρ the density) as the unit of length.

In the variables described and in the unit of length chosen, the equation of radiative transfer takes the form

$$\cos\theta \frac{\partial I}{\partial z} + \sin\theta\cos\phi \frac{\partial I}{\partial \varpi} + \frac{\sin\theta\sin\phi}{\varpi}\left(\frac{\partial I}{\partial \psi} - \frac{\partial I}{\partial \phi}\right) = I(z, \varpi, \psi;\ \mu, \phi) - \mathfrak{J}, \tag{2}$$

where $\mathfrak{J}$ denotes the source function.

In the problem of diffuse reflection and transmission, it is convenient to separate the incident radiation which prevails at all levels (with intensities reduced by determinate amounts) from the diffuse radiation which has resulted from one or more scattering processes in the medium. With this separation of the two fields and reserving I to refer to the intensity in the diffuse radiation field only, we can write for the source function (on the assumption of isotropic scattering) the expression

$$\mathfrak{J}(z, \varpi, \psi) = \frac{\lambda}{4\pi} \int_{-1}^{+1} \int_{0}^{2\pi} I(z, \varpi\, \psi;\ \mu'', \phi'')d\phi''d\mu'' + \frac{\lambda}{4} Fe^{-z/\mu_0}\ \Delta(\varpi - z \tan \theta_0;\ \psi), \tag{3}$$

where λ (≤ 1) is the albedo for single scattering and $\Delta(x - x_0;\ y - y_0)$ is a two-dimensional Dirac δ-function with the properties

$$\Delta(x - x_0;\ y - y_0) = 0 \qquad \text{if} \qquad x \neq x_0 \text{ and } y \neq y_0, \tag{4}$$

and

$$\iint \Delta(x - x_0;\ y - y_0)dxdy = 1, \tag{5}$$

if the domain of integration includes the point (x_0, y_0) and is zero otherwise.

The term in Δ $(\varpi - z \tan \theta_0;\ \psi)$ in the expression for the source function represents the scattering of radiation from the element of mass located at the point $(z, z \tan \theta_0, 0)$ where the incident pencil (reduced in intensity by the factor e^{-z/μ_0}) intersects the (ϖ, ψ)-plane at the level z.

Combining equations (2) and (3), we have

$$\cos \theta \frac{\partial I}{\partial z} + \sin \theta \cos \phi \frac{\partial I}{\partial \varpi} + \frac{\sin \theta \sin \phi}{\varpi} \left(\frac{\partial I}{\partial \psi} - \frac{\partial I}{\partial \phi}\right) = I(z, \varpi, \psi;\ \mu, \phi) - \frac{\lambda}{4\pi} \int_{-1}^{+1} \int_{0}^{2\pi} I(z, \varpi, \psi;\ \mu'', \phi'')d\phi''d\mu'' - \frac{1}{4} \lambda Fe^{-z/\mu_0}\ \Delta(\varpi - z \tan \theta_0;\ \psi). \tag{6}$$

A solution of this equation must be sought which satisfies the boundary conditions

$$I(0, \varpi, \psi;\ -\mu, \phi) \equiv \qquad \text{for} \qquad 0 < \mu \leq 1$$

and

$$e^{-z}I(z, \varpi, \psi;\ \mu, \phi) \rightarrow 0 \qquad \text{for} \qquad z \rightarrow \infty. \tag{7}$$

And the angular distribution of the emergent radiation,

$$I(0, \varpi, \psi;\ +\mu, \phi) \qquad (0 \leq \mu \leq 1), \tag{8}$$

will specify the required law of diffuse reflection. We shall express this law in terms of a *scattering function*, S, related to $I(0, \varpi, \psi;\ +\mu, \phi)$ by

$$I(0, \varpi, \psi;\ +\mu, \phi) = \frac{F}{4\mu} S\ (\varpi;\ \mu, \phi;\ \mu_0, \psi). \tag{9}$$

The reason for expressing the law of diffuse reflection in this manner is that the principle of reciprocity (*R.T.*, p. 172) requires that

$$S(\varpi;\ \mu, \phi;\ \mu_0, \psi) = S(\varpi;\ \mu_0, \psi;\ \mu, \phi). \tag{10}$$

3. *The Principle of Invariance.*—In analogy with the problem of diffuse reflection of a parallel beam by a semi-infinite atmosphere (*R.T.*, p. 91), we shall argue as follows:

Considering the radiation field in the atmosphere, we first distinguish between the incident pencil with the reduced flux

$$\pi Fe^{-z/\mu_0} \tag{11}$$

at the point $(z, z \tan \theta_0, 0)$, in the direction $(-\mu_0, 0)$, and the diffuse radiation field $I(z, \varpi, \psi; \mu, \phi)$. To distinguish further between the outward $(0 \leq \mu \leq 1)$ and the inward $(-1 < \mu < 0)$ directed radiations, we shall write

$$I(z, \varpi, \psi; +\mu, \phi) \qquad (0 \leq \mu \leq 1) \tag{12}$$

and

$$I(z, \varpi, \psi; -\mu, \phi) \qquad (0 < \mu < 1). \tag{13}$$

With these definitions we can formulate the following general principle:

The intensity $I(z, \varpi, \psi; +\mu, \phi)$ *in the outward direction at any level z results from the reflection of the pencil with the flux* $\pi Fe^{-z/\mu_0}$ *at* $(z, z \tan \theta_0, 0)$ *and the diffuse radiation* $I(z, \varpi', \psi'; -\mu', \phi')$ $(0 < \mu' \leq 1;\ 0 \leq \phi' \leq 2\pi;\ 0 \leq \varpi' < \infty;\ 0 \leq \psi' \leq 2\pi)$, *incident on the surface z, by the semi-infinite atmosphere below z.*

To obtain the mathematical expression for this principle, we first observe that the contribution to the outward intensity, $I(z, \varpi, \psi; +\mu, \phi)$ by the reflection of the reduced incident pencil at $(z, z \tan \theta_0, 0)$ is given by (cf. Fig. 3)

$$\frac{F}{4\mu} e^{-z/\mu_0} S(R; \mu, \phi - \chi; \mu_0, \psi + \chi), \tag{14}$$

where

$$R = (\varpi^2 + z^2 \tan^2 \theta_0 + 2\varpi z \tan \theta_0 \cos \psi)^{1/2} \tag{15}$$

and

$$\sin (\psi + \chi) = \frac{\varpi}{R} \sin \psi. \tag{16}$$

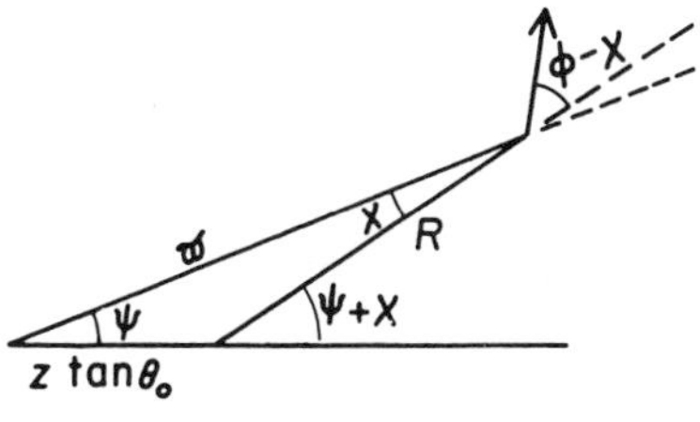

Fig. 3.

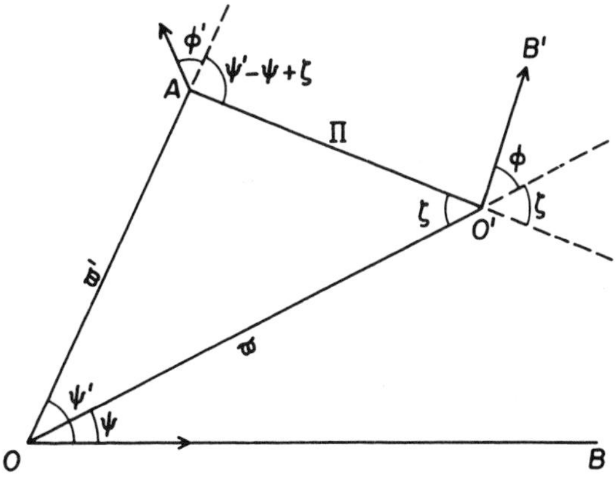

Fig. 4.

Similarly, the contribution to $I(z, \varpi, \psi; +\mu, \phi)$ by the diffuse reflection of the inward-direction radiation prevailing at the level z, is given by (cf. Fig. 4)

$$\frac{1}{4\pi\mu}\int_0^\infty\int_0^{2\pi}\int_0^1\int_0^{2\pi} S(\Pi;\ \mu, \phi + \zeta;\ \mu', 2\pi - [\psi' - \psi + \zeta + \phi'])\ \times$$
$$I(z, \varpi', \psi';\ -\mu', \phi')d\phi' d\mu' \varpi' d\psi' d\varpi', \quad (17)$$

where

$$\Pi = [\varpi^2 + \varpi'^2 + 2\varpi\varpi' \cos(\psi' - \psi)]^{1/2}, \quad (18)$$

and

$$\sin\zeta = \frac{\varpi'}{\Pi}\sin(\psi' - \psi). \quad (19)$$

It may be noted here that Π and ζ depend only on ϖ, ϖ' and $\psi' - \psi$; they do not depend on the other variables.

Combining the contributions (14) and (17) to $I(z, \varpi, \psi; +\mu, \phi)$, we have

$$I(z, \varpi, \psi;\ +\mu, \phi) = \frac{F}{4\mu}e^{-z/\mu_0}\ S(R;\ \mu, \phi - \chi;\ \mu_0, \psi + \chi)\ +$$
$$\frac{1}{4\pi\mu}\int_0^\infty\int_0^{2\pi}\int_0^1\int_0^{2\pi} S(\Pi;\ \mu, \phi + \zeta;\ \mu', 2\pi - [\psi' - \psi + \zeta + \phi'])\ \times$$
$$I(z, \varpi', \psi';\ -\mu', \phi')d\phi' d\mu' \varpi' d\psi' d\varpi'. \quad (20)$$

This equation expresses the invariance of the scattering function, S, to the addition (or, subtraction) of layers of arbitrary thicknesses to (or from) the semi-infinite atmosphere.

4. *The Integrodifferential Equation Governing S.*—As in the theory of the diffuse reflection of parallel beams (*R.T.*, pp. 91–94) we can obtain an equation for S by differentiating equation (20) with respect to z and passing to the limit $z = 0$. Thus we obtain

$$\left[\frac{\partial}{\partial z} I(z, \varpi, \psi;\ +\mu, \phi)\right]_{z=0} = -\frac{F}{4\mu\mu_0}S(\varpi;\ \mu, \phi;\ \mu_0, \psi)\ +$$
$$\frac{F}{4\mu}\left\{\frac{\partial S}{\partial\varpi}\left(\frac{\partial R}{\partial z}\right)_{z=0} + \left(\frac{\partial S}{\partial\psi} - \frac{\partial S}{\partial\phi}\right)\left(\frac{\partial\chi}{\partial z}\right)_{z=0}\right\} +$$
$$\frac{1}{4\pi\mu}\int_0^\infty\int_0^{2\pi}\int_0^1\int_0^{2\pi} S(\Pi;\ \mu, \phi + \zeta;\ \mu', 2\pi - [\psi' - \psi + \zeta + \phi'])\ \times$$
$$\left[\frac{\partial}{\partial z} I(z, \varpi', \psi';\ -\mu', \phi')\right]_{z=0} d\phi' d\mu' \varpi' d\psi' d\varpi'. \quad (21)$$

On the other hand, from equations (9), (15), and (16) and the equation of transfer (2), we derive

$$\left(\frac{\partial R}{\partial z}\right)_{z=0} = \tan\theta_0\cos\psi; \quad \left(\frac{\partial\chi}{\partial z}\right)_{z=0} = -\frac{\tan\theta_0\sin\psi}{\varpi}, \quad (22)$$

$$\left[\mu \frac{\partial}{\partial z} I(z, \varpi, \psi; +\mu, \phi)\right]_{z=0} + \mathfrak{J}(0, \varpi, \psi) =$$

$$\left[-\sin\theta\cos\phi \frac{\partial}{\partial\varpi} - \frac{\sin\theta\sin\phi}{\varpi}\left(\frac{\partial}{\partial\psi} \quad \frac{\partial}{\partial\phi}\right) + 1\right] I(0, \varpi, \psi; +\mu, \phi) = \tag{23}$$

$$\frac{F}{4\mu}\left[-\sin\theta\cos\phi \frac{\partial}{\partial\varpi} - \frac{\sin\theta\sin\phi}{\varpi}\left(\frac{\partial}{\partial\psi} - \frac{\partial}{\partial\phi}\right) + 1\right] S(\varpi; \mu, \phi; \mu_0, \psi),$$

and

$$\left[\frac{\partial}{\partial z} I(z, \varpi', \psi'; -\mu', \phi')\right]_{z=0} = \frac{1}{\mu'} \mathfrak{J}(0, \varpi', \psi'). \tag{24}$$

Making use of these relations in equation (21), we find after some rearrangement of the terms,

$$\frac{1}{4} F \left\{\left(\frac{1}{\mu_0} + \frac{1}{\mu}\right) S - (\tan\theta_0 \cos\psi + \tan\theta\cos\phi)\frac{\partial S}{\partial\varpi} + \right.$$

$$\left.\frac{1}{\varpi}(\tan\theta_0 \sin\psi - \tan\theta\sin\phi)\left(\frac{\partial S}{\partial\psi} - \frac{\partial S}{\partial\phi}\right)\right\} =$$

$$\mathfrak{J}(0, \varpi, \psi) + \frac{1}{4\pi}\int_0^\infty \int_0^{2\pi} \int_0^1 \int_0^{2\pi} \mathfrak{J}(0, \varpi', \psi') \times$$

$$S(\Pi; \mu, \phi + \zeta; \mu', 2\pi - [\psi' - \psi + \zeta + \phi'])d\phi' \frac{d\mu'}{\mu'} \varpi' d\psi' d\varpi' =$$

$$\mathfrak{J}(0, \varpi, \psi) + \frac{1}{4\pi}\int_0^\infty \int_0^{2\pi} \int_0^1 \int_0^{2\pi} \mathfrak{J}(0, \varpi', \psi') \times$$

$$S(\Pi; \mu, \phi + \zeta; \mu', \phi')d\phi' \frac{d\mu'}{\mu'} \varpi' d\psi' d\varpi'; \tag{25}$$

the alternative form of the integral over $\mathfrak{J}(0, \varpi', \psi')$ in the last line results from the independence of the source function on μ' and ϕ'.

According to equations (3) and (9),

$$\mathfrak{J}(0, \varpi, \psi) = \frac{\lambda}{4} F \left[\Delta(\varpi; \psi) + \frac{1}{4\pi}\int_0^1 \int_0^{2\pi} S(\varpi; \mu'', \phi''; \mu_0, \psi)d\phi'' \frac{d\mu''}{\mu''}\right]. \tag{26}$$

Using this result in equation (25), we finally obtain

$$\left\{\left(\frac{1}{\mu_0} + \frac{1}{\mu}\right) - (\tan\theta_0 \cos\psi + \tan\theta\cos\phi)\frac{\partial}{\partial\varpi} + \right.$$

$$\left.\frac{1}{\varpi}(\tan\theta_0 \sin\psi - \tan\theta\sin\phi)\left(\frac{\partial}{\partial\psi} - \frac{\partial}{\partial\phi}\right)\right\} S(\varpi; \mu, \phi\ \mu_0, \psi) =$$

$$\lambda \left\{\Delta(\varpi; \psi) + \frac{1}{4\pi}\int_0^1 \int_0^{2\pi} S(\varpi; \mu'', \phi''; \mu_0, \psi)d\phi'' \frac{d\mu''}{\mu''} + \right.$$

$$\frac{1}{4\pi}\int_0^1 \int_0^{2\pi} S(\varpi; \mu, \phi; \mu', \phi')\, d\phi' \frac{d\mu'}{\mu'} + \tag{27}$$

$$\frac{1}{16\pi^2}\int_0^\infty\int_0^{2\pi}\int_0^1\int_0^{2\pi}\int_0^1\int_0^{2\pi} S(\Pi;\ \mu,\ \phi+\zeta;\ \mu',\ \phi')\ \times$$

$$S\ (\varpi';\ \mu'',\ \phi'';\ \mu_0,\ \psi')\ d\phi''\ \frac{d\mu''}{\mu''}\ d\phi'\ \frac{d\mu'}{\mu'}\ \varpi' d\psi' d\varpi'\Bigg\}.$$

This is the required integrodifferential equation for $S\ (\varpi;\ \mu,\ \phi;\ \mu_0,\ \psi)$.

The structure of equation (27) becomes clearer if we rewrite the integrals which occur on the right-hand side of the equation in terms of the functions

$$P(\varpi;\ \mu,\ \phi) = \frac{1}{4\pi}\int_0^1\int_0^{2\pi} S(\varpi;\ \mu',\ \phi';\ \mu,\ \phi)d\phi'\ \frac{d\mu'}{\mu'} \tag{28}$$

and

$$\tilde{P}(\varpi;\ \mu,\ \phi) = \frac{1}{4\pi}\int_0^1\int_0^{2\pi} S(\varpi;\ \mu,\ \phi;\ \mu',\ \phi')d\phi'\ \frac{d\mu'}{\mu'} \tag{29}$$

we have

$$\Bigg\{\left(\frac{1}{\mu_0}+\frac{1}{\mu}\right) - (\tan\theta_0\cos\psi + \tan\theta\cos\phi)\ \frac{\partial}{\partial\varpi} +$$

$$\frac{1}{\varpi}\ (\tan\theta_0\sin\psi - \tan\theta\sin\phi)\left(\frac{\partial}{\partial\psi}-\frac{\partial}{\partial\phi}\right)\Bigg\}\ S(\varpi;\ \mu,\ \phi;\ \mu_0,\ \psi) =$$

$$\lambda\ \left\{\Delta(\varpi;\ \psi) + P(\varpi;\ \mu_0,\ \psi) + \tilde{P}(\varpi;\ \mu,\ \phi) + Q(\varpi;\ \mu,\ \phi;\ \mu_0,\ \psi)\right\}, \tag{30}$$

where

$$Q(\varpi;\ \mu,\ \phi;\ \mu_0,\ \psi) = \int_0^\infty\int_0^{2\pi} \tilde{P}(\Pi;\ \mu,\ \phi+\zeta)P(\varpi';\ \mu_0,\ \psi')\varpi' d\psi' d\varpi'. \tag{31}$$

The function $Q\ (\varpi;\ \mu,\ \phi;\ \mu_0,\ \psi)$ has a simple geometrical relationship with the directions of incidence $(\mu_0,\ \psi)$ and observation $(\mu,\ \phi)$. It is a surface integral for all positions of the vertex A of the triangle OAO' in Figure 4; the base, OO', of this triangle is fixed, and so are the directions of incidence and observation represented by OB and $O'B'$; the function $P\ (\varpi';\ \mu_0,\ \psi')$ is defined at O, while the function $\tilde{P}(\Pi;\ \mu,\ \phi+\zeta)$ is defined at O', and they are defined in a manner which is clearly symmetric with respect to the vertex, A, of the triangle. Accordingly, we may write (symbolically)

$$Q(\varpi;\ \mu,\ \phi;\ \mu_0,\ \psi) = \int \tilde{P}(\overrightarrow{AO'};\ \mu,\ \phi)P(\overrightarrow{AO};\ \mu_0,\ \psi)dA, \tag{32}$$

where dA is an element area at A.

5. *The Principle of Reciprocity.*—We shall now verify that if $S(\varpi;\ \mu,\ \phi;\ \mu_0,\ \psi)$ is a solution of equation (30), then so is

$$\tilde{S}(\varpi;\ \mu,\ \phi;\ \mu_0,\ \psi) = S(\varpi;\ \mu_0,\ \psi;\ \mu,\ \phi), \tag{33}$$

obtained by transposing the variables $(\mu,\ \phi)$ and $(\mu_0,\ \psi)$.

The effect of the transposition on the left-hand side of equation (30) is to replace S by $\tilde{S}$ but otherwise leave it unaltered, since the operator acting on S is symmetrical in the pair of variables $(\mu,\ \phi)$ and $(\mu_0,\ \psi)$.

The effect of the transposition on the right-hand side is to change it into

$$\lambda\{\Delta(\varpi;\ \phi) + P_S(\varpi;\ \mu, \phi) + \tilde{P}_S(\varpi;\ \mu_0, \psi) + Q_S(\varpi;\ \mu_0, \psi;\ \mu, \phi)\}, \tag{34}$$

where the subscript S signifies that the function in question is evaluated in terms of S. Clearly,

$$P_S(\varpi;\ \mu, \phi) = \tilde{P}_{\tilde{S}}(\varpi;\ \mu, \phi) \quad \text{and} \quad \tilde{P}_S(\varpi;\ \mu_0, \psi) = P_{\tilde{S}}(\varpi;\ \mu_0, \psi). \tag{35}$$

Also,

$$\begin{aligned} Q_S(\varpi;\ \mu_0, \psi;\ \mu, \phi) &= \int \tilde{P}_S(\overrightarrow{AO'};\ \mu_0, \psi) P_S(\overrightarrow{AO};\ \mu, \phi) dA \\ &= \int P_{\tilde{S}}(\overrightarrow{AO'};\ \mu_0, \psi) \tilde{P}_{\tilde{S}}(\overrightarrow{AO};\ \mu, \phi) dA \\ &= Q_{\tilde{S}}(\varpi;\ \mu, \phi;\ \mu_0, \psi). \end{aligned} \tag{36}$$

Hence (34) is the same as

$$\lambda\{\Delta(\varpi;\ \phi) + \tilde{P}_{\tilde{S}}(\varpi;\ \mu, \phi) + P_{\tilde{S}}(\varpi;\ \mu_0, \psi) + Q_{\tilde{S}}(\varpi;\ \mu, \phi;\ \mu_0, \psi)\}; \tag{37}$$

but this *is* the right-hand side of equation (30) for $\tilde{S}$. This completes the proof that if S *is a solution, then so is* $\tilde{S}$. The equation we have derived for S is therefore consistent with the principle of reciprocity. This latter principle requires that

$$S = \tilde{S}, \tag{38}$$

in which case,

$$P = \tilde{P}. \tag{39}$$

6. *Concluding Remarks.*—The principal object of this paper was to derive, in one simple case, an equation for the scattering function, S, which expresses a basic invariance of the problem. In a later paper, we shall show how this equation can be solved by a simple iteration procedure.

It is clear that the methods of this paper admit ready generalizations to less idealized situations: for example, to the problem of the diffuse reflection and transmission by atmospheres of finite optical thicknesses. But these can await the completion of the solution of the equation derived in this paper.

[1] S. Chandrasekhar, *Radiative Transfer* (Oxford: Clarendon Press; 1950); references to this book will be prefaced by "*R.T.*"

[2] Paul I. Richards, "Scattering from a Point Source in Plane Clouds," *J. Opt. Soc. Amer.*, **46**, 927–934, 1956.

DIFFUSE REFLECTION BY A SEMI-INFINITE ATMOSPHERE

HENRY G. HORAK
University of Kansas
AND
S. CHANDRASEKHAR
University of Chicago

Received January 10, 1961

ABSTRACT

Parallel light of flux density πF_0 is incident on a plane-parallel, semi-infinite atmosphere which scatters light in accordance with the phase function $p(\cos\Theta) = \varpi_0 + \varpi_1 P_1(\cos\Theta) + \varpi_2 P_2(\cos\Theta)$. The exact solution for the emergent radiation field is found by using the invariance-principle method.

I. INTRODUCTION

We shall consider the following problem in radiative transfer: parallel light of flux density πF_0 is incident on a plane-parallel, semi-infinite atmosphere which scatters light in accordance with the phase function

$$p(\cos\Theta) = \varpi_0 + \varpi_1 P_1(\cos\Theta) + \varpi_2 P_2(\cos\Theta) \qquad (\varpi_0 \leq 1), \tag{1}$$

where ϖ_0 (the albedo) ϖ_1, ϖ_2 are constants, and P_1 and P_2 are Legendre polynomials. We are interested in finding the intensity distribution of the light diffusely reflected by this atmosphere.

The conservation case ($\varpi_0 = 1$) has already been solved by Chandrasekhar (1946, 1950), although the details of the solution using the invariance-principle method have not been published. We therefore present the complete derivation here with $\varpi_0 \leq 1$.

II. THE EQUATION OF TRANSFER

The equation of transfer for this problem may be written as follows (*Radiative Transfer* [hereinafter cited as "*R.T.*"], chap. i, eqs. [41] and [63]):

$$\mu \frac{dI(\tau, \mu, \phi)}{d\tau} = I(\tau, \mu, \phi) - \mathfrak{J}(\tau, \mu, \phi), \tag{2}$$

where

$$\mathfrak{J}(\tau, \mu, \phi) = \frac{1}{4\pi}\int_0^{2\pi}\int_{-1}^{+1} I(\tau, \mu', \phi')\, p(\mu, \phi; \mu', \phi')\, d\mu' d\phi'. \tag{3}$$

It is convenient to separate explicitly the diffuse and direct-reduced fields; thus we let

$$I(\tau, +\mu, \phi) = I_D(\tau, +\mu, \phi) \qquad (0 < \mu \leq 1),$$

and

$$I(\tau, -\mu, \phi) = I_D(\tau, -\mu, \phi) + \pi F_0\, \delta(\mu - \mu_0)\, \delta(\phi - \phi_0) \quad (0 < \mu \leq 1), \tag{4}$$

where δ is the Dirac delta function, and the subscript D refers to the diffuse field. Equation (2) becomes

$$\mu \frac{dI_D(\tau, \mu, \phi)}{d\tau} = I_D(\tau, \mu, \phi) - \mathfrak{J}(\tau, \mu, \phi), \tag{5}$$

where

$$\mathfrak{J}(\tau, \mu, \phi) = \frac{1}{4\pi}\int_0^{2\pi}\int_{-1}^{+1} I_D(\tau, \mu', \phi')\, p(\mu, \phi; \mu', \phi')\, d\mu' d\phi' + \frac{F_0}{4} e^{-\tau/\mu_0} p(\mu, \phi; -\mu_0, \phi_0). \tag{6}$$

The boundary conditions are

$$I_D(0,-\mu,\phi)=0 \quad (0<\mu\leq 1) \qquad \text{at} \qquad \tau=0\,, \tag{7}$$

and

$$I_D(\tau,\mu,\phi)\to 0 \qquad \text{as} \qquad \tau\to\infty\,. \tag{8}$$

We shall suppress the subscript D in the remainder of the discussion.

III. THE SCATTERING FUNCTION

The intensity of the light diffusely reflected from the atmosphere can be expressed in terms of a scattering function thus:

$$I(0,+\mu,\phi)=\frac{F_0}{4\mu}S(\mu,\phi;\mu_0,\phi_0)\,. \tag{9}$$

One invariance principle (*R.T.*, p. 91, eq. [9]) will suffice for our purposes, viz.,

$$\begin{aligned} I(\tau,+\mu,\phi)=&\frac{F_0}{4\mu}e^{-\tau/\mu_0}S(\mu,\phi;\mu_0,\phi_0)\\ &+\frac{1}{4\pi\mu}\int_0^{2\pi}\int_0^1 S(\mu,\phi;\mu',\phi')\,I(\tau,-\mu',\phi')\,d\mu' d\phi'\,. \end{aligned} \tag{10}$$

Differentiating this equation with respect to τ, letting $\tau\to 0$, and evaluating the derivatives of the intensity by the equation of transfer enables us to get an integral equation (*R.T.*, p. 94, eq. [28]) for the scattering function,

$$\begin{aligned} \left(\frac{1}{\mu}+\frac{1}{\mu_0}\right)S(\mu,\phi;\mu_0,\phi_0)=&\,p(\mu,\phi;-\mu_0,\phi_0)\\ &+\frac{1}{4\pi}\int_0^{2\pi}\int_0^1 p(\mu,\phi;\mu'',\phi'')\,S(\mu'',\phi'';\mu_0,\phi_0)\frac{d\mu''}{\mu''}\,d\phi''\\ &+\frac{1}{4\pi}\int_0^{2\pi}\int_0^1 S(\mu,\phi;\mu',\phi')\,p(-\mu',\phi';-\mu_0,\phi_0)\frac{d\mu'}{\mu'}\,d\phi'\\ &+\frac{1}{16\pi^2}\int_0^{2\pi}\int_0^1\int_0^{2\pi}\int_0^1 S(\mu,\phi;\mu',\phi')\,p(-\mu',\phi';\mu'',\phi'')\\ &\times S(\mu'',\phi'';\mu_0,\phi_0)\frac{d\mu'}{\mu'}\,d\phi'\,\frac{d\mu''}{\mu''}\,d\phi''\,. \end{aligned} \tag{11}$$

The scattering function obeys the principle of reciprocity, i.e., it is symmetric in the pair of variables (μ,ϕ) and (μ_0,ϕ_0). The detailed proof of this is given in *R.T.*, page 171.

IV. FACTORIZATION OF THE INTEGRAL EQUATION

It is possible to separate the variables μ and ϕ in the scattering function, by expressing S in the following form:

$$\begin{aligned} S(\mu,\phi;\mu_0,\phi_0)=&\,S^{(0)}(\mu,\mu_0)+\sqrt{(1-\mu^2)}\sqrt{(1-\mu_0^2)}\,S^{(1)}(\mu,\mu_0)\\ &\cos(\phi-\phi_0)+(1-\mu^2)(1-\mu_0^2)\,S^{(2)}(\mu,\mu_0)\cos 2(\phi-\phi_0)\,, \end{aligned} \tag{12}$$

where the equations satisfied by $S^{(0)}$, $S^{(1)}$, and $S^{(2)}$ are to be determined.

The phase function given by equation (1) may be written

$$\begin{aligned} p(\mu,\phi;\mu',\phi')=&\,p^{(0)}(\mu,\mu')+(\varpi_1+3\varpi_2\mu\mu')[\sqrt{(1-\mu^2)}][\sqrt{(1-\mu'^2)}]\\ &\cos(\phi'-\phi)+\tfrac{3}{4}\varpi_2(1-\mu^2)(1-\mu'^2)\cos 2(\phi'-\phi)\,, \end{aligned} \tag{13}$$

where it may be readily verified that

$$p^{(0)}(\mu, \mu') = \frac{1}{2\pi}\int_0^{2\pi} p(\mu, \phi; \mu', \phi')\, d\phi'$$
$$= \frac{3\varpi_2}{4}\left(\frac{3\varpi_2}{4\varpi_0+\varpi_2}\right)\left[\left(\frac{4\varpi_0+\varpi_2}{3\varpi_2} - \mu^2\right)\left(\frac{4\varpi_0+\varpi_2}{3\varpi_2} - \mu'^2\right) + \frac{4\varpi_0}{\varpi_2}\mu^2\mu'^2\right] + \varpi_1\mu\mu' . \tag{14}$$

We shall substitute equation (12) in equation (11) and make use of equation (13). Also we need the relations

$$\begin{aligned}\frac{1}{2\pi}\int_0^{2\pi} \cos m(\phi' - \phi_0)\cos n(\phi - \phi')\, d\phi' &= 0 \text{ if } m \neq n ,\\ &= 1 \text{ if } m = n = 0 ,\\ &= \tfrac{1}{2}\cos m(\phi - \phi_0)\\ &\quad \text{if } m = n \neq 0 .\end{aligned} \tag{15}$$

The resulting expression is factorable, and it may be shown that $S^{(0)}$, $S^{(1)}$, and $S^{(2)}$ satisfy the following integral equations:

$$\left(\frac{1}{\mu}+\frac{1}{\mu_0}\right)S^{(0)}(\mu, \mu_0) = -\varpi_1\chi(\mu)\chi(\mu_0) + \frac{3\varpi_0}{\xi}\phi(\mu)\phi(\mu_0) + \frac{3\varpi_2}{4\xi}\psi(\mu)\psi(\mu_0), \tag{16}$$

$$\left(\frac{1}{\mu}+\frac{1}{\mu_0}\right)S^{(1)}(\mu, \mu_0) = \varpi_1\theta(\mu)\theta(\mu_0) - 3\varpi_2\sigma(\mu)\sigma(\mu_0), \tag{17}$$

$$\left(\frac{1}{\mu}+\frac{1}{\mu_0}\right)S^{(2)}(\mu, \mu_0) = \tfrac{3}{4}\varpi_2 H^{(2)}(\mu) H^{(2)}(\mu_0), \tag{18}$$

where

$$\xi = \frac{(4\varpi_0+\varpi_2)}{3\varpi_2}, \tag{19}$$

$$\chi(\mu) = \mu - \frac{1}{2}\int_0^1 S^{(0)}(\mu, \mu')\, d\mu' , \tag{20}$$

$$\phi(\mu) = \mu^2 + \frac{1}{2}\int_0^1 S^{(0)}(\mu, \mu')\mu'\, d\mu' , \tag{21}$$

$$\psi(\mu) = (\xi - \mu^2) + \frac{1}{2}\int_0^1 S^{(0)}(\mu, \mu')(\xi - \mu'^2)\frac{d\mu'}{\mu'}, \tag{22}$$

$$\theta(\mu) = 1 + \frac{1}{4}\int_0^1 S^{(1)}(\mu, \mu')(1 - \mu'^2)\frac{d\mu'}{\mu'}, \tag{23}$$

$$\sigma(\mu) = \mu - \frac{1}{4}\int_0^1 S^{(1)}(\mu, \mu')(1 - \mu'^2)\, d\mu' , \tag{24}$$

$$H^{(2)}(\mu) = 1 + \frac{1}{4}\int_0^1 S^{(2)}(\mu, \mu')(1 - \mu'^2)^2\frac{d\mu'}{\mu'}. \tag{25}$$

We shall next investigate these integral equations.

V. THE INTEGRAL EQUATION FOR $S^{(2)}$

Equations (18) and (25) may be easily combined, giving

$$H^{(2)}(\mu) = 1 + \mu H^{(2)}(\mu) \int_0^1 \frac{\Psi^{(2)}(\mu')}{\mu + \mu'} H^{(2)}(\mu')\, d\mu' , \tag{26}$$

where $\Psi^{(2)}$, the characteristic function, has the form

$$\Psi^{(2)}(\mu') = \frac{3\varpi_2}{16}(1 - \mu'^2)^2 . \tag{27}$$

This completes the solution for $S^{(2)}$.

VI. THE INTEGRAL EQUATION FOR $S^{(1)}$

Let us assume that $S^{(1)}$ is expressible in terms of a single $H^{(1)}$-function by an equation of the form

$$S^{(1)}(\mu, \mu_0) = \frac{\mu\mu_0}{\mu + \mu_0} K H^{(1)}(\mu) H^{(1)}(\mu_0) [A + B(\mu + \mu_0) + C\mu\mu_0] , \tag{28}$$

where K, A, B, and C are constants to be determined, and $H^{(1)}$ satisfies the equation

$$H^{(1)}(\mu) = 1 + \mu H^{(1)}(\mu) \int_0^1 \frac{\Psi^{(1)}(\mu')}{\mu + \mu'} H^{(1)}(\mu')\, d\mu' . \tag{29}$$

The characteristic function $\Psi^{(1)}$ may be written

$$\Psi^{(1)}(\mu) = a + b\mu^2 + c\mu^4 , \tag{30}$$

where a, b, and c are constants unspecified as yet.

We shall next substitute equation (28) in equations (23) and (24). In order to simplify the resulting expressions, it is necessary to evaluate integrals of the form

$$\Lambda^{(n)} = \int_0^1 \frac{\mu'^n}{\mu + \mu'} H^{(1)}(\mu')\, d\mu' \quad (n = 0, 1, 2, \ldots) . \tag{31}$$

A simple recurrence relation exists, viz.,

$$\begin{aligned} \Lambda^{(n)} &= \int_0^1 \left(1 - \frac{\mu}{\mu + \mu'}\right) \mu'^{n-1} H^{(1)}(\mu')\, d\mu' \\ &= \alpha_{n-1} - \mu\Lambda^{(n-1)} , \end{aligned} \tag{32}$$

where

$$\alpha_i = \int_0^1 \mu^i H^{(1)}(\mu)\, d\mu . \tag{33}$$

It remains to evaluate $\Lambda^{(0)}$. From the definition of the $H^{(1)}$-function (eqs. [29] and [30]) it is apparent that

$$\frac{H^{(1)}(\mu) - 1}{\mu H^{(1)}(\mu)} = a\Lambda^{(0)} + b\Lambda^{(2)} + c\Lambda^{(4)} \tag{34}$$

$$\begin{aligned} &= (b\alpha_1 + c\alpha_3) - (b\alpha_0 + c\alpha_2)\mu + (c\alpha_1)\mu^2 \\ &\quad - (c\alpha_0)\mu^3 + \Psi^{(1)}(\mu)\Lambda^{(0)} . \end{aligned} \tag{35}$$

The expressions for $\theta(\mu)$ and $\sigma(\mu)$ (eqs. [23] and [24]) become

$$\begin{aligned}\theta(\mu) &= 1+\frac{1}{4}\left[\int_0^1 S^{(1)}(\mu,\mu')\frac{d\mu'}{\mu'}-\int_0^1 S^{(1)}(\mu,\mu')\mu' d\mu'\right]\\ &= 1+\tfrac{1}{4}K\mu H(\mu)\{[A-(C+A)\mu^2+C\mu^4]\Lambda^{(0)}\\ &\quad +B(\alpha_0-\alpha_2)-A\alpha_1+\mu(A\alpha_0-C\alpha_2)\\ &\quad +\mu^2(C\alpha_1)-\mu^3(C\alpha_0)\};\end{aligned} \tag{36}$$

$$\begin{aligned}\sigma(\mu) &= \mu+\frac{1}{4}\left[\int_0^1 S^{(1)}(\mu,\mu')\mu'^2 d\mu'-\int_0^1 S^{(1)}(\mu,\mu')d\mu'\right]\\ &= \mu+\tfrac{1}{4}K\mu H(\mu)\{\mu[A-(C+A)\mu^2+C\mu^4]\Lambda^{(0)}\\ &\quad -A(\alpha_0-a_2)-B(\alpha_1-\alpha_3)-\mu[(C+A)\alpha_1-C\alpha_3]\\ &\quad +\mu^2[(C+A)\alpha_0-C\alpha_2]+\mu^3(C\alpha_1)-\mu^4(C\alpha_0)\}.\end{aligned} \tag{37}$$

In each of these we see that the coefficient of $\Lambda^{(0)}$ has the same form as the characteristic function, so that it is reasonable to make the following identifications:

$$A=a\,,\qquad -(C+A)=b\,,\qquad C=c\,. \tag{38}$$

Using these relations, setting $K = 4$, and introducing equation (35), it follows directly that

$$\theta(\mu)=H^{(1)}(\mu)(1+l\mu)\,, \tag{39}$$

where

$$l=c(\alpha_1-\alpha_3)+B(\alpha_0-\alpha_2)\,, \tag{40}$$

and

$$\sigma(\mu)=m\mu H^{(1)}(\mu)\,, \tag{41}$$

where

$$m=1-a(\alpha_0-\alpha_2)-B(\alpha_1-\alpha_3)\,. \tag{42}$$

Substituting equations (39) and (41) in equation (17) gives

$$\left(\frac{1}{\mu}+\frac{1}{\mu_0}\right)S^{(1)}(\mu,\mu_0)=H^{(1)}(\mu)H^{(1)}(\mu_0)\,[\varpi_1+\varpi_1 l(\mu+\mu_0)+(\varpi_1 l^2-3\varpi_2 m^2)\mu\mu_0]\,. \tag{43}$$

This must be consistent with equation (28), viz.,

$$\left(\frac{1}{\mu}+\frac{1}{\mu_0}\right)S^{(1)}(\mu,\mu_0)=H^{(1)}(\mu)H^{(1)}(\mu_0)[4a+4B(\mu+\mu_0)+4c\mu\mu_0]\,. \tag{44}$$

Hence we require that

$$a=\frac{\varpi_1}{4}\,, \tag{45}$$

$$\varpi_1 l=4B\,, \tag{46}$$

$$\varpi_1 l^2-3\varpi_2 m^2=4c\,. \tag{47}$$

We shall solve equations (46) and (47) for B and c. Combining (40) and (46) gives

$$B=\frac{acA_1}{1-aA_0}\,, \tag{48}$$

where $A_0 = \alpha_0 - \alpha_2$ and $A_1 = \alpha_1 - \alpha_3$. Substituting this result back in equations (40) and (42), we obtain

$$l = \frac{c A_1}{1 - a A_0}, \tag{49}$$

$$m = \frac{(1 - a A_0)^2 - a c A_1^2}{(1 - a A_0)}. \tag{50}$$

Inserting these into equation (47), we have

$$c = a \left(\frac{c A_1}{1 - a A_0} \right)^2 - \frac{3\varpi_2}{4} \left[\frac{(1 - a A_0)^2 - a c A_1^2}{1 - a A_0} \right]^2,$$

or

$$c [(1 - a A_0)^2 - a c A_1^2] = -\frac{3\varpi_2}{4} [(1 - a A_0)^2 - a c A_1^2]^2.$$

Therefore,

$$c = \frac{3\varpi_2}{4} [(1 - a A_0)^2 - a (c A_1^2)]. \tag{51}$$

We can obtain an expression for cA_1^2 by applying an H-function theorem, viz.:

When the characteristic function $\Psi(\mu)$ has the form $\Psi(\mu) = a + b\mu^2 + c\mu^4$, where a, b, and c are constants, then

$$A_0 = \tfrac{2}{3} + \tfrac{1}{2} (a A_0^2 - c A_1^2) + \tfrac{1}{2} (a + b + c)(\alpha_1^2 - \alpha_2^2), \tag{52}$$

where

$$A_0 = \alpha_0 - \alpha_2, \qquad A_1 = \alpha_1 - \alpha_3,$$

$$\alpha_i = \int_0^1 \mu^i H(\mu) \, d\mu \qquad (i = 0, 1, 2, 3).$$

This may be easily proved by virtue of the definition of an H-function (eq. [29]):

$$\begin{aligned} A_0 &= \int_0^1 H(\mu)(1 - \mu^2) \, d\mu \\ &= \tfrac{2}{3} + \int_0^1 \int_0^1 \frac{(1 - \mu^2) \mu \Psi(\mu')}{\mu + \mu'} H(\mu) H(\mu') \, d\mu d\mu' \\ &= \tfrac{2}{3} + I_1 + I_2 + I_3, \end{aligned} \tag{53}$$

where

$$I_1 = \int_0^1 \int_0^1 \frac{a \mu' (1 - \mu'^2)}{\mu + \mu'} H(\mu) H(\mu') \, d\mu d\mu',$$

$$I_2 = \int_0^1 \int_0^1 \frac{b \mu' \mu^2 (1 - \mu'^2)}{\mu + \mu'} H(\mu) H(\mu') \, d\mu d\mu',$$

$$I_3 = \int_0^1 \int_0^1 \frac{c \mu' \mu^4 (1 - \mu'^2)}{\mu + \mu'} H(\mu) H(\mu') \, d\mu d\mu'.$$

Let us consider I_3; it may be written

$$I_3 = \tfrac{1}{2} c \int_0^1 \int_0^1 \left(\frac{\mu^4 \mu' + \mu'^4 \mu}{\mu + \mu'} - \frac{\mu^4 \mu'^3 + \mu'^4 \mu^3}{\mu + \mu'} \right) H(\mu) H(\mu') \, d\mu d\mu'$$

$$= \tfrac{1}{2} c \int_0^1 \int_0^1 [\mu\mu'(\mu^2 - \mu\mu' + \mu'^2) - \mu^3\mu'^3]\, H(\mu) H(\mu')\, d\mu\, d\mu'$$

$$= \tfrac{1}{2} c (2\alpha_1\alpha_3 - \alpha_2^2 - \alpha_3^2) = \tfrac{1}{2} c [(\alpha_1^2 - \alpha_2^2) - (\alpha_1 - \alpha_3)^2] \ .$$

Similarly we can show that

$$I_2 = \tfrac{1}{2} b (\alpha_1^2 - \alpha_2^2),$$

and

$$I_1 = \tfrac{1}{2} a (\alpha_0^2 - 2\alpha_0\alpha_2 + \alpha_1^2) .$$

The theorem follows at once by substituting these results in equation (53).

In the case which we are considering, $c = -(a + b)$ (eq. [38]), so that the theorem reduces to

$$A_0 = \tfrac{2}{3} + \tfrac{1}{2} (a A_0^2 - c A_1^2),$$

or

$$c A_1^2 = a A_0^2 - 2 A_0 + \tfrac{4}{3} . \tag{54}$$

Inserting this in equation (51) gives

$$c = \frac{3\varpi_2}{4}\left[\frac{4a}{3} - 1\right] = \frac{\varpi_2}{4}(\varpi_1 - 3) . \tag{55}$$

It follows that

$$b = \tfrac{1}{4} [\varpi_2 (3 - \varpi_1) - \varpi_1] . \tag{56}$$

This completes the solution for $S^{(1)}$.

VII. THE INTEGRAL EQUATION FOR $S^{(0)}$

We shall assume that $S^{(0)}$ is expressible in terms of a single $H^{(0)}$-function by an equation of the form

$$S^{(0)}(\mu, \mu_0) = \frac{\mu\mu_0}{\mu + \mu_0} 2 H^{(0)}(\mu) H^{(0)}(\mu_0)$$

$$[A + B(\mu + \mu_0) + C\mu\mu_0 + D\mu\mu_0(\mu + \mu_0) + E\mu^2\mu_0^2 + F(\mu^2 + \mu_0^2)] , \tag{57}$$

where A, B, C, D, E, and F are constants to be determined and $H^{(0)}$ satisfies the equation

$$H^{(0)}(\mu) = 1 + \mu H^{(0)}(\mu) \int_0^1 \frac{\Psi^{(0)}(\mu')}{\mu + \mu'} H^{(0)}(\mu')\, d\mu' . \tag{58}$$

The characteristic function, $\Psi^{(0)}$, may be written

$$\Psi^{(0)}(\mu) = a + b\mu^2 + c\mu^4 , \tag{59}$$

where a, b, and c are constants unspecified as yet.

We shall substitute the expression for $S^{(0)}$ given by equation (57) in equations (20), (21), and (22) and introduce $\Lambda^{(n)}$ ($n = 0, 2, 4$) defined by equation (31). We obtain the following:

$$\begin{aligned} \chi(\mu) = \mu - H^{(0)}(\mu) \{ & -\mu^2 [A\Lambda^{(0)} + (2F - C)\Lambda^{(2)} + E\Lambda^{(4)}] \\ & + \mu [A\alpha_0 + B\alpha_1 + F\alpha_2] + \mu^2 [F\alpha_1 + D\alpha_2 + E\alpha_3] \} ; \end{aligned} \tag{60}$$

$$\begin{aligned} \phi(\mu) = \mu^2 + \mu H^{(0)}(\mu) \{ & \mu^2 [A\Lambda^{(0)} + (2F - C)\Lambda^{(2)} + E\Lambda^{(4)}] \\ & + [A\alpha_1 + B\alpha_2 + F\alpha_3] + \mu [-A\alpha_0 + (C - F)\alpha_2 + D\alpha_3] \} ; \end{aligned} \tag{61}$$

$$\phi(\mu)+\psi(\mu)=\xi\{1+\mu H^{(0)}(\mu)[A\Lambda^{(0)}+(2F-C)\Lambda^{(2)}+E\Lambda^{(4)}] + [B\alpha_0+(C-F)\alpha_1-E\alpha_3]+\mu[F\alpha_0+D\alpha_1+E\alpha_2]\}. \quad (62)$$

It seems reasonable to make the following identifications:

$$A=a, \qquad 2F-C=b, \qquad E=c. \quad (63)$$

Using these relations and introducing equation (34), it follows that

$$\chi(\mu)=\mu H^{(0)}(\mu)(p+q\mu), \quad (64)$$

$$\phi(\mu)=\mu H^{(0)}(\mu)(r+s\mu), \quad (65)$$

$$\psi(\mu)=H^{(0)}(\mu)(\xi+t\mu+v\mu^2), \quad (66)$$

where

$$\begin{aligned} p&=1-a\alpha_0-B\alpha_1-F\alpha_2, \qquad q=-F\alpha_1-D\alpha_2-c\alpha_3, \\ r&=a\alpha_1+B\alpha_2+F\alpha_3, \qquad s=1-a\alpha_0+(C-F)\alpha_2+D\alpha_3, \\ t&=\xi B\alpha_0+\xi(C-F)\alpha_1-\xi c\alpha_3-r, \\ v&=\xi F\alpha_0+\xi D\alpha_1+\xi c\alpha_2-s. \end{aligned} \quad (67)$$

Substituting the expressions for $\chi(\mu)$, $\phi(\mu)$, and $\psi(\mu)$ given by equations (64), (65), and (66) in equation (16) gives

$$\begin{aligned} S^{(0)}(\mu,\mu_0)=\frac{\mu\mu_0}{\mu+\mu_0}H^{(0)}(\mu)H^{(0)}(\mu_0) \times\Big\{&\frac{3\varpi_0}{\xi}[r^2\mu\mu_0+rs\mu\mu_0(\mu+\mu_0)+s^2\mu^2\mu_0^2] \\ &-\varpi_1[p^2\mu\mu_0+pq\mu\mu_0(\mu+\mu_0)+q^2\mu^2\mu_0^2] \\ &+\frac{3\varpi_2}{4\xi}[t^2\mu\mu_0+tv\mu\mu_0(\mu+\mu_0)+v^2\mu^2\mu_0^2+\xi v(\mu^2+\mu_0^2) \\ &+\xi t(\mu+\mu_0)+\xi^2]\Big\}. \end{aligned} \quad (68)$$

Also we require that (refer to eqs. [57] and [63])

$$S^{(0)}(\mu,\mu_0)=\frac{\mu\mu_0}{\mu+\mu_0}2H^{(0)}(\mu)H^{(0)}(\mu_0) [a+B(\mu+\mu_0)+C\mu\mu_0+D\mu\mu_0(\mu+\mu_0)+c\mu^2\mu_0^2+F(\mu^2+\mu_0^2)]. \quad (69)$$

Consistency between these last two equations requires that the coefficients of like terms be equal, viz.,

$$\frac{3\varpi_2}{8}\xi=a, \quad (70)$$

$$\frac{3\varpi_2}{8}t=B, \quad (71)$$

$$\frac{3\varpi_0}{2\xi}r^2-\frac{\varpi_1}{2}p^2+\frac{3\varpi_2}{8\xi}t^2=C=2F-b, \quad (72$$

$$\frac{3\varpi_0}{2\xi}\, r s - \frac{\varpi_1}{2}\, p q + \frac{3\varpi_2}{8\xi}\, t v = D\,, \tag{73}$$

$$\frac{3\varpi_0}{2\xi}\, s^2 - \frac{\varpi_1}{2}\, q^2 + \frac{3\varpi_2}{8\xi}\, v^2 = E = c\,, \tag{74}$$

$$\frac{3\varpi_2}{8}\, v = F\,. \tag{75}$$

From equation (70),

$$a = \frac{4\varpi_0 + \varpi_2}{8}\,. \tag{76}$$

It remains to determine b, c, B, D, and F. In principle, this can be done from equations (71)–(75) combined with equation (52), which gives a relation between the moments of $H^{(0)}$. The algebra becomes very intricate. It is preferable, therefore, to derive b and c by obtaining the characteristic function directly from the equation of transfer. The characteristic equation has the form (*R.T.*, p. 153, eq. [113])

$$1 = \frac{1}{2}\sum_{j=1}^{n} \frac{a_j}{1 - \mu_j^2 k^2} \Big\{ 2\varpi_0 + 2(1-\varpi_0)\varpi_1 \mu_j^2 + \frac{\varpi_2}{2}\Big[\frac{(1-\varpi_0)(3-\varpi_1)}{k^2} - 1\Big](3\mu_j^2 - 1) \Big\}\,, \tag{77}$$

and this may be rewritten

$$1 = \frac{1}{2}\sum_{j=1}^{n} \frac{a_j}{1 - \mu_j^2 k^2} \Big\{ \Big[2\varpi_0 + \frac{2(1-\varpi_0)\varpi_1}{3}\Big] - \Big[\frac{\varpi_2}{2} - \frac{2(1-\varpi_0)\varpi_1}{3}\Big](3\mu_j^2 - 1) + \Big[\frac{\varpi_2}{2}(1-\varpi_0)(3-\varpi_1)\Big]\mu_j^2(3\mu_j^2 - 1) \Big\}\,. \tag{78}$$

Thus the characteristic function is given by

$$\Psi^{(0)}(\mu) = a + b\mu^2 + c\mu^4\,, \tag{79}$$

where

$$a = \frac{4\varpi_0 + \varpi_2}{8}\,, \tag{80}$$

$$b = \frac{(1-\varpi_0)}{2}\,\varpi_1 - \frac{\varpi_2}{8}\,[3 + (1-\varpi_0)(3-\varpi_1)]\,, \tag{81}$$

and

$$c = \tfrac{3}{8}\varpi_2(1-\varpi_0)(3-\varpi_1)\,. \tag{82}$$

Assuming, then, that a, b, and c are correctly given by equations (80)–(82), we can now derive the values of B, D, and F. It will be observed that equations (71) and (72) involve only B and F, while equations (74) and (75) involve only D and F. Considering equations (71) and (72) first, we have

$$B = \boldsymbol{A} - \boldsymbol{B}F\,, \tag{83}$$

and

$$\boldsymbol{E} + \boldsymbol{F}B^2 + \boldsymbol{G}B + \boldsymbol{H}BF + \boldsymbol{I}F + \boldsymbol{J}G^2 = 0\,, \tag{84}$$

where

$$\boldsymbol{A} = \frac{(a + \xi b)\alpha_1 + \xi c \alpha_3}{\xi\alpha_0 - \alpha_2 - (8/3\varpi^2)}\,, \qquad \boldsymbol{B} = \frac{\xi\alpha_1 - \alpha_3}{\xi\alpha_0 - \alpha_2 - (8/3\varpi_2)}\,,$$

$$E=\frac{3\varpi_0}{2\xi}a^2\alpha_1^2-\frac{\varpi_1}{2}(1-a\alpha_0)^2+b\,,\qquad F=\frac{3\varpi_0}{2\xi}\alpha_2^2-\frac{\varpi_1}{2}\alpha_1^2+\frac{8}{4\varpi_0+\varpi_2}\,,$$

$$G=\frac{3\varpi_0}{\xi}a\alpha_1\alpha_2+\varpi_1(1-a\alpha_0)\alpha_1\,,\qquad H=\frac{3\varpi_0}{\xi}\alpha_2\alpha_3-\varpi_1\alpha_1\alpha_2\,, \tag{85}$$

$$I=\frac{3\varpi_0}{\xi}a\alpha_1\alpha_3+\varpi_1(1-a\alpha_0)\alpha_2-2\,,\qquad J=\frac{3\varpi_0}{2\xi}\alpha_3^2-\frac{\varpi_1}{2}\alpha_2^2\,.$$

Combining equations (83) and (84), we obtain a quadratic in F:

$$(E+A^2F+GA)+(AH+I-2ABF-BG)F+(J+B^2F-BH)F^2=0\,. \tag{86}$$

Considering, next, equations (74) and (75), we have

$$D=C-DF\,, \tag{87}$$

and

$$K+LD^2+MD+NDF+PF+QF^2=0\,, \tag{88}$$

where

$$C=\frac{1-a\alpha_0-(b+\xi c)\alpha_2}{\xi\alpha_1-\alpha_3}\,,\qquad D=\frac{1}{B}$$

$$K=\frac{3\varpi_0}{2\xi}(1-a\alpha_0-b\alpha_2)^2-\frac{\varpi_1}{2}c^2\alpha_3^2-c\,,\qquad L=J$$

$$M=\frac{3\varpi_0}{\xi}(1-a\alpha_0-b\alpha_2)\alpha_3-\varpi_1 c\alpha_2\alpha_3\,,\qquad N=H \tag{89}$$

$$P=\frac{3\varpi_0}{\xi}(1-a\alpha_0-b\alpha_2)\alpha_2-\varpi_1 c\alpha_1\alpha_3\,,\qquad Q=F\,.$$

Combining equations (87) and (88), we obtain a second quadratic in F:

$$B^2(K+C^2J+MC)+(B^2HC+B^2P-2BCJ-BM)F+(J+B^2F-BH)F^2=0\,. \tag{90}$$

Equations (86) and (90) have a common root given by

$$F=\frac{B^2(K+C^2J+MC)-(E+A^2F+GA)}{(AH+I-2ABF-BG)-(B^2HC+B^2P-2BCJ-BM)} \tag{91}$$

corresponding to which we have values for B and D given by equations (83) and (87). Furthermore, these resulting values of F, B, and D, together with the values of a, b, and c given by equations (80)–(82), must satisfy equation (73). It has not been possible to complete this last verification algebraically, since the algebra is so very involved; therefore, in any particular application of the formulae, it is essential that the verification be carried out numerically. An extensive series of calculations have been carried out at the Kansas University Computation Center using an IBM 650 digital computer. In every case so far examined the verification has been accomplished.

VIII. SUMMARY OF RESULTS

Parallel light of flux density πF_0 is incident on a plane-parallel, semi-infinite atmosphere which scatters light in accordance with the phase function $p(\cos\Theta)=\varpi_0+\varpi_1P_1(\cos\Theta)+\varpi_2P_2(\cos\Theta)$. The intensity of the light diffusely reflected by this atmosphere is given by the following formulae:

$$I(0, \mu, \phi) = \frac{F_0}{4\mu} S(\mu, \phi; \mu_0, \phi_0),$$

$$S(\mu, \phi; \mu_0, \phi_0) = S^{(0)}(\mu, \mu_0) + \sqrt{(1-\mu^2)}\sqrt{(1-\mu_0^2)} S^{(1)}(\mu, \mu_0)\cos(\phi - \phi_0)$$
$$+ (1-\mu^2)(1-\mu_0^2) S^{(2)}(\mu, \mu_0)\cos 2(\phi - \phi_0),$$

$$\left(\frac{1}{\mu}+\frac{1}{\mu_0}\right) S^{(0)}(\mu, \mu_0) = -\varpi_1 \chi(\mu)\chi(\mu_0) + \frac{3\varpi_0}{\xi}\phi(\mu)\phi(\mu_0) + \frac{3\varpi_2}{4\xi}\psi(\mu)\psi(\mu_0),$$

$$\left(\frac{1}{\mu}+\frac{1}{\mu_0}\right) S^{(1)}(\mu, \mu_0) = \varpi_1 \theta(\mu)\theta(\mu_0) - 3\varpi_2 \sigma(\mu)\sigma(\mu_0),$$

$$\left(\frac{1}{\mu}+\frac{1}{\mu_0}\right) S^{(2)}(\mu, \mu_0) = \frac{3\varpi_2}{4} H^{(2)}(\mu) H^{(2)}(\mu_0),$$

$$\chi(\mu) = \mu H^{(0)}(\mu)(p + q\mu), \qquad \phi(\mu) = \mu H^{(0)}(\mu)(r + s\mu),$$
$$\psi(\mu) = H^{(0)}(\mu)(\xi + t\mu + v\mu^2), \qquad \theta(\mu) = H^{(1)}(\mu)(1 + l\mu),$$
$$\sigma(\mu) = m\mu H^{(1)}(\mu),$$

$$H^{(i)}(\mu) = 1 + \mu H^{(i)}(\mu)\int_0^1 \frac{\Psi^{(i)}(\mu')}{\mu+\mu'} H^{(i)}(\mu')\,d\mu' \qquad (i = 0, 1, 2),$$

$$\Psi^{(i)}(\mu) = a^{(i)} + b^{(i)}\mu^2 + c^{(i)}\mu^4 \qquad (i = 0, 1, 2),$$

$$\alpha_n^{(i)} = \int_0^1 \mu^n H^{(i)}(\mu)\,d\mu \qquad (i = 0, 1, 2;\ n = \text{integer}),$$

$$a^{(0)} = \frac{4\varpi_0 + \varpi_2}{8}, \qquad b^{(0)} = \frac{(1-\varpi_0)}{2}\varpi_1 - \frac{\varpi_2}{8}[3 + (1-\varpi_0)(3-\varpi_1)],$$

$$c^{(0)} = \frac{3}{8}\varpi_2(1-\varpi_0)(3-\varpi_1), \qquad \xi = \frac{4\varpi_0 + \varpi_2}{3\varpi_2},$$

$$a^{(1)} = \frac{\varpi_1}{4}, \qquad b^{(1)} = \frac{1}{4}[\varpi_2(3-\varpi_1) - \varpi_1] \qquad c^{(1)} = \frac{\varpi_2}{4}(\varpi_1 - 3),$$

$$a^{(2)} = c^{(2)} = \frac{3\varpi_2}{16}, \qquad b^{(2)} = -\frac{3\varpi_2}{8},$$

$$p = 1 - a^{(0)}\alpha_0^{(0)} - B\alpha_1^{(0)} - F\alpha_2^{(0)},$$

$$q = -F\alpha_1^{(0)} - D\alpha_2^{(0)} - c^{(0)}\alpha_3^{(0)},$$

$$r = a^{(0)}\alpha_1^{(0)} + B\alpha_2^{(0)} + F\alpha_3^{(0)},$$

$$s = [1 - a^{(0)}\alpha_0^{(0)} - b^{(0)}\alpha_2^{(0)}] + F\alpha_2^{(0)} + D\alpha_3^{(0)},$$

$$t = -[\xi b^{(0)} + a^{(0)}]\alpha_1^{(0)} - \xi c^{(0)}\alpha_3^{(0)} + [\xi\alpha_0^{(0)} - \alpha_2^{(0)}]B - [\xi\alpha_1^{(0)} - \alpha_3^{(0)}]F,$$

$$v = [\xi c^{(0)} + b^{(0)}]\, \alpha_2^{(0)} - 1 + a^{(0)} \alpha_0^{(0)} + [\xi \alpha_0^{(0)} - \alpha_2^{(0)}]\, F + [\xi \alpha_1^{(0)} - \alpha_3^{(0)}]\, D\,,$$

$$l = c^{(1)}\, [\alpha_1^{(1)} - \alpha_3^{(1)}] + \Delta\, [\alpha_0^{(1)} - \alpha_2^{(1)}]\,,$$

$$m = 1 - a^{(1)}\, [\alpha_0^{(1)} - \alpha_2^{(1)}] - \Delta\, [\alpha_1^{(1)} - \alpha_3^{(1)}]\,,$$

$$\Delta = \frac{a^{(1)} c^{(1)}\, [\alpha_1^{(1)} - \alpha_3^{(1)}]}{1 - a^{(1)}\, [\alpha_0^{(1)} - \alpha_2^{(1)}]}\,.$$

F, B, and D are given by equations (91), (83), and (87). Further, it must be verified numerically that the following relation is satisfied:

$$\frac{3\varpi_0}{2\xi}\, r s - \frac{\varpi_1}{2}\, p q + \frac{3\varpi_2}{8\xi}\, t v - D = 0\,.$$

When $\varpi_0 = 1$, it can be shown that D, F, $c^{(0)}$, and $\chi(\mu)$ vanish, whereupon one obtains the results given in *R.T.*, page 158.

This work was financed in part by National Science Foundation contracts Nos. 1077 and G13353.

REFERENCES

Chandrasekhar, S. 1946, *Ap. J.*, **104**, 200.
———. 1950, *Radiative Transfer* (Oxford: Clarendon Press).

4. *Miscellaneous Problems*

20

The Formation of Absorption Lines in a Moving Atmosphere

S. CHANDRASEKHAR
Yerkes Observatory of the University of Chicago, Williams Bay, Wisconsin

1. INTRODUCTION

THE phenomenon of the scattering of light in a moving atmosphere has considerable interest for astrophysics. It occurs in Novae, Wolf-Rayet stars, planetary nebulae, the solar prominences, and the Corona. And more recently Struve's studies[1] of the spectra of stars like 48 Librae and 17 Leporis have emphasized its importance for stellar spectroscopy in general. But on consideration one soon realizes the unusual difficulties which must confront a rigorous theoretical analysis of these problems. For, in atmospheres in which large scale motions are present, on account of Doppler effect, the radiation scattered in different directions will have different frequencies, and, as a result of this, the radiation field in the different frequencies will interact with each other in a manner which is not always easy to visualize. However, in the astrophysical contexts, two circumstances simplify the problem. First, the velocities which are involved are small compared to the velocity of light, c, and second, the only effects of consequence are those which arise from the sensitive dependence of the scattering coefficient $\sigma(\nu)$ on the frequency ν. This last circumstance in particular allows us to ignore all effects such as aberration etc., and concentrate only on the effects arising from the change of frequency on scattering. The equation of transfer appropriate to these conditions has been written down by W. H. McCrea and K. K. Mitra.[2] But these writers did not succeed in solving any specific problem. However, we shall show how with certain approximations explicit solutions can be found which illustrate the effects which may be expected in the contours of absorption lines formed in an atmosphere in which differential motions exist. On the mathematical side, the novelty of the problem arises from the very unusual type of boundary value problem in hyperbolic equations which it presents.

2. THE EQUATION OF TRANSFER AND ITS APPROXIMATE FORMS

We shall consider an atmosphere stratified in parallel planes and in which all the properties are assumed to be constant over the planes z = constant (see Fig. 1).

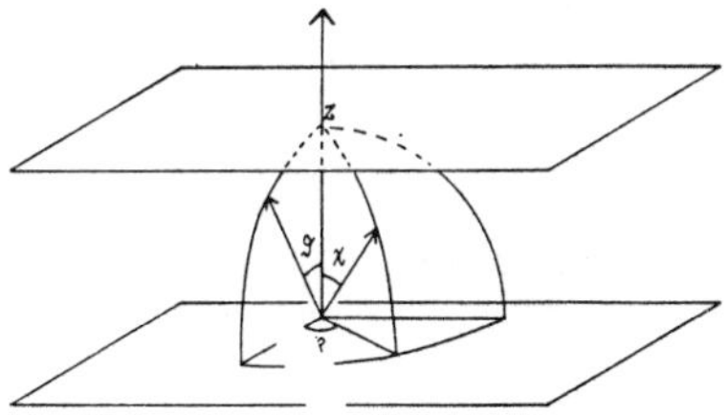

FIG. 1.

Let $\rho(z)$ be the density of the scattering material at height z and $w(z)$ the velocity of the material at the same height assumed parallel to the z direction. Further, let $\sigma(\nu)$ denote the mass scattering coefficient for the frequency ν as judged by an observer at rest with respect to the material. Since our principal interest is in the formation of absorption lines, we shall suppose that $\sigma(\nu)$ differs appreciably from zero only in a small range of ν. However, it is in the essence of the astrophysical problem that the "half-width" of $\sigma(\nu)$ is of the same order as the Doppler shifts in the frequency caused by the differential motions in the atmosphere. Indeed, it is this last

[1] O. Struve, Astrophys. J. **98**, 98 (1943). Also W. Hiltner, Astrophys. J. **99**, 103 (1944); P. W. Merrill and R. Sanford, Astrophys. J. **100**, 14 (1944).
[2] W. H. McCrea and K. K. Mitra, Zeits. f. Astrophys. **11**, 359 (1936).

circumstance which makes the change of frequency on scattering the only optical effect of the motion $w(z)$ which has any importance.

Consider then a pencil of radiation inclined at an angle ϑ to the positive normal and having a frequency ν as judged by a stationary observer. This radiation will appear to an observer at rest with respect to the material at z as having a frequency

$$\nu\left(1-\frac{w}{c}\cos\vartheta\right). \tag{1}$$

It will accordingly be scattered as such in all directions with a scattering coefficient

$$\sigma\left(\nu\left[1-\frac{w}{c}\cos\vartheta\right]\right). \tag{2}$$

We may, therefore, write the equation of transfer for the specific intensity $I(\nu, z, \vartheta)$ in the form

$$\cos\vartheta\frac{\partial I(\nu, z, \vartheta)}{\rho\partial z}=-\sigma\left(\nu\left[1-\frac{w}{c}\cos\vartheta\right]\right)I(\nu, z, \vartheta)+\mathfrak{J}(\nu, z, \vartheta), \tag{3}$$

where $\mathfrak{J}(\nu, z, \vartheta)$ denotes the emission per unit time and per unit solid angle in the frequency ν and in the direction ϑ. It is seen that this emission is given by

$$\mathfrak{J}(\nu, z, \vartheta)=\sigma\left(\nu\left[1-\frac{w}{c}\cos\vartheta\right]\right)\int_0^{2\pi}\int_0^{\pi}I\left(\nu\left[1-\frac{w}{c}\cos\vartheta+\frac{w}{c}\cos\chi\right], z, \chi\right)\sin\chi d\chi\frac{d\varphi}{4\pi}, \tag{4}$$

or in view of the symmetry about the z direction

$$\mathfrak{J}(\nu, z, \vartheta)=\tfrac{1}{2}\sigma\left(\nu\left[1-\frac{w}{c}\cos\vartheta\right]\right)\int_0^{\pi}I\left(\nu\left[1-\frac{w}{c}\cos\vartheta+\frac{w}{c}\cos\chi\right], z, \chi\right)\sin\chi d\chi. \tag{5}$$

To verify the foregoing expression for $\mathfrak{J}(\nu, z, \vartheta)$, we observe that the emission in the direction ϑ arises from the scattering into this direction of radiation from other directions. And, considering the contribution to $\mathfrak{J}$ from the scattering of the radiation in the direction specified by the polar angles χ and φ (see Fig. 1) into the direction $(\vartheta, 0)$, it is evident that the radiation must have the frequency

$$\nu\left(1-\frac{w}{c}\cos\vartheta+\frac{w}{c}\cos\chi\right), \tag{6}$$

as judged by a stationary observer; for, radiation of this frequency in the χ direction will appear to an observer at rest with respect to the material at z as having a frequency

$$\nu\left(1-\frac{w}{c}\cos\vartheta+\frac{w}{c}\cos\chi\right)\left(1-\frac{w}{c}\cos\chi\right)\simeq\nu\left(1-\frac{w}{c}\cos\vartheta\right), \tag{7}$$

which will accordingly be scattered uniformly in all directions with a scattering coefficient

$$\sigma\left(\nu\left[1-\frac{w}{c}\cos\vartheta\right]\right); \tag{8}$$

the radiation scattered into the ϑ-direction will have the same frequency (7) with respect to the material; to a stationary observer, it will appear as having a frequency ν. And summing over the contributions from all directions (χ, φ) we obtain (4).

Combining Eqs. (3) and (5) we have

$$\mu\frac{\partial I(\nu, z, \mu)}{\rho\partial z}=-\sigma\left(\nu\left[1-\frac{w}{c}\mu\right]\right)\left\{I(\nu, z, \mu)-\tfrac{1}{2}\int_{-1}^{+1}I\left(\nu\left[1-\frac{w}{c}\mu+\frac{w}{c}\mu'\right], z, \mu'\right)d\mu'\right\}, \tag{9}$$

where we have written μ and μ' for $\cos\vartheta$ and $\cos\chi$, respectively.

In solving Eq. (9) we shall adopt the method of approximation which has recently been developed in connection with the various problems of radiative transfer in the theory of stellar atmospheres.[3] The essence of this method is to replace the integrals which appear in the equation of transfer by sums according to Gauss's formula for numerical quadratures. Thus, considering Eq. (9) we replace it in the nth approximation by the system of $2n$ equations

$$\mu_i \frac{\partial I_i(\nu, z)}{\rho \partial z} = -\sigma\left(\nu\left[1-\frac{w}{c}\mu_i\right]\right)\left\{I_i(\nu, z) - \tfrac{1}{2}\sum a_j I_j\left(\nu\left[1-\frac{w}{c}\mu_i+\frac{w}{c}\mu_j\right], z\right)\right\}, \quad (i=\pm 1, \cdots, \pm n) \qquad (10)$$

where the μ_i's, $(i=\pm 1, \cdots, \pm n)$, are the zeros of the Legendre polynomial $P_{2n}(\mu)$, and the a_i's are the appropriate weights. Further, in Eq. (10) we have written $I_i(\nu, z)$ for $I(\nu, z, \mu_i)$.

At this stage one further simplification of Eq. (9) is possible. In evaluating the Doppler shifts, we need not distinguish between

$$\nu\left(1-\frac{w}{c}\mu\right) \quad \text{and} \quad \nu-\nu_0\frac{w}{c}\mu,$$

where ν_0 denotes the frequency of the center of the line. We may, therefore, replace Eq. (10) by the simpler one

$$\mu_i \frac{\partial I_i(\nu, z)}{\rho \partial z} = -\sigma\left(\nu-\nu_0\frac{w}{c}\mu_i\right)\left\{I_i(\nu, z) - \tfrac{1}{2}\sum a_j I_j\left(\nu-\nu_0\frac{w}{c}\mu_i+\nu_0\frac{w}{c}\mu_j, z\right)\right\}, \quad (i=\pm 1, \cdots, \pm n). \qquad (11)$$

The form of Eq. (11) suggests that instead of considering the intensities I_i, $(i=\pm 1, \cdots, \pm n)$, for some fixed frequency ν, we consider them for the frequencies

$$\nu_i = \nu + \nu_0\frac{w}{c}\mu_i \quad (i=\pm 1, \cdots, \pm n), \qquad (12)$$

which are functions of z. In Eq. (12), ν is a "fixed" frequency. If we now let

$$I_i(\nu_i, z) = \psi_i(\nu, z) \quad (i=\pm 1, \cdots, \pm n), \qquad (13)$$

we have

$$\frac{\partial \psi_i}{\partial z} = \left[\frac{\partial I_i(\nu, z)}{\partial z}\right]_{\nu=\nu_i} + \frac{\partial I_i(\nu_i, z)}{\partial \nu_i}\frac{\partial \nu_i}{\partial z}, \qquad (14)$$

or, according to Eqs. (12) and (13)

$$\frac{\partial \psi_i}{\partial z} = \left[\frac{\partial I_i(\nu, z)}{\partial z}\right]_{\nu=\nu_i} + \mu_i\frac{\nu_0}{c}\frac{dw}{dz}\frac{\partial \psi_i}{\partial \nu}. \qquad (15)$$

Substituting for the first term on the right-hand side of the foregoing equation from Eq. (11), we obtain

$$\mu_i\frac{\partial \psi_i}{\partial z} - \mu_i^2\frac{\nu_0}{c}\frac{dw}{dz}\frac{\partial \psi_i}{\partial \nu} = -\rho\sigma(\nu)(\psi_i - \tfrac{1}{2}\sum a_j\psi_j) \quad (i=\pm 1, \cdots, \pm n), \qquad (16)$$

which is clearly the most convenient form in which to study the equation of transfer for a moving atmosphere.

In our subsequent work we shall restrict ourselves to the first approximation. In this approximation

$$\mu_1 = -\mu_{-1} = 1/\sqrt{3} \quad \text{and} \quad a_1 = a_{-1} = 1, \qquad (17)$$

and Eqs. (11) and (16) lead to the two pairs of equations

[3] S. Chandrasekhar, Astrophys. J. **100**, 76, 117 (1944) and **101**, 95, 328, 348 (1945).

$$\mu_1 \frac{\partial I_{+1}(\nu, z)}{\rho \partial z} = -\tfrac{1}{2}\sigma\left(\nu - \nu_0 \frac{w}{c}\mu_1\right)\left[I_{+1}(\nu, z) - I_{-1}\left(\nu - 2\nu_0 \frac{w}{c}\mu_1, z\right)\right] \tag{18}$$

$$\mu_1 \frac{\partial I_{-1}(\nu, z)}{\rho \partial z} = +\tfrac{1}{2}\sigma\left(\nu + \nu_0 \frac{w}{c}\mu_1\right)\left[I_{-1}(\nu, z) - I_{+1}\left(\nu + 2\nu_0 \frac{w}{c}\mu_1, z\right)\right], \tag{19}$$

and

$$\mu_1 \frac{\partial \psi_{+1}}{\partial z} - \mu_1^2 \frac{\nu_0}{c} \frac{dw}{dz} \frac{\partial \psi_{+1}}{\partial \nu} = -\tfrac{1}{2}\rho\sigma(\nu)(\psi_{+1} - \psi_{-1}), \tag{20}$$

$$\mu_1 \frac{\partial \psi_{-1}}{\partial z} + \mu_1^2 \frac{\nu_0}{c} \frac{dw}{dz} \frac{\partial \psi_{-1}}{\partial \nu} = -\tfrac{1}{2}\rho\sigma(\nu)(\psi_{+1} - \psi_{-1}), \tag{21}$$

where it may be recalled that

$$\psi_{+1}(\nu, z) = I_{+1}\left(\nu + \mu_1 \nu_0 \frac{w}{c}, z\right), \quad \psi_{-1}(\nu, z) = I_{-1}\left(\nu - \mu_1 \nu_0 \frac{w}{c}, z\right). \tag{22}$$

3. SCHUSTER'S PROBLEM FOR A MOVING ATMOSPHERE

A classical problem first formulated by Schuster[4] provides the simplest model in terms of which the formation of absorption lines in a stellar atmosphere can be analyzed. In this model we consider a plane-stratified scattering atmosphere lying above a plane surface which radiates in a known manner and absorbs all radiation falling on it. The problem is to determine the radiation field in the atmosphere and in particular to relate the distribution in intensity of the emergent radiation with that radiated by the surface below. The appropriateness of this model for a first analysis of stellar absorption lines consists in the suitable idealization which it provides of the notions of a *photospheric surface* and the *reversing layers*. Consequently, when considering moving atmospheres it would seem proper that we retain the essentials of the Schuster model and generalize it only to the extent of admitting large scale motions. More particularly, we shall suppose that the photospheric surface is at $z=0$, and that it radiates uniformly in all outward directions $(0 \leqslant \vartheta < \pi/2)$ and in all frequencies. In other words, we suppose that

$$I(\nu, z, \vartheta) = \text{constant at } z=0 \text{ for } 0 \leqslant \vartheta < \pi/2 \text{ and for all frequencies.} \tag{23}$$

The state of motions in the atmosphere will be specified by the function $w(z)$ giving the velocity (assumed parallel to the z direction) at height z.

Finally, if $z=z_1$ defines the outer boundary of the atmosphere, we must require that here

$$I(\nu, z, \vartheta) \equiv 0, \quad \pi/2 < \vartheta < \pi \text{ at } z=z_1, \tag{24}$$

in accordance with the assumed non-existence of any radiation from the outside being incident on the atmosphere.

Schuster's problem for a moving atmosphere consists then in solving the equation of transfer (9), or the equivalent systems of equations in the various approximations, together with the boundary conditions (23) and (24). In the first approximation, the equivalent boundary conditions are that

$$I_{+1}(\nu, z) = \text{constant independent of } \nu \text{ at } z=0, \tag{25}$$

and

$$I_{-1}(\nu, z) \equiv 0 \text{ at } z=z_1. \tag{26}$$

4. THE REDUCTION TO A BOUNDARY VALUE PROBLEM FOR THE CASE $\sigma(\nu)$ = CONSTANT FOR $\nu_0 - \Delta\nu \leqslant \nu \leqslant \nu_0 + \Delta\nu$ AND ZERO OUTSIDE THIS INTERVAL AND FOR A LINEAR INCREASE OF w WITH THE OPTICAL DEPTH

In this paper we shall consider the solution to Schuster's problem formulated in the preceding

[4] A. Schuster, Astrophys. J. **21**, 1 (1905).

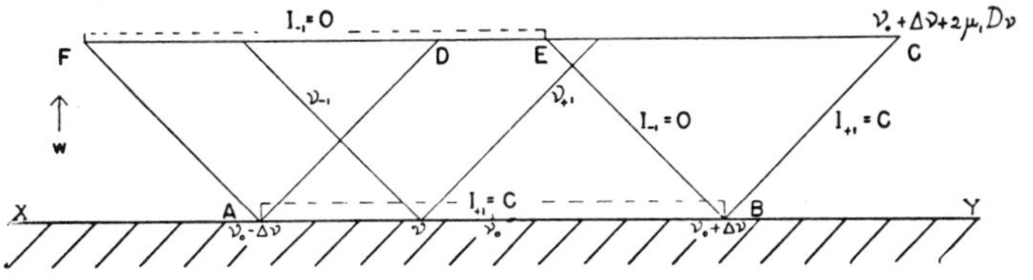

FIG. 2.

section for the case

$$\begin{aligned}\sigma(\nu) &= \text{constant} = \sigma_0 \text{ for } \nu_0-\Delta\nu \leqslant \nu \leqslant \nu_0+\Delta\nu, \\ &= 0 \text{ otherwise,}\end{aligned} \tag{27}$$

and

$$\frac{1}{\rho}\frac{dw}{dz} = \text{constant}. \tag{28}$$

Further, we shall restrict ourselves to the first approximation.

When $\sigma(\nu)$ has the form (27) some care is required in the formulation of the boundary conditions. For, according to Eqs. (18) and (19)

$$\partial I_{+1}/\partial z \neq 0 \text{ only if } \nu_0-\Delta\nu+\mu_1\nu_0\frac{w}{c} \leqslant \nu \leqslant \nu_0+\Delta\nu+\mu_1\nu_0\frac{w}{c}, \tag{29}$$

and

$$\partial I_{-1}/\partial z \neq 0 \text{ only if } \nu_0-\Delta\nu-\mu_1\nu_0\frac{w}{c} \leqslant \nu \leqslant \nu_0+\Delta\nu-\mu_1\nu_0\frac{w}{c}. \tag{30}$$

Accordingly, in the (ν, w) plane the lines

$$\nu = \nu_0-\Delta\nu+\mu_1\nu_0\frac{w}{c} \quad \text{and} \quad \nu = \nu_0+\Delta\nu+\mu_1\nu_0\frac{w}{c}, \tag{31}$$

demark the regions in which I_{+1} is different from a constant from the regions in which it is a constant for varying z. The situation is further clarified in Fig. 2 where AD and BC represent the lines (31). Similarly, the lines (AF and BE in Fig. 2)

$$\nu = \nu_0-\Delta\nu-\mu_1\nu_0\frac{w}{c} \quad \text{and} \quad \nu = \nu_0+\Delta\nu-\mu_1\nu_0\frac{w}{c}, \tag{32}$$

demark the regions in which I_{-1} is different from a constant from the regions in which it is a constant for varying z.

Now, since the outward intensity I_{+1} is a constant independent of ν on the photospheric surface (represented by the line $XABY$ in Fig. 2), it is clear that, we must, in accordance with our foregoing remarks, require that

$$I_{+1}(\nu, z) = \text{constant along } AB \text{ and } BC. \tag{33}$$

Similarly, the non-existence of any radiation incident on the atmosphere from the outside requires that

$$I_{-1}(\nu, z) = 0 \text{ along } BE \text{ and } EF. \tag{34}$$

When we pass to the intensities ψ_{+1} and ψ_{-1} defined as in Eq. (22), the boundary conditions (33) and (34) are equivalent to (cf. Fig. 3):

$$\begin{aligned}\psi_{+1} &= C = \text{constant on } AB\text{:} \quad z=0 \text{ and } \nu_0-\Delta\nu \leqslant \nu \leqslant \nu_0+\Delta\nu, \\ &= \text{the same constant on } BC\text{:} \quad \nu=\nu_0+\Delta\nu \text{ and } 0 \leqslant z \leqslant z_1,\end{aligned} \tag{35}$$

and

$$\begin{aligned}\psi_{-1} &= 0 \text{ on } CD\text{:} \quad z=z_1 \text{ and } \nu_0-\Delta\nu \leqslant \nu \leqslant \nu_0+\Delta\nu, \\ &= 0 \text{ on } BC\text{:} \quad \nu=\nu_0+\Delta\nu \text{ and } 0 \leqslant z \leqslant z_1.\end{aligned} \tag{36}$$

We now transform Eqs. (20) and (21) to forms which are more convenient for their solution:

Let t denote the optical depth of the atmosphere measured from the boundary inward in terms of σ_0. Then

$$\rho\sigma_0 dz = -dt. \tag{37}$$

In transforming Eqs. (20) and (21) it is, however, more convenient to use instead of the optical depth t the variable

$$x = \frac{1}{2\mu_1} t = \frac{\sqrt{3}}{2} t. \tag{38}$$

In terms of x Eqs. (20) and (21) are

$$\frac{\partial\psi_{+1}}{\partial x} - \mu_1 \frac{\nu_0}{c} \frac{dw}{dx} \frac{\partial\psi_{+1}}{\partial\nu} = \psi_{+1} - \psi_{-1}, \tag{39}$$

and

$$\frac{\partial\psi_{-1}}{\partial x} + \mu_1 \frac{\nu_0}{c} \frac{dw}{dx} \frac{\partial\psi_{-1}}{\partial\nu} = \psi_{+1} - \psi_{-1}. \tag{40}$$

Now the assumption (28) concerning the variation of w clearly implies that the velocity is a linear function of x. And as it entails no loss of generality, we shall suppose that $w=0$ at the base of the atmosphere. Further, let

$$w = w_1 \text{ at } t=0 \text{ and } x=0. \tag{41}$$

Under these circumstances we can write

$$w = w_1(1 - x/x_1) = w_1(1 - t/t_1), \tag{42}$$

where t_1 denotes the optical thickness in σ_0 of the entire atmosphere lying above the radiating surface. According to Eqs. (41) and (42) w_1 denotes the difference in velocity between the top and the bottom of the atmosphere. This velocity can be expressed in terms of a *Doppler width* $D\nu$ according to

$$D\nu = \tfrac{1}{2}\nu_0 w_1/c. \tag{43}$$

With these definitions

$$\mu_1 \frac{\nu_0}{c} \frac{dw}{dx} = -\mu_1 \frac{\nu_0}{c} \frac{w_1}{x_1} = -\frac{2}{\sqrt{3}} \frac{D\nu}{x_1} = -\frac{4}{3} \frac{D\nu}{t_1}, \tag{44}$$

and Eqs. (39) and (40) can be rewritten as

$$\frac{\partial\psi_{+1}}{\partial x} + 2\mu_1 \frac{D\nu}{x_1} \frac{\partial\psi_{+1}}{\partial\nu} = \psi_{+1} - \psi_{-1}, \tag{45}$$

and

$$\frac{\partial\psi_{-1}}{\partial x} - 2\mu_1 \frac{D\nu}{x_1} \frac{\partial\psi_{-1}}{\partial\nu} = \psi_{+1} - \psi_{-1}. \tag{46}$$

We now introduce the variable y defined by

$$(\nu_0 + \Delta\nu) - \nu = 2\mu_1 \frac{D\nu}{x_1} y. \tag{47}$$

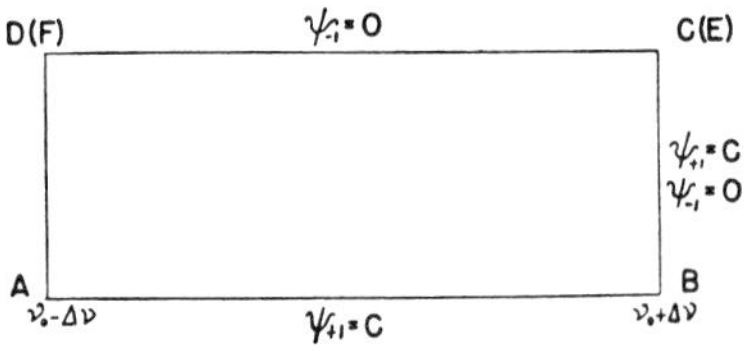

FIG. 3.

In other words, y measures the frequency shifts from the *violet edge* of $\sigma(\nu)$ in units of

$$2\mu_1 D\nu/x_1 \quad \text{(unit of frequency)}. \tag{48}$$

Equations (45) and (46) simplify to the forms

$$\partial\psi_{+1}/\partial x - \partial\psi_{+1}/\partial y = \psi_{+1} - \psi_{-1}, \tag{49}$$

and

$$\partial\psi_{-1}/\partial x + \partial\psi_{-1}/\partial y = \psi_{+1} - \psi_{-1}. \tag{50}$$

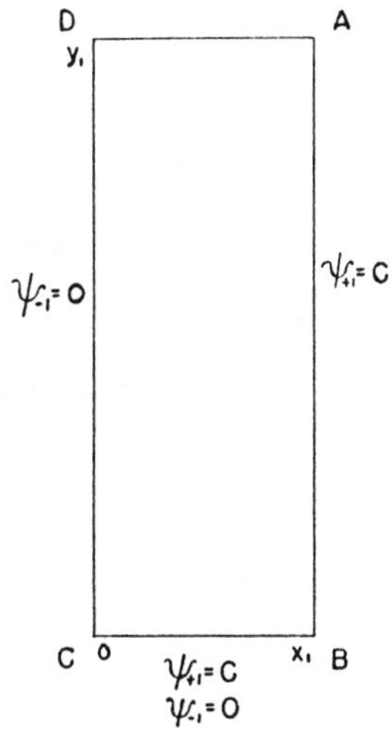

FIG. 4.

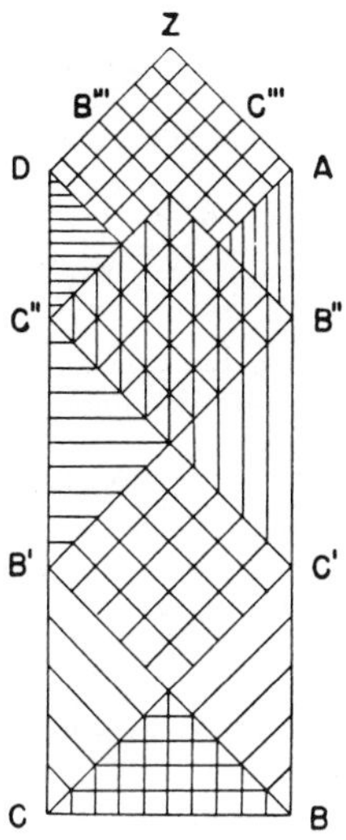

FIG. 5.

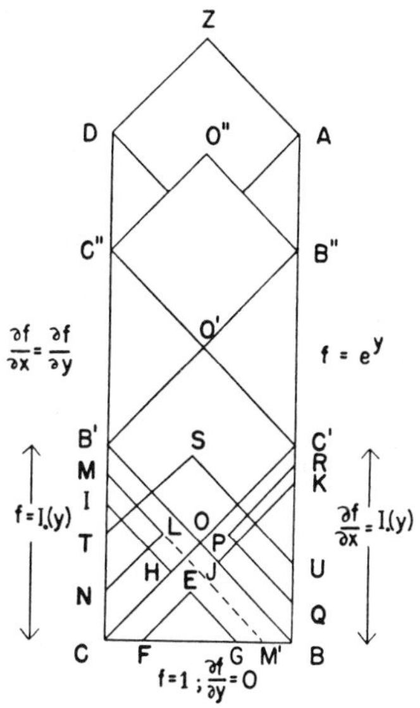

FIG. 6.

The range of the variables x and y in which the solution has to be sought is (Cf. Eq. [47])

$$0 \leqslant x \leqslant x_1 \quad \text{and} \quad 0 \leqslant y \leqslant y_1 = \frac{1}{\mu_1}\frac{\Delta\nu}{D\nu}x_1. \tag{51}$$

And the boundary conditions with respect to which Eqs. (49) and (50) have to be solved in the ranges (51) are (see Fig. 4):

$$\left.\begin{aligned} \psi_{+1} &= C \text{ on } BA: \quad x = x_1 \text{ and } 0 \leqslant y \leqslant y_1, \\ &= C \text{ on } CB: \quad y = 0 \text{ and } 0 \leqslant x \leqslant x_1. \end{aligned}\right\} \tag{52}$$

and

$$\left.\begin{aligned} \psi_{-1} &= 0 \text{ on } CD: \quad x = 0 \text{ and } 0 \leqslant y \leqslant y_1, \\ &= 0 \text{ on } CB: \quad y = 0 \text{ and } 0 \leqslant x \leqslant x_1. \end{aligned}\right\} \tag{53}$$

Since the Eqs. (49) and (50) are linear and homogeneous, there is no loss of generality if we set

$$C = 1. \tag{54}$$

We shall assume this normalization in our further work. Finally, we may note that in terms of the variables x and y Eq. (12) allowing the passage from the ψ's to the I's becomes

$$y_{\pm 1} = y \mp (x_1 - x). \tag{55}$$

It is convenient to introduce one further transformation of the variables. Let

$$\psi_{+1} = e^{-y}f \quad \text{and} \quad \psi_{-1} = e^{-y}g. \tag{56}$$

Equations (49) and (50) reduce to

$$\partial f/\partial x - \partial f/\partial y = -g, \tag{57}$$

and

$$\partial g/\partial x + \partial g/\partial y = +f. \tag{58}$$

Eliminating g between Eqs. (57) and (58), we obtain

$$\partial^2 f/\partial x^2 - \partial^2 f/\partial y^2 + f = 0. \tag{59}$$

We require to solve this hyperbolic equation with the boundary conditions (cf. Eqs. (52) and (53) and Fig. 5)

$$\left.\begin{aligned} &f = e^y \text{ on } AB: \quad x = x_1 \text{ and } 0 \leqslant y \leqslant y_1, \\ &f = 1 \text{ and } \partial f/\partial x = \partial f/\partial y = 0 \text{ on } BC: \quad y = 0 \text{ and } 0 \leqslant x \leqslant x_1, \\ &\partial f/\partial x = \partial f/\partial y \text{ on } CD: \quad x = 0 \text{ and } 0 \leqslant y \leqslant y_1, \end{aligned}\right\} \tag{60}$$

since $\psi_{-1} = 0$ implies that $g = 0$ and according to Eq. (57) this, in turn, implies that $\partial f/\partial x = \partial f/\partial y$.

Since, in the problem of the formation of absorption lines, our principal interest is on the ratio of the emergent intensity $I_{+1}(\nu, t)$ at $t = 0$ to the constant outward intensity on the radiating surface, we are most interested in the value of f on CD and DA. We may recall in this connection that $\Delta\nu$, $D\nu$ and x_1 are to be regarded as the parameters of the problem. Given these, y_1 is determined according to the relation (cf. Eq. (51))

$$y_1 = \sqrt{3}\, x_1 \frac{\Delta\nu}{D\nu} = \frac{3}{2} \text{ optical depth } \frac{\text{line width}}{\text{Doppler width}}. \tag{61}$$

Further, according to Eqs. (48) and (61), y measures the frequency as it enters in ψ_{+1} and ψ_{-1}, from the violet edge of $\sigma(\nu)$ and in the units

$$\frac{2}{y_1}\Delta\nu = \frac{1}{y_1} \text{ line width}. \tag{62}$$

Accordingly, the line contour will have a width

$$\frac{2}{y_1}(x_1 + y_1)\Delta\nu = \left(1 + \frac{x_1}{y_1}\right) \text{ line width}. \tag{63}$$

This is in agreement with what can be inferred directly from Fig. 2. As can be seen from this figure, the contour (on our present first approximation) must extend from

$$\nu_0 - \Delta\nu \text{ to } \nu_0 + \Delta\nu + \mu_1 \nu_0 \frac{w_1}{c}, \tag{64}$$

and must, therefore, have the width

$$2\Delta\nu + \mu_1 \nu_0 \frac{w_1}{c}, \tag{65}$$

or, according to Eqs. (43) and (61):

$$2\Delta\nu + 2\mu_1 D\nu = 2\Delta\nu\left(1 + \mu_1 \frac{D\nu}{\Delta\nu}\right) = 2\Delta\nu\left(1 + \frac{x_1}{y_1}\right). \tag{66}$$

It is evident that the line contour will itself be given by

$$r = e^{-y} f, \quad x = 0, \quad 0 \leqslant y \leqslant y_1, \tag{67}$$

and

$$r = e^{-y_1} f, \quad y = y_1, \quad 0 \leqslant x \leqslant x_1. \tag{68}$$

Equation (67) refers to the part of the contour which extends from

$$\nu = \nu_0 - \Delta\nu + \mu_1 \nu_0 \frac{w_1}{c} \text{ to } \nu_0 + \Delta\nu + \mu_1 \nu_0 \frac{w_1}{c} \tag{69}$$

while Eq. (68) refers to the part

$$\nu_0 - \Delta\nu + \mu_1 \nu_0 \frac{w_1}{c} \geqslant \nu \geqslant \nu_0 - \Delta\nu. \tag{70}$$

Finally, it may be noted that according to Eq. (55) the scale of frequency is the same for both the x and the y axis.

5. THE SOLUTION OF THE BOUNDARY VALUE PROBLEM

In the preceding section we have seen how the determination of the radiation field in a scattering atmosphere in which differential motions are present can be reduced to a boundary value problem in partial differential equations of the hyperbolic type. Under the conditions (27) and (28), the hyperbolic equation is one with constant coefficients and is of the simplest kind; indeed it is of the same form as the well-known equation of telegraphy.[5] But where our problem differs from the standard ones is in the boundary conditions. And it is the nature of our boundary conditions which prevents a direct application of the methods of Cauchy or Riemann.[6] For in these latter methods, only those situations in which the function and its derivatives are assigned along curves which do not intersect any characteristic more than once are contemplated. Our boundary conditions (60) are not as simple, the "supporting curve" $DCBA$ in fact intersecting every characteristic through a point inside the fundamental rectangle twice. Moreover, the function and its derivatives are assigned only on a part of the contour namely CB, while on the rest of the contour either the function alone, or a relation between its derivatives is specified. We shall, however, show how the boundary conditions (60) just suffice to determine f uniquely in the region $ZDCBA$. The method of solution we are going to describe is an adaptation of Riemann's method and is based on Green's theorem.

Now Green's theorem as applied to Eq. (59) is that the integral

$$\oint P dy - Q dx \tag{71}$$

where

$$P = v\frac{\partial f}{\partial x} - f\frac{\partial v}{\partial x} \quad \text{and} \quad Q = f\frac{\partial v}{\partial y} - v\frac{\partial f}{\partial y}, \tag{72}$$

around a closed contour vanishes if f and v are any two functions which satisfy Eq. (59) on and inside the contour.

As in Riemann's method we shall apply Green's theorem to contours which in parts are the characteristics $x-\xi=\pm(y-\eta)$ passing through some selected point (ξ, η) and choose for v a solution which is constant along the characteristics through (ξ, η). For Eq. (59) such a "Riemann function" $v(x, y; \xi, \eta)$ is known and depending on the quadrant in which the contour lies is

$$v(x, y; \xi, \eta) = I_0([(y-\eta)^2-(x-\xi)^2]^{\frac{1}{2}}), \tag{73}$$

or

$$v(x, y; \xi, \eta) = J_0([(x-\xi)^2-(y-\eta)^2]^{\frac{1}{2}}), \tag{74}$$

where J_0 and I_0 are the Bessel functions of order zero for real and imaginary arguments, respectively.

With the choice of the Riemann function for v, it is readily verified that

$$\int_{x-\xi=\pm(y-\eta)} P dy - Q dx = \pm f \tag{75}$$

if the integral on the right-hand side is a line integral along the characteristic $x-\xi=\pm(y-\eta)$.

In solving Eq. (59) consistent with the boundary conditions (60), we shall find it necessary to treat the various regions distinguished in Fig. 5 separately.

---

[5] Cf. A. G. Webster, *Partial differential equations of mathematical physics* (1933), Section 46, p. 173.

[6] For a general exposition of these classical methods see Webster, reference 5, pp. 160–188 and 239–255; or P. Frank and R. von Mises, *Die Differential und Integralgleichungen der Mechanik und Physik* (Rosenberg, New York, 1943), Vol. I, pp. 779–817.

(a) **The Solution in the Region *OCB***

Let the characteristic $x=y$ through C intersect AB at C' and the characteristic $x_1-x=y$ through B intersect CD at B'. Further, let CC' and BB' intersect at O. (See Fig. 6.)

Now, since the function and its derivatives are specified along CB, the solution inside the region OCB (including the sides OC and OB) can be found directly by Riemann's method. Thus applying Green's theorem to a contour such as $EFGE$ where EF and EG are the characteristics through $E=(\xi,\eta)$, and using Eq. (75) to evaluate the integrals along the characteristics we readily find that

$$f(\xi,\eta)=1-\tfrac{1}{2}\int_{\xi-\eta}^{\xi+\eta}\left(f\frac{\partial v}{\partial y}-v\frac{\partial f}{\partial y}\right)_{y=0}dx, \tag{76}$$

or, remembering that along CB

$$f=1 \quad \text{and} \quad \frac{\partial f}{\partial x}=\frac{\partial f}{\partial y}=0, \tag{77}$$

we have

$$f(\xi,\eta)=1-\tfrac{1}{2}\int_{\xi-\eta}^{\xi+\eta}\left(\frac{\partial v}{\partial y}\right)_{y=0}dx. \tag{78}$$

Now the Riemann function appropriate to our present contour is (73). Accordingly,

$$\begin{aligned}\left(\frac{\partial v}{\partial y}\right)_{y=0}&=\left[I_1([(y-\eta)^2-(x-\xi)^2]^{\frac{1}{2}})\frac{y-\eta}{[(y-\eta)^2-(x-\xi)^2]^{\frac{1}{2}}}\right]_{y=0}\\&=-I_1([\eta^2-(x-\xi)^2]^{\frac{1}{2}})\frac{\eta}{[\eta^2-(x-\xi)^2]^{\frac{1}{2}}}.\end{aligned} \tag{79}$$

Hence,

$$f(\xi,\eta)=1+\tfrac{1}{2}\eta\int_{\xi-\eta}^{\xi+\eta}I_1([\eta^2-(x-\xi)^2]^{\frac{1}{2}})\frac{dx}{[\eta^2-(x-\xi)^2]^{\frac{1}{2}}}. \tag{80}$$

Equation (80) determines f in the region OCB.

To evaluate the integral on the right-hand side of Eq. (80) we let

$$x-\xi=\eta\cos\vartheta, \tag{81}$$

and obtain

$$f(\xi,\eta)=1+\tfrac{1}{2}\eta\int_0^{\pi}I_1(\eta\sin\vartheta)d\vartheta. \tag{82}$$

Replacing I_1 in the foregoing equation by its equivalent series expansion and integrating term by term we find

$$\begin{aligned}f(\xi,\eta)&=1+\tfrac{1}{2}\eta\sum_{m=0}^{\infty}\int_0^{\pi}\frac{(\frac{1}{2}\eta\sin\vartheta)^{2m+1}}{m!\Gamma(m+2)}d\vartheta,\\&=1+\eta\sum_{m=0}^{\infty}(\tfrac{1}{2}\eta)^{2m+1}\frac{1}{m!\Gamma(m+2)}\int_0^{\pi/2}\sin^{2m+1}\vartheta d\vartheta,\\&=1+\eta\sum_{m=0}^{\infty}(\tfrac{1}{2}\eta)^{2m+1}\frac{1}{m!\Gamma(m+2)}\frac{2^{2m}(m!)^2}{(2m+1)!}\\&=1+\sum_{m=0}^{\infty}\frac{\eta^{2m+2}}{(2m+2)!}\\&=\sum_{m=0}^{\infty}\frac{\eta^{2m}}{(2m)!}.\end{aligned} \tag{83}$$

Thus, $$f(\xi,\eta)=\cosh\eta \tag{84}$$

inside and on the triangular contour OCB.

(b) The Integral Equation which Ensures the Continuity of the Solution Along OC

We have seen how the boundary conditions along CB determine the solution in the region OCB and on the sides OC and OB. We shall now show how this knowledge of the function along OC and OB together with the boundary conditions on CB' and BC' enables us to continue the solution into the region $O'B'COBC'O'$ (including the sides $B'O'$ and $O'C'$).

Thus, applying Green's theorem to contours such as $ICHI$ and $KJBK$ we shall obtain integral equations relating the values which the function takes along CO and OB with the values which the function and its derivatives take along CB' and BC'. And, as we shall see presently, these integral equations suffice to determine f along CB' and $\partial f/\partial x$ along BC' uniquely and secure at the same time the continuity of the solutions along OC and OB.

Considering first the condition which ensures the continuity of the solution along CO apply Green's theorem to a contour such as $ICHI$ where $H=(\eta,\eta)$ is a point on CO and HI is the characteristic $\eta-x=y-\eta$ through H. Using Eq. (75) to evaluate the integrals along the characteristics HI and CH and remembering that f takes the values 1 and $\cosh\eta$ at C and H, respectively, we find that

$$2\cosh\eta-1=f(2\eta,0)+\int_0^{2\eta}\left(v\frac{\partial f}{\partial x}-f\frac{\partial v}{\partial x}\right)_{x=0}dy, \tag{85}$$

or, since $\partial f/\partial x=\partial f/\partial y$ along CB', we have

$$2\cosh\eta-1=f(0,2\eta)+\int_0^{2\eta}\left(v\frac{\partial f}{\partial y}\right)_{x=0}dy-\int_0^{2\eta}\left(f\frac{\partial v}{\partial x}\right)_{x=0}dy. \tag{86}$$

Integrating by parts the first of the two integrals on the right-hand side of Eq. (86) we obtain

$$2\cosh\eta=2f(0,2\eta)-\int_0^{2\eta}f(0,y)\left(\frac{\partial v}{\partial x}+\frac{\partial v}{\partial y}\right)_{x=0}dy. \tag{87}$$

The Riemann function appropriate to our present contour is (74) with $\xi=\eta$. With this choice of v we find after some minor reductions that

$$\cosh\eta=f(0,2\eta)-\tfrac{1}{2}\int_0^{2\eta}f(0,y)J_1([\eta^2-(y-\eta)^2]^{\frac{1}{2}})\frac{y\,dy}{[\eta^2-(y-\eta)^2]^{\frac{1}{2}}}, \tag{88}$$

which is seen to be an integral equation for f along CB'. It is seen that Eq. (88) is equivalent to an integral equation of Volterra's type. For, by differentiating the equation

$$\sinh\eta=\tfrac{1}{2}\int_0^{2\eta}f(0,y)J_0([\eta^2-(y-\eta)^2]^{\frac{1}{2}})dy \tag{89}$$

with respect to η we may readily verify that we recover Eq. (88).

(c) The Solution of the Integral Eq. (89)

To solve Eq. (89) we apply a Laplace transformation to this equation. Thus multiplying both sides of Eq. (89) by $e^{-s\eta}$ and integrating over η from 0 to ∞ we obtain

$$\frac{2}{s^2-1}=\int_0^{\infty}d\eta e^{-s\eta}\int_0^{2\eta}dyf(0,y)J_0([\eta^2-(y-\eta)^2]^{\frac{1}{2}}), \tag{90}$$

or inverting the order of the integration on the right-hand side we have

$$\frac{2}{s^2-1}=\int_0^\infty dyf(0,y)\int_{y/2}^\infty d\eta e^{-s\eta}J_0([2\eta y-y^2]^{\frac{1}{2}}). \tag{91}$$

Introducing the variable

$$t=(2\eta y-y^2)^{\frac{1}{2}} \tag{92}$$

instead of η we find

$$\frac{2}{s^2-1}=\int_0^\infty \frac{dy}{y}e^{-sy/2}f(0,y)\int_0^\infty dt\, t\exp[-st^2/2y]J_0(t). \tag{93}$$

In Eq. (93) the integral over t is equivalent to the so-called Weber's first exponential integral in the theory of Bessel functions[7] and its value is given by

$$\int_0^\infty \exp[-st^2/2y]J_0(t)t\,dt=\frac{y}{s}e^{-y/2s}. \tag{94}$$

Using this result Eq. (93) reduces to

$$\frac{2s}{s^2-1}=\int_0^\infty f(0,y)\exp[-y(s+s^{-1})/2]dy. \tag{95}$$

If we now let

$$s+s^{-1}=2u, \tag{96}$$

Eq. (95) becomes

$$\frac{1}{(u^2-1)^{\frac{1}{2}}}=\int_0^\infty f(0,y)e^{-yu}dy. \tag{97}$$

In other words, we have shown that the Laplace transform of $f(0, y)$ is $(u^2-1)^{\frac{1}{2}}$. But it is known that the Laplace transform of $I_0(y)$ is exactly this. Hence

$$f(0,y)=I_0(y)\quad (0<y\leqslant x_1). \tag{98}$$

Thus, the requirement of the continuity of the solution along CO has determined f along CB'. Its derivatives along CB' are also deducible. We have

$$\left(\frac{\partial f}{dx}\right)_{x=0}=\left(\frac{\partial f}{\partial y}\right)_{x=0}=I_1(y)\quad (0<y\leqslant x_1). \tag{99}$$

(d) The Integral Equation Ensuring the Continuity of the Solution Along *OB* and Its Solution

Along BA we know f and its derivative with respect to y. But we do not know $\partial f/\partial x$ along this line. However, as the solution along OB is known the requirement that the solution be continuous on this line will determine $\partial f/\partial x$ along BC'. Thus, applying Green's theorem to a contour such as $JKBJ$ where $J=(x_1-\eta, \eta)$ is a point on OB and JK the characteristic $x-x_1+\eta=y-\eta$ through J we find in the usual manner that

$$2\cosh\eta=1+e^{2\eta}-\int_0^{2\eta}\left(v\frac{\partial f}{\partial x}-e^y\frac{\partial v}{\partial x}\right)_{x=x_1}dy. \tag{100}$$

The Riemann function appropriate to our present contour is

$$v=J_0([(x-x_1+\eta)^2-(y-\eta)^2]^{\frac{1}{2}}). \tag{101}$$

With v given by Eq. (101), Eq. (100) becomes

$$2\cosh\eta=1+e^{2\eta}-\int_0^{2\eta}e^yJ_1([\eta^2-(y-\eta)^2]^{\frac{1}{2}})\frac{\eta dy}{[\eta^2-(y-\eta)^2]^{\frac{1}{2}}}-\int_0^{2\eta}J_0([\eta^2-(y-\eta)^2]^{\frac{1}{2}})\left(\frac{\partial f}{\partial x}\right)_{x=x_1}dy. \tag{102}$$

[7] Cf. G. N. Watson, *Theory of Bessel Functions* (Cambridge University Press, England, 1944), p. 393.

Putting
$$y-\eta=\eta\cos\vartheta \tag{103}$$
in the first of the two integrals on the right-hand side of Eq. (102), it can be expressed in the form
$$2\cosh\eta=1+e^{2\eta}-e^{\eta}G(\eta)-\int_0^{2\eta}J_0([\eta^2-(y-\eta)^2]^{\frac{1}{2}})\left(\frac{\partial f}{\partial x}\right)_{x=x_1}dy, \tag{104}$$
where
$$G(\eta)=\eta\int_0^{\pi}e^{\eta\cos\vartheta}J_1(\eta\sin\vartheta)d\vartheta. \tag{105}$$

To evaluate $G(\eta)$ we replace $e^{\eta\cos\vartheta}$ and $J_1(\eta\sin\vartheta)$ by their respective series expansions and integrate term by term. In this manner we find that
$$\begin{aligned}
G(\eta)&=\eta\int_0^{\pi}\sum_{n=0}^{\infty}\frac{(\eta\cos\vartheta)^n}{n!}\sum_{m=0}^{\infty}(-1)^m\frac{(\frac{1}{2}\eta\sin\vartheta)^{2m+1}}{m!\Gamma(m+2)}d\vartheta,\\
&=\sum_{n=0}^{\infty}\frac{\eta^{2n+1}}{(2n)!}\Gamma(n+\tfrac{1}{2})\sum_{m=0}^{\infty}(-1)^m\frac{(\frac{1}{2}\eta)^{2m+1}}{\Gamma(m+2)\Gamma(m+n+\frac{3}{2})},\\
&=-\sum_{n=0}^{\infty}\frac{(2\eta)^{n+\frac{1}{2}}\Gamma(n+\frac{1}{2})}{(2n)!}\sum_{m=1}^{\infty}(-1)^m\frac{(\frac{1}{2}\eta)^{2m+n-\frac{1}{2}}}{\Gamma(m+1)\Gamma(m+n+\frac{1}{2})},\\
&=\sum_{n=0}^{\infty}\frac{(2\eta)^{n+\frac{1}{2}}\Gamma(n+\frac{1}{2})}{(2n)!}\left[\frac{(\frac{1}{2}\eta)^{n-\frac{1}{2}}}{\Gamma(n+\frac{1}{2})}-J_{n-\frac{1}{2}}(\eta)\right],\\
&=2\sum_{n=0}^{\infty}\frac{\eta^{2n}}{(2n)!}-\sum_{n=0}^{\infty}\frac{(2\eta)^{n+\frac{1}{2}}}{(2n)!}\Gamma(n+\tfrac{1}{2})J_{n-\frac{1}{2}}(\eta),\\
&=2\cosh\eta-(2\pi\eta)^{\frac{1}{2}}\sum_{n=0}^{\infty}\left(\frac{\eta}{2}\right)^n\frac{1}{n!}J_{n-\frac{1}{2}}(\eta).
\end{aligned} \tag{106}$$
But according to a formula of Lommel[8]
$$\sum_{n=0}^{\infty}(\tfrac{1}{2}\eta)^n\frac{1}{n!}J_{n-\frac{1}{2}}(\eta)=\left(\frac{2}{\pi\eta}\right)^{\frac{1}{2}}. \tag{107}$$
Hence
$$G(\eta)=2(\cosh\eta-1). \tag{108}$$
Substituting this result in Eq. (104) we find
$$\sinh\eta=\tfrac{1}{2}\int_0^{2\eta}J_0([\eta^2-(y-\eta)^2]^{\frac{1}{2}})\left(\frac{\partial f}{\partial x}\right)_{x=x_1}dy, \tag{109}$$
which is a Volterra integral equation for $(\partial f/\partial x)_{x=x_1}$.

It is seen that Eq. (109) is of the same form as Eq. (89). Accordingly
$$(\partial f/\partial x)_{x=x_1}=I_0(y)\quad(0\leqslant y\leqslant x_1). \tag{110}$$

(e) The Solution in the Region $O'B'COBC'O'$

With the determination of f along CB' and of $\partial f/\partial x$ along BC' our knowledge of the function and its derivatives along $B'CBC'$ is complete, and in the region $O'B'COBC'O'$ the solution becomes determinate. Thus, as in Riemann's method, applying Green's theorem to contours such as $LMNL$, $PQRP$, and $STCBUS$ we find that we can express f in the regions $OB'C$, OBC', and $O'B'OC'$ as follows:[9]

[8] See G. N. Watson, *Theory of Bessel Functions* (Cambridge University Press, England, 1944), p. 141, Eq. (7).

[9] For a point such as L, Green's theorem can be applied to either of the two contours $LMNL$ and $LNCM'L$. But clearly no ambiguity is implied as the continuity of the solution along OC (and OB) has already been ensured.

$$f(\xi, \eta)=I_0(\xi+\eta)-\tfrac{1}{2}\xi\int_0^{\pi} I_0(\eta+\xi \cos \vartheta)J_1(\xi \sin \vartheta)(1+\cos \vartheta)d\vartheta, \quad [(\xi, \eta) \text{ in } OB'C], \tag{111}$$

$$f(\xi, \eta)=e^{\eta}-\tfrac{1}{2}(x_1-\xi)\int_0^{\pi} I_0(\eta+[x_1-\xi]\cos \vartheta)J_0([x_1-\xi] \sin \vartheta) \sin \vartheta d\vartheta, \quad [(\xi, \eta) \text{ in } OBC']; \tag{112}$$

and

$$\begin{aligned} f(\xi, \eta)=&\, I_0([\eta^2-\xi^2]^{\frac{1}{2}})+e^{\eta-x_1+\xi}+\eta\int_{\cos^{-1}(x_1-\xi)/\eta}^{\cos\ \xi/\eta} I_1(\eta \sin \vartheta)d\vartheta \\ &-\xi\int_0^{\cosh^{-1}\eta/\xi} I_0(\eta-\xi \cosh \vartheta)I_1(\xi \sinh \vartheta)(\cosh \vartheta-1)d\vartheta \\ &+(x_1-\xi)\int_0^{\cosh^{-1}\eta/(x_1-\xi)} I_0([x_1-\xi] \sinh \vartheta)I_0(\eta-[x_1-\xi] \cosh \vartheta) \sinh \vartheta d\vartheta \\ &+(x_1-\xi)\int_0^{\cosh^{-1}\eta/(x_1-\xi)} e^{\eta-(x_1-\xi)\cosh\vartheta} I_1([x_1-\xi]\sinh \vartheta)d\vartheta, \quad [(\xi, \eta) \text{ in } O'B'OC']. \end{aligned} \tag{113}$$

In particular we may note that Eq. (113) will enable us to determine the solution along the sides $B'O'$ and $O'C'$.

(f) Further Continuation of the Solution

In the preceding paragraphs we have seen how the knowledge of the function along COB, together with the boundary conditions on CB' and BC' enables us to determine f in the region $O'B'COBC'$ including the sides $B'O'$ and $O'C'$. It is now apparent that in the same way we can utilize our present knowledge of the function along $B'O'C'$ to extend the solution further into the region $O'B'C''O''B''C'$. And this process can be continued until the solution inside the entire rectangle $DCBA$ is completed. However, in this paper we shall not consider these further extensions but content ourselves with the solution which has been completed in the first square $B'CBC'$. According to Eq. (61) this will suffice to determine the radiation field in all cases in which the ratio $D\nu:\Delta\nu$ exceeds $\sqrt{3}$.

6. THE CONTOURS OF THE ABSORPTION LINES FORMED. NUMERICAL ILLUSTRATIONS

In the preceding section we have seen how the boundary value problem formulated in Section 4 can, in principle, be solved. In terms of the solution thus found, we can specify the radiation field in an atmosphere with differential motions and under the conditions prescribed in Sections 3 and 4. While the determination of the radiation field in the entire atmosphere is necessary to answer all questions relating to the formation of the absorption lines (see Section 7 below), greatest interest is, however, attached to the contour of the resulting line. In the first approximation in which we have studied the problem, this is given in terms of the emergent value of the outward intensity $I_{+1}(\nu)$. More specifically the form of the line is given by Eqs. (67) and (68) where f is the solution of the boundary value problem. We shall now consider in some detail the predicted nature of these contours.

Now along CB' the solution is given by (cf. Eq. (98))

$$f=I_0(y) \quad (x=0,\ 0\leqslant y\leqslant x_1). \tag{114}$$

According to Eq. (68) we may, therefore, write down a formula for the *residual intensity* r which will be valid for a part of the line contour. Thus

$$r=e^{-y}I_0(y), \tag{115}$$

will describe the line in the frequency interval

$$\nu_0+\Delta\nu+2\mu_1 D\nu \geqslant \nu \geqslant \nu_0-\Delta\nu+2\mu_1 D\nu, \tag{116}$$

or

$$\nu_0+\Delta\nu+2\mu_1 D\nu \geqslant \nu \geqslant \nu_0+\Delta\nu, \tag{116'}$$

TABLE I. The function $r = e^{-y}I_0(y)$.

y	$e^{-y}I_0(y)$	y	$e^{-y}I_0(y)$	y	$e^{-y}I_0(y)$	y	$e^{-y}I_0(y)$
0	1.0000	0.7	0.5593	1.40	0.3831	3.0	0.2430
0.1	0.9071	0.8	0.5241	1.50	0.3674	3.5	0.2228
0.2	0.8269	0.9	0.4932	1.75	0.3346	4.0	0.2070
0.3	0.7576	1.0	0.4658	2.00	0.3085	4.5	0.1942
0.4	0.6974	1.1	0.4414	2.25	0.2874	5.0	0.1835
0.5	0.6450	1.2	0.4198	2.50	0.2700	6.0	0.1667
0.6	0.5993	1.3	0.4004	2.75	0.2555	8.0	0.1434
						10.0	0.1278

TABLE II. $f(\xi, \eta)$.

ξ \ η	1.0	2.0	2.5	3.0	4.0	5.0
0	1.2661	2.2796	3.2898	4.8808	11.302	27.240
0.5	1.4762	2.9697	4.4134	6.6792	15.847	39.601
1.0	1.5431	3.4374	5.2449	8.0836	19.628	51.362
1.5	1.5431	3.6897	5.7814	9.0767	23.354	62.659
2.0	1.5431	3.7622	6.0569	9.6891	26.931	73.746
2.5	1.5431	3.7622	6.1323	10.6881	30.579	84.930
3.0	1.5431	3.7622	6.7927	11.8944	34.506	96.502
3.5	1.5431	4.3857	7.7767	13.4600	38.859	108.657
4.0	1.5431	5.2378	9.0554	15.4183	43.693	121.460
4.5	2.0969	6.2657	10.5584	17.6724	48.998	134.802
5.0	2.7183	7.3891	12.1825	20.0855	54.598	148.413

TABLE III. Line contours of absorption lines formed in a moving atmosphere ($x_1 = 5$; $y_1 = 1, 2, 2.5, 3, 4, 5$).

$\frac{\nu-\nu_0+\Delta\nu}{2\Delta\nu}$	r	$\frac{\nu-\nu_0+\Delta\nu}{2\Delta\nu}$	r	$\frac{\nu-\nu_0+\Delta\nu}{2\Delta\nu}$	r	$\frac{\nu-\nu_0+\Delta\nu}{2\Delta\nu}$	r	$\frac{\nu-\nu_0+\Delta\nu}{2\Delta\nu}$	r	$\frac{\nu-\nu_0+\Delta\nu}{2\Delta\nu}$	r
0.0	1.0000	0.0	1.0000	0.0	1.0000	0.0	1.0000	0.0	1.0000	0.0	1.0000
0.5	0.7714	0.25	0.8480	0.2	0.8667	$0.1\dot{6}$	0.8799	0.125	0.8974	0.1	0.9083
1.0	0.5677	0.50	0.7089	0.4	0.7433	$0.3\dot{3}$	0.7676	0.250	0.8003	0.2	0.8184
1.5	0.5677	0.75	0.5935	0.6	0.6384	0.50	0.6701	0.375	0.7117	0.3	0.7321
2.0	0.5677	1.00	0.5092	0.8	0.5576	$0.6\dot{6}$	0.5922	0.500	0.6320	0.4	0.6502
2.5	0.5677	1.25	0.5092	1.0	0.5034	$0.8\dot{3}$	0.5321	0.625	0.5601	0.5	0.5723
3.0	0.5677	1.50	0.5092	1.2	0.4972	1.00	0.4824	0.750	0.4933	0.6	0.4969
3.5	0.5677	1.75	0.4993	1.4	0.4746	$1.1\dot{6}$	0.4519	0.875	0.4277	0.7	0.4222
4.0	0.5677	2.00	0.4652	1.6	0.4305	$1.3\dot{3}$	0.4025	1.000	0.3595	0.8	0.3461
4.5	0.5431	2.25	0.4019	1.8	0.3623	1.50	0.3325	1.125	0.2902	0.9	0.2668
5.0	0.4658	2.50	0.3085	2.0	0.2700	$1.6\dot{6}$	0.2430	1.250	0.2070	1.0	0.1835
5.5	0.6450	2.75	0.3674	2.2	0.3085	$1.8\dot{3}$	0.2700	1.375	0.2228	1.1	0.1942
6.0	1.0000	3.00	0.4658	2.4	0.3674	2.00	0.3085	1.500	0.2430	1.2	0.2070
		3.25	0.6450	2.6	0.4658	$2.1\dot{6}$	0.3674	1.625	0.2700	1.3	0.2228
		3.50	1.0000	2.8	0.6450	$2.3\dot{3}$	0.4658	1.750	0.3085	1.4	0.2430
				3.0	1.0000	2.50	0.6450	1.875	0.3674	1.5	0.2700
						$2.6\dot{6}$	1.0000	2.000	0.4658	1.6	0.3085
								2.125	0.6450	1.7	0.3674
								2.250	1.0000	1.8	0.4658
										1.9	0.6450
										2.0	1.0000

depending on whether

$$\mu_1 D\nu \gtrless \Delta\nu \quad \text{or} \quad \mu D\nu \lessgtr \Delta\nu. \tag{117}$$

It should be noted in this connection that in our present context y measures the frequency shifts from the violet edge $\nu_0+\Delta\nu+2\mu_1 D\nu$ of the *line contour* in the unit $2\Delta\nu/y_1$.

For convenience we have provided a brief table of the function on the right-hand side of Eq. (115) (see Table I).

Again, according to Eqs. (68) and (84), when

$$D\nu > (2/\mu_1)\Delta\nu = 2\sqrt{3}\Delta\nu, \tag{118}$$

in the frequency interval

$$\nu_0 - 3\Delta\nu + 2\mu_1 D\nu \geqslant \nu \geqslant \nu_0 + \Delta\nu, \tag{119}$$

the contour is flat, the residual intensity having the constant value

$$r = e^{-y_1} \cosh y_1 \quad (y_1 \leqslant x \leqslant x_1 - y_1). \tag{120}$$

This flat portion occupies a fraction

$$(x_1 - 2y_1)/(x_1 + y_1) \tag{121}$$

of the entire contour. As $D\nu/\Delta\nu \to \infty$ and $y_1 \to 0$, the fraction (121) tends to unity: the line accordingly becomes very shallow and very broad. More specifically, as $y_1 \to 0$

$$1 - r \to y_1 \quad (y_1 \to 0), \tag{122}$$

and

$$\text{the width of the line contour} \to 2\Delta\nu x_1/y_1 \quad (y_1 \to 0). \tag{123}$$

The equivalent width, therefore, tends to the limiting value

$$\text{Equivalent width} \to 2\Delta\nu x_1 = \sqrt{3}\Delta\nu t_1 = \sqrt{3}\Delta\nu N\sigma_0 m^{-1} \quad (y_1 \to 0), \tag{124}$$

where N denotes the number of scattering atoms in a column of unit cross section in the atmosphere and m the mass of the atom.

Returning to the general case, it is seen that the specification of the line contour over its entire range

$$\nu_0 - \Delta\nu \leqslant \nu \leqslant \nu_0 + \Delta\nu + 2\mu_1 D\nu, \tag{125}$$

requires a knowledge of the function f along the line $y = y_1$ and for $0 \leqslant x \leqslant x_1$. However, since the solution for f has been found in an explicit form only in the first square, $B'CBC'$, complete contours can be given only for those cases in which $D\nu \geqslant \sqrt{3}\Delta\nu$. And even then, the part of the solution not included in the triangle OBC and the sides CB' and BC' can be found only after several numerical quadratures. For, in these regions the solution is given by the formulae (111)–(113), and it does not seem that the various integrals occurring in these formulae can be evaluated explicitly.

As illustrating the solution found in Section 5 we have considered in detail the case

$$x_1 = 5 \tag{126}$$

and determined the line contours for the following ratios of the Doppler width to the line width:

$$\mu_1 D\nu/\Delta\nu = 5,\ 2.5,\ 2,\ 1\tfrac{2}{3},\ 1.25,\ \text{and}\ 1. \tag{127}$$

According to Eq. (61) the specification of the contours for these ratios of $\mu_1 D\nu : \Delta\nu$, requires the evaluation of f along the lines

$$y = 1,\ 2,\ 2.5,\ 3,\ 4,\ \text{and}\ 5, \tag{128}$$

for $0 \leqslant x \leqslant 5$. The values of f for several points along these lines and intercepted in the region $O'B'COBC'$ were determined according to Eqs. (111)–(113). The various integrals occurring in these equations were evaluated numerically.[10] The results of these calculations are included in Table II. In Table III, the values of f given in Table II are converted into residual intensities according to

[10] The carrying out of the numerical quadratures were immensely facilitated by the *British Association Mathematical Tables*, VI: *Bessel functions of order zero and unity* (Cambridge University Press, England, 1937). I should record here my indebtedness to Mrs. Frances Herman Breen for assistance with these calculations.

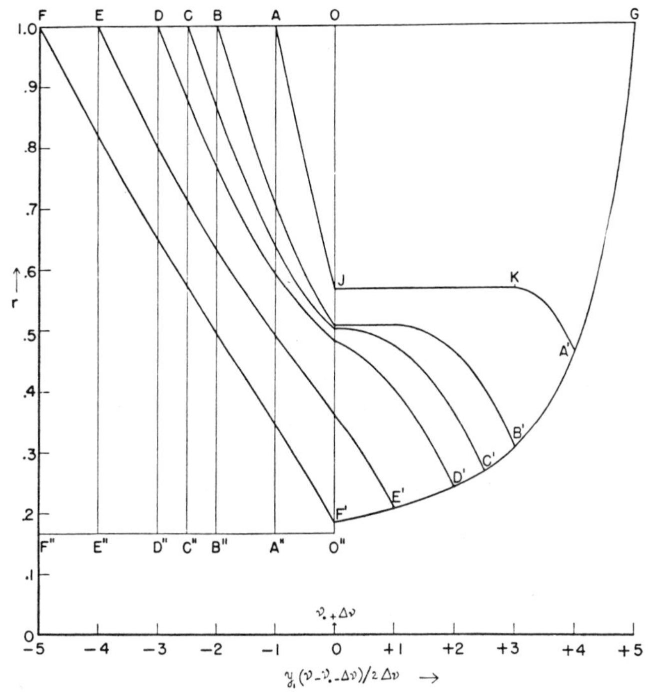

FIG. 7.

Eq. (68) and are tabulated together with the values of r for the remaining parts of the contours given by Eq. (115). The arguments in Table III are the frequency shifts measured from the red end $\nu_0-\Delta\nu$ of the contour in the unit $2\Delta\nu$.

The residual intensities tabulated in Table III are further illustrated as line contours in Fig. 7. In this figure the various contours are plotted on different frequency scales, the width $2\Delta\nu$ of $\sigma(\nu)$ always extending from the red end of the contour to O. Thus the contour $BB'G$ corresponds to a case in which the line formed under the same conditions in a static atmosphere would extend from B to O.

From Fig. 7 it is apparent that in all cases in which $D\nu > 2\sqrt{3}\Delta\nu$ the contour consists of four distinct parts, namely,

$$\left.\begin{aligned} \nu_0-\Delta\nu \leqslant \nu \leqslant \nu_0+\Delta\nu, &\quad \text{(i)} \\ \nu_0+\Delta\nu \leqslant \nu \leqslant \nu_0-3\Delta\nu+2\mu_1 D\nu, &\quad \text{(ii)} \\ \nu_0-3\Delta\nu+2\mu_1 D\nu \leqslant \nu \leqslant \nu_0-\Delta\nu+2\mu_1 D\nu, &\quad \text{(iii)} \\ \nu_0-\Delta\nu+2\mu_1 D\nu \leqslant \nu \leqslant \nu_0+\Delta\nu+2\mu_1 D\nu. &\quad \text{(iv)} \end{aligned}\right\} \tag{129}$$

In each of these parts r is given by a different analytical expression. It decreases from 1 in the first interval, remains constant in the second, and decreases some more in the third attaining its minimum at $\nu=\nu_0-\Delta\nu+2\mu_1 D\nu$. In the last interval it increases again to 1. It is in this fourth interval that the line contour is described by Eq. (115). In Fig. 7 we have indicated these four parts on the contour $AA'G$. The parts are respectively, AJ, JK, KA', and $A'G$. The reason for the existence of these four parts can be understood from a reference to Fig. 8. In this figure, which is similar to Fig. 2, the regions in which I_{+1} and I_{-1}, respectively, are different from constants (for varying z) are marked. We have further

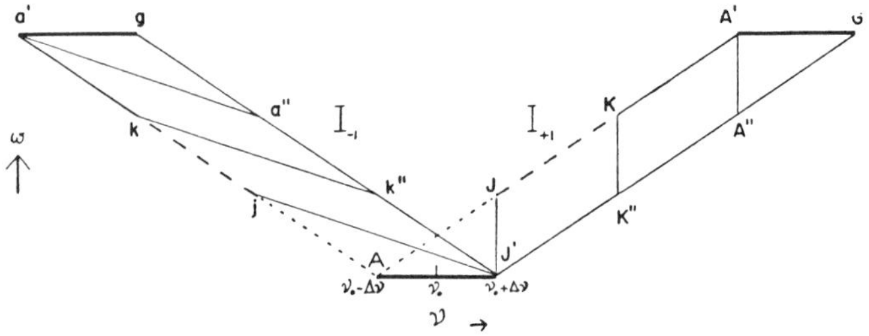

FIG. 8.

indicated the frequency intervals in which the different parts of the contour arise. (The lettering in Figs. 7 and 8 correspond). Now, according to our discussion in Sections 3 and 4, the outward intensity I_{+1} for a frequency ν interacts with the inward intensity I_{-1} for a frequency $\nu-2\mu_1\nu_0(w/c)$. Hence I_{+1} for the frequencies in the intervals AJ, JK, KA', and $A'G$ in their transfer through the atmosphere have interacted with I_{-1} for the frequencies in the regions AjJ', $jkk''J'$, $ka'a''k''$, and $a'ga''$, respectively. The reason for the existence of the four distinct parts in the line contour now becomes apparent. Moreover, this discussion makes it clear why it is that the problem increases in complexity as $\mu_1 D\nu/\Delta\nu$ decreases below unity.

Finally, it is of interest to compare the contours we have obtained with those which would be expected in an atmosphere in which no gradient of velocity exists. To discuss this case we have to go back to Eqs. (39) and (40). Setting $dw/dx=0$ in these equations and solving them with the boundary conditions appropriate to Schuster's problem we readily find that

$$\left.\begin{aligned} r &= \frac{1}{1+x_1} \qquad (\nu_0-\Delta\nu \leqslant \nu \leqslant \nu_0+\Delta\nu) \\ &= 0 \qquad\qquad\qquad \text{otherwise.} \end{aligned}\right\} \tag{130}$$

The contours are therefore rectangular. For $x_1=5$, $r=\frac{1}{6}$. These rectangular contours which will be obtained in the limit $D\nu=0$ are also shown in Fig. 7. Thus the contour $BB'G$ should be compared with $BB''O''O$; and similarly for the others.

7. REMARKS ON FUTURE WORK

The successful solution of a specific problem in the theory of moving atmospheres which we have presented in the preceding sections justifies the hope that it will be possible to solve problems more general and less idealized than the one considered in this paper. Indeed, there are several problems in the theory of moving atmospheres which come already within the scope of the methods developed in this paper. For example, there is the problem of the variation of line contours with the angle of emergence from the atmosphere. The solution to this problem will depend on the radiation field in the entire atmosphere. For, the intensity $I(\nu, z_1, \mu)$ of the radiation of frequency ν (as judged by an observer at rest with respect to the radiating surface at $z=0$) emergent in a direction with a direction cosine μ with respect to the positive normal can be expressed as an integral in the form (cf., Eq. (9))

$$I(\nu, z_1, \mu)=\int_0^{z_1} J(\nu, z, \mu)\exp\left\{-\int_z^{z_1}\rho\sigma(\nu-[\nu_0/c]w\mu)/\mu\right\}\frac{dz}{\mu}, \tag{131}$$

where

$$J(\nu, z, \mu)=\tfrac{1}{2}\int_{-1}^{+1} I\left(\nu-\nu_0\frac{w}{c}\mu+\nu_0\frac{w}{c}\mu', z, \mu'\right)d\mu'. \tag{132}$$

In the first approximation we can express the integral on the right-hand side of Eq. (132) as a Gauss sum with two terms. Thus,

$$J(\nu, z, \mu) = \tfrac{1}{2}\left\{ I_{+1}\left(\nu - \nu_0\frac{w}{c}\mu + \nu_0\frac{w}{c}\mu_1\right) + I_{-1}\left(\nu - \nu_0\frac{w}{c}\mu - \nu_0\frac{w}{c}\mu_1, z\right)\right\}. \tag{133}$$

The source function J can, therefore, be expressed in terms of the solutions for $I_{+1}(\nu, z)$ and $I_{-1}(\nu, z)$ which we have found in Section 5 and there will be no formal difficulty in solving for $I(\nu, z_1, \mu)$ according to Eq. (131).

Again, the determination of $I(\nu, z_1, \mu)$ by the procedure we have outlined above will be of particular importance for deriving contours comparable to those *observed* in cases in which the photospheric surface is itself moving with a velocity w_0. It will be recalled in this connection that our discussion of the equation of transfer involves no assumption concerning w_0 since everything was referred to an observer at rest with respect to the surface at $z=0$ and $t=t_1$. However, the line contour as seen by an observer outside the star will not be given by $F(\nu, z_1)$, as allowance will have to be made for the fact that the photospheric surface from which the radiation is emerging at an angle ϑ has a motion $w_0 \cos\vartheta$ towards the observer. Accordingly, the contour as judged by an external observer at a great distance from the star will be determined by

$$\mathfrak{F}(\nu) = 2\int_0^1 I\left(\nu + \nu_0\frac{w_0}{c}\mu, z_1, \mu\right)\mu d\mu, \tag{134}$$

where $I(\nu, z_1, \mu)$ has the same meaning as in Eq. (131).

Another problem which can be solved by the methods of the present paper is the radiative equilibrium of a planetary nebula. It is known that large differential motions are present in planetary nebulae and a problem of considerable interest relates to the question of the radiation pressure in the Lyman α-radiation. It can be shown that with the same assumptions (27) and (28) concerning $\sigma(\nu)$ and the variation of w through the atmosphere, the problem can be reduced to a boundary value problem very similar to the one considered in Sections 4 and 5 and the solution can also be found by similar methods.

And finally there is the general problem of line formation in moving atmospheres in which $\sigma(\nu)$ is allowed to be more general than the rectangular form considered in this paper. It can be shown that under these more general conditions the problem can still be reduced to a boundary value problem in hyperbolic equations. If the velocity be assumed to vary linearly with the optical depth, the equation we have to consider differs from (59) only in the occurrence of a factor depending on y in front of f. It does not seem impossible that with suitable simplifications, progress toward the solution of these more difficult problems can be made.

21

THE RADIATIVE EQUILIBRIUM OF AN EXPANDING PLANETARY NEBULA

I. RADIATION PRESSURE IN LYMAN-α

S. CHANDRASEKHAR
Yerkes Observatory

Received September 16, 1945

ABSTRACT

In this paper the problem of the radiative equilibrium in Lyman-α of a differentially expanding planetary nebula is formulated *de novo*. Particular attention has been given to the formulation of the boundary conditions. It is shown that the transfer problem reduces to a novel boundary-value problem in hyperbolic partial differential equations. Explicit solutions have been found for the case in which the line absorption coefficient $\sigma(\nu)$ has a rectangular form. On the basis of the solutions obtained, the question of the radiation pressure in Lyman-α has been re-examined. It is shown that when the Doppler shift due to the difference in velocities between the inner and the outer boundaries of the nebula exceeds the line width $2\Delta\nu$ of $\sigma(\nu)$ by a factor $2\sqrt{3}$, the nebula can be divided into three parts: an inner, a central, and an outer part. In the central part the radiation pressure in Lyman-α is of the same order as that in the continuum, while in the inner and the outer parts it is not inappreciable. The bearing of this manner in which the radiation pressure in Lyman-α operates in an expanding atmosphere on the dynamics of a planetary nebula and also on other astrophysical problems is briefly indicated.

1. *Introduction.*—As is well known on Zanstra's theory[1] of nebular luminosity, the hydrogen emission in a planetary nebula is traced to the conversion in the nebula of the incident ultraviolet radiation of the central star beyond the head of the Lyman series into radiation principally in the Lyman and the Balmer series. More specifically, if the optical thickness of the nebula for the ultraviolet radiation[2] is sufficiently large, this conversion will be so nearly complete that for every ultraviolet quantum in the incident starlight, the nebula will emit a quantum in the first member of the Lyman series, namely, Lyman-α. This conversion of the radiation arises from the fact that, whenever an ultraviolet quantum is scattered, there is only a probability p of order $\frac{1}{2}$ that it will be re-emitted as such.[3] And on the occasions when the ultraviolet quantum is not re-emitted, a chain of absorptions and emissions takes place which so rapidly leads to the emission of a Lyman-α quantum that we may almost say that every time an ultraviolet quantum is scattered, there is a definite probability $(1-p) \sim \frac{1}{2}$ that a Lyman-α quantum is emitted.[4] The theory of radiative transfer of the ultraviolet radiation through the nebula has been developed along these lines by Ambarzumian and others[5] and shows that the conversion of the ultraviolet radiation into line radiation is so far advanced through most of the nebula that, if nothing else intervened, the nebula would be subject to enormous radiation pressure because the absorption coefficient in Lyman-α is several thousand times larger than in the continuum. The magnitude of the radiation pressure which may thus act on the nebula is so large that it will seriously endanger its permanence even over relatively limited periods of time. The consequences of this radiation pressure in Lyman-α

[1] *Ap. J.*, **65**, 50, 1927; *Pub. Dom. Ap. Obs.*, **4**, 209, 1931.

[2] In this paper we shall use the term "ultraviolet radiation" to denote all radiation beyond the head of the Lyman series.

[3] G. Cillié, *M.N.*, **92**, 820, 1932; **96**, 777, 1936.

[4] Cf. V. A. Ambarzumian, *M.N.*, **93**, 50, 1932; also *Bull. de l'Obs. centr. à Poulkovo*, **13**, No. 14, 3, 1933.

[5] *Ibid.* More accurate solutions of the transfer problem have been given by S. Chandrasekhar, *Zs. f. Ap.*, **9**, 266, 1935, *Ap. J.*, **100**, 76, 1944; and C. U. Cesco, S. Chandrasekhar, and J. Sahade, *Ap. J.*, **101**, 320, 1945.

acting on the nebula are, indeed, such as to lead one to re-examine whether some factor has not been overlooked, which, when properly allowed for, will cut down the radiation pressure. In this connection, Zanstra[6] has suggested that the very likely presence of differential motions in the nebula may very effectively reduce the radiation pressure which may act on the nebula. Unfortunately, Zanstra's particular considerations relating to this problem are vitiated by the use of an equation of transfer which, to put it bluntly, is incorrect under the circumstances envisaged. However, it appears that the methods which have been recently developed to treat an analogous problem in theory of the formation of absorption lines in a moving atmosphere[7] are sufficiently general to enable one to solve the problem of radiative transfer in an expanding nebula. In this paper we therefore propose to study the problem of an expanding planetary nebula, particularly with a view to estimating the selective radiation pressure in Lyman-α. In later papers it is our intention to extend the methods of this investigation to the consideration of a variety of related questions.

2. *The equation of transfer for Lyman-α radiation and its approximate forms.*—As was first pointed out by E. A. Milne,[8] in studying the transfer of radiation in a planetary nebula we can ignore the curvature of the layers except when formulating the boundary conditions. We shall accordingly consider the atmosphere as stratified in parallel planes in which all properties are assumed to be constant over the planes z = constant. Let $\rho(z)$ denote the density of the scattering material (neutral hydrogen atoms in our present context) at height z, and $w(z)$ the velocity of the material at the same height assumed parallel to the z-direction. Further, let $\sigma(\nu)$ denote the mass-scattering coefficient in Lyman-α for the frequency ν as judged by an observer at rest with respect to the material. We shall suppose that $\sigma(\nu)$ differs appreciably from zero only in a small range of ν. However, it is in the essence of the problem that the half-width of $\sigma(\nu)$ is of the same order as the Doppler shifts in frequency caused by the differential motions in the nebula. This last circumstance makes the change of frequency on scattering the only optical effect of the motion $w(z)$ which has any importance and allows us to ignore all such effects as aberration, etc.

In writing down the equation of transfer for the problem of an expanding nebula, it is especially important that we be careful in referring all the frequencies to some chosen fixed observer. As will become apparent when we come to formulating our boundary conditions (§3), it is most convenient to choose our fixed observer as at rest with respect to the central star. Let $I(\nu, z, \vartheta)$ then denote the specific intensity of the radiation at height z inclined at angle ϑ to the positive normal and in the frequency ν as judged by our fixed observer. This radiation will appear to an observer at rest with respect to the material at z as having a frequency

$$\nu\left(1-\frac{w}{c}\cos\vartheta\right). \tag{1}$$

It will accordingly be scattered as such in all directions with a scattering coefficient of

$$\sigma\left(\nu\left[1-\frac{w}{c}\cos\vartheta\right]\right). \tag{2}$$

The equation of transfer will accordingly take the form

$$\cos\vartheta\,\frac{\partial I(\nu, z, \vartheta)}{\rho\,\partial z}=-\sigma\left(\nu\left[1-\frac{w}{c}\cos\vartheta\right]\right)I(\nu, z, \vartheta)+j(\nu, z, \vartheta)\,, \tag{3}$$

[6] *M.N.* **95**, 16, 1934.

[7] S. Chandrasekhar, *Rev. Mod. Phys.*, **17**, 138, 1945. This paper will be referred to as "Moving Atmospheres."

[8] *Zs. f. Ap.*, **1**, 98, 1930.

where $j(\nu, z, \vartheta)$ denotes the emission per unit time and per unit solid angle in the frequency ν and in the direction ϑ. This emission will consist of two parts: that derived from the scattering of the radiation of appropriate frequencies from other directions into the direction considered and that derived from the conversion of the ultraviolet radiation at z into Lyman-α. The former is given by (cf. "Moving Atmospheres," eq. [5])

$$\tfrac{1}{2}\sigma\left(\nu\left[1-\frac{w}{c}\cos\vartheta\right]\right)\int_0^{\pi} I\left(\nu\left[1-\frac{w}{c}\cos\vartheta+\frac{w}{c}\cos\chi\right], z, \chi\right)\sin\chi d\chi\,. \tag{4}$$

As for the latter, it will depend, among other things, on the probability $(1-p)$ with which an ultraviolet quantum is converted into a Lyman-α quantum on scattering and on the source function characteristic of the ultraviolet radiation. Thus, if πS_c denotes the flux of ultraviolet radiation incident per unit area of the inner boundary of the planetary nebula, the source function for the ultraviolet radiation is (cf. the references in n. 5),

$$p\left\{\frac{1}{2}\int_0^{\pi} I_c \sin\vartheta d\vartheta+\tfrac{1}{4}S_c e^{-(\tau_1-\tau)}\right\}, \tag{5}$$

where I_c denotes the intensity of the diffuse ultraviolet light, τ the optical depth for the ultraviolet radiation measured from the outer boundary inward, and τ_1 the total optical thickness of the nebula (also in ultraviolet light). To find the emission in Lyman-α at frequency ν, consider the number of ultraviolet quanta absorbed in a slab of material of unit cross-section and height dz at z. This is clearly given by

$$\frac{1}{h\nu_c}\{J_c(\tau)+\tfrac{1}{4}S_c e^{-(\tau_1-\tau)}\}\,\kappa_c\rho dz\,, \tag{6}$$

where κ_c denotes the absorption coefficient in the continuum, ν_c a suitably averaged frequency to represent the ultraviolet radiation, and

$$J_c=\frac{1}{2}\int_0^{\pi} I_c\sin\vartheta d\vartheta\,. \tag{7}$$

Since a Lyman-α quantum is emitted with a probability $(1-p)$ every time an ultraviolet quantum is scattered, the total energy emitted per unit time in Lyman-α (i.e., integrated over all the frequencies in the line) by the slab of material considered is

$$(1-p)\frac{\nu_0}{\nu_c}\{J_c(\tau)+\tfrac{1}{4}S_c e^{-(\tau_1-\tau)}\}\,\kappa_c\rho dz\,, \tag{8}$$

where ν_0 denotes the frequency of Lyman-α. The fraction of the energy (8) which will appear in the frequency ν, as judged by the fixed observer, is

$$\frac{\sigma\left(\nu\left[1-\frac{w}{c}\cos\vartheta\right]\right)}{\int_0^{\infty}\sigma(\nu)\,d\nu}\,. \tag{9}$$

Hence the contribution to the emission $j(\nu, z, \vartheta)$ by the conversion of the ultraviolet radiation is

$$(1-p)\frac{\nu_0}{\nu_c}\frac{\kappa_c}{2\sigma_0\Delta\nu}\{J_c(\tau)+\tfrac{1}{4}S_c e^{-(\tau_1-\tau)}\}\,\sigma\left(\nu\left[1-\frac{w}{c}\cos\vartheta\right]\right), \tag{10}$$

where we have written

$$2\sigma_0\Delta\nu=\int_0^{\infty}\sigma(\nu)\,d\nu\,. \tag{11}$$

The equation of transfer for the radiation of frequency ν (as judged by the fixed observer) can accordingly be written in the form

$$\left.\begin{aligned} \mu \frac{\partial I(\nu, \mu, z)}{\rho \partial z} &= -\sigma\left(\nu\left[1-\frac{w}{c}\mu\right]\right)\Big\{ I(\nu, \mu, z) \\ &\quad -\frac{1}{2}\int_{-1}^{+1} I\left(\nu\left[1-\frac{w}{c}\mu+\frac{w}{c}\mu'\right], z, \mu'\right) d\mu' - \tfrac{1}{2}E(z)\Big\}, \end{aligned}\right\} \quad (12)$$

where, for the sake of brevity, we have written

$$E(z) = (1-p)\frac{\nu_0}{\nu_c}\frac{\kappa_c}{\sigma_0 \Delta\nu}\{J_c(\tau) + \tfrac{1}{4}S_c e^{-(\tau_1-\tau)}\}. \quad (13)$$

Further, in equation (12) we have written μ and μ' for cos ϑ and cos χ, respectively.

In solving the equation of transfer (12), we shall adopt the method of approximation which has recently been developed in connection with the various problems of radiative transfer in the theory of stellar atmospheres.[9] In this method we replace the integrals which appear in the equation of transfer by sums according to Gauss's formula for numerical quadratures and replace the integrodifferential equation by a system of linear equations. Thus, considering equation (12), we replace it in the nth approximation by the system of $2n$ equations,

$$\left.\begin{aligned} \mu_i \frac{\partial I_i(\nu, z)}{\rho \partial z} &= -\sigma\left(\nu\left[1-\frac{w}{c}\mu_i\right]\right)\Big\{ I_i(\nu, z) \\ &\quad -\tfrac{1}{2}\Sigma a_j I_j\left(\nu\left[1-\frac{w}{c}\mu_i+\frac{w}{c}\mu_j\right], z\right) - \tfrac{1}{2}E(z)\Big\} \qquad (i=\pm 1, \ldots, \pm n), \end{aligned}\right\} \quad (14)$$

where the μ_i's ($i = \pm 1, \ldots, \pm n$) are the zeros of the Legendre polynomial $P_{2n}(\mu)$ and the a_i's are the appropriate Gaussian weights. Further, in equation (14) we have written $I_i(\nu, z)$ for $I(\nu, z, \mu_i)$.

At this stage one further simplification of equation (14) is possible. In evaluating the Doppler shifts we need not distinguish between

$$\nu\left(1-\frac{w}{c}\mu\right) \quad \text{and} \quad \nu - \nu_0 \frac{w}{c}\mu, \quad (15)$$

where ν_0 denotes the frequency of the center of the line. We may therefore replace equation (14) by the simpler one,

$$\left.\begin{aligned} \mu_i \frac{\partial I_i(\nu, z)}{\rho \partial z} &= -\sigma\left(\nu-\nu_0\frac{w}{c}\mu_i\right)\Big\{ I_i(\nu, z) \\ &\quad -\tfrac{1}{2}\Sigma a_j I_j\left(\nu-\nu_0\frac{w}{c}\mu_i+\nu_0\frac{w}{c}\mu_j, z\right) - \tfrac{1}{2}E(z)\Big\} \qquad (i=\pm 1, \ldots, \pm n). \end{aligned}\right\} \quad (16)$$

As in "Moving Atmospheres," instead of considering the intensities I_i ($i = \pm 1, \ldots, \pm n$) for some fixed frequency ν, we shall consider them for the frequencies

$$\nu_i = \nu + \nu_0 \frac{w}{c}\mu_i \qquad (i = \pm 1, \ldots, \pm n), \quad (17)$$

which are functions of z. If we now let

$$I_i(\nu_i, z) = \psi_i(\nu, z) \qquad (i = \pm 1, \ldots, \pm n), \quad (18)$$

[9] S. Chandrasekhar, *Ap. J.*, **100,** 76, 117, 1944; **101,** 95, 328, 348, 1945 (see particularly the first of these papers).

the differential equations for the ψ_i's become (cf. "Moving Atmospheres," eq. [16])

$$\mu_i \frac{\partial \psi_i}{\partial z} - \mu_i^2 \frac{\nu_0}{c} \frac{dw}{dz} \frac{\partial \psi_i}{\partial \nu} = -\rho\sigma(\nu)\left(\psi_i - \tfrac{1}{2}\Sigma a_j \psi_j - \tfrac{1}{2}E\right). \tag{19}$$

In our subsequent work we shall restrict ourselves to the first approximation. In this approximation

$$\mu_1 = -\mu_{-1} = \frac{1}{\sqrt{3}} \quad \text{and} \quad a_1 = a_{-1} = 1\,, \tag{20}$$

and equations (16) and (19) become

$$\mu_1 \frac{\partial I_{+1}(\nu, z)}{\rho \partial z} = -\tfrac{1}{2}\sigma\left(\nu - \nu_0 \frac{w}{c}\mu_1\right)\left[I_{+1}(\nu, z) - I_{-1}\left(\nu - 2\nu_0 \frac{w}{c}\mu_1, z\right) - E(z)\right], \tag{21}$$

$$\mu_1 \frac{\partial I_{-1}(\nu, z)}{\rho \partial z} = +\tfrac{1}{2}\sigma\left(\nu + \nu_0 \frac{w}{c}\mu_1\right)\left[I_{-1}(\nu, z) - I_{+1}\left(\nu + 2\nu_0 \frac{w}{c}\mu_1, z\right) - E(z)\right], \tag{22}$$

and

$$\mu_1 \frac{\partial \psi_{+1}}{\partial z} - \mu_1^2 \frac{\nu_0}{c} \frac{dw}{dz} \frac{\partial \psi_{+1}}{\partial \nu} = -\tfrac{1}{2}\rho\sigma(\nu)\left(\psi_{+1} - \psi_{-1} - E\right), \tag{23}$$

$$\mu_1 \frac{\partial \psi_{-1}}{\partial z} + \mu_1^2 \frac{\nu_0}{c} \frac{dw}{dz} \frac{\partial \psi_{-1}}{\partial \nu} = +\tfrac{1}{2}\rho\sigma(\nu)\left(\psi_{-1} - \psi_{+1} - E\right), \tag{24}$$

where it may be recalled that

$$\psi_{+1}(\nu, z) = I_{+1}\left(\nu + \mu_1\nu_0 \frac{w}{c}, z\right) \quad \text{and} \quad \psi_{-1}(\nu, z) = I_{-1}\left(\nu - \mu_1\nu_0 \frac{w}{c}, z\right). \tag{25}$$

3. *The reduction to a boundary-value problem for the case $\sigma(\nu)$ = constant for $\nu_0 - \Delta\nu \leqslant \nu \leqslant \nu_0 + \Delta\nu$ and zero outside this interval and for a linear increase of* w *with the optical depth.*—In this paper we shall consider the problem of the expanding nebula for the case

$$\begin{aligned} \sigma(\nu) &= \sigma_0 = \text{constant for } \nu_0 - \Delta\nu \leqslant \nu \leqslant \nu_0 + \Delta\nu \\ &= 0 \quad \text{otherwise}\,, \end{aligned} \tag{26}$$

and

$$\frac{1}{\rho}\frac{dw}{dz} = \text{constant}\,. \tag{27}$$

Further, we shall restrict ourselves to the first approximation.

When $\sigma(\nu)$ has the form (26), some care is required in the formulation of the boundary conditions; for, according to equations (21) and (22),

$$\frac{\partial I_{+1}}{\partial z} \neq 0 \quad \text{only if} \quad \nu_0 - \Delta\nu + \mu_1\nu_0 \frac{w}{c} \leqslant \nu \leqslant \nu_0 + \Delta\nu + \mu_1\nu_0 \frac{w}{c}, \tag{28}$$

and

$$\frac{\partial I_{-1}}{\partial z} \neq 0 \quad \text{only if} \quad \nu_0 - \Delta\nu - \mu_1\nu_0 \frac{w}{c} \leqslant \nu \leqslant \nu_0 + \Delta\nu - \mu_1\nu_0 \frac{w}{c}. \tag{29}$$

Accordingly, in the (ν, w)-plane the lines

$$\nu = \nu_0 - \Delta\nu + \mu_1\nu_0 \frac{w}{c} \quad \text{and} \quad \nu = \nu_0 + \Delta\nu + \mu_1\nu_0 \frac{w}{c} \tag{30}$$

delimit the regions in which I_{+1} is different from a constant from the regions in which it is a constant for varying z. The situation is further clarified in Figure 1, where AD and BC represent the lines (30). Similarly, the lines (AF and BE in Fig. 1)

$$\nu = \nu_0 - \Delta\nu - \mu_1\nu_0 \frac{w}{c} \quad \text{and} \quad \nu = \nu_0 + \Delta\nu - \mu_1\nu_0 \frac{w}{c} \tag{31}$$

delimit the regions in which I_{-1} is different from a constant from the regions in which it is a constant for varying z.

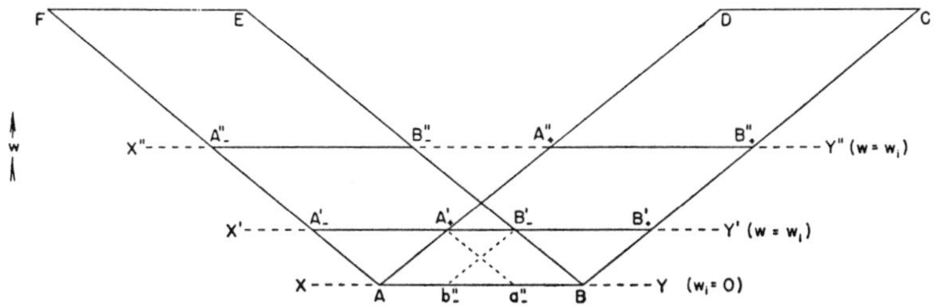

FIG. 1

In formulating the boundary conditions we must allow for the possibility that the inner boundary of the nebula may itself have a motion relative to our fixed observer, who, it will be remembered, is assumed to be at rest only with respect to the central star. Let w_i then denote the velocity of the nebula at its inner boundary. From Figure 1 it is apparent that we will have to distinguish three cases:

$$\left.\begin{aligned} &\text{case I:} \quad w_i = 0 \,, \\ &\text{case II:} \quad \nu_0 - \Delta\nu + \mu_1\nu_0 \frac{w_i}{c} < \nu_0 \,, \\ &\text{case III:} \quad \nu_0 - \Delta\nu + \mu_1\nu_0 \frac{w_i}{c} \geqslant \nu_0 \,. \end{aligned}\right\} \tag{32}$$

These cases are, however, more conveniently distinguished in terms of the Doppler shift, $2D\nu^{(i)}$, due to the velocity w_i:

$$2\,D\nu^{(i)} = \nu_0 \frac{w_i}{c} \,. \tag{33}$$

We have

$$\left.\begin{aligned} &\text{case I:} \quad D\nu^{(i)} = 0 \\ &\text{case II:} \quad 0 < D\nu^{(i)} < \frac{\Delta\nu}{2\mu_1} = \frac{\sqrt{3}}{2}\Delta\nu \,, \\ &\text{case III:} \quad D\nu^{(i)} \geqslant \frac{\Delta\nu}{2\mu_1} = \frac{\sqrt{3}}{2}\Delta\nu \,. \end{aligned}\right\} \tag{34}$$

These three cases correspond to situations in which the inner boundary of the nebula is at XY, $X'Y'$, or $X''Y''$, respectively, as shown in Figure 1.

Now the physical boundary conditions which have to be translated into mathematical terms are, first, that there is no radiation incident on the nebula from the outside and, second, that

$$I^*(\nu, \vartheta) = I^*(\nu, \pi - \vartheta) \tag{35}$$

whenever $I^*(\nu, \vartheta)$ refers to the intensity of a ray which, as judged by the fixed observer at the central star, has not suffered any change by absorptions or emissions by the intervening medium (see Fig. 2). This latter boundary condition arises from the geometry of

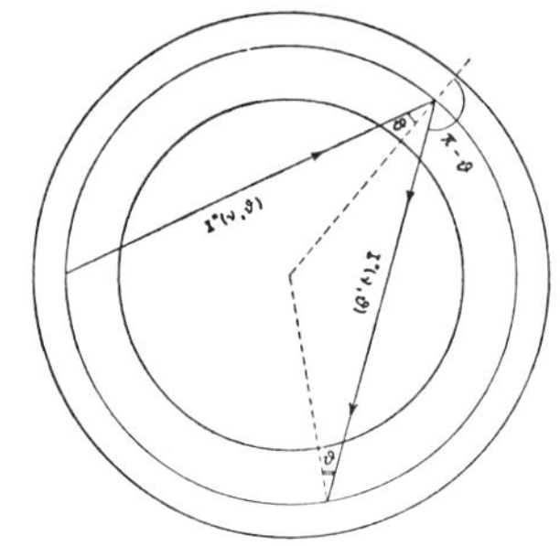

FIG. 2

the problem and the fact that the specific intensity along a ray does not change if no absorption or emission takes place.[10] Accordingly, the boundary conditions with respect to which equations (21) and (22) have to be solved are (see Fig. 1):

$$\text{case I} \quad \left\{ \begin{array}{ll} I_{-1}(\nu) = 0 \text{ along } BE \text{ and } EF, \\ I_{+1}(\nu) = 0 \text{ along } BC, \\ I_{+1}(\nu) = I_{-1}(\nu) \text{ along } AB; \end{array} \right\} \tag{36}$$

$$\text{case II} \quad \left\{ \begin{array}{ll} I_{-1}(\nu) = 0 \text{ along } B'_-E \text{ and } EF \\ I_{+1}(\nu) = 0 \text{ along } B'_-B'_+ \text{ and } B'_{+1}C, \\ I_{+1}(\nu) = I_{-1}(\nu) \text{ along } A'_+B'_-; \end{array} \right\} \tag{37}$$

and

$$\text{case III} \quad \left\{ \begin{array}{ll} I_{-1}(\nu) = 0 \text{ along } B''_-E \text{ and } EF, \\ I_{+1}(\nu) = 0 \text{ along } A''_+B''_+ \text{ and } B''_+C. \end{array} \right\} \tag{38}$$

When expressed for the intensities ψ_{+1} and ψ_{-1}, the foregoing boundary conditions become

$$\left. \begin{array}{lll} \psi_{-1}(\nu, z_1) = 0 & \text{for} & \nu_0 - \Delta\nu \leqslant \nu \leqslant \nu_0 + \Delta\nu, \\ \psi_{-1}(\nu_0 + \Delta\nu, z) = 0 & \text{for} & 0 \leqslant z \leqslant z_1, \\ \psi_{+1}(\nu_0 + \Delta\nu, z) = 0 & \text{for} & 0 \leqslant z \leqslant z_1, \end{array} \right\} \tag{39}$$

for all three cases, while

$$\psi_{-1}(\nu, 0) = \psi_{+1}(\nu, 0) \quad \text{in case I}. \tag{40}$$

$$\left. \begin{array}{l} \psi_{+1}(\nu, 0) = 0 \quad \text{for} \quad \nu_0 + \Delta\nu - 4D\nu^{(i)}\mu_1 \leqslant \nu \leqslant \nu_0 + \Delta\nu, \\ \psi_{+1}(\nu, 0) = \psi_{-1}(\nu + 4D\nu^{(i)}\mu_1, 0) \quad \text{for} \quad \nu_0 - \Delta\nu \leqslant \nu < \nu_0 + \Delta\nu - 4D\nu^{(i)}\mu_1 \end{array} \right\} \text{in case II}. \tag{41}$$

and

$$\psi_{+1}(\nu, 0) = 0 \quad \text{for} \quad \nu_0 - \Delta\nu \leqslant \nu \leqslant \nu_0 + \Delta\nu \quad \text{in case III}. \tag{42}$$

In the forgoing equations, $z = 0$ refers to the inner boundary of the nebula, while $z = z_1$ refers to the outer boundary.

[10] Cf. Milne, *op. cit.* (n. 8.)

We shall now transform equations (23) and (24) to forms more convenient for their solution.

Let t denote the optical depth of the atmosphere measured from the outer boundary inward in terms of σ_0. Then

$$dt = -\rho\sigma_0 dz \,. \tag{43}$$

It is, however, more convenient to use the variable

$$x = \frac{1}{2\mu_1} t = \frac{\sqrt{3}}{2} t \tag{44}$$

instead of t. In terms of x, equations (23) and 24) become

$$\frac{\partial\psi_{+1}}{\partial x} - \mu_1 \frac{\nu_0}{c} \frac{dw}{dx} \frac{\partial\psi_{+1}}{d\nu} = \psi_{+1} - \psi_{-1} - E(x)\,, \tag{45}$$

and

$$\frac{\partial\psi_{-1}}{\partial x} + \mu_1 \frac{\nu_0}{c} \frac{dw}{dx} \frac{\partial\psi_{-1}}{\partial\nu} = \psi_{+1} - \psi_{-1} + E(x)\,. \tag{46}$$

Now the assumption (27) concerning the variation of w clearly implies that the velocity is a linear function of x. Accordingly, we may write

$$w = w_1 - (w_1 - w_i)\frac{x}{x_1}, \tag{47}$$

where w_1 and w_i denote the velocities at the outer ($x = 0$) and the inner ($x = x_1$) boundaries of the planetary nebula, respectively. As in "Moving Atmospheres," we shall express the difference in velocities between the inner and the outer boundaries of the nebula in terms of a *Doppler width*, $D\nu$, according to

$$2\,D\nu = \frac{\nu_0}{c}(w_1 - w_i)\,. \tag{48}$$

With these definitions,

$$\mu_1 \frac{\nu_0}{c} \frac{dw}{dx} = -2\mu_1 \frac{D\nu}{x_1} = -\frac{4}{3} \frac{D\nu}{t_1}\,. \tag{49}$$

Equations (45) and (46) become

$$\frac{\partial\psi_{+1}}{\partial x} + 2\mu_1 \frac{D\nu}{x_1} \frac{\partial\psi_{+1}}{\partial\nu} = \psi_{+1} - \psi_{-1} - E(x) \tag{50}$$

and

$$\frac{\partial\psi_{-1}}{\partial x} - 2\mu_1 \frac{D\nu}{x_1} \frac{\partial\psi_{-1}}{\partial\nu} = \psi_{+1} - \psi_{-1} + E(x)\,. \tag{51}$$

We now introduce the variable y, defined by

$$(\nu_0 + \Delta\nu) - \nu = 2\mu_1 \frac{D\nu}{x_1} y\,; \tag{52}$$

y therefore measures the frequency shifts from the violet edge of $\sigma(\nu)$ (as they enter the intensities ψ_{+1} and ψ_{-1}) in units of

$$2\mu_1 \frac{D\nu}{x_1}\,. \tag{53}$$

Equations (50) and (51) simplify to the forms

$$\frac{\partial \psi_{+1}}{\partial x} - \frac{\partial \psi_{+1}}{\partial y} = \psi_{+1} - \psi_{-1} - E(x) \tag{54}$$

and

$$\frac{\partial \psi_{-1}}{\partial x} + \frac{\partial \psi_{-1}}{\partial y} = \psi_{+1} - \psi_{-1} + E(x)\,. \tag{55}$$

The range of the variables x and y in which the solution has to be sought is (cf. eq. [52])

$$0 \leqslant x \leqslant x_1 \quad \text{and} \quad 0 \leqslant y \leqslant y_1 = \frac{1}{\mu_1}\frac{\Delta\nu}{D\nu}\,x_1\,; \tag{56}$$

and the boundary conditions with respect to which equations (54) and (55) have to be solved are (see Fig. 3):

$$\left.\begin{aligned} \psi_{-1} &= 0 \quad \text{on} \quad CD: \quad x = 0 \quad \text{and} \quad 0 \leqslant y \leqslant y_1\,, \\ &= 0 \quad \text{on} \quad CB: \quad y = 0 \quad \text{and} \quad 0 \leqslant x \leqslant x_1\,, \\ \psi_{+1} &= 0 \quad \text{on} \quad CB: \quad y = 0 \quad \text{and} \quad 0 \leqslant x \leqslant x_1 \end{aligned}\right\} \text{in all three cases}\,; \tag{57}$$

and on BA

$$\psi_{+1} = \psi_{-1} \quad \text{for} \quad x = x_1 \quad \text{and} \quad 0 \leqslant y \leqslant y_1 \quad (\text{case I})\,. \tag{58}$$

$$\left.\begin{aligned} \psi_{+1}(y, x_1) &= 0 \quad \text{for} \quad 0 \leqslant y \leqslant y^* = 2\,\frac{D\nu^{(i)}}{D\nu}\,x_1\,, \\ \psi_{+1}(y, x_1) &= \psi_{-1}(y - y^*, x_1) \quad \text{for} \quad y^* < y \leqslant y_1 \end{aligned}\right\} (\text{case II})\,. \tag{59}$$

and

$$\psi_{+1} = 0 \quad \text{for} \quad x = x_1 \quad \text{and} \quad 0 \leqslant y \leqslant y_1 \quad (\text{case III})\,. \tag{60}$$

It is convenient to introduce one further transformation of the variable. Let

$$\psi_{+1} = e^{-y}F \quad \text{and} \quad \psi_{-1} = e^{-y}G\,. \tag{61}$$

Equations (54) and (55) become

$$\frac{\partial F}{\partial x} - \frac{\partial F}{\partial y} = -G - e^{y}E(x) \tag{62}$$

and

$$\frac{\partial G}{\partial x} + \frac{\partial G}{\partial y} = +F + e^{y}E(x)\,. \tag{63}$$

The corresponding boundary conditions follow from equations (57)-(60). Thus,

$$\left.\begin{aligned} G &= 0 \quad \text{on} \quad CD: \quad x = 0 \quad \text{and} \quad 0 \leqslant y \leqslant y_1 \\ &= 0 \quad \text{on} \quad CB: \quad y = 0 \quad \text{and} \quad 0 \leqslant x \leqslant x_1\,, \\ F &= 0 \quad \text{on} \quad CB: \quad y = 0 \quad \text{and} \quad 0 \leqslant x \leqslant x_1 \end{aligned}\right\} \text{in all three cases}\,; \tag{64}$$

and on BA

$$F = G \quad \text{for} \quad x = x_1 \quad \text{and} \quad 0 \leqslant y \leqslant y_1 \quad (\text{case I})\,. \tag{65}$$

$$\left.\begin{aligned} F(y, x_1) &= 0 \quad \text{for} \quad 0 \leqslant y \leqslant y^*\,. \\ F(y, x_1) &= e^{+y^*}G(y - y^*, x_1) \quad \text{for} \quad y^* < y \leqslant y_1 \end{aligned}\right\} (\text{case II})\,. \tag{66}$$

and

$$F = 0 \quad \text{for} \quad x = x_1 \quad \text{and} \quad 0 \leqslant y \leqslant y_1 \quad (\text{case III})\,. \tag{67}$$

Finally, we may note that $E(x)$ can be written in the form (cf. eq. [13])

$$E(x) = \frac{1}{2\mu_1}(1-p)\frac{\nu_0}{\nu_c\Delta\nu}\,a\{J_c(ax) + \tfrac{1}{4}S_c e^{-\tau_1}e^{ax}\}. \tag{68}$$

where

$$ax = \tau \quad \text{and} \quad a = 2\mu_1\frac{\kappa_c}{\sigma_0}. \tag{69}$$

The quantity a introduced in equation (68) is essentially the ratio of the absorption coefficients in the continuum and in the line; this ratio is generally of the order of 10^{-4}, so that, if the need should arise, we may treat this as a small quantity. However, it should be remembered that the optical depth t_1 in Lyman-α, under the circumstances of our problem, is so large that $at_1(=\tau_1)$ is of the order of unity.

4. *The solution of the boundary-value problem for the case* p = 0.—Before we can proceed with the solution of the boundary-value problem formulated in the preceding section, it is necessary for us to have an expression for the density of the diffuse ultraviolet radiation. This is given by the theory of radiative transfer of the ultraviolet radiation, and in a first approximation the solution for J_c has the form[11]

$$J_c(\tau) = Ae^{-\lambda\tau} + Be^{+\lambda\tau} + \frac{3p}{4(2-3p)}S_c e^{-(\tau_1-\tau)}, \tag{70}$$

where A and B are certain constants and $\lambda^2 = 3(1-p)$. While the solution of the boundary-value problem with this general form for $J_c(\tau)$ is entirely feasible, we shall be avoiding a great deal of formal complexity without losing, at the same time, any of the essential features of the problem by considering the case $p = 0$. Indeed, it is known that the flux in the diffuse ultraviolet radiation is generally so small that the case $p = 0$ provides ample accuracy for most problems. Moreover, since in this investigation our prime interest is to evaluate the selective radiation pressure in Lyman-α, we shall not be restricting our treatment in any essential way if we put $p = 0$. It is evident that, when this is the case,

$$J_c(\tau) \equiv 0 \qquad (p = 0). \tag{71}$$

The expression for $E(x)$ now reduces to (cf. eq. [68])

$$E(x) = \left(\frac{\nu_0}{8\mu_1\nu_c\Delta\nu}S_c e^{-\tau_1}\right)a\,e^{ax}; \tag{72}$$

and equations (62) and (63) become

$$\frac{\partial F}{\partial x} - \frac{\partial F}{\partial y} = -G - Qa\,e^{ax+y} \tag{73}$$

and

$$\frac{\partial G}{\partial x} + \frac{\partial G}{\partial y} = +F + Qa\,e^{ax+y}, \tag{74}$$

where we have written

$$Q = \frac{\nu_0}{8\mu_1\nu_c\Delta\nu}S_c e^{-\tau_1}. \tag{75}$$

Equations (73) and (74) are nonhomogeneous. But they can be reduced to homogeneous forms, since a particular integral is readily found. Thus

$$F = -Q\frac{2+a}{a}e^{ax+y} \quad \text{and} \quad G = -Q\frac{2-a}{a}e^{ax+y} \tag{76}$$

[11] Cf. S. Chandrasekhar, *Zs. f. Ap.*, **9**, 266, 1935 (see eqs. [24] and [28]).

are seen to satisfy equations (73) and (74). We accordingly write

$$F = Q\left(f - \frac{2+a}{a}\, e^{ax+y}\right) \tag{77}$$

and

$$G = Q\left(g - \frac{2-a}{a}\, e^{ax+y}\right) \tag{78}$$

and obtain for f and g the homogeneous equations

$$\frac{\partial f}{\partial x} - \frac{\partial f}{\partial y} = -g\,. \tag{79}$$

and

$$\frac{\partial g}{\partial x} + \frac{\partial g}{\partial y} = +f\,. \tag{80}$$

The boundary conditions with respect to which the foregoing equations have to be solved are (cf. eqs. [64]-[67] and [77] and [78]):

$$\left.\begin{aligned} &g = \frac{2-a}{a}\, e^{y} \quad \text{on} \quad CD\colon x = 0 \quad \text{and} \quad 0 \leqslant y \leqslant y_1\,, \\ &f = \frac{2+a}{a}\, e^{ax} \quad \text{and} \quad g = \frac{2-a}{a}\, e^{ax} \quad \text{on} \quad CB\colon y = 0 \quad \text{and} \quad 0 \leqslant x \leqslant x_1\,, \end{aligned}\right\} \tag{81}$$

in all three cases, and on AB ($x = x_1$ and $0 \leqslant y \leqslant y_1$)

$$f = g + 2e^{ax_1+y} \qquad \text{(case I)}\,. \tag{82}$$

and

$$f = \frac{2+a}{a}\, e^{ax_1+y} \qquad \text{(case III)}\,. \tag{83}$$

For case II there is a similar boundary condition on AB. But as we shall not be obtaining explicitly the solution for this case, we shall not continue to write down the boundary conditions for this case.

Eliminating g between equations (79) and (80), we obtain

$$\frac{\partial^2 f}{\partial x^2} - \frac{\partial^2 f}{\partial y^2} + f = 0\,. \tag{84}$$

We require to solve this hyperbolic equation with the boundary conditions (cf. eqs. [79]-[83])

$$\left.\begin{aligned} &f = \frac{2+a}{a}\, e^{ax}, \quad \frac{\partial f}{\partial x} = (2+a)\, e^{ax}, \quad \text{and} \quad \frac{\partial f}{\partial y} = \frac{a^2+a+2}{a}\, e^{ax} \\ &\qquad\qquad \text{on} \quad BC\colon y = 0 \quad \text{and} \quad 0 \leqslant x \leqslant x_1\,, \\ &\frac{\partial f}{\partial x} \mp \frac{\partial f}{\partial y} - \frac{2-a}{a}\, e^{y} \quad \text{on} \quad CD\colon x = 0 \quad \text{and} \quad 0 \leqslant y \leqslant y_1\,, \end{aligned}\right\} \tag{85}$$

in all three cases, and on AB ($x = x_1$ and $0 \leqslant y \leqslant y_1$)

$$f + \frac{\partial f}{\partial x} - \frac{\partial f}{\partial y} = 2e^{ax_1+y} \qquad \text{(case I)}\,. \tag{86}$$

and

$$f = \frac{2+a}{a}\, e^{ax_1+y} \qquad \text{(case III)}\,. \tag{87}$$

The solution to the boundary-value problem we have just formulated can be carried out in a manner quite analogous to that which was followed in the solution of a similar boundary-value problem in "Moving Atmospheres." Briefly, the method consists in applying Green's theorem to contours which are, in parts, the characteristics $x - \xi = \pm (y - \eta)$, passing through suitably selected points (ξ, η) and with the further choice of the Riemann function $v(x, y; \xi, \eta)$ for the solution of the adjoint equation.[12]

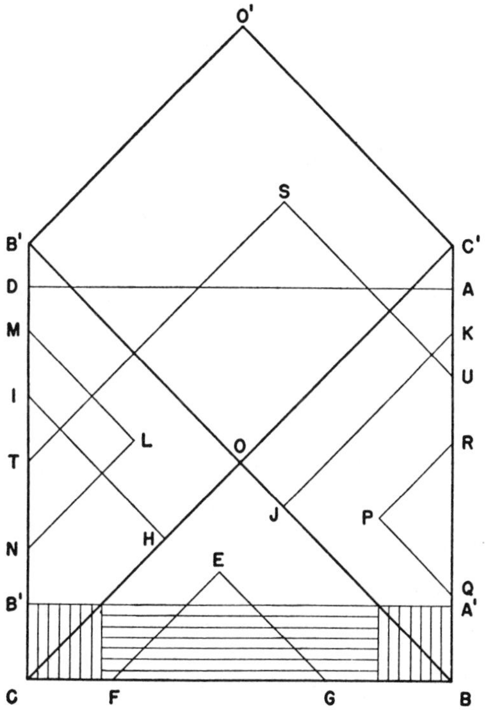

FIG. 3

a) The solution in the region OCB.—Let the characteristic $x = y$ through C intersect AB at C' and the characteristic $x_1 - x = y$ through B intersect CD at B'. Further, let CC' and BB' intersect at O (see Fig. 3).

Now, since the function and its derivatives are specified along CB, the solution in the region OCB (including the sides OC and OB) can be found directly by Riemann's method. Thus, applying Green's theorem to a contour such as $EFGE$ where EF and EG are the characteristics through $E = (\xi, \eta)$, we find that

$$f(\xi, \eta) = \frac{2+a}{a} e^{a\xi} \cosh a\eta - \frac{1}{2}\int_{\xi-\eta}^{\xi+\eta} \left(f \frac{\partial v}{\partial y} - v \frac{\partial f}{\partial y} \right)_{y=0} dx . \tag{88}$$

The Reimann function appropriate for the contour $EFGE$ is

$$v(x, y; \xi, \eta) = I_0([(y-\eta)^2 - (x-\xi)^2]^{\frac{1}{2}}) . \tag{89}$$

[12] A more detailed description of the method will be found in "Moving Atmospheres," § 5.

Using this v in equation (88) and substituting also for f and $\partial f/\partial y$ according to equation (85), we find

$$\left.\begin{aligned} f(\xi,\eta) = \frac{2+a}{a} e^{a\xi}\cosh a\eta + \frac{a^2+a+2}{2a}\int_{\xi-\eta}^{\xi+\eta} e^{ax} I_0([\eta^2-(x-\xi)^2]^{\frac{1}{2}})\,dx \\ + \frac{2+a}{2a}\,\eta\int_{\xi-\eta}^{\xi+\eta} e^{ax}\frac{I_1([\eta^2-(x-\xi)^2]^{\frac{1}{2}})\,dx}{[\eta^2-(x-\xi)^2]^{\frac{1}{2}}}. \end{aligned}\right\} \quad (90)$$

Putting

$$x - \xi = \eta\cos\vartheta \quad (91)$$

into the two integrals on the right-hand side of equation (90), we obtain

$$\left.\begin{aligned} f(\xi,\eta) = \frac{2+a}{a} e^{a\xi}\cosh a\eta + \frac{a^2+a+2}{2a}\,\eta e^{a\xi}\int_0^{\pi} e^{a\eta\cos\vartheta} I_0(\eta\sin\vartheta)\sin\vartheta d\vartheta \\ + \frac{2+a}{2a}\,\eta e^{a\xi}\int_0^{\pi} e^{a\eta\cos\vartheta} I_1(\eta\sin\vartheta)\,d\vartheta. \end{aligned}\right\} \quad (92)$$

The integrals occurring in the foregoing equation can be evaluated in the following manner:

Considering, first, the integral

$$a\eta\int_0^{\pi} e^{a\eta\cos\vartheta} I_0(\eta\sin\vartheta)\sin\vartheta d\vartheta, \quad (93)$$

we replace $e^{a\eta\cos\vartheta}$ and $I_0(\eta\sin\vartheta)$ by their respective series expansions and integrate term by term. We find

$$\left.\begin{aligned} a\eta\int_0^{\pi} & e^{a\eta\cos\vartheta} I_0(\eta\sin\vartheta)\sin\vartheta d\vartheta \\ &= 2a\eta\sum_{n=0}^{\infty}\frac{(a\eta)^{2n}}{(2n)!}\sum_{m=0}^{\infty}\frac{(\frac{1}{2}\eta)^{2m}}{m!\Gamma(m+1)}\int_0^{\pi/2}\cos^{2n}\vartheta\sin^{2m+1}\vartheta d\vartheta \\ &= \sum_{n=0}^{\infty}\Gamma(n+\tfrac{1}{2})\frac{(a\eta)^{2n+1}}{(2n)!}\sum_{m=0}^{\infty}\frac{(\frac{1}{2}\eta)^{2m}}{m!\Gamma(m+n+\frac{3}{2})} \\ &= \sum_{n=0}^{\infty}\Gamma(n+\tfrac{1}{2})\frac{(2a^2\eta)^{n+\frac{1}{2}}}{(2n)!}\sum_{m=0}^{\infty}\frac{(\frac{1}{2}\eta)^{2m+n+\frac{1}{2}}}{m!\Gamma(m+n+\frac{3}{2})} \\ &= \sum_{n=0}^{\infty}\Gamma(n+\tfrac{1}{2})\frac{(2a^2\eta)^{n+\frac{1}{2}}}{(2n)!} I_{n+\frac{1}{2}}(\eta) \\ &= (2\pi\eta)^{\frac{1}{2}}a\sum_{n=0}^{\infty}(\tfrac{1}{2}a^2\eta)^n\frac{1}{n!}I_{n+\frac{1}{2}}(\eta). \end{aligned}\right\} \quad (94)$$

But

$$I_\nu\{\eta\sqrt{(1+a^2)}\} = (1+a^2)^{\frac{1}{2}\nu}\sum_{n=0}^{\infty}(\tfrac{1}{2}a^2\eta)^n\frac{1}{n!}I_{\nu+n}(\eta). \quad (95)$$ [13]

[13] This formula can be established by following the method used by G. N. Watson in his *Theory of Bessel Functions*, p. 141, Cambridge University Press, 1944, in proving a similar relation involving the Bessel functions with real arguments.

The last summation which occurs in equation (94) is therefore a special case of equation (95). We have

$$\sum_{n=0}^{\infty} (\tfrac{1}{2}a^2\eta)^n \frac{1}{n!} I_{n+\frac{1}{2}}(\eta) = \frac{1}{(1+a^2)^{1/4}} I_{\frac{1}{2}}\{\eta\sqrt{(1+a^2)}\}. \tag{96}$$

Hence,

$$\eta\int_0^{\pi} e^{a\eta\cos\vartheta} I_0(\eta\sin\vartheta)\sin\vartheta d\vartheta = \frac{(2\pi\eta)^{\frac{1}{2}}}{(1+a^2)^{1/4}} I_{\frac{1}{2}}\{\eta\sqrt{(1+a^2)}\}; \tag{97}$$

or, substituting the explicit expression for $I_{\frac{1}{2}}$, we have

$$\eta\int_0^{\pi} e^{a\eta\cos\vartheta} I_0(\eta\sin\vartheta)\sin\vartheta d\vartheta = \frac{2}{\sqrt{(1+a^2)}}\sinh\{\eta\sqrt{(1+a^2)}\}. \tag{98}$$

Similarly,

$$\left.\begin{aligned}
a\eta&\int_0^{\pi} e^{a\eta\cos\vartheta} I_1(\eta\sin\vartheta)\, d\vartheta \\
&= 2\sum_{n=0}^{\infty}\frac{(a\eta)^{2n+1}}{(2n)!}\sum_{m=0}^{\infty}\frac{(\frac{1}{2}\eta)^{2m+1}}{m!\Gamma(m+2)}\int_0^{\pi/2}\cos^{2n}\vartheta\sin^{2m+1}\vartheta d\vartheta \\
&= \sum_{n=0}^{\infty}\Gamma(n+\tfrac{1}{2})\frac{(a\eta)^{2n+1}}{(2n)!}\sum_{m=0}^{\infty}\frac{(\frac{1}{2}\eta)^{2m+1}}{(m+1)!\Gamma(m+n+\frac{3}{2})} \\
&= \sum_{n=0}^{\infty}\Gamma(n+\tfrac{1}{2})\frac{(2a^2\eta)^{n+\frac{1}{2}}}{(2n)!}\sum_{m=1}^{\infty}\frac{(\frac{1}{2}\eta)^{2m+n-\frac{1}{2}}}{m!\Gamma(m+n+\frac{1}{2})} \\
&= \sum_{n=0}^{\infty}\Gamma(n+\tfrac{1}{2})\frac{(2a^2\eta)^{n+\frac{1}{2}}}{(2n)!}\left[I_{n-\frac{1}{2}}(\eta)-\frac{(\frac{1}{2}\eta)^{n-\frac{1}{2}}}{\Gamma(n+\frac{1}{2})}\right] \\
&= a(2\pi\eta)^{\frac{1}{2}}\sum_{n=0}^{\infty}(\tfrac{1}{2}a^2\eta)^n\frac{1}{n!}I_{n-\frac{1}{2}}(\eta) - 2a\sum_{n=0}^{\infty}\frac{(a\eta)^{2n}}{(2n)!} \\
&= a(2\pi\eta)^{\frac{1}{2}}(1+a^2)^{1/4}I_{-\frac{1}{2}}\{\eta\sqrt{(1+a^2)}\} - 2a\cosh a\eta\,.
\end{aligned}\right\} \tag{99}$$

Hence,

$$\eta\int_0^{\pi} e^{a\eta\cos\vartheta} I_1(\eta\sin\vartheta)\, d\vartheta = 2\cosh\{\eta\sqrt{(1+a^2)}\} - 2\cosh a\eta\,. \tag{100}$$

Substituting from equations (98) and (100) in equation (92), we obtain

$$\left.\begin{aligned}
f(\xi,\eta) = \frac{1}{a}e^{a\xi}\Bigg[&\frac{a^2+a+2}{\sqrt{(1+a^2)}}\sinh\{\eta\sqrt{(1+a^2)}\} \\
&+(2+a)\cosh\{\eta\sqrt{(1+a^2)}\}\Bigg].
\end{aligned}\right\} \tag{101}$$

This solves explicitly for f in the region OCB. With this solution for f, g can also be found in this region; for (cf. eq. [79])

$$g(\xi,\eta) = -\frac{\partial f}{\partial\xi}+\frac{\partial f}{\partial\eta}, \tag{102}$$

and equation (101) yields

$$g(\xi, \eta) = \frac{1}{\alpha} e^{\alpha\xi} \left[\frac{\alpha^2 - \alpha + 2}{\sqrt{(1+\alpha^2)}} \sinh\{\eta \sqrt{(1+\alpha^2)}\} + (2-\alpha) \cosh\{\eta \sqrt{(1+\alpha^2)}\} \right]. \tag{103}$$

The corresponding solutions for ψ_{+1} and ψ_{-1} are readily found. We have

$$\psi_{+1} = \frac{Q}{\alpha} e^{\alpha x - y} \left[\frac{\alpha^2 + \alpha + 2}{\sqrt{(1+\alpha^2)}} \sinh\{y \sqrt{(1+\alpha^2)}\} + (2+\alpha) \cosh\{y \sqrt{(1+\alpha^2)}\} - (2+\alpha) e^y \right] \tag{104}$$

and

$$\psi_{-1} = \frac{Q}{\alpha} e^{\alpha x - y} \left[\frac{\alpha^2 - \alpha + 2}{\sqrt{(1+\alpha^2)}} \sinh\{y \sqrt{(1+\alpha^2)}\} + (2-\alpha) \cosh\{y \sqrt{(1+\alpha^2)}\} - (2-\alpha) e^y \right]. \tag{105}$$

b) The integral equation which insures the continuity of the solution along CO.—Along *CD* neither the function nor its derivatives are known: only a relation between them is given (cf. eq. [85]). But our knowledge of the solution along *CO* and the requirement that the solution be continuous along this line suffice to determine $f(0, y)$ as the solution of an integral equation of Volterra's type. To obtain this integral equation we apply Green's theorem to a contour such as *CIHI*, where $H = (\eta, \eta)$ is a point on *OC* and *HI* is the characteristic $\eta - x = y - \eta$ through *H*. We find

$$2f(\eta, \eta) = f(0, 2\eta) + \frac{2+\alpha}{\alpha} + \int_0^{2\eta} \left(v \frac{\partial f}{\partial x} - f \frac{\partial v}{\partial x} \right)_{x=0} dy. \tag{106}$$

On the other hand, since (cf. eq. [85])

$$\frac{\partial f}{\partial x} = \frac{\partial f}{\partial y} - \frac{2-\alpha}{\alpha} e^y \tag{107}$$

along *CD*, we have

$$2f(\eta, \eta) = f(0\ 2\eta) + \frac{2+\alpha}{\alpha} - \frac{2-\alpha}{\alpha} \int_0^{2\eta} e^y [v]_{x=0}\, dy + \int_0^{2\eta} \left(v \frac{\partial f}{\partial y} \right)_{x=0} dy - \int_0^{2\eta} f(0, y) \left(\frac{\partial v}{\partial x} \right)_{x=0} dy. \tag{108}$$

Integrating by parts the second of the three integrals occurring on the right-hand side of equation (108) and remembering that the Riemann function for this problem is always unity along the characteristics, we find that

$$2f(\eta, \eta) = 2f(0, 2\eta) - \frac{2-\alpha}{\alpha} \int_0^{2\eta} e^y [v]_{x=0}\, dy - \int_0^{2\eta} f(0, y) \left(\frac{\partial v}{\partial x} + \frac{\partial v}{\partial y} \right)_{x=0} dy. \tag{109}$$

The Riemann function appropriate for our present contour is

$$v(x, y; \xi, \eta) = J_0([(\eta - x)^2 - (y - \eta)^2]^{\frac{1}{2}}). \tag{110}$$

With this choice of v, equation (109) reduces to the form

$$\left.\begin{aligned} f(\eta,\eta) &= f(0,2\eta) - \frac{2-a}{2a}\,\eta e^{\eta}\int_0^{\pi} e^{\eta\cos\vartheta} J_0(\eta\sin\vartheta)\sin\vartheta\, d\vartheta \\ &\quad -\frac{1}{2}\int_0^{2\eta} f(0,y)\, J_1([\eta^2-(y-\eta)^2]^{\frac{1}{2}})\frac{y\,dy}{[\eta^2-(y-\eta)^2]^{\frac{1}{2}}}. \end{aligned}\right\} \quad (111)$$

The definite integral on the right-hand side of equation (111) can be evaluated. We have

$$\left.\begin{aligned} \eta\int_0^{\pi} & e^{\eta\cos\vartheta} J_0(\eta\sin\vartheta)\sin\vartheta\, d\vartheta \\ &= \sum_{n=0}^{\infty}\frac{\eta^{2n+1}}{(2n)!}\sum_{m=0}^{\infty}(-1)^m\frac{(\frac{1}{2}\eta)^{2m}}{m!\Gamma(m+1)}\int_0^{\pi/2}\cos^{2n}\vartheta\,\sin^{2m+1}\vartheta\, d\vartheta \\ &= \sum_{n=0}^{\infty}\Gamma(n+\tfrac{1}{2})\frac{\eta^{2n+1}}{(2n)!}\sum_{m=0}^{\infty}(-1)^m\frac{(\frac{1}{2}\eta)^{2m}}{m!\Gamma(m+n+\frac{3}{2})} \\ &= \sum_{n=0}^{\infty}\Gamma(n+\tfrac{1}{2})\frac{(2\eta)^{n+\frac{1}{2}}}{(2n)!}\sum_{m=0}^{\infty}(-1)^m\frac{(\frac{1}{2}\eta)^{2m+n+\frac{1}{2}}}{m!\Gamma(m+n+\frac{3}{2})} \\ &= (2\pi\eta)^{\frac{1}{2}}\sum_{n=0}^{\infty}(\tfrac{1}{2}\eta)^n\frac{1}{n!}J_{n+\frac{1}{2}}(\eta) \\ &= (2\pi\eta)^{\frac{1}{2}}\frac{(\frac{1}{2}\eta)^{\frac{1}{2}}}{\Gamma(1+\frac{1}{2})} = 2\eta\,. \end{aligned}\right\} \quad (112)$$

Equation (111) thus becomes

$$\left.\begin{aligned} \frac{1}{a}e^{a\eta}&\left[\frac{a^2+a+2}{\sqrt{(1+a^2)}}\sinh\{\eta\sqrt{(1+a^2)}\}+(2+a)\cosh\{\eta\sqrt{(1+a^2)}\}\right] \\ &+\frac{2-a}{a}\eta e^{\eta} = f(0,2\eta) - \frac{1}{2}\int_0^{2\eta} f(0,y)\frac{J_1([\eta^2-(y-\eta)^2]^{\frac{1}{2}})}{[\eta^2-(y-\eta)^2]^{\frac{1}{2}}}\,y\,dy, \end{aligned}\right\} \quad (113)$$

where we have also substituted for $f(\eta, \eta)$ according to equation (101).

It is now seen that the right-hand side of equation (113) is simply the derivative of

$$\frac{1}{2}\int^{2\eta} f(0,y)\, J_0([\eta^2-(y-\eta)^2]^{\frac{1}{2}})\, dy \quad (114)$$

with respect to η. We can accordingly reduce equation (113) to an integral equation of Volterra's type. To perform this reduction we need only know the integral of the left-hand side of equation (113). We thus find that

$$\left.\begin{aligned} \frac{1}{a}e^{a\eta}&\left[(2-a)\cosh\{\eta\sqrt{(1+a^2)}\}+\frac{a^2-a+2}{\sqrt{(1+a^2)}}\sinh\{\eta\sqrt{(1+a^2)}\}\right] \\ &+\frac{2-a}{a}(\eta-1)e^{\eta} = \frac{1}{2}\int_0^{2\eta} f(0,y)\, J_0([\eta^2-(y-\eta)^2]^{\frac{1}{2}})\, dy, \end{aligned}\right\} \quad (115)$$

which is the required equation for $f(0, y)$.

c) The solution of the integral equation (115).—To solve equation (115) we apply a Laplace transformation to this equation, i.e., we multiply both sides of the equation by $e^{-s\eta}$ and integrate over η from 0 to ∞. The right-hand side then becomes (cf. "Moving Atmospheres," eqs. [90]-[95])

$$\frac{1}{2s}\int_0^\infty f(0, y)\exp[-y(s+s^{-1})/2]\,dy\,, \tag{116}$$

while the evaluation of the Laplace transform of the left-hand side requires only elementary integrals. We find

$$\left.\begin{aligned}&\frac{1}{\alpha}\left[\frac{(2-\alpha)(s-\alpha)}{(s-\alpha)^2-(1+\alpha^2)}+\frac{\alpha^2-\alpha+2}{(s-\alpha)^2-(1+\alpha^2)}\right]+\frac{(2-\alpha)(2-s)}{\alpha(s-1)^2}\\&\qquad=\frac{1}{2s}\int_0^\infty f(0, y)\exp[-y(s+s^{-1})/2]\,dy\,.\end{aligned}\right\} \tag{117}$$

We can re-write the foregoing equation in the form

$$\left.\begin{aligned}&\frac{2(s+1)-\alpha(s+3)+2\alpha^2}{\alpha(s^2-2s\alpha-1)}+\frac{(2-\alpha)(2-s)}{\alpha(s-1)^2}\\&\qquad=\frac{1}{2s}\int_0^\infty f(0, y)\exp[-y(s+s^{-1})/2]\,dy\,.\end{aligned}\right\} \tag{118}$$

If we now let

$$s+s^{-1}=2u \tag{119}$$

or, equivalently,

$$s=u+\sqrt{(u^2-1)}\,, \tag{120}$$

equation (118) provides the simple Laplace transform of $f(0, y)$, and the solution can, in principle, be found. However, since α is a small quantity of the order of 10^{-4}, the solution in the form of a series expansion in α will suffice for most purposes. Accordingly, we expand the left-hand side of equation (118) in powers of α. We thus obtain

$$\left.\begin{aligned}&\frac{2+\alpha}{\alpha}\,\frac{2s}{(s-1)^2}+\alpha\,\frac{4s}{(s-1)^3}+O(\alpha^2)\\&\qquad=\int_0^\infty f(0, y)\exp[-y(s+s^{-1})/2]\,dy\,.\end{aligned}\right\} \tag{121}$$

With the substitutions (119) and (120), equation (121) becomes

$$\left.\begin{aligned}&\frac{2+\alpha}{\alpha}\,\frac{1}{u-1}+\frac{2\alpha}{(u-1)^{3/2}[\sqrt{(u+1)}+\sqrt{(u-1)}]}+O(\alpha^2)\\&\qquad=\int_0^\infty f(0, y)\,e^{-yu}dy\,.\end{aligned}\right\} \tag{122}$$

Now the inverse Laplace transform of the first term on the left-hand side of equation (122) is

$$\frac{2+\alpha}{\alpha}\,e^{y}\,; \tag{123}$$

while writing the second term in the form

$$\alpha\left[\frac{1}{(u^2-1)^{\frac{1}{2}}}+\frac{2}{(u-1)^{3/2}(u+1)^{\frac{1}{2}}}-\frac{1}{u-1}\right], \tag{124}$$

we see that its inverse Laplace transform is

$$\alpha\{I_0(y)+2y[I_0(y)+I_1(y)]-e^y\}. \tag{125}$$

Hence,

$$f(0,y)=\frac{2+\alpha}{\alpha}e^y+\alpha\{I_0(y)+2y[I_0(y)+I_1(y)]-e^y\}+O(\alpha^2). \tag{126}$$

The corresponding solution for $\psi_{+1}(0, y)$ is

$$\left.\begin{aligned}\psi_{+1}(0,y)=Q\alpha e^{-y}\{I_0(y)+2y[I_0(y)+I_1(y)]-e^y\}+O(\alpha^2)\\(0\leqslant y\leqslant x_1),\end{aligned}\right\} \tag{127}$$

while $\psi_{-1}(0, y)$, of course, vanishes along this line, as required by the boundary conditions.

d) The integral equation insuring the continuity of the solution along OB *and its solution (case I).*—Applying, next, Green's theorem to a contour such as $JKBJ$, where $J = (x_1-\eta, \eta)$ is a point on OB, and JK is the characteristic $x-x_1+\eta=y-\eta$ through J, we find in the usual manner that

$$2f(x_1-\eta,\eta)=\frac{2+\alpha}{\alpha}e^{\alpha x_1}+f(x_1,2\eta)-\int_0^{2\eta}\left(v\frac{\partial f}{\partial x}-f\frac{\partial v}{\partial x}\right)_{x=x_1}dy. \tag{128}$$

In case I,

$$\frac{\partial f}{\partial x}=-f+\frac{\partial f}{\partial y}+2e^{\alpha x_1+y} \tag{129}$$

along AB (cf. eq. [86]). Using this in equation (128), we find, after some minor reductions, that

$$\left.\begin{aligned}f(x_1-\eta,\eta)=\frac{2+\alpha}{\alpha}e^{\alpha x_1}-e^{\alpha x_1}\int_0^{2\eta}e^y[v]_{x=x_1}dy\\+\frac{1}{2}\int_0^{2\eta}f(x_1,y)\left(v+\frac{\partial v}{\partial x}+\frac{\partial v}{\partial y}\right)_{x=x_1}dy.\end{aligned}\right\} \tag{130}$$

The Riemann function appropriate to our present contour is

$$v(x,y;x_1-\eta,\eta)=J_0([(x-x_1+\eta)^2-(y-\eta)^2]^{\frac{1}{2}}). \tag{131}$$

With v given by equation (131), equation (130) becomes

$$\left.\begin{aligned}f(x_1-\eta,\eta)=\frac{2+\alpha}{\alpha}e^{\alpha x_1}-\eta e^{\alpha x_1+\eta}\int_0^{\pi}e^{\eta\cos\vartheta}J_0(\eta\sin\vartheta)\sin\vartheta d\vartheta\\+\frac{1}{2}\int_0^{2\eta}f(x_1,y)J_0([\eta^2-(y-\eta)^2]^{\frac{1}{2}})dy\\-\frac{1}{2}\int_0^{2\eta}f(x_1,y)J_1([\eta^2-(y-\eta)^2]^{\frac{1}{2}})\frac{2\eta-y}{[\eta^2-(y-\eta)^2]^{\frac{1}{2}}}dy.\end{aligned}\right\} \tag{132}$$

But (cf. eq. [112])

$$\int_0^{\pi}e^{\eta\cos\vartheta}J_0(\eta\sin\vartheta)\sin\vartheta d\vartheta=2. \tag{133}$$

Using this result in equation (132) and substituting also for $f(x_1 - \eta, \eta)$ according to equation (101), we obtain

$$\left.\begin{aligned} &\frac{1}{a} e^{a(x_1-\eta)}\left[\frac{a^2+a+2}{\sqrt{(1+a^2)}}\sinh\{\eta\sqrt{(1+a^2)}\}+(2+a)\cosh\{\eta\sqrt{(1+a^2)}\}\right] \\ &\qquad -\frac{2+a}{a}e^{ax_1}+2\eta e^{ax_1+\eta}=\frac{1}{2}\int_0^{2\eta} f(x_1, y)\, J_0([\eta^2-(y-\eta)^2]^{\frac{1}{2}})\, dy \\ &\qquad -\frac{1}{2}\int_0^{2\eta}\frac{f(x_1, y)}{y} J_1\{\sqrt{(2\eta y-y^2)}\}\sqrt{(2\eta y-y^2)}\, dy\,, \end{aligned}\right\} \tag{134}$$

which is again an integral equation of Volterra's type for $f(x_1, y)$.

To solve equation (134) we apply, as before, a Laplace transformation to this equation. We find (cf. eq. [116])

$$\left.\begin{aligned} &\frac{1}{a} e^{ax_1}\left[\frac{a^2+a+2}{(s+a)^2-(1+a^2)}+\frac{(2+a)(s+a)}{(s+a)^2-(1+a^2)}\right]-\frac{1}{a}e^{ax_1}\frac{2+a}{s} \\ &\qquad +\frac{2e^{ax_1}}{(s-1)^2}=\frac{1}{2s}\int_0^{\infty} f(x_1, y)\exp[-y(s+s^{-1})/2]\, dy \\ &\qquad -\frac{1}{2}\int_0^{\infty} d\eta\, e^{-s\eta}\int_0^{2\eta}\frac{f(x_1, y)}{y}J_1\{\sqrt{(2\eta y-y^2)}\}\sqrt{(2\eta y-y^2)}\, dy\,. \end{aligned}\right\} \tag{135}$$

Inverting the order of the integration in the double integral in equation (135) and introducing the variable

$$2\eta y - y^2 = t^2 \tag{136}$$

we find that it reduces to

$$\int_0^{\infty}\frac{dy}{y^2} f(x_1, y)\, e^{-sy/2}\int_0^{\infty} dt\, t^2 J_1(t)\, e^{-st^2/2y}\,. \tag{137}$$

But the integral over t in (137) has the value[14]

$$\left(\frac{y}{s}\right)^2 e^{-y/2s}\,. \tag{138}$$

Accordingly, the

$$\left.\begin{aligned} &\text{Laplace transform of}\int_0^{2\eta}\frac{f(x_1, y)}{y}J_1\{\sqrt{(2\eta y-y^2)}\}\sqrt{(2\eta y-y^2)}\, dy \\ &\qquad =\frac{1}{s^2}\int_0^{\infty} f(x_1, y)\exp[-y(s+s^{-1})/2]\, d^{y}\,, \end{aligned}\right\} \tag{139}$$

and equation (135) becomes

$$\left.\begin{aligned} &\frac{1}{a}e^{ax_1}\left[\frac{2(s+1)+a(s+3)+2a^2}{s^2+2sa-1}-\frac{2+a}{s}+\frac{2a}{(s-1)^2}\right] \\ &\qquad =\frac{s-1}{2s^2}\int_0^{\infty} f(x_1, y)\exp[-y(s+s^{-1})/2]\, dy\,. \end{aligned}\right\} \tag{140}$$

With the substitution of equation (119) and (120) in equation (140), we shall obtain the Laplace transform of $f(x_1, y)$, and the solution beomes determinate. But again, as in

[14] Cf. Watson, *op. cit.*, p. 394, eq. (4).

the preceding subsection, we shall content ourselves with finding a solution in the form of a series expansion in a. Thus, expanding the left-hand side of equation (140) in powers of a and after some rearranging of the terms, we find

$$\left.\begin{aligned}\frac{2}{a}e^{ax_1}\left[\frac{(2+a)\,s}{(s-1)^2}+2a^2\frac{s^2}{(s-1)^4}+O(a^3)\right]\\ =\int_0^\infty f(x_1,y)\exp\left[-y(s+s^{-1})/2\right]dy\,.\end{aligned}\right\}\qquad(141)$$

In terms of the variable u defined as in equation (119) or equation (120), the foregoing equation reduces to

$$e^{ax_1}\left[\frac{2+a}{a}\,\frac{1}{(u-1)}+\frac{a}{(u-1)^2}+O(a^2)\right]=\int_0^\infty f(x_1,y)\,e^{-yu}dy\,.\qquad(142)$$

Hence,

$$f(x_1,y)=\frac{2+a}{a}\,e^{ax_1+y}+a\,y\,e^{ax_1+y}+O(a^2)\,.\qquad(143)$$

The corresponding solutions for ψ_{+1} and ψ_{-1} are (cf. eq. [58])

$$\psi_{+1}(x_1,y)=Q\,a\,y\,e^{ax_1}+O(a^2)\qquad(144)$$

and

$$\psi_{-1}(x_1,y)=Q\,a\,y\,e^{ax_1}+O(a^2)\,.\qquad(145)$$

e) The integral equation insuring the continuity of the solution along OB *and its solution (case III).*—In case III the boundary condition specified along AB is that (cf. eq. [87])

$$f=\frac{2+a}{a}\,e^{ax_1+y}\,,\qquad(146)$$

and we require to find $(\partial f/\partial x)_{x=x_1}$. The solution proceeds along lines now familiar. Equation (128), which is valid also under our present conditions, gives

$$\left.\begin{aligned}2f(x_1-\eta,\eta)=\frac{2+a}{a}\,e^{ax_1}(1+e^{2\eta})+\frac{2+a}{a}\,e^{ax_1}\int_0^{2\eta}e^{y}\left(\frac{\partial v}{\partial x}\right)_{x=x_1}dy\\ -\int_0^{2\eta}\left(v\,\frac{\partial f}{\partial x}\right)_{x=x_1}dy\,.\end{aligned}\right\}\qquad(147)$$

The Riemann function is the same as in case I and is given in equation (131). Using this in equation (147), we find after some minor transformations that

$$\left.\begin{aligned}2f(x_1-\eta,\eta)=\frac{2+a}{a}\,e^{ax_1}(1+e^{2\eta})-\frac{2+a}{a}\,\eta\,e^{ax_1+\eta}\int_0^{\pi}e^{\eta\cos\vartheta}J_1(\sin\vartheta)\,d\vartheta\\ -\int_0^{2\eta}\left(\frac{\partial f}{\partial x}\right)_{x=x_1}J_0([\eta^2-(y-\eta)^2]^{\frac{1}{2}})\,dy\,.\end{aligned}\right\}\qquad(148)$$

But (cf. "Moving Atmospheres," eqs. [105] and [106])

$$\eta\int_0^{\pi}e^{\eta\cos\vartheta}J_1(\eta\sin\vartheta)\,d\vartheta=2(\cosh\eta-1)\,.\qquad(149)$$

Using this result in equation (148) and substituting also for $f(x_1 - \eta, \eta)$ according to equation (101), we obtain

$$\left.\begin{aligned}&\frac{2+a}{a} e^{ax_1+\eta} - \frac{1}{a} e^{a(x_1-\eta)} \left[\frac{a^2+a+2}{\sqrt{(1+a^2)}} \sinh \{ \eta \sqrt{(1+a^2)} \} + (2+a) \right. \\ &\qquad \left. \times \cosh \{ \eta \sqrt{(1+a^2)} \} \right] = \frac{1}{2} \int_0^{2\eta} \left(\frac{\partial f}{\partial x} \right)_{x=x_1} J_0 ([\eta^2 - (y-\eta)^2]^{\frac{1}{2}}) \, dy .\end{aligned}\right\} \quad (150)$$

This is the required Volterra integral equation for $(\partial f/\partial x)_{x=x_1}$.

The solution of the integral equation (150) proceeds as in the other cases. We apply a Laplace transformation to the equation and obtain

$$\left.\begin{aligned}&\frac{1}{a} e^{ax_1} \left[\frac{2+a}{s-1} - \frac{2(s+1) + a(s+3) + 2a^2}{s^2 + 2sa - 1} \right] \\ &\qquad = \frac{1}{2s} \int_0^{\infty} \left(\frac{\partial f}{\partial x} \right)_{x=x_1} \exp [-y (s + s^{-1}) / 2] \, dy .\end{aligned}\right\} \quad (151)$$

And again, as in the earlier cases, we shall obtain a solution of this equation in the form of a series expansion in a by expanding the left-hand side of the equation in powers of a. We thus find

$$\left.\begin{aligned}&e^{ax_1} \left[\frac{4s}{(s-1)^2} - a \frac{4s}{(s-1)^3} + O(a^2) \right] \\ &\qquad = \int_0^{\infty} \left(\frac{\partial f}{\partial x} \right)_{x=x_1} \exp [-y (s + s^{-1}) / 2] \, dy .\end{aligned}\right\} \quad (152)$$

With the substitutions (119) and (120), equation (152) becomes

$$\left.\begin{aligned}&e^{ax_1} \left[\frac{2}{u-1} - \frac{2a}{(u-1)^{3/2} [\sqrt{(u+1)} + \sqrt{(u-1)}]} + O(a^2) \right] \\ &\qquad = \int_0^{\infty} \left(\frac{\partial f}{\partial x} \right)_{x=x_1} e^{-yu} dy .\end{aligned}\right\} \quad (153)$$

Equation (153) is seen to be similar in form to equation (122). We can accordingly write (cf. eqs. [124] and [125])

$$\left(\frac{\partial f}{\partial x} \right)_{x=x_1} = 2 e^{ax_1+y} - a e^{ax_1} \{ I_0(y) + 2y [I_0(y) + I_1(y)] - e^y \} + O(a^2) . \quad (154)$$

Since (cf. eq. [146])

$$\left(\frac{\partial f}{\partial y} \right)_{x=x_1} = \frac{2+a}{a} e^{ax_1+y} , \quad (155)$$

we have

$$\left.\begin{aligned}g(x_1, y) = \frac{2-a}{a} e^{ax_1+y} + a e^{ax_1} \{ I_0(y) + 2y [I_0(y) + I_1(y)] \\ - e^y \} + O(a^2) .\end{aligned}\right\} \quad (156)$$

The corresponding solution for $\psi_{-1}(x_1, y)$ is

$$\left.\begin{aligned}\psi_{-1}(x_1, y) = Q a e^{ax_1 - y} \{ I_0(y) + 2y [I_0(y) + I_1(y)] - e^y \} + O(a^2) \\ (0 \leqslant y \leqslant x_1) ,\end{aligned}\right\} \quad (157)$$

while $\psi_{+1}(x_1, y)$ vanishes in accordance with the boundary conditions.

f) The solution in the region O′B′COBC′O′ *and its further continuation.*—With the determination of f along CB' and f and its derivatives along BC', our knowledge of the function and its derivatives along $B'CBC'$ is complete, and the solution in the region $O'B'COBC'O'$ becomes determinate; for, as in Riemann's method, by applying Green's theorem to contours such as $LMNL$, $PQRP$, $STCBUS$, we can find the solution in the regions $OB'C$, OBC', and $O'B'OC'$. It is seen how a knowledge of the function along COB, together with the boundary conditions on CB' and BC', enables us to determine f in the region $O'B'COBC'O'$, including the sides $B'O'$ and $O'C'$. It is now apparent that, in the same way, we can utilize our present knowledge of the function along $B'O'C'$ to extend the solution still further. We shall not, however, consider these further extensions of the solution in this paper but content ourselves with the solution which has been completed in the first square, $B'CBC'$. According to equation (56), this will suffice to determine the radiation field in all cases in which the ratio $D\nu$:$\Delta\nu$ exceeds $\sqrt{3}$.

5. *Formulae for the radiation pressure in an expanding atmosphere.*—Consider a slab of material of unit cross-section and height dz. The normal force acting on this slab due to the absorption of radiation is

$$\frac{1}{c}\,2\pi\rho\,d\,z\int_0^\infty d\nu\sigma\,(\nu)\int_0^\pi d\vartheta I\left(\nu+\nu_0\frac{w}{c}\cos\vartheta,\ z,\ \vartheta\right)\cos\vartheta\sin\vartheta\,, \tag{158}$$

since the radiation in the direction ϑ, which will appear to our fixed observer as having a frequency $\nu+\nu_0(w/c)\cos\vartheta$, will be judged by an observer at rest with respect to the material as having a frequency ν: it will accordingly be absorbed only as such.

The pressure Π exerted by the radiation is therefore given by

$$\Pi\,(z)=\frac{1}{c}\,2\pi\rho\int_0^\infty d\nu\sigma\,(\nu)\int_{-1}^{+1}d\mu I\left(\nu+\nu_0\frac{w}{c}\mu,\ z,\ \mu\right)\mu\,. \tag{159}$$

Equation (159) is perfectly general. We shall now consider certain alternative forms of this equation suitable under various conditions and approximations.

First, replacing the integral over μ in equation (159) by a sum according to Gauss's formula for numerical quadratures, we obtain

$$\Pi\,(z)=\frac{1}{c}\,2\pi\rho\int_0^\infty\sigma\,(\nu)\sum_i a_i\mu_i I_i\left(\nu+\nu_0\frac{w}{c}\mu_i,\ z\right)d\nu\,. \tag{160}$$

The intensities which appear in equation (160) are exactly the intensities $\psi_i(\nu, z)$ as we have defined them in equations (17) and (18). We may therefore write

$$\Pi\,(z)=\frac{1}{c}\,2\pi\rho\int_0^\infty\sigma\,(\nu)\sum_i a_i\mu_i\psi_i\,(\nu,\ z)\,d\nu\,. \tag{161}$$

In the first approximation, equation (160) reduces to

$$\Pi\,(z)=\frac{1}{c}\,2\pi\mu_1\rho\int_0^\infty\sigma\,(\nu)\,[\psi_{+1}\,(\nu,\ z)-\psi_{-1}\,(\nu,\ z)]\,d\nu\qquad\left(\mu_1=\frac{1}{\sqrt{3}}\right), \tag{162}$$

an equation which has an obvious physical interpretation.

For the particular form (26) for $\sigma(\nu)$, equation (162) further simplifies to

$$\Pi\,(z)=\frac{1}{c}\,2\pi\mu_1\rho\sigma_0\int_{\nu_0-\Delta\nu}^{\nu_0+\Delta\nu}[\psi_{+1}\,(\nu,\ z)-\psi_{-1}\,(\nu,\ z)]\,d\nu\,, \tag{163}$$

or, expressing ν in the unit (53), we have

$$\Pi\,(x)=\frac{1}{c}\,4\pi\mu_1^2\frac{D\nu}{x_1}\,\rho\sigma_0\int_0^{y_1}[\psi_{+1}\,(x,\ y)-\psi_{-1}\,(x,\ y)]\,dy\,. \tag{164}$$

On the other hand, since (cf. eq. [44])

$$x_1 = \frac{1}{2\mu_1} t_1 , \tag{165}$$

where t_1 is the optical depth of the atmosphere in σ_0, we can re-write equation (164) in the form

$$\Pi(x) = \frac{1}{c} 8\pi\mu_1^3 D\nu \frac{\rho\sigma_0}{t_1} \int_0^{y_1} [\psi_{+1}(x, y) - \psi_{-1}(x, y)]\, dy . \tag{166}$$

For the particular model considered in this paper,

$$\frac{\rho\sigma_0}{t_1} = \frac{\rho\kappa_c}{\tau_1} , \tag{167}$$

where κ_c and τ_1 refer to the ultraviolet continuum. Hence, in this case we may also write

$$\Pi(x) = \frac{1}{c} 8\pi\mu_1^3 D\nu \frac{\rho\kappa_c}{\tau_1} \int_0^{y_1} [\psi_{+1}(x\ \ y) - \psi_{-1}(x\ \ y)]\, dy . \tag{168}$$

6. *The radiation pressure in Lyman-α in an expanding planetary nebula.*—With the solution of the transfer problem completed in the preceding sections, we are now in a

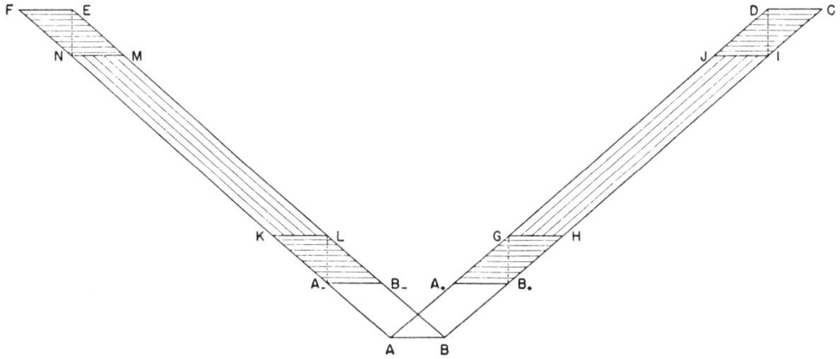

FIG. 4

position to answer the question raised in the introductory section, namely, as to how effective differential expansion can be in reducing the magnitude of the radiation pressure in Lyman-α which will otherwise act. As we shall presently see, the case of greatest interest in this connection arises when

$$D\nu > \frac{2}{\mu_1} \Delta\nu \tag{169}$$

or, alternatively, when (cf. eq. [56])

$$y_1 < \tfrac{1}{2} x_1 . \tag{170}$$

When (170) is the case, for

$$y_1 \leqslant x \leqslant x_1 - y_1 , \tag{171}$$

the solution for the radiation field is known explicitly and is given by equations (104) and (105) (see Figs. 3 and 4, where the regions so defined are indicated). The range (171) for x corresponds to the range

$$\frac{y_1}{x_1} \leqslant \frac{t}{t_1} \leqslant 1 - \frac{y_1}{x_1} \tag{172}$$

for the optical depth t. Since (cf. eq. [56])

$$\frac{y_1}{x_1} = \frac{1}{\mu_1} \frac{\Delta\nu}{D\nu}, \tag{173}$$

we can re-write (172) in the form

$$t_1 \frac{1}{\mu_1} \frac{\Delta\nu}{D\nu} \leqslant t \leqslant t_1 \left(1 - \frac{1}{\mu_1} \frac{\Delta\nu}{D\nu}\right). \tag{174}$$

We shall refer to the part of the nebula included in the range of optical depths specified by (174) as the *central part* of the nebula. Similarly, we shall refer to the parts included in

$$0 < t < t_1 \frac{1}{\mu_1} \frac{\Delta\nu}{D\nu} \tag{175}$$

and

$$t_1 \left(1 - \frac{1}{\mu_1} \frac{\Delta\nu}{D\nu}\right) < t < t_1 \tag{176}$$

as the *outer* and the *inner* parts of the nebula, respectively. It should be emphasized at this point that this distinction between the outer, central, and inner parts of a nebula applies only to those cases in which the Doppler width, $2D\nu$, exceeds the line width, $2\Delta\nu$, by a factor $2\sqrt{3}$.

We shall now show how the selective radiation pressure in the central part of a nebula can be estimated. From equations (104) and (105) we find that

$$\left.\begin{aligned} \psi_{+1} - \psi_{-1} = 2Q e^{ax} \bigg[\frac{e^{-y}}{\sqrt{(1+a^2)}} \sinh\{y\sqrt{(1+a^2)}\} \\ + e^{+y} \cosh\{y\sqrt{(1+a^2)}\} - 1 \bigg]; \end{aligned}\right\} \tag{177}$$

and, since we have assumed that $y_1 < \frac{1}{2}x_1$, this solution is valid for the entire range of y for x in the range (171).

Now, according to equation (168), the radiation pressure in Lyman-α is related directly to the integral of $\psi_{+1} - \psi_{-1}$ over the line; and this can be readily found from the solution (177). We have

$$\left.\begin{aligned} \int_0^{y_1} (\psi_{+1} - \psi_{-1})\, dy = 2Q e^{ax} \bigg\{ \frac{e^{-y_1}}{a^2} \bigg[2\cosh\{y_1\sqrt{(1+a^2)}\} \\ + \frac{2+a^2}{\sqrt{(1+a^2)}} \sinh\{y_1\sqrt{(1+a^2)}\} \bigg] - \frac{2}{a^2} - y_1 \bigg\} \quad (y_1 \leqslant x \leqslant x_1 - y_1). \end{aligned}\right\} \tag{178}$$

Inserting this value in equation (168) and substituting also for Q according to equation (75), we obtain

$$\left.\begin{aligned} \Pi(x) = \left(\frac{\pi\rho\kappa_c}{c} \frac{\nu_0}{\nu_c} S_c\right) 2\mu_1^2 \frac{D\nu}{\Delta\nu} \frac{e^{-(\tau_1-\tau)}}{\tau_1} \bigg\{ \frac{e^{-y_1}}{a^2} \bigg[2\cosh\{y_1\sqrt{(1+a^2)}\} \\ + \frac{2+a^2}{\sqrt{(1+a^2)}} \sinh\{y_1\sqrt{(1+a^2)}\} \bigg] - \frac{2}{a^2} - y_1 \bigg\}, \end{aligned}\right\} \tag{179}$$

where we have further used the relation $ax = \tau$ (eq. [69]).

The full implications of equation (179) are best understood when we expand the quantity in braces in powers of α and retain the first nonvanishing term. We find in this manner that

$$\Pi=\left(\frac{\pi\rho\kappa_c\nu_0}{c}\frac{\nu_0}{\nu_c}S_c\right)2\mu_1^2\frac{D\nu}{\Delta\nu}\frac{e^{-(\tau_1-\tau)}}{\tau_1}\left\{\frac{\alpha^2}{4}\left[y_1^2-y_1+\tfrac{1}{2}(1-e^{-2y_1})\right]+O(\alpha^4)\right\}. \tag{180}$$

A further simplification of this equation is possible. Since y_1 is a large quantity of the order of 10^3 or 10^4, we may write, to a sufficient accuracy,

$$\Pi=\left(\frac{\pi\rho\kappa_c\nu_0}{c}\frac{\nu_0}{\nu_c}S_c\right)\tfrac{1}{2}\mu_1^2\frac{D\nu}{\Delta\nu}\frac{e^{-(\tau_1-\tau)}}{\tau_1}\alpha^2y_1^2. \tag{181}$$

But (cf. eqs. [69] and [173])

$$\alpha y_1=\frac{1}{\mu_1}\frac{\Delta\nu}{D\nu}\alpha x_1=\frac{1}{\mu_1}\frac{\Delta\nu}{D\nu}\tau_1. \tag{182}$$

Equation (181) therefore reduces to

$$\Pi(x)=\left(\frac{\pi\rho\kappa_c\nu_0}{c}\frac{\nu_0}{\nu_c}S_c\right)\frac{1}{2}\frac{\Delta\nu}{D\nu}\tau_1e^{-(\tau_1-\tau)}\qquad(y_1\leqslant x\leqslant x_1-y_1). \tag{183}$$

In other words, *in the central part of a nebula the radiation pressures due to Lyman-α and the Lyman continuum are of the same order of magnitude.*

Now for the case $p=0$ (no re-emission in the Lyman continuum), the selective radiation pressure which will act in a static nebula can be readily written down. We must clearly have

$$\Pi_{\text{static}}=\left(\frac{\pi\rho\sigma_0\nu_0}{c}\frac{\nu_0}{\nu_c}S_c\right)\left[1-e^{-(\tau_1-\tau)}\right]. \tag{184}$$

From a comparison of equations (183) and (184) the remarkable result emerges that if a differential expansion to the extent required by equation (169) exists, then in the central part of the nebula the radiation pressure in Lyman-α is effectively cut down by a factor of the order of 10^4.

We cannot, of course, expect that the large reduction in the radiation pressure achieved in the central part will be maintained throughout the nebula. We should rather expect certain "edge effects." To investigate the nature of these edge effects in detail, we need the radiation field in regions where the solution can be found only by quadratures (cf. §4, subsec. *f*). However, we can estimate the radiation pressure acting on the outer and the inner boundaries of the nebula, and these might provide some indications.

At $\tau=0$ the inward intensity $\psi_{-1}(0,y)$ vanishes, and the radiation pressure depends only on the integral of the outward intensity $\psi_{+1}(0,y)$ over the line. Using the solution (127) for $\psi_{+1}(0,y)$, we have, for the radiation pressure acting at $\tau=0$,

$$\left.\begin{aligned}\Pi(0)&=\left(\frac{\pi\rho\kappa_c\nu_0}{c}\frac{\nu_0}{\nu_c}S_c\right)\mu_1^2\frac{D\nu}{\Delta\nu}\frac{e^{-\tau_1}}{\tau_1}\\&\times\left[\alpha\int_0^{y_1}e^{-y}\{I_0(y)+2y[I_0(y)+I_1(y)]-e^y\}dy+O(\alpha^2)\right].\end{aligned}\right\} \tag{185}$$

It does not appear that the integrals over the Bessel functions occurring in equation (185) can be evaluated explicitly. However, its asymptotic value for large y_1 can be readily found. We have

$$\left.\begin{aligned}&\int_0^{y_1}e^{-y}\{I_0(y)+2y[I_0(y)+I_1(y)]-e^y\}dy\\&\qquad\to\frac{4}{(2\pi)^{1/2}}\int_0^{y_1}y^{1/2}dy=\frac{8}{3(2\pi)^{1/2}}y_1^{3/2}.\end{aligned}\right\} \tag{186}$$

Using this result in equation (185), we obtain

$$\Pi(0) \simeq \left(\frac{\pi\rho\kappa_c\nu_0}{c}\frac{\nu_0}{\nu_c}S_c\right)\frac{8}{3(2\pi a)^{\frac{1}{2}}}\mu_1^2\frac{D\nu}{\Delta\nu}\frac{e^{-\tau_1}}{\tau_1}(ay_1)^{3/2}. \tag{187}$$

Substituting for ay_1 from equation (182), we finally have

$$\Pi(0) \simeq \left(\frac{\pi\rho\kappa_c\nu_0}{c}\frac{\nu_0}{\nu_c}S_c\right)\frac{8}{3}\left(\frac{\mu_1\tau_1}{2\pi a}\frac{\Delta\nu}{D\nu}\right)^{\frac{1}{2}}e^{-\tau_1}. \tag{188}$$

Accordingly, at $\tau = 0$, the radiation pressure due to Lyman-α is cut down only by a factor of the order of $\sqrt{(D\nu/a\Delta\nu)}$. We may, therefore, expect that in the outer parts of a nebula the radiation pressure due to Lyman-α will be appreciable, though it is not likely to exceed the radiation pressure due to the Lyman continuum by any very large factor.

The radiation pressure acting at $\tau = \tau_1$ will depend on the boundary conditions here. If we suppose that the inner boundary is at rest with respect to the central star (case I), the radiation pressure at $\tau = \tau_1$ vanishes identically, simply in virtue of the boundary conditions. But this will not clearly be true of the rest of the inner part, and it is likely that a truer estimate of the radiation pressure which may act in these regions is to be found from that acting at $\tau = \tau_1$ in our case III. It will be recalled that this case arises when the Doppler shift, owing to the velocity at the inner boundary, exceeds the line width by a factor $\sqrt{3}/2$ (eq. [34]). When this happens, ψ_{+1} vanishes at $\tau = \tau_1$, but there is a net flux directed *inward* owing to the nonvanishing of ψ_{-1}. Since the solution (157) for $\psi_{-1}(x_1, y)$ in case III is similar in form to the solution (127) for $\psi_{+1}(0, y)$, we can at once write down (cf. eq. [188])

$$\Pi(\tau_1) \simeq \left(\frac{\pi\rho\kappa_c\nu_0}{c}\frac{\nu_0}{\nu_c}S_c\right)\frac{8}{3}\left(\frac{\mu_1\tau_1}{2\pi a}\frac{\Delta\nu}{D\nu}\right)^{\frac{1}{2}} \qquad \text{(case III)}. \tag{189}$$

which has the same validity as equation (188). A comparison of equations (188) and (189) suggests that, in general, we may expect that the order of magnitude of the radiation pressure acting in the outer and the inner parts will be about the same.

We may summarize the results of our discussion so far in the following general terms.

In a static nebula, selective radiation pressure in Lyman-α is so large that we may expect a differential expansion to set in. This will have a tendency to reduce the magnitude of the radiation pressure acting. But the reduction, while it would set in only gradually, becomes suddenly very effective when the differential expansion present exceeds a certain critical value. On the basis of our calculations we expect reduction factors of the order of 10^4 over at least parts of the nebula when the Doppler shift due to the difference in velocities at the inner and outer voundaries exceeds the undisplaced line width by a factor of the order of 3.5. When this happens, the nebula may be divided into a central part and the edges. In the central part, selective radiation pressure is cut down effectively to zero, while at the edges it may still be appreciable.

An exact discussion of the dynamics of a planetary nebula, which will properly take into account the rather complex manner in which radiation pressure in Lyman-α acts, is likely to be a problem of considerable difficulty. But we may expect that a nebula, initially static, will "feel its way" to a state in which the parts we have described as central are fairly extensive; for such a state will have the character of quasi-stationariness, and dissipation, to the extent that it is present, will be confined only to the inner and the outer edges. We can expect such a quasi-stationary state to be reached, since the effectiveness with which selective radiation pressure acts is controlled very sensitively when the differential expansion has reached a certain stage.

7. *The role of radiation pressure in Lyman-α in other astrophysical problems.*—The manner and effectiveness of operation of the radiation pressure in Lyman-α which we have described in the preceding section is likely to have a bearing on other astrophysical problems besides that of emission nebulae. Here we shall make reference to only two such groups of problems.

The first of these relates to the problem of gaseous shells surrounding Be stars. As Struve[15] has pointed out, the clue to the variety of questions which a study of these interesting stars raises is probably to be found in the manner in which the radiation pressure in Lyman-α acts on these shells. While a detailed discussion of these questions is beyond the scope of this investigation, we may refer particularly to the fact to which Struve has called attention, namely, that the atmospheres of these stars appear to consist of three separate layers: a normal reversing layer, a relatively stationary but extensive shell, and an extreme outer part which is expanding. This division of the atmosphere into three parts is strongly reminiscent of our own distinction of the inner, the central, and the outer parts of a planetary nebula. It is, moreover, not inconceivable that the very capriciousness of the phenomena exhibited by the Be stars is to be understood in terms of the extreme sensitiveness with which the effectiveness of the radiation pressure in Lyman-α is controlled by the extent of the differential expansion present.

A second group of problems to which we wish to make reference relates to the spheres of ionized hydrogen surrounding luminous early-type stars. As B. Strömgren[16] has particularly called attention, the conversion of the radiation in the Lyman continuum into radiation in Lyman-α (in the manner of Zanstra's theory) is a principal feature of this problem. We cannot, therefore, escape the conclusion that something of the sort which we expect to happen in a planetary nebula also happens in these ionized hydrogen regions. And it is interesting to speculate on the bearing of this, in turn, on the still larger problems of the interstellar clouds.

From the foregoing brief discussion it is apparent that the clue to the understanding of a great many astrophysical problems may lie in the very remarkable manner in which the radiation pressure in Lyman-α operates in an expanding atmosphere.

15 *Ap. J.*, **95**, 134, 1942. The writer is indebted to Dr. O. Struve for illuminating discussions of these and other stellar spectroscopic problems.

16 *Ap. J.*, **89**, 526, 1939.

22

The softening of radiation by multiple Compton scattering

By S. Chandrasekhar, F.R.S., *Yerkes Observatory*

(*Received* 26 *August* 1947)

The problem of the softening of radiation by multiple Compton scattering in an atmosphere of free electrons is considered. An idealized problem in plane-parallel atmospheres is formulated and the appropriate equation of transfer is approximately solved. The modified distributions with wave-length of the radiation emergent after transmission through various optical thicknesses of an incident monochromatic flux of radiation are tabulated. The calculations show that there is a relatively high probability for quite large shifts after transmission through optical thicknesses of order even unity.

1. Introduction

A problem of some interest which arises in certain astrophysical and other contexts relates to the manner in which radiation of a particular wave-length gets modified by multiple Compton scattering in transmission through an atmosphere of free electrons. In considering this problem we shall suppose that an infinite plane surface radiates uniformly in the outward directions with a known spectral distribution and that above such a radiating surface there is an atmosphere of free electrons. It is required to find the modified distribution in wave-length of the emergent radiation. In this paper will be shown how, with certain simplifying assumptions, this problem can be solved. On the mathematical side the novelty of the problem arises from the somewhat unusual type of boundary value problem in elliptic equations to which it leads.

2. The equation of transfer and its approximate forms

As has already been indicated, we shall consider an atmosphere of free electrons stratified in parallel planes, in which all the properties are constant over the planes $z =$ constant. We shall further suppose that, for the wave-lengths which come under discussion, it is an adequate approximation to consider the coefficient of scattering as independent of wave-length and as having the classical (Thomson) value

$$\sigma = \frac{8\pi}{3}\frac{e^4}{m^2c^4}N_e, \tag{1}$$

where e denotes the charge on the electron, m its mass, c the velocity of light and N_e the number of electrons per unit mass. However, it is in the essence of this problem that we do not ignore the change of wave-length, $\delta\lambda$, on scattering. We shall suppose that this is given by

$$\delta\lambda = \frac{h}{mc}(1-\cos\Theta), \tag{2}$$

where Θ is the angle of scattering and h is Planck's constant. This last assumption implies that the thermal motions of the electrons can be neglected (cf. Dirac 1925).

Under the assumptions stated in the preceding paragraph, the equation of transfer governing the radiation field is

$$\cos\vartheta\frac{\partial I(z,\vartheta,\lambda)}{\rho\sigma\partial z} = -I(z,\vartheta,\lambda)+\frac{1}{4\pi}\int_0^\pi\int_0^{2\pi} I(z,\vartheta',\lambda-\frac{h}{mc}[1-\cos\Theta])\sin\vartheta'\,d\vartheta'\,d\phi', \quad (3)$$

where ρ is the density and $I(z,\vartheta,\lambda)$ is the specific intensity of the radiation of wave-length λ at height z and in the direction ϑ to the outward normal. Also in equation (3)

$$\cos\Theta = \cos\vartheta\,\cos\vartheta'+\sin\vartheta\,\sin\vartheta'\,\cos\phi'. \quad (4)$$

The form for the source function represented by the second term on the right-hand side of equation (3) arises from the fact that in accordance with equation (2) we must consider radiation of wave-length

$$\lambda-\frac{h}{mc}(1-\cos\Theta)$$

in the direction (ϑ',ϕ') in order that when scattered in the direction $(\vartheta, 0)$ it may have the wave-length λ considered.

It will be noticed that in writing the equation of transfer in the form (3) we have assumed that the radiation is scattered isotropically by the electrons. It might be considered an improvement to allow for an anisotropy of the scattered radiation according to the 'phase function'

$$\tfrac{3}{4}(1+\cos^2\Theta)\frac{d\omega'}{4\pi}.$$

But to allow in this manner for the anisotropy of the scattered radiation without at the same time allowing for the partial polarization of the scattered radiation is not a strictly justifiable procedure (cf. Chandrasekhar 1946). However, these are refinements which do not make any difference in the 'first approximation' in which we shall solve the equation of transfer.

Writing $\mu = \cos\vartheta$ and introducing the optical thickness

$$\tau = \int_z^\infty \rho\sigma\,dz, \quad (5)$$

the equation of transfer becomes

$$\mu\frac{\partial I(\tau,\mu,\lambda)}{\partial\tau} = I(\tau,\mu,\lambda)-\frac{1}{4\pi}\int_0^{2\pi}\int_{-1}^{+1} I(\tau,\mu',\lambda-\gamma[1-\cos\Theta])\,d\mu'\,d\phi', \quad (6)$$

where

$$\gamma = \frac{h}{mc} = 0{\cdot}024\,\text{A} \quad (7)$$

is the Compton wave-length.

We shall now suppose that $I(\tau,\mu',\lambda-\gamma[1-\cos\Theta])$ can be expanded as a Taylor series in the form

$$I(\tau,\mu',\lambda-\gamma[1-\cos\Theta]) = I(\tau,\mu',\lambda)-\gamma(1-\cos\Theta)\frac{\partial I(\tau,\mu',\lambda)}{\partial\lambda}+\ldots, \quad (8)$$

and that it is sufficient to retain only the first two terms in the expansion. (The limitations on the solution implied by this assumption will become apparent later.) We therefore replace equation (6) by

$$\mu \frac{\partial I(\tau,\mu,\lambda)}{\partial \tau} = I(\tau,\mu,\lambda) - \frac{1}{2}\int_{-1}^{+1}\left[I(\tau,\mu',\lambda) - \gamma(1-\mu\mu')\frac{\partial I(\tau,\mu',\lambda)}{\partial \lambda}\right]d\mu'. \tag{9}$$

In solving equation (9) we shall adopt the method of approximation which the writer has developed in recent years for solving the various problems of radiative transfer in the theory of stellar atmospheres. (For a general account of these investigations see Chandrasekhar 1947.) The essence of this method is to replace the integrals which occur in the equations of transfer by sums according to Gauss's formula for numerical quadratures. Thus, in the nth approximation, we replace equation (9) by the system of $2n$ equations

$$\mu_i \frac{\partial I_i(\tau,\lambda)}{\partial \tau} = I_i(\tau,\lambda) - \tfrac{1}{2}\sum_j a_j I_j(\tau,\lambda) + \tfrac{1}{2}\gamma \sum_j a_j(1-\mu_i\mu_j)\frac{\partial I_j(\tau,\lambda)}{\partial \lambda} \quad (i = \pm 1, \ldots, \pm n), \tag{10}$$

where the μ_i's are the zeroes of the Legendre polynomial $P_{2n}(\mu)$ and the a_j's ($a_j = a_{-j}$ and $j = \pm 1, \ldots, \pm n$) are the appropriate Gaussian weights. Further, in equation (10), we have written $I_i(\tau, \lambda)$ for $I(\tau, \mu_i, \lambda)$.

In subsequent work we shall restrict ourselves to the first approximation. In this approximation

$$a_{+1} = a_{-1} = 1 \quad \text{and} \quad \mu_{+1} = -\mu_{-1} = \frac{1}{\sqrt{3}}. \tag{11}$$

Equation (10) now leads to the following pair of equations:

$$\frac{1}{\sqrt{3}}\frac{\partial I_{+1}}{\partial \tau} - \tfrac{1}{3}\gamma\frac{\partial I_{+1}}{\partial \lambda} - \tfrac{2}{3}\gamma\frac{\partial I_{-1}}{\partial \lambda} = \tfrac{1}{2}(I_{+1} - I_{-1}), \tag{12}$$

and

$$\frac{1}{\sqrt{3}}\frac{\partial I_{-1}}{\partial \tau} + \tfrac{2}{3}\gamma\frac{\partial I_{+1}}{\partial \lambda} + \tfrac{1}{3}\gamma\frac{\partial I_{-1}}{\partial \lambda} = \tfrac{1}{2}(I_{+1} - I_{-1}). \tag{13}$$

It is convenient to introduce the variables

$$x = \tfrac{3}{2}\tau \quad \text{and} \quad y = \frac{3}{2\gamma}(\lambda - \lambda_0), \tag{14}$$

where λ_0 is some suitably chosen constant wave-length. In terms of the variables x and y equations (12) and (13) become

$$\frac{\sqrt{3}}{2}\frac{\partial I_{+1}}{\partial x} - \frac{1}{2}\frac{\partial I_{+1}}{\partial y} - \frac{\partial I_{-1}}{\partial y} = \tfrac{1}{2}(I_{+1} - I_{-1}), \tag{15}$$

and

$$\frac{\sqrt{3}}{2}\frac{\partial I_{-1}}{\partial x} + \frac{\partial I_{+1}}{\partial y} + \frac{1}{2}\frac{\partial I_{-1}}{\partial y} = \tfrac{1}{2}(I_{+1} - I_{-1}). \tag{16}$$

According to the remarks in § 1 it is necessary to solve equations (15) and (16) with the boundary conditions

$$I_{+1}(x_1, y) = \text{a known function of } y$$
$$= \psi(y) \quad \text{(say)} \tag{17}$$

and
$$I_{-1}(0, y) \equiv 0, \tag{18}$$

specifying respectively the known spectral distribution of the outward radiation at the base of the atmosphere at $x = x_1$ and the absence of any radiation incident on the atmosphere at $x = 0$.

3. The reduction to a boundary-value problem

Subtracting equation (16) from (15) we have

$$\frac{\sqrt{3}}{2}\frac{\partial}{\partial x}(I_{+1}-I_{-1}) = \frac{3}{2}\frac{\partial}{\partial y}(I_{+1}+I_{-1}). \tag{19}$$

We can therefore write

$$I_{+1}-I_{-1} = \sqrt{3}\frac{\partial S(x,y)}{\partial y}, \tag{20}$$

and
$$I_{+1}+I_{-1} = \frac{\partial S(x,y)}{\partial x}. \tag{21}$$

The conservation of the net integrated flux of radiation readily follows from equation (20).

Next, adding equations (15) and (16), we have

$$\frac{\sqrt{3}}{2}\frac{\partial}{\partial x}(I_{+1}+I_{-1})+\frac{1}{2}\frac{\partial}{\partial y}(I_{+1}-I_{-1}) = I_{+1}-I_{-1}, \tag{22}$$

or, substituting for $I_{+1}+I_{-1}$ and $I_{+1}-I_{-1}$ according to equations (20) and (21) we have

$$\frac{\partial^2 S}{\partial x^2}+\frac{\partial^2 S}{\partial y^2} = 2\frac{\partial S}{\partial y}. \tag{23}$$

Now put
$$S = e^y f(x,y). \tag{24}$$

Equation (22) then reduces to the standard *elliptic equation* in two variables. We have

$$\frac{\partial^2 f}{\partial x^2}+\frac{\partial^2 f}{\partial y^2}-f = 0. \tag{25}$$

Returning to equations (20), (21) and (24), we find that

$$I_{+1}(x,y) = \tfrac{1}{2}e^y\left[\frac{\partial f}{\partial x}+\sqrt{3}\left(f+\frac{\partial f}{\partial y}\right)\right], \tag{26}$$

and
$$I_{-1}(x,y) = \tfrac{1}{2}e^y\left[\frac{\partial f}{\partial x}-\sqrt{3}\left(f+\frac{\partial f}{\partial y}\right)\right]. \tag{27}$$

The boundary conditions (17) and (18) are therefore equivalent to

$$\left[\frac{\partial f}{\partial x}+\sqrt{3}\left(f+\frac{\partial f}{\partial y}\right)\right]_{x=x_1} = 2e^{-y}\psi(y), \tag{28}$$

and

$$\left[\frac{\partial f}{\partial x}-\sqrt{3}\left(f+\frac{\partial f}{\partial y}\right)\right]_{x=0} \equiv 0. \tag{29}$$

Our problem then is to solve the elliptic equation (25) with the boundary conditions (28) and (29). It is of interest to recall in this connexion that in an analogous problem in the theory of differentially moving atmospheres, we are led to a similar boundary value problem in *hyperbolic equations* (cf. Chandrasekhar 1945).

4. The solution of the boundary-value problem

It will now be shown how the boundary-value problem formulated in § 3 can be solved. The method of solution is based on Green's theorem.

Now Green's theorem as applied to equation (25) is that the integral

$$\int\left(v\frac{\partial f}{\partial x}-f\frac{\partial v}{\partial x}\right)dy-\left(v\frac{\partial f}{\partial y}-f\frac{\partial v}{\partial y}\right)dx$$

around any closed contour vanishes if f and v are any two functions which satisfy equation (25). In applying this theorem to our problem we shall let f stand for the solution satisfying the boundary conditions (28) and (29) and choose v to satisfy certain other conditions to be specified later. We shall further suppose that f tends to zero sufficiently rapidly for $y \to \pm\infty$, for the various integrals which occur in Green's theorem to converge when taken round the closed contour consisting of the pair of infinite lines $x = 0$ and $x = x_1$ extending from $-\infty$ to $+\infty$ and described in the opposite directions. (It will later become apparent that for $\psi(y)$ tending to zero sufficiently rapidly for $y \to \pm\infty$, this will indeed be the case.) Green's theorem applied to such a contour gives

$$\int_{-\infty}^{+\infty}\left(v\frac{\partial f}{\partial x}-f\frac{\partial v}{\partial x}\right)_{x=0}dy = \int_{-\infty}^{+\infty}\left(v\frac{\partial f}{\partial x}-f\frac{\partial v}{\partial x}\right)_{x=x_1}dy. \tag{30}$$

On the other hand, according to equation (29)

$$\left(\frac{\partial f}{\partial x}\right)_{x=0} = \sqrt{3}\left(f+\frac{\partial f}{\partial y}\right)_{x=0}. \tag{31}$$

Inserting this value for $(\partial f/\partial x)_{x=0}$ in the integral on the left-hand side of equation (30), we find, after an integration by parts, that

$$\int_{-\infty}^{+\infty}\left(v\frac{\partial f}{\partial x}-f\frac{\partial v}{\partial x}\right)_{x=0}dy = \int_{-\infty}^{+\infty}f(0,y)\left[\sqrt{3}\left(v-\frac{\partial v}{\partial y}\right)-\frac{\partial v}{\partial x}\right]_{x=0}dy. \tag{32}$$

Similarly, using equation (28) we find that

$$\int_{-\infty}^{+\infty}\left(v\frac{\partial f}{\partial x}-f\frac{\partial v}{\partial x}\right)_{x=x_1}dy = 2\int_{-\infty}^{+\infty}v(x_1,y)\,e^{-y}\,\psi(y)\,dy$$
$$-\int_{-\infty}^{+\infty}f(x_1,y)\left[\sqrt{3}\left(v-\frac{\partial v}{\partial y}\right)+\frac{\partial v}{\partial x}\right]_{x=x_1}dy. \tag{33}$$

Combining equations (30), (32) and (33) we have

$$\int_{-\infty}^{+\infty} f(0,y)\left[\sqrt{3}\left(v-\frac{\partial v}{\partial y}\right)-\frac{\partial v}{\partial x}\right]_{x=0} dy+\int_{-\infty}^{+\infty} f(x_1,y)\left[\sqrt{3}\left(v-\frac{\partial v}{\partial y}\right)+\frac{\partial v}{\partial x}\right]_{x=x_1} dy$$

$$= 2\int_{-\infty}^{+\infty} v(x_1,y)\,e^{-y}\,\psi(y)\,dy. \quad (34)$$

This is our basic formula.

(a) Determination of $f(0,y)$

Let v be so chosen that

$$\left[\sqrt{3}\left(v-\frac{\partial v}{\partial y}\right)+\frac{\partial v}{\partial x}\right]_{x=x_1} = 0. \quad (35)$$

Then equation (34) becomes

$$\int_{-\infty}^{+\infty} f(0,y)\left[\sqrt{3}\left(v-\frac{\partial v}{\partial y}\right)-\frac{\partial v}{\partial x}\right]_{x=0} dy = 2\int_{-\infty}^{+\infty} v(x_1,y)\,e^{-y}\,\psi(y)\,dy. \quad (36)$$

To determine v satisfying the condition (35), we first observe that

$$e^{i\beta y+(1+\beta^2)^{\frac{1}{2}}x},$$

where β is an arbitrary constant, satisfies equation (25). We shall accordingly assume that

$$v = e^{i\beta y}\,[A\cosh x\,\sqrt{(1+\beta^2)}+B\sinh x\,\sqrt{(1+\beta^2)}], \quad (37)$$

where A and B are two constants. For v of this form

$$\left[\sqrt{3}\left(v-\frac{\partial v}{\partial y}\right)+\frac{\partial v}{\partial x}\right]_{x=x_1} = e^{i\beta y}[A\{\sqrt{3}\,.\,(1-i\beta)\cosh\alpha x_1+\alpha\sinh\alpha x_1\}$$

$$+B\{\sqrt{3}\,.\,(1-i\beta)\sinh\alpha x_1+\alpha\cosh\alpha x_1\}], \quad (38)$$

where for the sake of brevity we have written

$$\alpha = \sqrt{(1+\beta^2)}. \quad (39)$$

From equation (38) it follows that with the choice

$$\left.\begin{aligned} A &= \sqrt{3}\,.\,(1-i\beta)\sinh\alpha x_1+\alpha\cosh\alpha x_1 \\ \text{and}\quad B &= -\sqrt{3}\,.\,(1-i\beta)\cosh\alpha x_1-\alpha\sinh\alpha x_1. \end{aligned}\right\} \quad (40)$$

v given by equation (37) will satisfy the required condition (35).

For v chosen in the manner described, it is readily verified that

$$v(x_1,y) = \alpha e^{i\beta y}, \quad (41)$$

and

$$\left[\sqrt{3}\left(v-\frac{\partial v}{\partial y}\right)-\frac{\partial v}{\partial x}\right]_{x=0} = 2(1-i\beta)\,e^{i\beta y}[(2-i\beta)\sinh\alpha x_1+\sqrt{3}\,.\,\alpha\cosh\alpha x_1]. \quad (42)$$

Equation (36) therefore becomes

$$(1-i\beta)\,[(2-i\beta)\sinh\alpha x_1+\sqrt{3}\,.\,\alpha\cosh\alpha x_1]\int_{-\infty}^{+\infty} e^{i\beta y}f(0,y)\,dy$$
$$=\alpha\int_{-\infty}^{+\infty} e^{i\beta y}e^{-y}\psi(y)\,dy. \tag{43}$$

Letting
$$\Psi(\beta)=\frac{1}{\sqrt{(2\pi)}}\int_{-\infty}^{+\infty} e^{i\beta y}e^{-y}\psi(y)\,dy, \tag{44}$$

we have
$$\frac{1}{\sqrt{(2\pi)}}\int_{-\infty}^{+\infty} e^{i\beta y}f(0,y)\,dy=\frac{\Psi(\beta)\sqrt{(1+\beta^2)}}{(1-i\beta)\,(p-i\beta q)}, \tag{45}$$

where
$$\left.\begin{aligned} p&=\sqrt{\{3(1+\beta^2)\}}\cosh x_1\sqrt{(1+\beta^2)}+2\sinh x_1\sqrt{(1+\beta^2)}\\ q&=\sinh x_1\sqrt{(1+\beta^2)}.\end{aligned}\right\} \tag{46}$$
and

Finally, inverting equation (45) we have

$$f(0,y)=\frac{1}{\sqrt{(2\pi)}}\int_{-\infty}^{+\infty}\frac{e^{-i\beta y}\Psi(\beta)\sqrt{(1+\beta^2)}}{(1-i\beta)\,(p-i\beta q)}\,d\beta. \tag{47}$$

Alternatively, we can also write

$$f(0,y)=\frac{1}{\sqrt{(2\pi)}}\int_{-\infty}^{+\infty}\frac{e^{-i\beta y}\Psi(\beta)\,[(p-\beta^2 q)+i\beta(p+q)]}{(p^2+\beta^2q^2)\sqrt{(1+\beta^2)}}\,d\beta. \tag{48}$$

Equation (48) determines f along the line $x=0$. The derivative $(\partial f/\partial x)_{x=0}$ then follows from equation (29).

(b) The determination of $f(x_1,y)$

Let v be now so chosen that

$$\left[\sqrt{3}\left(v-\frac{\partial v}{\partial y}\right)-\frac{\partial v}{\partial x}\right]_{x=0}=0. \tag{49}$$

Equation (34) then gives

$$\int_{-\infty}^{+\infty} f(x_1,y)\left[\sqrt{3}\left(v-\frac{\partial v}{\partial y}\right)+\frac{\partial v}{\partial x}\right]_{x=x_1}dy=2\int_{-\infty}^{+\infty} v(x_1,y)\,e^{-y}\psi(y)\,dy. \tag{50}$$

A solution of equation (25) which satisfies the condition (49) is seen to be

$$v(x,y)=e^{i\beta y}[\alpha\cosh\alpha x+\sqrt{3}\,.\,(1-i\beta)\sinh\alpha x]. \tag{51}$$

For v given by this equation

$$\left[\sqrt{3}\left(v-\frac{\partial v}{\partial y}\right)+\frac{\partial v}{\partial x}\right]_{x=x_1}=2(1-i\beta)\,(p-i\beta q)\,e^{i\beta y}, \tag{52}$$

and
$$v(x_1,y)=\frac{1}{\sqrt{3}}[p+(1-3i\beta)\,q]\,e^{i\beta y}, \tag{53}$$

where p and q have the same meanings as in equations (46). Combining equations (50), (52) and (53) we therefore have

$$\frac{1}{\sqrt{(2\pi)}}\int_{-\infty}^{+\infty} e^{i\beta y} f(x_1, y)\,dy = \frac{1}{\sqrt{3}}\frac{p+(1-3i\beta)\,q}{(1-i\beta)\,(p-i\beta q)}\Psi(\beta), \tag{54}$$

where $\Psi(\beta)$ is defined as in equation (44).

And now inverting equation (54) we have

$$f(x_1, y) = \frac{1}{\sqrt{(6\pi)}}\int_{-\infty}^{+\infty}\frac{e^{-i\beta y}\Psi(\beta)\,[p+(1-3i\beta)\,q]}{(1-i\beta)\,(p-i\beta q)}\,d\beta. \tag{55}$$

Equation (55) determines $f(x_1, y)$ and the derivative $(\partial f/\partial x)_{x=x_1}$ then follows from equation (28).

With our knowledge of f and its derivatives along $x = 0$ and $x = x_1$ completed in this manner, the value of f at any interior point of the infinite strip bounded by the lines $x = 0$ and $x = x_1$ can be determined by a simple application of Green's *formula* (cf. Frank & von Mises 1943).

5. The spectral distribution of the emergent radiation

From the linearity of the equation of transfer and more particularly of the reduced equation (25) it follows that the solution for an arbitrary continuous function $\psi(y)$ can be derived simply from the solution for the case

$$\psi(y) = \delta(y), \tag{56}$$

where $\delta(y)$ is the δ-function of Dirac. Thus, if $I_{+1}(0, y;\, \delta)$ denotes the solution for the emergent radiation for the case when the surface at $x = x_1$ radiates *monochromatically* at a wave-length λ_0 (cf. equation (14)), the solution when the surface radiates with a spectral distribution corresponding to an assigned $\psi(y)$, is given by

$$I_{+1}(0, y;\, \psi) = \int_{-\infty}^{+\infty} I_{+1}(0, y-\eta;\, \delta)\,\psi(\eta)\,d\eta. \tag{57}$$

It might be thought that the assumption of monochromatic emission by the radiating surface is incompatible with our earlier expansion of $I(\tau, \mu, \lambda)$ in a Taylor series (cf. equation (8)). But this is not the case. For, the manner of deriving the solution for an arbitrary ψ in terms of the solution for the case (56) is a mathematical statement strictly concerning the solution of the boundary value problem formulated in § 3 which has no bearing on what has gone before. However, our earlier assumption concerning $I(\tau, \mu, \lambda)$ means now, that physically significant solutions are obtained only after 'smearing' $I_{+1}(0, y;\, \delta)$ by a relatively smooth function $\psi(y)$ according to equation (57). With this understanding we proceed to consider the basic solution $I_{+1}(0, y;\, \delta)$ of our problem.

For $\psi(y) = \delta(y)$ it is apparent that (cf. equation (44))

$$\Psi(\beta) = \frac{1}{\sqrt{(2\pi)}} \tag{58}$$

and equation (48) reduces to

$$f(0,y) = \frac{1}{2\pi}\int_{-\infty}^{+\infty}\frac{e^{-i\beta y}[(p-\beta^2q)+i\beta(p+q)]}{(p^2+\beta^2q^2)\sqrt{(1+\beta^2)}}d\beta, \tag{59}$$

or, equivalently

$$f(0,y) = \frac{1}{\pi}\int_{0}^{+\infty}\frac{(p-\beta^2q)\cos\beta y+(p+q)\beta\sin\beta y}{(p^2+q^2\beta^2)\sqrt{(1+\beta^2)}}d\beta. \tag{60}$$

We are, of course, particularly interested in the distribution with wave-length of the emergent radiation. According to equation (26)

$$I_{+1}(0,y) = \tfrac{1}{2}e^{y}\left[\frac{\partial f}{\partial x}+\sqrt{3}\left(f+\frac{\partial f}{\partial y}\right)\right]_{x=0}, \tag{61}$$

or, using equation (29) we have

$$I_{+1}(0,y) = \sqrt{3}\,.\,e^{y}\left(f+\frac{\partial f}{\partial y}\right)_{x=0}. \tag{62}$$

Now substituting for $f(0,y)$ according to equation (60) we find, after some minor rearranging of the terms, that

$$I_{+1}(0,y;\,\delta) = \frac{\sqrt{3}}{\pi}e^{y}\int_{0}^{\infty}\frac{(p\cos\beta y+\beta q\sin\beta y)\sqrt{(1+\beta^2)}}{(p^2+\beta^2q^2)}d\beta, \tag{63}$$

where it may be recalled that p and q are given by equations (46).

TABLE 1. THE SPECTRAL DISTRIBUTION $I_{+1}(0,y;\,\delta)$ OF THE RADIATION EMERGENT AFTER TRANSMISSION THROUGH VARIOUS OPTICAL THICKNESSES OF AN INCIDENT MONOCHROMATIC FLUX

	$I_{+1}(0,y;\,\delta)$		
y	$\tau_1=\frac{2}{3}$	$\tau_1=\frac{4}{3}$	$\tau_1=2$
− 1·00	0·0141	—	—
− 0·50	0·0685	0·0218	—
0	0·2041	0·0448	0·01257
+ 0·25	0·2635	—	—
+ 0·50	0·2840	—	—
+ 0·75	0·2693	—	—
+ 1·00	0·2361	0·0900	0·02906
+ 1·5	0·1626	0·0957	—
+ 2·0	0·1061	0·0914	0·04210
+ 2·5	0·0681	—	—
+ 3·0	0·0437	0·0707	0·04565
+ 3·5	0·0287	—	—
+ 4·0	0·0181	0·0501	0·04245
+ 4·5	0·0116	—	—
+ 5·0	0·0069	0·0346	—
+ 6·0	—	0·0239	0·03035
+ 7·0	—	0·0173	—
+ 8·0	—	0·0125	0·02002
+10·0	—	—	0·01333
+12·5	—	—	0·00678

For any given x_1 and y, the emergent intensity can therefore be found by evaluating the integral (63). In this manner Mrs Frances H. Breen and the writer have evaluated $I_{+1}(0, y; \delta)$ for various values of y and for values of $x_1 = 1$, 2 and 3. The results of the calculations are summarized in table 1 and further illustrated in figure 1. In examining these results it should be remembered that y measures the wave-length shifts in units of $\frac{2}{3}$ the Compton wave-length ($= 0{\cdot}016$ A). And the values of x_1 for which the calculations have been made correspond to optical thicknesses equal to $\frac{2}{3}$, $\frac{4}{3}$ and 2, respectively.

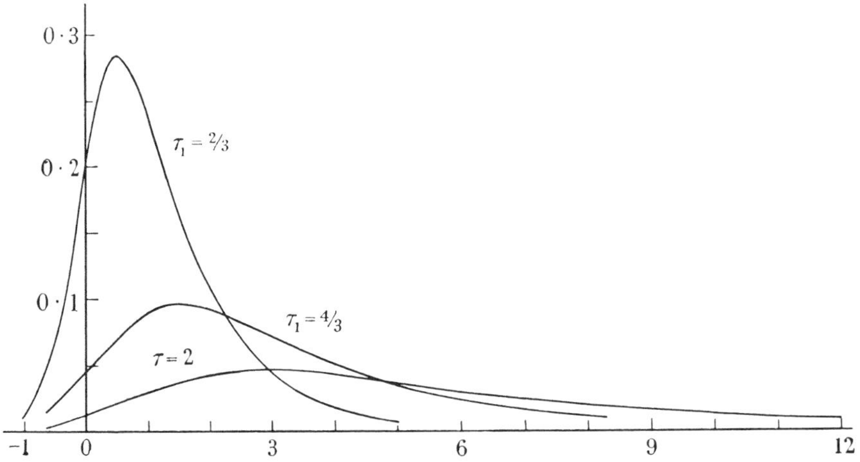

FIGURE 1. The spectral distribution, $I_{+1}(0, y; \delta)$, of the radiation emergent after transmission through various optical thicknesses of an incident monochromatic flux of radiation. The abscissae denote the wave-length shifts in units of $\frac{2}{3}$ Compton wave-length ($= 0{\cdot}016$ A) and the ordinates the emergent intensities in units of the integrated outward intensity at the base of the atmosphere.

From table 1 and figure 1, we observe that the calculated distributions predict finite intensities to the violet as well. An exact solution of the equation of transfer (3) will not, of course, predict this. The error in our treatment was clearly introduced in passing from equation (3) to (9). Indeed, the area to the left of $y = 0$ may be regarded as a measure of the inaccuracy introduced by our approximative treatment of the equation of transfer (3). It is therefore gratifying to note that even for $\tau_1 = \frac{2}{3}$, the contribution to the violet in the total intensity is less than 15 %. A further point to which attention may be drawn in this connexion is that the derived spectral distributions (table 1 and figure 1) are probably less affected by the passage from the exact equation (3) to the approximate equation (9), in the region of large wave-length shifts than in the region of small shifts; for, as τ becomes different from τ_1, the circumstances become more favourable for the use of the Taylor expansion (8) when $\delta\lambda/\gamma$ is large than when $\delta\lambda/\gamma$ is small.

One somewhat unexpected result which emerges from the calculations is the relatively high probability of quite large shifts after transmission through optical thicknesses of order even unity. (According to the remarks in the preceding paragraph, this result is probably not dependent on our approximate method of solution of the equation of transfer.) This result has some interesting astrophysical applications. But it will take us too far outside the scope of this paper to go into them here.

The writer is indebted to Mrs Frances H. Breen for valuable assistance with the numerical work.

References

Chandrasekhar, S. 1945 *Rev. Mod. Phys.* **17**, 138.
Chandrasekhar, S. 1946 *Astrophys. J.* **103**, 351 and **104**, 110.
Chandrasekhar, S. 1947 *Bull. Amer. Math. Soc.* **53**, 640.
Dirac, P. A. M. 1925 *Mon. Not. R. Astr. Soc.* **85**, 825.
Frank, P. & von Mises, R. 1943 *Die Differentialgleichungen der Mechanik und Physik*, **1**, pp. 578–581. New York: Rosenberg.

23

A NEW TYPE OF BOUNDARY-VALUE PROBLEM IN HYPERBOLIC EQUATIONS

By S. CHANDRASEKHAR

Received 10 December 1945

1. *Introduction*. In developing the theory of radiative transfer in expanding atmospheres (Chandrasekhar, 1945*a*, *b*) the author has recently encountered certain novel types of boundary-value problems in hyperbolic equations which appear to merit consideration for their own sake. As related to the equation

$$\frac{\partial^2 f}{\partial x^2}-\frac{\partial^2 f}{\partial y^2}+f=0, \tag{1}$$

the boundary-value problems which occur are of the following general type:

The value of f and its derivatives are assigned for

$$y=0 \quad \text{and} \quad 0\leqslant x\leqslant l_1, \tag{2}$$

i.e. along AB in Fig. 1. Along AD ($x=0$ and $0\leqslant y\leqslant l_2$) and BC ($x=l_1$, $0\leqslant y\leqslant l_2$) we are further given that

$$\left(\frac{\partial f}{\partial x}\right)_{x=0}=\left(\frac{\partial f}{\partial y}\right)_{x=0}+\phi(y) \quad (0\leqslant y\leqslant l_2), \tag{3}$$

and

$$f(l_1,y)=\psi(y) \quad (0\leqslant y\leqslant l_2), \tag{4}$$

where $\phi(y)$ and $\psi(y)$ are two known functions. The problem is to solve equation (1) in the rectangular strip $ABCD$ satisfying the stated boundary conditions. In the papers referred to, a systematic method of solving such boundary-value problems has been outlined and the solutions for the particular problems appearing in the astrophysical contexts obtained. Briefly, the method is the following:

Let the characteristics $x=y$ and $l_1-x=y$ through A and B intersect BC and AD at A' and B' respectively. Further, let AA' and BB' intersect at O. First, since the function and its derivatives are known along AB, we can find the solution inside and on the sides of the triangle OAB directly by Riemann's well-known method (cf. Webster, 1933; or Frank and von Mises, 1943). Next we use the requirement that the solution be unambiguously defined along AO and OB together with the boundary conditions specified along AD and BC to determine f along AB' and $\partial f/\partial x$ along BA' as solutions of certain integral equations of Volterra's type. With the knowledge of the function and its derivatives thus completed along the part $B'ABA'$ of the 'supporting curve' $ABCD$, the solution inside the entire region $O'B'ABA'$ becomes determinate by Riemann's method. In particular, the function and its derivatives along $B'A'$ can be found and the continuation of the solution into the second square $B'A'B''A''$ follows along similar lines.

While the method outlined above shows how the solution satisfying the given boundary conditions can be found in principle, it suffers from the disadvantage that the method of solution depends on solving a succession of Volterra integral equations;

and unless the boundary conditions specified along $DABC$ are specially simple we should not expect to go very far in the explicit carrying out of the solution by this method. It would, therefore, be useful if an alternative method of solution can be devised which will eliminate the need of solving integral equations and reduce the practical problem to one involving (at most!) only quadratures. It is the object of this paper to show how this can be accomplished in terms of certain appropriately constructed Green's functions.

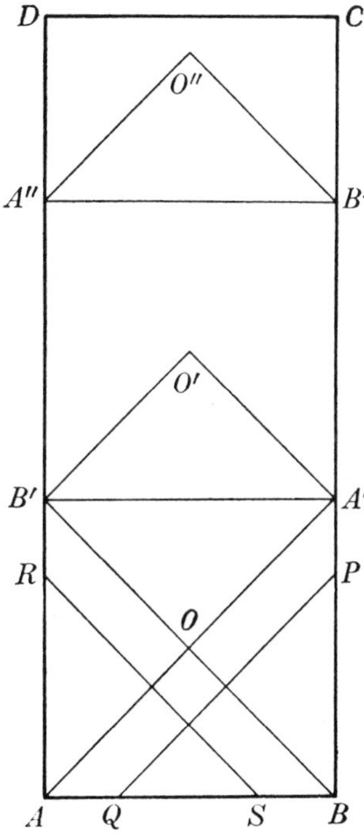

Fig. 1.

2. *The Green's functions* $C(x, y;\ x_1, y_1)$ *and* $C'(x, y;\ x_1, y_1)$. Let $C(x, y;\ x_1, y_1)$ be a solution of the hyperbolic equation (1) which satisfies the following boundary conditions:

$$C(x_1, y;\ x_1, y_1) = 1 \quad (y \leqslant y_1), \tag{5}$$

and

$$C(x, x_1+y_1-x;\ x_1, y_1) = 1 \quad (x \geqslant x_1). \tag{6}$$

In other words, if O represents the point (x_1, y_1), OB the characteristic $x+y = x_1+y_1$ through O and OA the line through O parallel to the y-axis, then the boundary conditions require that C take the value 1 along both OA and OB (see Fig. 2). We shall first show how such a solution can be constructed.

Consider a point $D = (x_1+y_1-\eta, \eta)$ on OB and apply Green's theorem to a contour such as DOE where DE is the characteristic $x-y = x_1+y_1-2\eta$ through D. Now Green's theorem as applied to equation (1) is that the integral

$$\int P\,dy - Q\,dx, \tag{7}$$

where

$$P = v\frac{\partial f}{\partial x} - f\frac{\partial v}{\partial x} \quad \text{and} \quad Q = f\frac{\partial v}{\partial y} - v\frac{\partial f}{\partial y}, \tag{8}$$

around any closed contour vanishes if f and v are any two functions which satisfy equation (1) on and inside the contour. In our particular context we shall let $f = C(x, y;\ x_1, y_1)$ and choose for v the Riemann function which is unity along the two characteristics through D. It is known, quite generally, that

$$v(x, y;\ \xi, \eta) = J_0([(x-\xi)^2-(y-\eta)^2]^{\frac{1}{2}}), \tag{9}$$

where J_0 denotes the Bessel function of order zero.

Remembering that with the choice of the Riemann function for v

$$\int_{x-\xi=\pm(y-\eta)} P\,dy - Q\,dx = \pm C, \tag{10}$$

if the integral on the left-hand side is a line integral along $x-\xi = \pm(y-\eta)$, we find that the application of Green's theorem to the contour DOE gives

$$\int_{2\eta-y_1}^{y_1}\left(v\frac{\partial C}{\partial x} - C\frac{\partial v}{\partial x}\right)_{x=x_1} dy = 0. \tag{11}$$

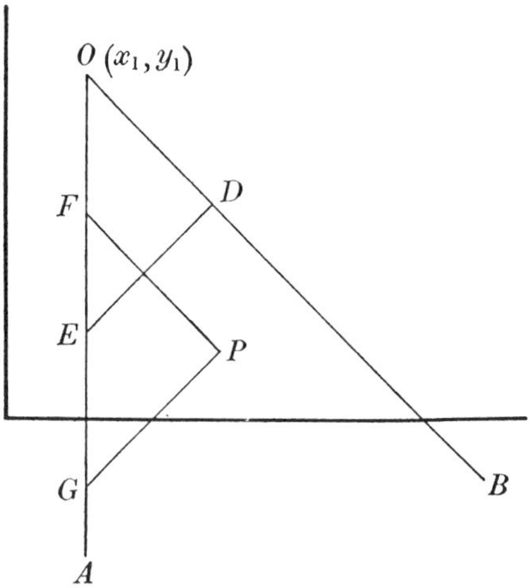

Fig. 2.

Since $C = 1$ along AO and

$$\left(\frac{\partial v}{\partial x}\right)_{x=x_1} = J_1([(y_1-\eta)^2-(y-\eta)^2]^{\frac{1}{2}})\frac{y_1-\eta}{[(y_1-\eta)^2-(y-\eta)^2]^{\frac{1}{2}}}, \tag{12}$$

equation (11) reduces to

$$\int_{2\eta-y_1}^{y_1} J_0([(y_1-\eta)^2-(y-\eta)^2]^{\frac{1}{2}})\,F(y)\,dy$$
$$= \int_{2\eta-y_1}^{y_1} J_1([(y_1-\eta)^2-(y-\eta)^2]^{\frac{1}{2}})\frac{(y_1-\eta)\,dy}{[(y_1-\eta)^2-(y-\eta)^2]^{\frac{1}{2}}}, \tag{13}$$

where we have written

$$\left(\frac{\partial C}{\partial x}\right)_{x=x_1} = F(y). \tag{14}$$

If we now write
$$\zeta = y_1 - y, \quad z = y_1 - \eta, \tag{15}$$
and
$$\chi(\zeta) = F(y_1 - \zeta), \tag{16}$$
equation (13) becomes
$$\int_0^{2z} J_0([z^2-(z-\zeta)^2]^{\frac{1}{2}})\,\chi(\zeta)\,d\zeta = \int_0^{2z} \frac{J_1([z^2-(z-\zeta)^2]^{\frac{1}{2}})}{[z^2-(z-\zeta)^2]^{\frac{1}{2}}}\,z\,d\zeta. \tag{17}$$

To evaluate the integral on the right-hand side of equation (17) we set
$$z - \zeta = z\cos\vartheta, \tag{18}$$
and obtain
$$z\int_0^{\pi} J_1(z\sin\vartheta)\,d\vartheta. \tag{19}$$
Replacing $J_1(z\sin\vartheta)$ in (19) by its equivalent series expansion and integrating term by term we obtain
$$\begin{aligned} z\int_0^{\pi} J_1(z\sin\vartheta)\,d\vartheta &= 2z\sum_{m=0}^{\infty}(-1)^m\frac{(\frac{1}{2}z)^{2m+1}}{m!\,\Gamma(m+2)}\int_0^{\frac{1}{2}\pi}\sin^{2m+1}\vartheta\,d\vartheta \\ &= 2\sum_{m=0}^{\infty}(-1)^m\frac{z^{2m+2}}{(2m+2)!} \\ &= 2-2\sum_{m=0}^{\infty}(-1)^m\frac{z^{2m}}{(2m)!} \\ &= 2(1-\cos z). \end{aligned} \tag{20}$$
Equation (17) thus reduces to the form
$$1-\cos z = \frac{1}{2}\int_0^{2z} J_0([z^2-(z-\zeta)^2]^{\frac{1}{2}})\,\chi(\zeta)\,d\zeta, \tag{21}$$
which is seen to be an integral equation of Volterra's type for $\chi(\zeta)$.

To solve the integral equation (21) we apply a Laplace transformation to this equation. Thus multiplying equation (21) by e^{-sz} and integrating over z from 0 to ∞ we obtain (Chandrasekhar, 1945*a*, eqs. [90]–[98])
$$\frac{2}{s^2+1} = \int_0^{\infty}\chi(\zeta)\exp\left[-\tfrac{1}{2}\zeta(s+s^{-1})\right]d\zeta. \tag{22}$$
If we now let
$$s + s^{-1} = 2u, \tag{23}$$
or equivalently
$$s = u + \sqrt{(u^2-1)}, \tag{24}$$
equation (22) becomes
$$\int_0^{\infty} e^{-\zeta u}\,\chi(\zeta)\,d\zeta = \frac{u-\sqrt{(u^2-1)}}{u}. \tag{25}$$
But it can be readily shown that the Laplace transform of the Bessel integral
$$Ii_1(\zeta) = \int_0^{\zeta}\frac{I_1(t)}{t}\,dt \tag{26}$$
is precisely the quantity on the right-hand side of equation (25). Accordingly
$$\chi(\zeta) = Ii_1(\zeta). \tag{27}$$
In other words, we have shown that (cf. eq. (16))
$$\left(\frac{\partial C}{\partial x}\right)_{x=x_1} = Ii_1(y_1-y) = \int_0^{y_1-y}\frac{I_1(t)}{t}\,dt. \tag{28}$$

With $(\partial C/\partial x)$ thus determined along AO the solution in the entire triangular area AOB becomes determinate. Thus, applying Green's theorem to a contour such as PFG, where $P = (x_1+\xi, \eta)$ and PF and PG are the two characteristics through P, and choosing for v the Riemann function $J_0([(x-x_1-\xi)^2-(y-\eta)^2]^{\frac{1}{2}})$, we readily find that

$$C(x_1+\xi, \eta;\ x_1, y_1)-1 = \frac{1}{2}\int_{\eta-\xi}^{\eta+\xi} J_0([\xi^2-(y-\eta)^2]^{\frac{1}{2}})\, Ii_1(y_1-y)\, dy - \tfrac{1}{2}\xi\int_{\eta-\xi}^{\eta+\xi}\frac{J_1([\xi^2-(y-\eta)^2]^{\frac{1}{2}})}{[\xi^2-(y-\eta)^2]^{\frac{1}{2}}}\,dy. \tag{29}$$

The second term on the right-hand side has the value $1-\cos\xi$ (cf. eq. (20)). Hence

$$C(x_1+\xi, \eta;\ x_1, y_1) = \cos\xi+\frac{1}{2}\int_{\eta-\xi}^{\eta+\xi} J_0([\xi^2-(y-\eta)^2]^{\frac{1}{2}})\, Ii_1(y_1-y)\, dy. \tag{30}$$

Thus the solution $C(x, y;\ x_1, y_1)$ satisfying the boundary conditions (5) and (6) can be expressed in the form

$$C(x, y;\ x_1, y_1) = \cos(x-x_1)+\tfrac{1}{2}(x-x_1)\int_0^{\pi} J_0([x-x_1]\sin\vartheta) \times Ii_1(y_1-y-[x-x_1]\cos\vartheta)\sin\vartheta\, d\vartheta. \tag{31}$$

To facilitate the numerical evaluation of $C(x, y;\ x_1, y_1)$ for practical purposes we have provided a table of the Bessel integral Ii_1 in the form of an appendix to this paper.

For later reference we may note here that according to equations (28) and (31)

$$\frac{\partial C}{\partial y} = -\tfrac{1}{2}(x-x_1)\int_0^{\pi}\frac{J_0([x-x_1]\sin\vartheta)\, I_1(y_1-y-[x-x_1]\cos\vartheta)}{y_1-y-(x-x_1)\cos\vartheta}\sin\vartheta\, d\vartheta. \tag{32}$$

A related solution $C'(x, y;\ x_1, y_1)$ of the hyperbolic equation which we shall find useful in our subsequent work is the one which satisfies the boundary conditions

$$C'(x_1, x;\ x_1, y_1) = 1 \quad (y \leqslant y_1), \tag{33}$$

and

$$C'(x, x-x_1+y_1;\ x_1, y_1) = 1 \quad (x \leqslant x_1). \tag{34}$$

It is evident that

$$C'(x_1-x, y;\ x_1, y_1) = C(x-x_1, y;\ x_1, y_1). \tag{35}$$

3. *The Green's function* $\Gamma(x, y;\ x_1, y_1)$ *and* $\Gamma'(x, y;\ x_1, y_1)$. It is evident that the function $\Gamma(x, y;\ x_1, y_1)$ defined in terms of $C(x, y;\ x_1, y_1)$, according to the formula

$$\Gamma(x, y;\ x_1, y_1) = \frac{\partial C}{\partial x}-\frac{\partial C}{\partial y}, \tag{36}$$

is a solution of the hyperbolic equation (1) which vanishes along the characteristic $x+y = x_1+y_1$ through (x_1, y_1). For the quantity on the right-hand side of equation (36) is proportional to the total derivative of C along $x+y = x_1+y_1$, and since C is constant along this line it must vanish. Thus

$$\Gamma(x, x_1+y_1-x;\ x_1, y_1) = 0 \quad (x \geqslant x_1). \tag{37}$$

Moreover, along $x = x_1$

$$\Gamma(x_1, y;\ x_1, y_1) = \left(\frac{\partial C}{\partial x}\right)_{x=x_1}-\left(\frac{\partial C}{\partial y}\right)_{x=x_1}. \tag{38}$$

But $(\partial C/\partial y)_{x=x_1} = 0$ (cf. eq. (5)). Hence

$$\Gamma(x_1, y;\ x_1, y_1) = \left(\frac{\partial C}{\partial x}\right)_{x=x_1} = Ii_1(y_1 - y). \tag{39}$$

Further,

$$\frac{\partial \Gamma}{\partial x} = \frac{\partial^2 C}{\partial x^2} - \frac{\partial^2 C}{\partial x\, \partial y}. \tag{40}$$

Since C satisfies equation (1) we can rewrite the foregoing equation in the form

$$\frac{\partial \Gamma}{\partial x} = \frac{\partial^2 C}{\partial y^2} - C - \frac{\partial^2 C}{\partial x\, \partial y} = -\frac{\partial}{\partial y}\left(\frac{\partial C}{\partial x} - \frac{\partial C}{\partial y}\right) - C, \tag{41}$$

or

$$\frac{\partial \Gamma}{\partial x} = -\frac{\partial \Gamma}{\partial y} - C. \tag{42}$$

In particular

$$\left(\frac{\partial \Gamma}{\partial x}\right)_{x=x_1} = -\left(\frac{\partial \Gamma}{\partial y}\right)_{x=x_1} - 1. \tag{43}$$

In an analogous relation to $C'(x, y;\ x_1, y_1)$ we can define the function

$$\Gamma'(x, y;\ x_1, y_1) = \frac{\partial C'}{\partial x} + \frac{\partial C'}{\partial y}, \tag{44}$$

which vanishes along the characteristic $x - y = x_1 - y_1$ through (x_1, y_1). For this function we have the relations

$$\Gamma(x_1, y;\ x_1, y_1) = -Ii_1(y_1 - y), \tag{45}$$

and

$$\left(\frac{\partial \Gamma'}{\partial x}\right)_{x=x_1} = \left(\frac{\partial \Gamma'}{\partial y}\right)_{x=x_1} - 1. \tag{46}$$

4. *The determination of f along AB'* ($x = 0$, $0 \leqslant y \leqslant l_1$). Returning to the boundary-value problem formulated in § 1 it is evident that if we can determine from the specified boundary conditions the values of f along AB' ($x = 0$ and $0 \leqslant y \leqslant l_1$) and of $\partial f/\partial x$ along BA' ($x = l_1$, $0 \leqslant y \leqslant l_1$), the solution in the first square $B'ABA'$ (see Fig. 1) will become determinate, and the further continuation of the solution into the succeeding squares will follow similarly. We shall now show how in terms of the Green's functions constructed in the preceding sections we can determine f along AB' and $\partial f/\partial x$ along BA' by processes which involve only quadratures.

Considering first the determination of f along AB' we shall apply Green's theorem to a contour such as RAS where $R = (0, \eta)$ is a point on AB' and RS is the characteristic $x + y = \eta$ through R. Further, in applying Green's theorem to this contour we shall choose $\Gamma(x, y;\ 0, \eta)$ as the Green's function v.

Since $\Gamma(x, y;\ 0, \eta)$ vanishes identically along RS (cf. eq. (37)), it is readily seen that in this case Green's theorem gives

$$\int_0^\eta \left(\Gamma\frac{\partial f}{\partial x} - f\frac{\partial \Gamma}{\partial x}\right)_{x=0} dy + \int_0^\eta \left(f\frac{\partial \Gamma}{\partial y} - \Gamma\frac{\partial f}{\partial y}\right)_{y=0} dx = 0. \tag{47}$$

Now, according to equations (3) and (43),

$$\begin{aligned}\left(\Gamma\frac{\partial f}{\partial x} - f\frac{\partial \Gamma}{\partial x}\right)_{x=0} &= \Gamma(0, y;\ 0, \eta)\left[\left(\frac{\partial f}{\partial y}\right)_{x=0} + \phi(y)\right] + f(0, y)\left[\left(\frac{\partial \Gamma}{\partial y}\right)_{x=0} + 1\right] \\ &= \frac{\partial}{\partial y}[\Gamma(0, y;\ 0, \eta) f(0, y)] + \Gamma(0, y;\ 0, \eta)\phi(y) + f(0, y).\end{aligned} \tag{48}$$

Hence

$$\int_0^\eta [P]_{x=0}\,dy = -\Gamma(0,0;\ 0,\eta)f(0,0)+\int_0^\eta \Gamma(0,y;\ 0,\eta)\phi(y)\,dy+\int_0^\eta f(0,y)\,dy, \tag{49}$$

and equation (47) reduces to the form

$$\int_0^\eta f(0,y)\,dy = Ii_1(\eta)f(0,0)-\int_0^\eta \Gamma(0,y;\ 0,\eta)\phi(y)\,dy$$

$$-\int_0^\eta \left[f(x,0)\left(\frac{\partial\Gamma}{\partial y}\right)_{y=0}-\Gamma(x,0;\ 0,\eta)\left(\frac{\partial f}{\partial y}\right)_{y=0}\right]dx. \tag{50}$$

It is now seen that all the integrals on the right-hand side of this equation can be evaluated in terms of known quantities. Accordingly, by differentiating equation (50) with respect to η we can obtain $f(0,y)$ for $0\leqslant y\leqslant l_1$.

5. *The determination of* $\partial f/\partial x$ *along* BA' ($x = l_1$, $0\leqslant y\leqslant l_1$). Turning next to the determination of $\partial f/\partial x$ along BA' we shall apply Green's theorem to a contour such as PQB where $P = (l_1,\eta)$ is a point on BA' and PQ is the characteristic $x-y = l_1-\eta$ through P. But we shall now use as the Green's function $C'(x,y;\ x_1,y_1)$ which is unity along PQ and PB. Since C' is unity along PQ it is apparent that

$$\int_{PQ} P\,dy-Q\,dx = f. \tag{51}$$

Green's theorem therefore gives

$$f(l_1-\eta,0)-f(l_1,\eta)-\int_{l_1-\eta}^{l_1}\left(f\frac{\partial C'}{\partial y}-C'\frac{\partial f}{\partial y}\right)_{y=0}dx+\int_0^\eta\left(C'\frac{\partial f}{\partial x}-f\frac{\partial C'}{\partial x}\right)_{x=l_1}dy = 0. \tag{52}$$

Since C' is unity along PB and

$$\left(\frac{\partial C'}{\partial x}\right)_{x=l_1} = -Ii_1(\eta-y), \tag{53}$$

equation (52) can be rewritten in the form

$$\int_0^\eta\left(\frac{\partial f}{\partial x}\right)_{x=l_1}dy = \psi(\eta)-f(l_1-\eta,0)-\int_0^\eta \psi(y)\,Ii_1(\eta-y)\,dy$$

$$+\int_{l_1-\eta}^{l_1}\left[f(x,0)\left(\frac{\partial C'}{\partial y}\right)_{y=0}-C'(x,0;\ l_1,\eta)\left(\frac{\partial f}{\partial y}\right)_{y=0}\right]dx. \tag{54}$$

Again, all the quantities on the right-hand side of equation (54) can be determined in terms of the specified boundary conditions. And differentiating this equation with respect to η we can obtain $(\partial f/\partial x)_{x=l_1}$ for $0\leqslant y\leqslant l_1$.

According to our remarks in § 4, with the determination of f along AB' and $\partial f/\partial x$ along BA' we have formally solved the boundary-value problem formulated in § 1. It will be further noted that we have reduced the problem to one involving only quadratures.

6. *Some related questions.* In some ways the method of solution described in the preceding sections is unexpected in so far as it has been possible to avoid the solution of a series of Volterra integral equations which would otherwise be necessary. Conversely, it follows that *all* the Volterra integral equations which may arise in following the systematic method of solution of the earlier papers (Chandrasekhar, 1945*a*, *b*) can

in principle be solved by quadratures only involving the Green's functions constructed in §§ 2 and 3. It is, moreover, possible that in this manner we may be able to establish integral relations involving the Bessel functions which it may be difficult to prove in any other way. We may give two examples of such relations.

In one of the papers we have referred to (Chandrasekhar, 1945*a*) it has been shown that for the boundary conditions

$$\left.\begin{aligned} &f = 1, \quad \frac{\partial f}{\partial x} = \frac{\partial f}{\partial y} = 0 && \text{for} \quad y = 0 \quad \text{and} \quad 0 \leqslant x \leqslant l_1, \\ &\frac{\partial f}{\partial x} = \frac{\partial f}{\partial y} && \text{for} \quad x = 0 \quad \text{and} \quad 0 \leqslant y \leqslant l_2, \\ &f = e^y && \text{for} \quad x = l_1 \quad \text{and} \quad 0 \leqslant y \leqslant l_2, \end{aligned}\right\} \tag{55}$$

we have

$$f(0,y) = I_0(y) \quad \text{and} \quad \left(\frac{\partial f}{\partial x}\right)_{x=x_1} = I_0(y) \quad (0 \leqslant y \leqslant l_1). \tag{56}$$

With the foregoing substitutions equations (50) and (54) become

$$\int_0^\eta I_0(y)\,dy = \int_0^\eta C(x,0;\ 0,\eta)\,dx, \tag{57}$$

and

$$\int_0^\eta I_0(y)\,dy = e^\eta - 1 - \int_0^\eta e^y Ii_1(\eta - y)\,dy + \int_{l_1-\eta}^{l_1} \left(\frac{\partial C'}{\partial y}\right)_{y=0} dx, \tag{58}$$

which are essentially integral relations involving the Bessel functions.

Finally, some remarks may be made on the possibilities of further extensions of the method described in this paper. One generalization is fairly obvious, namely when the function and its derivatives are specified on a part of the y-axis, say, for $x = 0$ and $0 \leqslant y \leqslant l_2$ while along the lines $y = 0$ and $y = l_2$ we are given either the function alone or a relation between its derivatives. Under these conditions the solution can be found in an analogous manner in terms of Green's functions similar to those constructed in §§ 2 and 3. Thus the Green's function which will play the same role as $C(x,y;\ x_1,y_1)$ is the solution $D(x,y;\ x_1,y_1)$ of equation (1) which satisfies the boundary conditions

$$D(x,y_1;\ x_1,y_1) = 1 \quad (x \geqslant x_1), \tag{59}$$

and

$$D(x,x_1+y_1-x;\ x_1,y_1) = 1 \quad (x \geqslant x_1). \tag{60}$$

Such a solution can be readily found. For, it can be shown that

$$\left(\frac{\partial D}{\partial y}\right)_{y=y_1} = -\int_0^{x-x_1} \frac{J_1(t)}{t}\,dt. \tag{61}$$

However, the question as to whether the method can be extended for more general types of hyperbolic equations than the one considered in this paper is not so easily answered. But we hope to return to this and related matters on a later occasion.

In conclusion I wish to record my indebtedness to Professor Fritz John who first suggested the investigation of the boundary value problem formulated in § 1 along the lines of this paper.

A table of the Bessel integral $Ii_1(x)$

The following table of the Bessel integral $Ii_1(x)$ was computed by Mrs Frances Herman Breen and the writer.

x	Ii_1	x	Ii_1	x	Ii_1
0	0	5·05	7·76082	7·60	47·3133
0·1	0·050021	5·10	8·01771	7·65	49·1544
0·2	0·100167	5·15	8·28409	7·70	51·0718
0·3	0·150564	5·20	8·56032	7·75	53·0687
0·4	0·201339	5·25	8·84681	7·80	55·1485
0·5	0·252621	5·30	9·14397	7·85	57·3148
0·6	0·304541	5·35	9·45223	7·90	59·5713
0·7	0·357234	5·40	9·77206	7·95	61·9219
0·8	0·410839	5·45	10·10391	8·00	64·3706
0·9	0·465499	5·50	10·4483	8·05	66·9217
1·0	0·521362	5·55	10·8057	8·10	69·5796
1·1	0·578583	5·60	11·1767	8·15	72·3490
1·2	0·637324	5·65	11·5617	8·20	75·2346
1·3	0·697754	5·70	11·9615	8·25	78·2416
1·4	0·760051	5·75	12·3766	8·30	81·3750
1·5	0·824403	5·80	12·8076	8·35	84·6405
1·6	0·891008	5·85	13·2552	8·40	88·0438
1·7	0·960076	5·90	13·7201	8·45	91·5909
1·8	1·031831	5·95	14·2029	8·50	95·2879
1·9	1·10651	6·00	14·7045	8·55	99·1416
2·0	1·18437	6·05	15·2256	8·60	103·1586
2·1	1·26567	6·10	15·7669	8·65	107·346
2·2	1·35070	6·15	16·3294	8·70	111·712
2·3	1·43978	6·20	16·9140	8·75	116·263
2·4	1·53323	6·25	17·5215	8·80	121·008
2·5	1·63141	6·30	18·1528	8·85	125·955
2·6	1·73470	6·35	18·8091	8·90	131·113
2·7	1·84351	6·40	19·4913	8·95	136·492
2·8	1·95827	6·45	20·2005	9·00	142·101
2·9	2·07946	6·50	20·9378	9·05	147·950
3·0	2·20759	6·55	21·7045	9·10	154·050
3·1	2·34321	6·60	22·5017	9·15	160·412
3·2	2·48691	6·65	23·3308	9·20	167·046
3·3	2·63932	6·70	24·1930	9·25	173·967
3·4	2·80114	6·75	25·0898	9·30	181·185
3·5	2·97311	6·80	26·0226	9·35	188·713
3·6	3·15604	6·85	26·9929	9·40	196·567
3·7	3·35079	6·90	28·0024	9·45	204·760
3·8	3·55830	6·95	29·0526	9·50	213·306
3·9	3·77958	7·00	30·1453	9·55	222·222
4·0	4·01574	7·05	31·2823	9·60	231·524
4·1	4·26797	7·10	32·4655	9·65	241·230
4·2	4·53754	7·15	33·6968	9·70	251·356
4·3	4·82586	7·20	34·9783	9·75	261·921
4·4	5·13443	7·25	36·3122	9·80	272·946
4·5	5·46489	7·30	37·7005	9·85	284·450
4·6	5·81901	7·35	39·1457	9·90	296·455
4·7	6·19871	7·40	40·6502	9·95	308·982
4·8	6·60608	7·45	42·2165	10·00	322·056
4·9	7·04337	7·50	43·8473		
5·0	7·51303	7·55	45·5452		

x	Ii_1	$e^{-x}Ii_1$	x	Ii_1	$e^{-x}Ii_1$
10·00	322·056	0·0146213	15·1	26784·8	0·00741380
10·1	349·941	0·0143754	15·2	29290·1	0·00733577
10·2	380·316	0·0141365	15·3	32032·3	0·00725909
10·3	413·409	0·0139042	15·4	35033·8	0·00718377
10·4	449·468	0·0136784	15·5	38319·4	0·00710975
10·5	488·764	0·0134588	15·6	41916·2	0·00703700
10·6	531·593	0·0132452	15·7	45853·8	0·00696550
10·7	578·278	0·0130373	15·8	50164·9	0·00689521
10·8	629·174	0·0128348	15·9	54885·2	0·00682610
10·9	684·666	0·0126377	16·0	60053·7	0·00675815
11·0	745·178	0·0124457	16·1	65713·4	0·00669133
11·1	811·170	0·0122587	16·2	71911·2	0·00662561
11·2	883·148	0·0120763	16·3	78698·9	0·00656098
11·3	961·664	0·0118986	16·4	86132·7	0·00649739
11·4	1047·320	0·0117253	16·5	94274·9	0·00643483
11·5	1140·78	0·0115562	16·6	103193·2	0·00637327
11·6	1242·76	0·0113912	16·7	112962	0·00631270
11·7	1354·05	0·0112302	16·8	123664	0·00625309
11·8	1475·52	0·0110731	16·9	135387	0·00619442
11·9	1608·10	0·0109197	17·0	148231	0·00613666
12·0	1752·83	0·0107698	17·1	162303	0·00607981
12·1	1910·85	0·0106234	17·2	177721	0·00602383
12·2	2083·38	0·0104803	17·3	194614	0·00596871
12·3	2271·78	0·0103405	17·4	213126	0·00591443
12·4	2477·52	0·0102039	17·5	233412	0·00586098
12·5	2702·23	0·0100703	17·6	255643	0·00580833
12·6	2947·67	0·00993958	17·7	280007	0·00575647
12·7	3215·78	0·00981175	17·8	306709	0·00570538
12·8	3508·69	0·00968669	17·9	335975	0·00565504
12·9	3828·71	0·00956431	18·0	368054	0·00560545
13·0	4178·39	0·00944453	18·1	403216	0·00555658
13·1	4560·50	0·00932727	18·2	441760	0·00550842
13·2	4978·08	0·00921244	18·3	484014	0·00546095
13·3	5434·48	0·00910000	18·4	530335	0·00541417
13·4	5933·34	0·00898986	18·5	581119	0·00536806
13·5	6478·64	0·00888195	18·6	636797	0·00532260
13·6	7074·77	0·00877622	18·7	697843	0·00527777
13·7	7726·52	0·00867260	18·8	764779	0·00523359
13·8	8439·12	0·00857103	18·9	838174	0·00519001
13·9	9218·31	0·00847146	19·0	918656	0·00514704
14·0	10070·40	0·00837383	19·1	1006913	0·00510466
14·1	11002·2	0·00827807	19·2	1103699	0·00506286
14·2	12021·4	0·00818415	19·3	1209843	0·00502164
14·3	13136·2	0·00809203	19·4	1326255	0·00498098
14·4	14355·5	0·00800162	19·5	1453933	0·00494086
14·5	15689·4	0·00791293	19·6	1593971	0·00490127
14·6	17148·7	0·00782587	19·7	1747574	0·00486222
14·7	18745·3	0·00774041	19·8	1916061	0·00482368
14·8	20492·2	0·00765652	19·9	2100881	0·00478566
14·9	22403·8	0·00757413	20·0	2303632	0·00474814
15·0	24495·6	0·00749324			

REFERENCES

CHANDRASEKHAR, S. *Rev. Mod. Phys.* 17 (1945*a*), 138.
CHANDRASEKHAR, S. *Astrophys. J.* 102 (1945*b*), 402.
FRANK, P. and VON MISES, R. *Die Differential Gleichungen der Mechanik und Physik* (New York: Rosenberg (1943)), pp. 779–817.
WEBSTER, A. G. *Partial Differential Equations of Mathematical Physics* (New York: Stechert (1933)), pp. 160–88.

YERKES OBSERVATORY
WILLIAMS BAY, WISCONSIN

5. *Review Articles*

24

THE TRANSFER OF RADIATION IN STELLAR ATMOSPHERES

S. CHANDRASEKHAR

1. **Introductory remarks.** The advances in the various branches of theoretical physics have often resulted in the creation of new mathematical disciplines: disciplines which, in their way, are as characteristic of the subjects as the physical phenomena they are devoted to. That several of the earlier Gibbs lectures should illustrate this point is not surprising since the example provided by the scientific writings of Gibbs is indeed the most striking in this connection. With your permission, I would like to illustrate the same thing this evening, on a more modest level, by considering the recent advances in theoretical astrophysics. More specifically, I should like to show how astrophysical studies relating to the transfer of radiation in stellar atmospheres have led to some characteristic developments in the theory of integro-differential and functional equations.

2. **The physical problem.** In a general way, the preoccupation of the astrophysicist with the transport of radiant energy in an atmosphere which absorbs, emits and scatters radiation is not difficult to understand. For, it is in the characteristics of the radiation emergent from a star—in its variation over the stellar disc, and in its distribution with wavelength through the spectrum—that he has to seek for information concerning the constitution and structure of the stellar atmosphere.

3. **Definitions. Intensity.** Now, the basic equation which governs the transfer of radiation in any medium is the so-called *equation of transfer*. This is the equation which governs the variation of radiant intensity in terms of the local properties of the medium. But, first, I should explain that for most purposes it is sufficient to characterize the radiation field by the specific intensity I; in terms of this quantity the amount of radiant energy crossing an element of area $d\sigma$, in a direction inclined at an angle ϑ to the normal, and confined to an element of solid angle $d\omega$, and in time dt can be expressed in the form

$$I \cos \vartheta d\sigma d\omega dt. \tag{1}$$

Accordingly, to characterize a radiation field completely, the specific

The twentieth Josiah Willard Gibbs lecture delivered at Swarthmore, Pennsylvania, December 26, 1946, under the auspices of the American Mathematical Society; received by the editors January 13, 1947.

Reprinted from "The Transfer of Radiation in Stellar Atmospheres" *Bulletin of the American Mathematical Society,* (1947), Volume 53, Pages 641–711, by permission of the American Mathematical Society.

intensity will have to be defined at every point and for every direction through a point.

4. **The equation of transfer.** To establish the equation which the intensity must satisfy in a stationary radiation field (that is, in a radiation field which is constant with time) we consider the equilibrium with respect to the transport of energy of a small cylindrical element of cross section $d\sigma$ and height ds in the medium. From the definition of intensity, it follows that the difference in the radiant energy crossing the two faces normally, in a time dt and confined to an element of solid angle $d\omega$, is

$$\frac{dI}{ds}\, ds\, d\sigma\, d\omega\, dt. \tag{2}$$

Here I denotes the specific intensity in the direction of s. This difference (2) must arise from the excess of emission over absorption of radiant energy in time dt and the element of solid angle $d\omega$ considered. Now the absorption of radiation by an element of mass is expressed in terms of the *mass absorption coefficient* denoted by κ, such that, of the energy $I d\sigma d\omega dt$ normally incident on the face $d\sigma$, the amount absorbed in time dt is

$$\kappa \rho ds \times I d\sigma d\omega dt, \tag{3}$$

where ρ denotes the density of the material. Similarly, the emission coefficient j is defined in such a way that the element of mass $\rho d\sigma ds$ emits in directions confined to $d\omega$ and in time dt an amount of radiant energy given by

$$j \rho d\sigma ds d\omega dt. \tag{4}$$

It should be noted that, according to our definition, the emission coefficient j can very well depend on the direction at each point, differing in this respect from the absorption coefficient which is a function of position only.

From our earlier remarks and the expressions (3) and (4) for absorption and emission, we have the equation of transfer

$$\frac{dI}{\rho ds} = -\kappa I + j, \tag{5}$$

expressing the conservation of radiant energy.

5. **The equation of transfer for plane-parallel atmospheres. The**

source function. In our further discussion we shall restrict ourselves to transfer problems in a semi-infinite plane-parallel atmosphere (Fig. 1). Let z measure distances normal to the plane of stratification and ϑ and φ the corresponding polar angles. The specific intensity I in such an atmosphere will, in general, depend on all three variables, z, ϑ

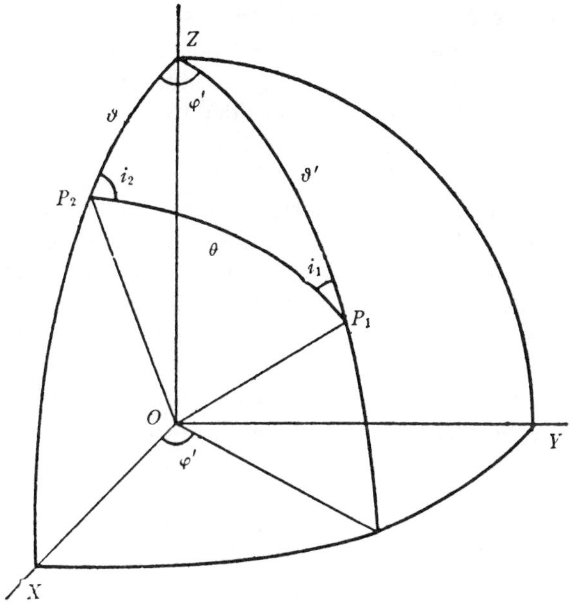

FIG. 1.

and φ. For such an atmosphere the equation of transfer can be written in the form

$$\cos\vartheta \frac{dI}{\rho dz} = -\kappa I + j. \tag{6}$$

It is convenient at this stage to introduce the *optical thickness*

$$\tau = \int_z^\infty \kappa\rho dz, \tag{7}$$

measured from the boundary *inward* as our variable instead of z. In terms of τ, the equation of transfer becomes

$$\cos\vartheta \frac{dI}{d\tau} = I(\tau, \vartheta, \varphi) - \mathfrak{J}(\tau, \vartheta, \varphi), \tag{8}$$

where we have written

$$\mathfrak{J} = \frac{j}{\kappa}. \tag{9}$$

This is the *source function.*

6. **The source function for a scattering atmosphere. The phase function.** Different physical problems lead to different *functional* dependences of the source function on I. To illustrate the nature of such functional dependences we shall consider the case of a *scattering atmosphere.* In such an atmosphere, absorption arises simply on account of the fact that when a pencil of radiation is incident on an element of gas, a determinate part of the radiant energy is scattered in other directions. In the same way, the radiation in any given direction can be reinforced by the scattering of radiation from other directions into the particular direction considered. It is now evident that, to formulate quantitatively the concept of scattering, we must specify the angular distribution of the scattered radiation when a pencil of radiation is incident on an element of gas. This angular distribution is generally given by a *phase function* $p(\cos \Theta)$ in such a way that

$$p(\cos \Theta) \frac{d\omega}{4\pi} \tag{10}$$

governs the probability that radiation will be scattered in a direction inclined at an angle Θ with the direction of incidence. If, as we have assumed, scattering is the only process by which radiation and matter interact with each other, then it is evident that

$$\int p(\cos \Theta) \frac{d\omega}{4\pi} = 1 \tag{11}$$

when the integration is extended over the complete sphere. The corresponding source function is

$$\begin{aligned} \mathfrak{J}(\tau, \vartheta, \varphi) = \frac{1}{4\pi} \int_0^{\pi} \int_0^{2\pi} & I(\tau, \vartheta', \varphi') p(\cos \vartheta \cos \vartheta' \\ & + \sin \vartheta \sin \vartheta' \cos [\varphi' - \varphi]) \sin \vartheta' d\vartheta' d\varphi'. \end{aligned} \tag{12}$$

When the scattering is *isotropic*

$$p(\cos \Theta) \equiv 1 \tag{13}$$

and the source function reduces to

(14) $$\mathfrak{J}(\tau) = \frac{1}{4\pi}\int_0^{\pi}\int_0^{2\pi} I(\tau, \vartheta', \varphi') \sin \vartheta' d\vartheta' d\varphi'.$$

Next to this isotropic case, greatest interest is attached to *Rayleigh's phase function*

(15) $$p(\cos \Theta) = \frac{3}{4}(1 + \cos^2 \Theta).$$

In general, we may, however, suppose that the phase function can be expanded as a series in Legendre polynomials of the form

(16) $$p(\cos \Theta) = \sum \varpi_l P_l(\cos \Theta)$$

where $\varpi_0 = 1$ and ϖ_l ($l \neq 0$) are some constants. (The condition $\varpi_0 = 1$ follows from equation (11).)

7. **Phase-functions with absorption.** Our foregoing remarks strictly apply to the case of perfect scattering; that is, to cases in which the interaction between matter and radiation results only in the transformation of radiation in one direction into radiation in other directions. However, cases are known in which, in the process of scattering, a certain amount of the radiation is consumptively transformed into other forms of energy. In such cases, we can still describe the scattered radiation by a phase function $p(\cos \Theta)$; but the "normalizing condition" (11) will no longer be satisfied and we shall have instead

(17) $$\int p(\cos \Theta) \frac{d\omega}{4\pi} = \lambda \qquad (\lambda \leqq 1).$$

The quantity λ defined in this way is called the *albedo*. Formally, this does not introduce anything very essential, as the source function will continue to be given by equation (12).

The case

(18) $$p(\cos \Theta) = \lambda(1 + x \cos \Theta) \qquad (-1 < x < 1)$$

is of particular interest for the analysis of planetary illumination.

8. **The problem of a semi-infinite plane-parallel atmosphere with a constant net flux.** Returning to the case of a perfectly scattering atmosphere, we observe that the equation of transfer

(19) $$\cos \vartheta \frac{dI(\tau, \vartheta, \varphi)}{d\tau} = I(\tau, \vartheta, \varphi) - \frac{1}{4\pi}\int_0^{\pi}\int_0^{2\pi} I(\tau, \vartheta', \varphi') p(\cos \Theta) d\omega',$$

where

(20) $$\cos \Theta = \cos \vartheta \cos \vartheta' + \sin \vartheta \sin \vartheta' \cos (\varphi' - \varphi),$$

admits the first integral

$$\frac{d}{d\tau}\int_0^{\pi}\int_0^{2\pi} I(\tau, \vartheta, \varphi) \cos\vartheta \sin\vartheta d\vartheta d\varphi = 0, \tag{21}$$

or

$$\pi F = \int_0^{\pi}\int_0^{2\pi} I(\tau, \vartheta, \varphi) \cos\vartheta \sin\vartheta d\vartheta d\varphi = \text{constant}. \tag{22}$$

The integral representing F is proportional to the net flux of radiant energy crossing unit area in the plane of stratification; and this is a constant for a perfectly scattering atmosphere. Because of this constancy of the net flux, a type of transfer problem which arises in connection with such perfectly scattering atmsopheres is that of a plane-parallel semi-infinite atmosphere with no incident radiation, and a constant net flux πF through the atmosphere normal to the plane of stratification. It is evident that for these problems, solutions of the appropriate equations of transfer must be sought which depend only on τ and ϑ, that is, solutions which exhibit rotational symmetry about the z-axis. Further, in these problems, the greatest interest is attached to the angular distribution of the *emergent radiation* $I(0, \mu)$ at $\tau=0$ and for $\mu>0$.

Finally, we may note that for the cases of isotropic scattering and scattering in accordance with Rayleigh's phase function, the equations of transfer have explicitly the forms

$$\text{I} \qquad \boxed{\mu\frac{dI(\tau, \mu)}{d\tau} = I(\tau, \mu) - \frac{1}{2}\int_{-1}^{+1} I(\tau, \mu')d\mu',}$$

and

$$\text{II} \qquad \boxed{\begin{aligned}\mu\frac{dI(\tau, \mu)}{d\tau} = I(\tau, \mu) - \frac{3}{16}\Big[(3-\mu^2)\int_{-1}^{+1} I(\tau, \mu')d\mu' \\ + (3\mu^2-1)\int_{-1}^{+1} I(\tau, \mu')\mu'^2 d\mu'\Big].\end{aligned}}$$

In the foregoing equations we have written μ for $\cos\vartheta$.

We require solutions of the integro-differential equations (I) and (II) which satisfy the boundary condition

$$I(0, \mu) = 0 \qquad \text{for } -1 < \mu < 0. \tag{23}$$

9. **The problem of diffuse reflection.** Another type of transfer problem which arises quite generally (that is, also in cases when λ is different from unity) is that relating to the phenomenon of *diffuse reflection.* In these contexts we are principally interested in the angular distribution $I(\vartheta, \varphi; \vartheta_0, \varphi_0)$ of the radiation diffusely reflected when a parallel beam of radiation of net flux πF per unit area normal to itself is incident on the atmosphere in some specified direction $(\pi - \vartheta_0, \varphi_0)$.

One general remark relating to problems of diffuse reflection may be made here. It is that, at any level, we may distinguish between the reduced incident radiation $\pi F e^{-\tau \sec \vartheta_0}$ which penetrates to the depth τ without suffering any scattering or absorbing mechanism, and the *diffuse radiation field* which arises in consequence of multiple scattering. We shall characterize this diffuse radiation field by $I(\tau, \mu, \varphi)$. Making this distinction between these two fields of radiation, we may write the equation of transfer for problems of diffuse reflection in the form

$$\begin{aligned} \cos\vartheta \frac{dI(\tau, \vartheta, \varphi)}{d\tau} &= I(\tau, \vartheta, \varphi) - \frac{1}{4\pi}\int_0^{\pi}\int_0^{2\pi} I(\tau, \vartheta', \varphi') \\ &\times p(\cos\vartheta\cos\vartheta' + \sin\vartheta\sin\vartheta'\cos[\varphi' - \varphi]) \sin\vartheta' d\vartheta' d\varphi' \\ &- \frac{1}{4} F e^{-\tau\sec\vartheta_0} p(-\cos\vartheta\cos\vartheta_0 + \sin\vartheta\sin\vartheta_0\cos[\varphi_0 - \varphi]). \end{aligned} \tag{24}$$

For the cases of isotropic scattering, and scattering in accordance with the phase functions (15) and (18), we have the equations

$$\mu \frac{dI(\tau, \mu)}{d\tau} = I(\tau, \mu) - \frac{1}{2}\int_{-1}^{+1} I(\tau, \mu') d\mu' - \frac{1}{4} F e^{-\tau/\mu_0}, \tag{III}$$

$$\begin{aligned} \mu \frac{dI(\tau, \vartheta, \varphi)}{d\tau} &= I(\tau, \vartheta, \varphi) \\ &- \frac{\lambda}{4\pi}\int_0^{\pi}\int_0^{2\pi} I(\tau, \vartheta', \varphi')[1 + x(\cos\vartheta\cos\vartheta' \\ &\qquad + \sin\vartheta\sin\vartheta'\cos[\varphi' - \varphi])] \sin\vartheta' d\vartheta' d\varphi' \\ &- \frac{1}{4}\lambda F e^{-\tau\sec\vartheta_0}[1 + x(-\cos\vartheta\cos\vartheta_0 + \sin\vartheta\sin\vartheta_0\cos\varphi)] \end{aligned} \tag{IV}$$

and

V

$$\cos\vartheta \frac{dI(\tau,\vartheta,\varphi)}{d\tau} = I(\tau,\vartheta,\varphi)$$
$$- \frac{3}{16\pi}\int_0^{\pi}\int_0^{2\pi} I(\tau,\vartheta',\varphi')\Big[1+\cos^2\vartheta\cos^2\vartheta'$$
$$+\frac{1}{2}\sin^2\vartheta\sin^2\vartheta' + 2\cos\vartheta\sin\vartheta\cos\vartheta'\sin\vartheta'\cos(\varphi'-\varphi)$$
$$+\frac{1}{2}\sin^2\vartheta\sin^2\vartheta'\cos 2(\varphi'-\varphi)\Big]\sin\vartheta' d\vartheta' d\varphi'$$
$$-\frac{3}{16}Fe^{-\tau\sec\vartheta_0}\Big[1+\cos^2\vartheta\cos^2\vartheta_0$$
$$+\frac{1}{2}\sin^2\vartheta\sin^2\vartheta_0 - 2\cos\vartheta\sin\vartheta\cos\vartheta_0\sin\vartheta_0\cos\varphi$$
$$+\frac{1}{2}\sin^2\vartheta\sin^2\vartheta_0\cos 2\varphi\Big].$$

And again we require solutions of these equations which satisfy the boundary condition (23).

It appears from the foregoing equations that for problems of diffuse reflection which arise in practice, we can expand $I(\tau, \mu, \varphi)$ in a finite Fourier series of the form

$$I(\tau,\mu,\varphi) = \sum_{m=0}^{N} I^{(m)}(\tau,\mu)\cos m\varphi. \tag{25}$$

And finally, it should be remarked that in the problem of diffuse reflection by a semi-infinite atmosphere, greatest interest is attached to the *scattering function* $S(\mu, \varphi; \mu_0, \varphi_0)$ which is related to the reflected intensity $I(0, \mu, \varphi)$ according to the equation

$$I(0,\mu,\varphi) = \frac{1}{4\mu}FS(\mu,\varphi;\mu_0,\varphi_0). \tag{26}$$

Defined in this manner $S(\mu, \varphi; \mu_0, \varphi_0)$ *must be symmetrical in the pair of variables* (μ, φ) *and* (μ_0, φ_0). This is required by a *principle of reciprocity* due to Helmholtz.

10. **The equations of transfer incorporating the polarization of the scattered radiation in accordance with Rayleigh's law. The axially symmetric case.** The transfer problems which we have described in

the preceding sections do not exhaust all the physical possibilities that may arise. For example, an important factor which we have not included in our discussion so far concerns the *state of polarization* of the radiation field. For many problems, this must be taken into account before we may be said to have an adequate physical description of the phenomenon under consideration, since, on scattering, light in general gets polarized. On Rayleigh's classical laws for instance, when an initially unpolarized beam is scattered in a direction inclined at an angle Θ with the direction of incidence, it becomes partially plane polarized with a ratio of intensities $1:\cos^2\Theta$ in directions perpendicular, respectively, parallel to the *plane of scattering*. (This is the plane which contains the directions of the incident and the scattered light.) It is, therefore, apparent that the diffuse radiation field in a scattering atmosphere must be partially polarized. The question immediately arises as to how best we can characterize the radiation field under these conditions in order that the relevant equations of transfer may be most conveniently formulated. This is a fundamental question: on its answer will depend the solution of a variety of problems, including the important one of the illumination and the polarization of the sunlit sky—a problem which was in fact first considered in a general way by Rayleigh in 1871. It is, therefore, somewhat surprising to find that even the basic equations of the problem should not have been written down before the problem arose in an astrophysical context earlier this year. Perhaps I may briefly explain the particular astrophysical problem which gave the final impetus for formulating and solving transfer problems in which the polarization characteristics of the radiation field are incorporated and allowance for the polarization of the scattered radiation is made:

It is now believed that in the atmospheres of stars with surface temperatures exceeding 15,000°K, the transfer of radiation must be predominantly controlled by the scattering by free electrons. And, according to J. J. Thomson's laws for this process, scattering by free electrons must result in the partial polarization of the scattered radiation. Thomson's laws agree, in fact, with Rayleigh's as regards the angular distribution and the state of polarization of the scattered radiation.

Now, if our belief in the important role played by the Thomson scattering in the atmospheres of hot stars is correct, we should expect that the radiation emergent from the atmospheres of such stars should be partially polarized. The question of the *degree of polarization* to be expected under such conditions therefore becomes one of exceptional astrophysical interest. But, before we can answer this question, we must formulate the relevant equations of transfer, distinguishing the different states of polarization, and solve them! One circumstance, however, simplifies this particular problem. It is, that in a plane-parallel atmosphere with no incident radiation, the radiation field at any point must exhibit rotational symmetry about the z-axis; accordingly, the plane of polarization must be along the principal meridian (or, at right angles to it). In other words, under the conditions of our present problem, the specific intensities I_l and I_r, referring respectively to the two states of polarization in which the electric vector vibrates in the meridian plane and at right angles to it, are sufficient to characterize the radiation field completely. And the relevant equations of transfer appropriate for this problem are found to be:

$$\text{VI} \qquad \begin{aligned} \mu \frac{dI_l}{d\tau} &= I_l - \frac{3}{8}\left\{\int_{-1}^{+1} I_l(\tau, \mu')[2(1-\mu'^2) + \mu^2(3\mu'^2 - 2)]d\mu' + \mu^2 \int_{-1}^{+1} I_r(\tau, \mu')d\mu'\right\}, \\ \mu \frac{dI_r}{d\tau} &= I_r - \frac{3}{8}\left\{\int_{-1}^{+1} I_l(\tau, \mu')\mu'^2 d\mu' + \int_{-1}^{+1} I_r(\tau, \mu')d\mu'\right\}. \end{aligned}$$

In other words, we now have a pair of simultaneous integro-differential equations to solve with the boundary conditions

$$I_l(0, \mu) = I_r(0, \mu) = 0 \qquad (-1 < \mu < 0). \tag{27}$$

11. **The parametric representation of partially plane-polarized light. The vector form of the equation of transfer.** Returning to the general problem of a partially polarized radiation field, we at once realize that the intensities I_l and I_r are not sufficient to characterize the radiation field completely. A third quantity is needed which will determine the plane of polarization. But the inclination of the plane of polarization itself to the meridian plane (for example) would be a most unsuitable parameter to choose: for, a proper parametric representation of partially polarized light should not only provide a complete set of variables to describe the radiation field, but also enable a convenient formulation of the rule of composition of a mixture of *independent* polarized streams in a given direction into a single partially polarized stream. On consideration it appears that the most convenient third variable to choose is a quantity U also of the dimensions of intensity and in terms of which the inclination, χ, of the plane of polarization to the direction to which I_l refers is given by

$$\chi = \frac{1}{2} \tan^{-1} \frac{U}{I_l - I_r}. \tag{28}$$

The rule of composition of a number of independent plane-polarized streams can be formulated as follows: If we have a number of independent plane-polarized streams of intensities $I^{(n)}$ (say) in a given direction, the resulting mixture will be partially plane-polarized and will be described by the parameters

$$I_l = \sum I^{(n)} \cos^2 \chi_n; \qquad I_r = \sum I^{(n)} \sin^2 \chi_n, \tag{29}$$

and

$$U = \sum I^{(n)} \sin 2\chi_n \tag{29'}$$

where χ_n denotes the inclination of the plane of polarization of the stream $I^{(n)}$ to the direction (in the transverse plane of the electric and the magnetic vectors) to which I_l refers.

With the rule of composition, as we have just stated, the equations of transfer for I_l, I_r and U can be formulated in accordance, for example, with Rayleigh's laws. For the problem of diffuse reflection of a partially plane-polarized parallel beam of radiation (characterized by the parameters F_l, F_r and $U^{(0)}$) incident on a plane-parallel atmosphere in the direction $(-\mu_0, \varphi_0)$, the three equations of transfer for the intensities I_l, I_r and U of the diffuse radiation field can be combined into a single vector equation of the form

$$\begin{aligned} \mu \frac{d\boldsymbol{I}}{d\tau} = \boldsymbol{I}(\tau, \mu, \varphi) & - \frac{3}{16\pi} \boldsymbol{Q} \int_{-1}^{+1} \int_0^{2\pi} \boldsymbol{J}(\mu, \varphi; -\mu', \varphi') \boldsymbol{I}(\tau, \mu', \varphi') d\mu' d\varphi' \\ & - \frac{3}{16} \boldsymbol{Q} \boldsymbol{J}(\mu, \varphi; \mu_0, \varphi_0) \boldsymbol{F} e^{-\tau/\mu_0}, \end{aligned} \tag{VII}$$

where

$$\boldsymbol{I} = (I_l, I_r, U), \qquad \boldsymbol{F} = (F_l, F_r, U^{(0)}), \tag{30}$$

$$\boldsymbol{Q} = \begin{pmatrix} 1 & 0 & 0 \\ 0 & 1 & 0 \\ 0 & 0 & 2 \end{pmatrix}, \tag{31}$$

and

$$J(\mu, \varphi; \mu_0, \varphi_0)$$

$$(32)\quad = \begin{pmatrix} 2(1-\mu^2)(1-\mu_0^2)+\mu^2\mu_0^2 & \mu^2 & \\ -4\mu\mu_0(1-\mu^2)^{1/2}(1-\mu_0^2)^{1/2}\cos(\varphi-\varphi_0) & & -2\mu(1-\mu^2)^{1/2}(1-\mu_0^2)^{1/2}\sin(\varphi-\varphi_0) \\ +\mu^2\mu_0^2\cos 2(\varphi-\varphi_0) & -\mu^2\cos 2(\varphi-\varphi_0) & +\mu^2\mu_0\sin 2(\varphi-\varphi_0) \\ & & \\ \mu_0^2-\mu_0^2\cos 2(\varphi-\varphi_0) & 1+\cos 2(\varphi-\varphi_0) & -\mu_0\sin 2(\varphi-\varphi_0) \\ & & \\ -2\mu_0(1-\mu^2)^{1/2}(1-\mu_0^2)^{1/2}\sin(\varphi-\varphi_0) & & (1-\mu^2)^{1/2}(1-\mu_0^2)^{1/2}\cos(\varphi-\varphi_0) \\ +\mu\mu_0^2\sin 2(\varphi-\varphi_0) & -\mu\sin 2(\varphi-\varphi_0) & -\mu\mu_0\cos 2(\varphi-\varphi_0) \end{pmatrix}$$

The matrix $\boldsymbol{QJ}$ therefore plays the same role for this problem as the phase function $p(\cos\Theta)$ does in the simpler problems in which polarization is not allowed for.

We require to solve (VII) which satisfies the boundary condition

$$(33)\qquad \boldsymbol{I}(0, \mu, \varphi) \equiv 0 \qquad (-1 < \mu < 0).$$

Since the equation of transfer is linear in $\boldsymbol{F}$, it is clear that the intensity $\boldsymbol{I}(\mu, \varphi; \mu_0, \varphi_0)$ diffusely reflected by the atmosphere in the outward direction (μ, φ) must be expressible in the form

$$(34)\qquad \boldsymbol{I}(\mu, \varphi; \mu_0, \varphi_0) = \frac{3}{16\mu}\boldsymbol{QS}(\mu, \varphi; \mu_0, \varphi_0)\boldsymbol{F}.$$

And our principal interest in this problem is in the determination of this *scattering matrix* $\boldsymbol{S}$.

12. **The Schwarzschild-Milne integral equation.** In the preceding sections we have formulated several typical problems in the theory of radiative transfer and we have seen how all of them lead to integro-differential equations of varying degrees of complexity. The question now arises as to whether there are any general methods for solving such equations. Until recently, the only example of such integro-differential equations which had been considered to any extent is the simplest we have written down, namely equation (I). And here the method used was one of reducing the integro-differential equation to an integral equation. This reduction to an integral equation was achieved in the following way:

Writing equation (I) in the form

$$(35)\qquad \mu\frac{dI}{d\tau} = I - J(\tau),$$

where

$$(36)\qquad J(\tau) = \frac{1}{2}\int_{-1}^{+1} I(\tau, \mu)d\mu,$$

we observe that its formal solution is

(37)
$$I(\tau, \mu) = \int_\tau^\infty e^{-(t-\tau)/\mu} J(t)\,\frac{dt}{\mu} \qquad (0 < \mu \leqq 1)$$
$$= -\int_0^\tau e^{-(t-\tau)/\mu} J(t)\,\frac{dt}{\mu} \qquad (-1 \leqq \mu < 0).$$

Substituting this solution for $I(\tau, \mu)$ back into the equation defining $J(\tau)$ (equation (36)) we find, after some elementary reductions, that

(38)
$$J(\tau) = \frac{1}{2}\int_0^\infty E_1(|t-\tau|)J(t)dt,$$

where

(39)
$$E_1(x) = \int_x^\infty e^{-y}\frac{dy}{y},$$

is the "first exponential integral." This is the *Schwarzschild-Milne integral equation* for $J(\tau)$.

13. **The method of Hopf and Wiener.** The integral equation (38) for $J(\tau)$ is seen to be homogeneous with a symmetric kernel. However, the kernel has a logarithmic singularity at $t=\tau$. Integral equations of this general structure have therefore a definite mathematical interest and have, in fact, been considered by a number of mathematicians including E. Hopf, J. Brönstein and N. Wiener. But this method based on the reduction of the equation of transfer to a linear integral equation for the source function has not so far been extended successfully to the solution of the other more interesting equations of transfer we have formulated. I shall therefore pass on to a different method of attack on these equations which has been developed during the past three years and which allows the solution of practically all the different types of transfer problems which arise in practice, by a systematic method of approximation. Also, for the particular problems enumerated, *exact solutions* for the angular distribution of the emergent (or the diffusely reflected) radiation can be found explicitly by passing to the limit of "infinite" approximation.

14. **The method of replacing the integro-differential equations by systems of linear equations.** The fundamental idea in this new method of solving an integro-differential equation is to replace it in a certain approximation by a system of ordinary linear equations. This reduction to an equivalent linear system is made by approximating the integrals over μ which occur in the equation by a weighted

average of the values which the various integrands take at a suitably selected set of points in the interval $(-1, +1)$. It is evident that in making the necessary division of the interval $(-1, +1)$, we must be guided by reasons of "economy" in the sense that with a given number of divisions we must try to achieve the maximum accuracy in the evaluation of the integrals. The problem which we encounter here is therefore the same as that considered by Gauss in 1814 in deriving his formula for numerical quadratures. In this formula of Gauss, the interval $(-1, +1)$ is divided according to the zeros of a Legendre polynomial, $P_m(\mu)$ (say), and the integral of a function $f(\mu)$ over the interval $(-1, +1)$ is expressed as a sum in the form

$$\int_{-1}^{+1} f(\mu)d\mu \simeq \sum a_j f(\mu_j) \tag{40}$$

where the *weights* a_j are given by

$$a_j = \frac{1}{P'_m(\mu_j)} \int_{-1}^{+1} \frac{P_m(\mu)}{\mu - \mu_j} d\mu. \tag{41}$$

The reason Gauss's formula is superior to other formulae for quadratures in the interval $(-1, +1)$ is that for a given m it evalues the integrals *exactly* for *all* polynomial of degree less than, or equal to, $2m-1$; in other words, it is almost "twice as accurate" as we should expect a formula which uses only m values of the function in the interval to be.

For certain formal reasons, we shall not enter into here, in our further work we shall use divisions of the interval $(-1, +1)$ only according to the zeros of the even Legendre polynomials. Further, when the division according to the zeros of $P_{2n}(\mu)$ is selected, we shall say that we are working in the nth *approximation*. It is, therefore, apparent that when an equation of transfer is reduced to an equivalent linear system in the nth approximation, we are effectively using a polynomial representation of order $4n-1$ (in μ) for the various intensities which occur in our problem.

Finally, it may be observed that, when working in the nth approximation, we may use the following relations between the points of the Gaussian division, μ_j, and the Gaussian weights a_j:

$$a_j = a_{-j}; \qquad \mu_j = -\mu_{-j} \qquad (j = \pm 1, \cdots, \pm n), \tag{42}$$

and

$$\sum_{j=1}^{n} a_j \mu_j^{2m} \equiv \frac{1}{2m+1} \qquad (m = 1, \cdots, 2n-1). \tag{43}$$

The identity (43) arises from the fact that in the nth approximation Gauss's formula evaluates exactly integrals of all polynomials of degree less than or equal to $4n-1$.

15. **The solution of equation** (I) **by the new method.** We shall illustrate the method of solving linear integro-differential equations by reducing them to equivalent linear systems in the various approximations by considering equation I.

In our nth approximation we replace equation (I) by the system of $2n$ equations

$$\mu_i \frac{dI_i}{d\tau} = I_i - \frac{1}{2} \sum a_j I_j \qquad (i = \pm 1, \cdots, \pm n), \tag{44}$$

where, for the sake of brevity, we have written $I_i(\tau)$ for $I(\tau, \mu_i)$.

In solving the system of equations represented by equation (44), we shall first obtain the different linearly independent solutions and later, by combining them, obtain the general solution.

As our first trial, we seek a solution of the form

$$I_i = g_i e^{-k\tau} \qquad (i = \pm 1, \cdots, \pm n), \tag{45}$$

where the g_i's and k are constants, for the present unspecified. Introducing equation (45) in equation (44), we obtain the relation

$$g_i(1 + \mu_i k) = \frac{1}{2} \sum a_j g_j. \tag{46}$$

Hence,

$$g_i = \frac{\text{constant}}{1 + \mu_i k} \qquad (i = \pm 1, \cdots, \pm n), \tag{47}$$

where the "constant" is independent of i. Substituting the foregoing form for g_i in equation (46), we obtain the *characteristic equation* for k:

$$1 = \frac{1}{2} \sum \frac{a_j}{1 + \mu_j k}. \tag{48}$$

Remembering that $a_{-j} = a_j$ and $\mu_{-j} = -\mu_j$, we can rewrite the characteristic equation in the form

$$1 = \sum_{j=1}^{n} \frac{a_j}{1 - \mu_j^2 k^2}. \tag{49}$$

It is, therefore, seen that k^2 must satisfy an algebraic equation of order n. However, since

(50) $$\sum_{j=1}^{n} a_j = 1,$$

$k^2=0$ is a root of equation (49). Equation (48) accordingly admits only $2n-2$ distinct nonzero roots which must occur in pairs as

(51) $$\pm k_\alpha \qquad (\alpha = 1, \cdots, n-1).$$

And corresponding to these $2n-2$ roots, we have $2n-2$ independent solutions of equation (44). The solution is completed by observing that

(52) $$I_i = b(\tau + \mu_i + Q) \qquad (i = \pm 1, \cdots, \pm n),$$

where b and Q are two arbitrary constants, also satisfies equation (44). The general solution of equation (44) can therefore be written in the form

(53) $$I_i = b\left\{\sum_{\alpha=1}^{n-1} \frac{L_\alpha e^{-k_\alpha \tau}}{1+\mu_i k_\alpha} + \sum_{\alpha=1}^{n-1} \frac{L_{-\alpha} e^{+k_\alpha \tau}}{1-\mu_i k_\alpha} + \tau + \mu_i + Q\right\} \qquad (i = \pm 1, \cdots, \pm n),$$

where b, $L_{\pm\alpha}$ $(\alpha=1, \cdots, n-1)$ and Q are the $2n$ arbitrary constants of integration.

For the astrophysical problem on hand, the boundary conditions are that none of the I_i's increase exponentially as $\tau \to \infty$ and that, further, there is no radiation incident on $\tau=0$. The first of these conditions requires that in the general solution (53) we omit all terms in $\exp(+k_\alpha \tau)$, thus leaving

(54) $$I_i = b\left\{\sum_{\alpha=1}^{n-1} \frac{L_\alpha e^{-k_\alpha \tau}}{1+\mu_i k_\alpha} + \tau + \mu_i + Q\right\} \qquad (i = \pm 1, \cdots, \pm n).$$

Next, the absence of any radiation in the directions $-1<\mu<0$ at $\tau=0$ implies that in our approximation we must require

(55) $$I_{-i} = 0 \qquad \text{at } \tau = 0 \text{ for } i = 1, \cdots, n.$$

Hence, according to equation (54),

(56) $$\sum_{\alpha=1}^{n-1} \frac{L_\alpha}{1-\mu_i k_\alpha} - \mu_i + Q = 0 \qquad (i = 1, \cdots, n).$$

These are the n equations which determine the n constants of integration L_α $(\alpha=1, \cdots, n-1)$ and Q. The constant b is left arbitrary, and this, as we shall presently see, is related to the assigned constant net flux of radiation through the atmosphere.

16. **Some elementary identities.** The further discussion of the solution obtained in §15 requires certain relations which we shall now establish.

Let

$$D_m(x) = \sum_i \frac{a_i \mu_i^m}{1 + \mu_i x} = (-1)^m \sum_i \frac{a_i \mu_i^m}{1 - \mu_i x} \qquad (m = 0, \cdots, 4n). \tag{57}$$

There is a simple recursion formula which $D_m(x)$ defined in this manner satisfies. We have:

$$D_m(x) = \frac{1}{x} \sum_i a_i \mu_i^{m-1} \left(1 - \frac{1}{1 + \mu_i x}\right), \tag{58}$$

or, using equations (42) and (43),

$$D_m(x) = \frac{1}{x} \left[\frac{2}{m} \epsilon_{m,\text{odd}} - D_{m-1}(x)\right], \tag{59}$$

where

$$\epsilon_{m,\text{odd}} = \begin{cases} 1 & \text{if } m \text{ is odd,} \\ 0 & \text{if } m \text{ is even.} \end{cases} \tag{60}$$

For odd, respectively, even values of m, equation (59) takes the form

$$D_{2j-1}(x) = \frac{1}{x} \left[\frac{2}{2j - 1} - D_{2j-2}(x)\right], \tag{61}$$

and

$$D_{2j}(x) = -\frac{1}{x} D_{2j-1}(x). \tag{62}$$

Combining these relations, we have

$$D_{2j-1}(x) = \frac{1}{x} \left[\frac{2}{2j - 1} + \frac{1}{x} D_{2j-3}(x)\right] = -x D_{2j}(x). \tag{63}$$

From this formula, we readily deduce that

$$\begin{aligned} D_{2j-1}(x) = & \frac{2}{(2j-1)x} + \frac{2}{(2j-3)x^2} + \cdots \\ & + \frac{2}{3x^{2j-3}} + \frac{2}{3x^{2j-1}} [2 - D_0(x)], \end{aligned} \tag{64}$$

and

$$D_{2j}(x) = -\frac{2}{(2j-1)x^2} - \frac{2}{(2j-3)x^4} - \cdots - \frac{2}{3x^{2j-2}} - \frac{2}{3x^{2j}}[2 - D_0(x)]. \tag{65}$$

If we now let x be a root k of the characteristic equation (48),

$$D_0(k) = 2, \tag{66}$$

and we find from equations (64) and (65) that

$$D_1(k) = D_2(k) = 0, \tag{67}$$

$$D_{2j}(k) = -\frac{2}{(2j-1)k^2} - \frac{2}{(2j-3)k^4} - \cdots - \frac{2}{3k^{2j-2}} \quad (j = 2, \cdots, n), \tag{68}$$

$$D_{2j-1}(k) = \frac{2}{(2j-1)k} + \frac{2}{(2j-3)k^3} + \cdots + \frac{2}{3k^{2j-3}} \quad (j = 2, \cdots, n). \tag{69}$$

17. **A relation between the characteristic roots and the zeros of the Legendre polynomial.** In terms of the $D_{2j}(k)$'s introduced in §16, we can express the characteristic equation for k in a form which does not explicitly involve the Gaussian weights and divisions: Let p_{2j} be the coefficient of μ^{2j} in the polynomial representation of the Legendre polynomial $P_{2n}(\mu)$, so that

$$P_{2n}(\mu) = \sum_{j=0}^{n} p_{2j}\mu^{2j}. \tag{70}$$

Now consider

$$\sum_{j=0}^{n} p_{2j}D_{2j}(k) = \sum_i \frac{a_i}{1+\mu_i k}\left(\sum_{j=0}^{n} p_{2j}\mu_i^{2j}\right). \tag{71}$$

Since the μ_i's are the zeros of $P_{2n}(\mu)$

$$\sum_{j=0}^{n} p_{2j}\mu_i^{2j} \equiv 0 \tag{72}$$

and the characteristic equation can be expressed in the form

$$\sum_{j=0}^{n} p_{2j}D_{2j}(k) = 0, \tag{73}$$

where the D_{2j}'s are defined as in equation (68). Substituting in particular for D_{2n} and D_0, we have

(74) $$\frac{2}{3}\frac{p_{2n}}{k^{2n-2}} + \cdots + 2p_0 = 0.$$

From this equation it follows that

(75) $$\frac{1}{(k_1 \cdots k_{n-1})^2} = (-1)^{n-1}\frac{3p_0}{p_{2n}} = 3\mu_1^2 \cdots \mu_n^2,$$

or

(76) $$k_1 \cdots k_{n-1}\mu_1 \cdots \mu_n = 1/3^{1/2}\,.$$

18. **The law of darkening.** Returning to the solution (54) of the transfer problem considered in §15, we evaluate the net flux πF according to the formula

(77) $$F = 2\int_{-1}^{+1} I\mu d\mu.$$

In our present scheme of approximation, we can write

(78) $$F = 2\sum a_i I_i \mu_i.$$

Evaluating this sum with the I_i's given by equation (54), we obtain

(79) $$F = 2b\left\{\sum_{\alpha=1}^{n-1} L_\alpha e^{-k_\alpha \tau} D_1(k_\alpha) + \sum_i a_i \mu_i^2\right\}.$$

Using equations (43) and (67), we have

(80) $$F = 4b/3.$$

We can now rewrite the solution (54) in the form

(81) $$I_i = \frac{3}{4}F\left\{\sum_{\alpha=1}^{n-1}\frac{L_\alpha e^{-k_\alpha \tau}}{1+\mu_i k_\alpha} + \tau + \mu_i + Q\right\} \quad (i = \pm 1, \cdots, \pm n).$$

In terms of this solution, the source function for the problem under consideration can be obtained in the following manner. We have

(82) $$\begin{aligned} J &= \frac{1}{2}\int_{-1}^{+1} I d\mu \simeq \frac{1}{2}\sum a_j I_j \\ &= \frac{3}{8}F\left\{\sum_{\alpha=1}^{n-1} L_\alpha e^{-k_\alpha \tau} D_0(k_\alpha) + 2(\tau + Q)\right\}, \end{aligned}$$

or, using equation (66),

$$J = \frac{3}{4} F \left\{ \sum_{\alpha=1}^{n-1} L_\alpha e^{-k_\alpha \tau} + \tau + Q \right\}. \tag{83}$$

The angular distribution $I(0, \mu)$ of the emergent radiation can be found from the source function (83) in accordance with the formula (cf. equation (37))

$$I(0, \mu) = \int_0^\infty J(\tau) e^{-\tau/\mu} \frac{d\tau}{\mu}. \tag{84}$$

We find

$$I(0, \mu) = \frac{3}{4} F \left\{ \sum_{\alpha=1}^{n-1} \frac{L_\alpha}{1 + \mu k_\alpha} + \mu + Q \right\}. \tag{85}$$

It is to be particularly noted that the foregoing expression for $I(0, \mu)$ is in agreement with the solution (81) for $\tau = 0$ and at the points of the Gaussian division $\mu = \mu_i$.

Comparing equation (85) with equation (56) which determines the constants L_α and Q, we observe that the angular distribution of the emergent radiation defined for the interval $0 \leqq \mu \leqq 1$ is determined in terms of a function which has zeros assigned in the complementary interval $-1 < \mu < 0$. Thus, letting

$$S(\mu) = \sum_{\alpha=1}^{n-1} \frac{L_\alpha}{1 - \mu k_\alpha} - \mu + Q, \tag{86}$$

the boundary conditions require that

$$S(\mu_i) = 0 \qquad (i = 1, \cdots, n), \tag{87}$$

while the angular distribution of the emergent radiation is given by

$$I(0, \mu) = \frac{3}{4} F S(-\mu). \tag{88}$$

19. **The elimination of the constants and the expression of the solution in closed form.** We shall now show how an explicit formula for $S(\mu)$ can be found without solving explicitly for the constants L_α and Q:

Consider the function

$$\prod_{\alpha=1}^{n-1} (1 - \mu k_\alpha) S(\mu). \tag{89}$$

This is a polynomial of degree n in μ which vanishes for $\mu=\mu_i$, $i=1, \cdots, n$. Consequently there must exist a proportionality of the form

$$\prod_{\alpha=1}^{n-1}(1-k_\alpha\mu)S(\mu)\propto\prod_{i=1}^{n-1}(\mu-\mu_i). \tag{90}$$

The constant of proportionality can be found from a comparison of the coefficients of the highest powers of μ on either side. In this manner we find that

$$S(\mu)=(-1)^n k_1\cdots k_{n-1}\frac{\prod_{i=1}^{n}(\mu-\mu_i)}{\prod_{\alpha=1}^{n-1}(1-k_\alpha\mu)}. \tag{91}$$

This is the required formula.

According to equation (91)

$$S(-\mu)=k_1\cdots k_{n-1}\frac{\prod_{i=1}^{n}(\mu+\mu_i)}{\prod_{\alpha=1}^{n-1}(1+k_\alpha\mu)}, \tag{92}$$

or, using the result (76), we can write

$$S(-\mu)=\frac{1}{3^{1/2}}H(\mu), \tag{93}$$

where

$$H(\mu)=\frac{1}{\mu_1\cdots\mu_n}\frac{\prod_{i=1}^{n}(\mu+\mu_i)}{\prod_{\alpha=1}^{n-1}(1+k_\alpha\mu)}. \tag{94}$$

In terms of the function $H(\mu)$ defined in this manner, the angular distribution of the emergent radiation can be expressed in the form

$$I(0,\mu)=\frac{3^{1/2}}{4}FH(\mu). \tag{95}$$

20. **The solution of equation (III) by the new method. A particular integral.** As a further illustration of our method of solving equations

of transfer, we shall next consider equation III appropriate for the problem of diffuse reflection by an isotropically scattering atmosphere. The equivalent system of linear equations in the nth approximation is

$$\mu_i \frac{dI_i}{d\tau} = I_i - \frac{1}{2}\sum a_j I_j - \frac{1}{4} F e^{-\tau/\mu_0} \qquad (i = \pm 1, \cdots, \pm n). \tag{96}$$

It is seen that the homogeneous system associated with equation (96) is the same as equation (44). Accordingly, the complimentary function for the solution of equation (96) is the same as the general solution (53) of the homogeneous system. To complete the solution we require only a particular integral. This can be found in the following manner: Setting

$$I_i = \frac{1}{4} F h_i e^{-\tau/\mu_0} \qquad (i = \pm 1, \cdots, \pm n) \tag{97}$$

in equation (96) (the h_i's are constants unspecified for the present) we verify that we must have

$$h_i(1 + \mu_i/\mu_0) = \frac{1}{2}\sum a_j h_j + 1. \tag{98}$$

Equation (98) implies that the constants h_i must be expressible in the form

$$h_i = \frac{\gamma}{1 + \mu_i/\mu_0} \qquad (i = \pm 1, \cdots, \pm n), \tag{99}$$

where the constant γ has to be determined from the condition (cf. equation (98))

$$\gamma = \frac{1}{2}\gamma \sum \frac{a_i}{1 + \mu_i/\mu_0} + 1. \tag{100}$$

In other words

$$I_i = \frac{1}{4} F \frac{\gamma e^{-\tau/\mu_0}}{1 + \mu_i/\mu_0} \qquad (i = \pm 1, \cdots, \pm n), \tag{101}$$

with

$$\gamma = \frac{1}{1 - \sum_{j=1}^{n} \frac{a_j}{1 - \mu_j^2/\mu_0^2}}, \tag{102}$$

represents the required particular integral of equation (96).

The constant γ defined as in equation (102) can be expressed in terms of the H-function introduced in §19 (equation (94)). For this purpose consider the function

(103) $$T(x) = 1 - \sum_{j=1}^{n} \frac{a_j}{1 - \mu_j^2 x}.$$

This clearly vanishes for

(104) $$x = 0 \quad \text{and} \quad x = k_\alpha^2 \qquad (\alpha = 1, \cdots, n-1).$$

Accordingly

(105) $$T(x) \prod_{j=1}^{n} (1 - \mu_j^2 x)$$

cannot differ from

(106) $$x \prod_{\alpha=1}^{n-1} (x - k_\alpha^2)$$

by more than a constant factor since (105) represents a polynomial of degree n in x. The constant of proportionality can be determined by comparing the coefficients of the highest powers. In this manner we find that

(107) $$T(x) \equiv (-1)^n \mu_1^2 \cdots \mu_n^2 \frac{x \prod_{\alpha=1}^{n-1} (x - k_\alpha^2)}{\prod_{j=1}^{n} (1 - \mu_j^2 x)}.$$

But

(108) $$\gamma = \frac{1}{T(\mu_0^{-2})}.$$

Hence

(109) $$\gamma = (-1)^n \frac{1}{\mu_1^2 \cdots \mu_n^2} \frac{\prod_{i=1}^{n} (\mu_0^2 - \mu_i^2)}{\prod_{\alpha=1}^{n-1} (1 - k_\alpha^2 \mu_0^2)}$$

or according to our definition of $H(\mu)$

$$\gamma = H(\mu_0)H(-\mu_0). \tag{110}$$

21. **The law of diffuse reflection for the case of isotropic scattering.** Now, adding to the particular integral (101) the general solution of the homogeneous system compatible with our present requirement of boundedness of the solution for $\tau \to \infty$, we have

$$I_i = \frac{1}{4} F \left\{ \sum_{\alpha=1}^{n-1} \frac{L_\alpha e^{-k_\alpha \tau}}{1+\mu_i k_\alpha} + Q + \frac{H(\mu_0)H(-\mu_0)}{1+\mu_i/\mu_0} e^{-\tau/\mu_0} \right\} \tag{111}$$

$$(i = 1, \cdots, n),$$

where the constants L_α $(\alpha = 1, \cdots, n-1)$ and Q are to be determined by the boundary conditions at $\tau = 0$.

At $\tau = 0$ we have no diffuse radiation directed inward. The conditions which determine the constants of integration are therefore

$$\sum_{\alpha=1}^{n-1} \frac{L_\alpha}{1-\mu_i k_\alpha} + Q + \frac{H(\mu_0)H(-\mu_0)}{1-\mu_i/\mu_0} = 0 \quad (i = \pm 1, \cdots, \pm n). \tag{112}$$

The angular distribution of the diffusely reflected radiation can be found in terms of the source function

$$\mathfrak{J}(\tau) = \frac{1}{2} \sum a_i I_i + \frac{1}{4} F e^{-\tau/\mu_0}, \tag{113}$$

according to the formula

$$I(0, \mu) = \int_0^\infty \mathfrak{J}(\tau) e^{-\tau/\mu} \frac{d\tau}{\mu}. \tag{114}$$

We find

$$I(0, \mu; \mu_0) = \frac{1}{4} F \left\{ \sum_{\alpha=1}^{n-1} \frac{L_\alpha}{1+\mu k_\alpha} + Q + \frac{H(\mu_0)H(-\mu_0)}{1+\mu/\mu_0} \right\}. \tag{115}$$

As in §18, we again observe that the angular distribution of the diffusely reflected radiation defined in the interval $0 \leqq \mu \leqq 1$ is described in terms of a function which has zeros assigned in the interval $0 > \mu > -1$. Thus, letting

$$S(\mu) = \sum_{\alpha=1}^{n-1} \frac{L_\alpha}{1-\mu k_\alpha} + Q + \frac{H(\mu_0)H(-\mu_0)}{1-\mu/\mu_0}, \tag{116}$$

we have

$$S(\mu_i) = 0 \qquad (i = 1, \cdots, n), \tag{117}$$

and

$$I(0, \mu; \mu_0) = \frac{1}{4} FS(-\mu). \tag{118}$$

And, again as in the earlier problem, we can find an explicit formula for $S(\mu)$ without having to solve explicitly for the constants L_α and Q.

Thus, considering the function

$$(1 - \mu/\mu_0)\prod_{\alpha=1}^{n-1}(1 - k_\alpha \mu)S(\mu) \tag{119}$$

we observe that it is a polynomial of degree n in μ which vanishes for $\mu = \mu_i$, $i = 1, \cdots, n$. There must, therefore, exist a relation of the form

$$S(\mu) = (-1)^n \frac{X}{\mu_1 \cdots \mu_n} \frac{\prod_{i=1}^{n}(\mu - \mu_i)}{\prod_{\alpha=1}^{n-1}(1 - k_\alpha \mu)} \frac{1}{(1 - \mu/\mu_0)}, \tag{120}$$

where X is a constant. In terms of $H(\mu)$ we can rewrite the foregoing equation as

$$S(\mu) = X \frac{H(-\mu)}{1 - \mu/\mu_0}. \tag{121}$$

The constant X appearing in equation (121) can be determined from the relation (cf. equation (116))

$$\lim_{\mu \to \mu_0} (1 - \mu/\mu_0)S(\mu) = H(\mu_0)H(-\mu_0). \tag{122}$$

According to equation (121) the left-hand side of equation (122) is $XH(-\mu_0)$. Hence

$$X = H(\mu_0) \tag{123}$$

and

$$S(\mu) = H(\mu_0)H(-\mu) \frac{\mu_0}{\mu_0 - \mu}. \tag{124}$$

The expression (118) for the angular distribution of the reflected radiation therefore becomes

$$I(0, \mu; \mu_0) = \frac{1}{4} FH(\mu)H(\mu_0) \frac{\mu_0}{\mu + \mu_0}. \tag{125}$$

22. **Tabulation of the solutions of equations** (I)–(VII). The method of solution of transfer problems which we have illustrated in the preceding sections by considering the two simplest problems is, on examination, found to be sufficiently general for adaptation to the solution of the more difficult problems presented by the other equations of transfer formulated in §§8–11. While the details of the solution of these other equations (particularly equations VI and VII) are considerably more elaborate and complex, the analysis nevertheless shows similarities with the simple problems we have considered in the broad features. Thus, it is found that, in all cases, the angular distribution of the emergent (equivalently, reflected) radiation is described by a function for the argument in the range (0, 1) while the boundary conditions specify the zeros of the same function in the interval (0, −1). This remarkable *reciprocity*, which exists in all the problems, enables the *elimination of the constants* and allows the reduction of the solutions to *closed forms* in the general nth approximation. And finally, these solutions, apart from certain constants, involve only H-functions of the form

$$H(\mu) = \frac{1}{\mu_1 \cdots \mu_n} \frac{\prod_{i=1}^{n} (\mu + \mu_i)}{\prod_{\alpha=1}^{n} (1 + k_\alpha \mu)}, \tag{126}$$

where the μ_i's are the zeros of the Legendre polynomial $P_{2n}(\mu)$ and the k_α's are the positive (or zero) roots of a characteristic equation of the form

$$1 = 2\sum_{j=1}^{n} \frac{a_j \Psi(\mu_j)}{1 - k^2 \mu_j^2}, \tag{127}$$

where $\Psi(\mu)$ is an even polynomial in μ satisfying the condition

$$\int_0^1 \Psi(\mu) d\mu \leqq \frac{1}{2}. \tag{128}$$

(This last condition is necessary for $H(\mu)$ to be real.)

The different physical problems naturally lead to different characteristic equations and therefore to different H-functions. However, as the H-functions differ from one another only through the characteristic equations which define the roots k_α, we may properly call $\Psi(\mu)$ the *characteristic function* in terms of which H is defined.

We tabulate below the solutions for the various transfer problems (equations I–VII) obtained in the *n*th approximation of our method of solutions.

Solutions for the emergent and the diffusely reflected radiation for various transfer problems

A. *Isotropic scattering*

Problem with constant net flux: the law of darkening:

$$I(\mu) = \frac{3^{1/2}}{4} F H(\mu).$$

Law of diffuse reflection:

$$I(\mu; \mu_0) = \frac{1}{4} F H(\mu) H(\mu_0) \frac{\mu_0}{\mu + \mu_0} \cdot$$

The characteristic function in terms of which $H(\mu)$ is defined is

$$\Psi(\mu) = 1/2.$$

B. *Scattering in accordance with Rayleigh's phase function* $3(1+\cos^2\Theta)/4$

Problem with constant net flux: the law of darkening:

$$I(\mu) = \frac{3}{4} q F H^{(0)}(\mu) \cdot$$

Law of diffuse reflection:

$$\begin{aligned} I(\mu, \varphi; \mu_0, \varphi_0) = \frac{3}{32} F \{ & H^{(0)}(\mu) H^{(0)}(\mu_0) [3 - (3 - 8q^2)^{1/2}(\mu + \mu_0) + \mu\mu_0] \\ & - 4\mu\mu_0(1 - \mu^2)^{1/2}(1 - \mu_0^2)^{1/2} H^{(1)}(\mu) H^{(1)}(\mu_0) \cos(\varphi - \varphi_0) \\ & + (1 - \mu^2)(1 - \mu_0^2) H^{(2)}(\mu) H^{(2)}(\mu_0) \cos 2(\varphi - \varphi_0) \} \frac{\mu_0}{\mu + \mu_0} \cdot \end{aligned}$$

The characteristic functions in terms of which $H^{(0)}(\mu)$, $H^{(1)}(\mu)$ and $H^{(2)}(\mu)$ are defined are respectively

$$\Psi^{(0)}(\mu) = \frac{3}{16}(3 - \mu^2),$$

$$\Psi^{(1)}(\mu) = \frac{3}{8}\mu^2(1 - \mu^2),$$

$$\Psi^{(2)}(\mu) = \frac{3}{32}(1 - \mu^2)^2.$$

And finally,

$$q = \frac{2(3^{1/2})}{H^{(0)}(+3^{1/2}) - H^{(0)}(-3^{1/2})}.$$

C. *Scattering in accordance with the phase function* $\lambda(1+x\cos\Theta)$

Law of diffuse reflection:

$$I(\mu, \varphi; \mu_0, \varphi_0)$$
$$= \frac{1}{4}\lambda F\{H^{(0)}(\mu)H^{(0)}(\mu_0)[1 - c(\mu+\mu_0) - x(1-\lambda)\mu\mu_0]$$
$$+ x(1-\mu^2)^{1/2}(1-\mu_0^2)^{1/2}H^{(1)}(\mu)H^{(1)}(\mu_0)\cos(\varphi-\varphi_0)\}\frac{\mu_0}{\mu+\mu_0}.$$

The characteristic functions in terms of which $H^{(0)}(\mu)$ and $H^{(1)}(\mu)$ are defined are respectively

$$\Psi^{(0)}(\mu) = \frac{1}{2}\lambda[1 + x(1-\lambda)\mu^2],$$

$$\Psi^{(1)}(\mu) = \frac{1}{4}x\lambda(1-\mu^2).$$

The constant c depends in a somewhat complicated manner on the characteristic roots defining $H^{(0)}(\mu)$ (cf. S. Chandrasekhar, Astrophysical Journal vol. 103 (1946) p. 165, equation [108]).

D. *Scattering in accordance with Rayleigh's law and allowing for the polarization of the scattered radiation*

Problem with constant net flux: the law of darkening in the two states of polarization:

$$I_l(\mu) = \frac{3}{8}F(1-c^2)^{1/2}H_l(\mu),$$

$$I_r(\mu) = \frac{3}{8}F\frac{1}{2^{1/2}}H_r(\mu)(\mu+c).$$

Problem of diffuse reflection: the scattering matrix:

$$\boldsymbol{I}(\mu, \varphi; \mu_0, \varphi_0) = \frac{3}{16\mu}\boldsymbol{QS}(\mu, \varphi; \mu_0, \varphi_0)\boldsymbol{F}$$

where

$$
\left(\frac{1}{\mu}+\frac{1}{\mu_0}\right)S(\mu,\varphi;\mu_0,\varphi_0)
$$

$$
=\left\{\begin{array}{lll}
2H_l(\mu)H_l(\mu_0)[1+\mu\mu_0-c(\mu+\mu_0) & (2(1-c^2))^{1/2}H_r(\mu_0)H_l(\mu)(\mu+\mu_0) & \\
-4\mu\mu_0(1-\mu^2)^{1/2}(1-\mu_0^2)^{1/2}H^{(1)}(\mu)H^{(1)}(\mu_0)\cos(\varphi-\varphi_0) & & -2\mu(1-\mu^2)^{1/2}(1-\mu_0^2)^{1/2}H^{(1)}(\mu)H^{(1)}(\mu_0)\sin(\varphi-\varphi_0) \\
+\mu^2\mu_0^2H^{(2)}(\mu)H^{(2)}(\mu_0)\cos 2(\varphi-\varphi_0) & -\mu^2H^{(2)}(\mu)H^{(2)}(\mu_0)\cos 2(\varphi-\varphi_0) & +\mu^2\mu_0H^{(2)}(\mu)H^{(2)}(\mu_0)\sin 2(\varphi-\varphi_0) \\
 & & \\
(2(1-c^2))^{1/2}H_r(\mu)H_l(\mu_0)(\mu+\mu_0) & H_r(\mu)H_r(\mu_0)[1+\mu\mu_0+c(\mu+\mu_0) & \\
-\mu_0^2H^{(2)}(\mu)H^{(2)}(\mu_0)\cos 2(\varphi-\varphi_0) & +H^{(2)}(\mu)H^{(2)}(\mu_0)\cos 2(\varphi-\varphi_0) & -\mu_0H^{(2)}(\mu)H^{(2)}(\mu_0)\sin 2(\varphi-\varphi_0) \\
 & & \\
-2\mu_0(1-\mu_0^2)^{1/2}(1-\mu^2)^{1/2}H^{(1)}(\mu)H^{(1)}(\mu_0)\sin(\varphi-\varphi_0) & & (1-\mu^2)^{1/2}(1-\mu_0^2)^{1/2}H^{(1)}(\mu)H^{(1)}(\mu_0)\cos(\varphi-\varphi_0) \\
+\mu\mu_0^2H^{(2)}(\mu)H^{(2)}(\mu_0)\sin 2(\varphi-\varphi_0) & -\mu H^{(2)}(\mu)H^{(2)}(\mu_0)\sin 2(\varphi-\varphi_0) & -\mu\mu_0H^{(2)}(\mu)H^{(2)}(\mu_0)\cos 2(\varphi-\varphi_0)
\end{array}\right.
$$

The characteristic functions in terms of which $H_l(\mu)$, $H_r(\mu)$, $H^{(1)}(\mu)$ and $H^{(2)}(\mu)$ are defined are respectively

$$\Psi_l(\mu) = \frac{3}{4}(1-\mu^2);$$

$$\Psi^{(1)}(\mu) = \frac{3}{8}(1-\mu^2)(1+2\mu^2);$$

$$\Psi_r(\mu) = \frac{3}{8}(1-\mu^2),$$

$$\Psi^{(2)}(\mu) = \frac{3}{16}(1+\mu^2)^2.$$

And finally, the constant c is given by

$$c = \frac{H_l(+1)H_r(-1) + H_l(-1)H_r(+1)}{H_l(+1)H_r(-1) - H_l(-1)H_r(+1)}\,.$$

23. **Remarks on the tabulated solutions. The possibility of passage to the "infinite approximation."** An examination of the solutions of the various transfer problems given in the preceding section discloses remarkable relationships between the *laws of darkening* of the emergent radiation from a semi-infinite plane-parallel atmosphere with constant net flux (and no incident radiation) and the *laws of diffuse reflection* by the same atmosphere. The relationship is naturally the simplest for the case of an isotropically scattering atmosphere and can indeed be established directly from the equations of transfer. However, in the other cases, the relationship is of a more complex nature and has to be sought between the darkening function for the problem with a constant net flux and the *azimuth independent* term in the law of diffuse reflection. It is seen that both these functions involve the same H-functions and the same constants.

While the relationship between the two problems has been established only in a particular scheme of approximation, it is apparent that the *relationship itself* must be an *exact one* since it is present in every approximation and must consequently be also present in the *limit of infinite approximation* when the solutions will become the exact ones of the problem. A further result of this train of thought is the realization that if we can solve the problem of diffuse reflection exactly, we shall, at the same time, have also solved exactly the axially symmetric problem with a constant net flux and no incident radiation.

Having thus been led to conceive the passage to the limit of infinite approximation, we naturally ask ourselves: *Can we in fact perform this limiting process and thus obtain the exact solutions for the various problems?* The answer to this question must clearly depend on our ability to pass to the limit of the H-functions as we have defined them, as $n \to \infty$. We shall now indicate how this limiting process can be achieved in practice.

24. **The equation satisfied by** $H(\mu)$. For the purposes of passing to the limit of infinite approximations of the solutions of the various transfer problems, we shall first establish the following basic theorem relating to the H-functions.

THEOREM 1. *Let* $\Psi(\mu)$ *be an even polynomial of degree* $2m$ *in* μ *such that*

$$\int_0^1 \Psi(\mu) d\mu \leqq \frac{1}{2} .$$

Let μ_j $(j = \pm 1, \cdots, \pm n)$ *denote the division of the interval* $(-1, +1)$ *according to the zeros of the Legendre polynomial of order* $2n$ $(>m)$; *further, let* a_j $(=a_{-j})$ *denote the corresponding Gaussian weights. Finally, let* k_α $(\alpha = 1, \cdots, n)$ *denote the distinct positive* (*or zero*) *roots of the characteristic equation*

$$1 = \sum_j \frac{a_j \Psi(\mu_j)}{1 + k\mu_j} = 2 \sum_{j=1}^{n} \frac{a_j \Psi(\mu_j)}{1 - k^2 \mu_j^2} .$$

Then the function

$$H(\mu) = \frac{1}{\mu_1 \cdots \mu_n} \frac{\prod_{i=1}^{n} (\mu + \mu_i)}{\prod_{\alpha=1}^{n} (1 + k_\alpha \mu)}$$

satisfies identically the equation

$$H(\mu) \equiv 1 + \mu H(\mu) \sum_{j=1}^{n} \frac{a_j H(\mu_j) \Psi(\mu_j)}{\mu + \mu_j} .$$

PROOF. We shall first consider the case

$$\int_0^1 \Psi(\mu) d\mu < \frac{1}{2} . \tag{129}$$

In this case the characteristic equation admits n distinct nonvanishing positive roots and we consider the function

$$S(\mu) = \sum_{\alpha=1}^{n} \frac{L_\alpha}{1 - k_\alpha \mu} + 1, \tag{130}$$

where L_α $(\alpha = 1, \cdots, n)$ are certain constants to be determined from the equations

$$\sum_{\alpha=1}^{n} \frac{L_\alpha}{1 - k_\alpha \mu_i} + 1 = 0 \qquad (i = 1, \cdots, n), \tag{131}$$

or equivalently

$$S(\mu_i) = 0 \qquad (i = 1, \cdots, n). \tag{132}$$

By considerations of the type we are now familiar with, it can be shown that

$$S(\mu) = k_1 \cdots k_n \mu_1 \cdots \mu_n \frac{(-1)^n}{\mu_1 \cdots \mu_n} \frac{\prod_i (\mu - \mu_i)}{\prod_\alpha (1 - k_\alpha \mu)} \cdot \tag{133}$$

In other words,

$$S(\mu) = k_1 \cdots k_n \mu_1 \cdots \mu_n H(-\mu). \tag{134}$$

Since $H(0) = 1$, we can rewrite equation (134) alternatively in the form

$$S(\mu) = S(0) H(-\mu), \tag{135}$$

where

$$S(0) = \sum_{\alpha=1}^{n} L_\alpha + 1 = k_1 \cdots k_n \mu_1 \cdots \mu_n. \tag{136}$$

Now, since $\Psi(\mu)$ is even in μ, we can rewrite the equation which a characteristic root satisfies in either of the forms

$$1 = \sum_j \frac{a_j \Psi(\mu_j)}{1 + k\mu_j} \tag{137}$$

or

$$1 = \sum_j \frac{a_j \Psi(\mu_j)}{1 - k\mu_j} \cdot \tag{138}$$

Let k_α denote a particular characteristic root. Then, on account of equations (137) and (138), which are satisfied by any of the characteristic roots, we can clearly write

$$S(0) = \sum_{\beta=1}^{n} L_\beta + 1$$

$$(139) \qquad = \sum_{\beta=1}^{n} \frac{L_\beta}{k_\alpha + k_\beta}\left[\sum_j a_j\Psi(\mu_j)\left\{\frac{k_\alpha}{1+k_\alpha\mu_j} + \frac{k_\beta}{1-k_\beta\mu_j}\right\}\right]$$

$$+ \sum_j \frac{a_j\Psi(\mu_j)}{1+k_\alpha\mu_j} \, .$$

Simplifying the quantity in brackets in the foregoing equation, we have

$$(140) \qquad S(0) = \sum_{\beta=1}^{n} L_\beta\left[\sum_j \frac{a_j\Psi(\mu_j)}{(1+k_\alpha\mu_j)(1-k_\beta\mu_j)}\right] + \sum_j \frac{a_j\Psi(\mu_j)}{1+k_\alpha\mu_j}$$

or, inverting the order of the summation,

$$(141) \qquad S(0) = \sum \frac{a_j\Psi(\mu_j)}{1+k_\alpha\mu_j}\left[\sum_{\beta=1}^{n} \frac{L_\beta}{1-k_\beta\mu_j} + 1\right].$$

But the quantity in brackets in equation (141) is $S(\mu_j)$. Hence

$$(142) \qquad S(0) = \sum_j \frac{a_j S(\mu_j)\Psi(\mu_j)}{1+k_\alpha\mu_j} \, .$$

In equation (142) (as in equations (137)–(141)) the summation is, of course, extended over all values of j, positive and negative. However, since $S(+\mu_j)=0$ (equation (132)), in equation (142) only the terms with negative j make a nonzero contribution. We can, therefore, write

$$(143) \qquad S(0) = \sum_{j=1}^{n} \frac{a_j S(-\mu_j)\Psi(\mu_j)}{1-k_\alpha\mu_j} \qquad (\alpha = 1, \cdots, n)$$

or, in view of equation (135),

$$(144) \qquad 1 = \sum_{j=1}^{n} \frac{a_j H(\mu_j)\Psi(\mu_j)}{1-k_\alpha\mu_j} \qquad (\alpha = 1, \cdots, n).$$

Now, consider the function

$$(145) \qquad 1 - \mu\sum_{j=1}^{n} \frac{a_j H(\mu_j)\Psi(\mu_j)}{\mu+\mu_j} \, .$$

According to equation (144), this vanishes for $\mu=-1/k_\alpha$ ($\alpha=1, \cdots, n$); for

$$1+\frac{1}{k_\alpha}\sum_{j=1}^{n}\frac{a_jH(\mu_j)\Psi(\mu_j)}{(-1/k_\alpha)+\mu_j}=1-\sum_{j=1}^{n}\frac{a_jH(\mu_j)\Psi(\mu_j)}{1-k_\alpha\mu_j}=0. \tag{146}$$

Hence

$$\prod_{j=1}^{n}(\mu+\mu_j)-\mu\sum_{j=1}^{n}a_jH(\mu_j)\Psi(\mu_j)\prod_{i\neq j}(\mu+\mu_i) \tag{147}$$

also vanishes for

$$\mu=-1/k_\alpha \qquad (\alpha=1,\cdots,n). \tag{148}$$

But the expression (147) is a polynomial of degree n in μ. It cannot, therefore, differ from

$$\prod_{\alpha=1}^{n}(1+k_\alpha\mu) \tag{149}$$

by more than a constant factor; and the constant of proportionality is seen to be

$$\mu_1\cdots\mu_n \tag{150}$$

from a comparison of the two functions at $\mu=0$. It therefore follows that

$$1-\mu\sum_{j=1}^{n}\frac{a_jH(\mu_j)\Psi(\mu_j)}{\mu+\mu_j}\equiv\mu_1\cdots\mu_n\frac{\prod_{\alpha=1}^{n}(1+k_\alpha\mu)}{\prod_{j=1}^{n}(\mu+\mu_j)}=\frac{1}{H(\mu)}. \tag{151}$$

Hence

$$H(\mu)\equiv1+\mu H(\mu)\sum_{j=1}^{n}\frac{a_jH(\mu_j)\Psi(\mu_j)}{\mu+\mu_j}. \tag{152}$$

This proves the theorem for the case (129).

Turning now to the case

$$\int_0^1\Psi(\mu)d\mu=\frac{1}{2}, \tag{153}$$

we observe that, in this case, $k=0$ is a root of the characteristic equation, and we are left with only $(n-1)$ positive roots. We, therefore, consider in this case the function

$$S(\mu) = \sum_{\alpha=1}^{n-1} \frac{L_\alpha}{1 - k_\alpha \mu} + L_0 - \tag{154}$$

in place of equation (130). However, the constants L_0 and L_α ($\alpha = 1, \cdots, n-1$) are again to be determined by the conditions

$$S(\mu_i) = 0 \qquad (i = 1, \cdots, n). \tag{155}$$

With this definition of $S(\mu)$, equation (135) continues to be valid and the rest of the proof follows on similar lines. The only essential point of departure that needs to be noted is that, at the stage of the proof corresponding to equation (140) and before inverting the order of the summation, we must add the extra term

$$-\sum_j \frac{a_j \Psi(\mu_j)\mu_j}{1 + k_\alpha \mu_j} \tag{156}$$

to the right-hand side of the equation. We can do this without altering anything, since the quantity we thus add is zero; for

$$\begin{aligned}\sum_j \frac{a_j \Psi(\mu_j)\mu_j}{1 + k_\alpha \mu_j} &= \frac{1}{k_\alpha}\left[\sum_j a_j \Psi(\mu_j)\left\{1 - \frac{1}{1 + k_\alpha \mu_j}\right\}\right] \\ &= \frac{1}{k_\alpha}\left[1 - \sum_j \frac{a_j \Psi(\mu_j)}{1 + k_\alpha \mu_j}\right] = 0.\end{aligned} \tag{157}$$

(Note that we are permitted to set $\sum a_j \Psi(\mu_j) = 1$, since the Gauss sum in the nth approximation evaluates the integrals exactly for all polynomials of degree less than or equal to $4n-1$; and we have assumed $2n > m$.)

This completes the proof of the theorem.

25. **The limit of $H(\mu)$ as $n \to \infty$. The basic functional equation.** The theorem proved in §24 suggests how the limit of the H-function as $n \to \infty$ can be obtained. We shall state this in the form of a theorem.

THEOREM 2. *The solution of the functional equation*

$$H(\mu) = 1 + \mu H(\mu) \int_0^1 \frac{H(\mu')\Psi(\mu')}{\mu + \mu'} d\mu',$$

where $\Psi(\mu)$ is an even polynomial satisfying the condition

$$\int_0^1 \Psi(\mu) d\mu \leqq \frac{1}{2},$$

is the limit function

$$\lim_{n\to\infty} \frac{1}{\mu_1 \cdots \mu_n} \frac{\prod_{i=1}^{n} (\mu + \mu_i)}{\prod_{\alpha=1}^{n} (1 + k_\alpha \mu)}$$

where the μ_i's and the k_α's have the same meanings as in Theorem 1.

SKETCH OF PROOF. The theorem arises in the following way: It is known that the integral of a bounded function over the interval (0, 1) can be approximated by a Gauss sum with any desired degree of accuracy by choosing a division of the interval according to the zeros of a Legendre polynomial of a sufficiently high degree. The integral which occurs on the right-hand side of the functional equation for $H(\mu)$ can, therefore, be replaced by the Gauss sum

$$\sum_{j=1}^{n} \frac{a_j H(\mu_j) \Psi(\mu_j)}{\mu + \mu_j} \tag{158}$$

to any desired accuracy by choosing a sufficiently large n. But by Theorem 1, for a finite n, no matter how large, the unique solution of the equation

$$H(\mu) = 1 + \mu H(\mu) \sum_{j=1}^{n} \frac{a_j H(\mu_j) \Psi(\mu_j)}{\mu + \mu_j} \tag{159}$$

is

$$\frac{1}{\mu_1 \cdots \mu_n} \frac{\prod_{i=1}^{n} (\mu + \mu_i)}{\prod_{\alpha=1}^{n} (1 + k_\alpha \mu)} . \tag{160}$$

If we now let $n \to \infty$, equation (159) becomes

$$H(\mu) = 1 + \mu H(\mu) \int_0^1 \frac{H(\mu')\Psi(\mu')}{\mu + \mu'} d\mu' . \tag{161}$$

The solution of this functional equation is, therefore, seen to be the limit of the function (160) as $n \to \infty$.

It is realized that the sketch of the proof of Theorem 2 we have just outlined does not meet the full demands of a rigorous mathematical demonstration. It is, moreover, probable that precisely the questions of uniqueness and existence which we have ignored will cause the principal difficulties in the constructions of a rigorous mathematical proof. However, as the equation arises in a physical context, and the physical situations are such as to leave no room for ambiguity, it is hardly to be doubted that the theorem is true. Indeed, as we shall presently show quite rigorously by an entirely different line of argument, the exact solutions for the various transfer problems are of the forms tabulated in §22 with the H-functions redefined in terms of functional equations of the form (161) instead of in terms of the Gaussian division and characteristic roots. Nevertheless, it may be of interest to pursue further the purely mathematical questions raised by Theorem 2.

26. **A practical method of determining the exact H-functions as solutions of the functional equations they satisfy.** Assuming for the present the indications of the preceding sections that the exact solutions for the various transfer problems are of the *forms* found in our method of solution and that the H-functions which occur in them have to be redefined as solutions of functional equations of the form (161), we may observe that Theorem 1 of §24 suggests a simple practical method for determining the exact H-functions numerically. For, starting with an approximate solution for $H(\mu)$ (in the third approximation, for example) we can determine the exact H-functions by a process of iteration using for this purpose the functional equation which it satisfies. In this manner the exact H-functions which occur in the solutions of the various transfer problems involving isotropic scattering with an albedo $\lambda \leqq 1$, Rayleigh phase function and Rayleigh scattering (including the state of polarization of the scattered radiation) have all been numerically evaluated. We have therefore now available exact numerical solutions for the cases A, B, C (for $x=0$) and D tabulated in §22.

27. **The constants in the solution.** It is to be noted that the determination of the limit to which the H-functions which occur in the solutions for the emergent (or diffusely reflected) radiation in the nth approximation tend, as $n \rightarrow \infty$, still leaves open the question of the exact limiting values of the constants which occur in these solutions. It does not seem that any direct or simple limiting process can be applied to the formulae which define them in the nth approximation. Attention may be particularly drawn in this connection to the fact that the formulae which define these constants (in the nth approximation) often involve the values of H-functions (now defined as rational functions in terms of the Gaussian division and the characteristic

roots) *outside* the interval (0, 1) (sometimes even for complex values of the argument!), whereas it would seem that in the limit of infinite approximation $H(\mu)$ has a meaning only in the interval (0, 1). (See, for example, the definitions of the constants c and q under the headings B and D in the tabulation of §22.) However, it appears that these constants can be determined *indirectly* by appealing to certain other identical relations which the problems must satisfy.

Thus, considering the transfer problems involving the Rayleigh phase function, we have

$$I(\mu) = \frac{3}{4} qFH^{(0)}(\mu), \tag{162}$$

where $H^{(0)}(\mu)$ is defined as the solution of the functional equation

$$H^{(0)}(\mu) = 1 + \frac{3}{16} \mu H^{(0)}(\mu) \int_0^1 \frac{(3 - \mu'^2)H(\mu')}{\mu + \mu'} d\mu'. \tag{163}$$

Now, as equation (162) gives the angular distribution of the emergent radiation for the axially symmetric problem with constant net flux, it follows that the outward flux of the emergent radiation must also equal πF. In other words, we must have

$$F = 2 \int_0^1 I(\mu) \mu d\mu = \frac{3}{2} qF\alpha_1, \tag{164}$$

where α_1 denotes the first moment of $H^{(0)}(\mu)$. Hence,

$$q = \frac{2}{3\alpha_1} \cdot \tag{165}$$

Thus, once the solution of equation (163) has been determined (by iteration based on an approximate $H(\mu)$ as suggested in §26) the constant q can be determined directly in terms of its first moment. Since q is the only constant which occurs in the solutions, they become determinate in this fashion.

Similarly, in the transfer problems in which the polarization of the scattered radiation in accordance with Rayleigh's law has to be properly allowed for, the solutions again involve a constant c. It does not seem possible to pass directly to the limit of infinite approximation in the formula defining this constant in the nth approximation. But we can determine it by appealing to the flux condition in the problem with the constant net flux. Thus, with the solutions $I_l(\mu)$ and $I_r(\mu)$ for the emergent radiations in the two states of polarization as given in §22 (with the functions $H_l(\mu)$ and $H_r(\mu)$ now defined properly in terms of functional equations of the form (161)), we must have

$$
\begin{aligned}
F &= 2\int_0^1 [I_l(\mu) + I_r(\mu)]\mu d\mu \\
&= \frac{3}{4} F\left[(1 - c^2)^{1/2}\alpha_1 + \frac{1}{2^{1/2}}(A_2 + cA_1)\right]
\end{aligned}
\tag{166}
$$

where α_1 denotes the first moment of $H_l(\mu)$ and A_1 and A_2 are the first and the second moments, respectively, of $H_r(\mu)$. Hence, we can determine c from the equation

$$
(2(1 - c^2))^{1/2}\alpha_1 + A_2 + cA_1 = \frac{4}{3} 2^{1/2}. \tag{167}
$$

28. **The functional equation for the problem of diffuse reflection.** The discussion in the preceding sections has shown how we can obtain in *practice* the exact solutions for the angular distribution of the emergent and the reflected radiations from a semi-infinite plane-parallel atmosphere for a wide variety of scattering laws. From a strictly mathematical point of view, the limiting process by which the passage to the limit of infinite approximation was achieved may not have been as rigorously justified as one might have wished. We shall, therefore, now show how the exact solutions obtained in the manner of the preceding sections (by redefining, for example, the H-functions which occur in the solutions in the general nth approximation, in terms of functional equations of the form (161)) can be justified by following an entirely different line of argument. The basic idea in the development we are now going to describe is due to the Armenian astrophysicist, V. A. Ambarzumian.

In the problem of diffuse reflection, we are interested in the solution of the relevant equations of transfer principally, only to the extent that we want to establish the law of diffuse reflection as specified by the $\sigma(\mu, \varphi; \mu_0, \varphi_0)$ which gives the intensity reflected in the direction (μ, φ) when a parallel beam of radiation of unit flux normal to itself is incident on the atmosphere in the direction $(-\mu_0, \varphi_0)$. Now, Ambarzumian starts with the almost trivial observation that the intensity $\sigma(\mu, \varphi; \mu_0, \varphi_0)$ *must be invariant to the addition (or removal) of layers of arbitrary optical thickness to (from) the atmosphere* and shows (and this is really the point of the observation) how this invariance can be used to derive a *functional equation* for the *scattering function* $\sigma(\mu, \varphi; \mu_0, \varphi_0)\mu$. Ambarzumian has explicitly derived the form of this functional equation for the law of diffuse reflection from an atmosphere scattering radiation in accordance with a general phase function, that is, the functional equation associated with the equation of transfer (24). However, when one proceeds to solve the resulting functional equation, one is soon led to simultaneous systems of nonlinear, nonhomogeneous functional equations of such a highly complex nature that one might almost despair of solving them! But, on examination, it soon appears that a knowledge of the *forms* of the solution obtained by the method described in the earlier parts of this lecture enables us, in all cases, to reduce the Ambarzumian type of functional equations to equations of the following standard form

$$H(\mu) = 1 + \mu H(\mu) \int_0^1 \frac{H(\mu')\Psi(\mu')}{\mu + \mu'} d\mu', \tag{168}$$

and helps us, moreover, to confirm the results obtained by our method of passage to the limit of infinite approximation.

29. **The functional equation for the scattering matrix.** In this section we shall derive, following Ambarzumian's general ideas, the functional equation for the scattering matrix $\boldsymbol{S}$ introduced in §11 (equation (34)). This problem is, therefore, more advanced than the ones concered by Ambarzumian; but it serves to illustrate the power of his idea.

To obtain the functional equation for S, we first rewrite the equation of transfer VII in the form

$$\mu \frac{d\boldsymbol{I}}{d\tau} = \boldsymbol{I}(\tau, \mu, \varphi) - \boldsymbol{B}(\tau, \mu, \varphi), \tag{169}$$

where

$$\begin{aligned} \boldsymbol{B}(\tau, \mu, \varphi) = {} & \frac{3}{16\pi} \boldsymbol{Q} \int_{-1}^{+1} \int_0^{2\pi} \boldsymbol{J}(\mu, \varphi; -\mu', \varphi') \boldsymbol{I}(\tau, \mu', \varphi') d\mu' d\varphi' \\ & + \frac{3}{16} \boldsymbol{Q} \boldsymbol{J}(\mu, \varphi; \mu_0, \varphi_0) \boldsymbol{F} e^{-\tau/\mu_0}, \end{aligned} \tag{170}$$

and express the intensity $\boldsymbol{I}(\mu, \varphi; \mu_0, \varphi_0)$ reflected in the direction (μ, φ), where radiation with a net flux $\pi \boldsymbol{F}$ is incident in the direction $(-\mu_0, \varphi_0)$, in the form

$$\boldsymbol{I}(\mu, \varphi; \mu_0, \varphi_0) = \frac{3}{16\mu} \boldsymbol{Q} \boldsymbol{S}(\mu, \varphi; \mu_0, \varphi_0) \boldsymbol{F}. \tag{171}$$

Now, consider a level at a depth $d\tau$ below the boundary of the atmosphere at $\tau=0$. At this level the radiation field present can be decomposed into two parts: first, there is the reduced incident flux of amount

$$\pi F\left(1-\frac{d\tau}{\mu_0}\right) \tag{172}$$

and, second, there is a diffuse radiation field. The amount of this diffuse radiation field which is directed inward can be inferred from the equation of transfer: for, since at $\tau=0$ there is no inward intensity, at the level $d\tau$, we must have an inward intensity

$$I(d\tau,\,-\mu',\,\varphi') = B(0,\,-\mu',\,\varphi')\,\frac{d\tau}{\mu'} \tag{173}$$

in the direction $(-\mu',\,\varphi')$. Both of these radiation fields will be reflected by the atmosphere below $d\tau$ by the *same laws* as those by which the atmosphere below $\tau=0$ reflects. *This invariance is due to the fact that the removal of a layer of arbitrary thickness from a semi-infinite atmosphere cannot alter its reflecting power.* This is Ambarzumian's basic idea. Accordingly, the reflection of the radiations (172) and (173) by the atmosphere below $d\tau$ will contribute to an *outward* intensity, in the direction $(\mu,\,\varphi)$, the amount

$$\begin{aligned} I(d\tau,\,\mu,\,\varphi) &= \frac{3}{16\mu}\left(1-\frac{d\tau}{\mu_0}\right)QS(\mu,\,\varphi;\,\mu_0,\,\varphi_0)F \\ &\quad+\frac{3}{16\pi\mu}\,d\tau Q\int_0^1\int_0^{2\pi} S(\mu,\,\varphi;\,\mu',\,\varphi')B(0,\,-\mu',\,\varphi')\,\frac{d\mu}{\mu'}\,d\varphi'. \end{aligned} \tag{174}$$

On the other hand, from the equation of transfer, we conclude that

$$\begin{aligned} I(d\tau,\,\mu,\,\varphi) &= I(0,\,\mu,\,\varphi)+\frac{d\tau}{\mu}\,[I(0,\,\mu,\,\varphi)-B(0,\,\mu,\,\varphi)] \\ &= \frac{3}{16\mu}\left(1+\frac{d\tau}{\mu}\right)QS(\mu,\,\varphi;\,\mu_0,\,\varphi_0)F-\frac{d\tau}{\mu}\,B(0,\,\mu,\,\varphi). \end{aligned} \tag{175}$$

Combining equations (174) and (175) and passing to the limit $d\tau=0$, we have

$$\begin{aligned} \frac{3}{16}\left(\frac{1}{\mu}+\frac{1}{\mu_0}\right)QS(\mu,\,\varphi;\,\mu_0,\,\varphi_0)F &= B(0,\,\mu,\,\varphi) \\ &\quad+\frac{3}{16\pi}\,Q\int_0^1\int_0^{2\pi} S(\mu,\,\varphi;\,\mu',\,\varphi')B(0,\,-\mu',\,\varphi')\,\frac{d\mu'}{\mu'}\,d\varphi'. \end{aligned} \tag{176}$$

But, according to equations (170) and (171)

$$
\begin{aligned}
B(0, \mu, \varphi) &= \frac{3}{16} Q \Bigg[J(\mu, \varphi; \mu_0, \varphi_0) \\
&+ \frac{3}{16\pi} \int_0^1 \int_0^{2\pi} J(\mu, \varphi; -\mu'', \varphi'') Q S(\mu'', \varphi''; \mu_0, \varphi_0) \frac{d\mu''}{\mu''} d\varphi'' \Bigg] F.
\end{aligned} \tag{177}
$$

Substituting for $B(0, \mu, \varphi)$ from the foregoing equation in equation (176) and remembering that F can be an arbitrary vector, we find, after some minor reductions, that

$$
\begin{aligned}
\left(\frac{1}{\mu} + \frac{1}{\mu_0}\right) S(\mu, \varphi; \mu_0, \varphi_0) &= J(\mu, \varphi; \mu_0, \varphi_0) \\
&+ \frac{3}{16\pi} \int_0^1 \int_0^{2\pi} J(\mu, \varphi; -\mu'', \varphi'') Q S(\mu'', \varphi''; \mu_0, \varphi_0) \frac{d\mu''}{\mu''} d\varphi'' \\
&+ \frac{3}{16\pi} \int_0^1 \int_0^{2\pi} S(\mu, \varphi; \mu', \varphi') Q J(-\mu', \varphi'; \mu_0, \varphi_0) \frac{d\mu'}{\mu'} d\varphi' \\
&+ \frac{9}{256\pi^2} \int_0^1 \int_0^{2\pi} \int_0^1 \int_0^{2\pi} S(\mu, \varphi; \mu', \varphi') Q J(-\mu', \varphi'; -\mu'', \varphi'') Q \\
&\qquad \times S(\mu'', \varphi''; \mu_0, \varphi_0) \frac{d\mu'}{\mu'} d\varphi' \frac{d\mu''}{\mu''} d\varphi''.
\end{aligned} \tag{178}
$$

This is the required functional equation for the scattering matrix.

Now, the matrix J (see equation (32)) has the property

$$
J_{ik}(\mu, \varphi; \mu_0, \varphi_0) = J_{ki}(\mu_0, \varphi; \mu, \varphi_0). \tag{179}
$$

From this property of J for transposition, it follows from equation (178) that S has also the same property:

$$
S_{ik}(\mu, \varphi; \mu_0, \varphi_0) = S_{ki}(\mu_0, \varphi; \mu, \varphi_0). \tag{180}
$$

It can be verified that equation (180) is equivalent to *Helmholtz' principle of reciprocity* for the problem under consideration.

30. **The reduction of the functional equation for S.** We shall now indicate how the functional equation for S derived in the preceding section can be solved.

From the form of the equation for S and the manner of its relation to J, it is evident that in a Fourier analysis of the elements of S in $(\varphi - \varphi_0)$ we must have the same nonvanishing components as in the corresponding elements of J. We may accordingly assume without loss of generality that S has the form

$$
\begin{aligned}
&S(\mu, \varphi; \mu_0, \varphi) \\
&\text{(181)} \\
&= \begin{pmatrix}
\begin{aligned} &S_{11}^{(0)}(\mu, \mu_0) \\ &- 4\mu\mu_0(1-\mu^2)^{1/2}(1-\mu_0^2)^{1/2}S_{11}^{(1)}(\mu, \mu_0)\cos(\varphi-\varphi_0) \\ &+ \mu^2\mu_0^2 S_{11}^{(2)}(\mu, \mu_0)\cos 2(\varphi-\varphi_0) \end{aligned} &
\begin{aligned} &S_{12}^{(0)}(\mu, \mu_0) \\ & \\ &- \mu^2 S_{12}^{(2)}(\mu, \mu_0)\cos 2(\varphi-\varphi_0) \end{aligned} &
\begin{aligned} & \\ &- 2\mu(1-\mu^2)^{1/2}(1-\mu_0^2)^{1/2}S_{13}^{(1)}(\mu, \mu_0)\sin(\varphi-\varphi_0) \\ &+ \mu^2\mu_0 S_{13}^{(2)}(\mu, \mu_0)\sin 2(\varphi-\varphi_0) \end{aligned} \\
\begin{aligned} &S_{21}^{(0)}(\mu, \mu_0) \\ &- \mu_0^2 S_{21}^{(2)}(\mu, \mu_0)\cos 2(\varphi-\varphi_0) \end{aligned} &
\begin{aligned} &S_{22}^{(0)}(\mu, \mu_0) \\ &+ S_{22}^{(2)}(\mu, \mu_0)\cos 2(\varphi-\varphi_0) \end{aligned} &
\begin{aligned} & \\ &- \mu_0 S_{23}^{(2)}(\mu, \mu_0)\sin 2(\varphi-\varphi_0) \end{aligned} \\
\begin{aligned} &- 2\mu_0(1-\mu_0^2)^{1/2}(1-\mu^2)^{1/2}S_{31}^{(1)}(\mu, \mu_0)\sin(\varphi-\varphi_0) \\ &+ \mu_0^2\mu S_{31}^{(2)}(\mu, \mu_0)\sin 2(\varphi-\varphi_0) \end{aligned} &
\begin{aligned} & \\ &- \mu S_{32}^{(2)}(\mu, \mu_0)\sin 2(\varphi-\varphi_0) \end{aligned} &
\begin{aligned} &(1-\mu^2)^{1/2}(1-\mu_0^2)^{1/2}S_{33}^{(1)}(\mu, \mu_0)\cos(\varphi-\varphi_0) \\ &- \mu\mu_0 S_{33}^{(2)}(\mu, \mu_0)\cos 2(\varphi-\varphi_0) \end{aligned}
\end{pmatrix}
\end{aligned}
$$

where, as the notation indicates, $S_{11}^{(0)}$, and so on, are all functions of μ and μ_0 only. From the property (180) of $\boldsymbol{S}$ for transposition, we now conclude that

$$S_{jk}^{(i)}(\mu, \mu_0) = S_{kj}^{(i)}(\mu_0, \mu). \tag{182}$$

If we now substitute the form (181) for $\boldsymbol{S}$ in equation (178) and equate the different Fourier components of the various elements, we shall clearly obtain three systems of functional equations governing the functions of the different orders, distinguished by their superscripts. Of these systems, the first, involving the zero-order functions $S_{11}^{(0)}$, $S_{12}^{(0)}$, $S_{21}^{(0)}$ and $S_{22}^{(0)}$, is the most important and, at the same time, the most difficult. We shall accordingly consider this system briefly.

First, we may write down the equations which are found for $S_{11}^{(0)}$, $S_{12}^{(0)}$, $S_{21}^{(0)}$, and $S_{22}^{(0)}$. The equations are

$$\begin{aligned}\left(\frac{1}{\mu}+\frac{1}{\mu_0}\right)&S_{11}^{(0)}(\mu,\mu_0)\\ &= \left\{\mu^2+\frac{3}{8}\int_0^1\frac{d\mu'}{\mu'}\left[\mu'^2S_{11}^{(0)}(\mu,\mu')+S_{12}^{(0)}(\mu,\mu')\right]\right\}\\ &\quad\times\left\{\mu_0^2+\frac{3}{8}\int_0^1\frac{d\mu'}{\mu'}\left[\mu'^2S_{11}^{(0)}(\mu',\mu_0)+S_{21}^{(0)}(\mu',\mu_0)\right]\right\}\\ &\quad+2\left\{1-\mu^2+\frac{3}{8}\int_0^1\frac{d\mu'}{\mu'}(1-\mu'^2)S_{11}^{(0)}(\mu,\mu')\right\}\\ &\quad\times\left\{1-\mu_0^2+\frac{3}{8}\int_0^1\frac{d\mu'}{\mu'}(1-\mu'^2)S_{11}^{(0)}(\mu',\mu_0)\right\};\end{aligned} \tag{183}$$

$$\begin{aligned}\left(\frac{1}{\mu}+\frac{1}{\mu_0}\right)&S_{12}^{(0)}(\mu,\mu_0)\\ &= \left\{\mu^2+\frac{3}{8}\int_0^1\frac{d\mu'}{\mu'}\left[\mu'^2S_{11}^{(0)}(\mu,\mu')+S_{12}^{(0)}(\mu,\mu')\right]\right\}\\ &\quad\times\left\{1+\frac{3}{8}\int_0^1\frac{d\mu'}{\mu'}\left[\mu'^2S_{12}^{(0)}(\mu',\mu_0)+S_{22}^{(0)}(\mu',\mu_0)\right]\right\}\\ &\quad+2\left\{1-\mu^2+\frac{3}{8}\int_0^1\frac{d\mu'}{\mu'}(1-\mu'^2)S_{11}^{(0)}(\mu,\mu')\right\}\\ &\quad\times\left\{\frac{3}{8}\int_0^1\frac{d\mu'}{\mu'}(1-\mu'^2)S_{12}^{(0)}(\mu',\mu_0)\right\};\end{aligned} \tag{184}$$

$$\left(\frac{1}{\mu}+\frac{1}{\mu_0}\right)S_{21}^{(0)}(\mu,\mu_0)$$

$$=\left\{\mu_0^2+\frac{3}{8}\int_0^1\frac{d\mu'}{\mu'}\left[\mu'^2S_{11}^{(0)}(\mu',\mu_0)+S_{21}^{(0)}(\mu',\mu_0)\right]\right\}$$

$$\times\left\{1+\frac{3}{8}\int_0^1\frac{d\mu'}{\mu'}\left[\mu'^2S_{21}^{(0)}(\mu,\mu')+S_{22}^{(0)}(\mu,\mu')\right]\right\} \tag{185}$$

$$+2\left\{1-\mu_0^2+\frac{3}{8}\int_0^1\frac{d\mu'}{\mu'}(1-\mu'^2)S_{11}^{(0)}(\mu',\mu_0)\right\}$$

$$\times\left\{\frac{3}{8}\int_0^1\frac{d\mu'}{\mu'}(1-\mu'^2)S_{21}^{(0)}(\mu,\mu')\right\};$$

$$\left(\frac{1}{\mu}+\frac{1}{\mu_0}\right)S_{22}^{(0)}(\mu,\mu_0)$$

$$=\left\{1+\frac{3}{8}\int_0^1\frac{d\mu'}{\mu'}\left[\mu'^2S_{12}^{(0)}(\mu',\mu_0)+S_{22}^{(0)}(\mu',\mu_0)\right]\right\}$$

$$\times\left\{1+\frac{3}{8}\int_0^1\frac{d\mu'}{\mu'}\left[\mu'^2S_{21}^{(0)}(\mu,\mu')+S_{22}^{(0)}(\mu,\mu')\right]\right\} \tag{186}$$

$$+\frac{9}{32}\int_0^1\frac{d\mu'}{\mu'}(1-\mu'^2)S_{21}^{(0)}(\mu,\mu')$$

$$\times\int_0^1\frac{d\mu'}{\mu'}(1-\mu'^2)S_{12}^{(0)}(\mu',\mu_0).$$

An inspection of these equations shows what we have already seen from the functional equation for **S**, that among these functions of zero-order the relation (182) must hold in particular. In other words

$$S_{11}^{(0)}(\mu,\mu_0)=S_{11}^{(0)}(\mu_0,\mu);\qquad S_{12}^{(0)}(\mu,\mu_0)=S_{21}^{(0)}(\mu_0,\mu);$$
$$S_{22}^{(0)}(\mu,\mu_0)=S_{22}^{(0)}(\mu_0,\mu). \tag{187}$$

In view of these relations, it follows from equations (183)–(186) that we can express the functions $S_{11}^{(0)}$, $S_{12}^{(0)}$, $S_{21}^{(0)}$ and $S_{22}^{(0)}$ in the forms

$$\left(\frac{1}{\mu}+\frac{1}{\mu_0}\right)S_{11}^{(0)}(\mu,\mu_0)=\psi(\mu)\psi(\mu_0)+2\phi(\mu)\phi(\mu_0), \tag{188}$$

$$\left(\frac{1}{\mu}+\frac{1}{\mu_0}\right)S_{12}^{(0)}(\mu,\mu_0)=\psi(\mu)\chi(\mu_0)+2\phi(\mu)\zeta(\mu_0), \tag{189}$$

$$\left(\frac{1}{\mu}+\frac{1}{\mu_0}\right)S_{21}^{(0)}(\mu,\mu_0)=\psi(\mu_0)\chi(\mu)+2\phi(\mu_0)\zeta(\mu), \tag{190}$$

and

$$\left(\frac{1}{\mu}+\frac{1}{\mu_0}\right)S_{22}^{(0)}(\mu,\mu_0)=\chi(\mu)\chi(\mu_0)+2\zeta(\mu)\zeta(\mu_0), \tag{191}$$

where

$$\psi(\mu)=\mu^2+\frac{3}{8}\int_0^1\frac{d\mu'}{\mu'}\left[\mu'^2S_{11}^{(0)}(\mu,\mu')+S_{12}^{(0)}(\mu,\mu')\right], \tag{192}$$

$$\phi(\mu)=1-\mu^2+\frac{3}{8}\int_0^1\frac{d\mu'}{\mu'}(1-\mu'^2)S_{11}^{(0)}(\mu,\mu'), \tag{193}$$

$$\chi(\mu)=1+\frac{3}{8}\int_0^1\frac{d\mu'}{\mu'}\left[\mu'^2S_{21}^{(0)}(\mu,\mu')+S_{22}^{(0)}(\mu,\mu')\right], \tag{194}$$

and

$$\zeta(\mu)=\frac{3}{8}\int_0^1\frac{d\mu'}{\mu'}(1-\mu'^2)S_{21}^{(0)}(\mu,\mu'). \tag{195}$$

Substituting for $S_{11}^{(0)}$, and so on, from equations (188)–(191) back into equations (192)–(195) we obtain the functional equations for the problem in their normal forms. We have

$$\begin{aligned}\psi(\mu)&=\mu^2+\frac{3}{8}\mu\psi(\mu)\int_0^1\frac{d\mu'}{\mu+\mu'}\left[\mu'^2\psi(\mu')+\chi(\mu')\right]\\&\quad+\frac{3}{4}\mu\phi(\mu)\int_0^1\frac{d\mu'}{\mu+\mu'}\left[\mu'^2\phi(\mu')+\zeta(\mu')\right],\end{aligned} \tag{196}$$

$$\begin{aligned}\phi(\mu)&=1-\mu^2+\frac{3}{8}\mu\psi(\mu)\int_0^1\frac{d\mu'}{\mu+\mu'}(1-\mu'^2)\psi(\mu')\\&\quad+\frac{3}{4}\mu\phi(\mu)\int_0^1\frac{d\mu'}{\mu+\mu'}(1-\mu'^2)\phi(\mu'),\end{aligned} \tag{197}$$

$$\begin{aligned}\chi(\mu)&=1+\frac{3}{8}\mu\chi(\mu)\int_0^1\frac{d\mu'}{\mu+\mu'}\left[\mu'^2\psi(\mu')+\chi(\mu')\right]\\&\quad+\frac{3}{4}\mu\zeta(\mu)\int_0^1\frac{d\mu'}{\mu+\mu'}\left[\mu'^2\phi(\mu')+\zeta(\mu')\right],\end{aligned} \tag{198}$$

(199)
$$\zeta(\mu) = \frac{3}{8}\mu\chi(\mu)\int_0^1 \frac{d\mu'}{\mu+\mu'}(1-\mu'^2)\psi(\mu') + \frac{3}{4}\mu\zeta(\mu)\int_0^1 \frac{d\mu'}{\mu+\mu'}(1-\mu'^2)\phi(\mu').$$

31. **The solution of the functional equations** (196)–(199). Equations (196)–(199) represent a nonlinear, nonhomogeneous system of four simultaneous functional equations which one might despair of even attempting to solve. However, with the guidance provided by the form of the solution obtained in our general nth approximation, it is possible to reduce the solution of equations (196)–(199) to two simple functional equations, each of form (168). Thus, it can be verified by direct substitution that the solution of the system of equations (196)–(199) is given by

(200) $$\psi(\mu) = (2(1-c^2))^{1/2}\mu H_l(\mu),$$

(201) $$\phi(\mu) = H_l(\mu)(1-c\mu),$$

(202) $$\chi(\mu) = H_r(\mu)(1+c\mu),$$

and

(203) $$\zeta(\mu) = \frac{(1-c^2)^{1/2}}{2^{1/2}}\mu H_r(\mu),$$

where $H_l(\mu)$ and $H_r(\mu)$ are defined in terms of the functional equations

(204) $$H_l(\mu) = 1 + \frac{3}{4}\mu H_l(\mu)\int_0^1 \frac{H_l(\mu')}{\mu+\mu'}(1-\mu'^2)d\mu'$$

and

(205) $$H_r(\mu) = 1 + \frac{3}{8}\mu H_r(\mu)\int_0^1 \frac{H_r(\mu')}{\mu+\mu'}(1-\mu'^2)d\mu',$$

and c is a constant related in a determinate way with the moments of $H_l(\mu)$ and $H_r(\mu)$. We find

(206) $$c = \frac{8(A_1-\alpha_1) + 3(2\alpha_1\alpha_0 - A_1A_0)}{3(A_1^2 + 2\alpha_1^2)},$$

where α_0, A_0 and α_1, A_1 are the moments of order zero and one of $H_l(\mu)$ and $H_r(\mu)$, respectively.

With $\psi(\mu)$, $\phi(\mu)$, and so on, defined as in equations (200)–(203) we can verify that the solutions for $S_{11}^{(0)}$, and so on, are in entire agreement with our earlier results obtained by passing to the limit of infinite approximation.

32. **The completion of the solution for *S*.** The discussion of the other two systems for the functions of order one and two turns out to be very simple, as it appears that all the functions $S^{(1)}$ are equal to each other and similarly all the functions $S^{(2)}$ are equal to each other. Therefore, writing

$$S_{i,j}^{(1)}(\mu, \mu_0) = S^{(1)}(\mu, \mu_0), \tag{207}$$

and

$$S_{i,j}^{(2)}(\mu, \mu_0) = S^{(2)}(\mu, \mu_0), \tag{208}$$

it is found that the equations governing $S^{(1)}$ and $S^{(2)}$ are

$$\begin{aligned}\left(\frac{1}{\mu}+\frac{1}{\mu_0}\right)&S^{(1)}(\mu, \mu_0)\\ &= \left\{1+\frac{3}{8}\int_0^1 \frac{d\mu'}{\mu'}(1-\mu'^2)(1+2\mu'^2)S^{(1)}(\mu', \mu_0)\right\}\\ &\quad\times\left\{1+\frac{3}{8}\int_0^1 \frac{d\mu'}{\mu'}(1-\mu'^2)(1+2\mu'^2)S^{(1)}(\mu, \mu')\right\},\end{aligned} \tag{209}$$

and

$$\begin{aligned}\left(\frac{1}{\mu}+\frac{1}{\mu_0}\right)S^{(2)}(\mu, \mu_0) &= \left\{1+\frac{3}{16}\int_0^1 \frac{d\mu'}{\mu'}(1+\mu'^2)^2S^{(2)}(\mu', \mu_0)\right\}\\ &\quad\times\left\{1+\frac{3}{16}\int_0^1 \frac{d\mu'}{\mu'}(1+\mu'^2)^2S^{(2)}(\mu, \mu')\right\}.\end{aligned} \tag{210}$$

From these equations it follows that the functions $S^{(1)}(\mu, \mu_0)$ and $S^{(2)}(\mu, \mu_0)$ are symmetrical in the variables μ and μ_0 and that they are expressible in the forms

$$\left(\frac{1}{\mu}+\frac{1}{\mu_0}\right)S^{(1)}(\mu, \mu_0) = H^{(1)}(\mu)H^{(1)}(\mu_0), \tag{211}$$

and

$$\left(\frac{1}{\mu}+\frac{1}{\mu_0}\right)S^{(2)}(\mu, \mu_0) = H^{(2)}(\mu)H^{(2)}(\mu_0) \tag{212}$$

where $H^{(1)}(\mu)$ and $H^{(2)}(\mu)$ are solutions of the functional equations

$$(213) \quad H^{(1)}(\mu) = 1 + \frac{3}{8}\mu H^{(1)}(\mu)\int_0^1 \frac{H^{(1)}(\mu')}{\mu+\mu'}(1-\mu'^2)(1+2\mu'^2)d\mu',$$

and

$$(214) \quad H^{(2)}(\mu) = 1 + \frac{3}{16}\mu H^{(2)}(\mu)\int_0^1 \frac{H^{(2)}(\mu')}{\mu+\mu'}(1+\mu'^2)^2 d\mu'.$$

With this, the solution of the functional equation for $\boldsymbol{S}$ is completed and it will be observed that the solution for $\boldsymbol{S}$ which we have now obtained is of exactly the same form as that given in §22 (under D) with the only difference that the H-functions which appear in the solution are now defined in terms of the exact functional equations which they satisfy; further, the constant c is shown to be related in a definite way with the moments of $H_l(\mu)$ and $H_r(\mu)$.

33. **A class of functional equations and their solution.** The solution of the functional equations for the problem of diffuse reflection for laws of scattering, other than the one we have described, can be carried out in an analogous manner. It is not our intention to go into the details of the solution of these other cases here, but it may be of some mathematical interest to see the type of functional equations which these problems lead us to consider.

The problem of diffuse reflection in accordance with Rayleigh's phase function leads to the following simultaneous pair of functional equations:

$$(215) \quad \begin{aligned} \psi(\mu) = 3 - \mu^2 &+ \frac{1}{16}\mu\psi(\mu)\int_0^1 \frac{\psi(\mu')}{\mu+\mu'}(3-\mu'^2)d\mu' \\ &+ \frac{1}{2}\mu\phi(\mu)\int_0^1 \frac{\phi(\mu')}{\mu+\mu'}(3-\mu'^2)d\mu' \end{aligned}$$

and

$$(216) \quad \begin{aligned} \phi(\mu) = \mu^2 &+ \frac{1}{16}\mu\psi(\mu)\int_0^1 \frac{\psi(\mu')}{\mu+\mu'}\mu'^2 d\mu' \\ &+ \frac{1}{2}\mu\phi(\mu)\int_0^1 \frac{\phi(\mu')}{\mu+\mu'}\mu'^2 d\mu'. \end{aligned}$$

Again, guided by the form of the solution obtained by the direct solution of the equation of transfer (see the tabulation in §22, under B), we are led to surmise that the solutions of equations (215) and

(216) must be of the forms

$$\psi(\mu) = H^{(0)}(\mu)(3 - c\mu) \tag{217}$$

and

$$\phi(\mu) = q\mu H^{(0)}(\mu), \tag{218}$$

where $H^{(0)}(\mu)$ satisfies the functional equation

$$H^{(0)}(\mu) = 1 + \frac{3}{16}\mu H^{(0)}(\mu)\int_0^1 \frac{H^{(0)}(\mu')}{\mu + \mu'}(3 - \mu'^2)d\mu', \tag{219}$$

and q and c are two constants related in the manner

$$8q^2 = 3 - c^2. \tag{220}$$

Direct substitution confirms that the solutions of equations (215) and (216) are indeed of the forms surmised and shows further that, in agreement with equation (165),

$$q = \frac{2}{3\alpha_1} \tag{221}$$

where α_1 is the first moment of $H^{(0)}(\mu)$.

Similarly, the problem of diffuse reflection in accordance with the phase function $\lambda(1+x \cos \Theta)$ leads to the following pair of functional equations:

$$\begin{aligned}\psi(\mu) = 1 &+ \frac{1}{2}\lambda\mu\psi(\mu)\int_0^1 \frac{\psi(\mu')}{\mu + \mu'}d\mu' \\ &- \frac{1}{2}x\lambda\mu\phi(\mu)\int_0^1 \frac{\phi(\mu')}{\mu + \mu'}d\mu',\end{aligned} \tag{222}$$

and

$$\begin{aligned}\phi(\mu) = \mu &- \frac{1}{2}\lambda\mu\psi(\mu)\int_0^1 \frac{\psi(\mu')}{\mu + \mu'}\mu' d\mu' \\ &+ \frac{1}{2}x\lambda\mu\phi(\mu)\int_0^1 \frac{\phi(\mu')}{\mu + \mu'}\mu' d\mu'.\end{aligned} \tag{223}$$

And, it is found by direct verification that the solution of these equations can be expressed in the form

$$\psi(\mu) = H^{(0)}(\mu)(1 - c\mu), \tag{224}$$

and

$$\phi(\mu) = q\mu H^{(0)}(\mu), \tag{225}$$

where $H^{(0)}(\mu)$ now satisfies the functional equation

$$H^{(0)}(\mu) = 1 + \frac{1}{2}\lambda\mu H^{(0)}(\mu)\int_0^1 \frac{H^{(0)}(\mu')}{\mu + \mu'}[1 + x(1 - \lambda)\mu'^2]d\mu', \tag{226}$$

and q and c are two constants related in the manner

$$xq^2 = c^2 + x(1 - \lambda), \tag{227}$$

and given explicitly by the formulae

$$q = \frac{2(1 - \lambda)}{2 - \lambda\alpha_0} \quad \text{and} \quad c = x\lambda(1 - \lambda)\frac{\alpha_1}{2 - \lambda\alpha_0}, \tag{228}$$

α_0 and α_1 being the moments of order zero and one of $H^{(0)}(\mu)$.

34. **Some general remarks.** In some ways it is remarkable that systems of functional equations as complex in appearance as equations (196)–(199), (215)–(216) and (222)–(223) are should be capable of being reduced to single functional equations of the form (168). There must clearly be something in the structure of these equations which makes this reduction possible. But, as to what it precisely is, is at present shrouded in mystery!

35. **Some integral properties of the functions** $H(\mu)$. The discussion in the preceding sections has disclosed the important role which functional equations of the form

$$H(\mu) = 1 + \mu H(\mu)\int_0^1 \frac{H(\mu')}{\mu + \mu'}\Psi(\mu')d\mu' \tag{229}$$

play in the theory of radiative transfer. The investigation of the properties of these equations is, therefore, a matter of considerable interest. Of course, from the practical standpoint of solving such equations numerically, the most important property is that derived from Theorem 1 (§24), namely, that when we replace the integral on the right-hand side by a Gauss sum, the solution can be explicitly written down as a rational function involving the points of the Gaussian division and the roots of the associated characteristic equation

$$1 = 2\sum_{j=1}^{n} \frac{a_j\Psi(\mu_j)}{1 - k^2\mu_j^2}. \tag{230}$$

However, in addition to this property, there are a number of integral theorems (of an essentially elementary kind) which can be proved for functions satisfying equations of the form (229). We shall give two examples.

THEOREM 3. $\int_0^1 H(\mu)\Psi(\mu)d\mu = 1 - [1 - 2\int_0^1 \Psi(\mu)d\mu]^{1/2}$.

PROOF. Multiplying the equation satisfied by $H(\mu)$ by $\Psi(\mu)$ and integrating over the range of μ, we have

$$\int_0^1 H(\mu)\Psi(\mu)d\mu = \int_0^1 \Psi(\mu)d\mu + \int_0^1 \int_0^1 \frac{\mu}{\mu + \mu'} H(\mu)\Psi(\mu)H(\mu')\Psi(\mu')d\mu d\mu'. \tag{231}$$

Interchanging μ and μ' in the double integral in equation (231) and taking the average of the two equations, we obtain

$$\int_0^1 H(\mu)\Psi(\mu)d\mu = \int_0^1 \Psi(\mu)d\mu + \frac{1}{2}\int_0^1 \int_0^1 H(\mu)\Psi(\mu)H(\mu')\Psi(\mu')d\mu d\mu' \tag{232}$$

or, alternatively,

$$\frac{1}{2}\left[\int_0^1 H(\mu)\Psi(\mu)d\mu\right]^2 - \int_0^1 H(\mu)\Psi(\mu)d\mu + \int_0^1 \Psi(\mu)d\mu = 0. \tag{233}$$

Solving this equation for the integral in question, we have

$$\int_0^1 H(\mu)\psi(\mu)d\mu = 1 \pm \left[1 - 2\int_0^1 \Psi(\mu)d\mu\right]^{1/2}. \tag{234}$$

The ambiguity in the sign in equation (234) can be removed by the consideration that the integral on the left-hand side must uniformly converge to zero when $\Psi(\mu)$ tends to zero uniformly in the interval (0, 1). This requires us to choose the negative sign in equation (234) and the result stated follows.

COROLLARY. *A necessary and sufficient condition that $H(\mu)$ be real is*

$$\int_0^1 \Psi(\mu)d\mu \leqq \frac{1}{2}\cdot$$

This is, of course, an immediate consequence of the theorem.

The physical meaning of the limitation on $\Psi(\mu)$ implied by this corollary is interesting: it is really equivalent to the condition that, on each scattering, more radiation should not be emitted than was incident; further, the equality sign is admissible only in the case of perfect scattering in the sense of equation (11).

THEOREM 4. $[1-2\int_0^1\Psi(\mu)d\mu]^{1/2}\int_0^1 H(\mu)\Psi(\mu)\mu^2 d\mu+[\int_0^1 H(\mu)\Psi(\mu)\mu d\mu]^2/2 =\int_0^1\Psi(\mu)\mu^2 d\mu.$

PROOF. To prove this theorem, we multiply the equation defining $H(\mu)$ by $\Psi(\mu)\mu^2$ and integrate over the range of μ. We find

$$\begin{aligned}
&\int_0^1 H(\mu)\Psi(\mu)\mu^2 d\mu \\
&= \int_0^1 \Psi(\mu)\mu^2 d\mu + \int_0^1\int_0^1 \frac{H(\mu)H(\mu')\Psi(\mu)\Psi(\mu')}{\mu+\mu'}\mu^3 d\mu d\mu' \\
&= \int_0^1 \Psi(\mu)\mu^2 d\mu \\
&\quad + \frac{1}{2}\int_0^1\int_0^1 \frac{H(\mu)H(\mu')\Psi(\mu)\Psi(\mu')}{\mu+\mu'}(\mu^3+\mu'^3) d\mu d\mu' \\
&= \int_0^1 \Psi(\mu)\mu^2 d\mu \\
&\quad + \frac{1}{2}\int_0^1\int_0^1 H(\mu)H(\mu')\Psi(\mu)\Psi(\mu')(\mu^2-\mu\mu'+\mu'^2) d\mu d\mu' \\
&= \int_0^1 \Psi(\mu)\mu^2 d\mu + \left[\int_0^1 H(\mu)\Psi(\mu)\mu^2 d\mu\right]\left[\int_0^1 H(\mu)\Psi(\mu)d\mu\right] \\
&\quad - \frac{1}{2}\left[\int_0^1 H(\mu)\Psi(\mu)\mu d\mu\right]^2 .
\end{aligned} \tag{235}$$

Using Theorem 3 we obtain, after some minor reductions,

$$\begin{aligned}
&\left[1-2\int_0^1 \Psi(\mu)d\mu\right]^{1/2}\int_0^1 H(\mu)\Psi(\mu)\mu^2 d\mu \\
&\qquad + \frac{1}{2}\left[\int_0^1 H(\mu)\Psi(\mu)\mu d\mu\right]^2 = \int_0^1 \Psi(\mu)\mu^2 d\mu,
\end{aligned} \tag{236}$$

which is the required result.

COROLLARY. *For the case of a perfectly scattering atmosphere when*

$$\int_0^1 \Psi(\mu)d\mu = 1/2,$$

we have the further integral

$$\int_0^1 H(\mu)\Psi(\mu)\mu d\mu = \left[2\int_0^1 \Psi(\mu)\mu^2 d\mu \right]^{1/2}.$$

The corollary we have just stated generalizes a classical result of Hopf and Bronstein for the case of an isotropically scattering atmosphere. For, in this latter case

$$\Psi(\mu) = \text{constant} = 1/2, \tag{237}$$

and, according to Theorem 3 and the corollary of Theorem 4, we have

$$\int_0^1 H(\mu)d\mu = 2, \tag{238}$$

$$\int_0^1 H(\mu)\mu d\mu = \frac{2}{3^{1/2}}. \tag{239}$$

Hence

$$\frac{J(0)}{F} = \frac{\int_0^1 H(\mu)d\mu}{4\int_0^1 H(\mu)\mu d\mu} = \frac{3^{1/2}}{4}. \tag{240}$$

This is the Hopf-Bronstein relation. It, therefore, follows that for all cases of perfect scattering we have an integral of the Hopf-Bronstein type which is essentially that given by the corollary of Theorem 4.

36. **The equation of transfer in spherical atmospheres, and its reduction for the case $\kappa\rho\alpha r^{-n}$.** So far we have restricted ourselves to transfer problems in plane parallel atmospheres. We shall now briefly indicate how the methods we have described can be extended to treat transfer problems in spherically symmetric atmospheres. In such cases, the intensity in the radiation field will be a function of the distance r from the center of symmetry and the angle ϑ measured from the positive direction of the radius vector; and the equation of transfer will take the form

$$\mu\frac{\partial I}{\partial r} + \frac{1-\mu^2}{r}\frac{\partial I}{\partial \mu} = -\kappa\rho(I - \mathfrak{J}), \tag{241}$$

where $\mu = \cos\vartheta$ and $\mathfrak{J}$ denotes, as usual, the source function.

In outlining the manner in which equations of transfer of the form (241) can be solved, we shall restrict ourselves to an isotropically scattering atmosphere. In this case equation (241) becomes

$$\mu \frac{\partial I}{\partial r} + \frac{1 - \mu^2}{r} \frac{\partial I}{\partial \mu} = -\kappa\rho I + \frac{1}{2} \kappa\rho \int_{-1}^{+1} I(r, \mu')d\mu'. \tag{242}$$

According to the ideas developed in the earlier parts of this lecture, we shall replace the integral which occurs on the right-hand side of equation (242) by a sum according to Gauss's formula for numerical quadratures, and reduce the integrodifferential equation to the system

$$\mu_i \frac{dI_i}{dr} + \frac{1 - \mu_i^2}{r} \left(\frac{\partial I}{\partial \mu}\right)_{\mu=\mu_i} = -\kappa\rho I_i + \frac{1}{2} \kappa\rho \sum a_j I_j \tag{243}$$

$$(i = \pm 1, \cdots, \pm n)$$

of $2n$ ordinary linear equations in the nth approximation. It is at once seen that our present system of equations differs, in an essential way, from those which we have considered so far: equation (243) now involves $(\partial I/\partial \mu)_{\mu=\mu_i}$ and, before we can proceed any further, we must know the values which we are to assign to $\partial I/\partial \mu$ at the points of the Gaussian division in our present scheme of approximation. At first sight it might be supposed that the assignment of values to $\partial I/\partial \mu$ at $\mu = \mu_i$, $i = \pm 1, \cdots, \pm n$, is largely arbitrary, particularly when n is small. However, on consideration, it appears that this assignment can be done in a satisfactory manner in only one way and, indeed, according to the following device:

Define the polynomials $Q_m(\mu)$ according to the formula

$$P_m(\mu) = -\frac{dQ_m}{d\mu} \qquad (m = 1, \cdots, 2n), \tag{244}$$

and adjust the constant of integration in Q_m by requiring that

$$Q_m = 0 \qquad \text{for } |\mu| = 1. \tag{245}$$

This can always be accomplished, since, when m is odd, Q_m is even and when m is even the indefinite integral of $P_m(\mu)$ already contains $(1 - \mu^2)$ as a factor. The first few polynomials $Q_m(\mu)$ are given below:

m	$P_m(\mu)$	$Q_m(\mu)$	$\mathfrak{Q}_m(\mu)$
1	μ	$(1 - \mu^2)/2$	$1/2$
2	$(3\mu^2 - 1)/2$	$\mu(1 - \mu^2)/2$	$\mu/2$
3	$(5\mu^3 - 3\mu)/2$	$1/8(5\mu^2 - 1)(1 - \mu^2)$	$1/8(5\mu^2 - 1)$
4	$1/8(35\mu^4 - 30\mu^2 + 3)$	$1/8\mu(7\mu^2 - 3)(1 - \mu^2)$	$1/8\mu(7\mu^2 - 3)$
5	$1/8(63\mu^5 - 70\mu^3 + 15\mu)$	$1/16(21\mu^4 - 14\mu^2 + 1)(1 - \mu^2)$	$1/16(21\mu^4 - 14\mu^2 + 1)$

Now, by an integration by parts, we arrive at the identity

$$\int_{-1}^{+1} Q_m(\mu)\frac{\partial I}{\partial\mu}\,d\mu = -\int_{-1}^{+1} I\frac{dQ_m}{d\mu}\,d\mu = \int_{-1}^{+1} IP_m(\mu)d\mu. \tag{246}$$

Expressing the first and the last integrals in equation (246) as sums according to Gauss's formula, we have, in the nth approximation,

$$\sum a_i Q_m(\mu_i)\left(\frac{\partial I}{\partial\mu}\right)_{\mu=\mu_i} = \sum a_i I_i P_m(\mu_i) \qquad (m = 1, \cdots, 2n). \tag{247}$$

Equation (247) provides us with exactly the right number of equations to express $(\partial I/\partial\mu)_{\mu=\mu_i}$ $(i=\pm 1, \cdots, \pm n)$ as linear combinations of I_i. Essentially what equation (247) allows is to determine in a "best possible way" the derivatives of a function in terms of its values at the points of the Gaussian division. This problem has apparently not been considered before.

Returning to equation (243) we now observe that this equation, together with equation (246), provides the required reduction of the equation of transfer (242) to an equivalent system of linear equations.

For purposes of practical solution it appears most convenient to combine equations (243) and (247) in the following manner.

Since we have arranged $Q_m(\mu)$ to be divisible by $(1-\mu^2)$, we can clearly write

$$Q_m(\mu) = \mathfrak{Q}_m(\mu)(1 - \mu^2). \tag{248}$$

The first few of the polynomials $\mathfrak{Q}_m(\mu)$ are listed in the preceding tabulation.

Now, multiply equation (243) by $a_i\mathfrak{Q}_m(\mu_i)$ and sum over all i's. We obtain

$$\begin{aligned}\frac{d}{dr}\sum a_i\mu_i\mathfrak{Q}_m(\mu_i)I_i + \frac{1}{r}\sum a_i(1-\mu_i^2)\mathfrak{Q}_m(\mu_i)\left(\frac{\partial I}{\partial\mu}\right)_{\mu=\mu_i} \\ = -\kappa\rho\sum a_i\mathfrak{Q}_m(\mu_i)I_i + \frac{1}{2}\kappa\rho\left(\sum a_i I_i\right)\left[\sum a_i\mathfrak{Q}_m(\mu_i)\right]\end{aligned} \tag{249}$$

$$(m = 1, \cdots, 2n).$$

But, according to equations (247) and (248)

$$\begin{aligned}\sum a_i(1-\mu_i^2)\mathfrak{Q}_m(\mu_i)\left(\frac{\partial I}{\partial\mu}\right)_{\mu=\mu_i} &= \sum a_i Q_m(\mu_i)\left(\frac{\partial I}{\partial\mu}\right)_{\mu=\mu_i} \\ &= \sum a_i P_m(\mu_i) I_i.\end{aligned} \tag{250}$$

Equation (249), therefore, reduces to

$$\frac{d}{dr}\sum a_i\mu_i\mathfrak{Q}_m(\mu_i)I_i+\frac{1}{r}\sum a_iP_m(\mu_i)I_i$$

(251)

$$=-\kappa\rho\sum a_i\mathfrak{Q}_m(\mu_i)I_i+\frac{1}{2}\kappa\rho(\sum a_iI_i)[\sum a_i\mathfrak{Q}_m(\mu_i)]$$

$$(m=1,\cdots,2n).$$

This is the required system of linear equations in the nth approximation.

Equation (251) for the case $m=1$ admits of immediate integration. For, when $m=1$

(252) $$P_1(\mu)=\mu\quad\text{and}\quad\mathfrak{Q}_1(\mu)=1/2,$$

and equation (251) yields

(253) $$\frac{1}{2}\frac{d}{dr}\sum a_i\mu_iI_i+\frac{1}{r}\sum a_i\mu_iI_i=0$$

or

(254) $$\sum a_i\mu_iI_i=\frac{1}{2}\frac{F_0^1}{r^2},$$

where F_0 is a constant. This is the equivalent, in our approximation, of the flux integral

(255) $$F=2\int_{-1}^{+1}I\mu d\mu=\frac{F_0}{r^2},$$

which the equation of transfer (242) admits directly.

Again, since $\mathfrak{Q}_m(\mu)$ is odd when m is even, equation (251) reduces for even values of m to the form

(256) $$\frac{d}{dr}\sum a_i\mu_i\mathfrak{Q}_m(\mu_i)I_i+\frac{1}{r}\sum a_iP_m(\mu_i)I_i=-\kappa\rho\sum a_i\mathfrak{Q}_m(\mu_i)I_i$$

$$(m=2,4,\cdots,2n).$$

For $m=2n$, the foregoing equation further simplifies to

(257) $$\frac{d}{dr}\sum a_i\mu_i\mathfrak{Q}_{2n}(\mu_i)I_i=-\kappa\rho\sum a_i\mathfrak{Q}_{2n}(\mu_i)I_i.$$

Finally, we may note the explicit forms of the equations in the second approximation. They are

$$\sum a_i\mu_i I_i = \frac{1}{2}\frac{F_0}{r^2},$$

$$\frac{d}{dr}\sum a_i\mu_i^2 I_i + \frac{1}{r}\sum a_i(3\mu_i^2 - 1)I_i = -\kappa\rho\sum a_i\mu_i I_i,$$

$$\frac{d}{dr}\sum a_i\mu_i(5\mu_i^2 - 1)I_i + \frac{4}{r}\sum a_i\mu_i(5\mu_i^2 - 3)I_i = -\frac{5}{3}\kappa\rho\sum a_i(3\mu_i^2 - 1)I_i, \tag{258}$$

$$\frac{d}{dr}\sum a_i\mu_i^2(7\mu_i^2 - 3)I_i = -\kappa\rho\sum a_i\mu_i(7\mu_i^2 - 3)I_i.$$

From the point of view of astrophysical applications, greatest interest is attached to the case when $\kappa\rho$ varies as some inverse of power. And when

$$\kappa\rho = \frac{\text{constant}}{r^n} \qquad (n > 1), \tag{259}$$

the equations of the second approximation (258) can be solved explicitly and the various physical quantities expressed as integrals over the Bessel functions, I_ν and K_ν, of purely imaginary argument, and of order

$$\nu = \frac{n+5}{2(n-1)}. \tag{260}$$

It does not, however, seem that the passage to the limit of infinite approximation can be achieved as simply as in the case of transfer problems in plane parallel atmospheres.

37. **The equation of transfer in a differentially moving atmosphere. The influence of Doppler effect.** Finally, we shall turn to a class of transfer problems which is of an altogether different character from the ones we have considered so far. The problems in question arise in connection with the study of the transfer of radiation in atmospheres in which the different parts are in relative motion. To be specific, consider a plane-parallel atmosphere in which the material at height z has a velocity $w(z_1)$ with respect to a stationary observer. The novelty of the situation arises on account of Doppler effect which makes the radiation scattered in different directions have different frequencies

as judged by a stationary observer. Consequently, the radiation field in the different frequencies will interact with each other in a manner which is not always easy to visualize. However, in the astrophysical contexts, two circumstances simplify the problem. First, the velocities which are involved are small compared to the velocity of light, c, and second, the only effects of consequence are those which arise from the sensitive dependence of the scattering coefficient $\sigma(\nu)$ on frequency. This last circumstance, in particular, allows us to ignore all effects such as aberration and so on, and concentrate only on the effects arising from the change of frequency on scattering. Under these conditions, the equation of transfer can be shown to take the form

$$\mu \frac{\partial I(\nu, z, \mu)}{\rho \partial z} = -\sigma\left(\nu - \nu_0 \frac{w}{c}\mu\right)\left\{I(\nu, z, \mu) - \frac{1}{2}\int_{-1}^{+1} I\left(\nu - \nu_0 \frac{w}{c}\mu + \nu_0 \frac{w}{c}\mu', z, \mu'\right) d\mu'\right\}, \tag{261}$$

where ν_0 denotes the frequency of the "center of the line."

With suitable simplifying assumptions concerning $\sigma(\nu)$ and $w(z)$, the discussion of the equation of transfer (261), in the first approximation in our scheme of replacing integrals by Gauss sums, leads to a variety of novel types of boundary value problems in hyperbolic equations. It may be of some interest to specify the nature of these boundary value problems and indicate the methods which have been developed for their solution.

38. **A new class of boundary value problems in hyperbolic equations.** As related to the equation

$$\frac{\partial^2 f}{\partial x^2} - \frac{\partial^2 f}{\partial y^2} + f = 0 \tag{262}$$

the boundary value problems which are of most frequent occurrence are of the following general type:

The value of f and its derivatives are assigned for

$$y = 0 \quad \text{and} \quad 0 \leqq x \leqq l_1, \tag{263}$$

that is, along AB in Fig. 2. Along AD ($x=0$ and $0 \leqq y \leqq l_2$) and BC ($x=l_1$, $0 \leqq y \leqq l_2$), we are further given that

$$\left(\frac{\partial f}{\partial x}\right)_{x=0} = \left(\frac{\partial f}{\partial y}\right)_{x=0} + \phi(y) \qquad (0 \leqq y \leqq l_2), \tag{264}$$

$$f(l_1, y) = \psi(y) \qquad (0 \leqq y \leqq l_2) \tag{265}$$

where $\phi(y)$ and $\psi(y)$ are two known functions. The problem is to solve equation (1) in the rectangular strip $ABCD$ satisfying the stated boundary conditions. For the particular boundary value problems which occur in the astrophysical contexts, the following "systematic method" of solution has been found convenient.

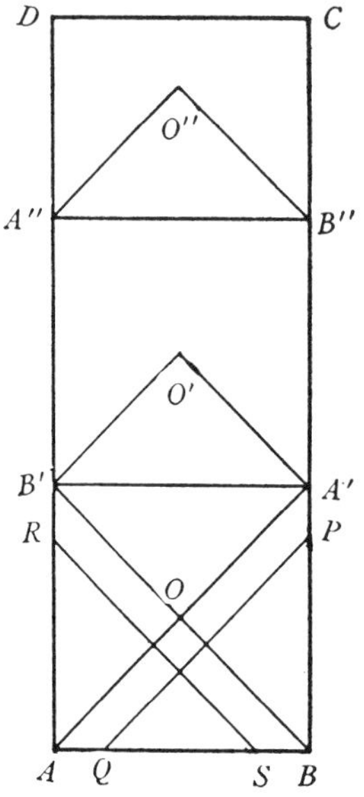

FIG. 2.

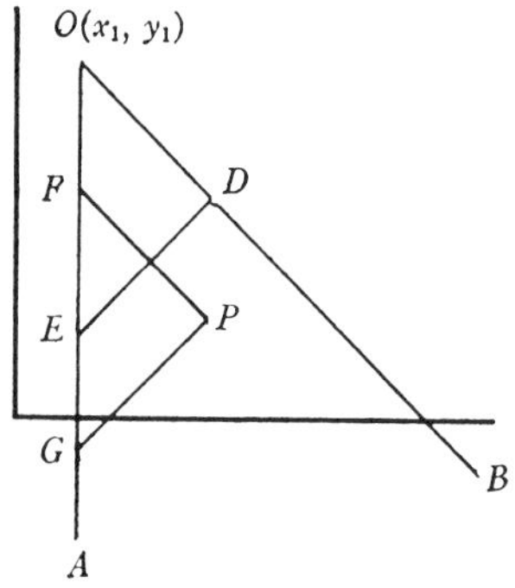

FIG. 3.

Let the characteristics $x=y$ and $l_1-x=y$ through A and B intersect BC and AD at A' and B', respectively. Further, let AA' and BB' intersect at O. First, since the function and its derivatives are known along AB, we can find the solution inside and on the sides of the triangle OAB directly by Riemann's well known method. Next we use the requirement that the solution be unambiguously defined along AO and OB together with the boundary conditions specified along AD and BC to determine f along AB' and $\partial f/\partial x$ along BA' as solutions of certain integral equations of Volterra's type. With the knowledge of the function and its derivatives thus completed along the part $B'ABA'$ of the "supporting curve" $ABCD$, the solution inside the entire region $O'B'ABA'$ becomes determinate by Riemann's method. In particular, the function and its derivatives along $B'A'$ can be found and the continuation of the solution in the second square $B'A'B''A''$ follows along similar lines.

While the method outlined above shows how solutions satisfying the given boundary conditions can be found in principle, it suffers from the disadvantage that the method of solution depends on solving a succession of Volterra integral equations; and, unless the boundary conditions specified along $DABC$ are specially simple, we should not expect to go very far in the explicit carrying out of the solution by this method. It would, therefore, be useful if an alternative method of solution can be devised which will eliminate the need of solving integral equations and reduce the practical problem to one involving (at most!) only quadratures. It is remarkable that this can actually be accomplished by constructing suitable Green's functions and applying Green's theorem to contours, such as RAS and PQB.

39. **The Green's functions $C(x, y; x_1, y_1)$ and $\Gamma(x, y; x_1, y_1)$.** It is found that Green's functions which are appropriate for the solution of the boundary value problems of the type formulated in §38 are $C(x, y; x_1, y_1)$ and $\Gamma(x, y; x_1, y_1)$, defined as follows:

$C(x, y; x_1, y_1)$ is a solution of the hyperbolic equation (262) which satisfies the boundary conditions

$$C(x_1, y; x_1, y_1) = 1 \qquad (y \leqq y_1) \tag{266}$$

and

$$C(x, x_1 + y_1 - x; x_1, y_1) = 1 \qquad (x \geqq x_1). \tag{267}$$

In other words, if O represents the point (x_1, y_1), OB the characteristic $x+y=x_1+y_1$ through O and OA the line through O parallel to the y-axis, then the boundary conditions require that C take the value 1 along both OA and OB (see Fig. 3). A solution satisfying these boundary conditions can be found explicitly. It is given by

$$\begin{aligned} C(x, y; x_1, y_1) &= \cos(x - x_1) + \frac{1}{2}(x - x_1)\int_0^\pi J_0([x - x_1]\sin\vartheta) \\ &\quad \times Ii_1(y_1 - y - [x - x_1]\cos\vartheta)\sin\vartheta d\vartheta, \end{aligned} \tag{268}$$

where J_0 denotes the Bessel function of order zero and $Ii_1(z)$ the "Bessel integral"

$$Ii_1(z) = \int_0^z \frac{I_1(t)}{t}\,dt. \tag{269}$$

($I_1(t)$ denotes the Bessel function of order 1 for a purely imaginary argument.)

The second function, $\Gamma(x, y; x_1, y_1)$, is defined in terms of

$C(x, y; x_1, y_1)$ according to the formula

(270) $$\Gamma(x, y; x_1, y_1) = \frac{\partial C}{\partial x} - \frac{\partial C}{\partial y}.$$

Continuity Conditions:

$$e^{-y}f = e^{y/\alpha}F$$

$$e^{-y}\left(\frac{df}{dx} - \frac{df}{dy}\right) = e^{y/\alpha}\left(\frac{dF}{dr} + \alpha\frac{dF}{dy}\right)$$

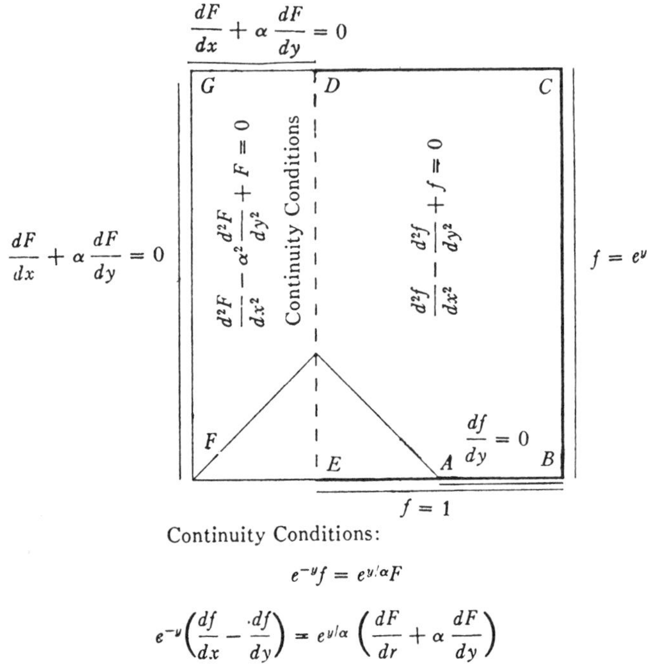

Fig. 4.

40. Some further boundary problems. The boundary value problem formulated in §38 does not exhaust the type of problems which occur in theory of radiative transfer in moving atmospheres. However, no progress has so far been made in the solution of these other problems. It may, therefore, be of particular interest to describe here the nature of these more complex boundary value problems.

A typical problem is to solve (see Fig. 4)

(271) $$\frac{\partial^2 F}{\partial x^2} - \alpha^2\frac{\partial^2 F}{\partial y^2} + F = 0 \qquad (\alpha = \text{constant}),$$

for F in the rectangular strip $FEDG$ $(0 \leqq x \leqq x_2,\ 0 \leqq y \leqq y_1)$, and

$$\frac{\partial^2 f}{\partial x^2} - \frac{\partial^2 f}{\partial y^2} + f = 0, \tag{272}$$

for f in the rectangular strip $EBCD$ ($x_2 \leqq x \leqq x_1$, $0 \leqq y \leqq y_1$) satisfying the following boundary conditions:

Along FG ($x=0$, $0 \leqq y \leqq y_1$) and GD ($y=y_1$, $0 \leqq x \leqq x_2$) relations of the form

$$\left(\frac{\partial F}{\partial x} + \alpha \frac{\partial F}{\partial y}\right)_{x=0} = \phi(y) \qquad (0 \leqq y \leqq y_1) \tag{273}$$

and

$$\left(\frac{\partial F}{\partial x} + \alpha \frac{\partial F}{\partial y}\right)_{y=y_1} = \psi(x) \qquad (0 \leqq x \leqq x_2) \tag{274}$$

are specified where $\phi(y)$ and $\psi(x)$ are two known functions. Along EB ($y=0$, $x_2 \leqq x \leqq x_1$) and BC ($x=x_1$, $0 \leqq y \leqq y_1$) f is given, while along the part AB ($x^* = x_2(1+\alpha) \leqq x \leqq x_1$, $y=0$) of the x-axis, the derivative $\partial f/\partial y$ is also given. And, finally, along ED ($x=x_2$, $0 \leqq y \leqq y_1$) certain "continuity conditions" of the type

$$f(x_2, y) = Q(y)F(x_2, y) \qquad (0 \leqq y \leqq y_1) \tag{275}$$

and

$$\left(\frac{\partial f}{\partial x} - \frac{\partial f}{\partial y}\right)_{x=x_2} = Q(y)\left(\frac{\partial F}{\partial x} + \alpha \frac{\partial F}{\partial y}\right)_{x=x_2} \qquad (0 \leqq y \leqq y_1) \tag{276}$$

are specified, where $Q(y)$ is a known function.

It would be of considerable interest to know how such boundary value problems can be solved. (In Fig. 4 the particular boundary conditions which occur in a specific problem are indicated.)

41. **Concluding remarks.** In concluding, I may recall what I said at the beginning, namely, that the advance of a branch of theoretical physics often leads to the creation of a new mathematical discipline. I think it may be conversely said, with almost equal truth, that the creation of a new mathematical discipline is often the sign that the particular branch of theoretical physics has reached maturity. I hope I have given you the impression that theoretical astrophysics has now come of age.

Since the lecture was given in December, the mathematical theory of radiative transfer has advanced along several directions and in this Addendum we shall briefly summarize the results of these newer investigations. The particular sections of the lecture to which these advances refer are indicated by the numbering of the paragraphs which follow; however, §42 breaks new ground not covered by the lecture.

11a. **Elliptically polarized radiation field.** In §11 we outlined how the equations of transfer for a partially plane-polarized radiation field can be formulated and gave the explicit form of these equations for the particular case of Rayleigh scattering. It is not difficult to extend this discussion to include the case of a general elliptic polarization of the radiation field. In this latter case, we must consider, in addition to the intensities I_l and I_r in two directions at right angles to each other in the plane of the electric and the magnetic vectors, the two further quantities

$$(277) \quad U = (I_l - I_r) \tan 2\chi \text{ and } V = (I_l - I_r) \tan 2\beta \sec 2\chi$$

where χ denotes the inclination of the plane of polarization to the direction to which l refers and $-\pi/2 \leqq \beta \leqq +\pi/2$ is an angle the tangent of which is equal to the ratio of axes of the ellipse characterizing the state of polarization. (The sign of β depends on whether the polarization is right-handed (+) or left-handed (−).) And the rule of composition (due to Stokes) is that a mixture of several *independent* streams of polarized light is characterized by values of the parameters I_l, I_r, U and V which are the sums of the respective parameters of the individual streams. With this rule of composition, the equations of transfer for I_l, I_r, U and V can be formulated in terms of the basic laws of single scattering.

For the case of Rayleigh scattering, it is found that the equations of transfer for I_l, I_r and U are of exactly the same forms as when, only, plane polarization is contemplated; that is, in the case of Rayleigh scattering, the equations are *reducible.* Moreover, it is found that for Rayleigh scattering, V is simply scattered in accordance with a phase function $3/2 \cos \Theta$; and the exact solution of the equation for V therefore presents no difficulty.

23a. **The functional equation relating the law of darkening and the scattering function for semi-infinite plane-parallel atmospheres.** In §23 we have remarked on the remarkable relationships between the angular distributions of the emergent radiation in the problem with a constant net flux and the law of diffuse reflection. It has since been possible to trace the origin of these relationships: they arise simply in consequence of the invariance of the emergent radiation $I(0, \mu)$ (in the problem with a constant net flux) to the addition (or removal) of layers of arbitrary thickness to (or from) the atmosphere. The mathematical expression of this invariance is that the outward radiation $I(\tau, +\mu)$ $(0<\mu<1)$ at any level τ differs from the emergent radiation $I(0, \mu)$ only on account of the fact that at τ there is an inward directed radiation field specified by $I(\tau, -\mu')$ $(0<\mu'<1)$ which will be reflected by the atmosphere below τ by the law of diffuse reflection of a semi-infinite atmosphere. In other words, we must have

$$I(\tau, +\mu) = I(0, \mu) + \frac{1}{2\mu}\int_0^1 S^{(0)}(\mu, \mu')I(\tau, -\mu')d\mu', \tag{278}$$

where

$$S^{(0)}(\mu, \mu') = \frac{1}{2\pi}\int_0^1 S(\mu, \phi; \mu', \phi')d\phi' \tag{279}$$

is the azimuth independent term in the scattering function $S(\mu, \phi; \mu', \phi')$ (cf. equation (26)).

Differentiating equation (278) with respect to τ and passing to the limit $\tau=0$, we obtain

$$\left[\frac{dI(\tau, +\mu)}{d\tau}\right]_{\tau=0} = \frac{1}{2\mu}\int_0^1 S^{(0)}(\mu, \mu')\left[\frac{dI(\tau, -\mu')}{d\tau}\right]_{\tau=0} d\mu'. \tag{280}$$

On the other hand, from the equation of transfer

$$\mu\frac{dI}{d\tau} = I(\tau, \mu) - B(\tau, \mu) \tag{281}$$

where $B(\tau, \mu)$ is the source function appropriate for the problem (see equations I, II and VI) we conclude that

$$\left[\frac{dI(\tau, +\mu)}{d\tau}\right]_{\tau=0} = \frac{1}{\mu}\left[I(0, \mu) - B(0, \mu)\right], \tag{282}$$

and

$$\left[\frac{dI(\tau, -\mu')}{d\tau}\right]_{\tau=0} = \frac{1}{\mu'}B(0, -\mu'). \tag{283}$$

Now combining equations (280), (282) and (283), we obtain

$$I(0, \mu) = B(0, \mu) + \frac{1}{2}\int_0^1 S^{(0)}(\mu, \mu')B(0, -\mu')\frac{d\mu'}{\mu'}, \tag{284}$$

which is a functional equation relating $I(0, \mu)$ and $S^{(0)}(\mu, \mu')$; it can be shown that it is precisely in consequence of this equation that the relationship between $I(0, \mu)$ and $S^{(0)}(\mu, \mu')$ noticed in §23 arises.

35a. **Representation of $H(\mu)$ as a complex integral.** According to equations (102) and (110)

$$1 - \sum_{j=1}^{n} \frac{a_j}{1 - \mu_j^2/z^2} \equiv \frac{1}{H(z)H(-z)} \tag{285}$$

where $H(z)$ is defined as usual in terms of the roots of the characteristic equation (see equation (49))

$$1 = \sum_{j=1}^{n} \frac{a_j}{1 - \mu_j^2 k^2} . \tag{286}$$

The arguments leading to equation (285) (§20) are seen to be sufficiently general to establish the identity

$$\begin{aligned} 1 - 2\sum_{j=1}^{n} \frac{a_j \Psi(\mu_j)}{1 - \mu_j^2/z^2} &\equiv \frac{1}{H(z)H(-z)} \\ &= \frac{\prod_\alpha (1 - k_\alpha^2 z^2)}{\prod_j (1 - z^2/\mu_j^2)}, \end{aligned} \tag{287}$$

where $H(z)$ is now defined as in equation (126) in terms of the roots of the characteristic equation

$$1 = 2\sum_{j=1}^{n} \frac{a_j \Psi(\mu_j)}{1 - k^2 \mu_j^2}, \tag{288}$$

and $\Psi(\mu)$ has the same meaning as in equation (127).

Now let

$$G(x) = \frac{1}{2\pi i}\int_{-i\infty}^{i\infty} \log T(z) \frac{x\,dz}{z^2 - x^2}, \tag{289}$$

where

$$T(z) = 1 - 2z^2 \sum_{j=1}^{n} \frac{a_j \Psi(\mu_j)}{z^2 - \mu_j^2} = \frac{\prod_\alpha (1 - k_\alpha^2 z^2)}{\prod_j (1 - z^2/\mu_j^2)} . \tag{290}$$

It is seen that defined in this manner $G(x)$ is regular for $R(x) > 0$.

By evaluating the residue at the pole on the right

$$\frac{1}{2\pi i}\int_{-i\infty}^{i\infty} \log\left(1 + \frac{z}{a}\right) \frac{x\,dz}{z^2 - x^2} = -\frac{1}{2}\log\left(1 + \frac{x}{a}\right), \tag{291}$$

if $R(x) > 0$ and $R(a) > 0$. Similarly, by evaluating the residue at the pole on the left, we have

$$\frac{1}{2\pi i}\int_{-i\infty}^{i\infty} \log\left(1 - \frac{z}{a}\right)\frac{xdz}{z^2 - x^2} = -\frac{1}{2}\log\left(1 + \frac{x}{a}\right). \tag{292}$$

Hence,

$$\frac{1}{2\pi i}\int_{-i\infty}^{i\infty} \log\left(1 - \frac{z^2}{a^2}\right)\frac{xdz}{z^2 - x^2} = -\log\left(1 + \frac{x}{a}\right). \tag{293}$$

Accordingly

$$\begin{aligned} G(x) &= \frac{1}{2\pi i}\int_{-i\infty}^{i\infty} \left\{\sum_{\alpha=1}^{n} \log\left(1 - k_\alpha^2 z^2\right) - \sum_{j=1}^{n} \log\left(1 - \frac{z^2}{\mu_j^2}\right)\right\}\frac{xdz}{z^2 - x^2} \\ &= -\sum_{\alpha=1}^{n} \log\left(1 + xk_\alpha\right) + \sum_{j=1}^{n} \log\left(1 + \frac{z}{\mu_j}\right) \\ &= \log H(z). \end{aligned} \tag{294}$$

We have thus shown that

$$\log H(x) = \frac{1}{2\pi i}\int_{-i\infty}^{i\infty} \log T(z)\,\frac{xdz}{z^2 - x^2}\,. \tag{295}$$

From the representation (295) of the H-function as a complex integral, it would appear that the solution of the *functional equation*

$$H(\mu) = 1 + \mu H(\mu)\int_0^1 \frac{H(\mu')}{\mu + \mu'}\,\Psi(\mu')d\mu' \tag{296}$$

has the representation (cf. Theorem 2, §25)

$$\log H(\mu) = \frac{1}{2\pi i}\int_{-i\infty}^{i\infty} \log T(z)\,\frac{\mu dz}{z^2 - \mu^2}\,, \tag{297}$$

where, now (see equation (290)),

$$T(z) = 1 - 2z^2\int_0^1 \frac{\Psi(\mu)d\mu}{z^2 - \mu^2}\,. \tag{298}$$

Our arguments do not of course establish rigorously the representation (297) of the solution of equation (296). However, Professor E. C. Titchmarsh, with whom I have corresponded, has kindly communicated to me a rigorous demonstration of the representation (297) by one of his colleagues, Mr. M. M. Crum.

42. **The theory of radiative transfer in atmospheres of finite optical thicknesses.** In the lecture attention was directed almost exclusively to transfer problems in semi-infinite plane-parallel atmospheres. The extension of this theory to the study of the transfer of radiation in plane-parallel atmospheres of finite optical thicknesses raises problems of a higher order of difficulties; these difficulties arise principally from the circumstance that boundary conditions have to be *explicitly* satisfied on both sides of the atmosphere. Thus, if we consider an atmosphere of optical thickness τ_1 ($< \infty$), we are interested, for example, in solutions of the equations of transfer I–VII which satisfy the boundary conditions

$$I(0, -\mu) = 0 \qquad (0 < \mu < 1), \tag{299}$$

and

$$I(\tau_1, +\mu) = 0 \qquad (0 < \mu < 1). \tag{300}$$

However, in analogy with the theory of semi-infinite atmospheres, we may expect that in the case of finite atmospheres, also, the angular distributions of the emergent radiations can be expressed in terms of functions (like $H(\mu)$) which will be explicitly known in any finite approximation and which, in the limit of infinite approximation, will become solutions of functional equations of a standard form. It now appears that this reduction can in fact be achieved and that the pair of functional equations

$$X(\mu) = 1 + \mu \int_0^1 \frac{X(\mu)X(\mu') - Y(\mu)Y(\mu')}{\mu + \mu'} \Psi(\mu')d\mu', \tag{301}$$

and

$$Y(\mu) = e^{-\tau_1/\mu} + \mu \int_0^1 \frac{Y(\mu)X(\mu') - X(\mu)Y(\mu')}{\mu - \mu'} \Psi(\mu')d\mu', \tag{302}$$

where $\Psi(\mu)$ is an even polynomial in μ satisfying the condition

$$\int_0^1 \Psi(\mu)d\mu < \frac{1}{2}, \tag{303}$$

plays the same basic role in the theory of atmospheres of finite optical thicknesses as the functional equation

$$H(\mu) = 1 + \mu H(\mu) \int_0^1 \frac{H(\mu')}{\mu + \mu'} \Psi(\mu')d\mu' \tag{304}$$

played in the theory of semi-infinite atmospheres. And just as

$$H(\mu) = \frac{1}{\mu_1 \cdots \mu_n} \frac{\prod_i (\mu + \mu_i)}{\prod_\alpha (1 + k_\alpha \mu)}, \tag{305}$$

where the k_α's are the positive roots of the characteristic equation

$$1 = 2\sum_{j=1}^{n} \frac{a_j \Psi(\mu_j)}{1 - k^2 \mu_j^2}, \tag{306}$$

provides a rational approximation to the solution of equation (304), so also the functions $X(\mu)$ and $Y(\mu)$ defined in the manner of the following equations provides an approximation to the solution of equations (301) and (302):

$$X(\mu) = \frac{(-1)^n}{\mu_1 \cdots \mu_n} \frac{1}{[C_0^2(0) - C_1^2(0)]^{1/2}} \frac{1}{W(\mu)} \cdot [P(-\mu)C_0(-\mu) - e^{-\tau_1/\mu}P(\mu)C_1(\mu)] \tag{307}$$

$$Y(\mu) = \frac{(-1)^n}{\mu_1 \cdots \mu_n} \frac{1}{[C_0^2(0) - C_1^2(0)]^{1/2}} \frac{1}{W(\mu)} \cdot [e^{-\tau_1/\mu}P(\mu)C_0(\mu) - P(-\mu)C_1(-\mu)] \tag{308}$$

where

$$P(\mu) = \prod_{i=1}^{n} (\mu - \mu_i), \qquad W(\mu) = \prod_{\alpha=1}^{n} (1 - k_\alpha^2 \mu^2), \tag{309}$$

$$C_0(\mu) = \sum_{2^{n-1}\text{ terms}} \epsilon_{2l}^{(0)} \frac{\prod_{i=1}^{2l} \prod_{m=1}^{n-2l} (k_{r_i} + k_{s_m})}{\prod_{i=1}^{2l} \prod_{m=1}^{n-2l} (k_{r_i} - k_{s_m})} \times \prod_{i=1}^{2l} \xi_{r_i}(1 + k_{r_i}\mu) \prod_{m=1}^{n-2l} \eta_{s_m}(1 - k_{s_m}\mu), \tag{310}$$

$$C_1(\mu) = \sum_{2^{n-1}\text{ terms}} \epsilon_{2l+1}^{(1)} \frac{\prod_{i=1}^{2l+1} \prod_{m=1}^{n-2l-1} (k_{r_i} + k_{s_m})}{\prod_{i=1}^{2l+1} \prod_{m=1}^{n-2l-1} (k_{r_i} - k_{s_m})} \times \prod_{i=1}^{2l+1} \xi_{r_i}(1 + k_{r_i}\mu) \prod_{m=1}^{n-2l-1} \eta_{r_i}(1 - k_{s_m}\mu). \tag{311}$$

In equations (310) and (311), $(r_1, \cdots, r_j)$ and $(s_1, \cdots, s_{n-j})$ are j, respectively, $n-j$ *distinct* integers from the set $(1, 2, \cdots, n)$,

$$\epsilon_{2l}^{(0)} = \begin{cases} +1 \text{ for even integers of the form } 4l, \\ -1 \text{ for even integers of the form } 4l+2, \end{cases} \tag{312}$$

$$\epsilon_{2l+1}^{(1)} = \begin{cases} +1 \text{ for odd integers of the form } 4l+1, \\ -1 \text{ for odd integers of the form } 4l+3, \end{cases} \tag{313}$$

and

$$\xi_\alpha = e^{k_\alpha \tau_1/2} P(-1/k_\alpha) \quad \text{and} \quad \eta_\alpha = e^{-k_\alpha \tau_1/2} P(+1/k_\alpha) \qquad (\alpha = 1, \cdots, n). \tag{314}$$

Finally, it should be noted that the definitions of $C_0(\mu)$ and $C_1(\mu)$ according to equations (307) and (308) are valid only in even orders of approximation; in odd orders the role of C_0 and C_1 should be interchanged.

Moreover, there exist also functional equations for the scattering and the transmission functions for the problem of diffuse reflection and transmission by atmospheres of finite optical thicknesses. These equations arise from general invariances of the type considered in §§28, 29 and 23a and lead to a whole new class of systems of functional equations which can all be reduced to the solution of pairs of functional equations of the form (301) and (302). It is therefore apparent that the study of the transfer of radiation in atmospheres of finite optical thicknesses will lead to the development of a mathematical theory at least as extensive as the one described in the lecture in the context of semi-infinite atmospheres.

References

The following papers and monographs substantially form the basis of the report given in the lecture. The list is not intended to be a complete bibliography of the subject of the transfer of radiation in stellar atmospheres.

V. A. Ambarzumian
C. R. (Doklady) Acad. Sci. URSS. vol. 38 (1943) p. 257.
Russian Astronomical Journal vol. 19 (1942).
Journal of Physics of the Academy of Sciences of the U.S.S.R. vol. 8 (1944) p. 65.
S. Chandrasekhar
Astrophysical Journal vol. 99 (1944) p. 180; vol. 100 (1944) pp. 76, 117, 355; vol. 101 (1945) pp. 95, 328, 348; vol. 102 (1945) p. 402; vol. 103 (1946) pp. 165, 351; vol. 104 (1946) pp. 110, 191; vol. 105 (1947) pp. 151, 164, 424, 435, 441, 461. (Further papers in this series are in course of publication in the same journal.)

Reviews of Modern Physics vol. 17 (1945) p. 138.
Proc. Chambridge Philos. Soc. vol. 42 (1946) p. 250.

E. Hopf
Mathematical problems of radiative equilibrium, Cambridge Mathematical Tracts, No. 31, Cambridge, England, 1934.

G. Münch
Astrophysical Journal vol. 104 (1946) p. 87.

Lord Rayleigh
Scientific papers of Lord Rayleigh, vol. 1, Cambridge, England, 1899, pp. 87, 104, 518.

Sir George Stokes
Trans. Cambridge Philos. Soc. vol. 9 (1852) p. 399.

M. Tuberg
Astrophysical Journal vol. 103 (1946) p. 145.

G. C. Wick
Zeitschrift für Physik vol. 120 (1943) p. 702.

Yerkes Observatory, University of Chicago

25

Radiative Transfer—A Personal Account

I had the pleasure of meeting Academician Viktor Amazaspovitch Ambartsumian in the company of Lev Davidovich Landau in the summer of 1934 in Leningrad; and the visit to the Hermitage with them is still very vivid in my memory. This is now my first opportunity, since that time, to visit the Soviet Union and to renew my personal acquaintance with Academician Ambartsumian. I feel greatly privileged to visit this Observatory, conceived and founded by his foresight and his efforts, and especially on this occasion when we are assembled to celebrate his innovative introduction of principles of invariance in the study of radiative transfer and in the fluctuations in brightness of the Milky Way.

Because of the exigencies of the Second World War, I first became aware of Academician Ambartsumian's first paper on principles of invariance[1] only in the summer of 1945; and I became aware of his second paper[2] very much later. At that time, I had already been involved for some two years in solving problems in radiative transfer by a different technique; and the impact of Ambartsumian's papers was immediate and profound. However, because of the particular circumstances in which I came to know of these innovative papers, their influence in midstream, so to say, in altering the course of my work was perhaps somewhat different from their influence on those of you who were more directly and immediately inspired by them. I hope, then, that you will forgive me if, on this occasion, I allow myself the license of giving an account of the evolution of my own investigations on radiative transfer during the years 1943–48 and how they were redirected by Ambartsumian's ideas. I am afraid that I must confine myself exclusively to those years, since, after the publication of my book on *Radiative Transfer* in 1950, my interests have strayed very far away and only rarely have I returned—I must confess with some nostalgia—to my interests of those youthful years.

Paper presented at the Symposium devoted to the 40th Anniversary of the Principle of Invariance: Introduction to the Radiation Transfer Theory, Byurakan, Armenia, USSR, 26–30 October 1981.

1. *C.R. (Doklady) Acad. URSS* 38 (1943): 257.
2. *J. Phys. Acad. Sci. USSR* 8 (1944): 65.

I. Preliminaries

I shall begin by formulating the problems I was concerned with when I became seriously interested in the theory of radiative transfer in the fall of 1943.

The problems relate to the transfer of radiation in atmospheres, stratified in parallel planes, which are either *semi-infinite* (normal to the plane of stratification) or *finite* (confined between two of the planes of stratification). The atmospheres are further characterized by a scattering coefficient, σ, so that a pencil of radiation of intensity I, incident on an element of mass dm, scatters the amount of radiation

$$\sigma\, dm\, I p(\cos\Theta)\, \frac{d\omega}{4\pi} \tag{1}$$

into an element of solid angle $d\omega$, in a direction inclined at an angle Θ to the direction of incidence, and $p(\cos\Theta)$ is the *phase function* for single scattering. If scattering is the only process by which matter and radiation interact with each other, then we must have

$$\int p(\cos\Theta)\, \frac{d\omega}{4\pi} = 1. \tag{2}$$

This is the *conservative case.* If, however, a certain fraction, $1 - \varpi_0$, of the incident radiant energy is consumptively absorbed, then we should have, instead,

$$\int p(\cos\Theta)\, \frac{d\omega}{4\pi} = \varpi_0 \quad (<1). \tag{3}$$

This is the *nonconservative case;* and ϖ_0 is the *albedo* for single scattering. In general we may suppose that the phase function can be expanded in a series of Legendre polynomials, $P_l(\cos\Theta)$, in the form

$$p(\cos\Theta) = \sum_{l=0}^{N} \varpi_l\, P_l(\cos\Theta), \tag{4}$$

where the ϖ_l's are constants and $\varpi_0 = 1$ in the conservative case.

In writing the equation of transfer, we shall introduce the normal *optical thickness*

$$\tau = \int_0^z \rho\sigma\, dz, \tag{5}$$

where ρ denotes the density and z is the normal distance from the boundary (at $\tau = 0$) measured *inward.* Also, we shall measure the polar angle θ with respect to the *outward* normal and let φ be the azimuthal angle referred to a suitably chosen x-axis in the plane of stratification. The equation of transfer then takes the form

$$\mu \frac{dI(\tau, \mu, \varphi)}{d\tau} = I(\tau, \mu, \varphi) - \frac{1}{4\pi} \int_{-1}^{+1} \int_{0}^{2\pi} p(\mu, \varphi; \mu', \varphi') I(\tau, \mu', \varphi') d\mu' d\varphi', \quad (6)$$

where $\mu = \cos\theta$.

In the conservative case ($\varpi_0 = 1$) equation (6) allows the *flux integral,*

$$\pi F = \int_{-1}^{+1} \int_{0}^{2\pi} \mu I(\tau, \mu, \varphi) d\mu d\varphi = \text{constant}. \quad (7)$$

There is a further integral which conservative problems admit: this is the *K-integral,*

$$K(\tau) = \frac{1}{4\pi} \int_{-1}^{+1} \int_{0}^{2\pi} I(\tau, \mu, \varphi)\mu^2 d\mu d\varphi$$
$$= \tfrac{1}{4}F[(1 - \tfrac{1}{3}\varpi_1) \tau + Q], \quad (8)$$

where Q is a constant.

In the foregoing framework there are two problems one generally considers: (1) the problem of a semi-infinite atmosphere with a constant net flux, πF, in the conservative case; and (2) the problem of diffuse reflection and transmission when a parallel beam of radiation of net flux πF per unit area normal to itself is incident in some specified direction, $(-\mu_0, \varphi_0)$.

In the first problem, the radiation field, at any point in the atmosphere, is clearly axisymmetric about the z-axis, and the intensity I, now a function only of τ and μ, satisfies the integrodifferential equation

$$\mu \frac{dI(\tau, \mu)}{d\tau} = I(\tau, \mu) - \frac{1}{2} \int_{-1}^{+1} p^{(0)}(\mu; \mu') I(\tau, \mu') d\mu', \quad (9)$$

where

$$p^{(0)}(\mu; \mu') = \frac{1}{2\pi} \int_{0}^{2\pi} p(\mu, \varphi; \mu', \varphi') d\varphi'. \quad (10)$$

And we require to solve equation (9) together with the boundary conditions

$$I(0, -\mu) = 0 \quad (0 < \mu \leq 1) \quad \text{and} \quad I(\tau, \mu) e^{-\tau} \to 0 \quad \text{as } \tau \to \infty. \quad (11)$$

In the problem of diffuse reflection and transmission, one considers, in general, an atmosphere of optical thickness τ_1 and asks for the intensity $I(0, +\mu, \varphi)$ $(0 < \mu \leq 1)$ *diffusely reflected* from the surface $\tau = 0$, and the intensity $I(\tau, -\mu, \varphi)$ $(0 < \mu \leq 1)$ *diffusely transmitted* below the surface $\tau = \tau_1$. The resulting laws of diffuse reflection and transmission are expressed in terms of a *scattering* and a *transmission* function,

$$S(\tau_1; \mu, \varphi; \mu_0, \varphi_0) \quad \text{and} \quad T(\tau_1; \mu, \varphi; \mu_0, \varphi_0), \quad (12)$$

defined in terms of the reflected and transmitted intensities by

$$I(0, +\mu, \varphi) = \frac{F}{4\mu} S(\tau_1; \mu, \varphi; \mu_0, \varphi_0)$$

and (13)

$$I(\tau_1, -\mu, \varphi) = \frac{F}{4\mu} T(\tau_1; \mu, \varphi; \mu_0, \varphi_0).$$

It is to be specifically noted that the reflected and the transmitted intensities refer only to the radiation that has suffered one or more scattering processes: $I(\tau_1; -\mu, \varphi)$ does not, for example, include the directly transmitted intensity, $\frac{1}{4}F\, e^{-\tau_1/\mu_0}$, in the direction $(-\mu_0, \varphi_0)$, which has not suffered any scattering process. More generally, in the treatment of the problem of diffuse reflection and transmission, we distinguish between the reduced incident radiation, $\pi F e^{-\tau/\mu_0}$, which penetrates to the depth τ without having suffered any scattering process, and the diffuse radiation field which has arisen from one or more scattering processes suffered by the incident beam. With this distinction, the equation of transfer takes the form

$$\mu \frac{dI(\tau, \mu, \varphi)}{d\tau} = I(\tau, \mu, \varphi) - \frac{1}{4\pi} \int_{-1}^{+1} \int_{0}^{2\pi} p(\mu, \varphi; \mu', \varphi')\, I(\tau, \mu', \varphi')\, d\mu' d\varphi' - \tfrac{1}{4}F\, e^{-\tau/\mu_0}\, p(\mu, \varphi; -\mu_0, \varphi_0); \quad (14)$$

and we require to solve equation (14) together with the boundary conditions

$$I(0, -\mu, \varphi) = 0 \quad (0 < \mu \leq 1) \quad \text{at} \quad \tau = 0$$

and (15)

$$I(\tau_1, +\mu, \varphi) = 0 \quad (0 < \mu \leq 1) \quad \text{at} \quad \tau = \tau_1.$$

II. The Solutions for Some Typical Problems by the Method of Discrete Ordinates

At the time I became interested in problems of radiative transfer, the two problems formulated in § I had been considered only for the case of conservative isotropic scattering when

$$p(\cos \Theta) = 1, \quad (16)$$

and the appropriate equations of transfer are

$$\mu \frac{dI(\tau, \mu)}{d\tau} = I(\tau, \mu) - \frac{1}{2} \int_{-1}^{+1} I(\tau, \mu')\, d\mu', \quad (17)$$

and

$$\mu \frac{dI(\tau, \mu)}{d\tau} = I(\tau, \mu) - \frac{1}{2} \int_{-1}^{+1} I(\tau, \mu')\, d\mu' - \tfrac{1}{4} F e^{-\tau/\mu_0}. \quad (18)$$

Also, in the context of equation (18) only the problem of diffuse reflection by a semi-infinite atmosphere had been considered. And through the combined researches of E. A. Milne, E. Hopf, and M. Bronstein, the solutions for the angular distributions of the emergent radiations, for the two problems, had been expressed in the forms

$$I(0, \mu) = \frac{\sqrt{3}}{4} F\, H(\mu), \tag{19}$$

and

$$I(0, \mu) = \tfrac{1}{4} F \frac{\mu_0}{\mu + \mu_0} H(\mu)\, H(\mu_0), \tag{20}$$

where $H(0) = 1$ and an integral representation for $H(\mu)$ had been given.

Since only the conservative case of isotropic scattering had been considered, I asked myself whether some systematic method could not be developed for solving the same problems for more general laws of scattering. I was particularly interested in solving the problem for *Rayleigh's phase function,*

$$p(\cos\Theta) = \tfrac{3}{4}(1 + \cos^2\Theta), \tag{21}$$

and the phase function,

$$p(\cos\Theta) = \varpi_0(1 + x\cos\Theta) \quad (0 \leq |x| \leq 1), \tag{22}$$

for the problem of diffuse reflection. A paper by G. C. Wick in which equation (17) had been treated by a novel method was most opportune for me. Wick's method consisted in replacing the integral over μ', which occurs in equation (17), by a sum using Gauss's quadrature formula. Thus,

$$\int_{-1}^{+1} I(\tau, \mu')\, d\mu' \quad \text{is replaced by} \quad \sum_{j=-n}^{+n} a_j\, I(\tau, \mu_j), \tag{23}$$

where $\pm\mu_j$ ($j = 1, \ldots, n$) are the zeros of the Legendre polynomial $P_{2n}(\mu)$ and $a_{+j} = a_{-j}$ ($j = 1, \ldots, n$) are the "Gaussian weights" which satisfy the conditions

$$\sum_{j=-n}^{+n} a_j \mu_j^{\,l} = \frac{2\delta_{l,\text{even}}}{l+1} \quad (l \leq 4n - 1), \tag{24}$$

where

$$\left.\begin{aligned} \delta_{l,\text{even}} &= 1 \quad \text{if } l \text{ is even} \\ &= 0 \quad \text{if } l \text{ is odd} \end{aligned}\right\}. \tag{25}$$

By this device, the equations of transfer are replaced by systems of linear equations with constant coefficients for the intensities, $I_{\pm j}(\tau) = I(\tau, \pm\mu_j)$, at the points of the Gaussian division; and the solution of these equations presents no difficulty of principle. The method could clearly be applied to the problems in which I was interested. But what was totally unexpected was that in the general "nth approximation" (in which the quadrature formula with $2n$ divisions of the interval $-1 \leq \mu \leq 1$ is used) the solutions for the emergent radiations, in all cases considered, could be obtained in closed forms and expressed in terms of *H-functions* defined in the manner

$$H(\mu) = \frac{1}{\mu_1 \ldots \mu_n} \frac{\prod_{i=1}^{n} (\mu + \mu_i)}{\prod_{\alpha=1}^{n} (1 + k_\alpha \mu)}, \tag{26}$$

where the μ_i's are the points of the Gaussian division in the positive half of the interval $(-1, +1)$, and the k_α's are the distinct positive (or zero) roots of a *characteristic equation,*

$$1 = 2 \sum_{j=1}^{n} \frac{a_j \Psi(\mu_j)}{1 - k^2 \mu^2}, \tag{27}$$

and $\Psi(\mu)$ is an even polynomial in μ satisfying the condition

$$\int_0^1 \Psi(\mu) = \sum_{j=0}^{n} a_j \Psi(\mu_j) = 1 \tag{28}$$

—the equality sign always obtaining for the conservative case.

Different physical problems lead to different characteristic equations and therefore to different H-functions. However, as the H-functions differ from one another only through the characteristic equations which determine the characteristic roots, k_α, the function $\Psi(\mu)$ is called the *characteristic function* which defines $H(\mu)$.

We tabulate below the solutions to the problems that were considered in the first instance. (A convenient reference to the historically minded reader is the author's Gibbs Lecture to the American Mathematical Society (given on 26 December 1946).[3]

A. *Isotropic Scattering*

Problem with constant flux: the law of darkening:

$$I(\mu) - \frac{\sqrt{3}}{4} F\, H(\mu). \tag{29}$$

Law of diffuse reflection:

$$I(\mu; \mu_0) = \tfrac{1}{4} F \frac{\mu_0}{\mu + \mu_0} H(\mu)\, H(\mu_0). \tag{30}$$

3. *Bull. Am. Math. Soc.* 53 (1947): 641–711 (paper 24 in this volume).

The characteristic function in terms of which $H(\mu)$ is defined is

$$\Psi(\mu) = \tfrac{1}{2}. \tag{31}$$

B. *Scattering in Accordance with Rayleigh's Phase Function* $\tfrac{3}{4}(1 + \cos^2 \Theta)$

Problem with constant net flux: the law of darkening:

$$I(\mu) = \tfrac{3}{4}qF\, H^{(0)}(\mu). \tag{32}$$

Law of diffuse reflection:

$$\begin{aligned} I(\mu, \varphi; \mu_0, \varphi_0) = \tfrac{3}{32} F\, \{&H^{(0)}(\mu)\, H^{(0)}(\mu_0)[3 - (3 - 8q^2)^{1/2}(\mu + \mu_0) + \mu\mu_0] \\ &-4\mu\mu_0\, (1 - \mu^2)^{1/2}(1 - \mu_0{}^2)^{1/2}\, H^{(1)}(\mu)\, H^{(1)}(\mu_0) \cos(\varphi - \varphi_0) \\ &+ (1 - \mu^2)(1 - \mu_0{}^2)\, H^{(2)}(\mu)\, H^{(2)}(\mu_0) \cos 2(\varphi - \varphi_0)\} \frac{\mu_0}{\mu + \mu_0}. \end{aligned} \tag{33}$$

The characteristic functions in terms of which $H^{(0)}(\mu)$, $H^{(1)}(\mu)$, and $H^{(2)}(\mu)$ are defined are, respectively,

$$\Psi^{(0)}(\mu) = \tfrac{3}{16}(3 - \mu^2), \qquad \Psi^{(1)}(\mu) = \tfrac{3}{8}\mu^2 (1 - \mu^2),$$
$$\text{and} \quad \Psi^{(2)}(\mu) = \tfrac{3}{32}(1 - \mu^2)^2. \tag{34}$$

And, finally,

$$q = \frac{2\sqrt{3}}{H^{(0)}(+\sqrt{3}) - H^{(0)}(-\sqrt{3})}. \tag{35}$$

C. *Scattering in Accordance with the Phase Function* $\varpi_0(1 + x \cos \Theta)$

Law of diffuse reflection:

$$\begin{aligned} I(\mu, \varphi; \mu_0, \varphi_0) = \tfrac{1}{4}\varpi_0\, F\{&H^{(0)}(\mu)\, H^{(0)}(\mu_0)\, [1 - c(\mu + \mu_0) - x(1 - \varpi_0)\mu\mu_0] \\ &+ x\, (1 - \mu^2)^{1/2}\, (1 - \mu_0{}^2)^{1/2}\, H^{(1)}(\mu)\, H^{(1)}(\mu_0) \cos(\varphi - \varphi_0)\} \frac{\mu_0}{\mu + \mu_0}. \end{aligned} \tag{36}$$

The characteristic functions in terms of which $H^{(0)}(\mu)$ and $H^{(1)}(\mu)$ are defined are, respectively,

$$\Psi^{(0)}(\mu) = \tfrac{1}{2}\varpi_0\, [1 + x(1 - \varpi_0)\mu^2] \quad \text{and} \quad \Psi^{(1)}(\mu) = \tfrac{1}{4}\, x\varpi_0(1 - \mu^2). \tag{37}$$

The constant c depends in a complicated manner on the characteristic roots defining $H^{(0)}(\mu)$.[4]

Law of diffuse reflection for the case of isotropic scattering with albedo ϖ_0:

$$I(\mu) = \tfrac{1}{4}\varpi_0\, F \frac{\mu_0}{\mu + \mu_0} H^{(0)}(\mu)\, H^{(0)}(\mu_0), \tag{38}$$

4. Cf. S. Chandrasekhar, *Ap. J.* 103 (1946): 165, eq. (108). See p. 148 in this volume.

where $H_0(\mu)$ is defined in terms of the characteristic function

$$\Psi(\mu) = \tfrac{1}{2}\,\varpi_0. \tag{39}$$

It was at this stage that I became aware of Ambartsumian's first paper on the principle of invariance for the problem of diffuse reflection by a semi-infinite atmosphere. An immediate inference that could be drawn from a comparison of Ambartsumian's solution for the isotropic case with the solution (38) was that in the limit of infinite approximation, i.e., in an *exact theory,* the H-functions, which appear in the solutions, derived by the method of discrete ordinates, must be redefined as solutions of the nonlinear integral equation

$$H(\mu) = 1 + \mu\, H(\mu) \int_0^1 \frac{\Psi(\mu')\, H(\mu')}{\mu + \mu'}\, d\mu'. \tag{40}$$

Once this became apparent, it was not difficult to prove that $H(\mu)$ *defined by equations* (26) *and* (27) *is indeed the unique solution of the finite-difference form of equation* (40), namely,

$$H(\mu) = 1 + \mu H(\mu) \sum_{j=1}^{n} \frac{a_j \Psi(\mu_j)\, H(\mu_j)}{\mu + \mu_j}. \tag{41}$$

At this point, it was abundantly clear to me that Ambartsumian's paper had opened an entirely original and novel approach to the problems of radiative transfer. But I had to put the matter aside without a closer examination for some months, since I was, at that time, engrossed in another aspect of the theory of radiative transfer. I consider this aspect, first, in §§ III and IV; and I return to the principles of invariance in § V.

III. The Equations of Transfer Incorporating the Polarization of the Radiation Field—The Problem with a Constant Net Flux

Early in 1946, I became interested in the question of the polarization of the continuous radiation of early-type stars. The reason for my interest was that, in the atmospheres of these stars, hydrogen will be almost completely ionized and the transfer of radiation will be controlled by the scattering by the free electrons. And according to J. J. Thomson's laws governing this process, an incident beam of unpolarized radiation will become partially polarized on scattering. On this account, one may expect the continuous radiation of early-type stars to exhibit partial polarization. To investigate this matter, one requires, of course, to formulate the equation of transfer correctly allowing for the state of polarization of the prevalent radiation field. But no such formulation existed in 1946.

In the context of the problem in plane-parallel atmospheres with no incident radiation, the formulation of the relevant equations of transfer is much simplified, since the axial symmetry of the radiation field, at each point, with respect to the outward normal, implies that the plane of polarization, at each point, coincides with the principal meridian. Therefore, the radiation field can be fully specified by the intensities I_l and I_r, in the two states of polarization, parallel and perpendicular, respectively, to the principal meridian. For Thomson scattering by free electrons, the equations of transfer governing the intensities I_l and I_r are

$$\mu \frac{dI_l(\tau, \mu)}{d\tau} = I_l(\tau, \mu) - \frac{3}{8}\left\{\int_{-1}^{+1} I_l(\tau, \mu')[2(1 - \mu'^2) + \mu^2 (3\mu'^2 - 2)]d\mu' + \mu^2 \int_{-1}^{+1} I_r(\tau, \mu')d\mu'\right\} \tag{42}$$

and

$$\mu \frac{dI_r(\tau, \mu')}{d\tau} = I_r(\tau, \mu) - \frac{3}{8}\left\{\int_{-1}^{+1} I_l(\tau, \mu')\mu'^2 d\mu' + \int_{-1}^{+1} I_r(\tau, \mu')d\mu'\right\}. \tag{43}$$

And solutions of these equations are required which satisfy the boundary conditions

$$I_l(0, \mu) = I_r(0, \mu) = 0 \quad (0 < \mu \leq 1),$$

and (44)

$$I_l(\tau, \mu)\, e^{-\tau} \to 0 \quad \text{and} \quad I_r(\tau, \mu)\, e^{-\tau} \to 0 \quad \text{for } \tau \to \infty.$$

Again, it was found that by the method of discrete ordinates the solutions of equations (42) and (43) for the emergent radiation in the two states of polarization can be found in the closed forms

$$I_l(0, \mu) = \tfrac{3}{8}F(1 - c^2)^{1/2}H_l(\mu)$$

and (45)

$$I_r(0, \mu) = \tfrac{3}{8}F \frac{1}{\sqrt{2}} H_r(\mu)(\mu + c),$$

where $H_l(\mu)$ and $H_r(\mu)$ are defined in terms of the characteristic functions

$$\Psi_l = \tfrac{3}{4}(1 - \mu^2) \quad \text{and} \quad \Psi_r(\mu) = \tfrac{3}{8}(1 - \mu^2), \tag{46}$$

and

$$c = \frac{H_l(+1)\, H_r(-1) + H_l(-1)\, H_r(+1)}{H_l(+1)\, H_r(-1) - H_l(-1)\, H_r(+1)}. \tag{47}$$

From the solutions for $I_l(0, \mu)$ and $I_r(0, \mu)$ given by equations (45)–(47), it was concluded that the radiation emerging tangentially at the "limb" ($\theta = \pi/2$

and $\mu = 0$) must be partially polarized to the extent of 11.4%. (The exact solution obtained a few months later gave 11.7%.) Figure 2 in paper 7 (p. 151) illustrates the predicted difference in the laws of darkening in the two states of polarization.

When the effect illustrated in this figure was found in 1946, it seemed to me that it would be worthwhile to look for it during the eclipse of binary stars, one component of which is expected to show the predicted polarization. And I suggested this observation to several leading photoelectric astronomers of the time. Dr. W. A. Hiltner (who was then a colleague of mine at the Yerkes Observatory) undertook, together with Dr. J. Hall, to look for the predicted effect. The first binary star they observed did indeed show a polarization during the last phases of eclipse; but contrary to prediction, the degree of polarization did not change with phase and it persisted even after the eclipse. Hiltner and Hall had discovered interstellar polarization instead!

By a curious coincidence, at the time of this writing (26 December 1982) I have in front of me a preprint of a paper by Dr. A. Kemp of the University of Oregon and his collaborators at Oregon and at the University of New Mexico which concludes with the statement: "We assert that the long-sought Chandrasekhar limb polarization in eclipsing binaries has now been discovered in the star system Algol."

IV. The General Vector Equation of Transfer in Terms of Stokes Parameters

As I stated at the conclusion of my paper written in February 1946 (in which the solution [45] is derived), "the successful solution of a specific problem in the theory of radiative transfer, distinguishing the different states of polarization, justifies the hope that it will be possible to solve other astrophysical problems in which polarization plays a significant part."[5] I was particularly anxious to solve the problem of diffuse reflection by a semi-infinite atmosphere on Rayleigh's laws of scattering. Indeed, a note added on 6 May 1946 to the same paper states: "The problem of diffuse reflection from a semi-infinite plane-parallel atmosphere, allowing for the partial polarization of the diffuse radiation, has now been solved." (The paper giving the solution[6] was communicated on 13 May 1946.) But before the solution was obtained, it was necessary to formulate the equations of radiative transfer in which the plane of polarization is also considered as one of the variables and not as something which is known from symmetry considerations (as in the problem considered in § III). The formulation was possible only after discovering a long-forgotten paper by Stokes published in 1852. Perhaps I may be allowed to go into some detail as to how I came to

5. *Ap. J.* 103 (1946): 351 (paper 7 in this volume).
6. *Ap. J.* 104 (1946): 110 (paper 8 in this volume).

discover Stokes's paper. The following is an abridged version of an account that I wrote more than thirty years ago, unpublished hitherto.

The fundamental question that confronted me was the following. We have an element of gas immersed in an anisotropic radiation field; and the anisotropy refers not only to the dependence of the intensity of the radiation on direction but also to the dependence of the polarization characteristics on direction. Clearly the first question concerns how one is to characterize a pencil of partially polarized radiation. All the standard and not so standard books on optics available until the early fifties contain only the obvious statements to the effect that in order to characterize partially polarized light we should specify the intensity, the plane of polarization, and the degree of polarization; and if the light is elliptically polarized, then we should in addition specify the ellipticity of the ellipse characterizing the polarized part of the radiation. In other words, the parameters of the problem are an intensity, a direction, a ratio, and a geometrical property of an ellipse. For the particular problem we have on hand, all these quantities of diverse dimensions should depend in turn on the level of the atmosphere at which we are and on the direction as well; and all these factors must be incorporated in a basic equation of transfer. As I said, none of the extant books of that time were helpful; and I also sought in vain the advice of several physicists (including G. Herzberg, G. Placzek, J. Wheeler, and G. Breit). Nevertheless, it did not seem to me that the basic question could have been overlooked by the great masters of the nineteenth century. And I recall how one afternoon I took down from the library shelves the collected papers of Rayleigh, Kelvin, and Stokes. I started with Stokes's papers; and as I was glancing down the table of contents of volume 3 of his *Collected Papers,* I came across the title "On the Composition and Resolution of Streams of Polarized Light from Different Sources." I was at once certain that this paper by Stokes ought to contain the answer to my question. This certain belief was confirmed when in the opening paragraph of the paper I read, "But when two polarized streams from different sources mix together, the mixture possesses properties intermediate between those of the original streams. . . ." And Stokes continues with characteristic modesty, "The properties of such mixtures form but an uninviting subject of investigation; and accordingly, though to a certain extent they are obvious, and must have forced themselves upon the attention of all who have paid any special attention to the physical theory of light, they do not seem hitherto to have been studied in detail." That was written in 1852; but no book on optics that had been written for one hundred years after Stokes (with one exception,[7] as I came to know later) had included an account of the topic, and no physicist whom I had consulted seemed even to be aware of the problem.

The meaning of the Stokes parameters for a partially plane-polarized beam

7. J. Walker, *The Analytical Theory of Light* (Cambridge: Cambridge University Press, 1904), §§ 20–22, pp. 28–32.

is particularly simple. If l and r refer to arbitrarily chosen directions at right angles to one another in the plane transverse to the direction of propagation of the beam, the intensity $I(\psi)$ in a direction at an angle ψ to the direction of l must go through two complete cycles as ψ goes through 360°. It must therefore be expressible in the form

$$I(\psi) = \tfrac{1}{2}(I + Q \cos 2\psi + U \sin 2\psi). \tag{48}$$

The coefficients I, Q, and U in this representation are the Stokes parameters. If one is dealing with an elliptically polarized beam, then we ask for the intensity in the direction ψ when we introduce a retardation ε in one component relative to the other; then the corresponding expression is

$$I(\psi) = \tfrac{1}{2}(I + Q \cos 2\psi + U \sin 2\psi \cos \varepsilon - V \sin 2\psi \sin \varepsilon). \tag{49}$$

In terms of the foregoing parameters the plane of polarization (referred to the direction $\psi = 0$) and the ellipticity ($= \tan \beta =$ ratio of axes of the ellipse) are given by

$$\tan 2\chi = \frac{U}{Q} \quad \text{and} \quad \sin 2\beta = \frac{V}{(Q^2 + U^2 + V^2)^{1/2}}. \tag{50}$$

The additive property of the Stokes parameters which makes them ideal for treating transfer problems is evident: *if two independent streams of polarized light are mixed, then the Stokes parameters characterizing the mixture are the sum of the Stokes parameters of the individual streams.*

In terms of the Stokes parameters, a law of scattering is specified by a matrix, $\boldsymbol{R}(\cos \Theta)$, since an elementary act of scattering results in a linear transformation of the parameters of the incident beam. Consequently, by considering the intensity as a vector $\boldsymbol{I}$, with components I_l, I_r, U, and V (where l and r refer to directions parallel and perpendicular, respectively, to the meridian through the point under consideration and the plane containing the directions of the beam and of the normal to the plane of stratification of the atmosphere) and by replacing the "phase function" commonly introduced to describe the angular distribution of the scattered radiation by a *phase matrix*, $\boldsymbol{P}$, we can formulate the basic equation of transfer without any difficulty of principle. The equation of transfer is

$$\begin{aligned} \mu \frac{d\boldsymbol{I}(\tau, \mu, \varphi)}{d\tau} &= \boldsymbol{I}(\tau, \mu, \varphi) - \frac{1}{4\pi} \int_{-1}^{+1} \int_0^{2\pi} \boldsymbol{P}(\mu, \varphi; \mu', \varphi')\, \boldsymbol{I}(\tau, \mu', \varphi') d\mu' d\varphi' \\ &\quad - \tfrac{1}{4} e^{-\tau/\mu_0} \boldsymbol{P}(\mu, \varphi; -\mu_0, \varphi_0)\, \boldsymbol{F}, \end{aligned} \tag{51}$$

where

$$\boldsymbol{F} = (F_l, F_r, F_U, F_V) \tag{52}$$

is the Stokes vector representing the parallel beam of radiation incident on the atmosphere in the direction $(-\mu_0, \varphi_0)$: πF_l, πF_r, πF_U, and πF_V denote the net fluxes per unit area normal to the beam in the four Stokes parameters; and the phase matrix $\boldsymbol{P}$ is obtained from $\boldsymbol{R}(\cos\Theta)$ by applying to it a linear transformation $\boldsymbol{L}(-i_1)$ on the *right*, to transform $\boldsymbol{I}(\tau, \mu', \varphi')$ to the orientation of the axes to which $\boldsymbol{R}(\cos\Theta)$ is generally specified, *and* a linear transformation $\boldsymbol{L}(\pi - i_2)$ on the *left* to transform the resulting scattered intensity to the orientation of the axes chosen at (μ, φ).[8]

For the particular case of Rayleigh scattering and an incident beam of partially plane-polarized light, the phase matrix is given by[9]

$$\boldsymbol{P}(\mu, \varphi; \mu', \varphi') = \boldsymbol{Q}[\boldsymbol{P}^{(0)}(\mu, \mu') + (1 - \mu^2)^{1/2}(1 - \mu'^2)^{1/2}\boldsymbol{P}^{(1)}(\mu, \varphi; \mu', \varphi') + \boldsymbol{P}^{(2)}(\mu, \varphi; \mu', \varphi')], \quad (53)$$

where

$$\boldsymbol{Q} = \begin{bmatrix} 1 & 0 & 0 \\ 0 & 1 & 0 \\ 0 & 0 & 2 \end{bmatrix}, \quad (54)$$

$$\boldsymbol{P}^{(0)}(\mu, \mu') = \frac{3}{4}\begin{bmatrix} 2(1 - \mu^2)(1 - \mu'^2) + \mu^2\mu'^2 & \mu^2 & 0 \\ \mu'^2 & 1 & 0 \\ 0 & 0 & 0 \end{bmatrix}, \quad (55)$$

$$\boldsymbol{P}^{(1)}(\mu, \varphi; \mu', \varphi') = \frac{3}{4}\begin{bmatrix} 4\mu\mu'\cos(\varphi' - \varphi) & 0 & 2\mu\sin(\varphi' - \varphi) \\ 0 & 0 & 0 \\ -2\mu'\sin(\varphi' - \varphi) & 0 & \cos(\varphi' - \varphi) \end{bmatrix}, \quad (56)$$

and

$$\boldsymbol{P}^{(2)}(\mu, \varphi; \mu', \varphi') = \frac{3}{4}\begin{bmatrix} \mu^2\mu'^2\cos 2(\varphi' - \varphi) & -\mu^2\cos 2(\varphi' - \varphi) & \mu^2\mu'\sin 2(\varphi' - \varphi) \\ -\mu'^2\cos 2(\varphi' - \varphi) & \cos 2(\varphi' - \varphi) & -\mu'\sin 2(\varphi' - \varphi) \\ -\mu\mu'^2\sin 2(\varphi' - \varphi) & \mu\sin 2(\varphi' - \varphi) & \mu\mu'\cos 2(\varphi' - \varphi) \end{bmatrix}. \quad (57)$$

The solution of equation (51) for the problem of diffuse reflection and transmission must, of course, satisfy the standard boundary conditions:

$$\boldsymbol{I}(0, -\mu, \varphi) = 0 \quad (0 < \mu \leq 1, \ 0 \leq \varphi \leq 2\pi)$$

and

$$\boldsymbol{I}(\tau_1, +\mu, \varphi) = 0 \quad (0 < \mu \leq 1, \ 0 \leq \varphi \leq 2\pi). \quad (58)$$

And the laws of diffuse reflection and transmission are expressed in terms of a *scattering matrix*, $\boldsymbol{S}(\tau_1; \mu, \varphi; \mu_0, \varphi_0)$, and a *transmission matrix*, $\boldsymbol{T}(\tau_1; \mu, \varphi; \mu_0, \varphi_0)$, such that the reflected and the transmitted intensities are given by

8. For a detailed derivation of eq. (51) see S. Chandrasekhar, *Radiative Transfer* (Oxford: Clarendon Press, 1950), §§ 16 and 17.

9. Cf. S. Chandrasekhar, *Ap. J.* 104 (1946): 110 (see eqs. [49]–[51]).

$$\boldsymbol{I}(0, +\mu, \varphi; \mu_0, \varphi_0) = \frac{1}{4\mu} \boldsymbol{QS}(\tau_1; \mu, \varphi; \mu_0, \varphi_0)\boldsymbol{F}$$

and (59)

$$\boldsymbol{I}(\tau_1; -\mu, \varphi; \mu_0, \varphi_0) = \frac{1}{4\mu} \boldsymbol{QT}(\tau_1; \mu, \varphi; \mu_0, \varphi_0)\boldsymbol{F}.$$

The solution for the problem of diffuse reflection by a semi-infinite atmosphere was found in closed form in the early summer of 1946; and the solution for the scattering matrix that was obtained is given in equation (60) on page 525, where $H_l(\mu)$ and $H_r(\mu)$ are the same H-functions introduced in § III and $H^{(1)}(\mu)$ and $H^{(2)}(\mu)$ are two additional H-functions defined in terms of the characteristic functions,

$$\Psi^{(1)}(\mu) = \tfrac{3}{8}(1 - \mu^2)(1 + 2\mu^2) \quad \text{and} \quad \Psi^{(2)}(\mu) = \tfrac{3}{16}(1 + \mu^2)^2, \tag{61}$$

and the constant c has the same value given in (47).

V. The Impact of Ambartsumian's Principles of Invariance

By the fall of 1946, when I had completed the solution of the problem of diffuse reflection by a semi-infinite Rayleigh-scattering atmosphere, it became imperative that I relate the solutions I had obtained for the various problems to the coupled systems of nonlinear integral equations that the application of Ambartsumian's principles of invariance would provide. Studying then his two papers carefully—his second paper devoted to the diffuse reflection and transmission by atmospheres of finite optical thicknesses had also become available meantime—I felt the need to supplement his ideas in several directions. But first let me state the principles of invariance as Ambartsumian stated them. They are the following:

I. *The law of diffuse reflection by a semi-infinite plane-parallel atmosphere must be invariant to the addition (or subtraction) of layers of arbitrary thicknesses to (or from) the atmosphere.*

II. *The laws of diffuse reflection and transmission by a plane-parallel atmosphere of a finite optical thickness, τ_1, must be invariant to the addition (or removal) of layers of arbitrary optical thicknesses to (or from) the atmosphere at the top (at $\tau = 0$) and the simultaneous removal (or addition) of layers of equal optical thicknesses from (or to) the atmosphere at the bottom (at $\tau = \tau_1$).*

Tracing the implications of these principles, when the layers added or subtracted are of infinitesimal thicknesses, Ambartsumian derived nonlinear integral equations for the scattering and the transmission functions defined in equation (13). These equations, when applied to specific examples of phase functions, lead to coupled systems of nonlinear equations of orders two, four, or eight even for the simplest cases. On the other hand, since the method of discrete ordinates had provided, for the cases considered, solutions in closed

$$\left(\frac{1}{\mu} + \frac{1}{\mu_0}\right)\boldsymbol{S}(\mu, \varphi; \mu_0, \varphi_0) =$$

$$\begin{bmatrix} \begin{array}{l} 2H_l(\mu)H_l(\mu_0)[1 + \mu\mu_0 - c(\mu + \mu_0)] \\ -4\mu\mu_0(1-\mu^2)^{1/2}(1-\mu_0^2)^{1/2}H^{(1)}(\mu)H^{(1)}(\mu_0)\cos(\varphi-\varphi_0) \\ +\mu^2\mu_0^2H^{(2)}(\mu)H^{(2)}(\mu_0)\cos 2(\varphi-\varphi_0) \end{array} & \begin{array}{l} [2(1-c^2)]^{1/2}H_r(\mu_0)H_l(\mu)(\mu+\mu_0) \\ \\ -\mu^2H^{(2)}(\mu)H^{(2)}(\mu_0)\cos 2(\varphi-\varphi_0) \end{array} & \begin{array}{l} \\ -2\mu(1-\mu^2)^{1/2}(1-\mu_0^2)^{1/2}H^{(1)}(\mu)H^{(1)}(\mu_0)\sin(\varphi-\varphi_0) \\ +\mu^2\mu_0H^{(2)}(\mu)H^{(2)}(\mu_0)\sin 2(\varphi-\varphi_0) \end{array} \\ \begin{array}{l} [2(1-c^2)]^{1/2}H_r(\mu)H_l(\mu_0)(\mu+\mu_0) \\ -\mu_0^2H^{(2)}(\mu)H^{(2)}(\mu_0)\cos 2(\varphi-\varphi_0) \end{array} & \begin{array}{l} H_r(\mu)H_r(\mu_0)[1+\mu\mu_0+c(\mu+\mu_0)] \\ +H^{(2)}(\mu)H^{(2)}(\mu_0)\cos 2(\varphi-\varphi_0) \end{array} & \begin{array}{l} \\ -\mu_0H^{(2)}(\mu)H^{(2)}(\mu_0)\sin 2(\varphi-\varphi_0) \end{array} \\ \begin{array}{l} -2\mu_0(1-\mu_0^2)^{1/2}(1-\mu^2)^{1/2}H^{(1)}(\mu)H^{(1)}(\mu_0)\sin(\varphi-\varphi_0) \\ +\mu\mu_0^2H^{(2)}(\mu)H^{(2)}(\mu_0)\sin 2(\varphi-\varphi_0) \end{array} & \begin{array}{l} \\ -\mu H^{(2)}(\mu)H^{(2)}(\mu_0)\sin 2(\varphi-\varphi_0) \end{array} & \begin{array}{l} (1-\mu^2)^{1/2}(1-\mu_0^2)^{1/2}H^{(1)}(\mu)H^{(1)}(\mu_0)\cos(\varphi-\varphi_0) \\ -\mu\mu_0H^{(2)}(\mu)H^{(2)}(\mu_0)\cos 2(\varphi-\varphi_0) \end{array} \end{bmatrix} \quad (60)$$

forms, it was clear that with the knowledge of the *forms* of the solutions, one should be able to reduce Ambartsumian's coupled systems of equations to one or more H-functions (for problems of reflection by semi-infinite atmospheres) or X- and Y-functions (for problems of reflection and transmission by atmospheres of finite optical thicknesses, as we shall see presently) defined with respect to suitable characteristic functions. Besides, Ambartsumian's principles required generalization to the case when the intensity is represented by a Stokes vector and scattering is governed by a phase matrix instead of a phase function. And finally, it appeared that Ambartsumian's principles could be formulated not so much in the sense of expressing invariance as of grasping the essential mathematical content of the geometrical scaffolding of the physical description that is sought.

Considering, then, the incidence of a parallel beam of radiation, of a net flux πF per unit area normal to itself, on a plane-parallel atmosphere of normal optical thickness τ_1, in a direction $(-\mu_0, \varphi_0)$ and distinguishing the outward, $(0 < \mu \leq 1)$, and the inward, $(-1 \leq \mu < 0)$, directed radiations,

$$I(\tau, +\mu, \varphi) \quad \text{and} \quad I(\tau, -\mu, \varphi) \quad (0 < \mu \leq 1), \tag{62}$$

prevailing at depth τ, we can give mathematical expression to the following four manifest consequences of the *definitions* of the scattering and the transmission functions.

I. *The intensity* $I(\tau_1 + \mu, \varphi)$ *in the outward direction at any level* τ *results from the reflection of the reduced incident flux* $\pi F e^{-\tau/\mu_0}$ *and the diffuse radiation* $I(\tau, -\mu', \varphi')$, $(0 < \mu' \leq 1)$, *incident on the surface* τ, *by the atmosphere of optical thickness* $(\tau_1 - \tau)$ *below* τ.

The mathematical expression of this principle is clearly (see fig. 1)

$$I(\tau, +\mu, \varphi) = \frac{F}{4\mu} e^{-\tau/\mu_0} S(\tau_1 - \tau; \mu, \varphi; \mu_0, \varphi_0) + \frac{1}{4\pi\mu} \int_0^1 \int_0^{2\pi} S(\tau_1 - \tau; \mu, \varphi; \mu', \varphi')\, I(\tau, -\mu', \varphi') d\mu' d\varphi'. \tag{63}$$

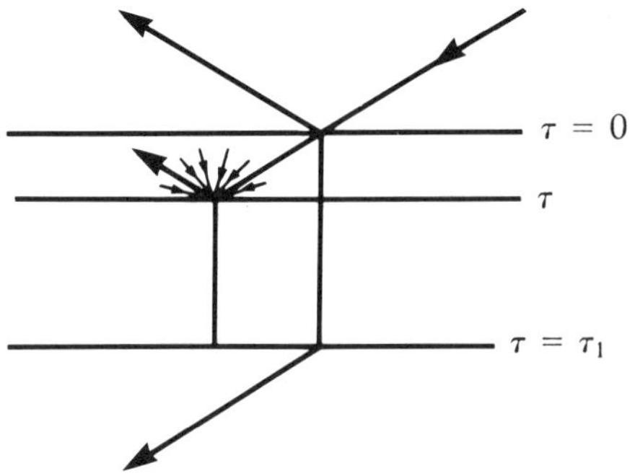

FIG. 1

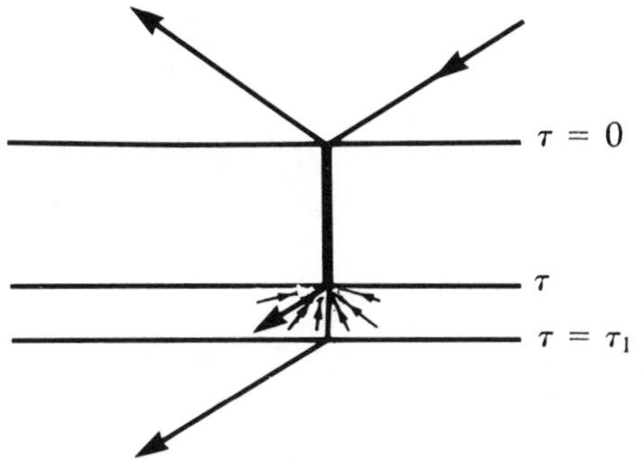

FIG. 2

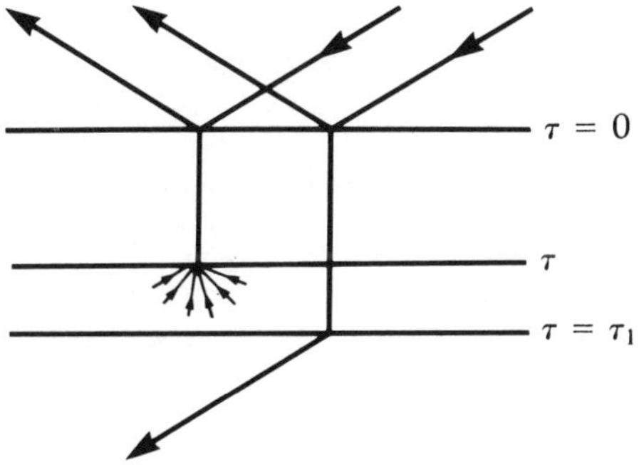

FIG. 3

II. *The intensity* $I(\tau, -\mu, \varphi)$ *in the inward direction at any level* τ *results from the transmission of the incident flux by the atmosphere of optical thickness* τ, *above the surface* τ, *and the reflection by this same surface of the diffuse radiation* $I(\tau, +\mu', \varphi')$, $(0 < \mu' \leq 1)$, *incident on it from below.*

The mathematical expression of this principle is (see fig. 2)

$$I(\tau, -\mu, \varphi) = \frac{F}{4\mu} T(\tau; \mu, \varphi; \mu_0, \varphi_0) + \frac{1}{4\pi\mu} \int_0^1 \int_0^{2\pi} S(\tau; \mu, \varphi; \mu', \varphi') \, I(\tau, +\mu', \varphi') d\mu' d\varphi'. \quad (64)$$

III. *The diffuse reflection of the incident light by the entire atmosphere is equivalent to the reflection by the part of the atmosphere of optical thickness* τ, *above the level* τ, *and the transmission by this same atmosphere of the diffuse radiation* $I(\tau, +\mu', \varphi')$, $(0 < \mu' \leq 1)$, *incident on the surface* τ *from below.*

The mathematical expression of this principle is (see fig. 3)

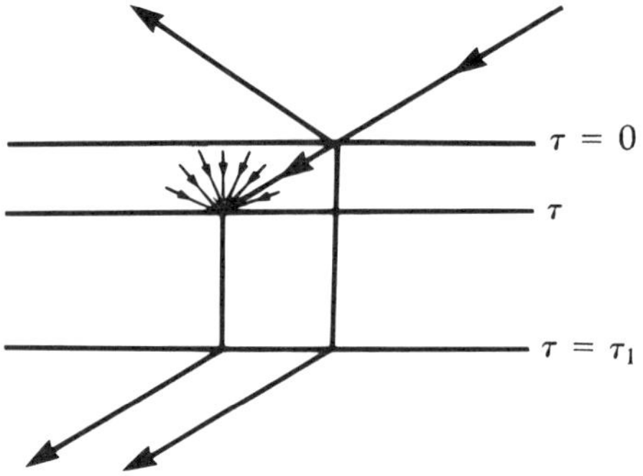

FIG. 4

$$\frac{F}{4\mu} S(\tau_1; \mu, \varphi; \mu_0, \varphi_0) = \frac{F}{4\mu} S(\tau; \mu, \varphi; \mu_0, \varphi_0) + e^{-\tau/\mu} I(\tau, +\mu, \varphi) + \frac{1}{4\pi\mu} \int_0^1 \int_0^{2\pi} T(\tau; \mu, \varphi; \mu', \varphi') I(\tau, +\mu', \varphi') d\mu' d\varphi'. \quad (65)$$

IV. *The diffuse transmission of the incident light by the entire atmosphere is equivalent to the transmission of the reduced incident flux* $\pi F e^{-\tau/\mu_0}$ *and the diffuse radiation* $I(\tau, -\mu', \varphi')$, $(0 < \mu' \leqslant 1)$, *incident on the surface* τ *by the atmosphere of optical thickness* $(\tau_1 - \tau)$ *below* τ.

The mathematical expression of this principle is (see fig. 4)

$$\frac{F}{4\mu} T(\tau_1; \mu, \varphi; \mu', \varphi') = \frac{F}{4\mu} e^{-\tau/\mu_0} T(\tau_1 - \tau; \mu, \varphi; \mu', \varphi') + e^{-(\tau_1 - \tau)/\mu} I(\tau, -\mu, \varphi) + \frac{1}{4\pi\mu} \int_0^1 \int_0^{2\pi} T(\tau_1 - \tau; \mu, \varphi; \mu', \varphi') I(\tau, -\mu', \varphi') d\mu' d\varphi'. \quad (66)$$

Similar expressions can be written down which relate the angular distribution of the emergent radiation, in the problem with the constant net flux, to the scattering and the transmission functions.

The basic integral equations which govern the scattering and the transmission functions can be obtained by differentiating equations (63)–(66) with respect to τ and passing to the limit $\tau = 0$ of expressions (63) and (66), and to the limit $\tau = \tau_1$ of expressions (64) and (65). This procedure leads to a set of four integrodifferential equations for the scattering and the transmission functions. Two linear combinations of these equations are equivalent to the integral equations derived by Ambartsumian from his principles of invariance; and we are left with two additional relations.

For the case of isotropic scattering with an albedo ϖ_0, one finds that with the definitions

$$X(\mu) = 1 + \frac{1}{2}\int_0^1 S(\tau_1;\, \mu, \mu')\,\frac{d\mu'}{\mu'} \tag{67}$$

and

$$Y(\mu) = e^{-\tau_1/\mu} + \frac{1}{2}\int_0^1 T(\tau_1;\, \mu, \mu')\,\frac{d\mu'}{\mu'}, \tag{68}$$

we obtain four equations,

$$\left(\frac{1}{\mu_0} + \frac{1}{\mu}\right) S(\tau_1;\, \mu, \mu_0) = \varpi_0[X(\mu)X(\mu_0) - Y(\mu)Y(\mu_0)], \tag{69}$$

$$\left(\frac{1}{\mu_0} - \frac{1}{\mu}\right) T(\tau_1;\, \mu, \mu_0) = \varpi_0[Y(\mu)X(\mu_0) - X(\mu)Y(\mu_0)], \tag{70}$$

$$\frac{\partial S(\tau_1;\, \mu, \mu_0)}{\partial \tau_1} = \varpi_0 Y(\mu)Y(\mu_0), \tag{71}$$

and

$$\left(\frac{1}{\mu_0} - \frac{1}{\mu}\right)\frac{\partial T(\tau_1;\, \mu, \mu_0)}{\partial \tau_1} = \varpi_0\left[\frac{1}{\mu_0}X(\mu)Y(\mu_0) - \frac{1}{\mu}Y(\mu)X(\mu_0)\right]. \tag{72}$$

Inserting expressions (69) and (70) for S and T in equations (67) and (68), we obtain the equations

$$X(\mu) = 1 + \tfrac{1}{2}\varpi_0\mu\int_0^1 \frac{d\mu'}{\mu + \mu'}\,[X(\mu)X(\mu') - Y(\mu)Y(\mu')] \tag{73}$$

and

$$Y(\mu) = e^{-\tau_1/\mu} + \tfrac{1}{2}\varpi_0\mu\int_0^1 \frac{d\mu'}{\mu - \mu'}\,[Y(\mu)X(\mu') - X(\mu)Y(\mu')], \tag{74}$$

and from equations (71) and (72) we obtain

$$\frac{\partial X(\mu, \tau_1)}{\partial \tau_1} = \tfrac{1}{2}\varpi_0 Y(\mu, \tau_1)\int_0^1 \frac{d\mu'}{\mu'}Y(\mu', \tau_1) \tag{75}$$

and

$$\frac{\partial Y(\mu, \tau_1)}{\partial \tau_1} + \frac{Y(\mu, \tau_1)}{\mu} = \tfrac{1}{2}\varpi_0 X(\mu, \tau_1)\int_0^1 \frac{d\mu'}{\mu'}\,Y(\mu', \tau_1). \tag{76}$$

From a comparison of equations (73) and (74) with equation (40), it would appear that the pair of coupled equations

$$X(\mu) = 1 + \mu\int_0^1 \frac{\Psi(\mu')}{\mu + \mu'}\,[X(\mu)X(\mu') - Y(\mu)Y(\mu')]d\mu' \tag{77}$$

and

$$Y(\mu) = e^{-\tau_1/\mu} + \mu \int_0^1 \frac{\Psi(\mu')}{\mu - \mu'} [Y(\mu)X(\mu') - X(\mu)Y(\mu')]d\mu', \tag{78}$$

where $\Psi(\mu)$ is a characteristic function, will play the same role in the theory of radiative transfer in atmospheres of finite optical thicknesses as equation (40) does in the theory of transfer in semi-infinite atmospheres. We may note parenthetically that in the context of equations (77) and (78), equations (75) and (76) are replaced by

$$\frac{\partial X(\mu, \tau_1)}{\partial \tau_1} = Y(\mu, \tau_1) \int_0^1 \frac{d\mu'}{\mu'} \Psi(\mu')Y(\mu', \tau_1) \tag{79}$$

and

$$\frac{\partial Y(\mu, \tau_1)}{\partial \tau_1} + \frac{Y(\mu, \tau_1)}{\mu} = X(\mu, \tau_1) \int_0^1 \frac{d\mu'}{\mu'} \Psi(\mu')Y(\mu', \tau_1). \tag{80}$$

These equations have played important roles in subsequent developments.

We shall consider in § VI below the manner of application of the coupled systems of equations, to which the principles of invariance lead, for laws of scattering more general than isotropic. But it is important to point out that even in the case of isotropic scattering the equations governing the functions $X(\mu)$ and $Y(\mu)$ do not, by themselves, suffice to specify the scattering and the transmission functions in the conservative case when $\varpi_0 = 1$. For, quite generally, when

$$\int_0^1 \Psi(\mu)d\mu = \tfrac{1}{2}, \tag{81}$$

it can be readily verified that if $X(\mu)$ and $Y(\mu)$ are solutions of equations (77) and (78), then so are

$$X(\mu) + Q\mu[X(\mu) + Y(\mu)]$$

and (82)

$$Y(\mu) - Q\mu\,[X(\mu) + Y(\mu)],$$

where Q is an arbitrary constant. On this account, in the conservative case (81), one defines *standard solutions* by the requirements

$$x_0 = \int_0^1 X(\mu)\Psi(\mu)d\mu = 1$$

and (83)[10]

$$y_0 = \int_0^1 Y(\mu)\Psi(\mu)d\mu = 0.$$

10. One can impose these conditions in the case (81), since, in general, we have identity,

$$x_0 = 1 - \left[1 - 2\int_0^1 \Psi(\mu)d\mu + y_0^2\right]^{1/2}.$$

The resulting ambiguity in the solution for the scattering and the transmisson functions can be removed by the consideration of the *K*-integral. In this manner, one finds that with the *X*- and *Y*-functions defined as the standard solutions, the required *X*- and *Y*-functions are given by equation (82) with

$$Q = -\frac{\alpha_1 - \beta_1}{(\alpha_1 + \beta_1)\tau_1 + 2(\alpha_2 + \beta_2)}, \tag{84}$$

where α_n and β_n denote the moments of order n of $X(\mu)$ and $Y(\mu)$:

$$\alpha_n = \int_0^1 X(\mu)\mu^n d\mu \quad \text{and} \quad \beta_n = \int_0^1 Y(\mu)\mu^n d\mu. \tag{85}$$

VI. Solutions of the Coupled Systems of Integral Equations which Follow from the Principles of Invariance

Considering first the problem of diffuse reflection by semi-infinite atmospheres for the scattering function, $S(\mu, \varphi; \mu_0, \varphi_0)$, we find that, in the isotropic case, the principle of invariance directly leads to the solution (as Ambartsumian first showed),

$$\left(\frac{1}{\mu_0} + \frac{1}{\mu}\right)S(\mu, \mu_0) = \varpi_0 H(\mu)H(\mu_0), \tag{86}$$

where $H(\mu)$ is defined in terms of the characteristic function, $\frac{1}{2}\varpi_0$. This solution confirms that $H(\mu)$, defined as the solution of the difference equation (41), becomes, in the limit of infinite approximation, the solution of the integral equation (40).

When we proceed to a consideration of problems with even slightly more general phase functions, the reduction of the integral equations, which follows from the principle of invariance, is not as straightforward. Thus, considering the case of Rayleigh's phase function (21), we find that the azimuth-dependent terms in the emergent intensity are, indeed, directly expressed in terms of the *H*-functions, $H^{(1)}(\mu)$ and $H^{(2)}(\mu)$, defined in equations (33) and (34). But the azimuth-independent term in the scattering function is expressed in the manner

$$\left(\frac{1}{\mu} + \frac{1}{\mu_0}\right)S^{(0)}(\mu, \mu_0) = \tfrac{1}{3}\psi(\mu)\psi(\mu_0) + \tfrac{8}{3}\phi(\mu)\phi(\mu_0), \tag{87}$$

where

$$\psi(\mu) = 3 - \mu^2 + \frac{3}{16}\int_0^1 (3 - \mu'^2)S^{(0)}(\mu, \mu')\frac{d\mu'}{\mu'} \tag{88}$$

and

$$\phi(\mu) = \mu^2 + \frac{3}{16}\int_0^1 \mu'^2\, S^{(0)}(\mu, \mu')\frac{d\mu'}{\mu'}. \tag{89}$$

The substitution of equation (87) in equations (88) and (89) provides the pair of coupled nonlinear integral equations

$$\psi(\mu) = 3 - \mu^2 + \tfrac{1}{16}\mu\psi(\mu)\int_0^1 \frac{3-\mu'^2}{\mu+\mu'}\psi(\mu')d\mu' + \tfrac{1}{2}\mu\phi(\mu)\int_0^1 \frac{3-\mu'^2}{\mu+\mu'}\phi(\mu')d\mu' \tag{90}$$

and

$$\phi(\mu) = \mu^2 + \tfrac{1}{16}\,\mu\psi(\mu)\int_0^1 \frac{\mu'^2}{\mu+\mu'}\psi(\mu')d\mu' + \tfrac{1}{2}\,\mu\phi(\mu)\int_0^1 \frac{\mu'^2}{\mu+\mu'}\phi(\mu')d\mu'. \tag{91}$$

I should not have known (in 1946) how to reduce these equations had it not been for the fact that the solution (33) (derived by the method of discrete ordinates) suggests that we seek solutions of the form

$$\psi(\mu) = (3 - c\mu)\,H^{(0)}(\mu) \quad \text{and} \quad \phi(\mu) = q\mu\,H^{(0)}(\mu), \tag{92}$$

where $H^{(0)}(\mu)$ is defined in terms of the characteristic function (cf. eq. [34])

$$\Psi^{(0)}(\mu) = \tfrac{3}{16}(3 - \mu^2), \tag{93}$$

and q and c are constants unspecified in the first instance. The substitution of solutions of the forms (92) in equations (90) and (91) determines the constants q and c. We find

$$q = \frac{2}{3\alpha_1} \quad \text{and} \quad c = \frac{\alpha_2}{\alpha_1}, \tag{94}$$

where α_1 and α_2 are the first and the second moments of $H^{(0)}(\mu)$.

Similarly, for scattering in accordance with the phase function (22), we find that the azimuth-independent term in the scattering function is expressed in the form

$$\left(\frac{1}{\mu} + \frac{1}{\mu_0}\right) S^{(0)}(\mu, \mu_0) = \psi(\mu)\psi(\mu_0) - x\phi(\mu)\phi(\mu_0), \tag{95}$$

where

$$\psi(\mu) = 1 + \tfrac{1}{2}\varpi_0 \int_0^1 S^{(0)}(\mu, \mu')\,\frac{d\mu'}{\mu'} \tag{96}$$

and

$$\phi(\mu) = \mu - \tfrac{1}{2}\varpi_0 \int_0^1 S^{(0)}(\mu, \mu')\,d\mu'. \tag{97}$$

The substitution of equation (95) in equations (96) and (97) now leads to the pair of equations

$$\psi(\mu) = 1 + \tfrac{1}{2}\varpi_0\mu\psi(\mu)\int_0^1 \frac{d\mu'}{\mu + \mu'}\psi(\mu') - \tfrac{1}{2}x\varpi_0\mu\phi(\mu)\int_0^1 \frac{d\mu'}{\mu + \mu'}\phi(\mu') \quad (98)$$

and

$$\phi(\mu) = \mu - \tfrac{1}{2}\varpi_0\mu\psi(\mu)\int_0^1 \frac{d\mu'}{\mu + \mu'}\mu'\psi(\mu') + \tfrac{1}{2}x\varpi_0\mu\phi(\mu)\int_0^1 \frac{d\mu'}{\mu + \mu'}\mu'\phi(\mu'). \quad (99)$$

Again from the knowledge of solution (36), obtained by the method of discrete ordinates, we find that the required solution of equations (98) and (99) is given by

$$\psi(\mu) = (1 - c\mu)H^{(0)}(\mu) \quad \text{and} \quad \phi(\mu) = q\mu H^{(0)}(\mu), \quad (100)$$

where $H^{(0)}(\mu)$ is defined in terms of the characteristic function $\Psi^{(0)}(\mu)$ (cf. eq. [37]):

$$\Psi^{(0)}(\mu) = \tfrac{1}{2}\varpi_0[1 + x(1 - \varpi_0)\mu^2], \quad (101)$$

and c and q are constants given by

$$c = x\varpi_0(1 - \varpi_0)\frac{\alpha_1}{2 - \varpi_0\alpha_0} \quad \text{and} \quad q = \frac{2(1 - \varpi_0)}{2 - \varpi_0\alpha_0}, \quad (102)$$

where α_n denotes the moment of order n of $H^{(0)}(\mu)$.

Finally, considering the case of diffuse reflection by a Rayleigh-scattering atmosphere, we find that the principle of invariance (now generalized to allow for scattering in accordance with a phase matrix) requires that the azimuth-independent part of the scattering matrix is of the form

$$\left(\frac{1}{\mu'} + \frac{1}{\mu}\right)S^{(0)}(\mu, \mu') = \begin{bmatrix} \psi(\mu) & \phi(\mu)\sqrt{2} \\ \chi(\mu) & \zeta(\mu)\sqrt{2} \end{bmatrix} \begin{bmatrix} \psi(\mu') & \chi(\mu') \\ \phi(\mu')\sqrt{2} & \zeta(\mu')\sqrt{2} \end{bmatrix}, \quad (103)$$

where

$$\psi(\mu) = \mu^2 + \frac{3}{8}\int_0^1 \frac{d\mu'}{\mu'}[\mu'^2 S_{ll}^{(0)}(\mu, \mu') + S_{lr}^{(0)}(\mu, \mu')], \quad (104)$$

$$\phi(\mu) = 1 - \mu^2 + \frac{3}{8}\int_0^1 \frac{d\mu'}{\mu'}(1 - \mu'^2)S_{ll}^{(0)}(\mu, \mu'), \quad (105)$$

$$\chi(\mu) = 1 + \frac{3}{8}\int_0^1 \frac{d\mu'}{\mu'}[\mu'^2 S_{rl}^{(0)}(\mu, \mu') + S_{rr}^{(0)}(\mu, \mu')], \quad (106)$$

and

$$\zeta(\mu) = \frac{3}{8}\int_0^1 \frac{d\mu'}{\mu'}(1 - \mu'^2)S_{rl}^{(0)}(\mu, \mu'). \quad (107)$$

And we are led to the following coupled system of equations of order 4:

$$\psi(\mu) = \mu^2 + \tfrac{3}{8}\mu\psi(\mu)\int_0^1 \frac{d\mu'}{\mu+\mu'}[\mu'^2\,\psi(\mu') + \chi(\mu')]$$
$$+ \tfrac{3}{4}\mu\phi(\mu)\int_0^1 \frac{d\mu'}{\mu+\mu'}[\mu'^2\phi(\mu') + \zeta(\mu')], \tag{108}$$

$$\phi(\mu) = 1 - \mu^2 + \tfrac{3}{8}\mu\psi(\mu)\int_0^1 \frac{d\mu'}{\mu+\mu'}(1-\mu'^2)\psi(\mu')$$
$$+ \tfrac{3}{4}\mu\phi(\mu)\int_0^1 \frac{d\mu'}{\mu+\mu'}(1-\mu'^2)\phi(\mu'), \tag{109}$$

$$\chi(\mu) = 1 + \tfrac{3}{8}\mu\chi(\mu)\int_0^1 \frac{d\mu'}{\mu+\mu'}[\mu'^2\,\psi(\mu') + \chi(\mu')]$$
$$+ \tfrac{3}{4}\mu\zeta(\mu)\int_0^1 \frac{d\mu'}{\mu+\mu'}[\mu'^2\phi(\mu') + \zeta(\mu')], \tag{110}$$

and

$$\zeta(\mu) = \tfrac{3}{8}\mu\chi(\mu)\int_0^1 \frac{d\mu'}{\mu+\mu'}(1-\mu'^2)\,\psi(\mu')$$
$$+ \tfrac{3}{4}\mu\zeta(\mu)\int_0^1 \frac{d\mu'}{\mu+\mu'}(1-\mu'^2)\phi(\mu'). \tag{111}$$

From the solution for the scattering matrix given in equation (60), we now deduce that the solution of the foregoing equations is given by

$$\psi(\mu) = q\mu\, H_l(\mu), \quad \phi(\mu) = (1 - c\mu)\, H_l(\mu),$$
$$\zeta(\mu) = \tfrac{1}{2}q\mu\, H_r(\mu), \quad \text{and} \quad \chi(\mu) = (1 + c\mu)\, H_r(\mu), \tag{112}$$

where $H_l(\mu)$ and $H_r(\mu)$ are defined in terms of the same characteristic functions, Ψ_l and Ψ_r, given in equations (46), and q and c are constants given by

$$q = \frac{8(A_1 + 2\alpha_1) - 6(A_0\alpha_1 + \alpha_0 A_1)}{3(A_1{}^2 + 2\alpha_1{}^2)} \tag{113}$$

and

$$c = \frac{8(A_1 - \alpha_1) + 3(2\alpha_1\alpha_0 - A_1A_0)}{3(A_1{}^2 + 2\alpha_1{}^2)}, \tag{114}$$

and α_n and A_n are the moments of order n of $H_l(\mu)$ and $H_r(\mu)$.

Turning next to the problem of diffuse reflection and transmission by atmospheres of finite optical thicknesses, we find that they are considerably more complex than the same problems in the context of semi-infinite atmospheres, at both levels: at the level of eliminating the constants and of obtaining solutions in closed forms in the method of discrete ordinates *and* at the level of reducing the coupled systems of equations, derived from the principles of invariance, to a few X- and Y-equations. The complexity of the analysis does not permit a review of even the most salient features. But I shall briefly dwell on an aspect of the problem of the elimination of the constants, in the method of discrete ordinates, which does not appear to have attracted any attention in the subsequent literature: it is the central role played by a problem which, in some sense, is a generalization of the problem underlying Lagrange's interpolation formula. The particular problem[11] is the determination of two polynomials, $C_0(\mu)$ and $C_1(\mu)$, both of degree n, which are related in the manner

$$C_0(1/k_\alpha) = e^{k_\alpha \tau_1} \frac{P(-1/k_\alpha)}{P(+1/k_\alpha)} C_1(-1/k_\alpha) \qquad (\alpha = \pm 1, \ldots, \pm n), \tag{115}$$

where $\pm k_\alpha$ ($\alpha = 1, \ldots, n$) are the distinct roots of the characteristic equation (27) and

$$P(\mu) = \prod_{i=1}^{n} (\mu - \mu_i). \tag{116}$$

(We are ignoring for the present the conservative case when the characteristic equation allows a pair of zero roots.) The solution of the problem (none too straightforward to obtain) is

$$C_0(\mu) = \sum_{2^{n-1}\text{ terms}}^{l=n,n-2,\ldots} \varepsilon_l^{(0)} \frac{\prod_{i=1}^{l}\prod_{m=1}^{n-l}(k_{r_i} + k_{s_m})}{\prod_{i=1}^{l}\prod_{m=1}^{n-l}(k_{r_i} - k_{s_m})} \prod_{i=1}^{l}(1 + k_{r_i}\mu) \prod_{m=1}^{n-l}\frac{1}{\lambda_{s_m}}(1 - k_{s_m}\mu) \tag{117}$$

and

$$C_1(\mu) = (-1)^{n-1} \sum_{2^{n-1}\text{ terms}}^{l=n-1,n-3,\ldots} \varepsilon_l^{(1)} \frac{\prod_{i=1}^{l}\prod_{m=1}^{n-l}(k_{r_i} + k_{s_m})}{\prod_{i=1}^{l}\prod_{m=1}^{n-l}(k_{r_i} - k_{s_m})}$$

$$\times \prod_{i=1}^{l}(1 + k_{r_i}\mu) \prod_{m=1}^{n-l}\frac{1}{\lambda_{s_m}}(1 - k_{s_m}\mu), \tag{118}$$

where $r_1, \ldots, r_l$ and $s_1, \ldots, s_{n-l}$ are selections of l and $n - l$ distinct integers from the set $(1, 2, \ldots, n)$,

11. The abstract mathematical problem that is posed is the determination of two polynomials, $F(x)$ and $G(x)$, both of degree n, which are related in the manner

$$F(x_j) = +\lambda_j F(-x_j) \qquad \text{and} \qquad G(x_j) = -\lambda_j G(-x_j),$$

where x_j ($j = 1, \ldots, n$) are n distinct values of the argument and λ_j ($j = 1, \ldots, n$) are n assigned numbers, all different from one another.

$$\left.\begin{aligned} \varepsilon_l^{(0)} &= +1 \quad \text{for integers of the form } n - 4m \\ &= -1 \quad \text{for integers of the form } n - 4m - 2 \\ &= 0 \quad\;\; \text{otherwise,} \\ \varepsilon_l^{(1)} &= +1 \quad \text{for integers of the form } n - 4m - 1 \\ &= -1 \quad \text{for integers of the form } n - 4m - 3 \\ &= 0 \quad\;\; \text{otherwise,} \\ \lambda_\alpha &= e^{k_\alpha \tau_1} \frac{P(-1/k_\alpha)}{P(+1/k_\alpha)} \qquad (\alpha = \pm 1, \ldots, \pm n). \end{aligned}\right\} \tag{119}$$

And the functions in terms of which the solutions to the various problems are obtained in closed forms are

$$X(\mu) = \frac{(-1)^n}{\mu_1 \ldots \mu_n} \frac{1}{[C_0^2(0) - C_1^2(0)]^{1/2}} \frac{1}{W(\mu)} [P(-\mu)C_0(-\mu) - e^{-\tau_1/\mu}P(\mu)C_1(\mu)]$$

and (120)

$$Y(\mu) = \frac{(-1)^n}{\mu_1 \ldots \mu_n} \frac{1}{[C_0^2(0) - C_1^2(0)]^{1/2}} \frac{1}{W(\mu)} [e^{-\tau_1/\mu}P(\mu)C_0(\mu) - P(-\mu)C_1(-\mu)],$$

where

$$W(\mu) = \sum_{\alpha=1}^{n} (1 - k_\alpha^2 \mu^2). \tag{121}$$

In the limit of infinite approximation, these rational functions are replaced, in the exact theory, by the X- and Y-functions defined by the integral equations (77) and (78).

In the conservative case, the characteristic equation allows a pair of zero roots. We then restrict relations (115) to the nonvanishing roots and consider polynomials, $C_0(\mu)$ and $C_1(\mu)$, of degree $(n - 1)$. The corresponding expressions for $X(\mu)$ and $Y(\mu)$, defined in the same manner as in equations (120), become, in the exact theory, the *standard solutions* defined in § V. The rigorous justification of these "empirical" facts remains to be established.

The facts then are that for the different laws of scattering considered in §§ II and IV, the solutions for the scattering and the transmission functions (or matrices) can be obtained in closed forms. With the forms of the solutions thus known, and with the correspondence enunciated between the rational representations (120) and the exact X- and Y-functions, we can complete the solution of the coupled systems of equations, derived from the principles of invariance, in the manner we have described earlier in the context of semi-infinite atmospheres. I shall leave the matter with this bare statement.

VII. The Polarization of the Sunlit Sky

I now turn to a problem in the theory of radiative transfer which bears on the polarization of the sunlit sky. I shall begin by tracing briefly the history of this problem.

In 1871, Lord Rayleigh at the age of twenty-nine and in his eighth published paper (his six volumes of *Scientific Papers* list four hundred and forty-six papers), entitled "On the Light from the Sky, Its Polarization and Color," accounted for the principal features of the phenomena to which his paper was addressed in terms of the law of scattering that has since come to be known under his name. In particular, he interpreted the blue color of the sky as a result of the inverse fourth power of the wavelength which appears in his law of scattering; and he interpreted the almost complete polarization of the sky in a direction at right angles to the sun in terms of the complete polarization predicted by his law of scattering in that direction. But of course at no point of the sky is the light completely polarized. Lord Rayleigh refers to this fact and interprets it correctly by attributing the incomplete polarization to multiple scattering. But a precise formulation and solution to Rayleigh's problem had to wait for nearly three-quarters of a century. In the meantime, however, Babinet, Brewster, and Arago had made interesting discoveries concerning the directions in which the sunlight is unpolarized. For angles of incidence not exceeding 70°, the *neutral points* (that is, the points of zero polarization) occur at about 10°–20° above and below the sun; these are the points of Babinet and Brewster. And when the sun is low, then near the horizon opposite the sun and about 20° above the antisolar point, a neutral point occurs: this is the Arago point. In fact, the setting of the Brewster point in the western sky coincides with the rising of the Arago point in the eastern sky. These facts should be contrasted with what should be expected on Rayleigh's laws of scattering, namely, that the polarization should be zero in the forward and in the backward directions.

The neutral points were observed and discussed a great deal during the nineteenth century, culminating in the earlier years of this century in the monumental work of the Swiss meteorologist C. Dorno. Dorno not only observed the neutral points on the principal meridian, but he also investigated in detail the continuation of these neutral points over the entire hemisphere along the so-called *neutral lines.* These neutral lines separate the regions of negative polarization from the regions of positive polarization and show a remarkable dependence on the direction of the sun, as can be seen in figure 4 of Paper 11 in this volume (p. 211). To explain these curves, one needs an exact solution of the underlying problem in radiative transfer.

It is clear that a rigorous theory allowing for all orders of scattering cannot be developed without deriving the equations governing the transfer of radiation in an atmosphere in which each element scatters in accordance with Rayleigh's laws. The need for a development along these lines was clearly stated by L. V. King in 1913.[12] But he also stated:

> The complete solution of the problem from this aspect would require us to split up the incident radiation into two components one of which is polarized in the principal plane and the other at right angles to it: the effect of self-illumination would lead to simultaneous integral equations in three variables the solution of which would be much too complicated to be useful.

With this comment, King proceeded to a highly approximative treatment of the problem; and so did others who followed him.[13]

With the equation of transfer now properly formulated in vector form for the Stokes parameters, the problem we are required to solve is no more than the extension of the solution (60), obtained in the context of semi-infinite atmospheres, to atmospheres of finite optical thicknesses. The required solution was obtained by the procedure described toward the end of § VI. Thus, the solutions for the scattering and the transmission matrices were first obtained in closed forms by the method of discrete ordinates.[14] With the forms of the solutions thus determined, the exact solutions were then obtained with the aid of the systems of equations derived from the proper vector generalizations of the principles of invariance. Without further ado, I shall simply transcribe the solution that was obtained in the fall of 1947.[15]

Writing the scattering and the transmission matrices in the forms

$$\begin{aligned} \boldsymbol{S}(\mu, \varphi; \mu_0, \varphi_0) = \boldsymbol{Q}[&\tfrac{3}{4}\boldsymbol{S}^{(0)}(\mu; \mu_0) \\ &+ (1 - \mu^2)^{1/2}(1 - \mu_0{}^2)^{1/2}\, \boldsymbol{S}^{(1)}(\mu, \varphi; \mu_0, \varphi_0) \\ &+ \boldsymbol{S}^{(2)}(\mu, \varphi; \mu_0, \varphi_0)] \end{aligned} \tag{122}$$

and

$$\begin{aligned} \boldsymbol{T}(\mu, \varphi; \mu_0, \varphi_0) = \boldsymbol{Q}[&\tfrac{3}{4}\boldsymbol{T}^{(0)}(\mu; \mu_0) \\ &+ (1 - \mu^2)^{1/2}(1 - \mu_0{}^2)^{1/2}\, \boldsymbol{T}^{(1)}(\mu, \varphi; \mu_0, \varphi_0) \\ &+ \boldsymbol{T}^{(2)}(\mu, \varphi; \mu_0, \varphi_0)], \end{aligned} \tag{123}$$

we find, directly, that the dependence of the azimuth-dependent terms ($\boldsymbol{S}^{(1)}$, $\boldsymbol{T}^{(1)}$) and ($\boldsymbol{S}^{(2)}$, $\boldsymbol{T}^{(2)}$) on $(\varphi - \varphi_0)$ is essentially the same as that of $\boldsymbol{P}^{(1)}$ and $\boldsymbol{P}^{(2)}$ in expression (53) for the phase matrix. Indeed, we find that

12. *Phil. Trans. Roy. Soc. London, A* 212 (1913): 375.
13. E.g., A. Hammad and S. Chapman, *Phil. Mag.*, ser. 7, 28 (1939): 99.
14. *Ap. J.* 106 (1947): 184–216.
15. *Ap. J.* 107 (1948): 199–214.

$$\left(\frac{1}{\mu_0}+\frac{1}{\mu}\right)\boldsymbol{S}^{(i)} = [X^{(i)}(\mu)X^{(i)}(\mu_0) - Y^{(i)}(\mu)Y^{(i)}(\mu_0)]\,\boldsymbol{P}^{(1)}(\mu, \varphi;\ -\mu_0, \varphi_0) \qquad (i = 1, 2), \quad (124)$$

and

$$\left(\frac{1}{\mu_0}-\frac{1}{\mu}\right)\boldsymbol{T}^{(i)} = [Y^{(i)}(\mu)X^{(i)}(\mu_0) - X^{(i)}(\mu)Y^{(i)}(\mu_0)]\,\boldsymbol{P}^{(i)}(-\mu, \varphi;\ -\mu_0, \varphi_0) \qquad (i = 1, 2), \quad (125)$$

where $(X^{(1)}, Y^{(1)})$ and $(X^{(2)}, Y^{(2)})$ are a pair of X- and Y-functions belonging to the characteristic functions $\Psi^{(1)}$ and $\Psi^{(2)}$ defined in equations (61).

In contrast to the azimuth-dependent terms, the azimuth-independent terms, $\boldsymbol{S}^{(0)}$ and $\boldsymbol{T}^{(0)}$, have a very complicated structure. They have the forms

$$\left(\frac{1}{\mu_0}+\frac{1}{\mu}\right)\boldsymbol{S}^{(0)}(\mu;\ \mu_0) = \begin{bmatrix}\psi(\mu) & \phi(\mu)\sqrt{2} & 0\\ \chi(\mu) & \zeta(\mu)\sqrt{2} & 0\\ 0 & 0 & 0\end{bmatrix}\begin{bmatrix}\psi(\mu_0) & \chi(\mu_0) & 0\\ \phi(\mu_0)\sqrt{2} & \zeta(\mu_0)\sqrt{2} & 0\\ 0 & 0 & 0\end{bmatrix} - \begin{bmatrix}\xi(\mu) & \eta(\mu)\sqrt{2} & 0\\ \sigma(\mu) & \theta(\mu)\sqrt{2} & 0\\ 0 & 0 & 0\end{bmatrix}\begin{bmatrix}\xi(\mu_0) & \sigma(\mu_0) & 0\\ \eta(\mu_0)\sqrt{2} & \theta(\mu_0)\sqrt{2} & 0\\ 0 & 0 & 0\end{bmatrix} \quad (126)$$

and

$$\left(\frac{1}{\mu_0}-\frac{1}{\mu}\right)\boldsymbol{T}^{(0)}(\mu;\ \mu_0) = \begin{bmatrix}\xi(\mu) & \eta(\mu)\sqrt{2} & 0\\ \sigma(\mu) & \theta(\mu)\sqrt{2} & 0\\ 0 & 0 & 0\end{bmatrix}\begin{bmatrix}\psi(\mu_0) & \chi(\mu_0) & 0\\ \phi(\mu_0)\sqrt{2} & \zeta(\mu_0)\sqrt{2} & 0\\ 0 & 0 & 0\end{bmatrix} - \begin{bmatrix}\psi(\mu) & \phi(\mu)\sqrt{2} & 0\\ \chi(\mu) & \zeta(\mu)\sqrt{2} & 0\\ 0 & 0 & 0\end{bmatrix}\begin{bmatrix}\xi(\mu_0) & \sigma(\mu_0) & 0\\ \eta(\mu_0)\sqrt{2} & \theta(\mu_0)\sqrt{2} & 0\\ 0 & 0 & 0\end{bmatrix}, \quad (127)$$

where ψ, ϕ, χ, etc., are eight functions which satisfy a coupled system of non-linear integral equations. By the procedure described earlier, we find that these eight functions are expressible in terms of a pair of X- and Y-functions, (X_l, Y_l) and (X_r, Y_r) belonging to the characteristic functions Ψ_l and Ψ_r defined in equations (46). However, since Ψ_l belongs to the conservative class, we choose for (X_l, Y_l) the standard solutions. In terms of (X_r, Y_r) and the standard solutions (X_l, Y_l), we find

$$\begin{aligned}
\psi(\mu) &= \mu[\nu_1 Y_l(\mu) - \nu_2 X_l(\mu)],\\
\xi(\mu) &= \mu[\nu_2 Y_l(\mu) - \nu_1 X_l(\mu)],\\
\phi(\mu) &= (1 + \nu_4\mu)X_l(\mu) - \nu_3\mu Y_l(\mu),\\
\eta(\mu) &= (1 - \nu_4\mu)Y_l(\mu) + \nu_3\mu X_l(\mu),\\
\chi(\mu) &= (1 - u_4\mu)X_r(\mu) + u_3\mu Y_r(\mu) + Q(u_4 - u_3)\mu^2[X_r(\mu) - Y_r(\mu)],
\end{aligned}$$

$$\sigma(\mu) = (1 + u_4\mu)Y_r(\mu) - u_3\mu X_r(\mu) - Q(u_4 - u_3)\mu^2[X_r(\mu) - Y_r(\mu)],$$
$$\zeta(\mu) = \tfrac{1}{2}\mu[\nu_1 Y_r(\mu) - \nu_2 X_r(\mu)] + \tfrac{1}{2}Q(\nu_2 - \nu_1)\mu^2[X_r(\mu) - Y_r(\mu)],$$
$$\theta(\mu) = \tfrac{1}{2}\mu[\nu_2 Y_r(\mu) - \nu_1 X_r(\mu)] - \tfrac{1}{2}Q(\nu_2 - \nu_1)\mu^2[X_r(\mu) - Y_r(\mu)], \tag{128}$$

where the constants $\nu_1, \nu_2, \nu_3, \nu_4, u_3, u_4$, and Q are determined by the following formulae:

$$\nu_2 + \nu_1 = 2\Delta_1(\kappa_1\delta_1 - \kappa_2\delta_2); \quad \nu_2 - \nu_1 = 2\Delta_2(\kappa_1\delta_1 - \kappa_2\delta_2),$$
$$\nu_4 + \nu_3 = \Delta_1(d_1\kappa_1 - d_0\kappa_2); \quad \nu_4 - \nu_3 = \Delta_2[c_1\delta_1 - c_0\delta_2 - 2Q(d_0\delta_1 - d_1\delta_2)],$$
$$u_4 + u_3 = \Delta_1(c_1\delta_1 - c_0\delta_2); \quad u_4 - u_3 = \Delta_2(d_1\kappa_1 - d_0\kappa_2),$$
$$\Delta_1 = (d_0\delta_1 - d_1\delta_2)^{-1}; \quad \Delta_2 = [c_0\kappa_1 - c_1\kappa_2 - 2Q(d_1\kappa_1 - d_0\kappa_2)]^{-1},$$
$$Q = (c_0 - c_2)[(d_0 - d_2)\tau_1 + 2(d_1 - d_3)]^{-1},$$
$$c_0 = A_0 + B_0 - \tfrac{8}{3}; \quad d_0 = A_0 - B_0 - \tfrac{8}{3},$$
$$c_n = A_n + B_n; \quad d_n = A_n - B_n; \quad \kappa_n = \alpha_n + \beta_n; \quad \delta_n = \alpha_n - \beta_n \quad (n = 1, 2, 3, \ldots), \tag{129}$$

and α_n, β_n, A_n, and B_n are the moments of order n of X_l, Y_l, X_r, and Y_r, respectively.

In the theory of the illumination of the sky we are interested in the transmitted light in the case of incident natural light. In this latter case $F_l = F_r = \frac{1}{2}F$ (where πF denotes the net flux of the incident natural light) and $F_U = 0$. The equations governing the intensity and polarization of the sky, as witnessed by an observer at $\tau = \tau_1$ in these circumstances, readily follow from the solutions already given; thus by setting $\boldsymbol{F} = \frac{1}{2}(F, F, 0)$ and combining equations (123), (125), and (127) appropriately, we find

$$\begin{aligned} I_l(\tau_1; -\mu, \varphi; \mu_0, \varphi_0) = \tfrac{3}{32}\{&[\psi(\mu_0) + \chi(\mu_0)]\,\xi(\mu) + 2[\phi(\mu_0) + \zeta(\mu_0)]\eta(\mu) \\ &- [\xi(\mu_0) + \sigma(\mu_0)]\psi(\mu) - 2[\theta(\mu_0) + \eta(\mu_0)]\phi(\mu) \\ &+ 4\mu\mu_0(1 - \mu^2)^{1/2}(1 - \mu_0^2)^{1/2}[X^{(1)}(\mu_0)Y^{(1)}(\mu) - Y^{(1)}(\mu_0)X^{(1)}(\mu)]\cos(\varphi_0 - \varphi) \\ &- \mu^2(1 - \mu_0^2)[X^{(2)}(\mu_0)Y^{(2)}(\mu) - Y^{(2)}(\mu_0)X^{(2)}(\mu)]\cos 2(\varphi_0 - \varphi)\}\,\frac{F\mu_0}{\mu - \mu_0}, \end{aligned} \tag{130}$$

$$\begin{aligned} I_r(\tau_1; -\mu, \varphi; \mu_0, \varphi_0) = \tfrac{3}{32}\{&[\psi(\mu_0) + \chi(\mu_0)]\,\sigma(\mu) + 2[\phi(\mu_0) + \zeta(\mu_0)]\theta(\mu) \\ &- [\xi(\mu_0) + \sigma(\mu_0)]\chi() - 2[\theta(\mu_0) + \eta(\mu_0)]\zeta(\mu) \\ &+ (1 - \mu_0^2)[X^{(2)}(\mu_0)Y^{(2)}(\mu) - Y^{(2)}(\mu_0)X^{(2)}(\mu)]\cos 2(\varphi_0 - \varphi)\}\,\frac{F\mu_0}{\mu - \mu_0}, \end{aligned} \tag{131}$$

and

$$U(\tau_1; -\mu, \varphi; \mu_0, \varphi_0) =$$

$$\tfrac{3}{16}\{2(1 - \mu^2)^{1/2}(1 - \mu_0^2)^{1/2}\mu_0[X^{(1)}(\mu_0)Y^{(1)}(\mu) - Y^{(1)}(\mu_0)X^{(1)}(\mu)]\sin(\varphi_0 - \varphi)$$

$$- \mu(1 - \mu_0^2)[X^{(2)}(\mu_0)Y^{(2)}(\mu) - Y^{(2)}(\mu_0)X^{(2)}(\mu)] \sin 2(\varphi_0 - \varphi)\} \frac{F\mu_0}{\mu - \mu_0}. \quad (132)$$

Equations (130)–(132) can be used to determine the distribution of the polarization over the sky for various zenith distances of the sun. Figure 5 in paper 11 (p. 212) illustrates the predicted neutral lines for an atmosphere of optical thickness 0.15. It is seen that the theoretically delineated neutral lines show a remarkable correspondence with Dorno's observations.

And so, at long last, the problem of the polarization of the sunlit sky reached its port.

May I, in concluding this personal account, express my deep-felt gratitude to Director Professor L. V. Mirzoyan, Professor M. A. Mnatsakanian, and the other organizers of the symposium for this opportunity to recall the happiest years of my scientific life? I could not have chosen a finer occasion than the present, since the incisive and original ideas of Academician Ambartsumian provided so much sustenance and inspiration to my efforts of those years.

Postscript

I greatly regret that in my account I have not acknowledged the important contributions to the theory of radiative transfer by the "Leningrad School," and most especially those of Professors V. V. Sobolev and V. V. Ivanov. Fortunately, an adequate account of these contributions has been given by Professor I. N. Minin (*Astrophysica* [Academy of Sciences of the Armenian Republic of USSR] 17 [1980]: 585–618).

I must also refer to the contributions of Professor T. W. Mullikin, who has, in particular, given an alternative (and a more rigorous) derivation of the solution, quoted in § VII, to the problem of diffuse reflection and transmission by a Rayleigh-scattering atmosphere; see in particular the papers by T. W. Mullikin, "Radiative Transfer in Finite Homogeneous Atmospheres with Anisotropic Scattering. I. Linear Singular Equations" (*Ap. J.* 139 [1964]: 379–95), and "The Complete Rayleigh-scattered Field within a Homogeneous Plane-parallel Atmosphere" (*Ap. J.* 145 [1966]: 886–931).

As I stated at the outset, my account has been confined exclusively to the years 1943–47, when I was engrossed in this subject: after writing my book on *Radiative Transfer* (Oxford: Clarendon Press, 1950) during the spring and summer of 1948, I have only on rare occasions returned to the subject. For an account which brings the subject to the present (with full references to the literature), the reader should consult the two magnificent volumes of H. C. van de Hulst (*Multiple Scattering: Tables, Formulas, and Applications,* vols. 1 and 2 [New York: Academic Press, 1980]).

PART TWO

Investigation on the Negative Ion of Hydrogen and of Two Electron Atoms

SOME REMARKS ON THE NEGATIVE HYDROGEN ION AND ITS ABSORPTION COEFFICIENT

S. Chandrasekhar
Yerkes Observatory
Received June 28, 1944

ABSTRACT

Some remarks on the quantum theory of the negative hydrogen ion are made, and attention is drawn to certain facts which make the evaluation of its continuous absorption coefficient a problem of extreme difficulty.

This paper will consist of a few disconnected remarks on the quantum theory of the negative hydrogen ion.

1. *The wave function for the ground state of* H^-.—Since the discovery of the stability of the negative ion of hydrogen by Bethe[1] and Hylleraas[2] and the recognition of its astrophysical importance by Wildt,[3] attempts[4] have been made to determine the electron affinity of hydrogen with as high a precision as possible. In these latter attempts the energy of the ground state is determined by applications of the Ritz principle, using forms for the wave functions suggested by Hylleraas' successful treatment of the ground state of the helium atom. Thus Williamson's six-parameter wave function is of exactly the same form as the "best" wave function of Hylleraas for helium. Similarly, Henrich's eleven-parameter wave function includes terms beyond those used by Williamson. While there can hardly be any doubt that Henrich's value for the electron affinity of 0.747 electron volts can be in error by more than a fraction of 1 per cent, the relatively weak convergence of the entire process (cf., e.g., Table 7 in Henrich's paper) leaves one with a suspicion that the formal analogy between the atomic configurations of H^- and He has perhaps been taken too literally. From one point of view it would seem that the structures of these two atoms must be very different indeed; for, while helium is a stable closed structure, the negative hydrogen ion is an open structure which exists principally on account of incomplete screening and polarization (see below). This suggests that it might be possible to obtain better representations of the true wave function by seeking forms which will explicitly take into account this difference. That such attempts may not prove unsuccessful is suggested by the following preliminary considerations.

As is well known, the success of Hylleraas' investigations on helium is due principally to the circumstance that a wave function of the form

$$\psi = e^{-a(r_1+r_2)} , \tag{1}$$

which ascribes a hydrogen-like wave function to each of the electrons in a suitably screened Coulomb field, already provides a good first approximation. More particularly the wave function of the form (1), which gives the lowest energy, is

$$\psi = e^{-(Z-[5/16])(r_1+r_2)} . \tag{2}$$

(In the foregoing equation r_1 and r_2 are measured in units of the Bohr radius. Similarly, in the rest of the paper we shall systematically use Hartree's atomic units.)

[1] *Zs. f. Phys.*, **57**, 815, 1929.

[2] *Zs. f. Phys.*, **60**, 624, 1930.

[3] *Ap. J.*, **89**, 295, 1939.

[4] R. E. Williamson, *Ap. J.*, **96**, 438, 1942; and L. R. Henrich, *Ap. J.*, **99**, 59, 1943.

When $Z = 1$, the wave function (2) predicts an energy $E = -0.473$, which actually makes H^- an unstable structure and is in error by fully 12 per cent. In other words, the first approximation, which is so satisfactory for *He*, fails completely for H^-. That this should happen is not surprising in view of our earlier remarks concerning the difference between the two atoms. On the other hand, it would appear that in contrast to *He* a natural first approximation for H^- is to ignore the screening of one of the electrons and adjust the screening constant for the second electron only. In other words, the starting-point for H^- should rather be a wave function of the form

$$\psi = e^{-r_1 - br_2} + e^{-r_2 - br_1}\,, \tag{3}$$

where b is the screening constant for the second electron. More generally, we may write

$$\psi = e^{-ar_1 - br_2} + e^{-ar_2 - br_1}\,, \tag{4}$$

where a and b are constants to be appropriately chosen. The Ritz principle applied to a wave function of the form (4) showed that the lowest value of energy is attained when

$$a = 1.03925 \qquad \text{and} \qquad b = 0.28309\,. \tag{5}$$

The corresponding value for the energy is

$$E_1 = -0.51330\,, \tag{6}$$

which predicts the stability of H^-. Moreover, in confirmation of our expectations it is seen that, while the inner electron is practically unscreened, the outer one is screened considerably and to the extent of 72 per cent. In view of this, it appears that a good second approximation may be provided by considering a wave function of the form

$$\psi = (e^{-ar_1 - br_2} + e^{-ar_2 - br_1})\,(1 + c\,r_{12})\,, \tag{7}$$

where a, b, and c are constants to be so chosen as to lead to a minimum value for the energy. It is found that with

$$a = 1.07478\,, \qquad b = 0.47758\,, \qquad \text{and} \qquad c = 0.31214 \tag{8}$$

we minimize the energy integral and obtain for it the value

$$E_2 = -0.52592\,. \tag{9}$$

This value for the energy, while inferior to those predicted by Williamson (0.5265) and Henrich (0.5276), is substantially better than the value 0.5253 given by the three-parameter wave function of Bethe and Hylleraas.

An interesting feature of the wave function (7) with the constants as given by equation (8) is that the inclusion of the term r_{12} reduces the screening of the outer electron from 0.72 to 0.52. This relatively large reduction in the screening is due to the strong polarizability of the hydrogen atom. Indeed, according to equations (6) and (9) we may say that the electron affinity of hydrogen is due about equally to the incomplete screening of the nucleus and to the polarization of the hydrogenic core.

The foregoing discussion suggests that it might be profitable to improve the wave function (7) by including further terms. This would be particularly useful for estimating the inherent uncertainty in the absorption cross-sections derived from different wave functions, all of which predict (within limits) the same value for energy. The practical importance for carrying out such a discussion will be apparent from our remarks in the following section.

2. *The absorption cross-sections for* H^-.—The calculations of the absorption cross-sections which have been carried out so far (Massey and Bates; Williamson; Henrich) are based on two approximations. The first consists in the use of the wave function for describing the bound state the ones derived from the minimal calculations and the second, in the use of a plane wave representation of the ejected outgoing electron. The validity or otherwise of these approximations will depend upon whether the principal contributions to the matrix element,

$$\mu = \int \Psi_d (r_1 + r_2) \Psi_c d\tau \,, \tag{10}$$

come from those regions of the configuration space in which the two approximations may be expected to be satisfactory. In equation (10) Ψ_d denotes the normalized wave function for the ground state of H^-, and Ψ_c the wave function of the continuous state normalized to correspond to an outgoing electron of unit density.

It appears that the use of the plane-wave representation for the free electron will not introduce any very serious error, since, as has been pointed out on an earlier occasion,[5] parts of the configuration space which are only relatively far from the hydrogenic core are relevant for the absorption process. But if this be admitted, the question immediately arises as to whether the wave function for the ground state derived from the Ritz principle can be trusted to these distances. It appears that the matter can be decided in the following manner.

First, we may observe that it might prove to be an adequate approximation to use for the continuous wave function that of an electron moving in the Hartree field of a hydrogen atom. In other words, it might be sufficient to use for Ψ_c the expression

$$\Psi_c = \frac{1}{\sqrt{2\pi}} \{ e^{-r_2} \phi(r_1) + e^{-r_1} \phi(r_2) \} \,, \tag{11}$$

where $\phi(r)$ satisfies the wave equation

$$\nabla^2 \phi + \left[k^2 + 2 \left(1 + \frac{1}{r} \right) e^{-2r} \right] \phi = 0 \tag{12}$$

and tends asymptotically at infinity to a plane wave of unit amplitude along some chosen direction. If this direction in which the ejected electron moves at infinity be chosen as the polar axis of a spherical system of co-ordinates, it is readily shown that the appropriate solution for ϕ can be expressed in the form

$$\phi = \sum_{l=0}^{\infty} \frac{1}{kr} (2l+1) P_l(\cos\vartheta) \chi_l(r) \,, \tag{13}$$

where the radial function χ_l is a solution of the equation

$$\frac{d^2 \chi_l}{dr^2} + \left\{ k^2 - \frac{l(l+1)}{r^2} + 2 \left(1 + \frac{1}{r} \right) e^{-2r} \right\} \chi_l = 0 \,, \tag{14}$$

which tends to a pure sinusoidal wave of unit amplitude at infinity. Thus, on our present approximation Ψ_c can be written in the form

$$\left. \begin{aligned} \Psi_c = \frac{1}{\sqrt{2\pi}} \Big\{ & e^{-r_2} \sum_{l=0}^{\infty} \frac{1}{kr_1} (2l+1) P_l(\cos\vartheta_1) \chi_l(r_1; k) \\ & + e^{-r_1} \sum_{l=0}^{\infty} \frac{1}{kr_2} (2l+1) P_l(\cos\vartheta_2) \chi_l(r_2; k) \Big\} \,. \end{aligned} \right\} \tag{15}$$

[5] S. Chandrasekhar and M. K. Krogdahl, *Ap. J.*, **98**, 205, 1943.

Using the foregoing form for Ψ_c, the standard formula for the absorption cross-section for a process in which a photoelectron with k atomic units of momentum is ejected can be reduced to the form

$$\kappa = 9.266 \times 10^{-19} \frac{k^2 + 0.05512}{k} \left| \int_0^\infty W(r)\chi_1(r)\,dr \right|^2 \text{cm}^2, \tag{16}$$

where $W(r)$ is a certain weight function which can be derived from and depends only on the wave function for the bound state. It is seen that, according to equation (16), the absorption cross-section depends only on the single radial function χ_1. This is to be expected, since the ground state, being an s-state, transitions can take place only to a p-state. It may be noted here that on the plane-wave representation of the free electron the appropriate form for χ_1 is

$$\chi_1(\text{plane wave}) = \frac{\sin kr}{kr} - \cos kr. \tag{17}$$

The function $W(r)$ corresponding to Henrich's eleven-parameter wave function has been computed and is tabulated in Table 1. The run of the function is further illustrated in Figure 1.

TABLE 1

THE WEIGHT FUNCTION $W(r)$

r	$W(r)$	r	$W(r)$	r	$W(r)$	r	$W(r)$
0	0	4.0	1.597	11.0	0.833	19.0	0.131
0.5	0.210	4.5	1.623	12.0	.703	20.0	.096
1.0	0.553	5.0	1.620	13.0	.585	21.0	.069
1.5	0.861	6.0	1.548	14.0	.478	22.0	.049
2.0	1.108	7.0	1.422	15.0	.383	23.0	.034
2.5	1.298	8.0	1.273	16.0	.301	24.0	.024
3.0	1.439	9.0	1.120	17.0	.233	25.0	0.016
3.5	1.538	10.0	0.972	18.0	0.177	∞	0

An examination of the values given in Table 1 discloses the somewhat disquieting fact that substantial contributions to the integral

$$\int_0^\infty W(r)\chi_1(r)\,dr \tag{18}$$

arise from values of r up to 25, while as much as 30–40 per cent of the entire value comes from $r \gtrsim 10$. This result has two consequences. The first is that the use of the p-spherical wave (17) instead of the solution derived from (cf. eq. [14])

$$\frac{d^2\chi_1}{dr^2} + \left\{ k^2 - \frac{2}{r^2} + 2\left(1 + \frac{1}{r}\right) e^{-2r} \right\} \chi_1 = 0, \tag{19}$$

will not lead to any serious error; for the solution of equation (19), which tends to a sine wave of unit amplitude at infinity, has the behavior

$$\chi_1 \to \frac{\sin(kr + \delta)}{kr} - \cos(kr + \delta) \qquad (r \to \infty), \tag{20}$$

and the "phase shift" δ may be taken as a measure of the distortion of the p-spherical wave by the hydrogen atom at the origin. Integrations of equation (19) for various values of k^2 have been carried out numerically, and the resulting phase shifts for some of them are given in Table 2. It is seen that the phase shifts are indeed quite small for values of k^2, which are of astrophysical interest.

The second consequence of the run of the function $W(r)$ is not so satisfactory; for an examination of the energy integral minimized in the Ritz principle reveals that over 95 per cent of the contribution to the integral arises from regions of the configuration space which correspond to $r < 10$. Accordingly, it would appear that the choice of the wave

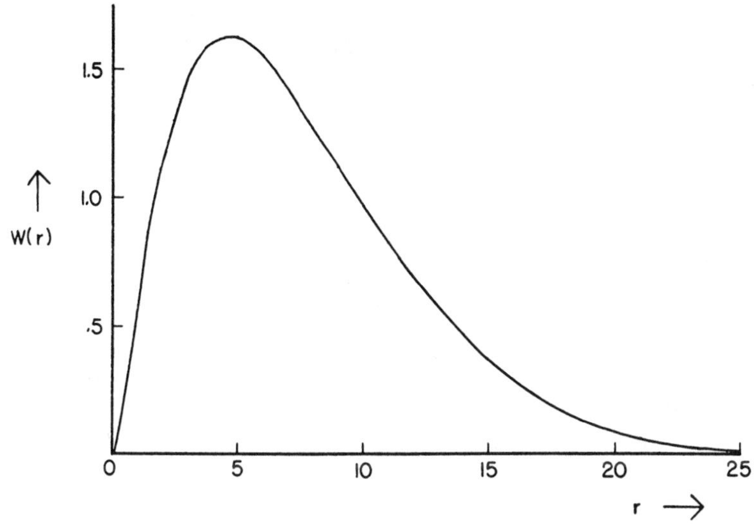

FIG. 1

TABLE 2

PHASE SHIFTS δ FOR THE p-SPHERICAL WAVES IN THE HARTREE FIELD OF A HYDROGEN ATOM

k^2	δ	k^2	δ	k^2	δ	k^2	δ
1.50.......	0.1486	0.80......	0.09244	0.25......	0.02605	0.100.....	0.007689
1.00.......	0.1115	0.50......	0.05838	0.125.....	0.01046	0.035.....	0.001709

function in accordance only with the Ritz principle cannot be expected to lead to values of $W(r)$ which are necessarily trustworthy for $r > 10$. Under these circumstances the best hope for improving the current wave functions would consist in first determining the true asymptotic forms of the wave function for large distances and later choosing functions which would lead not only to the best value for the energy but also to the correct asymptotic forms. However, such calculations are likely to be extremely laborious.

I am greatly indebted to Miss Frances Herman for valuable assistance in the numerical parts of the present investigation.

ON THE CONTINUOUS ABSORPTION COEFFICIENT OF THE NEGATIVE HYDROGEN ION

S. CHANDRASEKHAR
Yerkes Observatory
Received June 25, 1945

ABSTRACT

In this paper it is shown that the continuous absorption coefficient of the negative hydrogen ion is most reliably determined by a formula for the absorption cross-section which involves the matrix element of the momentum operator. A new absorption curve for H^- has been determined which places the maximum at λ 8500 A; at this wave length the atomic absorption coefficient has the value 4.37×10^{-17} cm^2.

1. *Introduction.*—In earlier discussions[1] by the writer attention has been drawn to the fact that the continuous absorption coefficient of the negative hydrogen ion, evaluated in terms of the matrix element

$$\mu = \int \Psi_d^* (r_1 + r_2) \Psi_c d\tau \tag{1}$$

(where Ψ_d denotes the wave function of the ground state of the ion and Ψ_c the wave function belonging to a continuous state normalized to correspond to an outgoing electron of unit density), depends very much on Ψ_d in regions of the configuration space which are relatively far from the hydrogenic core. This has the consequence that the absorption cross-sections are not trustworthily determined if wave functions derived by applications of the Ritz principle are used in the calculation of the matrix elements according to equation (1). This is evident, for example, from Figure 1, in which we have plotted the absorption coefficients as determined by Williamson[2] and Henrich,[3] using wave functions of the forms

$$\Psi_d = \mathfrak{N} e^{-\alpha s/2} (1 + \beta u + \gamma t^2 + \delta s + \epsilon s^2 + \zeta u^2) \tag{2}$$

and

$$\left.\begin{aligned}\Psi_d = \mathfrak{N} e^{-\alpha s/2} (1 + \beta u + \gamma t^2 + \delta s + \epsilon s^2 + \zeta u^2 + \chi_6 t^4 + \chi_7 t^6 \\ + \chi_8 t^4 u^2 + \chi_9 t^2 u^2 + \chi_{10} t^2 u^4) ,\end{aligned}\right\} \tag{3}$$

respectively. (In eqs. [2] and [3] $\mathfrak{N}$ is the normalizing factor; and α, β, γ, etc., are constants determined by the Ritz condition of minimum energy,

$$s = r_2 + r_1 , \quad t = r_2 - r_1 , \quad \text{and} \quad u = r_{12} , \tag{4}$$

where r_1, r_2, and r_{12} are the distances of the two electrons from the nucleus and from each other, respectively.) The wide divergence between the two curves in Figure 1 is too large to be explained in terms of only the improvement in energy effected by the wave function (3): it must arise principally from the fact that in the evaluation of the matrix elements according to equation (1) parts of the wave function are used which do not contribute appreciably to the energy integral and are therefore poorly determined. Indeed, this sen-

[1] *Ap. J.*, **100**, 176, 1944; also *Rev. Mod. Phys.*, **16**, 301, 1944.

[2] *Ap. J.*, **96**, 438, 1942.

[3] *Ap. J.*, **99**, 59, 1944.

sitiveness of the derived absorption coefficients to wave functions effecting only relatively slight improvements in the energy makes it difficult to assess the reliability of the computed absorption coefficients. However, in this paper we shall show how these difficulties can be avoided by using a somewhat different formula for the absorption cross-section.

2. *Alternative formulae for evaluating the absorption coefficient.*—It is well known that in the classical theory the radiative characteristics of an oscillating dipole can be expressed in terms of either its dipole moment, its momentum, or its acceleration. There are, of course, analogous formulations in the quantum theory, the matrix element

$$(a \mid z_j \mid b) = \int \psi_a^* z_j \psi_b d\tau \tag{5}$$

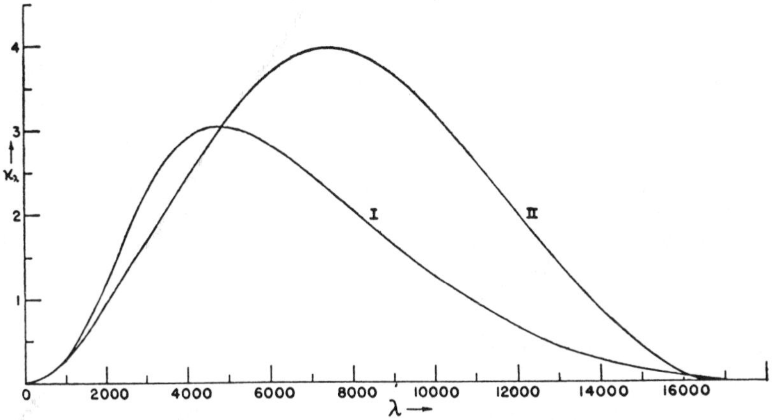

FIG. 1.—A comparison of the continuous absorption coefficient of H^- computed according to formula (I) and with wave functions of forms (2) (curve *I*) and (3) (curve *II*). The ordinates denote the absorption coefficients in units of 10^{-17} cm²; the abscissae, the wave length in angstroms.

for the co-ordinate z_j of the jth electron in an atom being simply related to the corresponding matrix element of the momentum operator or the acceleration. Thus, we have the relations

$$(a \mid z_j \mid b) = \frac{1}{(E_a - E_b)} \int \frac{\partial \psi_a^*}{\partial z_j} \psi_b d\tau = -\frac{1}{(E_a - E_b)} \int \psi_a^* \frac{\partial \psi_b}{\partial z_j} d\tau \tag{6}$$

and

$$(a \mid z_j \mid b) = \frac{1}{(E_a - E_b)^2} \int \psi_a^* \frac{\partial V}{\partial z_j} \psi_b d\tau \tag{7}$$

if all the quantities are measured in Hartree's atomic units and where E_a and E_b denote the energies of the states indicated by the letters a and b and where V denotes the potential energy arising from Coulomb interactions between the particles. More particularly for an atom (or ion) with two electrons, we have

$$\mu_z = \int \Psi_d^* (z_1 + z_2) \Psi_c d\tau \,, \tag{8}$$

$$\mu_z = -\frac{1}{(E_d - E_c)} \int \Psi_d^* \left(\frac{\partial}{\partial z_1} + \frac{\partial}{\partial z_2} \right) \Psi_c d\tau \,, \tag{9}$$

and

$$\mu_z = \frac{1}{(E_d - E_c)^2} \int \Psi_d^* \left(\frac{z_1}{r_1^3} + \frac{z_2}{r_2^3} \right) \Psi_c d\tau \,. \tag{10}$$

While the foregoing formulae are entirely equivalent to each other if Ψ_d abd Ψ_c are exact solutions of the wave equation, they are of different merits for the evaluation of μ_z if approximate wave functions are used. Thus, it is evident that formula (8) uses parts of the configuration space, which are more distant than relevant, for example, in the evaluation of the energy; similarly, formula (10) uses the wave functions in regions much nearer the origin. It would appear that formula (9) is the most suitable one for the evaluation of μ_z, particularly when wave functions derived by applications of the Ritz principle are used. The calculations which we shall present in the following sections confirm this anticipation; but before we proceed to such calculations, it is useful to have the explicit formulae for the absorption cross-sections on the basis of equations (8), (9), and (10).

In ordinary (c.g.s.) units the standard formula for the atomic absorption coefficient κ_ν for radiation of frequency ν, in which an electron with a velocity v is ejected, is

$$\kappa_\nu = \frac{32\pi^4 m^2 e^2}{3h^3 c} \nu v \left| \int \Psi_d^* (z_1 + z_2) \Psi_c d\tau \right|^2 , \tag{11}$$

where m, e, h, and c have their usual meanings. (In writing eq. [11] it has been assumed that the electron is ejected in the z-direction; see eq. [15] below.) By inserting the numerical values for the various atomic constants equation (11) can be expressed in the form

$$\kappa_\nu = 8.561 \times 10^{-19} (\nu_{\mathrm{at}} k \, |\mu_z|^2) \ \mathrm{cm}^2 , \tag{12}$$

where k denotes the momentum of the ejected electron and ν_{at} the frequency of the radiation absorbed, both measured in atomic units, and where, moreover, the matrix element μ_z has also to be evaluated in atomic units.

If I denotes the electron affinity (also expressed in atomic units)

$$4\pi\nu_{\mathrm{at}} = k^2 + 2I , \tag{13}$$

and depending on which of the formulae (8), (9), and (10) we use for evaluating κ_ν, we have

$$\kappa_\nu = 6.812 \times 10^{-20} k (k^2 + 2I) \left| \int \Psi_d^* (z_1 + z_2) \Psi_c d\tau \right|^2 . \tag{I}$$

$$\kappa_\nu = 2.725 \times 10^{-19} \frac{k}{(k^2 + 2I)} \left| \int \Psi_d^* \left(\frac{\partial}{\partial z_1} + \frac{\partial}{\partial z_2} \right) \Psi_c d\tau \right|^2 , \tag{II}$$

and

$$\kappa_\nu = 1.090 \times 10^{-18} \frac{k}{(k^2 + 2I)^3} \left| \int \Psi_d^* \left(\frac{z_1}{r_1^3} + \frac{z_2}{r_2^3} \right) \Psi_c d\tau \right|^2 . \tag{III}$$

Finally, we may note that if λ denotes the wave length of the radiation measured in angstroms, then

$$\lambda = \frac{911.3}{k^2 + 2I} \ \mathrm{A} . \tag{14}$$

3. *The continuous absorption coefficient of* H^- *evaluated according to formula* (*III*).—As we have already indicated, in the customary evaluations of κ_ν according to formula (I) the relatively more distant parts of the wave function are used. It is evident that we shall be going to the opposite extreme in using the wave function principally only near the origin if we evaluate κ_ν according to formula (III). For this reason it is of interest to consider first the absorption coefficient as determined by this formula.

In evaluating κ_ν according to formula (III), we shall use for Ψ_d a wave function of form (3) and for Ψ_c a plane wave representation of the outgoing electron:

$$\Psi_c = \frac{1}{\sqrt{2\pi}} \left(e^{-r_1 + ikz_2} + e^{-r_2 + ikz_1} \right) . \tag{15}$$

(In § 5 we refer to an improvement in Ψ_c which can be incorporated without much difficulty at this stage.) For Ψ_d and Ψ_c of forms (3) and (15) the evaluation of the matrix element

$$\int \Psi_d^* \left(\frac{z_1}{r_1^3} + \frac{z_2}{r_2^3} \right) \Psi_c d\tau \tag{16}$$

is straightforward, though it is somewhat involved. We find

$$\left.\begin{aligned}\int\Psi_d^*\left(\frac{z_1}{r_1^3}+\frac{z_2}{r_2^3}\right)\Psi_c d\tau = -(2048\pi^3)^{1/2}\frac{\mathfrak{N}}{(1+a)^3}\frac{i}{k^2}\Bigg[\sum_{j=-2}^{6} l_j\mathcal{L}_j^{(a)}\\ +\sum_{j=-2}^{3}\lambda_j\mathcal{L}_j^{(1+2a)}-\tfrac{1}{30}\beta(1+a)^4\left\{\sum_{j=-1}^{3}a_jS_j^{(1+2a)}+\sum_{j=-1}^{3}b_jC_j^{(1+2a)}\right\}\Bigg],\end{aligned}\right\} \quad (17)$$

where we have used the following abbreviations:

$$\left.\begin{aligned}\mathcal{L}_j^{(p)} &= \int_0^\infty e^{-py}\left(k\cos ky-\frac{\sin ky}{y}\right)y^jdy && (j=-2,-1,\ldots\ldots),\\ &=(j-1)!\rho^j\{j\rho k\cos[(j+1)\xi]-\sin j\xi\} && (j\geqslant 1),\\ &=\rho k\cos\xi-\xi && (j=0),\\ &=p\xi-k && (j=-1).\\ &=\frac{1}{2}\left(pk-\frac{\xi}{\rho^2}\right) && (j=-2).\end{aligned}\right\}\quad (18)$$

$$\left.\begin{aligned}S_j^{(p)} &= \int_0^\infty e^{-py}y^j\sin ky\,dy && (j=-1,0,\ldots\ldots).\\ &= j!\rho^{j+1}\sin[(j+1)\xi] && (j=0,1,\ldots\ldots).\\ &=\xi && (j=-1).\end{aligned}\right\}\quad (19)$$

and

$$\left.\begin{aligned}C_j^{(p)} &= \int_0^\infty e^{-py}y^j\cos ky\,dy = j!\rho^{j+1}\cos[(j+1)\xi] && (j=0,1,\ldots\ldots),\\ &=\int_0^\infty e^{-py}(e^{ay}-\cos ky)\frac{dy}{y}=\log\left(\frac{p}{p-a}\,|\sec\xi|\right) && (j=-1).\end{aligned}\right\}\quad (20)$$

where

$$\rho=\frac{1}{(k^2+p^2)^{1/2}}\quad\text{and}\quad\xi=\tan^{-1}\frac{k}{p};\quad (21)$$

and

$$\left.\begin{aligned}l_{-2}&=4q^2\beta,\\ l_{-1}&=1+q(\tfrac{1}{5}\beta+\delta)+12q^2(\gamma+\epsilon+\zeta)+360q^4(\chi_6+\chi_9)+20{,}160q^6(\chi_7+\chi_8+\chi_{10}),\\ l_0&=(\delta+\beta)-6q(\gamma-\epsilon)-120q^3(2\chi_6+\chi_9)-5040q^5(3\chi_7+2\chi_8+\chi_{10}),\\ l_1&=-\frac{\beta}{6q}+(\gamma+\epsilon+\zeta)+24q^2(3\chi_6+\chi_9)+120q^4(45\chi_7+21\chi_8+13\chi_{10}).\\ l_2&=-\frac{\zeta}{3q}-4q(3\chi_6+2\chi_9)-40q^3(30\chi_7+13\chi_8+12\chi_{10})\\ l_3&=(\chi_6+\tfrac{7}{3}\chi_9)+4q^2(45\chi_7+29\chi_8+21\chi_{10}),\\ l_4&=-\frac{\chi_9}{3q}-2q(9\chi_7+12\chi_8+7\chi_{10}).\\ l_5&=\chi_7+\tfrac{11}{3}(\chi_8+\chi_{10}).\\ l_6&=-\frac{1}{3q}(\chi_8+2\chi_{10}).\end{aligned}\right\}\quad (22)$$

$$\left.\begin{aligned} \lambda_{-2} &= -4\beta q^2 ; & \lambda_1 &= \frac{\beta}{15q}, \\ \lambda_{-1} &= -\tfrac{6}{5}\beta q ; & \lambda_2 &= -\frac{\beta}{30q^2}, \\ \lambda_0 &= -\tfrac{1}{5}\beta ; & \lambda_3 &= \frac{\beta}{30q^3}, \end{aligned}\right\} \tag{23}$$

where

$$q = \frac{1}{1+a}; \tag{24}$$

and

$$\left.\begin{aligned} a_{-1} &= 6\eta^4 [4\eta k^2 (5a^4 - 10a^2k^2 + k^4) + (a^4 - 6a^2k^2 + k^4)], \\ a_0 &= 6\eta^3 a [16\eta k^2 (a^2 - k^2) + (a^2 - 3k^2)], \\ a_1 &= 3\eta^2 [4\eta k^2 (3a^2 - k^2) + (a^2 - k^2)], \\ a_2 &= \eta a (8\eta k^2 + 1). \\ a_3 &= \eta k^2, \end{aligned}\right\} \tag{25}$$

$$\left.\begin{aligned} b_{-1} &= +24\eta^4 a k [\eta (a^4 - 10a^2k^2 + 5k^4) - (a^2 - k^2)]. \\ b_0 &= -6\eta^3 k [4\eta (a^4 - 6a^2k^2 + k^4) - (3a^2 - k^2)], \\ b_1 &= -6\eta^2 a k [2\eta (a^2 - 3k^2) - 1]. \\ b_2 &= -\eta k [4\eta (a^2 - k^2) - 1]. \\ b_3 &= -\eta a k . \end{aligned}\right\} \tag{26}$$

where

$$\eta = (a^2 + k^2)^{-1}. \tag{27}$$

Putting $\chi_6 = \chi_7 = \ldots = \chi_{10} = 0$ in the foregoing equations, we shall obtain the formulae which can be used with a wave function of form (2).

By using for the constants of wave functions (2) and (3) the values determined by Williamson and Henrich, the atomic absorption coefficient κ_ν has been computed according to the foregoing formulae for various wave lengths. The results of the calculations are given in Table 1 and are further illustrated in Figure 2. It is seen that, in contrast to what happened when formula (I) was used (cf. Fig. 1), wave function (2) now predicts systematically *larger* values for κ_ν than does wave function (3). The divergence between the two curves must now be attributed to the overweighting of the wave function near the origin, where it is again poorly determined by the Ritz method.

4. *The continuous absorption coefficient of* H^- *evaluated according to formula (II).*— Finally, returning to formula (II), which would appear to have the best chances for determining κ_ν most reliably, the calculations were again carried through for wave functions Ψ_d of forms (2) and (3) and for Ψ_c of form (15). Before we give the results of the calculations, we may note that for Ψ_d of form (3) and for Ψ_c of form (15)

$$\left.\begin{aligned} \int \Psi_d^* \left(\frac{\partial}{\partial z_1} + \frac{\partial}{\partial z_2}\right) \Psi_c d\tau = -(2048\pi^3)^{1/2} \frac{\mathfrak{N}}{(1+a)^3} \frac{i}{k^2} \Bigg[\sum_{j=-1}^{6} l_j \mathcal{L}_j^{(a)} \\ + \sum_{j=-1}^{+1} \lambda_j \mathcal{L}_j^{(1+2a)} + k^2 \left\{ \sum_{j=0}^{7} s_j S_j^{(a)} + \sum_{j=0}^{+1} \sigma_j S_j^{(1+2a)} \right\} \Bigg], \end{aligned}\right\} \tag{28}$$

TABLE 1

THE CONTINUOUS ABSORPTION COEFFICIENT OF H^- COMPUTED ACCORDING TO FORMULA III AND WITH WAVE FUNCTIONS OF FORMS (2) AND (3)

λ (A)	$\kappa_\lambda \times 10^{17}$ cm²		λ (A)	$\kappa_\lambda \times 10^{17}$ cm²	
	With Wave Function (3)	With Wave Function (2)		With Wave Function (3)	With Wave Function (2)
1000	0.225	0.241	7000	5.173	6.732
2000	0.955	1.010	7500	5.225	7.070
2500	1.459	1.538	8000	5.204	7.333
3000	2.010	2.125	8500	5.106	7.496
3500	2.580	2.752	9000	4.946	7.567
4000	3.139	3.400	9500	4.724	7.536
4500	3.657	4.046	10000	4.453	7.411
5000	4.118	4.676	12000	3.031	5.952
5500	4.505	5.271	14000	1.407	3.355
6000	4.812	5.820	16000	0.149	0.401
6500	5.036	6.310			

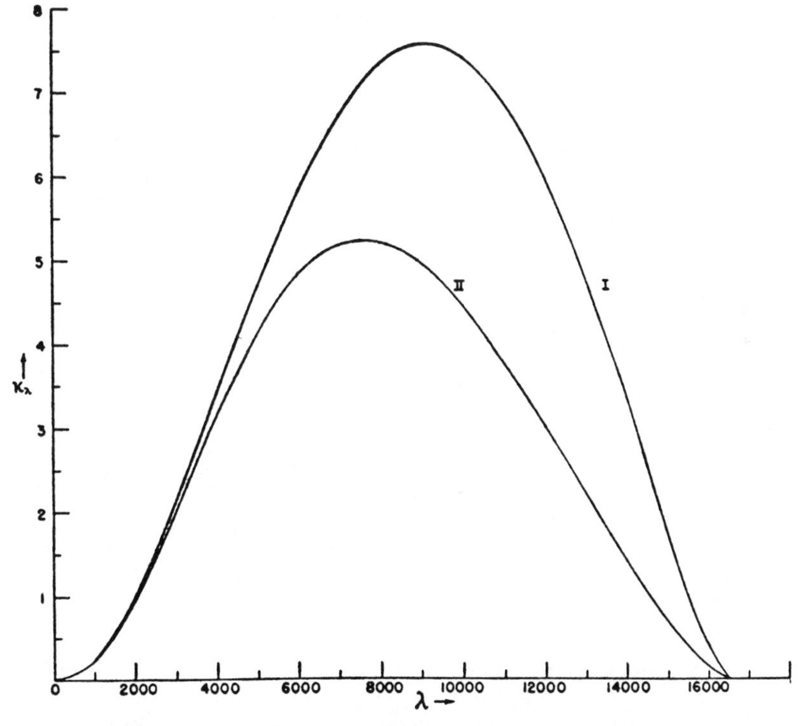

FIG. 2.—A comparison of the continuous absorption coefficient of H^- computed according to formula (III) and with wave functions of forms (2) (curve *I*) and (3) (curve *II*). The ordinates denote the absorption coefficients in units of 10^{-17} cm²; the abscissae, the wave length in angstroms.

where

$$\left.\begin{aligned} l_{-1} &= 4\beta q^3 ; \quad l_0 = 0 ; \quad l_1 = -\beta q . \\ l_2 &= -2q\zeta - 40q^3\chi_9 - 1680q^5(\chi_8 + 2\chi_{10}) . \\ l_3 &= 16q^2\chi_9 + 960q^4(\chi_8 + \chi_{10}) , \\ l_4 &= -2q\chi_9 - 80q^3(3\chi_8 + 2\chi_{10}) . \\ l_5 &= 32q^2(\chi_8 + \chi_{10}) ; \quad l_6 = -2q(\chi_8 + 2\chi_{10}) , \end{aligned}\right\} \tag{29}$$

$$\lambda_{-1} = -4\beta q^3 ; \quad \lambda_0 = -4\beta q^2 ; \quad \lambda_1 = -\beta q , \tag{30}$$

$$\left.\begin{aligned} s_0 &= 4\beta q^2 , \\ s_1 &= 1 + 3q\delta + 12q^2(\gamma+\epsilon+\zeta) + 360q^4(\chi_6+\chi_9) + 20{,}160q^6(\chi_7+\chi_8+\chi_{10}) , \\ s_2 &= (\delta+\beta) - 6q(\gamma-\epsilon) - 120q^2(2\chi_6+\chi_9) - 5040q^5(3\chi_7+2\chi_8+\chi_{10}) , \\ s_3 &= (\gamma+\epsilon+\zeta) + 24q^2(3\chi_6+\chi_9) + 120q^4(45\chi_7+21\chi_8+13\chi_{10}) , \\ s_4 &= -6q(2\chi_6+\chi_9) - 80q^3(15\chi_7+6\chi_8+5\chi_{10}) , \\ s_5 &= (\chi_6+\chi_9) + 4q^2(45\chi_7+21\chi_8+13\chi_{10}) . \\ s_6 &= -6q(3\chi_7+2\chi_8+\chi_{10}) , \\ s_7 &= \chi_7+\chi_8+\chi_{10} \end{aligned}\right\} \tag{31}$$

and

$$\sigma_0 = -4\beta q^2 ; \quad \sigma_1 = -\beta q . \tag{32}$$

Further, in equation (28) the quantities $\mathcal{L}_j^{(p)}$, $S_j^{(p)}$, and q have the same meanings as in equations (18), (19), (21), and (24).

TABLE 2

THE CONTINUOUS ABSORPTION COEFFICIENT OF H^- COMPUTED ACCORDING TO FORMULA II AND WITH WAVE FUNCTIONS OF FORMS (2) AND (3)

λ (A)	$\kappa_\lambda \times 10^{17}$ cm²		λ (A)	$\kappa_\lambda \times 10^{17}$ cm²	
	With Wave Function (3)	With Wave Function (2)		With Wave Function (3)	With Wave Function (2)
1000	0.271	0.270	7000	4.174	4.113
2000	0.945	0.991	7500	4.296	4.080
2500	1.335	1.461	8000	4.363	3.993
3000	1.730	1.955	8500	4.372	3.858
3500	2.119	2.437	9000	4.324	3.682
4000	2.498	2.880	9500	4.221	3.471
4500	2.860	3.265	10000	4.065	3.233
5000	3.197	3.581	12000	2.995	2.108
5500	3.504	3.822	14000	1.502	0.954
6000	3.773	3.989	16000	0.167	0.097
6500	3.998	4.084			

The absorption cross-sections, as calculated according to formula (II), and the foregoing equations are given in Table 2 and further illustrated in Figure 3. It is seen that, as anticipated, the two curves now do not diverge more than can be reasonably attributed to the betterment of the wave function in consequence of the increased number of parameters used in the Ritz method.

5. *Concluding remarks.*—A comparison of Figures 1, 2, and 3 clearly illustrates the superiority of formula (II) for the purposes of evaluating the continuous absorption coefficient of the negative hydrogen ion. The general reliability of the absorption cross-

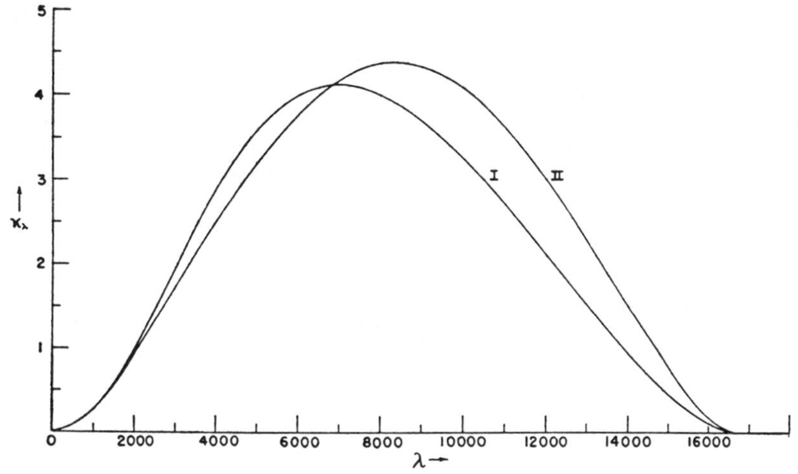

FIG. 3.—A comparison of the continuous absorption coefficient of H^- computed according to formula (II) and with wave functions of forms (2) (curve *I*) and (3) (curve *II*). The ordinates denote the absorption coefficients in units of 10^{-17} cm^2; the abscissae, the wave length in angstroms.

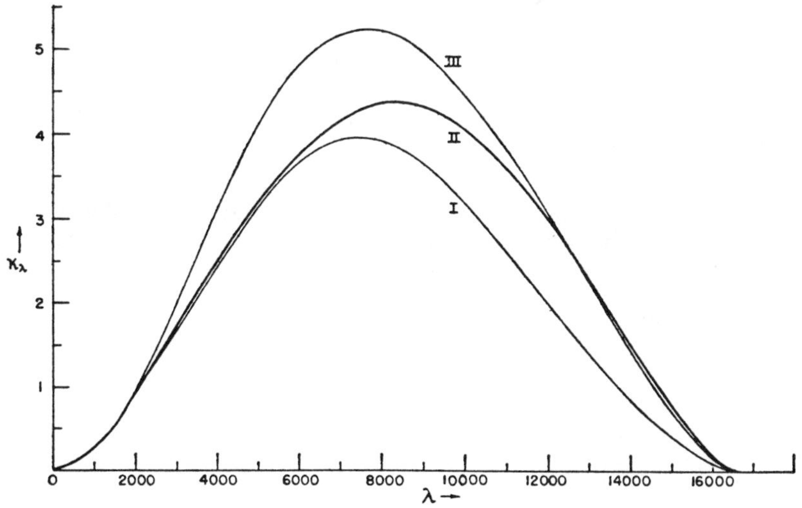

FIG. 4.—A comparison of the continuous absorption coefficient of H^- computed according to formulae (I) (curve *I*), (II) (curve *II*), and (III) (curve *III*) with a wave function of form (3). The ordinates denote the absorption coefficients in units of 10^{-17} cm^2; the abscissae, the wave length in angstroms.

sections derived on the basis of formula (II) and wave function (3) can be seen in another way. In Figure 4 we have plotted κ_ν as given by the three formulae and as obtained in each case with wave function (3). It is seen that, while the cross-sections given by formula (II) agree with those given by formula (I) in the visual and the violet part of the spectrum ($\lambda < 6000$ A), they agree with those given by formula (III) in the infrared ($\lambda > 12{,}000$ A). This is readily understood when it is remembered that on all the three formulae the absorption cross-sections in the infrared are relatively more dependent on the wave function at large distances than they are in the visual and the violet parts of the spectrum. Accordingly, it is to be expected that, as we approach the absorption limit of H^- at 16,550 A, formula (III) must give less unreliable values than it does at shorter wave lengths; formula (I), of course, ceases to be valid in the infrared. It is also clear that, as we go toward the violet, we have the converse situation.

Summarizing our conclusions so far, it may be said that in the framework of the approximation in which a plane-wave representation of the outgoing electron is used, formula (II), together with wave function (3), gives sufficiently reliable values for the absorption coefficient over the entire range of the spectrum. Attention may be particularly drawn to the fact that the maximum of the absorption-curve is now placed at λ 8500 A, where $\kappa_\lambda = 4.37 \times 10^{-17}\ \mathrm{cm}^2$.

The question still remains as to the improvements which can be effected in the choice of Ψ_c. As shown in an earlier paper,[4] it may be sufficient to use for Ψ_c the wave functions in the Hartree field of a hydrogen atom. On this approximation we should use (*op. cit.*, eq. [15])

$$\left.\begin{aligned}\Psi_c = \frac{1}{\sqrt{2\pi}}\Bigg\{ e^{-r_1}\sum_{l=0}^{\infty}\frac{i^l}{k r_2}(2l+1)\,P_l(\cos\vartheta_2)\,\chi_l(r_2;\,k)\\ + e^{-r_2}\sum_{l=0}^{\infty}\frac{i^l}{k r_1}(2l+1)\,P_l(\cos\vartheta_1)\,\chi_l(r_1;\,k)\Bigg\},\end{aligned}\right\} \qquad (33)$$

where χ_l is the solution of the equation

$$\frac{d^2\chi_l}{dr^2}+\left\{k^2-\frac{l(l+1)}{r^2}+2\left(1+\frac{1}{r}\right)e^{-2r}\right\}\chi_l=0\,, \qquad (34)$$

which tends to a pure sinusoidal wave of unit amplitude at infinity. We shall return to these further improvements in a later paper.

It is a pleasure to acknowledge my indebtedness to Professor E. P. Wigner for many helpful discussions and much valuable advice. My thanks are also due to Mrs. Frances Herman Breen for assistance with the numerical work.

[4] *Ap. J.*, **100**, 176, 1944.

ON THE CONTINUOUS ABSORPTION COEFFICIENT OF THE NEGATIVE HYDROGEN ION. II

S. Chandrasekhar
Yerkes Observatory

Received September 13, 1945

ABSTRACT

In this paper the continuous absorption coefficient of the negative hydrogen ion, determined in an earlier paper (*Ap. J.*, **102**, 223, 1945) in terms of the matrix elements of the momentum operator, is further improved to take into account the effect of the static field of the hydrogen atom on the motion of the ejected electron. It is shown that for wave functions for the ground state of H^- of the forms generally considered, the formula for the absorption cross-section can be reduced to the form

$$\kappa = \frac{3.7062}{k(k^2+0.05512)}\left|\int_0^\infty W_2(r)\chi_1(r)dr\right|^2 \times 10^{-18}\text{ cm}^2,$$

where k denotes the momentum of the ejected electron in atomic units, $W_2(r)$ a certain weight function which can be tabulated, and $\chi_1(r)$ the radial part of a p-spherical wave in the Hartree field of a hydrogen atom which tends to unit amplitude at infinity.

The absorption cross-sections of H^- have been evaluated according to the foregoing formula for various wave lengths. It is found that the new values are larger than those obtained with a plane-wave representation of the outgoing electron by about 5 per cent in the visual and the near infrared regions. The new absorption-curve places the maximum at about λ 8500 A; at this wave length the atomic absorption coefficient has the value 4.52×10^{-17} cm².

1. *Introduction.*—In an earlier paper[1] it was shown that the continuous absorption coefficient of the negative hydrogen ion is most reliably determined in terms of the matrix elements of the momentum operator. However, in the actual evaluation of the absorption cross-sections by this method, the plane-wave approximation for the ejected electron was used (see I, eq. [15]). In this paper we propose to consider certain refinements in this direction.

2. *Formulae for the evaluation of the continuous absorption coefficient of* H⁻ *using the wave functions of an electron in the Hartree field of a hydrogen atom.*—It would seem that, if we abandon the plane-wave approximation[2] for the ejected electron, the next simplest thing to do will be to use the wave functions of an electron moving in the static field of a hydrogen atom.[3] In other words, it would appear that in the "next approximation" we use for the wave functions Ψ_c describing the continuous states of H^- that of a hydrogen atom in its ground state, together with an electron moving in the Hartree field,

$$-\left(\frac{1}{r}+1\right)e^{-2r}. \qquad (1)^4$$

On this approximation Ψ_c will have the form

$$\Psi_c = \frac{1}{\sqrt{2\pi}}\{e^{-r_2}\phi(r_1) + e^{-r_1}\phi(r_2)\}, \qquad (2)$$

where $\phi(r)$ satisfies the wave equation

$$\nabla^2\phi + \left[k^2 + 2\left(1+\frac{1}{r}\right)e^{-2r}\right]\phi = 0 \qquad (3)$$

[1] S. Chandrasekhar, *Ap. J.*, **102**, 223, 1945. This paper will be referred to as "I."

[2] First suggested in this connection by H. S. W. Massey and D. R. Bates, *Ap. J.*, **91**, 202, 1940.

[3] A. Wheeler and R. Wildt, *Ap. J.*, **95**, 281, 1942; also S. Chandrasekhar, *Ap. J.*, **100**, 176, 1944.

[4] In writing this equation, we have adopted the atomic system of units. These units will be used in all our formal developments.

and tends asymptotically at infinity to a plane wave of unit amplitude along some chosen direction. If this direction, in which the ejected electron moves at infinity, be chosen as the polar axis of a spherical system of co-ordinates, the requirement is that

$$\phi(r) \to e^{ikr\cos\vartheta} \quad \text{as} \quad r \to \infty\,. \tag{4}$$

On the other hand, since

$$\left.\begin{aligned} e^{ikr\cos\vartheta} &= \left(\frac{\pi}{2kr}\right)^{1/2} \sum_{l=0}^{\infty} i^l (2l+1) P_l(\cos\vartheta) J_{l+1/2}(kr) \\ &\to \sum_{l=0}^{\infty} \frac{i^l(2l+1)}{kr} P_l(\cos\vartheta)\cos\left(kr - \tfrac{1}{2}l\pi - \tfrac{1}{2}\pi\right), \end{aligned}\right\} \tag{5}$$

it is evident that the solution for ϕ appropriate to our problem is

$$\phi = \sum_{l=0}^{\infty} \frac{i^l(2l+1)}{kr} P_l(\cos\vartheta)\chi_l(r)\,. \tag{6}$$

where the radial function $\chi_l(r)$ is a solution of the equation

$$\frac{d^2\chi_l}{dr^2} + \left\{k^2 - \frac{l(l+1)}{r^2} + 2\left(1+\frac{1}{r}\right)e^{-2r}\right\}\chi_l = 0\,, \tag{7}$$

which tends to a pure sinusoidal wave of unit amplitude at infinity.

Thus, on our present approximation, the wave function can be expressed in the form (cf. I, eq. [33])

$$\left.\begin{aligned} \Psi_c = \frac{1}{\sqrt{2\pi}} \Bigg\{ e^{-r_2} \sum_{l=0}^{\infty} \frac{i^l(2l+1)}{kr_1} P_l(\mu_1)\chi_l(r_1;k) \\ + e^{-r_1} \sum_{l=0}^{\infty} \frac{i^l(2l+1)}{kr_2} P_l(\mu_2)\chi_l(r_2;k) \Bigg\}, \end{aligned}\right\} \tag{8}$$

where we have used μ_1 and μ_2 to denote $\cos\vartheta_1$ and $\cos\vartheta_2$, respectively.

For the evaluation of the absorption cross-sections according to formula II of paper I, we need the quantity

$$\left(\frac{\partial}{\partial z_1} + \frac{\partial}{\partial z_2}\right)\Psi_c\,. \tag{9}$$

For Ψ_c given by equation (8) it can be readily shown that

$$\left.\begin{aligned} \left(\frac{\partial}{\partial z_1} + \frac{\partial}{\partial z_2}\right)\Psi_c = \frac{1}{\sqrt{2\pi}} \Bigg[&-\mu_1 e^{-r_1} \sum_{l=0}^{\infty} \frac{i^l(2l+1)}{kr_2} P_l(\mu_2)\chi_l(r_2) \\ &- \mu_2 e^{-r_2} \sum_{l=0}^{\infty} \frac{i^l(2l+1)}{kr_1} P_l(\mu_1)\chi_l(r_1) \\ &+ e^{-r_1} \sum_{l=0}^{\infty} \frac{i^l}{k} \{ lP_{l-1}(\mu_2) S_l(r_2) + (l+1) P_{l+1}(\mu_2) T_l(r_2) \} \\ &+ e^{-r_2} \sum_{l=0}^{\infty} \frac{i^l}{k} \{ lP_{l-1}(\mu_1) S_l(r_1) + (l+1) P_{l+1}(\mu_1) T_l(r_1) \} \Bigg], \end{aligned}\right\} \tag{10}$$

where

$$S_l(r) = \frac{\partial}{\partial r}\left(\frac{\chi_l}{r}\right) + (l+1)\frac{\chi_l}{r^2} \tag{11}$$

and

$$T_l(r) = \frac{\partial}{\partial r}\left(\frac{\chi_l}{r}\right) - l\frac{\chi_l}{r^2}. \tag{12}$$

For a wave function of the ground state of the form I, equation (3), we find after some lengthy but straightforward calculations that

$$\left.\begin{aligned} &\int \Psi_d\left(\frac{\partial}{\partial z_1}+\frac{\partial}{\partial z_2}\right)\Psi_c d\tau \\ &= -\frac{(2048\pi^3)^{1/2}\mathfrak{N}i}{k(1+a)^3}\left\{\int_0^\infty dr S_1(r)\left[e^{-ar}\sum_{j=0}^{7} s_j r^{j+1} + e^{-(1+2a)r}\sum_{j=0}^{+1}\sigma_j r^{j+1}\right]\right. \\ &\qquad \left. - \int_0^\infty dr \chi_1(r)\left[e^{-ar}\sum_{j=-1}^{6} l_j r^j + e^{-(1+2a)r}\sum_{j=-1}^{+1}\lambda_j r^j\right]\right\}, \end{aligned}\right\} \tag{13}$$

where s_j, σ_j, l_j, and λ_j have the same meanings as in I, equations (29)-(32). After some further reductions, equation (13) can be simplified to the form

$$\int \Psi_d\left(\frac{\partial}{\partial z_1}+\frac{\partial}{\partial z_2}\right)\Psi_c d\tau = -\frac{(2048\pi^3)^{1/2}\mathfrak{N}i}{k(1+a)^3}\int_0^\infty W_2(r)\chi_1(r)\,dr, \tag{14}$$

where

$$\left.\begin{aligned} W_2(r) &= e^{-ar}\sum_{j=-1}^{7}[a s_j - j s_{j+1} - l_j]\, r^j \\ &\quad + e^{-(1+2a)r}\sum_{j=-1}^{+1}[(1+2a)\sigma_j - j\sigma_{j+1} - \lambda_j]\, r^j, \end{aligned}\right\} \tag{15}$$

with the understanding that

$$s_{-1} = s_8 = \sigma_{-1} = \sigma_2 = l_7 = 0. \tag{16}$$

Formula II (paper I) for the absorption cross-section now becomes

$$\kappa = 2.725\times 10^{-19}\frac{2048\pi^3\mathfrak{N}^2}{(1+a)^6 k(k^2+2I)}\left|\int_0^\infty W_2(r)\chi_1(r)\,dr\right|^2. \tag{17}$$

Inserting the numerical values for the various constants in equations (15) and (17), we find:

$$\left.\begin{aligned} W_2(r) &= e^{-0.707735r}(0.1573261r^{-1} + 0.2686713 + 0.9780967r \\ &- 0.2397504r^2 + 0.06594301r^3 - 0.006107001r^4 + 0.0008050248r^5 \\ &- 0.00007231322r^6 + 0.000004366 32r^7) - e^{-2.415470r}(0.1573261r^{-1} \\ &\qquad + 0.5373426 + 0.2294096r) \end{aligned}\right\} \tag{18}$$

and

$$\kappa = \frac{3.7062}{k(k^2+0.055118)}\left|\int_0^\infty W_2(r)\chi_1(r)\,dr\right|^2 \times 10^{-18}\ \text{cm}^2. \tag{II$_1$}$$

For the purposes of evaluating the absorption cross-sections according to the foregoing formulae, it is convenient to have a fairly extensive table of the "weight function," $W_2(r)$. This is provided in Table 1.

TABLE 1

THE WEIGHT FUNCTION $W_2(r)$

r	$W_2(r)$	r	$W_2(r)$	r	$W_2(r)$	r	$W_2(r)$
0	0	4.4	0.18818	8.8	0.05010	16.4	0.00676
0.1	0.12932	4.5	.18220	8.9	.04872	16.6	.00638
0.2	.23292	4.6	.17644	9.0	.04738	16.8	.00601
0.3	.31486	4.7	.17089	9.1	.04609	17.0	.00567
0.4	.37865	4.8	.16554	9.2	.04484	17.2	.00530
0.5	.42731	4.9	.16039	9.3	.04363	17.4	.00502
0.6	.46340	5.0	.15541	9.4	.04245	17.6	.00473
0.7	.48911	5.1	.15061	9.5	.04132	17.8	.00443
0.8	.50628	5.2	.14598	9.6	.04021	18.0	.00417
0.9	.51649	5.3	.14150	9.7	.03915	18.2	.00391
1.0	.52105	5.4	.13718	9.8	.03811	18.4	.00366
1.1	.52106	5.5	.13300	9.9	.03711	18.6	.00342
1.2	.51744	5.6	.12896	10.0	.03614	18.8	.00321
1.3	.51097	5.7	.12506	10.2	.03428	19.0	.00302
1.4	.50229	5.8	.12128	10.4	.03253	19.2	.00281
1.5	.49192	5.9	.11763	10.6	.03089	19.4	.00264
1.6	.48030	6.0	.11409	10.8	.02934	19.6	.00246
1.7	.46778	6.1	.11067	11.0	.02787	19.8	.00230
1.8	.45467	6.2	.10736	11.2	.02649	20.0	.00214
1.9	.44118	6.3	.10415	11.4	.02518	20.2	.00200
2.0	.42753	6.4	.10105	11.6	.02394	20.4	.00187
2.1	.41384	6.5	.09804	11.8	.02277	20.6	.00174
2.2	.40026	6.6	.09514	12.0	.02165	20.8	.00162
2.3	.38686	6.7	.09232	12.2	.02059	21.0	.00151
2.4	.37373	6.8	.08959	12.4	.01959	21.2	.00140
2.5	.36090	6.9	.08695	12.6	.01863	21.4	.00130
2.6	.34844	7.0	.08440	12.8	.01771	21.6	.00121
2.7	.33635	7.1	.08192	13.0	.01684	21.8	.00112
2.8	.32465	7.2	.07952	13.2	.01601	22.0	.00104
2.9	.31338	7.3	.07720	13.4	.01522	22.2	.00097
3.0	.30250	7.4	.07496	13.6	.01446	22.4	.00090
3.1	.29204	7.5	.07278	13.8	.01373	22.6	.00083
3.2	.28197	7.6	.07068	14.0	.01304	22.8	.00077
3.3	.27230	7.7	.06864	14.2	.01239	23.0	.00071
3.4	.26301	7.8	.06666	14.4	.01175	23.2	.00066
3.5	.25410	7.9	.06475	14.6	.01115	23.4	.00061
3.6	.24553	8.0	.06291	14.8	.01057	23.6	.00056
3.7	.23732	8.1	.06112	15.0	.01001	23.8	.00052
3.8	.22943	8.2	.05938	15.2	.00949	24.0	.00048
3.9	.22185	8.3	.05771	15.4	.00899	24.2	.00044
4.0	.21457	8.4	.05608	15.6	.00849	24.4	.00041
4.1	.20758	8.5	.05451	15.8	.00803	24.6	.00038
4.2	.20086	8.6	.05299	16.0	.00758	24.8	.00035
4.3	0.19440	8.7	0.05152	16.2	0.00715	25.0	0.00032

In an analogous manner it is found that, under the same assumptions, formulae I and III of paper I become

$$\kappa = 9.266 \frac{k^2 + 0.055118}{k} \left| \int_0^\infty W_1(r)\, \chi_1(r)\, dr \right|^2 \times 10^{-19}\ \mathrm{cm}^2 \qquad (\mathrm{I}_1)$$

and

$$\kappa = 1.48\,25 \frac{1}{k(k^2+0.055118)^3} \left| \int_0^\infty W_3(r)\,\chi_1(r)\,dr \right|^2 \times 10^{-17}\ \text{cm}^2 , \quad (\text{III}_1)$$

where

$$\left.\begin{aligned} W_1(r) &= e^{-0.707735r}(0.7810175r + 1.0260897r^2 - 0.107743r^3 \\ &+ 0.0748147r^4 - 0.00432587r^5 + 0.000771209r^6 - 0.0000547250r^7 \\ &+ 0.00000616943r^8) - e^{-2.415407r}(0.7810175r + 1.333771r^2 \\ &\qquad + 1.138864r^3 + 0.3241463r^4) \end{aligned}\right\} \quad (19)$$

and

$$\left.\begin{aligned} W_3(r) &= e^{-0.707735r}(0.3796213r^{-2} + 1.0311631r^{-1} - 0.1091749 \\ &- 0.00361463r + 0.00176560r^2 + 0.000847872r^3 - 0.0000857360r^4 \\ &+ 0.00001105546r^5 - 0.000001669208r^6 - 0.07846795r^4 \\ &\times Ei[1.707735r]) - e^{-2.415470r}(0.3796213r^{-2} + 0.1944878r^{-1} \\ &\qquad + 0.0553556 - 0.03151090r + 0.02690613r^2 - 0.04594855r^3) . \end{aligned}\right\} \quad (20)$$

A brief table of the function $W_1(r)$ has been given in an earlier paper.[5] We now provide a similar tabulation of the function $W_3(r)$ (see Table 2). In Figure 1 we have further illustrated the dependence of the functions W_1, W_2, and W_3 on r. It is seen that the three weight functions are of entirely different orders of magnitude at large distances; it is in terms of these differences that the remarks made in paper I, §§ 2 and 5, have to be understood.

TABLE 2

THE WEIGHT FUNCTION $W_3(r)$

r	$W_3(r)$	r	$W_3(r)$	r	$W_3(r)$	r	$W_3(r)$
0.	∞	0.6	1.408	1.4	0.293	3.5	0.021
0.1	13.49	0.7	1.100	1.6	.215	4.0	.013
.2	6.13	0.8	0.879	1.8	.160	4.5	.009
.3	3.721	0.9	0.713	2.0	.121	5.0	.006
.4	2.526	1.0	0.586	2.5	.064	5.5	0.004
0.5	1.853	1.2	0.408	3.0	0.036		

3. *The continuous absorption coefficient of* H^- *evaluated according to formula* (II_1).—The evaluation of the absorption cross-sections according to formula (II_1) requires a knowledge of the p-spherical waves $\chi_1(r)$ of an electron in the Hartree field of a hydrogen atom. The writer has at his disposal a large number of tables of these functions for various values of k^2 in the range $1.75 \geq k^2 > 0$.[6] Using these tables and the table of the weight function $W_2(r)$, we have evaluated the absorption cross-sections according to formula (II_1) by straightforward numerical quadratures. The results of these integrations are given in Table 3.

[5] S. Chandrasekhar, *Ap. J.*, **100**, 176, 1944 (Table 1).

[6] These tables were computed by Mrs. Frances H. Breen and the writer. It is hoped to publish these (and similar tables of the s-waves, which have also been completed) in the near future.

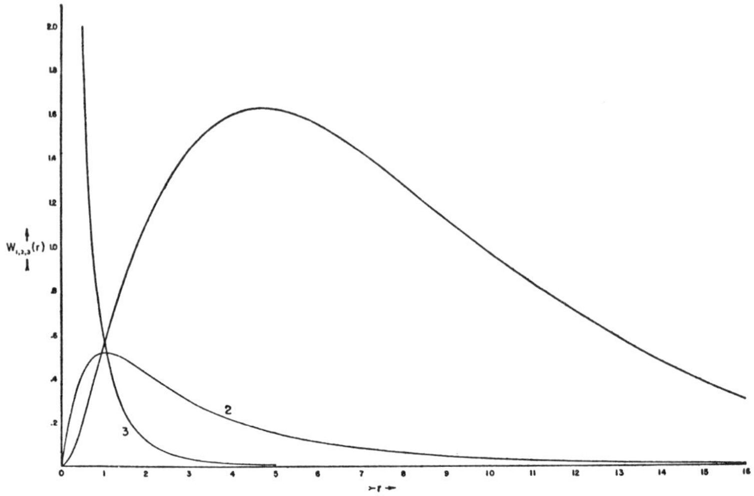

FIG. 1.—A comparison of the weight functions $W_1(r)$ (curve 1), $W_2(r)$ (curve 2), and $W_3(r)$ (curve 3), which occur in the formulae (I_1, II_1, and III_1) for the absorption cross-sections of H^- in terms of the matrix elements of the dipole moment, momentum, and acceleration, respectively.

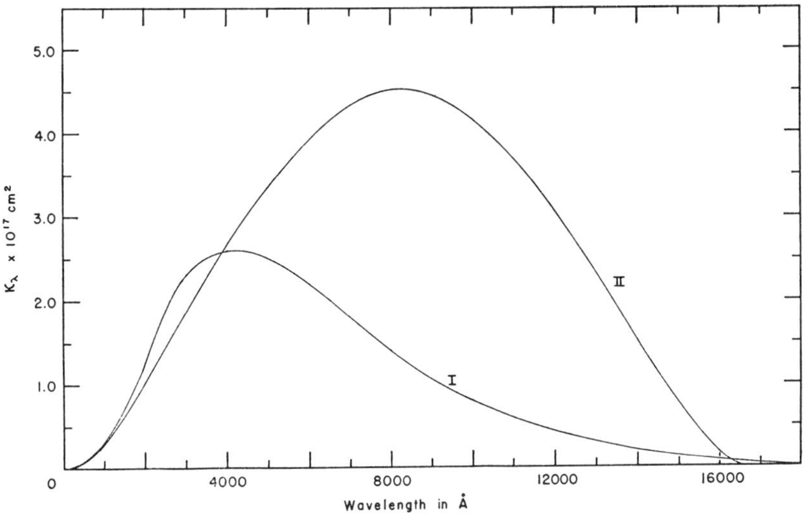

FIG. 2.—A comparison of the continuous absorption coefficient of H^- as determined by Massey and Bates (curve I) with the determination of the present paper (curve II).

It is seen that our present values differ from those obtained earlier on the plane-wave representation of the ejected electron by about 5 per cent in the entire spectral range of astrophysical interest. The maximum is, however, not appreciably shifted and is at the same place (λ 8500 A); here the atomic absorption coefficient has the value 4.52×10^{-17} cm.^{2} In view of the many determinations of the continuous absorption coefficient of H^- that have been made, it is perhaps of interest to compare the first of these determinations by Massey and Bates with that of the present paper. This is done in Figure 2.

TABLE 3

THE CONTINUOUS ABSORPTION COEFFICIENT OF H^- COMPUTED ACCORDING TO FORMULA (II_1)

k^2	λ(A)	$\kappa_\lambda \times 10^{17}$ Cm^2	k^2	λ(A)	$\kappa_\lambda \times 10^{17}$ Cm^2	k^2	λ(A)	$\kappa_\lambda \times 10^{17}$ Cm^2
1.75	505	0.0657	0.175	3960	2.62	0.055	8275	4.52
0.80	1066	0.333	.150	4443	2.97	.050	8669	4.50
0.50	1642	0.740	.125	5059	3.39	.045	9102	4.44
0.35	2249	1.231	.100	5875	3.87	.035	10,111	4.13
0.25	2987	1.84	.090	6280	4.06	.020	12,131	2.96
0.20	3572	2.32	0.070	7283	4.41	0.010	13,994	1.50

It is again a pleasure to record my indebtedness to Mrs. Frances H. Breen for valuable assistance with the numerical work.

ON THE CONTINUOUS ABSORPTION COEFFICIENT OF THE NEGATIVE HYDROGEN ION. III

S. Chandrasekhar and Frances Herman Breen
Yerkes Observatory
Received August 14, 1946

ABSTRACT

In this paper the contribution to the continuous absorption coefficient of the negative hydrogen ion by the free-free transitions is evaluated in terms of the matrix elements of the acceleration in the Hartree field of a hydrogen atom. The new coefficients are larger than the earlier determinations by factors exceeding 10 over the entire range of wave lengths of astrophysical interest.

Tables of the continuous absorption coefficient of H^-, including both the bound-free and the free-free transitions for various temperatures ($2520° \leqslant T \leqslant 10{,}080°$) and wave lengths ($\lambda > 4000$ A) are also provided. It is further shown that the new coefficients are sufficient to account for the solar continuous spectrum from λ 4000 A to λ 25,000 A.

1. *Introduction.*—In two earlier papers[1] the continuous absorption coefficient of the negative hydrogen ion has been considered, and its cross-sections for the radiative processes leading to its ionization have been determined with some degree of definitiveness. And recent astrophysical discussions[2] relating to the continuous spectrum of the sun have shown that the cross-sections which were derived for these "bound-free" transitions of H^- are adequate to account for the continuous absorption in the solar atmosphere between λ 4000 A and λ 10,000 A. Beyond λ 10,000 A, however, there appears to be an additional source of absorption, which it would be natural to suppose is due to the radiative transitions of free electrons in the field of neutral hydrogen atoms.[3] But the existing evaluations[4] of these "free-free" transitions make them insufficient to account for the observed amount of absorption beyond λ 10,000 A by factors exceeding 10. Indeed, on the strength of this discrepancy, the existence of a hitherto "unknown source of absorption" has been concluded.[5] However, as similar conclusions relating to the bound-free transitions have proved premature in the past, we have examined the earlier evaluations of the free-free transitions of H^- and have found, as we shall presently explain, that there are ample grounds for mistrusting them, even as to giving the correct orders of magnitude. We have, accordingly, made some further calculations to obtain estimates of the free-free transitions, on which at least some reliance could be placed. It is the object of this paper to present the results of such calculations and to show that these newly determined cross-sections for the free-free transitions, together with the cross-sections for the bound-free transitions given in Paper II, are sufficient to account for the continuous absorption in the solar and in the stellar atmospheres of neighboring spectral types in a manner which dispels any remaining belief in an "unknown source of absorption."

2. *The inadequacy of the Born approximation for the evaluation of the free-free transitions of* H$^-$.—The essential approximation which underlies all the earlier evaluations[4] is that of Born's for describing the motion of an electron in the field of a hydrogen atom.

[1] S. Chandrasekhar, *Ap. J.*, **102**, 223, 395, 1945. These papers will be referred to as Papers I and II, respectively.

[2] G. Münch, *Ap. J.*, **102**, 385, 1945; D. Chalonge and V. Kourganoff, *Ann. d' ap.* (in press).

[3] The possible astrophysical importance of this process was first suggested by A. Pannekoek, *M.N.*, **91**, 162, 1931.

[4] L. Nedelsky, *Phys. Rev.*, **42**, 641, 1932; D. H. Menzel and C. L. Pekeris, *M.N.*, **96**, 77, 1935; J. A. Wheeler and R. Wildt, *Ap. J.*, **95**, 281, 1942.

[5] See particularly the discussion of Chalonge and Kourganoff (*op. cit.*).

This is apparent, for example, from the agreement of the calculations of Menzel and Pekeris with those of Wheeler and Wildt, in which the Born approximation is explicitly made. However, on consideration it appears that for the range of energies which occur in stellar atmospheres the Born approximation for incident *s*-electrons must be a very bad one; for, as the mean energy of the electrons in a Maxwellian distribution at temperature T is

$$\frac{1}{2}\bar{k}^2 = 0.0243\left(\frac{T}{5040}\right) \tag{1}$$

when expressed in atomic units, it is evident that it is only electrons with $k^2 < 0.05$ that will principally contribute to the continuous absorption. And for *s*-electrons with as small energies as these, the Born approximation must fail. This is indeed well known from the work of P. M. Morse and W. P. Allis.[6] But it may be useful to illustrate this failure in a manner which will emphasize the magnitude of the errors to which the Born approximation may lead in the cross-sections for radiative transitions. For this purpose, we have compared in Figure 1 the radial *s*-waves

$$\chi_0\,(\text{Born}) = \sin kr \tag{2}$$

on the Born approximation with the properly normalized *s*-spherical waves in the Hartree field of a hydrogen atom, which the authors have recently tabulated.[7] It is seen that for $r < 2$ the Hartree waves have amplitudes which are larger than χ_0 (Born) by factors which, on the average, exceed 3. However, similar comparisons between the *p*-waves

$$\chi_1\,(\text{Born}) = \frac{\sin kr}{kr} - \cos kr \tag{3}$$

on the Born approximation and the corresponding Hartree waves show that equation (3) provides a satisfactory approximation for the energies in which we are interested. Remembering that, in the evaluation of the matrix elements of the acceleration, the wave functions for $r < 2$ are all that matters (cf. Fig. 2, in which we have plotted the acceleration $\ddot{r}$ in the static field of a hydrogen atom) and, further, that the contribution to the absorption coefficient arising from the $s \rightarrow p$ and $p \rightarrow s$ transitions must far outweigh all the others, it is apparent that in the Hartree approximation we shall obtain cross-sections which will be larger by ten or more times the values obtained with the Born approximation. In other words, we may expect a correct evaluation of the free-free transitions of H^- to bring about an agreement between physical theory and the demands of astrophysical data without the need of postulating a still unknown source of continuous absorption.

3. *Formula for the evaluation of the cross-sections for the free-free transitions of* H^- *on the Hartree approximation.*—Our remarks in the preceding section have shown the inadequacy of the Born approximation for the evaluation of the free-free transitions of H^-. The use of the Hartree approximation suggests itself as the next best, though the effects of exchange and polarization may very well be appreciable for the very slow *s*-electrons in which we are primarily interested. But it may be hoped that these latter effects will not, at any rate, affect the orders of magnitude of the derived quantities! In any case, to improve on the Hartree approximation would require an amount of numerical work which will be several fold; and the task is immense even as it is.[8] These considerations, to-

[6] *Phys. Rev.*, **44**, 269, 1933.

[7] S. Chandrasekhar and F. H. Breen, *Ap. J.*, **103**, 41, 1946.

[8] For example, the present work has required the numerical integration of 63 radial functions and the evaluation of 523 infinite integrals, not to mention the computation of numerous auxiliary functions and tables. (All this work was done with a Marchant.)

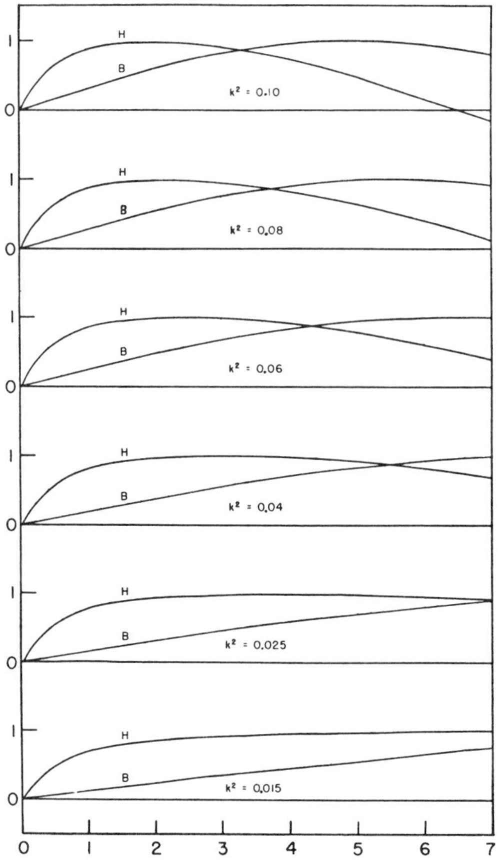

FIG. 1.—A comparison of the radial wave functions of *s*-electrons in the field of a hydrogen atom on the Born (B) and the Hartree (H) approximations for various energies of astrophysical interest. (The abscissa measures the distance from the center in units of the Bohr radius).

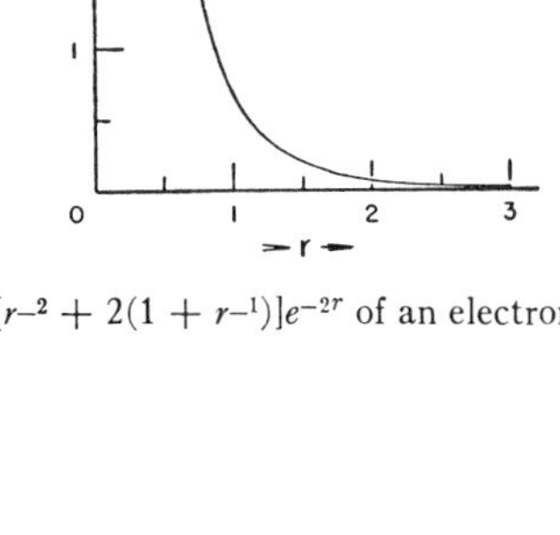

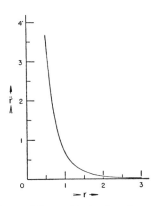

FIG. 2.—The acceleration $\ddot{r} = [r^{-2} + 2(1 + r^{-1})]e^{-2r}$ of an electron in the Hartree field of a hydrogen atom.

gether with the urgency of the astrophysical needs, have prompted us to undertake in some detail the evaluation of the cross-sections for the free-free transitions of H^-, using the *s*- and the *p*-spherical waves in the Hartree field of the hydrogen atom which we have tabulated in an earlier paper.[7]

Now, in the Hartree approximation a wave function representing a plane wave at infinity is given by

$$\Psi = \sum_{l=0}^{\infty} \frac{i^l (2l+1)}{k r} P_l (\mu) \chi_l (r; k^2) , \tag{4}$$

where the radial functions $\chi_l(r; k)$ are solutions of the equation

$$\frac{d^2\chi_l}{d r^2} + \left\{k^2 - \frac{l(l+1)}{r^2} + 2\left(1+\frac{1}{r}\right)e^{-2r}\right\}\chi_l = 0 , \tag{5}$$

which have unit amplitude at infinity. (In the foregoing equations we have adopted the atomic system of units). For wave functions of the form (4) the standard formula in the quantum theory, which gives the cross-section for a transition in which a free electron with an initial momentum k_0 (in atomic units) becomes an electron with a momentum k_1 by the absorption of a quantum of appropriate energy from an incident beam of unit specific intensity, can be written in the form[9]

$$\left.\begin{aligned} a(k_0^2; \Delta k^2) &= \frac{256\pi^2}{3}\left(\frac{2\pi e^2}{h c}\right)\left(\frac{h^2}{4\pi^2 m e^2}\right)^5 \frac{1}{k_0^2 k_1 (\Delta k^2)^3} \\ &\times \sum_{l=1}^{\infty} l\{ |(l, k_0^2| \ddot{r} | l-1, k_1^2)|^2 + |(l-1, k_0^2| \ddot{r} | l, k_1^2)|^2\} \text{cm}^5 , \end{aligned}\right\} \tag{6}$$

where the various matrix elements $(l, k_1^2|\ddot{r}|l-1, k_2^2)$ have to be evaluated in atomic units and

$$\Delta k^2 = k_1^2 - k_0^2 . \tag{7}$$

Equation (6) gives the cross-section for a single hydrogen atom in the ground state for the transition in question when there is one electron with momentum k_0 in unit volume. Moreover, the wave length, λ, of the radiation for which formula (6) gives the atomic absorption coefficient is

$$\lambda (\Delta k^2) = \frac{911.3}{\Delta k^2} A . \tag{8}$$

In the Hartree field of a hydrogen atom the acceleration is given by

$$\ddot{r} = \left[\frac{1}{r^2} + 2\left(1+\frac{1}{r}\right)\right] e^{-2r} , \tag{9}$$

and the matrix elements which have to be evaluated are of the type

$$(l, k_0^2| \ddot{r} | l-1, k_1^2) = \int_0^{\infty} \chi_l (r; k_0^2) \left[\frac{1}{r^2} + 2\left(1+\frac{1}{r}\right)\right] e^{-2r} \chi_{l-1} (r; k_1^2)\, d r . \tag{10}$$

To obtain the absorption coefficients appropriate for an assembly in which we have a Maxwellian distribution of electrons corresponding to a temperature T, we must average $a(k_0^2; \Delta k^2)$ over all initial k_0^2 (and for a fixed Δk^2) with the weight function

$$197.8\, \theta^{3/2} k_0 e^{-31.32\theta k_0^2} \tag{11}$$

[9] Cf. J. A. Gaunt, *Phil. Trans. R. Soc., London*, A, **229**, 163, 1930; see particularly formula (5.24) on p. 194; also Wheeler and Wildt, *op. cit.*, eq. (6) on p. 284.

where

$$\theta = \frac{5040}{T}. \tag{12}$$

In astrophysical applications it is convenient to express the free-free atomic absorption coefficients as per unit electron pressure. If we denote by $\kappa(\Delta k^2)$ the corresponding atomic absorption coefficient after averaging over a Maxwell distribution of initial velocities, we find that we can write our basic formula in the form

$$\left.\begin{aligned}\kappa(\Delta k^2) &= \frac{7.251\times 10^{-29}\theta^{5/2}}{(\Delta k^2)^3}\int_0^\infty d(k_0^2)\frac{f(k_0^2)}{k_0^2 k_1}\\ &\times\sum_{l=1}^{\infty} l\{|(l, k_0^2|\ddot{r}|l-1, k_1^2)|^2 + |(l-1, k_0^2|\ddot{r}|l, k_1^2)|^2\}\frac{\text{cm}^4}{\text{dyne}},\end{aligned}\right\} \tag{13}$$

where

$$f(k_0^2) = 100k_0 e^{-31.32\theta k_0^2}. \tag{14}$$

4. *Details of the evaluation: tables of the necessary matrix elements.*—The problem of numerically evaluating absorption coefficients for free-free transitions is a specially troublesome one, since for each wave length the coefficients must be computed for a range of initial velocities sufficient to allow for the averaging over Maxwell distributions for various temperatures. If the matrix elements for all the necessary values of k_0^2 and Δk^2 must be individually evaluated, then the number of wave functions which would be needed will be so many as to make the problem an impracticable one on this score alone. It may, therefore, be useful to place on record the details of our procedure in this instance.

First, we may observe that the infinite series in equation (13) is so rapidly convergent that we may ignore all terms except the first one. This may be verified in the following manner:

We have already indicated in § 2 that for $l \geqslant 1$ we can use the Born approximation without any serious error. We may therefore write

$$\chi_l(r; k^2) = \left(\frac{\pi k r}{2}\right)^{1/2} J_{l+\frac{1}{2}}(kr). \tag{15}$$

Accordingly, for $l \geqslant 2$,

$$\left.\begin{aligned}(l, k_1^2|\ddot{r}|l-1, k_2^2) &= \tfrac{1}{2}\pi(k_1k_2)^{1/2}\int_0^\infty rJ_{l+\frac{1}{2}}(k_1r)J_{l-\frac{1}{2}}(k_2r)\\ &\qquad\times\left[\frac{1}{r^2}+2\left(1+\frac{1}{r}\right)\right]e^{-2r}dr.\end{aligned}\right\} \tag{16}$$

Writing equation (16) in the form

$$\left.\begin{aligned}(l, k_1^2|\ddot{r}|l-1, k_2^2) &= -\tfrac{1}{2}\pi k_1k_2\int_0^\infty rJ_{l+\frac{1}{2}}(k_1r)J_{l-\frac{1}{2}}(k_2r)\\ &\qquad\times\frac{d}{dr}\left[\left(1+\frac{1}{r}\right)e^{-2r}\right]dr\end{aligned}\right\} \tag{17}$$

and integrating by parts, we find after some further reductions that

$$\left.\begin{aligned}(l, k_1^2|\ddot{r}|l-1, k_2^2) &= \tfrac{1}{2}\pi(k_1k_2)^{1/2}\int_0^\infty(1+r)e^{-2r}[k_1J_{l-\frac{1}{2}}(k_1r)J_{l-\frac{1}{2}}(k_2r)\\ &\qquad - k_2J_{l+\frac{1}{2}}(k_1r)J_{l+\frac{1}{2}}(k_2r)]\,dr.\end{aligned}\right\} \tag{18}$$

The integrals over the Bessel functions which occur in the foregoing equation can be expressed in terms of the Legendre functions of the second kind, $Q_l(x)$, and their derivatives, $Q_l'(x)$.[10] We find

$$(l,\, k_1^2 |\, \ddot{r}\, | \, l-1,\, k_2^2) = \tfrac{1}{2} k_1 Q_{l-1}(x) - \tfrac{1}{2} k_2 Q_l(x) - \frac{1}{k_2} Q_{l-1}'(x) + \frac{1}{k_1} Q_l'(x)\,, \quad (19)$$

where the argument for the Legendre functions is

$$x = \frac{k_1^2 + k_2^2 + 4}{2 k_1 k_2}\,. \quad (20)$$

Using equation (19), we can readily verify the fact that all transitions in which the incident electron is characterized by an l greater than 1 contribute less than a fraction of 1 per cent to the sum in equation (13). It is therefore sufficient to consider only the matrix elements $(0,\, k_0^2 |\ddot{r}| 1,\, k_1^2)$ and $(1,\, k_0^2 |\ddot{r}| 0,\, k_1^2)$. In evaluating these matrix elements it was found convenient to distinguish three cases, discussed below.

i) $k_0^2 < 0.015$ *and* $k_1^2 < 0.015$.—It has been shown[7] that, when $k^2 < 0.015$, the radial functions $\chi_0(r;\, k^2)$ and $\chi_1(r;\, k^2)$ can be expressed with sufficient accuracy in the form

$$\chi_0(r;\, k^2) = \frac{1}{A_0(k)} [X_0(r) - k^2 Y_0(r)] \quad (21)$$

and

$$\chi_1(r;\, k^2) = \frac{1}{A_1(k)} [X_1(r) - k^2 Y_1(r)]\,, \quad (22)$$

where $A_0(k)$ and $A_1(k)$ are the factors which normalize the functions in brackets to unit amplitude at infinity and X_0, Y_0, X_1, and Y_1 are certain functions which have been tabulated. With this representation of the wave functions, the matrix element $(0,\, k_0^2 |\ddot{r}| 1,\, k_1^2)$ can be expressed in the form

$$\left.\begin{aligned}(0,\, k_0^2 |\, \ddot{r}\, | \,1,\, k_1^2) = \frac{1}{A_0(k_0)\, A_1(k_1)} \Big[\int_0^\infty X_0 X_1 \ddot{r}\, dr - k_0^2 \int_0^\infty Y_0 X_1 \ddot{r}\, dr \\ - k_1^2 \int_0^\infty X_0 Y_1 \ddot{r}\, dr + k_0^2 k_1^2 \int_0^\infty Y_0 Y_1 \ddot{r}\, dr \Big]\,. \end{aligned}\right\} \quad (23)$$

The corresponding expression for $(1,\, k_0^2 |\ddot{r}| 0,\, k_1^2)$ can be obtained from equation (23) by simply interchanging k_0 and k_1.

Thus, when k_0^2 and k_1^2 are both less than 0.015, the necessary matrix elements can all be evaluated in terms of four definite integrals which are listed in Table 1.

TABLE 1

INTEGRALS FOR COMPUTING THE MATRIX ELEMENTS WHEN k_0^2 AND k_1^2 ARE BOTH LESS THAN 0.015

$\int_0^\infty X_0 X_1 \ddot{r} dr$	0.40044	$\int_0^\infty Y_0 X_1 \ddot{r} dr$	0.22641
$\int_0^\infty Y_0 Y_1 \ddot{r} dr$	0.16669	$\int_0^\infty X_0 Y_1 \ddot{r} dr$	0.09076

ii) $k_0^2 < 0.015$ *and* $k_1^2 \geqslant 0.015$.—In this case we can use the representation (20) or (21) only for the radial function describing the incident electron, and the required matrix elements can be expressed in the forms

$$(0,\, k_0^2 |\, \ddot{r}\, | \,1,\, k_1^2) = \frac{1}{A_0(k_0)} \Big[\int_0^\infty X_0 \chi_1(r;\, k_1)\, \ddot{r}\, dr - k_0^2 \int_0^\infty Y_0 \chi_1(r;\, k_1)\, \ddot{r}\, dr \Big] \quad (24)$$

[10] Cf. G. N. Watson, *Theory of Bessel Functions*, p. 389, Cambridge University Press, 1944.

and

$$(1,\ k_0^2|\ \ddot{r}\ |0,\ k_1^2) = \frac{1}{A_1(k_0)}\left[\int_0^\infty X_1\chi_0(r;\ k_1)\ \ddot{r}\,dr - k_0^2\int_0^\infty Y_1\chi_0(r;\ k_1)\ \ddot{r}\,dr\right]. \quad (25)$$

The integrals which occur on the right-hand side of equations (24) and (25) have been evaluated for various values of $k_1^2 \geqslant 0.015$ and are listed in Table 2; for these values of k_1^2 and for all $k_0^2 < 0.015$, the required matrix elements can be found without further numerical integrations according to equations (24) and (25).

iii) $k_0^2 \geqslant$ *0.015 and* $k_1^2 \geqslant$ *0.015*.—In this case there is no short cut to the evaluation of the matrix elements, and the integrals must all be evaluated individually. However, for this purpose we have only the radial functions tabulated in the paper by Chandrasekhar and Breen.

Our procedure, then, for determining the various matrix elements with assigned k_0^2 and Δk^2 was as follows:

TABLE 2

INTEGRALS FOR COMPUTING THE MATRIX ELEMENTS WHEN $k_0^2<0.015$

k_1^2	$\int_0^\infty X_1\chi_0(r;\ k_1^2)\ddot{r}dr$	$\int_0^\infty Y_1\chi_0(r;\ k_1^2)\ddot{r}dr$	$\int_0^\infty X_0\chi_1(r;\ k_1^2)\ddot{r}dr$	$\int_0^\infty Y_0\chi_1(r;\ k_1^2)\ddot{r}dr$
0.015	0.64185	0.14271	0.0033763	0.0018942
.030	.71329	.15557	.0067421	.0037565
.045	.74163	.15863	.0101066	.0055872
.060	.75618	.15858	.013458	.0073854
.080	.76622	.15650	.017920	.0097336
.100	.77079	.15328	.022366	.012025
.125	.77307	.14861	.027895	.014809
.150	.77310	.14360	.033392	.017504
.175	.77164	.13840	.038851	.020113
.200	.76954	.13170	.044272	.022636
0.250	0.76357	0.12293	0.054972	0.027421

First, the matrix elements listed in Table 3 were computed with the known radial functions. (This table also includes the matrix elements computed according to eqs. [23], [24], or [25]). The values given in Table 3 show sufficiently smooth differences with both arguments to enable satisfactory interpolation for any intermediate value. Using these directly computed values, we next found, by interpolation, all the matrix elements which were needed for the evaluation of $\kappa(\Delta k^2)$ according to equation (13); the matrix elements found in this manner are given in Table 4, which was then the basis for our further calculations.

5. *The continuous absorption coefficient of* H^-.—Even with as complete a table of matrix elements as Table 4, the evaluation of the continuous absorption coefficient for free-free transitions is a tiresome matter, since, for each wave length and temperature for which the coefficient is desired, a numerical integration must be performed. The coefficients tabulated in Table 5 were determined in this manner according to equation (13).

In Table 5 we have also entered the absorption coefficients due to the bound-free transitions for $\lambda < 16{,}500$ A. These latter coefficients were obtained by multiplying the cross-sections derived from those given in Paper II (Table 3) by the factor

$$\phi(\theta) = 4.158 \times 10^{-10}\theta^{5/2}e^{1.726\theta}, \quad (26)$$

which gives the number of H^- ions present per neutral hydrogen atom and unit electron pressure. The contributions to the continuous absorption coefficient of H^- by the

TABLE 3

THE MATRIX ELEMENTS $(0, k_0^2|\ddot{r}|1, k_1^2)$ AND $(1, k_0^2|\ddot{r}|0, k_1^2)$ EVALUATED BY EXACT NUMERICAL INTEGRATIONS

k^2	$(0, 0.0025\|\ddot{r}\|$ $1, k^2)$	$(1, 0.0025\|\ddot{r}\|$ $0, k^2)$	$(0, 0.005\|\ddot{r}\|$ $1, k^2)$	$(1, 0.005\|\ddot{r}\|$ $0, k^2)$	$(0, 0.0075\|\ddot{r}\|$ $1, k^2)$	$(1, 0.0075\|\ddot{r}\|$ $0, k^2)$
0.0025	0.0005256	0.0005257	....	....	....	....
.0050	.0010513	.0006778	0.001356	0.001356	....	....
.0075	.001577	.0007681	.002033	.001536	0.002304	0.002304
.0100	.002103	.0008293	.002711	.001659	.003073	.002488
.0125	.002629	.0008738	.003390	.001747	.003842	.002621
.0150	.003153	.0009022	.004066	.001804	.004608	.002707
.0300	.006297	.001003	.008119	.002005	.009201	.003008
.0450	.009439	.001043	.01217	.002085	.01379	.003128
.060	.01257	.001063	.01621	.002126	.01837	.003189
.080	.01674	.001077	.02158	.002154	.02446	.003232
.100	.02089	.001084	.02694	.002167	.03053	.003251
.125	.02605	.001087	.03360	.002174	.03808	.003261
.150	.03119	.001087	.04022	.002174	.04558	.003261
.175	.03629	.001085	.04680	.002170	.05304	.003255
.200	.04135	.001082	.05333	.002164	.06044	.003246
0.250	0.05135	0.001074	0.06622	0.002147	0.07506	0.003221

k^2	$(0, 0.01\|\ddot{r}\|$ $1, k^2)$	$(1, 0.01\|\ddot{r}\|$ $0, k^2)$	$(0, 0.0125\|\ddot{r}\|$ $1, k^2)$	$(1, 0.0125\|\ddot{r}\|$ $0, k^2)$	$(0, 0.015\|\ddot{r}\|$ $1, k^2)$	$(1, 0.015\|\ddot{r}\|$ $0, k^2)$
0.010	0.003318	0.003318	....	....	....	....
.0125	.004148	.003495	....	....	....	....
.015	.004975	.003609	0.005241	0.004513	0.005412	0.005412
.030	.009935	.004011	.010467	.005015	.010808	.006015
.045	.01489	.004171	.01569	.005215	.01620	.006254
.060	.01983	.004253	.02090	.005317	.02158	.006377
.080	.02641	.004309	.02783	.005388	.02873	.006463
.100	.03296	.004335	.03473	.005421	.03586	.006502
.125	.04112	.004348	.04332	.005437	.04473	.006522
.150	.04922	.004349	.05186	.005438	.05355	.006523
.175	.05727	.004341	.06035	.005428	.06232	.006516
.200	.06527	.004329	.06877	.005414	.07102	.006494
0.250	0.08105	0.004296	0.08541	0.005372	0.08820	0.006445

k^2	$(0, 0.02\|\ddot{r}\|$ $1, k^2)$	$(1, 0.02\|\ddot{r}\|$ $0, k^2)$	$(0, 0.025\|\ddot{r}\|$ $1, k^2)$	$(1, 0.025\|\ddot{r}\|$ $0, k^2)$	$(0, 0.03\|\ddot{r}\|$ $1, k^2)$	$(1, 0.03\|\ddot{r}\|$ $0, k^2)$
0.02	0.007594	0.007593	0.007842	0.009471	0.008021	0.01135
.04	.01517	.008260	.01567	.010303	.01603	.01237
.06	.02270	.008506	.02345	.010609	.02398	.01274
.08	.03023	.008619	.03123	.01075	.03194	.01291
.10	.03773	.008672	.03898	.01082	.03987	.01299
.125	.04707	.008699	.04862	.01085	.04974	.01303
.150	.05635	.008701	.05821	.01085	.05955	.01303
.175	.06557	.008685	.06774	.01083	.06930	.01301
.200	.07473	.008663	.07721	.01081	.07899	.01298
0.250	0.09277	0.008597	0.09590	0.01072	0.09811	0.01288

TABLE 3—*Continued*

k^2	$(0, 0.035 \| \ddot{r} \| 1, k^2)$	$(1, 0.035 \| \ddot{r} \| 0, k^2)$	$(0, 0.04 \| \ddot{r} \| 1, k^2)$	$(1, 0.04 \| \ddot{r} \| 0, k^2)$	$(0, 0.06 \| \ddot{r} \| 1, k^2)$	$(1, 0.06 \| \ddot{r} \| 0, k^2)$
0.04	0.01630	0.01444	0.01650	0.01651	...	...
.06	.02440	.01487	.02470	.01700	0.02544	0.02544
.08	.03249	.01507	.03290	.01723	.03388	.02578
.100	.04056	.01516	.04107	.01733	.04230	.02595
.125	.05060	.01521	.05123	.01739	.05279	.02603
.150	.06058	.01521	.06135	.01740	.06321	.02605
.175	.07050	.01519	.07139	.01737	.07357	.02600
.200	.08036	.01515	.08137	.01732	.08387	.02594
.250	0.09982	0.01504	0.10109	0.01720	.10422	.02576
0.350	...	...	...	...	0.1440	0.02528

k^2	$(0, 0.08 \| \ddot{r} \| 1, k^2)$	$(1, 0.08 \| \ddot{r} \| 0, k^2)$	$(0, 0.1 \| \ddot{r} \| 1, k^2)$	$(1, 0.1 \| \ddot{r} \| 0, k^2)$	$(0, 0.125 \| \ddot{r} \| 1, k^2)$	$(1, 0.125 \| \ddot{r} \| 0, k^2)$
0.080	0.03435	0.03435	...	...	...	...
.100	.04289	.03457	0.04317	0.04317	...	...
.125	.05352	.03469	.05388	.04332	0.05408	0.05408
.150	.06410	.03471	.06453	.04339	.06478	.05417
.175	.07462	.03466	.07513	.04330	.07543	.05406
.200	.08507	.03458	.08566	.04321	.08602	.05395
.250	.1057	.03434	.1065	.04292	.1070	.05361
.350	0.1461	0.03372	.1472	.04216	.1480	.05270
.450	...	...	.1866	.04130	0.1876	0.05165
.50	...	...	.2057	.04086	...	...
.60	...	...	.2427	.03996	...	...
.70	...	...	.2783	.03906	...	...
0.80	...	...	0.3123	0.03818	...	...

k^2	$(0, 0.15 \| \ddot{r} \| 1, k^2)$	$(1, 0.15 \| \ddot{r} \| 0, k^2)$	$(0, 0.175 \| \ddot{r} \| 1, k^2)$	$(1, 0.175 \| \ddot{r} \| 0, k^2)$	$(0, 0.2 \| \ddot{r} \| 1, k^2)$	$(1, 0.2 \| \ddot{r} \| 0, k^2)$
0.150	0.06485	0.06485	0.06478	0.07552	0.06467	0.08614
.175	.07552	.06478	...	...	...	...
.200	.08614	.06467	.08607	.07533	...	...
.250	.1071	.06428	.1071	.07490	.1070	.08547
.350	.1483	.06321	.1483	.07369	.1482	.08414
.450	.1881	.06198	0.1882	0.07230	.1882	.08259
.50	.2074	.06134	...	...	.2076	.08177
.60	.2451	.06005	...	...	.2455	.08010
.70	0.2812	0.05874	...	...	.2819	.07835
0.80	...	...	...	...	0.3169	0.07676

k^2	$(0, 0.25 \| \ddot{r} \| 1, k^2)$	$(1, 0.25 \| \ddot{r} \| 0, k^2)$	$(0, 0.35 \| \ddot{r} \| 1, k^2)$	$(1, 0.35 \| \ddot{r} \| 0, k^2)$	$(0, 0.45 \| \ddot{r} \| 1, k^2)$	$(1, 0.45 \| \ddot{r} \| 0, k^2)$
0.25	0.1064	0.1064	...	...	...	...
.35	.1476	.1049	0.1457	0.1457	...	...
.40	.1678	.1040	.1658	.1446	...	...
.45	.1876	.1030	.1856	.1434	0.1829	0.1829
.50	.2071	.1021	.2050	.1422	.2023	.1815
.60	.2451	.1001	.2430	.1396	.2402	.1785
.70	.2817	.09809	.2798	.1370	.2769	.1754
0.80	0.3169	0.09603	0.3152	0.1344	0.3124	0.1722

TABLE 3—*Continued*

k^2	$(0, 0.5 \mid \dot{r} \mid 1, k^2)$	$(1, 0.5 \mid \dot{r} \mid 0, k^2)$	$(0, 0.6 \mid \dot{r} \mid 1, k^2)$	$(1, 0.6 \mid \dot{r} \mid 0, k^2)$
0.50	0.2008	0.2008	. . . .	. . . .
.60	.2385	.1976	0.2351	0.2351
.70	.2752	.1942	.2716	.2314
0.80	0.3107	0.1909	0.3070	0.2277

bound-free and free-free transitions can thus be brought together in the same system (cm^4/dyne).

In Figure 3 we have illustrated the results of Table 5 for $T = 6300°$ ($\theta = 0.8$). For comparison we have also included the earlier evaluations.

In Table 6 the coefficients given in Table 5 have been reduced by the stimulated emission factor

$$(1 - e^{-31.32\theta\,\Delta k^2})\ , \tag{27}$$

which is quite appreciable in the infrared. The values given in Table 6 are such as to enable satisfactory interpolation for most astrophysical applications.

Finally, in Table 7 we give the net absorption coefficient of H^-, including both the bound-free and the free-free transitions.

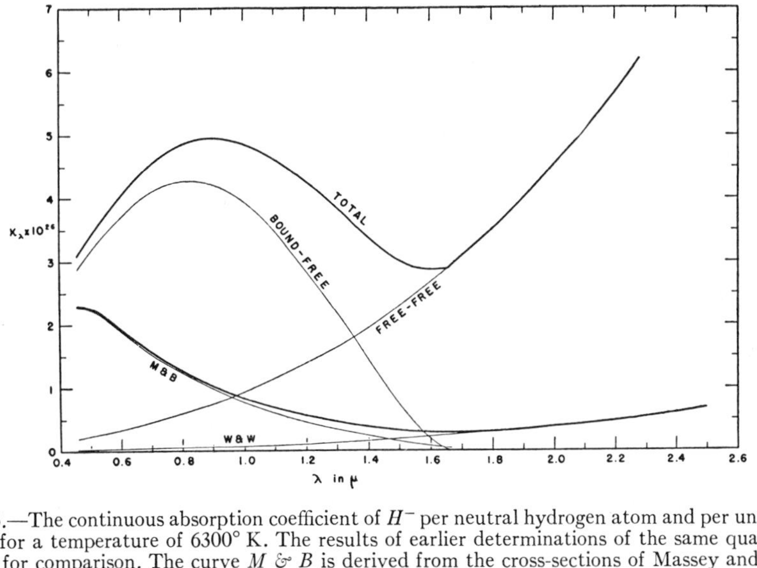

FIG. 3.—The continuous absorption coefficient of H^- per neutral hydrogen atom and per unit electron pressure for a temperature of 6300° K. The results of earlier determinations of the same quantities are included for comparison. The curve *M & B* is derived from the cross-sections of Massey and Bates for the bound-free transitions, and the curve *W & W* corresponds to the results of Wheeler and Wildt.

TABLE 4

THE MATRIX ELEMENTS $(0, k_0^2|\ddot{r}|1, k_0^2+\Delta k^2=k_1^2)$ AND $(1, k_0^2|\ddot{r}|0, k_0^2+\Delta k^2=k_1^2)$ FOR VARIOUS VALUES OF k_0^2 AND Δk^2 DERIVED BY INTERPOLATION FROM TABLE 3

Δk^2	$k_0^2=0.0025$		$k_0^2=0.0050$		$k_0^2=0.0075$		$k_0^2=0.0100$		$k_0^2=0.0125$		$k_0^2=0.0150$		$k_0^2=0.020$	
	$(0,k_0^2\|\ddot{r}\|1,k_1^2)$	$(1,k_0^2\|\ddot{r}\|0,k_1^2)$	$(0,k_0^2\|\ddot{r}\|1,k_1^2)$	$(1,k_0^2\|\ddot{r}\|0,k_1^2)$	$(0,k_0^2\|\ddot{r}\|1,k_1^2)$	$(1,k_0^2\|\ddot{r}\|0,k_1^2)$	$(0,k_0^2\|\ddot{r}\|1,k_1^2)$	$(1,k_0^2\|\ddot{r}\|0,k_1^2)$	$(0,k_0^2\|\ddot{r}\|1,k_1^2)$	$(1,k_0^2\|\ddot{r}\|0,k_1^2)$	$(0,k_0^2\|\ddot{r}\|1,k_1^2)$	$(1,k_0^2\|\ddot{r}\|0,k_1^2)$	$(0,k_0^2\|\ddot{r}\|1,k_1^2)$	$(1,k_0^2\|\ddot{r}\|0,k_1^2)$
0.	0.000526	0.000526	0.001356	0.001356	0.002304	0.002304	0.003318	0.003318	0.004370	0.004370	0.005412	0.005412	0.007594	0.007594
0.005. . . .	.001577	.000768	.002711	.001659	.003842	.002621	.004975	.003609	.006112	.004626	.007209	.005668	.009496	.007820
.010. . . .	.002629	.000874	.004069	.001815	.005373	.002774	.006628	.003780	.007854	.004814	.009008	.005866	.01139	.008003
.015. . . .	.003677	.000924	.005417	.001893	.006904	.002887	.008281	.003912	.009596	.004958	.010808	.006015	.01328	.008148
.020. . . .	.004725	.000961	.006769	.001958	.008436	.002974	.009935	.004011	.01134	.005064	.01261	.006123	.01517	.008260
.025. . . .	.005773	.000991	.008119	.002005	.009967	.003037	.01159	.004083	.01308	.005140	.01441	.006200	.01706	.008350
.030. . . .	.006821	.001012	.009470	.002041	.01150	.003083	.01324	.004134	.01482	.005194	.01620	.006254	.01894	.008418
.040. . . .	.008916	.001038	.01217	.002085	.01456	.003140	.01654	.004203	.01830	.005272	.01979	.006342	.02270	.008506
.050. . . .	.011006	.001054	.01487	.002114	.01761	.003180	.01983	.004253	.02176	.005328	.02337	.006400	.02647	.008575
.060. . . .	.01309	.001065	.01755	.002134	.02066	.003208	.02312	.004286	.02523	.005367	.02695	.006444	.03023	.008619
.070. . . .	.01518	.001073	.02024	.002148	.02370	.003228	.02641	.004309	.02869	.005393	.03052	.006473	.03399	.008649
.080. . . .	.01726	.001078	.02292	.002159	.02674	.003241	.02969	.004324	.03214	.005410	.03408	.006493	.03773	.008672
.090. . . .	.01933	.001082	.02560	.002166	.02977	.003249	.03296	.004335	.03559	.005423	.03764	.006507	.04147	.008686
.100. . . .	.02141	.001084	.02827	.002169	.03280	.003255	.03623	.004342	.03903	.005431	.04119	.006516	.04521	.008696
.120. . . .	.02554	.001087	.03360	.002174	.03883	.003261	.04274	.004349	.04589	.005439	.04827	.006524	.05265	.008703
.140. . . .	.02965	.001087	.03890	.002174	.04483	.003261	.04922	.004349	.05271	.005437	.05531	.006522	.06005	.008696
.160. . . .	.03374	.001086	.04417	.002172	.05081	.003257	.05566	.004343	.05950	.005429	.06232	.006516	.06741	.008681
.180. . . .	.03781	.001084	.04941	.002169	.05675	.003251	.06207	.004334	.06625	.005418	.06928	.006500	.07473	.008663
.200. . . .	.04186	.001082	.05463	.002162	.06265	.003243	.06844	.004323	.07296	.005404	.07621	.006480	.08200	.008638
0.225. . . .	0.04687	0.001078	0.06110	0.002154	0.06997	0.003231	0.07634	0.004307	0.08128	0.005383	0.08479	0.006455	0.09098	0.008604

Δk^2	$k_0^2=0.025$		$k_0^2=0.030$		$k_0^2=0.035$		$k_0^2=0.040$		$k_0^2=0.045$		$k_0^2=0.050$		$k_0^2=0.055$	
	$(0,k_0^2\|\ddot{r}\|1,k_1^2)$	$(1,k_0^2\|\ddot{r}\|0,k_1^2)$	$(0,k_0^2\|\ddot{r}\|1,k_1^2)$	$(1,k_0^2\|\ddot{r}\|0,k_1^2)$	$(0,k_0^2\|\ddot{r}\|1,k_1^2)$	$(1,k_0^2\|\ddot{r}\|0,k_1^2)$	$(0,k_0^2\|\ddot{r}\|1,k_1^2)$	$(1,k_0^2\|\ddot{r}\|0,k_1^2)$	$(0,k_0^2\|\ddot{r}\|1,k_1^2)$	$(1,k_0^2\|\ddot{r}\|0,k_1^2)$	$(0,k_0^2\|\ddot{r}\|1,k_1^2)$	$(1,k_0^2\|\ddot{r}\|0,k_1^2)$	$(0,k_0^2\|\ddot{r}\|1,k_1^2)$	$(1,k_0^2\|\ddot{r}\|0,k_1^2)$
0.	0.009805	0.009805	0.01204	0.01204	0.01428	0.01428	0.01651	0.01651	0.01874	0.01874	0.02097	0.02097	0.02321	0.02321
0.005. . . .	.01176	.009989	.01404	.01221	.01630	.01444	.01856	.01666	.02081	.01888	.02306	.02110	.02531	.02333
.010. . . .	.01372	.01017	.01603	.01237	.01833	.01458	.02060	.01679	.02288	.01900	.02515	.02121	.02741	.02343
.015. . . .	.01567	.01030	.01802	.01251	.02035	.01469	.02265	.01690	.02495	.01911	.02724	.02131	.02951	.02351
.020. . . .	.01762	.01042	.02000	.01261	.02237	.01479	.02470	.01700	.02702	.01919	.02932	.02139	.03161	.02359
.025. . . .	.01956	.01050	.02199	.01268	.02440	.01487	.02675	.01708	.02909	.01926	.03141	.02145	.03371	.02364
.030. . . .	.02151	.01056	.02398	.01274	.02642	.01494	.02880	.01714	.03117	.01932	.03350	.02151	.03581	.02369
.040. . . .	.02540	.01066	.02796	.01284	.03047	.01503	.03290	.01723	.03530	.01941	.03767	.02159	.04000	.02377
.050. . . .	.02929	.01073	.03194	.01291	.03451	.01510	.03699	.01729	.03944	.01947	.04183	.02164	.04418	.02382
.060. . . .	.03317	.01077	.03591	.01295	.03854	.01514	.04107	.01734	.04356	.01951	.04598	.02168	.04836	.02385
.070. . . .	.03704	.01080	.03987	.01299	.04257	.01518	.04514	.01736	.04767	.01953	.05013	.02170	.05253	.02387
.080. . . .	.04091	.01083	.04382	.01301	.04659	.01520	.04920	.01738	.05178	.01955	.05426	.02172	.05668	.02389
.090. . . .	.04477	.01084	.04777	.01302	.05060	.01521	.05326	.01739	.05587	.01956	.05839	.02173	.06083	.02389
.100. . . .	.04862	.01085	.05171	.01303	.05460	.01522	.05731	.01740	.05996	.01956	.06251	.02172	.06497	.02388
.120. . . .	.05630	.01085	.05955	.01303	.06257	.01521	.06537	.01739	.06811	.01954	.07071	.02170	.07321	.02384
.140. . . .	.06394	.01084	.06736	.01302	.07050	.01519	.07339	.01736	.07621	.01951	.07887	.02166	.08142	.02380
.160. . . .	.07153	.01082	.07512	.01299	.07839	.01516	.08137	.01732	.08427	.01947	.08699	.02161	.08957	.02374
.180. . . .	.07909	.01080	.08284	.01296	.08623	.01512	.08930	.01728	.09227	.01941	.09504	.02155	.09767	.02367
.200. . . .	.08659	.01077	.09050	.01292	.09402	.01507	.09718	.01722	.10022	.01935	.10305	.02148	.10571	.02360
0.225. . . .	0.09590	0.01073	0.10001	0.01287	0.10367	0.01502	0.10694	0.01716	0.11008	0.01928	0.11299	0.02139	0.11570	0.02350

TABLE 4—*Continued*

Δk^2	$k_0^2=0.06$		$k_0^2=0.07$		$k_0^2=0.08$		$k_0^2=0.09$		$k_0^2=0.100$		$k_0^2=0.125$		$k_0^2=0.150$	
	$(0,k_0^2\|\ddot{r}\|1,k_1^2)$	$(1,k_0^2\|\ddot{r}\|0,k_1^2)$	$(0,k_0^2\|\ddot{r}\|1,k_1^2)$	$(1,k_0^2\|\ddot{r}\|0,k_1^2)$	$(0,k_0^2\|\ddot{r}\|1,k_1^2)$	$(1,k_0^2\|\ddot{r}\|0,k_1^2)$	$(0,k_0^2\|\ddot{r}\|1,k_1^2)$	$(1,k_0^2\|\ddot{r}\|0,k_1^2)$	$(0,k_0^2\|\ddot{r}\|1,k_1^2)$	$(1,k_0^2\|\ddot{r}\|0,k_1^2)$	$(0,k_0^2\|\ddot{r}\|1,k_1^2)$	$(1,k_0^2\|\ddot{r}\|0,k_1^2)$	$(0,k_0^2\|\ddot{r}\|1,k_1^2)$	$(1,k_0^2\|\ddot{r}\|0,k_1^2)$
0.....	0.02544	0.02544	0.02990	0.02990	0.03435	0.03435	0.03876	0.03876	0.04317	0.04317	0.05408	0.05408	0.06485	0.06485
0.005....	.02755	.02555	.03203	.02998	.03648	.03440	.04091	.03881	.04531	.04320	.05622	.05411	.06699	.06484
.010....	.02966	.02564	.03416	.03005	.03862	.03446	.04305	.03885	.04746	.04323	.05837	.05413	.06912	.06483
.015....	.03177	.02572	.03628	.03012	.04076	.03451	.04519	.03889	.04960	.04326	.06051	.05415	.07126	.06482
.020....	.03388	.02578	.03841	.03018	.04289	.03457	.04733	.03893	.05174	.04329	.06265	.05416	.07339	.06480
.025....	.03599	.02583	.04053	.03022	.04502	.03460	.04947	.03896	.05388	.04332	.06478	.05417	.07552	.06478
.030....	.03810	.02588	.04265	.03026	.04715	.03463	.05160	.03899	.05601	.04334	.06692	.05416	.07766	.06476
.040....	.04230	.02595	.04689	.03031	.05140	.03467	.05586	.03903	.06028	.04338	.07118	.05411	.08190	.06472
.050....	.04650	.02599	.05111	.03035	.05564	.03470	.06011	.03905	.06453	.04339	.07543	.05406	.08614	.06467
.060....	.05069	.02602	.05533	.03037	.05988	.03471	.06435	.03905	.06878	.04336	.07968	.05400	.09036	.06460
.070....	.05488	.02604	.05954	.03038	.06410	.03471	.06859	.03902	.07301	.04332	.08391	.05396	.09457	.06453
.080....	.05905	.02605	.06374	.03038	.06831	.03469	.07281	.03899	.07724	.04327	.08812	.05392	.09877	.06445
.090....	.06321	.02605	.06793	.03036	.07252	.03467	.07703	.03896	.08146	.04323	.09233	.05386	.1030	.06437
.100....	.06736	.02603	.07211	.03034	.07672	.03464	.08123	.03894	.08566	.04321	.09653	.05379	.1071	.06428
.120....	.07564	.02599	.08044	.03029	.08507	.03458	.08959	.03884	.09402	.04310	.1049	.05365	.1155	.06409
.140....	.08387	.02594	.08872	.03022	.09337	.03448	.09791	.03874	.1023	.04298	.1132	.05349	.1238	.06389
.160....	.09205	.02587	.09694	.03014	.10163	.03439	.1062	.03863	.1106	.04285	.1215	.05332	.1320	.06368
.180....	.10017	.02580	.1051	.03005	.1098	.03429	.1144	.03851	.1189	.04271	.1297	.05314	.1402	.06345
.200....	.10825	.02572	.1132	.02996	.1180	.03418	.1226	.03838	.1270	.04256	.1379	.05295	.1483	.06321
.225....	.11827	.02561	0.1233	0.02984	.1281	.03405	0.1327	0.03821	.1372	.04237	.1480	.05270	.1584	.06291
.250....	.12821	.02549	....	....	0.1381	0.03387	....	....	.1472	.04216	.1580	.05244	.1684	.06261
0.300....	0.14786	0.02525	....	....	....	....	....	....	0.1671	0.04174	0.1778	0.05192	0.1881	0.06198

Δk^2	$k_0^2=0.175$		$k_0^2=0.20$		$k_0^2=0.25$		$k_0^2=0.35$		$k_0^2=0.45$		$k_0^2=0.50$	
	$(0,k_0^2\|\ddot{r}\|1,k_1^2)$	$(1,k_0^2\|\ddot{r}\|0,k_1^2)$	$(0,k_0^2\|\ddot{r}\|1,k_1^2)$	$(1,k_0^2\|\ddot{r}\|0,k_1^2)$	$(0,k_0^2\|\ddot{r}\|1,k_1^2)$	$(1,k_0^2\|\ddot{r}\|0,k_1^2)$	$(0,k_0^2\|\ddot{r}\|1,k_1^2)$	$(1,k_0^2\|\ddot{r}\|0,k_1^2)$	$(0,k_0^2\|\ddot{r}\|1,k_1^2)$	$(1,k_0^2\|\ddot{r}\|0,k_1^2)$	$(0,k_0^2\|\ddot{r}\|1,k_1^2)$	$(1,k_0^2\|\ddot{r}\|0,k_1^2)$
0.....	0.07546	0.07546	0.08594	0.08593	0.1064	0.1064	0.1457	0.1457	0.1829	0.1829	0.2008	0.2008
0.005....	.07759	.07544	.08805	.08589	.1085	.1064	.1478	.1456	.1849	.1828	.2027	.2006
.010....	.07971	.07541	.09016	.08586	.1106	.1063	.1498	.1455	.1868	.1826	.2046	.2004
.015....	.08183	.07539	.09227	.08582	.1127	.1062	.1518	.1454	.1888	.1825	.2065	.2003
.020....	.08395	.07536	.09438	.08577	.1148	.1062	.1538	.1453	.1907	.1824	.2084	.2001
.025....	.08607	.07533	.09648	.08573	.1168	.1061	.1558	.1452	.1926	.1822	.2103	.2000
.030....	.08818	.07529	.09858	.08568	.1189	.1060	.1578	.1451	.1946	.1821	.2122	.1998
.040....	.09240	.07522	.1028	.08558	.1230	.1059	.1618	.1449	.1984	.1818	.2160	.1995
.050....	.09661	.07514	.1070	.08547	.1272	.1057	.1658	.1446	.2023	.1815	.2198	.1992
.060....	.10081	.07504	.1111	.08536	.1313	.1056	.1698	.1444	.2061	.1812	.2236	.1989
.070....	.1050	.07495	.1153	.08524	.1354	.1054	.1737	.1442	.2099	.1809	.2273	.1986
.080....	.1092	.07485	.1195	.08512	.1395	.1052	.1777	.1439	.2138	.1806	.2311	.1982
.090....	.1134	.07474	.1236	.08499	.1436	.1051	.1816	.1437	.2176	.1803	.2348	.1979
.100....	.1175	.07463	.1277	.08486	.1476	.1049	.1856	.1434	.2214	.1800	.2385	.1976
.120....	.1258	.07440	.1360	.08458	.1557	.1045	.1934	.1430	.2289	.1794	.2460	.1969
.140....	.1340	.07415	.1441	.08429	.1638	.1042	.2011	.1425	.2364	.1788	.2533	.1963
.160....	.1422	.07389	.1523	.08399	.1718	.1038	.2089	.1420	.2439	.1782	.2607	.1956
.180....	.1503	.07362	.1603	.08369	.1797	.1034	.2166	.1415	.2513	.1776	.2680	.1949
.200....	.1584	.07335	.1684	.08338	.1876	.1030	.2242	.1410	.2587	.1769	.2752	.1942
.225....	.1684	.07301	.1783	.08299	.1974	.1026	.2336	.1403	.2678	.1761	.2842	.1934
.250....	0.1784	0.07266	.1882	.08259	.2071	.1021	.2430	.1396	.2769	.1754	.2931	.1926
0.300....	....	....	0.2076	0.08177	0.2262	0.1011	0.2616	0.1383	0.2948	0.1738	0.3107	0.1909

TABLE 5*

THE CONTINUOUS ABSORPTION COEFFICIENT OF THE NEGATIVE HYDROGEN ION PER NEUTRAL HYDROGEN ATOM AND PER UNIT ELECTRON PRESSURE FOR VARIOUS TEMPERATURES ($\theta=5040/T$) AND WAVE LENGTHS

Δk^2	λA	$\theta=0.5$	$\theta=0.6$	$\theta=0.7$	$\theta=0.8$	$\theta=0.9$	$\theta=1.0$	$\theta=1.2$	$\theta=1.4$	$\theta=1.6$	$\theta=1.8$	$\theta=2.0$
†......	....	1.86 (−30)	2.03 (−30)	2.17 (−30)	2.29 (−30)	2.40(−30)	2.49(−30)	2.66(−30)	2.79 (−30)	2.91(−30)	3.01(−30)	3.09(−30)
0.005....	182300	1.55 (−23)	1.70 (−23)	1.84 (−23)	1.95 (−23)	2.06(−23)	2.16(−23)	2.34(−23)	2.50 (−23)	2.64(−23)	2.77(−23)	2.89(−23)
.010....	91130	2.02 (−24)	2.24 (−24)	2.44 (−24)	2.61 (−24)	2.78(−24)	2.94(−24)	3.23(−24)	3.50 (−24)	3.76(−24)	4.00(−24)	4.23(−24)
.015....	60750	6.27 (−25)	7.00 (−25)	7.67 (−25)	8.30 (−25)	8.89(−25)	9.48(−25)	1.06(−24)	1.16 (−24)	1.27(−24)	1.37(−24)	1.46(−24)
.020....	45560	2.77 (−25)	3.12 (−25)	3.45 (−25)	3.76 (−25)	4.06(−25)	4.36(−25)	4.93(−25)	5.49 (−25)	6.05(−25)	6.60(−25)	7.14(−25)
.025....	36450	1.49 (−25)	1.69 (−25)	1.88 (−25)	2.06 (−25)	2.25(−25)	2.43(−25)	2.78(−25)	3.14 (−25)	3.49(−25)	3.84(−25)	4.20(−25)
.030....	30380	9.00 (−26)	1.03 (−25)	1.16 (−25)	1.28 (−25)	1.40(−25)	1.52(−25)	1.77(−25)	2.01 (−25)	2.26(−25)	2.51(−25)	2.77(−25)
.040....	22780	4.17 (−26)	4.84 (−26)	5.51 (−26)	6.18 (−26)	6.86(−26)	7.54(−26)	8.92(−26)	1.03 (−25)	1.18(−25)	1.33(−25)	1.48(−25)
.050....	18230	2.34 (−26)	2.76 (−26)	3.18 (−26)	3.60 (−26)	4.03(−26)	4.48(−26)	5.38(−26)	6.32 (−26)	7.29(−26)	8.29(−26)	9.31(−26)
.060....	15190	1.48 (−26)	1.77 (−26)	2.06 (−26)	2.36 (−26)	2.66(−26)	2.98(−26)	3.63(−26)	4.30 (−26)	5.01(−26)	5.74(−26)	6.48(−26)
		0.117(−26)	0.219(−26)	0.382(−26)	0.634(−26)	1.01(−26)	1.56(−26)	3.48(−26)	0.723(−25)	1.43(−25)	2.70(−25)	4.97(−25)
.070....	13020	1.02 (−26)	1.23 (−26)	1.44 (−26)	1.66 (−26)	1.89(−26)	2.13(−26)	2.63(−26)	3.14 (−26)	3.69(−26)	4.25(−26)	4.82(−26)
		0.398(−26)	0.746(−26)	1.30 (−26)	2.16 (−26)	3.45(−26)	5.33(−26)	11.9 (−26)	2.47 (−25)	4.87(−25)	9.22(−25)	17.0 (−25)
.080....	11390	7.41 (−27)	9.02 (−27)	1.07 (−26)	1.24 (−26)	1.42(−26)	1.61(−26)	2.00(−26)	2.41 (−26)	2.85(−26)	3.29(−26)	3.76(−26)
		0.603(−26)	1.13 (−26)	1.97 (−26)	3.28 (−26)	5.23(−26)	8.08(−26)	18.0 (−26)	3.74 (−25)	7.37(−25)	14.0 (−25)	25.7 (−25)
.090....	10130	5.64 (−27)	6.93 (−27)	8.27 (−27)	9.68 (−27)	1.11(−26)	1.27(−26)	1.59(−26)	1.92 (−26)	2.28(−26)	2.65(−26)	3.03(−26)
		0.719(−26)	1.35 (−26)	2.36 (−26)	3.91 (−26)	6.24(−26)	9.65(−26)	21.5 (−26)	4.46 (−25)	8.80(−25)	16.7 (−25)	30.7 (−25)
.100....	9113	4.45 (−27)	5.50 (−27)	6.61 (−27)	7.78 (−27)	8.99(−27)	1.03(−26)	1.29(−26)	1.58 (−26)	1.87(−26)	2.18(−26)	2.51(−26)
		0.774(−26)	1.45 (−26)	2.53 (−26)	4.20 (−26)	6.71(−26)	10.4 (−26)	23.1 (−26)	4.80 (−25)	9.46(−25)	17.9 (−25)	33.0 (−25)
.120....	7594	2.98 (−27)	3.73 (−27)	4.53 (−27)	5.38 (−27)	6.27(−27)	7.20(−27)	9.16(−27)	1.13 (−26)	1.35(−26)	1.58(−26)	1.82(−26)
		0.778(−26)	1.46 (−26)	2.55 (−26)	4.23 (−26)	6.75(−26)	10.4 (−26)	23.3 (−26)	4.83 (−25)	9.52(−25)	18.1 (−25)	33.2 (−25)
.140....	6509	2.15 (−27)	2.72 (−27)	3.33 (−27)	3.98 (−27)	4.66(−27)	5.38(−27)	6.89(−27)	8.52 (−27)	1.02(−26)	1.20(−26)	1.39(−26)
		0.725(−26)	1.36 (−26)	2.37 (−26)	3.94 (−26)	6.28(−26)	9.72(−26)	21.6 (−26)	4.49 (−25)	8.86(−25)	16.8 (−25)	30.9 (−25)
.160....	5695	1.63 (−27)	2.08 (−27)	2.57 (−27)	3.08 (−27)	3.63(−27)	4.20(−27)	5.42(−27)	6.72 (−27)	8.10(−27)	9.55(−27)	1.11(−26)
		0.657(−26)	1.23 (−26)	2.15 (−26)	3.57 (−26)	5.70(−26)	8.81(−26)	19.6 (−26)	4.08 (−25)	8.04(−25)	15.2 (−25)	28.0 (−25)
.180....	5063	1.29 (−27)	1.65 (−27)	2.05 (−27)	2.47 (−27)	2.92(−27)	3.39(−27)	4.39(−27)	5.47 (−27)	6.61(−27)	7.81(−27)	9.06(−27)
		0.591(−26)	1.11 (−26)	1.94 (−26)	3.21 (−26)	5.13(−26)	7.93(−26)	17.7 (−26)	3.67 (−25)	7.23(−25)	13.7 (−25)	25.2 (−25)
.200....	4556	1.04 (−27)	1.35 (−27)	1.68 (−27)	2.03 (−27)	2.41(−27)	2.81(−27)	3.65(−27)	4.56 (−27)	5.52(−27)	6.53(−27)	7.59(−27)
		0.532(−26)	0.997(−26)	1.74 (−26)	2.89 (−26)	4.61(−26)	7.13(−26)	15.9 (−26)	3.30 (−25)	6.51(−25)	12.3 (−25)	22.7 (−25)
0.225....	4050	8.30 (−28)	1.08 (−27)	1.35 (−27)	1.64 (−27)	1.95(−27)	2.28(−27)	2.97(−27)	3.72 (−27)	4.52(−27)	5.60(−27)	6.23 (−**27**)
		0.468(−26)	0.877(−26)	1.53 (−26)	2.54 (−26)	4.06(−26)	6.28(−26)	14.0 (−26)	2.90 (−25)	5.73(−25)	10.9 (−25)	20.0 (−25)

* The upper entries give the contributions to the absorption coefficients by the free-free transitions. The lower entries (for $\lambda > 16{,}500$ A) give the corresponding contributions due to the bound-free transitions derived from the cross-sections given in *Ap. J.*, **102**, 395, 1945. The numbers in parentheses are the powers of 10, by which the corresponding entries should be multiplied to get the absorption coefficients in the unit CM^4/dyne.

† The entries in this line when divided by $(\Delta k^2)^3$ will give the corresponding absorption coefficients to sufficient accuracy for all $\Delta k^2 < 0.005$.

TABLE 6*

THE CONTINUOUS ABSORPTION COEFFICIENT OF THE NEGATIVE HYDROGEN ION PER NEUTRAL HYDROGEN ATOM AND PER UNIT ELECTRON PRESSURE FOR VARIOUS TEMPERATURES AND WAVE LENGTHS AFTER ALLOWING FOR THE STIMULATED EMISSION FACTOR $(1-e^{-h\nu/kT})$

Δk^2	λA	$\theta=0.5$	$\theta=0.6$	$\theta=0.7$	$\theta=0.8$	$\theta=0.9$	$\theta=1.0$	$\theta=1.2$	$\theta=1.4$	$\theta=1.6$	$\theta=1.8$	$\theta=2.0$
†......	....	2.92 (−29)	3.81 (−29)	4.76 (−29)	5.74 (−29)	6.76 (−29)	7.81(−29)	9.98(−29)	1.22 (−28)	1.46(−28)	1.70(−28)	1.93(−28)
0.005...	182300	1.17 (−24)	1.53 (−24)	1.91 (−24)	2.30 (−24)	2.71 (−24)	3.13(−24)	4.01(−24)	4.91 (−24)	5.85(−24)	6.80(−24)	7.77(−24)
.010...	91130	2.93 (−25)	3.84 (−25)	4.80 (−25)	5.79 (−25)	6.83 (−25)	7.90(−25)	1.01(−24)	1.24 (−24)	1.48(−24)	1.72(−24)	1.97(−24)
.015...	60750	1.31 (−25)	1.72 (−25)	2.15 (−25)	2.60 (−25)	3.07 (−25)	3.55(−25)	4.56(−25)	5.61 (−25)	6.69(−25)	7.79(−25)	8.91(−25)
.020...	45560	7.45 (−26)	9.77 (−26)	1.22 (−25)	1.48 (−25)	1.75 (−25)	2.03(−25)	2.61(−25)	3.21 (−25)	3.83(−25)	4.46(−25)	5.10(−25)
.025...	36450	4.81 (−26)	6.32 (−26)	7.93 (−26)	9.61 (−26)	1.14 (−25)	1.32(−25)	1.69(−25)	2.09 (−25)	2.49(−25)	2.90(−25)	3.32(−25)
.030...	30380	3.38 (−26)	4.44 (−26)	5.57 (−26)	6.77 (−26)	8.01 (−26)	9.29(−26)	1.20(−25)	1.47 (−25)	1.76(−25)	2.05(−25)	2.34(−25)
.040...	22780	1.94 (−26)	2.56 (−26)	3.22 (−26)	3.91 (−26)	4.64 (−26)	5.38(−26)	6.94(−26)	8.54 (−26)	1.02(−25)	1.19(−25)	1.36(−25)
.050...	18230	1.27 (−26)	1.68 (−26)	2.11 (−26)	2.57 (−26)	3.05 (−26)	3.54(−26)	4.56(−26)	5.62 (−26)	6.70(−26)	7.79(−26)	8.91(−26)
.060...	15190	{9.02 (−27)	1.19 (−26)	1.50 (−26)	1.83 (−26)	2.17 (−26)	2.52(−26)	3.25(−26)	3.99 (−26)	4.76(−26)	5.54(−26)	6.33(−26)
		0.0710(−26)	0.148(−26)	0.279(−26)	0.493(−26)	0.825(−26)	1.32(−26)	3.12(−26)	0.671(−25)	1.36(−25)	2.61(−25)	4.86(−25)
.070...	13020	{6.77 (−27)	8.97 (−27)	1.13 (−26)	1.38 (−26)	1.63 (−26)	1.89(−26)	2.44(−26)	3.00 (−26)	3.58(−26)	4.16(−26)	4.76(−26)
		0.265 (−26)	0.545(−26)	1.02 (−26)	1.79 (−26)	2.97 (−26)	4.74(−26)	11.0 (−26)	2.35 (−25)	4.72(−25)	9.05(−25)	16.7 (−25)
.080...	11390	{5.29 (−27)	7.02 (−27)	8.84 (−27)	1.08 (−26)	1.27 (−26)	1.48(−26)	1.90(−26)	2.34 (−26)	2.79(−26)	3.26(−26)	3.73(−26)
		0.431 (−26)	0.879(−26)	1.63 (−26)	2.83 (−26)	4.68 (−26)	7.42(−26)	17.1 (−26)	3.63 (−25)	7.24(−25)	13.8 (−25)	25.5 (−25)
.090...	10130	{4.27 (−27)	5.65 (−27)	7.12 (−27)	8.66 (−27)	1.03 (−26)	1.19(−26)	1.53(−26)	1.89 (−26)	2.25(−26)	2.63(−26)	3.02(−26)
		0.544 (−26)	1.10 (−26)	2.03 (−26)	3.50 (−26)	5.74 (−26)	9.07(−26)	20.8 (−26)	4.38 (−25)	8.70(−25)	16.6 (−25)	30.6 (−25)
.100...	9113	{3.52 (−27)	4.66 (−27)	5.87 (−27)	7.14 (−27)	8.46 (−27)	9.81(−27)	1.26(−26)	1.56 (−26)	1.86(−26)	2.18(−26)	2.50(−26)
		0.612 (−26)	1.23 (−26)	2.25 (−26)	3.86 (−26)	6.31 (−26)	9.92(−26)	22.6 (−26)	4.74 (−25)	9.40(−25)	17.9 (−25)	32.9 (−25)
.120...	7594	{2.52 (−27)	3.34 (−27)	4.21 (−27)	5.11 (−27)	6.06 (−27)	7.03(−27)	9.06(−27)	1.12 (−26)	1.34(−26)	1.57(−26)	1.81(−26)
		0.659 (−26)	1.31 (−26)	2.37 (−26)	4.02 (−26)	6.52 (−26)	10.2 (−26)	23.0 (−26)	4.80 (−25)	9.50(−25)	18.0 (−25)	33.2 (−25)
.140...	6509	{1.91 (−27)	2.52 (−27)	3.18 (−27)	3.86 (−27)	4.57 (−27)	5.31(−27)	6.86(−27)	8.50 (−27)	1.02(−26)	1.20(−26)	1.39(−26)
		0.644 (−26)	1.26 (−26)	2.26 (−26)	3.82 (−26)	6.16 (−26)	9.60(−26)	21.5 (−26)	4.49 (−25)	8.86(−25)	16.8 (−25)	30.9 (−25)
.160...	5695	{1.50 (−27)	1.98 (−27)	2.49 (−27)	3.03 (−27)	3.59 (−27)	4.17(−27)	5.40(−27)	6.72 (−27)	8.10(−27)	9.55(−27)	1.11(−26)
		0.603 (−26)	1.17 (−26)	2.09 (−26)	3.51 (−26)	5.63 (−26)	8.75(−26)	19.6 (−26)	4.07 (−25)	8.04(−25)	15.2 (−25)	28.0 (−25)
.180...	5063	{1.21 (−27)	1.60 (−27)	2.01 (−27)	2.44 (−27)	2.90 (−27)	3.38(−27)	4.39(−27)	5.47 (−27)	6.61(−27)	7.81(−27)	9.06(−27)
		0.556 (−26)	1.07 (−26)	1.90 (−26)	3.18 (−26)	5.09 (−26)	7.90(−26)	17.6 (−26)	3.67 (−25)	7.23(−25)	13.7 (−25)	25.2 (−25)
.200...	4556	{9.98 (−28)	1.32 (−27)	1.66 (−27)	2.02 (−27)	2.40 (−27)	2.80(−27)	3.65(−27)	4.56 (−27)	5.52(−27)	6.53(−27)	7.59(−27)
		0.509 (−26)	0.974(−26)	1.72 (−26)	2.87 (−26)	4.60 (−26)	7.12(−26)	15.9 (−26)	3.30 (−25)	6.51(−25)	12.3 (−25)	22.7 (−25)
0.225...	4050	{8.05 (−28)	1.06 (−27)	1.34 (−27)	1.64 (−27)	1.95 (−27)	2.27(−27)	2.97(−27)	3.72 (−27)	4.52(−27)	5.60(−27)	6.23(−27)
		0.454 (−26)	0.865(−26)	1.52 (−26)	2.53 (−26)	4.05 (−26)	6.27(−26)	14.0 (−26)	2.90 (−25)	5.73(−25)	10.9 (−25)	20.0 (−25)

* The arrangement of this table is the same as Table 5.

† The entries in this line when divided by $(\Delta k^2)^2$ will give the corresponding absorption coefficients to a sufficient accuracy for all $\Delta k^2<0.005$.

TABLE 7*

THE CONTINUOUS ABSORPTION COEFFICIENT OF THE NEGATIVE HYDROGEN ION DUE TO THE FREE-FREE AND THE BOUND-FREE TRANSITIONS FOR VARIOUS TEMPERATURES AND WAVE LENGTHS AFTER ALLOWING FOR THE STIMULATED EMISSION FACTOR $(1-e^{-h\nu/kT})$

Δk^2	λA	θ=0.5	θ=0.6	θ=0.7	θ=0.8	θ=0.9	θ=1.0	θ=1.2	θ=1.4	θ=1.6	θ=1.8	θ=2.0
†		3.513 (−29)	4.593 (−29)	5.730(−29)	6.915(−29)	8.140(−29)	9.404(−29)	1.202(−28)	1.474 (−28)	1.754 (−28)	2.042 (−28)	2.330 (−28)
0.005	182300	1.168 (−24)	1.528 (−24)	1.906(−24)	2.301(−24)	2.710(−24)	3.132(−24)	4.006(−24)	4.913 (−24)	5.846 (−24)	6.795 (−24)	7.765 (−24)
.010	91130	2.934 (−25)	3.840 (−25)	4.796(−25)	5.794(−25)	6.829(−25)	7.897(−25)	1.012(−24)	1.243 (−24)	1.481 (−24)	1.724 (−24)	1.971 (−24)
.015	60750	1.313 (−25)	1.720 (−25)	2.151(−25)	2.602(−25)	3.066(−25)	3.554(−25)	4.560(−25)	5.610 (−25)	6.688 (−25)	7.793 (−25)	8.907 (−25)
.020	45560	7.447 (−26)	9.771 (−26)	1.223(−25)	1.482(−25)	1.750(−25)	2.028(−25)	2.606(−25)	3.209 (−25)	3.829 (−25)	4.463 (−25)	5.102 (−25)
.025	36450	4.811 (−26)	6.320 (−26)	7.925(−26)	9.607(−26)	1.136(−25)	1.317(−25)	1.695(−25)	2.088 (−25)	2.493 (−25)	2.904 (−25)	3.321 (−25)
.030	30380	3.375 (−26)	4.440 (−26)	5.575(−26)	6.765(−26)	8.006(−26)	9.289(−26)	1.196(−25)	1.474 (−25)	1.759 (−25)	2.049 (−25)	2.343 (−25)
.040	22780	1.941 (−26)	2.560 (−26)	3.220(−26)	3.913(−26)	4.636(−26)	5.383(−26)	6.935(−26)	0.8544(−25)	1.019 (−25)	1.186 (−25)	1.356 (−25)
.050	18230	1 270 (−26)	1.679 (−26)	2.115(−26)	2.572(−26)	3.049(−26)	3.541(−26)	4.561(−26)	0 5617(−25)	0.6695(−25)	0.7792(−25)	0.8901(−25)
.060	15190	0.9733(−26)	1.342 (−26)	1.784(−26)	2.324(−26)	2.996(−26)	3.846(−26)	6.363(−26)	1.071 (−25)	1.832 (−25)	3.167 (−25)	5.488 (−25)
.070	13020	0.9423(−26)	1.443 (−26)	2.153(−26)	3.164(−26)	4.600(−26)	6.631(−26)	13.46 (−26)	2.652 (−25)	5.077 (−25)	9.462 (−25)	17.22 (−25)
.080	11390	0.9599(−26)	1.580 (−26)	2.517(−26)	3.910(−26)	5.953(−26)	8.902(−26)	19.02 (−26)	3.861 (−25)	7.519 (−25)	14.15 (−25)	25.90 (−25)
.090	10130	0.9701(−26)	1.665 (−26)	2.741(−26)	4.366(−26)	6.770(−26)	10.261(−26)	22.29 (−26)	4.565 (−25)	8.928 (−25)	16.84 (−25)	30.86 (−25)
.100	9113	0.9638(−26)	1.695 (−26)	2.838(−26)	4.575(−26)	7.152(−26)	10.901(−26)	23.83 (−26)	4.894 (−25)	9.586 (−25)	18.10 (−25)	33.16 (−25)
.120	7594	0.9118(−26)	1.640 (−26)	2.786(−26)	4.532(−26)	7.124(−26)	10.895(−26)	23.90 (−26)	4.915 (−25)	9.632 (−25)	18.19 (−25)	33.34 (−25)
.140	6509	0.8344(−26)	1.513 (−26)	2.581(−26)	4.206(−26)	6.618(−26)	10.126(−26)	22.22 (−26)	4.570 (−25)	8.959 (−25)	16.92 (−25)	31.02 (−25)
.160	5695	0.7531(−26)	1.369 (−26)	2.337(−26)	3.809(−26)	5.993(−26)	9.169(−26)	20.12 (−26)	4.140 (−25)	8.117 (−25)	15.33 (−25)	28.12 (−25)
.180	5063	0.6768(−26)	1.230 (−26)	2.100(−26)	3.423(−26)	5.385(−26)	8.238(−26)	18.08 (−26)	3.721 (−25)	7.298 (−25)	13.79 (−25)	25.30 (−25)
.200	4556	0.6086(−26)	1.1058(−26)	1.887(−26)	3.074(−26)	4.837(−26)	7.401(−26)	16.25 (−26)	3.346 (−25)	6.564 (−25)	12.41 (−25)	22.76 (−25)
0.225	4050	0.5347(−26)	0.9710(−26)	1.656(−26)	2.698(−26)	4.246(−26)	6.499(−26)	14.28 (−26)	2.941 (−25)	5.772 (−25)	10.91 (−25)	20.02 (−25)

* The coefficients are per neutral hydrogen atom and per unit electron pressure. The numbers in parentheses give the powers of 10 by which the corresponding entries should be multiplied to get the coefficients in the unit CM⁴/dyne.

† The entries in this line when multiplied by $(\lambda\ A/1000)^2$ will give the absorption coefficients for the various θ-values for all wave lengths $\lambda > 180{,}000$ A.

6. *Concluding remarks.*—It is not our intention here to make detailed applications of our new absorption coefficients to various astrophysical problems. But one comparison is of interest. In Figure 4 we have compared the theoretical variation of the continuous absorption coefficient of H^- with wave length for $T = 6300°$ with that derived by Chalonge and Kourganoff in their recent discussion of the continuous spectrum of the sun for the same temperature. It is seen that the agreement is very satisfactory. In any event, there is no basis for the conclusion that the free-free transitions of H^- are not

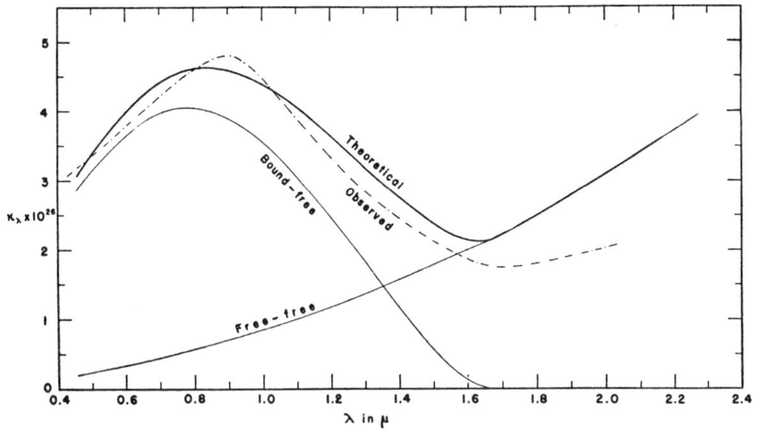

FIG. 4.—A comparison of the theoretical variation of the continuous absorption coefficient with wave length of H^- for $T = 6300°$, with the corresponding variation derived empirically from the solar continuous spectrum by Chalonge and Kourganoff.

adequate to account for the observed amount of absorption in the red beyond λ 10,000 A. Indeed, it would appear that the negative ion of hydrogen by itself is able to account quantitatively for the entire continuous spectrum of the sun over the range of wave lengths λλ 4000 A to 25,000 A.

On the physical side we should, however, emphasize that, while our new coefficients for the free-free transitions of H^- are probably to be trusted generally, the importance of exchange and polarization for the slow *s*-electrons may lead to further changes. It is likely that these effects will have a tendency to reduce our coefficients somewhat. We hope to return to these questions in the near future.

ON THE CONTINUOUS ABSORPTION COEFFICIENT OF THE NEGATIVE HYDROGEN ION. IV

S. CHANDRASEKHAR
Yerkes Observatory
Received April 24, 1958

ABSTRACT

The photoionization cross-sections of the negative ion of hydrogen are revised by making use of Hart and Herzberg's 20-parameter wave function for the ground state. The improvements over the earlier calculations based on Henrich's 11-parameter wave function are not very great.

I. INTRODUCTION AND RESULTS

During the early and the middle 1940's the writer and his associates (Williamson 1942; Chandrasekhar and Krogdahl 1943; Henrich 1944*a*, *b;* Chandrasekhar 1944*a*, *b*, 1945*a*, *b*, 1947; Chandrasekhar and Breen 1946; for brief accounts of these investigations from the point of view of atomic physics see Massey 1950, pp. 36–39, and Bethe and Salpeter 1957, pp. 151, 155, 252, 316–318) spent considerable time and effort on the determination of the continuous absorption coefficient of the negative hydrogen ion. The results of these calculations have since become incorporated in the developments in the theory of stellar atmospheres which have taken place in the intervening decade (see Unsöld 1955, p. 165, and Sec. 48, pp. 174–176); and they have further received strong confirmation in laboratory experiments (Lochte-Holtegreven 1951; Smith and Branscomb 1955*a*, *b*).

A re-examination of the photoionization cross-sections for H^- (Chandrasekhar 1945*a*, *b;* these papers will be referred to hereafter as "Paper I" and "Paper II," respectively) which are currently in use was suggested by the recent advances relative to the ground-state energies and wave functions of two-electron systems. These advances followed on the realization that, to attain spectroscopic precision in the theoretically predicted values for the ground-state energies, one must include a very large number of parameters in the basic variational calculations. This was a reversal of the views previously held; and it came about when it was pointed out (Chandrasekhar, Elbert, and Herzberg 1953) that the results of Hylleraas' calculations on an 8-parameter wave function for the ground state of helium (on the basis of which the earlier contrary view had been held) were unfortunately vitiated by an error. Once this was realized, results of calculations with increasing numbers of variational parameters quickly followed: Chandrasekhar, Elbert, and Herzberg (1953) with 10 parameters; Chandrasekhar and Herzberg (1955) with 18 parameters; Hart and Herzberg (1957) with 20 parameters; Hylleraas and Midtal (1956) with 24 parameters; and, finally, Kinoshita (1957) with 39 parameters.[1] The bearing of these advances on the problem of the photoionization cross-sections of H^- is the following:

The sensitiveness of the computed cross-sections on the accuracy of variationally determined wave functions for the ground state of H^- is well known (Chandrasekhar

[1] It was supposed for some time that Hylleraas and Midtal (1956), by their judicious selection of their 24 parameters, had actually succeeded in obtaining an energy value lower than even Kinoshita with 39 parameters. However, by repeating the calculations with the same 24 parameters, Herzberg discovered that Hylleraas and Midtal's calculation was in error; that the correct values are only very slightly different from what Hart and Herzberg had derived from a 20-parameter wave function; and that Kinoshita's value is, indeed, the best. These facts have since been confirmed by Hylleraas and Midtal (1958).

1944*a*, *b;* in this connection see also Geltman 1956). This had been taken into account in the earlier calculations by evaluating the required cross-sections not only in terms of the matrix elements of the dipole length but also in terms of the matrix elements of the dipole velocity and dipole acceleration (which weight different parts of the wave function differently) and comparing them. Nevertheless, in view of the central role which the continuous absorption coefficient of H^- plays in the theory of stellar atmospheres, it appeared worthwhile to repeat the 1945 calculations with the most accurate wave function now available for the ground state of H^-, namely, Hart and Herzberg's 20-parameter wave function. The results of the present calculations are given in Table 1 and illustrated in Figure 1; comparisons with the results of the earlier calculations are made in Figures 2 and 3.

TABLE 1

CONTINUOUS ABSORPTION COEFFICIENT OF NEGATIVE HYDROGEN ION COMPUTED ACCORDING TO DIPOLE-LENGTH AND DIPOLE-VELOCITY FORMULAE USING HART AND HERZBERG'S 20-PARAMETER WAVE FUNCTION FOR GROUND STATE

λ (A)	$\kappa_\lambda \times 10^{17}$ CM² FROM Dipole-Length Formula	$\kappa_\lambda \times 10^{17}$ CM² FROM Dipole-Velocity Formula	λ (A)	$\kappa_\lambda \times 10^{17}$ CM² FROM Dipole-Length Formula	$\kappa_\lambda \times 10^{17}$ CM² FROM Dipole-Velocity Formula
0	0	0	9000	3.421	4.263
500	0.07522	0.06315	9500	3.170	4.138
1000	0.2945	0.1747	10000	2.896	3.961
1500	0.5761	0.5902	10500	2.606	3.737
2000	0.8888	0.9463	11000	2.309	3.473
2500	1.251	1.324	11500	2.012	3.173
3000	1.666	1.713	12000	1.720	2.844
3500	2.115	2.106	12500	1.439	2.492
4000	2.567	2.495	13000	1.172	2.126
4500	2.989	2.870	13500	0.9226	1.753
5000	3.352	3.220	14000	0.6945	1.380
5500	3.638	3.535	14500	0.4903	1.019
6000	3.838	3.807	15000	0.3132	0.6804
6500	3.950	4.028	15500	0.1666	0.3781
7000	3.977	4.193	16000	0.05643	0.1336
7500	3.927	4.300	16250	0.0186	0.0456
8000	3.810	4.346	16450	0.001	0.00208
8500	3.638	4.334			

II. BASIC FORMULAE

The necessary formulae from atomic physics are given in Paper I, equations (11)–(14). Revising the numerical coefficients in them in accordance with the most recent values of the fundamental physical constants (Cohen and DuMond 1958), we have

$$\kappa_\nu = 6.8113_8 \times 10^{-20} k\,(k^2 + 2I)\,\left|\int \Psi_d^* (z_1 + z_2) \Psi_c d\tau\right|^2 \text{cm}^2 \tag{1}$$

and

$$\kappa_\nu = 2.7245_5 \times 10^{-19} \frac{k}{k^2 + 2I} \left|\int \Psi_d^* \left(\frac{\partial}{\partial z_1} + \frac{\partial}{\partial z_2}\right) \Psi_c d\tau\right|^2 \text{cm}^2, \tag{2}$$

where κ_ν is the atomic absorption coefficient for radiation of frequency ν in which an electron with a momentum k (in atomic units) is ejected; I is the electron affinity, also measured in atomic units; Ψ_d and Ψ_c are the wave functions describing the ground

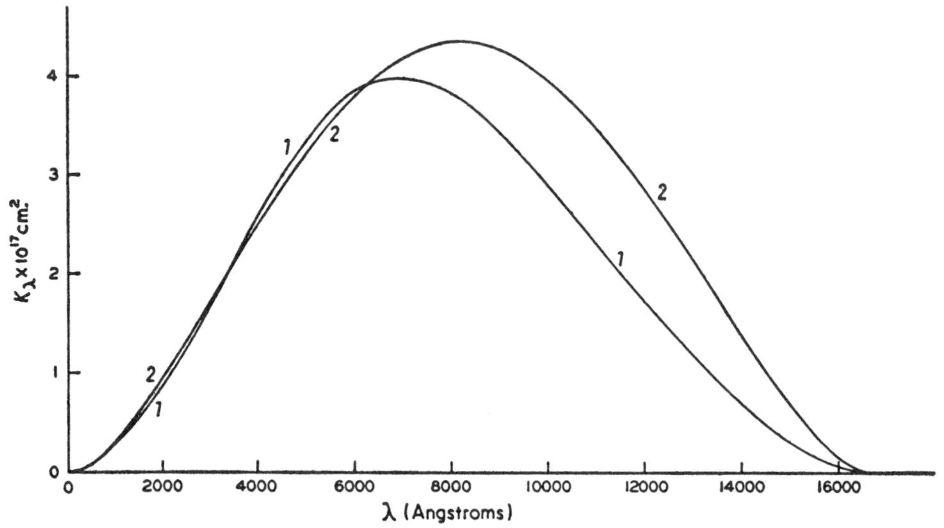

FIG. 1.—A comparison of the continuous absorption coefficient of H^- computed on the basis of the dipole-length (curve *1*) and the dipole-velocity (curve *2*) formulae, using Hart and Herzberg's 20-parameter wave function for the ground state. The ordinates denote the absorption coefficients in units of 10^{-17} cm^2; the abscissae, the wave length in angstroms.

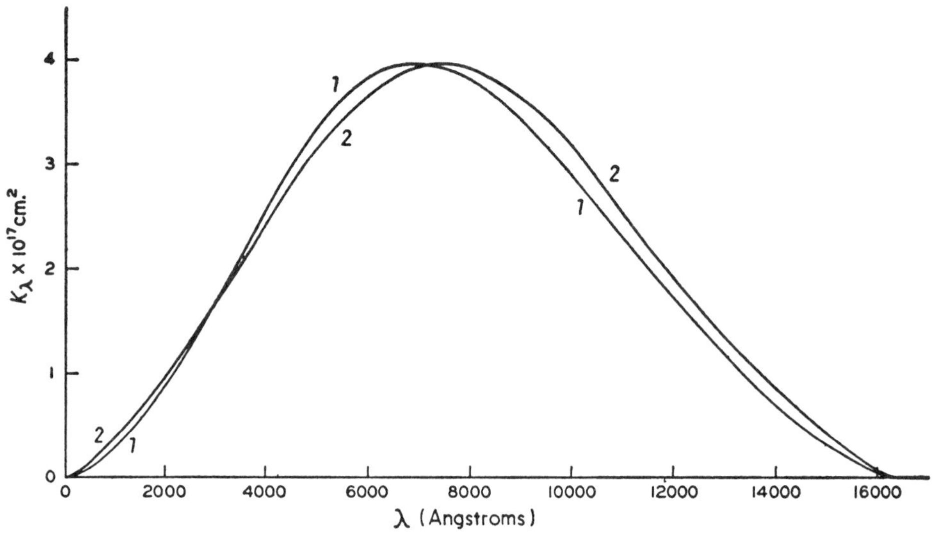

FIG. 2.—A comparison of the continuous absorption coefficient of H^- computed on the basis of the dipole-length formula with Hart and Herzberg's 20-parameter wave function (curve *1*) and Henrich's 11-parameter wave function (curve *2*). The relatively close agreement between the two curves suggests that the improvement in the relevant parts of the wave function by going to 20 parameters is not very significant. The ordinates denote the absorption coefficients in units of 10^{-17} cm^2; the abscissae, the wave length in angstroms.

state of H^- and the continuum, respectively. Equations (1) and (2) are, respectively, the dipole-length and the dipole-velocity formulae; in these formulae the matrix elements are to be evaluated in atomic units.

The wave length, λ, of the incident radiation measured in thousand-angstrom units is given by

$$\lambda \text{ (in 1000-A unit)} = \frac{0.9112671}{k^2 + 2I} . \tag{3}$$

In evaluating κ_ν in accordance with the foregoing formulae, we shall use for Ψ_d the

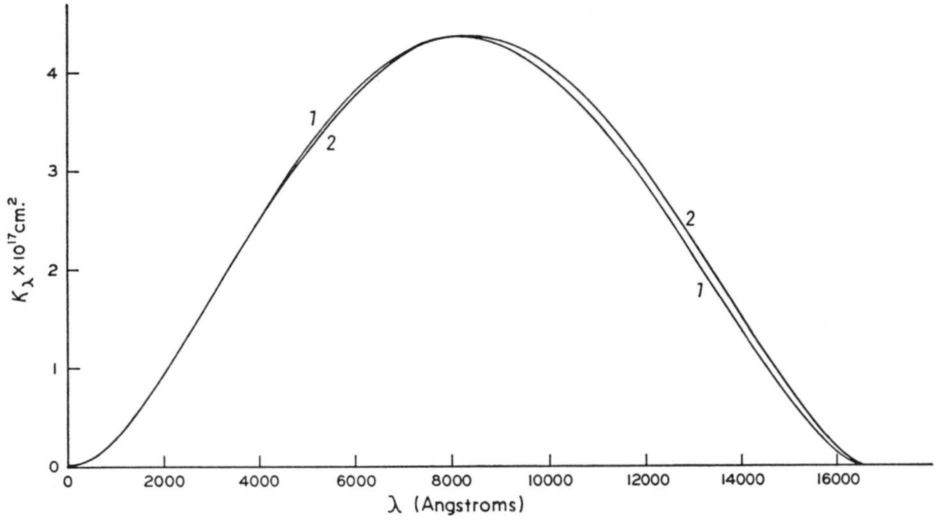

FIG. 3.—A comparison of the continuous absorption coefficient of H^- computed on the basis of the dipole-velocity formula with Hart and Herzberg's 20-parameter wave function (curve *1*) and Henrich's 11-parameter wave function (curve *2*). The relatively close agreement between the two curves suggests that the improvement in the relevant parts of the wave function by going to 20 parameters is not very significant. The ordinates denote the absorption coefficients in units of 10^{-17} cm^2; the abscissae, the wave length in angstroms.

20-parameter wave function of Hart and Herzberg (1955); their wave function is of the form

$$\begin{aligned}\Psi_d = \mathfrak{N} e^{-as/2}(1 &+ c_1 u + c_2 t^2 + c_3 s + c_4 s^2 + c_5 u^2 + c_6 su \\ &+ c_7 t^2 u + c_8 u^3 + c_9 t^2 u^2 + c_{10} st^2 + c_{11} s^3 + c_{12} t^2 u^4 \\ &+ c_{13} u^4 + c_{14} u^5 + c_{15} t^2 u^3 + c_{16} s^2 t^2 + c_{17} s^4 \\ &+ c_{18} st^2 u + c_{19} t^4) ,\end{aligned} \tag{4}$$

where $\mathfrak{N}$ is the normalization factor; $a, c_1, c_2, \ldots, c_{19}$ are the twenty constants determined by the Ritz condition of minimum energy:

$$s = r_1 + r_2 , \quad t = r_1 - r_2 , \quad \text{and} \quad u = r_{12} , \tag{5}$$

where r_1, r_2, and r_{12} are the distances of the two electrons from the nucleus and from each other, respectively. For convenience, we give in Table 2 the values of the various constants as determined by Hart and Herzberg.

The energy for the ground state which Hart and Herzberg's wave function gives is

$$E = -0.5276446692 \text{ atomic unit} . \tag{6}$$

This value should be contrasted with the values

$$-0.527559 \quad \text{and} \quad -0.527717 \tag{7}$$

obtained by Henrich (1944*a*) and Hylleraas and Midtal (1958) with 11 and 24 parameters, respectively. Hylleraas and Midtal have not published the coefficients of their wave function; moreover, since they have used fractional powers and logarithms of s, t, and u in their variational expansion, the evaluation of the cross-sections using their wave function would be excessively complicated. In any event, the improvement effected by going from Henrich's 11-parameter wave function to Hart and Herzberg's 20-parameter wave function is not sufficiently great to justify further efforts along these directions.

TABLE 2

VALUES OF COEFFICIENTS IN HART AND HERZBERG'S WAVE FUNCTION FOR GROUND STATE OF H^-

$E = -0.5276446692$

$\mathfrak{N} = +0.0716409212$

$a = +0.6750351$

$c_0 = 1$	$c_{10} = -1.77558678\times10^{-3}$
$c_1 = +0.337294236$	$c_{11} = -7.40841223\times10^{-4}$
$c_2 = +0.0808833954$	$c_{12} = +1.63058368\times10^{-6}$
$c_3 = -0.213129754$	$c_{13} = -2.73106176\times10^{-4}$
$c_4 = +0.0200385445$	$c_{14} = +6.27409440\times10^{-6}$
$c_5 = -0.0287160073$	$c_{15} = -6.38293145\times10^{-5}$
$c_6 = -0.0154381194$	$c_{16} = -1.84423133\times10^{-4}$
$c_7 = -9.21896695\times10^{-3}$	$c_{17} = +1.55857013\times10^{-5}$
$c_8 = +4.32904670\times10^{-3}$	$c_{18} = +6.48350632\times10^{-4}$
$c_9 = +7.86976451\times10^{-4}$	$c_{19} = +6.88602440\times10^{-4}$

Returning to equations (1) and (2), we shall use, for a plane-wave representation of the outgoing electron,

$$\Psi_c = \frac{1}{\sqrt{(2\pi)}}\left(e^{-r_1+ikz_2} + e^{-r_2+ikz_1}\right) . \tag{8}$$

In Section V we shall refer to an improvement in Ψ_c which we shall incorporate in a later paper.

III. DIPOLE-LENGTH MATRIX ELEMENT

For Ψ_d and Ψ_c of the forms given in Section II, we find, after some very lengthy calculations, that

$$\int \Psi_d^* (z_1 + z_2)\Psi_c d\tau = -i(2048\pi^3)^{1/2} \frac{\mathfrak{N} q^3}{k^2}\left[\sum_{j=-1}^{8} l_j \mathfrak{L}_j^{(a)} - \sum_{j=-1}^{3} \lambda_j \mathfrak{L}_j^{(1+2a)}\right] , \tag{9}$$

where

$$q = \frac{1}{1+\alpha} = 0.5970024 \,; \qquad (10)$$

$\mathfrak{L}_j^{(p)}$ are functions of k which are defined in Paper I, equation (18); and

$$\begin{aligned}
l_{-1} &= 24q^4c_1 + 168q^5c_6 + 1344q^6c_7 + 576q^6c_8 + 28800q^8c_{14} \\
&\quad + 51840q^8c_{15} + 12096q^7c_{18} \\
&= +0.54504286\,, \\
l_0 &= 24q^4c_6 - 336q^5c_7 - 10368q^7c_{15} - 1344q^6c_{18} \\
&= +0.166279695\,, \\
l_1 &= 24q^4c_7 - 72q^4c_8 - 5760q^6c_{14} - 3456q^6c_{15} - 168q^5c_{18} \\
&= -0.0676090978\,, \\
l_2 &= 1 + 12q^2c_2 + 3qc_3 + 12q^2c_4 + 4q^2c_5 + 120q^4c_9 + 60q^3c_{10} \\
&\quad + 60q^3c_{11} - 6720q^6c_{12} - 120q^4c_{13} + 1008\,q^5c_{15} + 360q^4c_{16} \\
&\quad + 360q^4c_{17} + 24q^4c_{18} + 360q^4c_{19} \\
&= +\,1.013384721\,, \\
l_3 &= c_1 - 6qc_2 + c_3 + 6qc_4 + 3qc_6 + 12q^2c_7 + 12q^2c_8 - 40q^3c_9 \\
&\quad - 12q^2c_{10} + 36q^2c_{11} + 1680q^5c_{12} + 288q^4c_{15} + 240q^3c_{17} \\
&\quad + 60q^3c_{18} - 240q^3c_{19} \\
&= -0.179174932\,, \\
l_4 &= c_2 + c_4 + c_5 + c_6 - 6qc_7 + 16q^2c_9 - 3qc_{10} + 9qc_{11} + 600q^4c_{12} \\
&\quad + 24q^2c_{13} - 120q^3c_{15} - 24q^2c_{16} + 72q^2c_{17} - 12q^2c_{18} \\
&\quad + 72q^2c_{19} \\
&= +0.109770777\,, \\
l_5 &= c_7 + c_8 - 6qc_9 + c_{10} + c_{11} - 240q^3c_{12} + 40q^2c_{14} + 24q^2c_{15} \\
&\quad + 12qc_{17} - 3qc_{18} - 12qc_{19} \\
&= -1.67478325 \times 10^{-2}\,, \\
l_6 &= c_9 + 36q^2c_{12} + c_{13} - 6qc_{15} + c_{16} + c_{17} + c_{18} + c_{19} \\
&= +1.93154519 \times 10^{-3}\,, \\
l_7 &= -6qc_{12} + c_{14} + c_{15} \\
&= -6.33959945 \times 10^{-5}\,, \\
l_8 &= c_{12} \\
&= +\,1.63058368 \times 10^{-6}\,,
\end{aligned}$$

$$\lambda_{-1} = 24q^4c_1 + 168q^5c_6 + 1344q^6c_7 + 576q^6c_8 + 28800q^8c_{14} + 51840q^8c_{15} + 12096q^7c_{18} = +0.54504286\,,$$

$$\lambda_0 = 24q^3c_1 + 192q^4c_6 + 1008q^5c_7 + 576q^5c_8 + 28800q^7c_{14} + 41472q^7c_{15} + 10752q^6c_{18} = +1.07924559\,,$$

$$\lambda_1 = 12q^2c_1 + 108q^3c_6 + 360q^4c_7 + 216q^4c_8 + 8640q^6c_{14} + 12096q^6c_{15} + 4536q^5c_{18} = +0.97554017\,,$$

$$\lambda_2 = 2qc_1 + 32q^2c_6 + 40q^3c_7 + 24q^3c_8 + 720q^5c_{14} + 1008q^5c_{15} + 960q^4c_{18} = +0.244827969\,,$$

$$\lambda_3 = 4qc_6 + 80q^3c_{18} = -0.0258299591\,.$$

Making use of equations (3) and (9), we find that equation (1) for κ_ν becomes

$$\kappa_\nu = \frac{9.1588}{\lambda k^3}\left|\sum\right|^2 \times 10^{-19}\ \mathrm{cm}^2\,, \tag{11}$$

where λ is the wave length in 1000-A units and Σ stands for the summations in brackets in equation (9). The results of the calculations based on the formulae of this section are given in Table 1.

IV. THE DIPOLE-VELOCITY MATRIX ELEMENT

We find:

$$\int \Psi_d^* \left(\frac{\partial}{\partial z_1} + \frac{\partial}{\partial z_2}\right)\Psi_c d\tau = -i(2048\pi^3)^{1/2}\frac{\mathfrak{N} q^3}{k^2}\left[\sum_{j=-1}^{7}(p_j + js_{j+1} - \alpha s_j)\mathfrak{L}_j^{(\alpha)} - \sum_{j=-1}^{2}(\varpi_j + j\sigma_{j+1} - [1+2\alpha]\sigma_j)\mathfrak{L}_j^{(1+2\alpha)}\right], \tag{12}$$

where

$$p_{-1} = 4q^3c_1 + 24q^4c_6 + 168q^5c_7 + 72q^5c_8 + 2880q^7c_{14} + 5184q^7c_{15} + 1344q^6c_{18} = +0.177188866\,,$$

$$p_0 = 4q^3c_6 - 48q^4c_7 - 1152q^6c_{15} - 168q^5c_{18} = +0.0381409776\,,$$

$$p_1 = -qc_1 - 4q^2c_6 - 16q^3c_7 - 24q^3c_8 - 1080q^5c_{14} - 936q^5c_{15}$$
$$\quad - 144q^4c_{18}$$
$$= -0.177920494\,,$$

$$p_2 = -2qc_5 - qc_6 + 8q^2c_7 - 40q^3c_9 - 3360q^5c_{12} - 80q^3c_{13}$$
$$\quad + 288q^4c_{15} + 24q^3c_{18}$$
$$= +0.0157287428\,,$$

$$p_3 = -qc_7 - 3qc_8 + 16q^2c_9 + 960q^4c_{12} - 180q^3c_{14} - 84q^3c_{15}$$
$$\quad + 4q^2c_{18}$$
$$= +4.26191106 \times 10^{-3}\,,$$

$$p_4 = -2qc_9 - 160q^3c_{12} - 4qc_{13} + 24q^2c_{15} - qc_{18}$$
$$= -1.27604161 \times 10^{-3}\,,$$

$$p_5 = 32q^2c_{12} - 5qc_{14} - 3qc_{15}$$
$$= +1.14187619 \times 10^{-4}\,,$$

$$p_6 = -4qc_{12}$$
$$= -3.89384963 \times 10^{-6}\,,$$

$$\varpi_{-1} = 4q^3c_1 + 24q^4c_6 + 168q^5c_7 + 72q^5c_8 + 2880q^7c_{14} + 5184q^7c_{15}$$
$$\quad + 1344q^6c_{18}$$
$$= +0.177188866\,,$$

$$\varpi_0 = 4q^2c_1 + 28q^3c_6 + 120q^4c_7 + 72q^4c_8 + 2880q^6c_{14} + 4032q^6c_{15}$$
$$\quad + 1176q^5c_{18}$$
$$= +0.334938544\,,$$

$$\varpi_1 = qc_1 + 12q^2c_6 + 20q^3c_7 + 12q^3c_8 + 360q^5c_{14} + 504q^5c_{15}$$
$$\quad + 360q^4c_{18}$$
$$= +0.134540148\,,$$

$$\varpi_2 = 2qc_6 + 40q^3c_{18}$$
$$= -0.0129149795\,,$$

$$s_0 = 4q^2c_1 + 20q^3c_6 + 120q^4c_7 + 72q^4c_8 + 2880q^6c_{14} + 4032q^6c_{15}$$
$$\quad + 840q^5c_{18}$$
$$= +0.344697028\,,$$

$$s_1 = 1 + 12q^2c_2 + 3qc_3 + 12q^2c_4 + 12q^2c_5 + 4q^2c_6 - 40q^3c_7$$
$$\quad + 360q^4c_9 + 60q^3c_{10} + 60q^3c_{11} + 20160q^6c_{12} + 360q^4c_{13}$$
$$\quad - 1008q^5c_{15} + 360q^4c_{16} + 360q^4c_{17} - 120q^4c_{18} + 360q^4c_{19}$$
$$= +0.995185094\,,$$

$$
\begin{aligned}
s_2 &= c_1 - 6qc_2 + c_3 + 6qc_4 + 3qc_6 + 16q^2c_7 + 24q^2c_8 - 120q^3c_9 \\
&\quad - 12q^2c_{10} + 36q^2c_{11} - 5040q^5c_{12} + 1080q^4c_{14} + 792q^4c_{15} \\
&\quad + 240q^3c_{17} + 40q^3c_{18} - 240q^3c_{19} \\
&= -0.194014864 ,
\end{aligned}
$$

$$
\begin{aligned}
s_3 &= c_2 + c_4 + c_5 + c_6 - 6qc_7 + 24q^2c_9 - 3qc_{10} + 9qc_{11} \\
&\quad + 1560q^4c_{12} + 40q^2c_{13} - 240q^3c_{15} - 24q^2c_{16} + 72q^2c_{17} \\
&\quad - 8q^2c_{18} + 72q^2c_{19} \\
&= +0.113210215 ,
\end{aligned}
$$

$$
\begin{aligned}
s_4 &= c_7 + c_8 - 6qc_9 + c_{10} + c_{11} - 400q^3c_{12} + 60q^2c_{14} + 36q^2c_{15} \\
&\quad + 12qc_{17} - 3qc_{18} - 12qc_{19} \\
&= -0.0170316162 ,
\end{aligned}
$$

$$
\begin{aligned}
s_5 &= c_9 + 52q^2c_{12} + c_{13} - 6qc_{15} + c_{16} + c_{17} + c_{18} + c_{19} \\
&= +1.94084374 \times 10^{-3} ,
\end{aligned}
$$

$$
\begin{aligned}
s_6 &= -6qc_{12} + c_{14} + c_{15} \\
&= -6.33959945 \times 10^{-5} ,
\end{aligned}
$$

$$
\begin{aligned}
s_7 &= c_{12} \\
&= +1.63058368 \times 10^{-6} ,
\end{aligned}
$$

$$
\begin{aligned}
\sigma_0 &= 4q^2c_1 + 20q^3c_6 + 120q^4c_7 + 72q^4c_8 + 2880q^6c_{14} + 4032q^6c_{15} \\
&\quad + 840q^5c_{18} \\
&= +0.344697028 ,
\end{aligned}
$$

$$
\begin{aligned}
\sigma_1 &= qc_1 + 12q^2c_6 + 20q^3c_7 + 12q^3c_8 + 360q^5c_{14} + 504q^5c_{15} \\
&\quad + 360q^4c_{18} \\
&= +0.134540148 ,
\end{aligned}
$$

$$
\begin{aligned}
\sigma_2 &= 2qc_6 + 40q^3c_{18} \\
&= -0.0129149795 .
\end{aligned}
$$

Making use of the foregoing equations, we find that equation (2) for κ_ν becomes

$$\kappa_\nu = 4.41172 \frac{\lambda}{k^3} \left| \sum \right|^2 \times 10^{-18}\ \text{cm}^2 , \tag{13}$$

where λ is again the wave length in 1000-A units and

$$\sum = \sum_{j=-1}^{7} L_j \mathfrak{L}_j^{(\alpha)} + \sum_{j=-1}^{2} S_j \mathfrak{L}_j^{(1+2\alpha)} , \tag{14}$$

where

$$L_{-1} = +0.167508162\,, \qquad L_6 = -4.86841726 \times 10^{-5}\,,$$
$$L_0 = +0.194541608\,, \qquad L_7 = +1.10070118 \times 10^{-6}\,,$$
$$L_1 = +1.043720208\,,$$
$$L_2 = -0.373116012\,, \qquad S_{-1} = -0.167508162\,,$$
$$L_3 = +0.123253804\,, \qquad S_0 = -0.475123656\,,$$
$$L_4 = -0.0179842718\,, \qquad S_1 = -0.194553619\,,$$
$$L_5 = +1.51292996 \times 10^{-3}\,, \qquad S_2 = +0.0174361284\,.$$

The results of the calculations based on the formulae of this section are given in Table 1; for reasons which have been elaborated in the earlier papers, the present results derived from the dipole-velocity matrix elements are to be preferred to those derived from either the dipole-length or the dipole-acceleration matrix elements.

V. CONCLUDING REMARKS

From an examination of Figures 2 and 3 it would appear that no further improvement in the computed cross-sections are to be expected by going to wave functions for the ground state more accurate than Hart and Herzberg's. An improvement in the representation of the wave function for the continuum could, of course, be considered, though it is unlikely that it will bring any substantial changes. In any event, it is hoped to publish soon the results of calculations based on the Hartree wave functions for the free electron in the field of a neutral hydrogen atom.

In conclusion I wish to record my indebtedness to Miss Donna Elbert for carrying out all the numerical calculations based on the formulae of Sections III and IV; to Dr. W. H. Reid for carefully checking the formulae for the matrix elements; and most of all to Dr. G. Herzberg, to whom we owe, more than to anyone else, the current renewed interest in the two-electron systems.

REFERENCES

Bethe, H. A., and Salpeter, E. E. 1957, *Quantum Mechanics of One and Two Electron Atoms* (New York: Academic Press).
Chandrasekhar, S. 1944*a*, *Ap. J.*, **100**, 176.
———. 1944*b*, *Rev. Mod. Phys.*, **16**, 301.
———. 1945*a*, *Ap. J.*, **102**, 223.
———. 1945*b*, *ibid.*, p. 395.
———. 1947, *Scient. Monthly*, **64**, 313.
Chandrasekhar, S., and Breen, F. H. 1946, *Ap. J.*, **104**, 430.
Chandrasekhar, S., Elbert, D. D., and Herzberg, G. 1953, *Phys. Rev.*, **91**, 1172.
Chandrasekhar, S., and Herzberg, G. 1955, *Phys. Rev.*, **98**, 1050.
Chandrasekhar, S., and Krogdahl, M. K. 1943, *Ap. J.*, **98**, 205.
Cohen, E. R., and DuMond, J. W. 1958, *Encyclopedia of Physics*, Vol. **35** (Berlin: Springer).
Geltman, S. 1955, *Phys. Rev.*, **104**, 346.
Hart, J. F., and Herzberg, G. 1957, *Phys. Rev.*, **106**, 79.
Henrich, L. F. 1944*a*, *Ap. J.*, **99**, 59.
———. 1944*b*, *ibid.*, p. 318.
Hylleraas. E. A., and Midtal, J. 1956, *Phys. Rev.*, **103**, 829.
———. 1958, *ibid.*, **109**, 1013.
Kinoshita, T. 1957, *Phys. Rev.*, **105**, 1490.
Lochte-Holtgreven, W. 1951, *Naturwiss.*, **38**, 258.
Massey, H. S. W. 1950, *Negative Ions* (Cambridge: Cambridge University Press).
Smith, S. J., and Branscomb, L. M. 1955*a*, *J. Research Nat. Bur. Standards*, **55**, 165.
———. 1955*b*, *Phys. Rev.*, **98**, 1028.
Unsöld, A. 1955, *Physik der Sternatmosphären* (Berlin: Springer).
Williamson, R. E. 1943, *Ap. J.*, **96**, 438.

ON THE CONTINUOUS ABSORPTION COEFFICIENT OF THE NEGATIVE HYDROGEN ION. V

S. CHANDRASEKHAR AND DONNA D. ELBERT
Yerkes Observatory

Received June 16, 1958

ABSTRACT

The photoionization cross-sections of the negative hydrogen ion derived in Paper IV, using Hart and Herzberg's 20-parameter wave function, are further improved by replacing the plane-wave approximation for the free electron by the Hartree approximation.

I. INTRODUCTION

In Paper IV (Chandrasekhar 1958) of this series, the photoionization cross-sections of the negative hydrogen ion were revised by making use of Hart and Herzberg's (1957) 20-parameter wave function for the ground state of the ion. However, the free electron was represented by a plane wave. In this paper we shall replace this representation by a Hartree approximation and consider the extent of the improvements effected.

II. THE BASIC FORMULA AND RESULTS

By following the procedure outlined in Paper II (Chandrasekhar 1945), we find that, with Hart and Herzberg's wave function for the ground state and the Hartree wave function for the free state, the basic formula for the dipole velocity matrix element becomes

$$\int \Psi_d \left(\frac{\partial}{\partial z_1} + \frac{\partial}{\partial z_2} \right) \Psi_c d\tau = -i (2048\pi^3)^{1/2} \frac{\mathfrak{N} q^3}{k} \int_0^\infty W_2(r) \chi_1(r)\, dr , \tag{1}$$

where $\chi_1(r)$ is the radial part of the p-spherical wave (as defined in Paper II, eqs. [6] and [7]) and

$$W_2(r) = \left(\sum_{j=-1}^{7} L_j r^j \right) e^{-ar} + \left(\sum_{j=-1}^{2} S_j r^j \right) e^{-(1+2a)r} . \tag{2}$$

In equation (2) the L_j's and the S_j's have the same meanings (and values) as in Paper IV, equation (14). The quantities k, $\mathfrak{N}$, q, and a also have the same meanings as in Paper IV.

The "weight function," $W_2(r)$, computed with the values for the constants given in Paper IV is listed in Table 1; this table should be contrasted with the corresponding table (Table 1) in Paper II derived from Henrich's 11-parameter wave function.

The corresponding formula for the absorption cross-section is given by

$$\kappa_\lambda = \frac{4.02024}{k(k^2+0.055289)} \left| \int_0^\infty W_2(r) \chi_1(r)\, dr \right|^2 \times 10^{-18}\ \text{cm}^2 ; \tag{3}$$

this cross-section refers to a wave length (cf. Paper IV, eq. [3])

$$\lambda = \frac{911.2671}{k^2+0.055289}\ \text{A} . \tag{4}$$

The values of κ_λ have been evaluated in accordance with equation (3) by making use of Chandrasekhar and Breen's (1946) tabulation of the functions $\chi_1(r)$. The results are given in Table 2. For comparison, the results of the earlier calculations are also included in this table. From this comparison it would appear that no substantial improvements in the deduced cross-sections are to be expected by going to more accurate wave functions

TABLE 1

THE WEIGHT FUNCTION $W_2(r)$

r	$W_2(r)$	r	$W_2(r)$	r	$W_2(r)$	r	$W_2(r)$
0	0	4.4	0.18140	8.8	0.04836	16.4	0.00558
0.1	0.12665	4.5	.17538	8.9	.04712	16.6	.00522
0.2	.22763	4.6	.16958	9.0	.04591	16.8	.00487
0.3	.30707	4.7	.16400	9.1	.04474	17.0	.00455
0.4	.36851	4.8	.15861	9.2	.04360	17.2	.00425
0.5	.41502	4.9	.15342	9.3	.04250	17.4	.00396
0.6	.44920	5.0	.14842	9.4	.04142	17.6	.00369
0.7	.47326	5.1	.14360	9.5	.04037	17.8	.00344
0.8	.48907	5.2	.13896	9.6	.03936	18.0	.00320
0.9	.49819	5.3	.13449	9.7	.03836	18.2	.00298
1.0	.50193	5.4	.13018	9.8	.03740	18.4	.00277
1.1	.50140	5.5	.12603	9.9	.03646	18.6	.00258
1.2	.49748	5.6	.12203	10.0	.03554	18.8	.00239
1.3	.49094	5.7	.11818	10.2	.03377	19.0	.00222
1.4	.48239	5.8	.11447	10.4	.03209	19.2	.00207
1.5	.47232	5.9	.11090	10.6	.03049	19.4	.00192
1.6	.46116	6.0	.10746	10.8	.02896	19.6	.00178
1.7	.44922	6.1	.10414	11.0	.02750	19.8	.00165
1.8	.43678	6.2	.10095	11.2	.02611	20.0	.00153
1.9	.42428	6.3	.09788	11.4	.02477	20.2	.00141
2.0	.41117	6.4	.09492	11.6	.02350	20.4	.00131
2.1	.39831	6.5	.09207	11.8	.02228	20.6	.00121
2.2	.38556	6.6	.08933	12.0	.02112	20.8	.00112
2.3	.37299	6.7	.08669	12.2	.02000	21.0	.00104
2.4	.36066	6.8	.08414	12.4	.01894	21.2	.00096
2.5	.34862	6.9	.08169	12.6	.01792	21.4	.00089
2.6	.33689	7.0	.07932	12.8	.01695	21.6	.00082
2.7	.32549	7.1	.07704	13.0	.01602	21.8	.00076
2.8	.31443	7.2	.07485	13.2	.01513	22.0	.00070
2.9	.30373	7.3	.07273	13.4	.01428	22.2	.00064
3.0	.29338	7.4	.07068	13.6	.01347	22.4	.00059
3.1	.28337	7.5	.06871	13.8	.01270	22.6	.00055
3.2	.27371	7.6	.06681	14.0	.01197	22.8	.00050
3.3	.26439	7.7	.06497	14.2	.01127	23.0	.00047
3.4	.25540	7.8	.06320	14.4	.01060	23.2	.00043
3.5	.24672	7.9	.06148	14.6	.00997	23.4	.00039
3.6	.23836	8.0	.05983	14.8	.00937	23.6	.00036
3.7	.23029	8.1	.05823	15.0	.00880	23.8	.00033
3.8	.22252	8.2	.05668	15.2	.00826	24.0	.00031
3.9	.21502	8.3	.05518	15.4	.00775	24.2	.00028
4.0	.20779	8.4	.05373	15.6	.00727	24.4	.00026
4.1	.20083	8.5	.05232	15.8	.00681	24.6	.00024
4.2	.19411	8.6	.05096	16.0	.00638	24.8	.00022
4.3	0.18764	8.7	0.04964	16.2	0.00597	25.0	0.00020

for the ground state of the ion. It is also unlikely that improvements will be affected by going to better representations of the free state.

TABLE 2

PHOTOIONIZATION CROSS-SECTIONS OF H^- COMPUTED WITH DIPOLE-VELOCITY FORMULA, USING HART AND HERZBERG'S 20-PARAMETER AND HENRICH'S 11-PARAMETER WAVE FUNCTIONS FOR GROUND STATE AND HARTREE WAVE FUNCTIONS FOR FREE STATE

k^2	WITH 11-PARAMETER WAVE FUNCTION		WITH 20-PARAMETER WAVE FUNCTION	
	λ (A)	$\kappa_\lambda \times 10^{17}$ cm^2	λ (A)	$\kappa_\lambda \times 10^{17}$ cm^2
1.75	505	0.0657	505	0.0656
0.80	1066	0.333	1065	0.342
0.50	1642	0.740	1641	0.751
0.35	2249	1.231	2248	1.226
0.25	2987	1.84	2985	1.83
0.20	3572	2.32	3570	2.30
0.175	3960	2.62	3957	2.61
0.150	4443	2.97	4439	2.98
0.125	5059	3.39	5054	3.41
0.100	5875	3.87	5868	3.90
0.090	6280	4.06	6272	4.09
0.070	7283	4.41	7273	4.41
0.055	8275	4.52	8263	4.49
0.050	8669	4.50	8655	4.45
0.045	9102	4.44	9086	4.37
0.035	10111	4.13	10093	4.03
0.020	12131	2.96	12104	2.83

Note added in proof: In a recent paper, Bransden, Dalgarno, John, and Seaton (1958) have published calculations for p-waves in the field of a neutral hydrogen atom allowing for exchange and polarization, and they have tabulated functions which are equivalent to our $\chi_1(r)$ in equation (1) for $k^2 = 0.25$. Using their functions (denoted by G_1^+ in Table 10 of their paper), together with the weight function $W_2(r)$ of the present paper, we find for κ_λ the values 1.50×10^{-17} cm^2 and 1.42×10^{-17} cm^2 for their functions with $d \neq 0$ and $d = 0$, respectively; these values should be contrasted with 1.83×10^{-17} cm^2 (see Table 1) obtained with the Hartree wave function. The difference arising from the use of a more accurate representation for the free state for $k^2 = 0.25$ (corresponding to $\lambda = 2985$ A) is therefore not inappreciable. However, the difference resulting from allowing for exchange and polarization will decrease for increasing λ, since the values of $W_2(r)$ for large values of r become increasingly important; and, as Bransden *et al.* point out, the difference between their wave functions and the Hartree wave functions (Chandrasekhar and Breen 1946) becomes negligible for $r > 4$.

REFERENCES

Bransden, B. H., Dalgarno, A., John, T. L., and Seaton, M. J. 1958, *Proc. Phys. Soc.*, **71**, 877.
Chandrasekhar, S. 1945, *Ap. J.*, **102**, 395.
———. 1958, *ibid.*, **128**, 114.
Chandrasekhar, S., and Breen, F. H. 1946, *Ap. J.*, **103**, 41.
Hart, J. F., and Herzberg, G. 1957, *Phys. Rev.*, **106**, 79.

THE CONTINUOUS SPECTRUM OF THE SUN AND THE STARS

S. Chandrasekhar and Guido Münch[1]
Yerkes Observatory
Received September 5, 1946

Using the recent determination of the continuous absorption coefficient of H^- by Chandrasekhar and Breen, we have shown that the dependence of the continuous absorption coefficient with wave length in the range 4000–24,000 A, which can be inferred from the intensity distribution in the continuous spectrum of the sun, can be quantitatively accounted for as due to H^-; and, further, that the color temperatures measured in the wave-length intervals 4100–6500 A (Greenwich) and 4000–4600 A (Barbier and Chalonge) for stars of the main sequence and of spectral types A0–G0 can also be interpreted in terms of the continuous absorption of H^- and neutral hydrogen atoms.

The problem of the discontinuities at the head of the Balmer and the Paschen series is also briefly considered on the revised physical theory of the continuous absorption coefficient.

1. *Introduction.*—The two principal problems in the theory of the continuous spectrum of the stars are, first, to identify the source of the continuous absorption in the solar atmosphere which will account for the intensity distribution in the continuous spectrum of the sun and the law of darkening in the different wave lengths and, second, to account for the observed relations between the color and the effective temperatures of the stars. In this paper we shall show that the major aspects of these two problems find their natural solution in terms of the continuous absorption coefficient of the negative hydrogen ion as recently determined by Chandrasekhar and Breen.[2] More particularly, we shall show that the dependence of the continuous absorption coefficient with wave length in the range 4000–24,000 A, which can be deduced from the solar data, can be quantitatively accounted for as due to H^-; and, further, that the color temperatures measured in the wave-length intervals 4100–6500 A (Greenwich[3]) and 4000–4600 A (Barbier and Chalonge[4]) for stars of the main sequence and of spectral types A0–G0 can also be interpreted in terms of the continuous absorption of H^- and neutral hydrogen.

In addition to the two problems we have mentioned, we shall also consider some related questions concerning the discontinuities at the head of the Balmer and the Paschen series.

2. *The mean absorption coefficients of* H$^-$ *and* H.—As is well known, the character of the emergent continuous radiation from a stellar atmosphere is determined in terms of the temperature distribution in the atmosphere; and, as has recently been shown,[5] the temperature distribution in a nongray atmosphere will be given approximately by a formula of the standard type

$$T^4 = \tfrac{3}{4} T_e^4 \left(\tau + q[\tau]\right), \tag{1}$$

where T_e denotes the effective temperature and $q(\tau)$ a certain monotonic increasing function of the optical depth τ, provided that the mean absorption coefficient κ, in terms of which τ is measured, is defined as a straight average of the monochromatic absorption coefficient weighted according to the net flux $F_\nu^{(1)}$ of radiation of frequency ν in a gray

[1] Fellow of the John Simon Guggenheim Memorial Foundation at the Yerkes Observatory.

[2] *Ap. J.*, **104**, 430, 1946.

[3] Sir Frank Dyson, *Observations of Color-Temperatures of Stars, 1926–1932*, London, 1932; also *M.N.*, **100**, 189, 1940.

[4] *Ann. d'ap.*, **4**, 30, Table 4, 1941.

[5] S. Chandrasekhar, *Ap. J.*, **101**, 328, 1945. This paper will be referred to as "Radiative Equilibrium VII."

atmosphere and if, further, $\kappa_\nu/\bar{\kappa}$ is independent of depth. On this approximation, then, the emergent intensity in a given frequency and in a given direction will depend not only on the continuous absorption coefficient at the frequency under consideration but also on the mean absorption coefficient $\bar{\kappa}$ over all frequencies.

As the discussion in this paper will establish for the stellar atmospheres considered, the contributions to $\bar{\kappa}$ in the visible and the infrared regions of the spectrum are essentially from only two sources: H^- and the neutral hydrogen atoms. The cross-sections for the absorption by H^- for various temperatures and wave lengths have been tabulated by Chandrasekhar and Breen in Table 7 of their paper, while those for hydrogen can be found from the formulae of Kramers and Gaunt, standardized, for example, by B. Strömgren.[6] However, the evaluation of the mean absorption coefficient for wave lengths shorter than 4000 A is made uncertain on two accounts: First, there is the absorption by the metals and the excessive crowding of the absorption lines toward the violet, which is particularly serious for spectral types later than F0; and, second, there is the absorption in the Lyman continuum. On both these accounts the true values of $\bar{\kappa}$ will be larger than those determined by ignoring them. But the exact amount by which they will be larger will be difficult to predict without a detailed theory of "blanketing,"[7] on the one hand, and without going into a more exact theory[8] of radiative transfer than represented by the approximations leading to equation (1), on the other. However, since in this paper our primary object is to establish only the adequacy of H^- as the source of absorption in the solar atmosphere over the entire visible and infrared regions of the spectrum and the corresponding role of H^- and H for stellar atmospheres with spectral types A2–G0, it appeared best to ignore the refinements indicated and simply determine $\bar{\kappa}$ by weighting κ_ν due to H^- and H (without the Lyman absorption) at the conditions prevailing at $\tau = 0.6$ by the flux $F_\nu^{(1)}$ at this level.[9] For only in this way can we use the solution to the transfer problem in the form of equation (1) in a consistent manner. It should, however, be remembered that the effects we have ignored may easily increase $\bar{\kappa}$, determined in terms of H^- and H (without the absorption in the Lyman continuum) by factors of the order of 1.5 and probably not exceeding 2.[10]

Turning our attention, next, to the evaluation of $\bar{\kappa}$, we may first observe that, since our present method of averaging is a straight one, the contributions to κ_ν from different sources are simply additive. We may, accordingly, consider the mean absorption coefficient of H^- and H separately.

Now the absorption coefficient of H^-, including both the bound-free and the free-free transitions, is most conveniently expressed as per neutral hydrogen atom and per unit electron pressure in the unit cm^4/dyne. The monochromatic coefficients κ'_ν, after allowing for the stimulated emission factor $(1 - e^{-h\nu/kT})$, are tabulated in Chandrasekhar and Breen's paper for various values of $\theta(=5040/T)$. If we now denote by $a(H^-)$ the average

[6] "Tables of Model Stellar Atmospheres," *Publ. mind. Meddel. Kobenhavns Obs.*, No. 138, 1944.

[7] Cf. G. Münch, *Ap. J.*, **104**, 87, 1946.

[8] Such as, e.g., the (2, 2) approximation given in "Radiative Equilibrium VII," § 6.

[9] The choice of $\tau = 0.6$ for the "representative point" was made after some preliminary trials (cf. G. Münch, *Ap. J.*, **102**, 385, 1945, esp. Table 3), though it is evident on general grounds that a level such as $\tau = 0.6$, where the local temperature is approximately the same as the effective temperature, would be the correct one in the scheme of approximations leading to eq. (1).

[10] In all earlier evaluations of $\bar{\kappa}$ the absorption in the Lyman continuum did not, indeed, play any role. This was due to the manner in which $\bar{\kappa}$ was defined in those investigations as the Rosseland mean. But in "Radiative Equilibrium VII" it has been shown that there is no justification for taking the Rosseland means as they have been hitherto. Since the method of averaging, by which we have now replaced the Rosseland mean, is a straight one, it is no longer permissible simply to ignore the absorption in the Lyman continuum. At the same time, it is not possible to take it into account satisfactorily in the (2, 1) approximation leading to eq. (1). We should have to go at least to the (2, 2) approximation of "Radiative Equilibrium VII."

value of the coefficients κ_ν', weighted according to the flux $F_\nu^{(1)}$ in a gray atmosphere at $\tau = 0.6$, where the temperature is approximately the effective temperature T_e, then the contribution $\bar{\kappa}(H^-)$ to the mass absorption coefficient $\bar{\kappa}$ by H^- is given by

$$\bar{\kappa}\,(H^-) = \frac{(1 - x_H)\,p_e}{m_H}\,a\,(H^-)\,, \tag{2}$$

where m_H is the mass of the hydrogen atom, p_e the electron pressure, and x_H the degree of ionization of hydrogen under the physical conditions represented by T_e and p_e.[11] The values of $a(H^-)$ found by graphical integration in accordance with the formula

$$a\,(H^-) = \frac{1}{F}\int_0^\infty \kappa_\nu' F^{(1)}\,(0.6)\,d\nu \tag{3}$$

for various values of $\theta = \theta_e$ are given in Table 1.

TABLE 1

THE MEAN ABSORPTION COEFFICIENTS $a(H^-)$ AND $a(H)$

θ	$a(H^-)$	$a(H)$	θ	$a(H^-)$	$a(H)$
0.5	0.563×10^{-26}	1.65×10^{-22}	0.9	6.08×10^{-26}	5.52×10^{-27}
0.6	1.145×10^{-26}	1.22×10^{-23}	1.0	9.32×10^{-26}	3.65×10^{-28}
0.7	2.25×10^{-26}	1.13×10^{-24}	1.2	2.00×10^{-25}	......
0.8	3.88×10^{-26}	8.00×10^{-26}	1.4	3.89×10^{-25}	......

Similarly, the contribution to $\bar{\kappa}$ by hydrogen can also be expressed in the form

$$\bar{\kappa}\,(H) = \frac{1 - x_H}{m_H}\,a\,(H)\,, \tag{4}$$

where

$$a\,(H) = \int_0^\infty \frac{fD}{a^3}\,(1 - e^{-a})\,\frac{F_a^{(1)}\,(0.6)}{F}\,da\,, \tag{5}$$

where $a = h\nu/kT_e$ and f and D are certain functions of temperature and frequency, respectively, which have been tabulated by Strömgren.[12] For reasons which we have already explained, we do not include the Lyman absorption in evaluating $a(H)$. The values of $a(H)$ for various temperatures are also listed in Table 1.

In terms of $a(H^-)$ and $a(H)$ given in Table 1, we can determine the combined mass absorption coefficient $\bar{\kappa}$ according to

$$\bar{\kappa} = \frac{1 - x_H}{m_H}\,[a\,(H^-)\,p_e + a\,(H)]\,. \tag{6}$$

Values of $\bar{\kappa}(H^-)$, $\bar{\kappa}(H)$, and $\bar{\kappa}$ determined in accordance with the foregoing equations for various temperatures and electron pressures are given in Table 2.

3. *The continuous absorption in the solar atmosphere*—As we have already stated in the introduction, one of the principal problems in the interpretation of the solar spectrum is the identification of the source of absorption which will predict the same dependence

[11] It will be noted that, in writing the mass absorption coefficient in the form (2), we have assumed the preponderant abundance of hydrogen in the stellar atmosphere.

[12] See the reference quoted in n. 6.

of the absorption coefficient with wave length in the range 4000–24,000 A, which can be inferred from the intensity distribution in the continuous spectrum and the law of darkening in the different wave lengths. While the amount and variation of the continuous absorption in the spectral region $\lambda\lambda$ 4000–24,000 A can be deduced in a variety of ways,[13] it appears that, for the purposes of the identification of the physical source of absorption, it is most direct to adopt the following procedure:

TABLE 2

THE MEAN MASS ABSORPTION COEFFICIENTS $\bar{\kappa}(H^-)$, $\bar{\kappa}(H)$, AND $\bar{\kappa}$ FOR VARIOUS TEMPERATURES AND ELECTRON PRESSURES

		$p_e=1$	$p_e=10$	$p_e=10^2$	$p_e=10^3$	$p_e=10^4$
$\theta_e=0.5$	$\bar{\kappa}(H^-)$	5.85×10^{-6}	3.70×10^{-4}	4.98×10^{-2}	2.12	32.1
	$\bar{\kappa}(H)$	1.71×10^{-1}	1.08	14.5	61.8	93.5
	$\bar{\kappa}$	1.71×10^{-1}	1.08	14.6	63.9	126
$\theta_e=0.6$	$\bar{\kappa}(H^-)$	3.93×10^{-4}	2.60×10^{-2}	5.82×10^{-1}	6.68	68.4
	$\bar{\kappa}(H)$	4.18×10^{-1}	2.76	6.19	7.10	7.27
	$\bar{\kappa}$	4.18×10^{-1}	2.79	6.77	13.8	75.7
$\theta_e=0.7$	$\bar{\kappa}(H^-)$	9.08×10^{-3}	1.28×10^{-1}	1.34	13.4	134
	$\bar{\kappa}(H)$	4.54×10^{-1}	6.41×10^{-1}	0.67	0.7	1
	$\bar{\kappa}$	4.63×10^{-1}	7.69×10^{-1}	2.01	14.1	135
$\theta_e=0.8$	$\bar{\kappa}(H^-)$	2.26×10^{-2}	2.32×10^{-1}	2.32	23.2	232
	$\bar{\kappa}(H)$	4.66×10^{-2}	0.48×10^{-1}	0.05	....	....
	$\bar{\kappa}$	6.92×10^{-2}	2.80×10^{-1}	2.37	23.2	232
$\theta_e=0.9$	$\bar{\kappa}(H^-)$	3.63×10^{-2}	3.63×10^{-1}	3.63	36.3	363
	$\bar{\kappa}(H)$	0.33×10^{-2}	0.03×10^{-1}	....	....	....
	$\bar{\kappa}$	3.96×10^{-2}	3.66×10^{-1}	3.63	36.3	363
$\theta_e=1.0$	$\bar{\kappa}(H^-)$	5.57×10^{-2}	5.57×10^{-1}	5.57	55.7	557
	$\bar{\kappa}(H)$	0.02×10^{-2}	....	....	....	....
	$\bar{\kappa}$	5.59×10^{-2}	5.57×10^{-1}	5.57	55.7	557
$\theta_e=1.2$	$\bar{\kappa}(H^-)$	1.20×10^{-1}	1.20	12.0	120	1200
$\theta_e=1.4$	$\bar{\kappa}(H^-)$	2.33×10^{-1}	2.33	23.3	233	2330

We compare the observed intensity distribution in the emergent solar flux F_λ (obs.) with the flux $F_\lambda^{(1)}$ (0) to be expected in a gray atmosphere.[14] It is evident that the *departures*,

$$\Delta \log F_\lambda = \log F_\lambda\,(\text{obs.}) - \log F_\lambda^{(1)}\,(0)\,, \tag{7}$$

must be related more or less directly with the dependence of the continuous absorption coefficient κ_ν' with wave length. Indeed, in the approximations leading to the temperature distribution (1) this relation must be one-one, since, with the adopted definition of $\bar{\kappa}$, the temperature distributions in the gray and the nongray atmospheres agree.[15] This suggests that, with the known value of κ_ν' due to H^-, we compare the predicted de-

[13] G. Mülders, *Zs. f. Ap.*, **11**, 132, 1935; G. Münch, *Ap. J.*, **102**, 385, 1945; D. Chalonge and V. Kourganoff, *Ann. d'ap.* (in press).

[14] The values of $F_\lambda^{(1)}$ (0) can be readily derived from the entries along the line $\tau = 0$ in Table 2 of "Radiative Equilibrium VII."

[15] Cf. the remarks in italics on p. 343 in "Radiative Equilibrium VII."

partures from $F_\lambda^{(1)}(0)$ with those observed. The only uncertainty in these predictions will be of the nature of a "zero-point" correction, since a value of $\bar{\kappa}$ different from the one adopted will lead to an approximately constant additive correction to $\Delta \log F_\lambda$.[16]

In order, then, to make the comparison suggested in the preceding paragraph, we need to determine $\Delta \log F_\lambda$ in terms of κ_λ' due to H^- and an adopted $\bar{\kappa}$. Assuming in the first instance that the contribution to $\bar{\kappa}$ is only H^-, we find that

$$a\,(H^-) = 5.62 \times 10^{-26}\ \text{cm}^4/\text{dyne} \tag{8}$$

for an adopted value of

$$\theta_e = 0.8822\ . \tag{9}$$

The value of κ_λ' for $\theta = 0.8822$ can be found by simple interpolation in Table 7 of Chandrasekhar and Breen's paper. The ratios $\kappa_\lambda'/a(H^-)$ derived in this manner are

TABLE 3

THE PREDICTED DEPARTURES [LOG F_λ (obs.) − LOG $F_\lambda^{(1)}(0)$] FROM GRAYNESS OF THE SOLAR ATMOSPHERE DUE TO THE ABSORPTION BY H^-

λA	$\dfrac{\kappa_\lambda'}{\bar{\kappa}(H^-)}$	$\dfrac{\kappa_\lambda'}{1.42\bar{\kappa}(H^-)}$	LOG F_λ (THEO.)		LOG $F_\lambda^{(1)}(0)$	Δ LOG F_λ	
			$\bar{\kappa}=\bar{\kappa}(H^-)$	$\bar{\kappa}=1.42\bar{\kappa}(H^-)$		$\bar{\kappa}=\bar{\kappa}(H^-)$	$=\bar{\kappa}1.42\bar{\kappa}(H^-)$
4000	0.686	0.483	14.489	14.532	14.358	+0.131	+0.174
4500	0.783	.551	14.468	....	14.390	+ .078	....
5000	0.881	.620	14.431	14.534	14.399	+ .032	+ .135
6000	1.029	.725	14.341	14.423	14.350	− .009	+ .073
7000	1.132	.797	14.259	14.329	14.285	− .026	+ .044
8000	1.188	.837	14.164	14.224	14.194	− .030	+ .030
9000	1.183	.833	14.070	14.128	14.098	− .028	+ .030
10,000	1.125	.792	13.982	14.034	13.999	− .017	+ .035
11,000	1.028	.724	13.897	13.946	13.901	− .004	+ .045
12,000	0.911	.642	13.817	13.862	13.805	+ .012	+ .057
13,000	0.788	.555	13.742	13.784	13.712	+ .030	+ .072
14,000	0.651	.458	13.674	13.713	13.624	+ .050	+ .089
15,000	0.523	.368	13.608	13.643	13.537	+ .071	+ .106
16,000	0.481	.339	13.532	13.566	13.454	+ .078	+ .112
17,000	0.486	.342	13.448	13.480	13.374	+ .074	+ .106
18,000	0.516	.363	13.362	13.395	13.297	+ .065	+ .098
19,000	0.562	.396	13.279	13.313	13.223	+ .056	+ .090
20,000	0.618	.435	13.197	13.230	13.152	+ .045	+ .078
21,000	0.679	.478	13.118	13.149	13.083	+ .035	+ .066
22,000	0.749	.527	13.042	13.073	13.016	+ .026	+ .057
23,000	0.820	0.577	12.971	13.003	13.952	+0.019	+0.051

given in Table 3 for various values of λ. With these values of $\kappa_\lambda'/a(H^-)$, the theoretical determination of $\Delta \log F_\lambda$ is straightforward with the help of the nomogram of Burkhardt's table,[17] which one of us has recently published.[18] The results of the determination are given in Table 3. In Figure 1 we have further compared the computed departures $\Delta \log F_\lambda$ with those observed.[19] It is seen that the predicted variation of the

[16] This is seen most directly in an approximation in which we expand the source function $B_\lambda(T)$ as a Taylor series about a suitable point and determine the emergent flux in terms of it (see, e.g., A. Unsöld, *Physik der Sternatmosphären*, p. 109. eq. [31.18], Berlin, 1938).

[17] *Zs.f.Ap.*, **13**, 56, 1936.

[18] G. Münch, *Ap. J.*, **102**, 385, Fig. 2, 1945.

[19] For $\lambda > 9000$ A the observed departures were obtained from a reduction of the solar data by M. Minnaert, *B.A.N.*, **2**, No. 51, 75, 1924; see also Unsöld, *op. cit.*, p. 32. For $\lambda < 9000$ A the reduction of G. Mülders (dissertation, Utrecht, 1934) was used.

departures runs remarkably parallel with the observed departures over the wavelength range 4000–20,000 A. (The observational data do not seem specially reliable for $\lambda > 20{,}000$ A.) However, the absolute values of the predicted departures are systematically less than the predicted departures by approximately a constant amount, indicating a zero-point correction in the sense that the adopted value of $\bar{\kappa}$ as due to H^- alone is somewhat too small. The calculations were accordingly repeated for other slightly larger values of $\bar{\kappa}$, and it was found that with $\bar{\kappa} = 1.42\ \kappa(H^-)$, the predicted and the observed departures agree entirely within the limits of the observational uncertainties over the whole region of the spectrum in which H^- contributes to the absorption. The

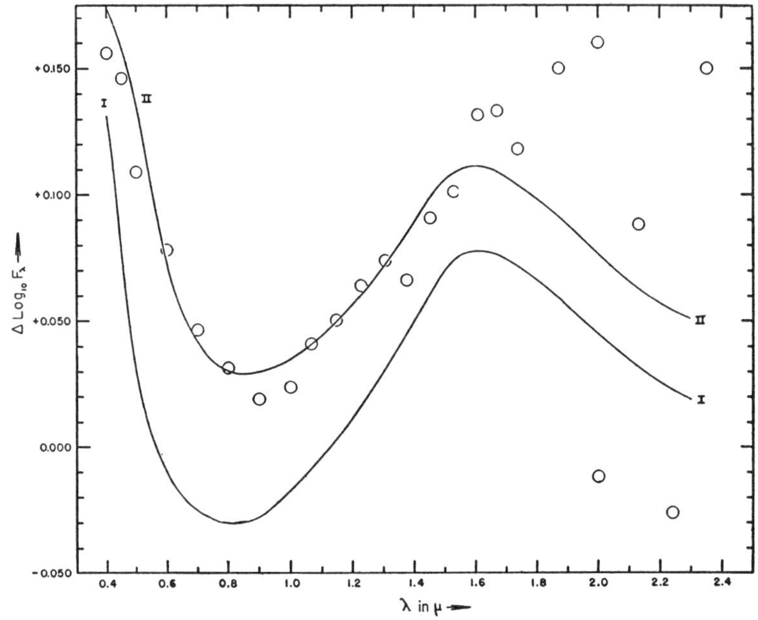

FIG. 1.—Comparison of the observed and the theoretically predicted departures [$\log F_\lambda - \log F_\lambda^1(0)$] from a gray atmosphere due to the absorption by H^-. The circles represent the observed departures of the solar emergent flux from that of a gray atmosphere, while Curves I and II are the theoretically derived departures on the two assumptions $\bar{\kappa} = \bar{\kappa}(H^-)$ and $\bar{\kappa} = 1.42\ \bar{\kappa}(H^-)$.

agreement is, in fact, so striking that we may say that H^- reveals its presence in the solar atmosphere by its absorption spectrum.

We may finally remark on the value of $\bar{\kappa} = 1.42\ \bar{\kappa}(H^-)$, indicated by the comparisons we have just made: it can, in fact, be deduced empirically from the solar data on the continuous spectrum that the absorption in the violet ($\lambda < 4000$ A) must increase the value of κ derived from the visible and the infrared regions of the spectrum by a factor of the order of 1.5.[20]

4. *The predicted color-effective temperature relations: comparison with observations.*—All earlier attempts[21] to predict the color temperatures in the region λλ 4000–6500 A for stars of spectral types A0–G0 in agreement with the observations have failed. This failure in the past has been due to the following circumstance: The observed relation between the color and the effective temperatures and, in particular, the fact that $T_c > T_e$

[20] Cf. G. Münch, *Ap. J.*, **102**, 385, 1945, esp. the remarks preceding eq. (19) on p. 394.

[21] R. Wildt, *Ap. J.*, **93**, 47, 1941, and *Observatory*, **64**, 195, 1942; R. E. Williamson, *Ap. J.*, **97**, 51, 1943.

implies that the continuous absorption coefficient is an increasing function of λ in the spectral region observed. But the physical theory on which the calculations were made placed the maximum of the absorption-curve in the region of λ 4500 A; this was incompatible with the observations and, moreover, predicted color temperatures less than the effective temperatures, contrary to all evidence. Indeed, on the strength of this discrepancy, it was concluded that H^- as a source of absorption was inadequate even in the region $\lambda\lambda$ 4500–6500 A, and the existence of an unknown source operative in this region was further inferred. However, later evaluations[22] of the bound-free transitions of H^- showed the unreliability of earlier determinations and placed the maximum of the absorption-curve in the neighborhood of λ 8500 A. The addition of the free-free transitions pushes this maximum only still further to the red. It is therefore evident that on the revised physical theory we should be able to remove the major discrepancies of the subject. We shall now show how complete the resolution of these past difficulties is.

From the point of view of establishing the adequacy of the physical theory in the region $\lambda\lambda$ 4000–6500 A, it is most instructive to consider the theoretical predictions for color temperatures which can be directly compared with the color determinations at Greenwich,[3] for the Greenwich measures are based on the mean gradients in the wavelength interval 4100–6500 A, and it is in the prediction of these colors that the earlier calculations were most discordant.[23]

Now, from the Planck formula in the form

$$i_\lambda = \frac{2hc^2}{\lambda^5} \frac{1}{e^{c_2/\lambda T} - 1}, \tag{10}$$

it readily follows that

$$\frac{1}{i_\lambda} \frac{d i_\lambda}{d(1/\lambda)} = 5\lambda - \frac{c_2}{T}\left(1 - e^{-c_2/\lambda T}\right)^{-1}. \tag{11}$$

Defining the gradient

$$\phi = \frac{c_2}{T}\left(1 - e^{-c_2/\lambda T}\right)^{-1} \tag{12}$$

in the usual manner, we can write

$$\frac{1}{M} \frac{d \log_{10} i_\lambda}{d(1/\lambda)} = 5\lambda - \phi \qquad \left(\frac{1}{M} = 2.303\right). \tag{13}$$

If F_{λ_1} and F_{λ_2} are the emergent fluxes at two wave lengths λ_1 and λ_2 and if ϕ is the mean gradient in this wave-length interval, then we can write, in accordance with equation (13)

$$\phi = 5\lambda_m - \frac{1}{M} \frac{\log_{10}(F_{\lambda_1}/F_{\lambda_2})}{\lambda_1^{-1} - \lambda_2^{-1}}, \tag{14}$$

where λ_m denotes an appropriate mean wave length for the interval to which the gradient ϕ refers. Equation (14) can be re-written in the following form:

$$\phi = 5\lambda_m - \frac{1}{M} \frac{\Delta \log_{10} F}{\Delta(1/\lambda)}. \tag{15}$$

According to equation (15), the theoretical determination of color temperatures will proceed by determining, first, the gradient ϕ from the values of F_λ at the end-points of the wave-length interval and then determining the temperature which will give this gradient.

[22] S. Chandrasekhar, *Ap. J.*, **102**, 223, 395, 1945

[23] Cf. Fig. 6 in Williamson's paper (*op. cit.*).

For the Greenwich measures $\lambda_m = 0.55\ \mu$, and equation (15) becomes

$$\phi\ (\text{Greenwich}) = 2.75 - 2.56 \log_{10}\left(\frac{F_{4100}}{F_{6500}}\right), \tag{16}$$

provided that, in determining the gradients, wave lengths are measured in microns.[24]

In Table 4 we have listed the values of κ'_λ/κ for the wave lengths 4100 A and 6500 A for various values of θ_e and p_e. In terms of these values the determination of the fluxes at the two wave lengths 4100 A and 6500 A is straightforward with the help of Burkhardt's table. The gradient ϕ then follows according to equation (16) and, from that, the color temperature. The reciprocal color temperatures $\theta_c = 5040/T_c$ derived in this manner are given in Table 5. The resulting color–effective temperature relations are illustrated

TABLE 4

$\kappa'_\lambda/\bar{\kappa}$ IN MODEL STELLAR ATMOSPHERES

$5040/T_e$	λA	$p_e=10$	$p_e=10^2$	$p_e=10^3$	$p_e=10^4$
	λ 3647	.....	3.45	3.22	2.68
		.....	0.161	0.181	0.332
	λ 4000	0.210	0.212	0.238	0.397
θ=0.5	λ 4600	0.316	0.318	0.346	0.513
	λ 6500	0.826	0.834	0.848	0.991
	λ 8203	.....	1.51	1.51	1.55
		.....	0.490	0.526	0.788
	λ 3647	.....	4.04	2.60	1.09
		.....	0.178	0.427	0.685
	λ 4000	0.148	0.205	0.481	0.770
θ=0.6	λ 4600	0.221	0.278	0.585	0.901
	λ 6500	0.579	0.637	0.935	1.25
	λ 8203	.....	1.27	1.36	1.45
		.....	0.424	0.885	1.36
	λ 3647	6.06	2.82	0.961	0.684
		0.176	0.461	0.625	0.645
	λ 4000	0.208	0.521	0.695	0.725
θ=0.7	λ 4600	0.276	0.617	0.810	0.848
	λ 6500	0.546	0.901	1.11	1.14
	λ 8203	0.889	1.105	1.22	1.24
		0.356	0.893	1.19	
	λ 3647	1.54	0.720	0.625	....
		0.518	0.601	0.614	0.611
	λ 4000	0.582	0.671	0.685	0.685
θ=0.8	λ 4600	0.681	0.781	0.800	0.800
	λ 6500	0.961	1.06	1.09	1.09
	λ 8203	1.11	1.178	1.19	1.19
		1.01	1.164		
	λ 3647	0.708	0.642	0.636	0.636
		0.631	0.636		
	λ 4000	0.690	0.690	0.690	0.690
θ=0.9	λ 4600	0.800	0.800	0.800	0.800
	λ 6500	1.09	1.09	1.09	1.09
	λ 8203	1.16	1.16	1.16	1.16
		.....	....	....	....

[24] The constant c_2 in eq. (12) then has the value 14,320.

in Figure 2. For comparison we have also plotted in this figure the Greenwich determinations for stars on the main sequence and of spectral types A0–G0 (reduced, however, to the Morgan, Keenan, and Kellman system of spectral classification). The color temperature of the sun for this wave-length interval is also plotted in Figure 1. It is seen from Figure 1 that the agreement between the observed and the theoretical color temperatures is entirely satisfactory, particularly when it is remembered that the earlier calculations failed even to predict the correct sign for $\theta_c - \theta_e$. It will, however, be noted that the observed values of θ_c for spectral types later than F0 are somewhat larger than the pre-

TABLE 5

THEORETICAL RECIPROCAL COLOR TEMPERATURES AND THE PREDICTED DISCONTINUITIES AT THE HEAD OF THE BALMER AND THE PASCHEN SERIES*

p_e		θ_e					
		0.5	0.6	0.7	0.8	0.9	1.0
10	$\theta_c(G)$	....	....	....	0.68	0.80	0.88
	$\theta_c(B$ and $C)$	....	....	....	.61	.69	.75
	D_B	....	....	....	.31	.015	....
10^2	$\theta_c(G)$	....	0.45	0.60	.73	.80	.88
	$\theta_c(B$ and $C)$	....	....	.47	.63	.69	.75
	D_B	....	....	[.50]	.07	....	....
	D_P	....	....	.030	.001	....	....
10^3	$\theta_c(G)$	[0.31]	.52	.65	.74	.80	.88
	$\theta_c(B$ and $C)$	....	.42	.56	.64	.69	.75
	D_B	[.58]	[.34]	.13	....	....	....
	D_P	.114	.051	.003	....	....	....
10^4	$\theta_c(G)$	[.39]	.59	.66	.74	.80	.88
	$\theta_c(B$ and $C)$	[.34]	.50	.58	.64	.69	.75
	D_B	[.40]	.12	.018	....	....	....
	D_P	[.071]	.006	....	....	....	....
Pure H^-	$\theta_c(G)$	.52	.59	.66	.74	.80	.88
	$\theta_c(B$ and $C)$	0.41	0.50	0.58	0.64	0.69	0.75

* $\theta_c(G)$ and $\theta_c(B$ and $C)$ are the reciprocal color temperatures, $5040/T_c$, appropriate for the wave-length intervals 4100–6500 A and 4000–4600 A, respectively; D_B and D_P, representing the logarithm of the ratio of the fluxes at the two sides of the series limits, are the expected Balmer and Paschen discontinuities, respectively.

dicted values, though the agreement is as good as can be expected in the case of the sun. The reason for this must undoubtedly be the crowding of the absorption lines toward the violet in the later spectral types and the consequent depression of the continuous spectrum in this region. The correctness of this explanation is apparent when it is noted that in the case of the sun, in which allowance has been made for this effect of the lines on the continuum, the discordance is not present.

Comparisons similar to those we have just made also can be made with the measurements of Barbier and Chalonge[4] on the color temperatures based on the observed gradients in the wave-length interval 4000–4600 A. The formula giving the theoretical gradient for this wave-length interval takes the form

$$\phi\ (B \text{ and } C) = 2.175 - 7.06 \log_{10}\left(\frac{F_{4000}}{F_{4600}}\right). \qquad (17)$$

The values of $\kappa'_\lambda/\bar{\kappa}$ at λ 4000 A and λ 4600 A are given in Table 4, and the reciprocal color temperatures derived from these in Table 5. The results are further illustrated in Figure 3, where the theoretical relations for various electron pressures are compared with the measures of Barbier and Chalonge (reduced also to the Morgan, Keenan, and Kellman system of spectral classification). It is seen that the general agreement is again good, though there are now somewhat larger differences between the computed and the observed color temperatures for spectral types later than F0 than were encountered in the comparison with the Greenwich colors. This must again be due to the crowding of the absorption lines toward the violet in the later spectral types and the further fact that

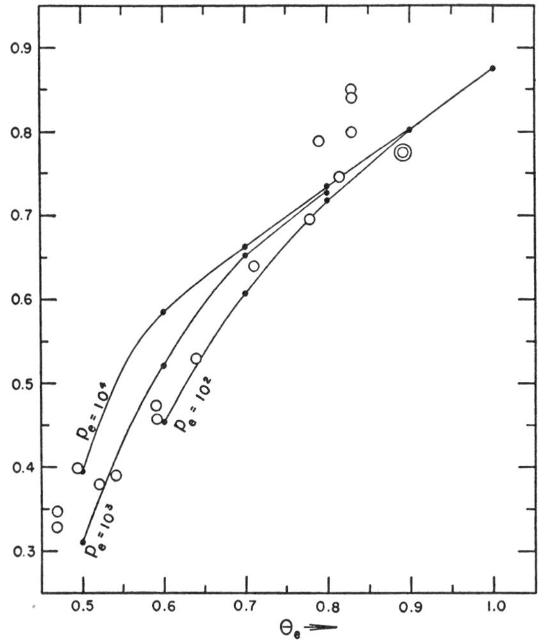

FIG. 2.—Comparison of the observed and the theoretical color effective-temperature relations for the wave-length interval 4100–6500 A. The ordinates denote the reciprocal color temperatures and the abscissae denote the reciprocal effective temperatures ($\theta = 5040/T$). The circles represent the Greenwich color determinations reduced to the Morgan, Keenan, and Kellman system of spectral classification. The double circle represents the sun.

the base line for the Barbier and Chalonge colors is much shorter than that for the Greenwich colors.

5. *The discontinuities at the head of the Balmer and the Paschen series.*—With the physical theory of the continuous absorption coefficient now available, we can also predict the extent of the discontinuities which we may expect at the head of the Balmer and the Paschen series of hydrogen. For this purpose the values of $\kappa'_\lambda/\bar{\kappa}$ on the two sides of the series limits are also given in Table 4. From these values it is a simple matter to estimate the discontinuities which will exist at the head of the Balmer and the Paschen series, and they are given in Table 5. The results for the Balmer discontinuities are further illustrated in Figure 4, in which the discontinuities measured by Barbier and Chalonge[4] for various stars are also plotted. The progressive increase of the electron pressure as we go from the later to the earlier spectral types is particularly apparent

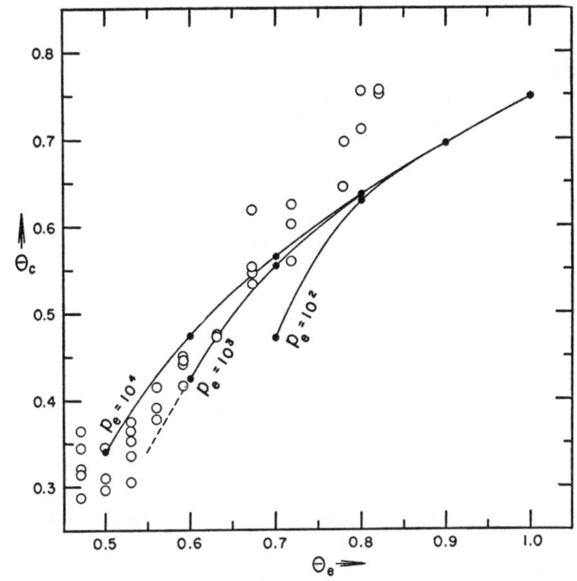

FIG. 3.—Comparison of the observed and the theoretical color effective-temperature relations for the wave-length interval 4000–4600 A. The ordinates denote the reciprocal color temperatures, and the abscissae denote the reciprocal effective temperatures $\theta = 5040/T$). The circles represent the color determinations of Barbier and Chalonge for the wave-length interval 4000–4600 A, reduced to the Morgan, Keenan, and Kellman system of spectral classification.

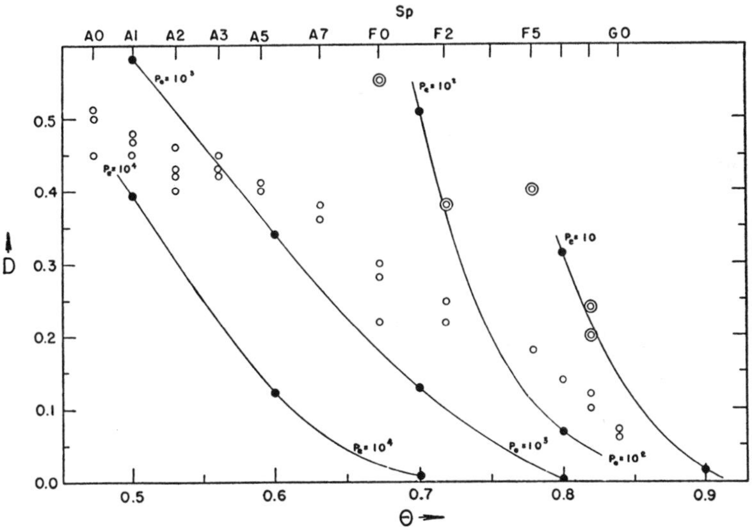

FIG. 4.—The predicted discontinuities (D) at the head of the Balmer series for various effective temperatures and electron pressures. The circles represent the discontinuities as measured by Barbier and Chalonge. (The double circles represent the observations for supergiants.)

from Figure 4; this progression is, moreover, in agreement with what is indicated by the color determinations (cf. Figs. 2 and 3). In Table 6 we give the electron pressures for the main-sequence stars of various spectral types estimated in this manner.

6. *Concluding remarks.*—While our discussion in the preceding sections has established the unique role which H^- plays in determining the character of the continuous spectrum of the sun and the stars, it should not be concluded that the various other

TABLE 6

ELECTRON PRESSURES FOR STARS ON THE MAIN SEQUENCE

Type	log p_e	Type	log p_e	Type	log p_e
A1	3.7	F0	2.4	F8	1.4
A2	3.3	F2	2.2	G0	1.2
A3	3.0	F4	2.0	G2	1.0
A5	2.8	F5	1.8		
A7	2.6	F6	1.6		

astrophysical elements of the theory are equally well established. Indeed, the theory of model stellar atmospheres as developed by Strömgren in recent years must not only be revised on the basis of the new absorption coefficients of H^- but also be advanced still further before we can be said to have a completely satisfactory account of all the classical problems of the theory of stellar atmospheres. But our discussion in this paper does give us the confidence that the continuous absorption by H^- discovered by Wildt must provide the key to the solution of many of these problems.

33

THE STORY OF TWO ATOMS

By S. CHANDRASEKHAR
Yerkes Observatory, Williams Bay, Wis.

IT IS now a matter of common knowledge that the nuclear model of the atom first proposed by Rutherford in 1911 is basic for all the natural sciences. On this model an atom is pictured as a minute positively charged nucleus, in which the greater part of its mass is concentrated, surrounded by a number of electrons. Under normal conditions, the number of electrons surrounding the nucleus is such as to make the whole atom electrically neutral. However, atoms with a net positive or negative charge are also known to exist. Thus, atoms with a positive charge can be obtained by simply removing one or more of the outer electrons of a neutral atom. Such *positive ions*—as they are sometimes called—can be produced in the laboratory under suitable conditions (usually in a spark discharge); they are also spectroscopically observed in the atmospheres of the sun and the stars where the high temperatures and low pressures prevailing are appropriate for their occurrence in the free state. On the other hand, *negative ions*, which in all cases known consist of one additional electron attached to a neutral atom, are of less common occurrence: indeed, whether a particular neutral atom can or cannot form a stable atomic configuration with an additional electron is a question not always easily answered. When the stability of a negative ion can be established, the corresponding neutral atom is said to have a positive *electron affinity*.

It is evident that on the Rutherford model, the simplest type of atomic structure will be represented by atoms (or ions) with one electron. The prototype of these is of course the hydrogen atom, which consists of an electron in the field of a proton. Singly ionized helium,[1] doubly ionized lithium, or triply ionized beryllium are also examples of one-electron systems. The next simplest type of atomic architecture will be represented by atoms (or ions) with two electrons: and the prototype of these is helium (named by Norman Lockyer after ἥλιος, "the sun"). Although other atomic configurations with two electrons (such as lithium, once ionized, or beryllium, twice ionized) are well known and have been produced in the laboratory, the case of the *negative hydrogen ion* consisting of an electron stably bound to a hydrogen atom is unique in that its stability has been established on purely theoretical grounds and it has been positively detected only in the sun's atmosphere. Indeed, it is remarkable that the story of the discovery and the identification of the two atoms, helium and negative hydrogen ion, should be so intimately connected with astrophysical studies relating to the constitution of the sun's atmosphere. And the story is worth telling since it illustrates once again the basic unity of all the sciences—in this instance, the unity of astronomical and laboratory physics.

IN THE year 1706 there was an eclipse of the sun visible in Switzerland, and there was one Stannyan who gave an account of what he saw at Bern. After describing the phenomena of the eclipse, he said, referring to the sun, "His getting out of eclipse was preceded by a blood-red streak of light." There are no photographic rec-

[1] An atom is said to be once ionized if one electron from the neutral atom has been removed, twice ionized if two electrons have been removed, and so on.

ords of the eclipse of 1706 (it was before the days of photography and astronomical spectroscopy), but we do have records of similar later eclipses, and in all of them Stannyan's observation has been confirmed. Indeed, it is now known that the origin of the blood-red streak which Stannyan first observed is to be found in one of the solar envelopes to which the name *chromosphere* was given by the English astronomer Norman Lockyer.

In the year 1868, during an eclipse of the sun visible in India, the spectroscope was first put into effective use for the study of the chromosphere by Lockyer, Pogson, and Janssen. In Figure 1 we have illustrated Pogson's original diagram of the spectrum of the chromosphere. It will be seen that in the spectrum there is a *bright line* appearing in the position of the dark (Fraunhofer) D line of the normal solar spectrum. Referring to this yellow line, Pogson said that it was "at D or near D." Almost the whole of the story of helium depends on this distinction.

After the eclipse of 1868 a method suggested much earlier for studying the chromospheric prominences without waiting for an eclipse was put into operation and has been extensively used ever since. The method consists in forming a telescopic image of the sun on the slit of a spectroscope so that the spectrum of the sun's edge and of the sun's surrounding envelope could be seen simultaneously. Consequently, exact coincidence or want of coincidence between the bright lines of the chromospheric prominences and the dark Fraunhofer lines of the normal solar spectrum could be ascertained at once. By using this method, it was shown that there was exact coincidence between the lines C and F (Hα and Hβ, respectively) appearing as dark lines in the solar spectrum and the strong bright lines in the red and the blue-green regions of the spectrum of the "blood-red streak" (Figs. 2 and 3). On the other hand, no such coincidence was found between the D lines originating from the sodium vapor present in the solar atmosphere and the bright yellow line of the prominences (Fig. 4). The prominence line in this position has no connection whatever with the dark lines of the ordinary solar spectrum. The considerable significance attached to this divergence by Lockyer, Janssen, and others who had examined the matter is therefore understandable. The new line was accordingly called D_3 to distinguish it from the sodium lines D_1 and D_2, and the presence in the solar atmosphere of a new element, "helium," not terrestrially known at the time, was concluded.

A considerable amount of laboratory work was done with regard to the D_3 line.

But the substance emitting the yellow ray lay outside the range of our acquaintanceship, and seemed unlikely to be brought within it. That contingency, nevertheless, came to pass. In the

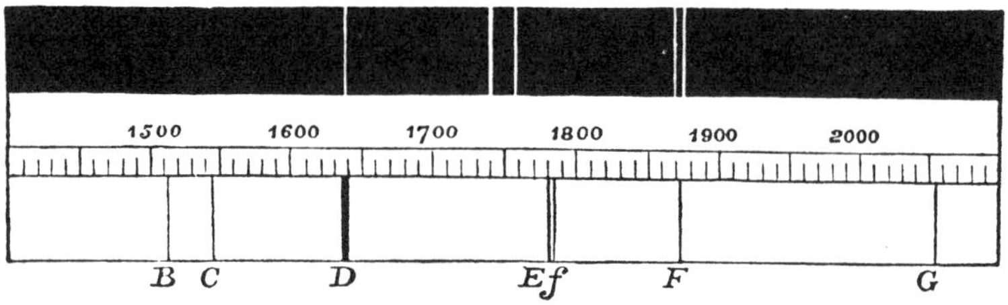

Fig. 1. Pogson's diagram of the spectrum of the sun's envelope during the eclipse of 1868. The bright lines which were seen are shown in the upper part of the diagram and the principal dark Fraunhofer lines in the normal solar spectrum are shown in the lower part.

Fig. 2. The coincidence of the red chromospheric-prominence line with the dark line $C(H_\alpha)$.

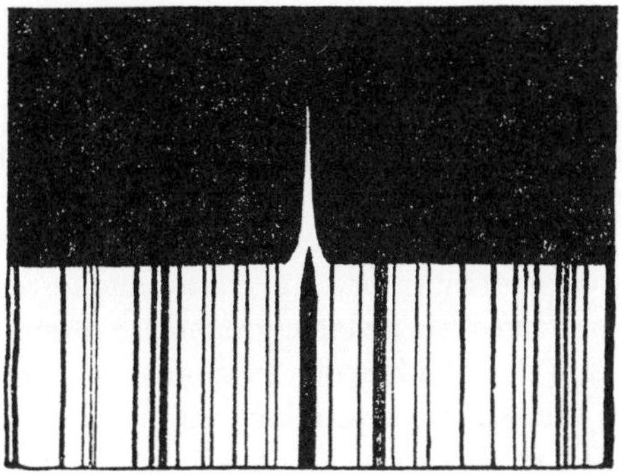

Fig. 3. The coincidence of the blue-green chromospheric-prominence line with the dark line $F(H_\beta)$.

course of a search for compounds of argon, Professor Ramsay, at the suggestion of Professor Miers,[2] fortunately examined the reputed nitrogen occluded by the Scandinavian mineral "clevite." This velvety-black stone, remarked as peculiar by Nordenskiöld and analyzed by Cleve, is a kind of pitchblende, composed of uranate of lead mixed with rare earths. The gas evolved from it at University College [London] gave a brilliant spectrum in which the prominence-line D_3 shone conspicuous. Helium was indeed captured! A beautiful confirmation of the identity was soon afterwards afforded. The golden line seen in the laboratory was perceived by Runge to have a faint close companion, and he declared that, unless the solar D_3 were also double, clevite-gas should be regarded as different from helium. The challenge was taken up on both sides of the Atlantic. Professor Hale on the 20th of June, 1895, and Sir William Huggins independently on July 10th, succeeded in resolving the prominence-ray into a delicate, unequal pair, and our possession of helium as a truly indigenous element was rendered incontrovertible.[3]

[2] Professor Miers, of the British Museum, recalled to Ramsay that W. F. Hildebrand, of the U. S. Geological Survey, had observed in 1889 that when certain uranium-containing minerals were boiled with sulphuric acid, a quantity of gas was evolved. The gas was, however, supposed to be nitrogen as its spectrum showed the characteristic fluting, and, consequently, it was not further investigated.

[3] Clerke, Agnes M. *Problems in Astrophysics.* London: Adam and Charles Black, 1903, 56.

THAT a neutral hydrogen atom, together with an electron, can form a stable atomic configuration was first conclusively proved by H. A. Bethe and E. A. Hylleraas, independently, in 1930. Since this is a unique instance in which the stability of an atom had been first established from theoretical considerations, it is of interest to see how Bethe and Hylleraas arrived at their conclusion.

On the quantum theory any arbitrary function of the coordinates of the electrons of an atomic system represents a possible state of the system; and the *probable*

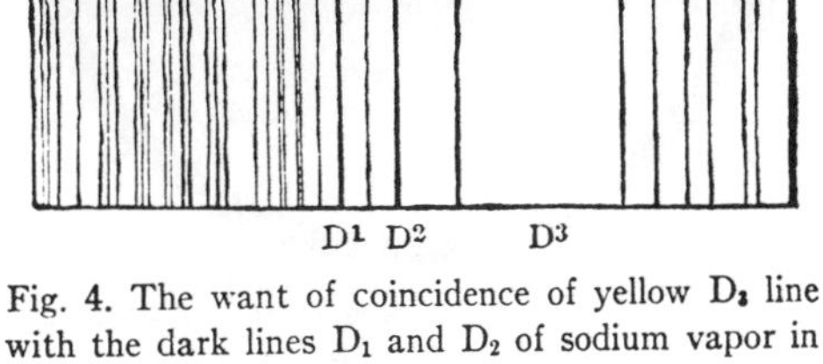

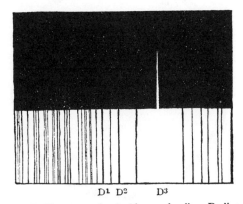

Fig. 4. The want of coincidence of yellow D_3 line with the dark lines D_1 and D_2 of sodium vapor in the sun's atmosphere.

energy (i.e., the energy which would be obtained as the average of many independent determinations on the system in the state considered) appropriate to the state in question can be found in a unique manner according to established rules of the quantum theory. If the atomic system considered has a stable ground (or normal) state then the function (*wave function*, as it is then called) describing such a state will lead to an energy which will be *less* than that given by *any* other function. This is one of the fundamental principles of the quantum theory. And we shall now indicate how this principle can be used to decide whether a given atomic system has or has not a stable ground state. We shall illustrate this by considering specifically the case of the negative hydrogen ion. The question is: Can an electron attach itself stably to a hydrogen atom?

We know that a proton and an electron can be stably bound together to form a hydrogen atom. And the energy of binding is 13.54 electron volts, meaning that an amount of work necessary for displacing an electron against a potential difference of 13.54 volts will be required for separating the constituents of a hydrogen atom, to rest, at infinite separation. We may accordingly say that the energy of the normal state of the hydrogen atom is −13.54 electron volts if we take as our zero of energy the state in which the charged particles constituting an atom are at rest at infinite separation. If the hydrogen atom together with an electron can form a stable atom, then its energy should be *less* than −13.54 electron volts: for, if a stable negative hydrogen ion exists, a finite amount of work should be necessary to separate the additional electron to infinity and leave a normal hydrogen atom behind with an energy of −13.54 electron volts. Consequently, *if* we can find *some* function which, according to the rules of the quantum theory, will lead to an energy less than −13.54 electron volts, we shall have proved the stability of the negative hydrogen ion: for the normal state must have an energy *lower* than any that we can find—unless we have been lucky enough to guess the correct wave function itself!

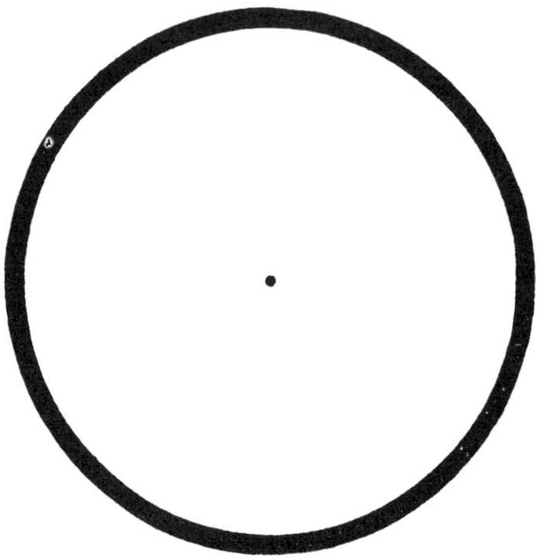

Fig. 5. A model of the hydrogen atom in its ground state. The electron will be found on the average at a distance of 0.79×10^{-8} cm. from the proton.

Bethe and Hylleraas were able to do precisely this: they were, in fact, able to isolate a function of the coordinates of the two electrons in the field of a proton which gave an energy of −14.24 electron volts indicating that the electron affinity of hydrogen is positive and must *exceed* 0.70 electron volts. A more recent theoretical determination has increased the value of the electron affinity of hydrogen to 0.75 electron volts; there are reasons to believe that this last value must be very near the true value.

Having seen that the negative hydrogen ion as a stable atom exists, we may ask, "What does it look like?" In one sense, it is not reasonable to ask such a question since on the quantum theory we cannot locate or individually identify the different electrons in an atom: we can only give the *probability*

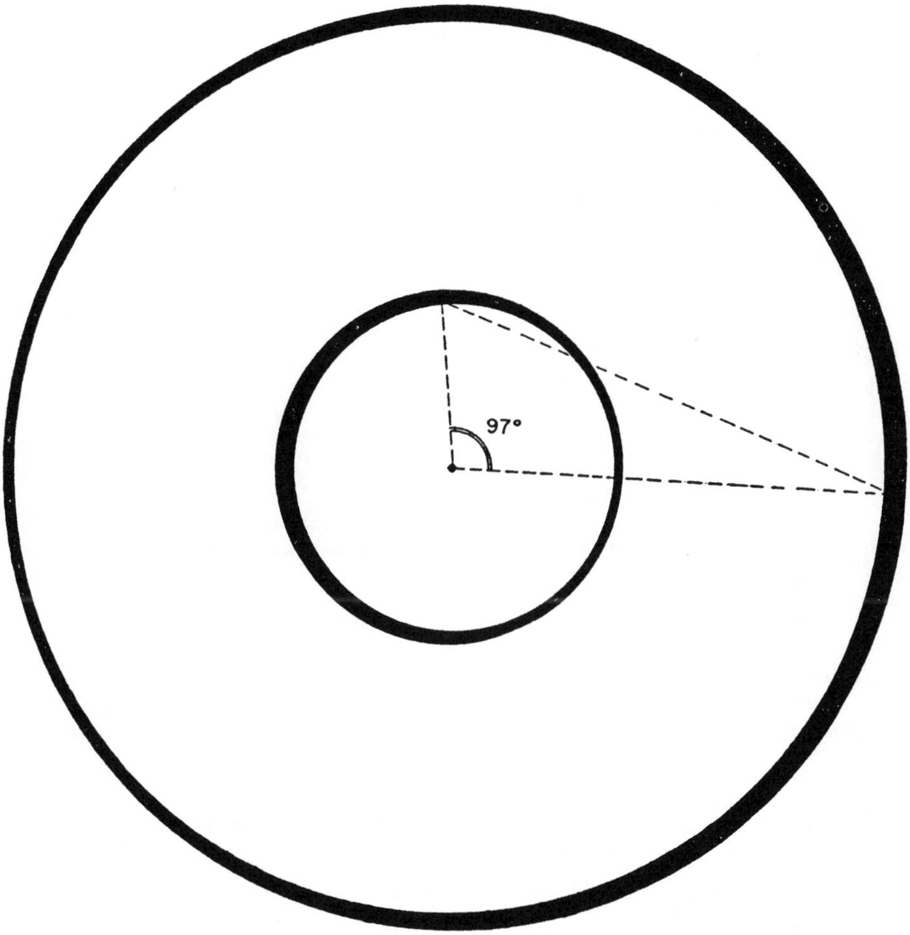

Fig. 6. A model of the negative hydrogen ion. The inner electron will be found on the average at a dis tance of 0.75×10^{-8} cm. from the proton while the outer electron will be found on the average at a distance of 2.02×10^{-8} cm. from the proton. The fact that the two electrons will be more often found on the opposite sides of the proton than on the same side is indicated by the unsymmetrical shading of the orbits. Calculations show that on the average the angle between the lines joining the proton to the two electrons will be about 97°.

that the electrons will be found in assigned positions. Nevertheless, we may ask where the electrons will be found on the *average.* In a hydrogen atom in its normal state, for example, the electron will be found at an average distance of 0.79×10^{-8} cm. from the proton (Fig. 5). With the same reservations, we may picture the negative hydrogen ion as in Figure 6, with the inner electron at a distance of 0.75×70^{-8}cm. from the proton—this is only slightly less than it is in the hydrogen atom—and the outer electron at a distance 2.02×10^{-8} cm. from the proton. However, the positions of the two electrons are correlated in the sense that the two electrons will be found more often on the opposite sides of the proton than on the same side: it is as though

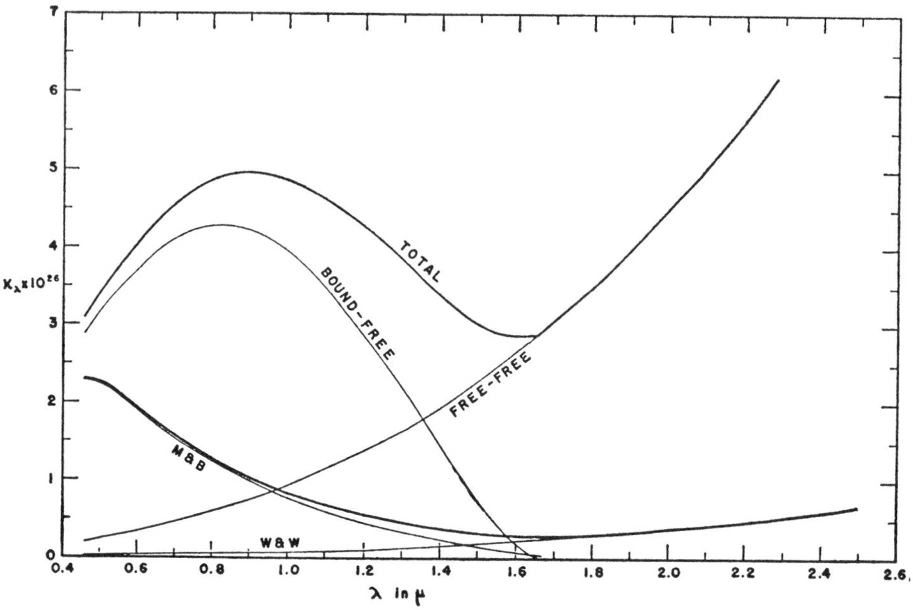

Fig. 7. The absorptive power of the negative hydrogen ion for light of various colors derived from physical theory. The curves are drawn appropriately for an atmosphere at a temperature of 6300° Kelvin and an electron pressure of one dyne/cm.^{2} and per neutral hydrogen atom. The curves *M* and *B*, *W* and *W*, and their sum are the results of earlier determinations by Massey and Bates, and Wheeler and Wildt. The results of more recent determination by Chandrasekhar and Breen are illustrated by the remaining curves.

the inner electron is accommodating enough to let the outer electron have an occasional glimpse of the proton to strengthen its binding!

We have indicated how the principles of the quantum theory enable us to determine the stability and structure of the negative hydrogen ion. But the atom has not so far been isolated with definiteness in the laboratory. We shall now show how its presence in the solar atmosphere has recently been inferred.

It is known from careful measurements of the *solar constant* (which gives the amount of radiant energy received from the sun, at the earth's distance, by a unit area normal to the line joining it to the sun) that the sun radiates 3.78×10^{33} ergs per second to the space outside. Since the radius of the sun is 6.95×10^{10} cm., it follows that each square centimeter of the sun radiates 6.23×10^{10} ergs per second. This *flux* of emergent radiant energy is not emitted in any particular wave length or color. The energy is actually emitted in all colors with a definite determinable distribution of intensity. This distribution of the emitted solar energy over the different colors has been determined with great precision by C. G. Abbot and his collaborators at the Smithsonian Institution. The experimentally derived distribution is illustrated in Figure 10 (curve 2). It is evident that we should be able to infer from this energy distribution something significant about the constitution of the solar atmosphere, particularly about the relative effectiveness with which the material composing the solar atmosphere absorbs light of various colors. For the radiation we receive at any particular wave length (or color) is the resultant of the radiation of this color emitted by the layers of the solar atmosphere at *all* depths: only the weight with which the deeper layers contribute to the emergent intensity will be progressively reduced on account of absorption by the intervening material. The effectiveness with which the deeper layers affect the emergent radiation in any particular color will therefore depend on the absorptive power of the solar material for radiation of this color. Thus, if the absorptive power of the solar material is less in blue-green than it is in yellow-orange, then in the blue-green region of the spectrum we should effectively see to greater depths in the solar atmosphere than we do in the yellow-orange region; consequently, the radiation in the blue-green region must be more characteristic of the deeper layers at a higher temperature than the radiation in the yellow-orange region, which will be more characteristic of the higher layers at a lower temperature. It is clear that in this manner we should be able to infer the relative absorptive power of the solar atmosphere for radiation of different colors. This was first done by E. A. Milne in 1922, and his results have been generally confirmed and extended by later investigators. These investigations show that the absorptive power of the solar atmosphere increases by a factor of the order of two as the wave length increases from 4,000 to 9,000A; beyond λ 9,000A the absorptive power decreases by about the same amount until we reach λ 16,000A in the infrared; and the indications are that, as we go further into the far infrared, the absorptive power again increases.

For many years one of the principal problems of astrophysics was to determine the source of continuous absorption in the solar atmosphere which will show the behavior we have described. All efforts in this direction failed until R. Wildt suggested in 1938 that we should perhaps look for the source of continuous absorption in the solar atmosphere in the presence of negative hydrogen ions in the atmosphere. The

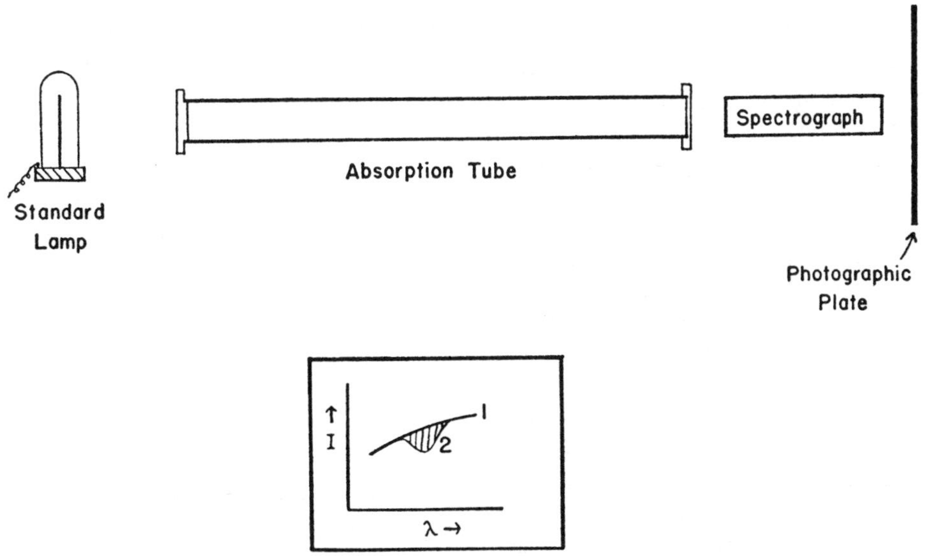

Fig. 8. The schematic experimental arrangement used in laboratory investigations for the spectroscopic identification of an unknown gas.

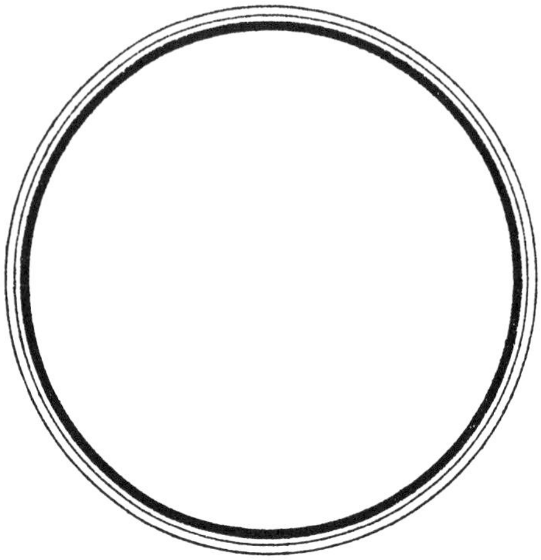

Fig. 9. A schematic drawing of the sun's photospheric layers.

grounds for the expectation were simply that in the solar atmosphere there is an abundance of neutral hydrogen atoms and there is also a supply of free electrons from easily ionized atoms such as sodium, calcium, magnesium, silicon, and the rest; and, in view of the positive electron affinity of hydrogen, some of the electrons will attach themselves to the neutral hydrogen atoms present. This suggestion of Wildt was taken up with great enthusiasm by other astrophysicists. But, before definite conclusions could be drawn, the absorptive power of negative hydrogen ions for radiation of different colors had to be derived, again from physical theory. The first determinations of the absorptive power of the negative hydrogen ion by Massey and Bates and others were disappointing. As will be seen from Figure 7, these early determinations were at complete variance with the astrophysical demands. However, it was soon realized that a reliable determination of the absorptive power of the negative hydrogen ion is an exceptionally difficult problem on account of its relatively very large size. The difficulties have now been overcome satisfactorily, and the new determinations are entirely in accord with the solar data. Indeed, we shall now show how, with the help of the newer determination of the absorptive power of the negative hydrogen ion, we can actually *see* the presences of these atoms in the solar atmosphere.

Let us recall how we normally identify any particular gaseous substance by its absorption spectrum in the laboratory (Fig. 8). First, we have a standard lamp which emits radiation in a known manner in all wave lengths (curve 1, Fig. 8). Next, we introduce the unknown gas in an absorption tube and we let the light from the standard lamp pass through the absorption tube and analyze the transmitted light spectroscopically. By comparing the intensity distribution in the transmitted light (curve 2, Fig. 8) with that emitted by the standard lamp (curve 1) we get the *absorption curve* of the gas. It is in terms of this *absorption* that the identification of the gas is made. Similarly, the radiation from the deeper layers of the sun passing through the outer (photospheric) layers gives rise to the observed emergent continuous spectrum of the sun (Fig. 9 and curve 2,

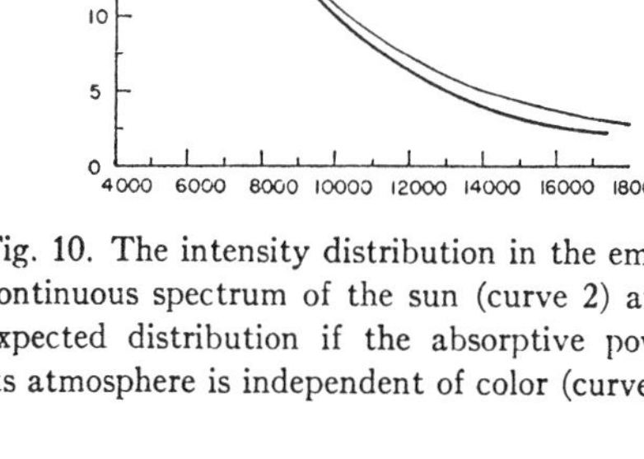

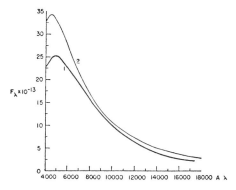

Fig. 10. The intensity distribution in the emergent continuous spectrum of the sun (curve 2) and the expected distribution if the absorptive power of its atmosphere is independent of color (curve 1).

Fig. 10). It would seem that the analogy with the laboratory experiment breaks down at this point since we do not seem to have the analogue of the standard lamp. However, we can construct a theoretical standard of reference in the following way: Suppose that the absorptive power of the solar atmosphere is the same in all colors; in other words, suppose it is *gray*. We can then predict in a relatively straightforward way what the intensity distribution in the emergent radiation should be. This is indicated by curve 1 in Figure 10. We use this as our standard of reference. The difference in curves 1 and 2 in Figure 10 must be due then to the *variation* of the absorptive power of the solar atmosphere with color. We may, if we like, call this difference the *solar absorption curve.* This is shown as a series of points in Figure 11. Now, we can compute what the solar absorption curve should be if the absorbing mechanism is provided by the negative hydrogen ions. This is shown as the full-line curve in Figure 11. It is seen that the agreement between the observed absorption curve and that to be expected from the negative hydrogen ion is very satisfactory; indeed, so satisfactory that we are justified in saying that this atom has been detected spectrophotometrically in the sun's atmosphere.

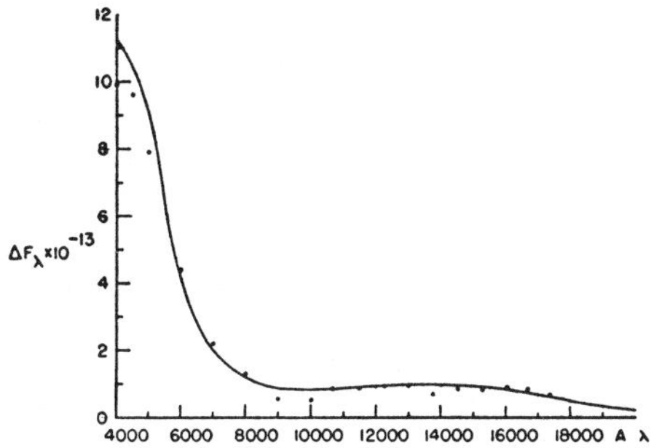

Fig. 11. A comparison of the observed "solar absorption curve"—the difference of curves 1 and 2 of Figure 10—and that to be expected on account of the presence of negative hydrogen ions in the solar atmosphere. The former is indicated by dots while the latter (derived from a recent investigation of Chandrasekhar and Münch) is shown by the full-line curve.

The story of the negative hydrogen ion we have just concluded has, in some respects, a plot which is strangely the reverse of that of helium. In the case of helium, the presence of a new element in the sun was first concluded on the strength of spectroscopic evidence, and it was only a quarter of a centry later that its terrestrial existence was established. In the case of the negative hydrogen ion, on the other hand, its stable existence was first established on theoretical grounds, and it was only some 15 years later that its presence in the solar atmosphere was concluded, on the strength of spectrophotometric evidence.

The stories of helium and negative hydrogen ion we have told have a moral. It is, however, the same as it was 50 years ago when Norman Lockyer, concluding the story of helium, wrote, "The more we can study the different branches of science in their relation to each other, the better for the progress of all sciences".

Shift of the 1^1S State of Helium

S. CHANDRASEKHAR AND DONNA ELBERT, *Yerkes Observatory, University of Chicago, Williams Bay, Wisconsin*

AND

G. HERZBERG, *Physics Division, National Research Council, Ottawa, Canada*

(Received May 22, 1953)

In this paper the shift of the 1^1S state of helium is reconsidered and it is shown that recent discussions of the problem are subject to considerable doubt. The doubt arises from the unreliability (to the required precision) of the current theoretical determinations of the energy of the ground state. The result of an improved calculation for the latter is presented; and if this is accepted as a sufficient approximation, an unexplained shift of 21.5 cm^{-1} (opposite in direction to that previously suggested) would result for He. Similar unexplained shifts (but in the same direction as those suggested earlier) are also predicted for the other He-like ions.

A NUMBER of investigators[1] have recently tried to determine the electromagnetic shift (Lamb shift) of the ground state of He and He-like ions. Unfortunately, the experimental value for the ionization potential of He is known[2] only to ± 15 cm^{-1}, or perhaps ± 5 cm^{-1}, which is of the order of the expected shift. In preparation for a new and more precise spectroscopic determination of the ionization potential of He at Ottawa, it appeared of interest to inquire how reliable the presently accepted theoretical value of I(He) is and whether its accuracy can be improved. It is generally assumed that the energy of the ground state of helium is known from theory with an accuracy (± 2 cm^{-1}) appreciably better than from experiment. But an examination of the calculations in the literature shows that the situation is by no means so favorable.

As is well known, Hylleraas[3] made the first successful attempt to reach a high degree of accuracy in the theoretical prediction of the ground state of two-electron systems, including helium and the negative hydrogen ion. His calculations were based on the variational principle; and in minimizing the energy given by wave functions of chosen forms, Hylleraas assumed that

$$\psi = e^{-\frac{1}{2}ks} \sum c_{lmn}(k)^{l+m+n} s^l t^m u^n, \qquad (1)$$

where s, t, and u are related to the distances r_1, r_2, and r_{12} (measured in atomic units) of the two electrons from the nucleus and each other, respectively, by

$$s = r_1 + r_2, \quad t = r_2 - r_1, \quad \text{and} \quad u = r_{12}. \qquad (2)$$

Further in Eq. (1), k and the c_{lmn}'s are constants with respect to which the energy is minimized. By choosing a wave function of the form

$$\psi = e^{-\frac{1}{2}ks}(1 + \beta u + \gamma t^2 + \delta s + \epsilon s^2 + \zeta u^2), \qquad (3)$$

with six parameters, Hylleraas[4] found for E the value

$$E(\text{helium}) = -2.90324 \text{ atomic units.} \qquad (4)$$

Also it has generally been stated that "an eighth approximation" gives the improved value[5,6]

$$E(\text{helium}) = -2.903745 \text{ atomic units.} \qquad (5)$$

Though this value has been widely quoted and used, it must be pointed out that it was not found by a strict application of the variational principle. The value (5) was derived by a semiempirical procedure based on an alternative method in which the energy of the ground state of a two-electron system is expressed as a series in the reciprocal of the nuclear charge Z. But this latter method cannot be relied upon to the same extent as the variational principle in that we cannot even be certain whether the calculated value is greater or less than the true value. That the value (5) should be suspect is apparent when we note that using the same method Hylleraas derived for the ground state of the negative hydrogen ion the value

$$E(H^-) = -0.52642 \text{ atomic unit,} \qquad (6)$$

whereas Henrich[7] has derived the value

$$E(H^-) = -0.52756 \text{ atomic unit} \qquad (7)$$

by a variational method using a wave function of the form (1) with eleven parameters.

For the reasons stated earlier, we have attempted an improved calculation of the energy of the ground state of helium by applying the variational principle to a wave function of the form

$$\psi = e^{-\frac{1}{2}ks}(1 + \beta u + \gamma t^2 + \delta s + \epsilon s^2 + \zeta u^2 + \chi_6 su + \chi_7 t^2 u + \chi_8 u^3 + \chi_9 t^2 u^2) \qquad (8)$$

[1] H. A. S. Eriksson, Nature **161**, 393 (1948); M. Günther, Physica **15**, 675 (1949); H. E. V. Håkansson, Arkiv. Fysik **1**, 555 (1950); B. Edlén, Arkiv. Fysik **4**, 441 (1951).

[2] C. E. Moore [*Atomic Energy Levels*, National Bureau of Standards Circular No. 467 (Government Printing Office, Washington, D. C. 1952)] gives the old Paschen value 198305±15 cm^{-1}, while J. J. Hopfield [Astrophys. J. **72**, 133 (1930)] and B. Edlén (reference 1) gives values of 198314 and 198312±5 cm^{-1}, respectively. *Note added in proof:*—Recent measurements by R. Zbinden and one of us (G. H.) at Ottawa confirm that the last-mentioned limit of error is a very conservative one.

[3] An account of Hylleraas's investigations will be found in H. Bethe, *Handbuch der Physik* (J. Springer, Berlin, 1933), Vol. 24, No. 1, pp. 353–363, see particularly pp. 358, 362, and 363.

[4] E. A. Hylleraas, Z. Physik **54**, 347 (1929), particularly p. 358.

[5] For example, Bethe (reference 3) states that "und nach einer noch genaueren Rechnung von Hylleraas wird schliesslich in achter Näherung."

[6] E. A. Hylleraas, Z. Physik **65**, 209 (1930).

[7] L. R. Henrich, Astrophys. J. **99**, 59 (1944).

with ten parameters. [It may be noted here that the terms in u, t^2, s, s^2, u^2, su, t^2u, and u^3 are the same ones which Hylleraas used in his calculations for determining the value (5).] The results of the calculations are summarized in Table I. This table gives the coefficients β, γ, etc., for three wave functions in the range in which E takes the minimum value. The energy of the ground state given by the calculations is therefore

$$E(\text{helium}) = -2.903603 \text{ atomic units.} \tag{9}$$

We conclude that there is no basis for supposing that the ground state of helium is given by the often quoted value (5).

The value (9), like Hylleraas's values (4) and (5), represents the total energy of He assuming infinite mass of the nucleus and is based on the nonrelativistic wave equation. Hylleraas and Bethe have shown that most of the effect of the motion of the nucleus can be taken into account by using the Rydberg constant R_{He} for He rather than that for infinite mass R_∞ in converting the atomic units to cm^{-1}. A small residual correction, called mass polarization by Bethe, amounts to only 5.2 cm^{-1}.

TABLE I. The constants of a ten-parameter wave function for the ground state of helium: $\psi = \mathfrak{N} e^{-\frac{1}{2}ks}(1+\beta u+\gamma t^2+\delta s+\epsilon s^2+\zeta u^2+\chi_6 su+\chi_7 t^2u+\chi_8 u^3+\chi_9 t^2u^2)$ ($\mathfrak{N}$ is the normalization factor).

k	3.5100255	3.5299360	3.5498639
β	+0.350563	+0.352547	+0.353024
γ	+0.157394	+0.157622	+0.160254
δ	−0.129341	−0.120909	−0.112542
ϵ	+0.0130191	+0.0126717	+0.0124213
ζ	−0.0681335	−0.0708333	−0.0722356
χ_6	+0.0192383	+0.0231770	+0.0272757
χ_7	−0.0338436	−0.0322601	−0.0333144
χ_8	+0.0055753	+0.0057980	+0.0056875
χ_9	+0.0053420	+0.0051526	+0.0056020
E	−2.9036027	−2.9036022	−2.9036014
$\mathfrak{N}$	0.37984145	0.37356893	0.36764807

The relativity correction has been subject to considerable change depending on the approximation used. According to a first approximation worked out by Bethe,[3] it was estimated to be -27 cm^{-1} with Hartree functions and -10 cm^{-1} using a still simpler eigenfunction. Eriksson,[8] on the basis of higher approximations, obtains $+2$ cm^{-1}. These numbers are to be understood as net corrections for the energy difference between the ground states of He and He^+. Including the mass and relativity corrections, one finds from (9) for the ionization potential of He, the value

$$I_{\text{calc}}(\text{He}) = 198287.7 \text{ cm}^{-1}, \tag{10}$$

while the observed value[2] is

$$I_{\text{obs}}(\text{He}) = 198313 \pm 5 \text{ cm}^{-1}. \tag{11}$$

In computing (10) from (9), the Lamb shift of the ground state of He^+ has been neglected. If it is assumed that this shift of He^+ is correctly given by the Bethe-Schwinger theory, i.e., is 3.8 cm^{-1}, one obtains

$$I_{\text{calc}}(\text{He}) = 198291.5 \text{ cm}^{-1}. \tag{12}$$

[8] H. A. S. Eriksson, Z. Physik **109**, 762 (1938).

TABLE II. Observed and calculated values of the ionization potentials of He-like ions.

Z	Ion	I.P.$_{\text{obs}}$, cm^{-1}	I.P.$_{\text{calc}}$, cm^{-1}	I.P.$_{\text{obs}}$ − I.P.$_{\text{calc}}$, cm^{-1}
2	He I	198313	198291.5	+21.5
3	Li II	610079	610049	+30
4	Be III	1241225	1241309	−84
5	B IV	209196_0	2092240	−280
6	C V	316245_0	3163009	−559
7	N VI	445280_0	4453848	−1048
8	O VII	596300_0	5964970	−1970

The difference between observed and calculated values is much greater than the estimated limit of error of the observed value. The question whether the difference (21.5 cm^{-1}) is due to an electromagnetic shift (opposite in direction to the Lamb shift), or to incorrect mass or relativistic corrections, or to a failure of the tenth-order approximation in approaching the correct value, must be left to future investigations. With regard to this last point, we are at the present time working on a wave function with 14 parameters.

Similar large discrepancies between observed and calculated values arise for the He-like ions. Hylleraas has derived a widely quoted general interpolation formula for the ionization potentials of these ions from the values for H^-, He, and a hypothetical ion with $Z=\infty$. Using the revised values for H^- and He and Hylleraas's value for $Z=\infty$, one finds the following modified formula

$$\text{I.P.} = R_Z\left(Z^2 - \frac{5}{4}Z + 0.31488 - \frac{0.020896}{Z} + \frac{0.011096}{Z^2}\right).$$

The values calculated from this formula, and corrected for relativity and mass polarization effects according to Eriksson[8] as well as for the Lamb shifts of the H-like ions, are compared with the observed values as given by Edlén[1] in Table II. The observed values are larger than the calculated ones for He and Li^+, but smaller for the other He-like ions and increasingly so with increasing Z. The difference (last column of Table II) represents a shift of the $1s^2\,{}^1S$ ground state referred to the state of the bare nucleus and the two electrons at infinite distance. For Li^+, as for He, this shift is opposite in direction to, and appreciably larger than, the "ordinary" Lamb shift, while for the other elements it is in the same direction and about twice as large as the Lamb shift for the corresponding one-electron systems. In agreement with expectation, the shift for $Z \gtrsim 4$ is roughly proportional to Z^4 as was already found by Edlén on the basis of the Hylleraas formula. Edlén's shifts are approximately half those of Table II, since he referred the ionization potential to the unshifted ground state of the corresponding H-like ions. It must be emphasized that for the higher He-like ions the probable error of the observed ionization potentials is a considerable fraction (about one-third) of the shift.

35

Energies of the Ground States of He, Li^+, and O^{6+}

S. Chandrasekhar, *University of Chicago, Chicago, Illinois*
AND
G. Herzberg, *Division of Physics, National Research Council, Ottawa, Canada*
(Received February 7, 1955)

A complete 14-parameter calculation for the ground state of He and similar 10-parameter calculations for the ground states of Li^+ and O^{6+} have been carried out. In addition, using the four most important of eleven terms, which were tried individually as fifteenth parameters, an 18-parameter calculation for He has been carried out but without minimizing against the scale parameter k. Similar 12-parameter calculations were carried out for Li^+ and O^{6+} using the two most important terms beyond the tenth in the He calculation. As a result, the 18-parameter nonrelativistic ionization potential of He is found to be 198311.4 cm^{-1}. The series of 6, 10, 14, 18 parameter values appears to converge to $198312._3$ cm^{-1} with an error of less than 2 cm^{-1}. Adding the relativistic corrections yields $198310._4$ cm^{-1} which agrees to 0.1 cm^{-1} with the latest experimental value. Considering the uncertainties of the theoretical and experimental values, the magnitude of the Lamb shift of the ground state of He (compared to He^+) must be less than 3 cm^{-1} which does not contradict present theoretical estimates. Similar agreements but within wider limits of error are found for the 12-parameter energy values of the ground states of Li^+ and O^{6+}.

A. INTRODUCTION

HYLLERAAS' well-known method of obtaining the energy of the ground state of helium has recently been carried to a higher (tenth) approximation than previously available.[1] This work showed that Hylleraas' earlier eighth approximation energy contained an error[2] and that the agreement between theory and experiment is not as good as previously believed. By assuming the mass polarization and relativistic corrections of Bethe[3] and Eriksson[4] respectively, a difference between observed and theoretical ionization potential of He of 25 cm^{-1} was found. It was suggested that this discrepancy might be due to an electromagnetic shift (opposite in direction to the Lamb shift) or to incomplete mass polarization or relativistic corrections or to a failure of the tenth approximation in approaching the correct nonrelativistic value. Since the publication of our previous paper, the relativistic corrections have been studied by Sucher and Foley.[5] They found that a term that may be interpreted as a spin-spin interaction had been neglected in the earlier treatments. It amounts to 4 cm^{-1} but is of such a sign that the discrepancy between theory and experiment is increased rather than decreased, i.e., is 29 rather than 25 cm^{-1}. The question of the mass polarization has been studied anew by Wilets[6] who obtained a slightly smaller correction than given by Bethe.[3]

Even before the new relativistic correction was known, work was started to carry the nonrelativistic Hylleraas calculation to still higher orders. When the preliminary results of these calculations were presented at the Rydberg Centennial Symposium at Lund, we learned that Hylleraas[2] had independently carried out similar calculations arriving at very similar results. In view of the importance of the subject and the ever-present possibility of numerical mistakes in the extensive calculations, it appeared worth while to complete and publish our calculations independently of Hylleraas' new work.

[1] Chandrasekhar, Elbert, and Herzberg, Phys. Rev. **91**, 1172 (1953), henceforth referred to as I.
[2] See also E. Hylleras, Proc. Rydberg Centennial Conference, Lund, Sweden, July, 1954, p. 83.
[3] H. A. Bethe in Geiger-Scheel's *Handbuch der Physik* (Verlag Julius Springer, Berlin, 1933), second edition, Vol. 24, Part 1.
[4] M. A. S. Eriksson, Z. Physik **109**, 762 (1938).
[5] J. Sucher and H. M. Foley, Phys. Rev. **95**, 966 (1954).
[6] L. Wilets (private communication).

B. HYLLERAAS FUNCTIONS AND CORRESPONDING ENERGIES OF THE GROUND STATE OF He

The Hylleraas type of wave function is of the form

$$\psi=\mathfrak{N}e^{-\frac{1}{2}ks}\sum c_{lmn}k^{l+m+n}s^l t^m u^n, \tag{1}$$

where s, t, and u are related to the distances r_1, r_2, and r_{12} (measured in atomic units) of the two electrons from the nucleus and from each other, respectively, by

$$s=r_1+r_2,\quad t=r_2-r_1,\quad u=r_{12}. \tag{2}$$

$\mathfrak{N}$ is a normalization constant and k and the c_{lmn} are constants which are to be adjusted so that the energy

Table I. Constants of 14-parameter wave function (14) for the ground state of He.

k (input) E (input)	3.85 −2.90370	3.75 −2.90370
k from (8)	3.8499301	3.7500555
E from (8)	−2.90370063	−2.90370089
β	+0.39836744	+0.39601198
γ	+0.17742685	+0.17483693
δ	+0.011878857	−0.041079523
ϵ	+0.020414801	+0.024913648
ζ	−0.11994054	−0.11315715
χ_6	+0.077281607	+0.054080045
χ_7	−0.084952179	−0.074771058
χ_8	+0.022483449	+0.021974359
χ_9	+0.014528286	+0.012947629
χ_{10}	+0.042902881	+0.030114033
χ_{11}	+0.0012248967	−0.0012596415
χ_{12}	−0.00010041525	−0.00010267038
χ_{13}	−0.0020615103	−0.0020306527
$\mathfrak{N}$	1.3617172	1.3633714

N.R.C. 3604

is minimized. In order to be minimized with regard to the c_{lmn}, the energy E must fulfill the determinantal equation (see Bethe[3]):

$$|k^2M_{ij}-kL_{ij}-EN_{ij}|=0. \tag{3}$$

Here i and j stand each for a set of numbers, $l_im_in_i$ and $l_jm_jn_j$ respectively, and M_{ij}, L_{ij}, and N_{ij} are the following sums of integrals:

$$\begin{aligned} M_{ij}=&+(l_il_j-m_im_j+l_in_j+n_il_j-m_in_j-n_im_j)\times[l_i+l_j\ ,m_i+m_j\ ,n_i+n_j+1] \\ &-0.5(l_i+l_j+n_i+n_j)\times[l_i+l_j+1,m_i+m_j\ ,n_i+n_j+1] \\ &-(n_in_j+l_in_j+n_il_j)\times[l_i+l_j\ ,m_i+m_j+2,n_i+n_j-1] \\ &+(n_in_j+m_in_j+n_im_j)\times[l_i+l_j+2,m_i+m_j\ ,n_i+n_j-1] \\ &+0.5(l_i+l_j)\times[l_i+l_j-1,m_i+m_j+2,n_i+n_j+1] \\ &+0.5(n_i+n_j)\times[l_i+l_j+1,m_i+m_j+2,n_i+n_j-1] \\ &-l_il_j\times[l_i+l_j-2,m_i+m_j+2,n_i+n_j+1] \\ &+m_im_j\times[l_i+l_j+2,m_i+m_j-2,n_i+n_j+1] \\ &+0.25\times[l_i+l_j+2,m_i+m_j\ ,n_i+n_j+1] \\ &-0.25\times[l_i+l_j\ ,m_i+m_j+2,n_i+n_j+1], \end{aligned} \tag{4}$$

$$N_{ij}=[l_i+l_j+2,\ m_i+m_j,\ n_i+n_j+1]-[l_i+l_j,\ m_i+m_j+2,\ n_i+n_j+1], \tag{5}$$

$$\begin{aligned} L_{ij}=&4Z[l_i+l_j+1,\ m_i+m_j,\ n_i+n_j+1]-[l_i+l_j+2,\ m_i+m_j\ ,\ n_i+n_j] \\ &+[l_i+l_j\ ,\ m_i+m_j+2,\ n_i+n_j], \end{aligned} \tag{6}$$

where Z is the nuclear charge ($=2$ for He). The brackets $[a,b,c]$ stand for the integrals

$$[a,b,c]=\int\int\int e^{-s}s^at^bu^cdsdudt=\frac{(a+b+c+2)!}{(b+1)(b+c+2)}. \tag{7}$$

The determinantal equation (3) must be solved for several values of k until E is minimized against k also. A check on the correctness of k and E is obtained from the formulas

$$k=L/2M,\quad E=L^2/4MN, \tag{8}$$

where

$$L=2\sum_{i,j}c_ic_jL_{ij},\quad M=2\sum_{i,j}c_ic_jM_{ij},\quad N=2\sum_{i,j}c_ic_jN_{ij}. \tag{9}$$

Here the c_i, c_j are the coefficients c_{lmn} derived from the secular determinant with the best E.

In I, the ten-parameter function

$$\begin{aligned}\psi=\mathfrak{N}e^{-\frac{1}{2}ks}(1+\beta u+\gamma t^2+\delta s+\epsilon s^2+\zeta u^2+\chi_6su \\ +\chi_7t^2u+\chi_8u^3+\chi_9t^2u^2)\end{aligned} \tag{10}$$

was used and an energy value for the ground state of He of

$$E=-2.903603 \text{ atomic units} \tag{11}$$

with $k=3.51$ was obtained.[7]

By adding the four terms

$$\chi_{10}t^4+\chi_{11}t^6+\chi_{12}t^4u^2+\chi_{13}t^2u^4 \tag{12}$$

to the bracket in (10), i.e., using a 14-parameter function and minimizing, an energy value of

$$E=-2.903629 \text{ atomic units} \tag{13}$$

was obtained. Here $k=3.53$ was assumed and no minimizing with regard to k was attempted (see, however, below).

In view of the smallness of the decrease of the energy in adding the terms (12), an attempt was made to ascertain whether perhaps other terms might have a larger effect. For this purpose, a fifteenth column was added to the secular determinant in turn corresponding to a term in u^4, or s^3, or t^2u^3, or st^2 and each time the energy was evaluated. The difference of the resulting E values from the value (13) indicated the relative importance of the terms considered. Similarly, the importance of the various terms (12) was ascertained by dropping the particular column of the secular determinant and finding the effect on the energy. In this way, it was found that the contributions of the terms in t^4, t^6, t^4u^2 are small[8] compared to those in t^2u^4, u^4, s^3, and st^2. Therefore, the following 14-parameter function was finally chosen:

$$\begin{aligned}\psi=\mathfrak{N}e^{-\frac{1}{2}ks}(1+\beta u+\gamma t^2+\delta s+\epsilon s^2+\zeta u^2+\chi_6su+\chi_7t^2u \\ +\chi_8u^3+\chi_9t^2u^2+\chi_{10}st^2+\chi_{11}s^3+\chi_{12}t^2u^4+\chi_{13}u^4.\end{aligned} \tag{14}$$

With $k=3.53$, this gave an energy value of

$$E=-2.903690 \text{ atomic units.} \tag{15}$$

The test of relation (8) showed that the assumed k value was not yet correct and several further (14×14) determinantal equations with different k values had to be solved. In Table I the coefficients, normalization constants and k and E values obtained from (8) are given for two sets of input values E and k near the

[7] The values for the normalization constant $\mathfrak{N}$ given in the last column of Table I of I are erroneous and should be replaced by 1.359625, 1.359841, and 1.360462.

[8] The term t^2u^3 was also found to be of little importance at this stage. However, later on it was found that a numerical error had occurred in the calculation for t^2u^3 and that actually t^2u^3 is more important than t^2u^4 as shown by the 18-parameter function given below.

TABLE II. Effect of adding various parameters to the 14-parameter Hylleraas function (14) with $k=3.85$.

Term added	E atomic units	ΔE 10^{-6} atomic units
...	-2.9037007	0
s^2u	-2.9037011	-0.4
su^2	-2.9037010	-0.3
s^4	-2.9037076	-6.9
s^2t^2	-2.9037036	-2.9
st^2u	-2.9037032	-2.5
t^4	-2.9037015	-0.8
t^2u^3	-2.9037070	-6.3
u^6	-2.9037076	-6.9
t^6	-2.9037014	-0.7
t^4u^2	-2.9037013	-0.6
$s^{-1}u$	-2.9037025	-1.8
$t^2u^3+u^6$	-2.9037098	-9.1
$s^4+s^2t^2+t^2u^3+u^6$	-2.9037162	-15.5

minimum. By interpolation from the corresponding values of the determinant (3), one obtains for the minimum:

$$k=3.80,\quad E=-2.903701 \text{ atomic units.} \tag{16}$$

By comparing with (15), it is seen that the effect on the energy of minimizing with respect to k is small.

In order to see what influence on the energy the addition of still higher terms in the series (1) might have, various trial terms were again added in turn to the 14-parameter function (14) as a fifteenth term and the energy determined in each case for $k=3.85$. This involved only the addition of a single column to the Gaussian algorithm by means of which the 14×14 determinants had been solved for $E=-2.90370$ and $E=-2.90371$. The results are shown in Table II. Only five terms give a noticeably different energy: u^6, s^4, t^2u^3, s^2t^2, and st^2u (in order of decreasing importance). The term $s^{-1}u$ was tried at the suggestion of Professor H. M. James (Purdue University) since he had found it of importance in a low-order approximation; but this term turned out to be unimportant in a 15-parameter function (see Table II).

If the four most important terms of Table II, *viz.*, u^6, s^4, t^2u^3, s^2t^2, are simultaneously added to the 14-parameter function (14), the following 18-parameter energy value is obtained:

$$E=-2.903716 \text{ atomic units.} \tag{17}$$

It should be noted that this value deviates from the 14-parameter value by much less than the sum of the four individual corrections taken from Table II (-15.5 against -23.0×10^{-6}). This gives one confidence that the other terms of Table II, if they were simultaneously added, would change the energy by less than the sum of the corrections of each term separately, that is, a 25-parameter value including all the terms in Table II would be between the value just given and

$$E=-2.903723 \text{ atomic units.} \tag{18}$$

It must be emphasized that the 18-parameter value (17) has been obtained with an assumed k-value, which is close to the minimum in the 14-parameter solution (16). From the change of E with k found there, it appears quite safe to conclude that minimizing of the 18-parameter value against k will change it by less than 0.000005. At the present stage, it did not seem worthwhile to carry through this minimizing process.

TABLE III. Constants of an 18-parameter wave function for the ground state of helium:

$$\psi=\mathfrak{N}e^{-\frac{1}{2}ks}(1+\beta u+\gamma t^2+\delta s+\epsilon s^2+\zeta u^2+\chi_6 su+\chi_7 t^2u+\chi_8 u^3+\chi_9 t^2u^2 +\chi_{10} st^2+\chi_{11} s^3+\chi_{12} t^2u^4+\chi_{13} u^4+\chi_{14} u^6+\chi_{15} t^2u^3+\chi_{16} s^2t^2+\chi_{17} s^4).$$

k (input)	3.85	χ_8	$+0.045441323$
E (input)	2.90371	χ_9	$+0.043516169$
k from (8)	3.8499613	χ_{10}	$+0.028227870$
E from (8)	2.9037063	χ_{11}	$+0.0071384413$
β	$+0.41389641$	χ_{12}	$+0.00050273143$
γ	$+0.21197114$	χ_{13}	-0.0099342061
δ	$+0.029010815$	χ_{14}	$+0.00093063179$
ϵ	$+0.0050395758$	χ_{15}	-0.0075260852
ζ	-0.14909338	χ_{16}	$+0.0030749706$
χ_6	$+0.079148647$	χ_{17}	-0.00080572559
χ_7	-0.12587484	$\mathfrak{N}$	1.3504631

TABLE IV. Constants of 10-parameter wave functions (10) of Li^+ and O^{6+}.

	Li^+		O^{6+}	
k (input)	5.60	5.70	16.40	16.48
k from (8)	5.600055	5.699877	16.400023	16.479994
E (input)	7.279760	7.279770	59.15640	59.156405
E from (8)	7.2797624	7.2797596	59.156413	59.156422
β	$+0.3598049$	$+0.3602588$	$+0.3771723$	$+0.3776096$
γ	$+0.2385639$	$+0.2569460$	$+0.6751016$	$+0.6935819$
δ	-0.08864538	-0.04417829	$+0.2922311$	$+0.3283520$
ϵ	$+0.01678235$	$+0.01818171$	$+0.1349207$	$+0.1569268$
ζ	-0.1303588	-0.1370898	-0.5061933	-0.5152307
χ_6	$+0.05504580$	$+0.07733141$	$+0.3317571$	$+0.3529975$
χ_7	-0.09060980	-0.1066555	-0.7846255	-0.8396899
χ_8	$+0.01606871$	$+0.01499941$	$+0.1682221$	$+0.1684930$
χ_9	$+0.01968222$	$+0.02688445$	$+0.4678856$	$+0.5306426$
$\mathfrak{N}$	5.750594	5.756579	141.2470	141.2869

The coefficients of the 18-parameter function nearest to the minimum are given in Table III. It should be noted that the k value from (8) is slightly less than the input value indicating that the best k value is less than 3.85.

C. HYLLERAAS FUNCTIONS AND CORRESPONDING ENERGIES OF THE GROUND STATES OF Li^+ AND O^{6+}

It appears of interest to carry through the Hylléraas calculation for some other two-electron systems. Li^+ and O^{6+} were chosen for this study and a 10-parameter calculation carried out similar to that for He in I. As may be seen from Eqs. (4)–(6), the quantities M_{ij} and N_{ij} in the determinantal equation (3) are the same as for He and only L_{ij} is different. Solving in the same way as for He, the sets of coefficients and corresponding energy values given in Table IV were found. By interpolation one obtains from the E and k values of Table IV and the values of 10×10 determinants calculated for several other E values:

$$\text{For Li}^+:\quad E=-\ 7.279763,\quad k=\ 5.63,$$
$$\text{For O}^{6+}:\quad E=-59.156404,\quad k=16.48. \tag{19}$$

The k value found for O^{6+} is remarkable since it is larger than $2Z$ which would be the k value in the absence of mutual interaction of the two electrons. It is also noteworthy that for high Z the coefficients $\beta, \gamma, \cdots, \chi_9$ decrease much less rapidly than for low Z.

If the two most important terms beyond the tenth, st^2 and s^3, as judged from the He calculation, are added and a 12-parameter calculation carried out for Li^+ with $k=5.60$ and for O^{6+} with 16.4, the following energy values are obtained:

$$\text{Li}^+:\quad E=-\ 7.27982_5 \text{ atomic units},$$
$$\text{O}^{6+}:\quad E=-59.15648_6 \text{ atomic units}. \tag{20}$$

Since the energy depends only slightly on k (see Table IV), it would appear that minimizing against k will only very slightly lower the energy values (20) which may therefore be safely considered as 12-parameter values.

After completing these calculations, a paper by Eriksson[9] came to our notice in which the author gives, as a 13-parameter value for Li^+, $E=-7.27984_4$ which agrees most satisfactorily with our value. The terms in (1) used by Eriksson are the same as those used by us except that he has added u^4. He does not give the coefficients in the eigenfunction.

D. DISCUSSION

In going from a Hylleraas wave function with 6 parameters to one with 10, 14, and finally 18 parameters, the nonrelativistic energy of the ground state of He changes from

$$-2.90324_0 \text{ to } -2.903603 \text{ to } -2.903701 \text{ to } -2.903716; \tag{21}$$

that is, there is a fairly good convergence. It will be remembered that the last four terms added have the largest effect on the energy among a considerable number of terms that have been tried. None of the other terms that have been tried give a contribution greater than 0.000002_5 and most of them much less. To be sure, not all of the fifty terms which are possible up to sixth order have been tried, but only those have not been tried for which a similar *lower* order term gave a contribution of less than 0.00001 to the energy, e.g., since s^2t^2 gives only a contribution of 0.0000029, s^2t^3, s^2t^4, $\cdots$, and s^3t^2, s^4t^2, $\cdots$ were not tried since they would be expected to give a contribution much smaller than s^2t^2. As shown previously, the resultant effect of a number of terms is less than additive. Since the terms that have not been used in the 18-parameter function give individual contributions less than 0.000002_5 and, with one exception, less than 0.0000008, it appears probable, as is also suggested by the series of numbers (21), that the energy of the ground state of He converges to -2.90372 atomic units and is almost certainly not below -2.90373. This assumes, of course, that the variation method does converge. Some doubt has been expressed with regard to this by various authors.[10] Even if these doubts were justified, the value (18) still remains an upper limit to the energy.

The energy value obtained in this way refers to a fixed nucleus. The motion of the nucleus is largely taken into account by multiplying by $2R_{He}$ rather than by $2R_\infty$ when converting to wave number units. A small correction, the mass polarization, first discussed by Bethe, has recently been re-evaluated by Wilets[6] to be $+4.1\ cm^{-1}$ (compared to Bethe's $5.2\ cm^{-1}$). In this way, subtracting the energy of the ground state of He^+, one obtains from (18) the following *lower* limit for the nonrelativistic ionization potential of He:

$$\text{I.P.}_{\text{n.r.}} \geqslant 198311.4\ \text{cm}^{-1}. \tag{22}$$

This number is $25.9\ cm^{-1}$ higher than the previous lower limit based on the 10-parameter approximation and the old mass polarization.

Assuming the extrapolated convergence suggested above, the energy value would be

$$\text{I.P.}_{\text{n.r.}}^{(\text{extrapol.})} = 198312._3\ \text{cm}^{-1}, \tag{23}$$

with an error probably not greater than $\pm 2\ cm^{-1}$ and more likely positive than negative.

If one were to use the old relativity correction of Eriksson, i.e., $+2._2\ cm^{-1}$, one would obtain

Eriksson:

$$\text{I.P.}_{\text{rel}}(\text{He}) \geqslant 198313.6\ \text{cm}^{-1},$$
$$\text{I.P.}_{\text{rel}}^{(\text{extrapol.})}(\text{He}) = 198314._5 \pm 2\ \text{cm}^{-1}. \tag{24}$$

However, if Sucher and Foley's relativistic correction is used, i.e., $-1.9\ cm^{-1}$, one obtains[11]

Sucher-Foley:

$$\text{I.P.}_{\text{rel}}(\text{He}) \geqslant 198309.5\ \text{cm}^{-1},$$
$$\text{I.P.}_{\text{rel}}^{(\text{extrapol.})}(\text{He}) = 198310._4 \pm 2\ \text{cm}^{-1}. \tag{25}$$

A provisional experimental value obtained by Zbinden and one of us[12] is

$$\text{I.P.}_{\text{exp}}(\text{He}) = 198310._5 \pm 1\ \text{cm}^{-1}. \tag{26}$$

[9] H. A. S. Eriksson, Arkiv. Mat. Astron. Fysik **B30**, No. 6 (1944).

[10] See, for example, Bartlett, Gibbons, and Dunn, Phys. Rev. **47**, 679 (1935); also V. A. Fock, Izvest. Akad. Nauk. S. S. S. R Ser. Fiz. **18**(2), 161 (1954). The objections of Bartlett *et al.* have been refuted by Coolidge and James [Phys. Rev. **51**, 855 (1937)] who have shown that the Hylleraas method does converge to the correct energy value. However, according to Kato [Trans. Am. Math. Soc. **70**, 212 (1951)], Coolidge and James' proof "is not complete from a mathematical standpoint." Kato has established the convergence of the variation method. He considers it as "highly plausible" but not as proven that it converges to the correct energy value.

[11] Here account has been taken of the note added in proof in Sucher and Foley's paper in which one part of the correction (E_3') is doubled.

[12] G. Herzberg and R. Zbinden (unpublished).

TABLE V. Ionization potentials and Lamb shifts of He and He-like ions.

	Ionization potential (cm^{-1}) nonrelativistic	relativistic	observed	Lamb shift (cm^{-1}) obs	calc
He	198312.₃	198310.₄	198310.₅	$+0.1\pm3$	-1.4
Li^+	610049	610087	610079	-8 ± 25	-8.5
O^{6+}	5959957	5963266	5963000	-266 ± 600	-460

The agreement between the experimental value and the extrapolated theoretical value using the Sucher-Foley relativistic correction is surprisingly close, much closer than one would have expected from the combined uncertainty of the theoretical and experimental values. At any rate, the large discrepancy found in I is entirely due to poor convergence of the tenth approximation.

Since the theoretical value (25) is based on the Dirac theory, the difference between the observed and the calculated value would represent an observed value for the electrodynamic (Lamb) shift of the ground state of He beyond that of He^+. This shift comes out to be

$$-4._0\pm3\ cm^{-1} \quad\text{and}\quad +0.1\pm3\ cm^{-1}$$

for the extrapolated values (24) and (25) based on the Eriksson and Sucher-Foley corrections, respectively. There appears to be general agreement that the Sucher-Foley correction is the correct one. Thus, we conclude that the Lamb shift of the ground state of He lies very probably between $+3$ cm^{-1} and -3 cm^{-1}. Günther[13] and Håkansson[14] have made rough theoretical estimates of the Lamb shift of He, obtaining -1.6 and -1.2 cm^{-1}, respectively. It is seen that these values are entirely compatible with the observed value. Conversely, if one considers the theoretical Lamb shift as correct, it would indicate that in extrapolating the convergence of the various approximations to the nonrelativistic energy of the He ground state, the effect of the neglected terms has been slightly underestimated (i.e., by 1.5 cm^{-1} or 0.000007 atomic units), which does not seem unreasonable. The extreme limit suggested above (-2.90373 atomic units) would give an "observed" Lamb shift of $-2._1$ cm^{-1}, that is, the predicted Lamb shift is well bracketed by the observed values following from the extrapolated and the extreme nonrelativistic energies.

For Li^+, the energy value (20) based on a 12-parameter Hylleraas function together with Bethe's mass polarization leads to a nonrelativistic ionization potential:

$$\text{I.P.}_{\text{n.r.}}(Li^+)\gtrsim 610049\ cm^{-1}. \tag{27}$$

Sucher and Foley have not calculated the relativistic correction including the spin-spin interaction and we are dependent on Eriksson's old value[15] of $+38$ cm^{-1}, yielding

$$\text{I.P.}_{\text{rel}}(Li^+)\gtrsim 610087\ cm^{-1}. \tag{28}$$

This value agrees remarkably well with Robinson's[16] experimental value of

$$\text{I.P.}_{\text{exp}}(Li^+)=610079\pm25\ cm^{-1}. \tag{29}$$

The difference obs-calc of -8 ± 25 cm^{-1} would represent an observed value of the Lamb shift of Li^+ (beyond that of Li^{++}). This value agrees with Håkansson's predicted value of -8.5 cm^{-1} far better than the accuracy of the data warrants. If one uses Eriksson's[10] 13-parameter value, the observed Lamb shift becomes -12 ± 25 cm^{-1}.

For O^{6+}, one finds from the 12-parameter energy value (20) the nonrelativistic ionization potential:

$$\text{I.P.}_{\text{n.r.}}(O^{6+})\gtrsim 5959957\ cm^{-1}. \tag{30}$$

In this case, Sucher and Foley[5] have calculated their relativistic correction, obtaining $+3309$ cm^{-1}, which may be compared with Eriksson's $+4110$ cm^{-1}. Using the former yields

$$\text{I.P.}_{\text{rel}}(O^{6+})\gtrsim 5963266\ cm^{-1}. \tag{31}$$

The experimental value of Tyrén[17] is

$$\text{I.P.}_{\text{exp}}(O^{6+})=5963000\pm600\ cm^{-1}. \tag{32}$$

From these two figures, an observed shift of -266 ± 600 cm^{-1} results which may be compared with Håkansson's predicted value of -460 cm^{-1}.

The results just presented are summarized in Table V. Neither the theoretical nor the experimental values for the ionization potentials of He, Li^+, and O^{6+} are as yet sufficiently precise to obtain reliable values for the Lamb shifts of the ground states of these systems, but the precision is now approaching that required for such a determination. Preparations are being made for an attempt to increase the accuracy of both theoretical and experimental values still further so that a determination of the Lamb shift will become possible.

The extensive computations underlying the present work, all done by desk machines, were carried out by Miss Alma Marcus, Miss Cecile DeChantigny, and Mrs. Sarah Segall at the National Research Council of Canada. Preparatory computations were done by Miss Donna Elbert at the Yerkes Observatory. We are very much indebted to all of them for their care and perseverance in carrying out these long and tedious calculations.

[13] M. Günther, Physica 15, 675 (1949).
[14] H. E. V. Håkansson, Arkiv. Fysik 1, 555 (1950).
[15] *Note added in proof.*—According to E. E. Salpeter (private communication), the Sucher-Foley correction for Li^+ amounts to $+14$ cm^{-1} leading to $\text{I.P.}_{\text{rel.}}(Li^+)\geq 610\,063$ cm^{-1}.
[16] H. A. Robinson, Phys. Rev. 51, 14 (1937).
[17] F. Tyrén, Nova Acta Reg. Soc. Sci. Ups. 12, nr 1 (1940).

Acknowledgements

The author and publisher are grateful to the following societies and publishers for their permission to reprint in this volume papers that originally appeared in print under their auspices.

American Association for the Advancement of Science (*The Scientific Monthly*)
American Mathematical Society
American Philosophical Society
The American Physical Society (*The Physical Review, Reviews of Modern Physics*)
Cambridge University Press
Canadian Journal of Physics
National Academy of Sciences
The Royal Society